Lecture Notes in Computer Science 6795

Commenced Publication in 1973
Founding and Former Series Editors:
Gerhard Goos, Juris Hartmanis, and Jan van Leeuwen

Giancarlo Mauri Alberto Leporati (Eds.)

Developments in Language Theory

15th International Conference, DLT 2011
Milan, Italy, July 19-22, 2011
Proceedings

Springer

Volume Editors

Giancarlo Mauri
Università degli Studi di Milano-Bicocca
Dipartimento di Informatica, Sistemistica e Comunicazione
Viale Sarca 336, Edificio U14
20126 Milano, Italy
E-mail: mauri@disco.unimib.it

Alberto Leporati
Università degli Studi di Milano-Bicocca
Dipartimento di Informatica, Sistemistica e Comunicazione
Viale Sarca 336, Edificio U14
20126 Milano, Italy
E-mail: leporati@disco.unimib.it

ISSN 0302-9743 e-ISSN 1611-3349
ISBN 978-3-642-22320-4 ISBN 978-3-642-22321-1 (eBook)
DOI 10.1007/978-3-642-22321-1
Springer Heidelberg Dordrecht London New York

Library of Congress Control Number: 2011930930

CR Subject Classification (1998): F.2, F.1, G.2, E.1, F.3, F.4

LNCS Sublibrary: SL 1 – Theoretical Computer Science and General Issues

Typesetting: Camera-ready by author, data conversion by Scientific Publishing Services, Chennai, India

Printed on acid-free paper

Springer is part of Springer Science+Business Media (www.springer.com)

Preface

The 15th International Conference on Developments in Language Theory (DLT 2011) was held in Milan, Italy, on the beautiful campus of the Milano–Bicocca University. It was a four-day conference starting July 19 and ending July 22, 2011.

The DLT conference series is one of the major international conference series in language theory. It started in Turku, Finland, in 1993. Initially, it was held once every two years. Since 2001, it has been held every year, odd years in Europe and even years in other continents.

The papers submitted to DLT 2011 were from 28 countries including Belgium, Canada, Czech Republic, Estonia, Finland, France, Germany, Hungary, Iceland, India, Iran, Israel, Italy, Japan, Malaysia, Poland, Portugal, Russian Federation, Serbia, Slovakia, South Korea, Spain, Switzerland, Taiwan, Turkey, UK, and the USA. Each paper was reviewed by three referees and discussed by the members of the Program Committee. Finally, 34 regular papers were selected by the Program Committee for presentation at the conference. There were five invited talks given at the conference. They were given by (in alphabetic order) Maxime Crochemore (Marne-la-Vallée), Antonio Restivo (Palermo), Arseny Shur (Ekaterinburg), Thomas Wilke (Kiel), and Sheng Yu (London, Ontario). In addition, there were seven posters on display at the conference. This volume includes all the 34 contributed papers, the papers or abstracts from the 5 invited speakers, and a 2-page abstract for each of the 7 poster papers.

We warmly thank all the invited speakers and all the authors of the submitted papers. Their efforts were the basis of the success of the conference.

We would like to thank all the members of the Program Committee and the external referees. Their work in evaluating the papers and comments during the discussions were essential to the decisions on the contributed papers. We would also like to thank all the members of the DLT Steering Committee, for their ideas and efforts in forming the Program Committee and selecting the invited speakers.

We wish to thank the conference sponsors: the University of Milano–Bicocca and the European Association for Theoretical Computer Science.

We would also like to thank the staff of the Computer Science Editorial at Springer, for their help in making this volume available before the conference. Their timely instructions were very helpful to our preparation of this volume.

July 2011

Alberto Leporati
Giancarlo Mauri

Organization

DLT 2011 was organized by the Department of Informatics, Systems and Communication, University of Milano–Bicocca.

Program Committee

Alberto Bertoni	Milan, Italy
Christian Choffrut	Paris, France
Stefano Crespi Reghizzi	Milan, Italy
Aldo de Luca	Naples, Italy
Manfred Droste	Leipzig, Germany
Zoltán Ésik	Szeged, Hungary
Paul Gastin	Cachan, France
Hendrik Jan Hoogeboom	Leiden, The Netherlands
Marcus Holzer	Giessen, Germany
Oscar H. Ibarra	Santa Barbara, USA
Lila Kari	London, Ontario, Canada
Natasha Jonoska	Tampa, USA
Giancarlo Mauri	Milan, Italy, Chair
Anca Muscholl	Bordeaux, France
Alexander Okhotin	Turku, Finland
Michel Rigo	Liège, Belgium
Wojciech Rytter	Warsaw, Poland
Mikhail V. Volkov	Ekaterinbug, Russia
Hsu-Chun Yen	Taipei, Taiwan
Takashi Yokomori	Tokyo, Japan

Organizing Committee

Paola Bonizzoni	Milan, Italy
Paolo Cazzaniga	Milan, Italy
Claudio Ferretti	Milan, Italy
Alberto Leporati	Milan, Italy, Chair
Dario Pescini	Milan, Italy
Antonio E. Porreca	Milan, Italy
Claudio Zandron	Milan, Italy

External Referees

Sayem Abu Sadat
Angela Angeleska
Marcella Anselmo
Vince Barany
Nicolas Bedon
Marcello Bersani
Valerie Berthe
Dietmar Berwanger
Daniela Besozzi
Maria Bianchi
Jean-Camille Birget
Henrik Björklund
Alexandre Blondin Massé
Bernard Boigelot
Benedikt Bollig
Henning Bordihn
Laurent Braud
Luca Breveglieri
Michelangelo Bucci
Marie-Pierre Béal
Arnaud Carayol
Arturo Carpi
Olivier Carton
Alessandra Cherubini
Alexander Clark
Thomas Colcombet
Bo Cui
James Currie
Elena Czeizler
Flavio D'Alessandro
Jurgen Dassow
Alessandro De Luca
Rodrigo De Souza
Alberto Dennunzio
Egor Dolzhenko
Mike Domaratzki
Omer Egecioglu
Duchene Eric
Claudio Ferretti
Kaoru Fujioka
Jean-Marc Fédou
Julia Gamzova
Yuan Gao
Silvio Ghilardi
Amy Glen
Serge Grigorieff
Giuliano Grossi
Pierre Guillon
Tero Harju
Niko Haubold
Fritz Henglein
Sebastian Jakobi
Artur Jeź
Galina Jiraskova
Christos Kapoutsis
Ines Klimann
Satoshi Kobayashi
Hans-Joerg Kreowski
Dietrich Kuske
Martin Kutrib
Zbynek Krivka
Ruggero Lanotte
Tommi Lehtinen
Julien Leroy
Anthony Widjaja Lin
Chun-Cheng Lin
Markus Lohrey
Violetta Lonati
Kalpana Mahalingam
Andreas Malcher
Andreas Maletti
Roberto Mantaci
Conrado Martinez
Oliver Matz
Katja Meckel
Carlo Mereghetti
Martin Middendorf
Lukasz Mikulski
Benjamin Monmege
Aniello Murano
Filip Murlak
Judit Nagy-György
Chrystopher L. Nehaniv
Cyril Nicaud
Lasse Nielsen

Fumiya Okubo
Friedrich Otto
Paritosh Pandya
Matthieu Picantin
Giovanni Pighizzini
Jean-Éric Pin
Wojciech Plandowski
Libor Polak
Antonio E. Porreca
Matteo Pradella
Jakub Radoszewski
Michael Rao
Bala Ravikumar
Klaus Reinhardt
Antonio Restivo
Emanuele Rodaro
Jacques Sakarovitch
Pierluigi San Pietro
Sylvain Schmitz
Stefan Schwoon
Shinnosuke Seki
Carla Selmi
José M. Sempere
Arseny Shur
Giulia Simi
Amir Simjour
Ludwig Staiger
Štěpán Starosta
Camille Vacher
Leonardo Vanneschi
Bow-Yaw Wang
Claudio Zandron
Marc Zeitoun

Sponsoring Institutions

The University of Milano–Bicocca
European Association for Theoretical Computer Science

Table of Contents

Invited Talks

Regular Papers

Short Papers

Hunting Redundancies in Strings

Golnaz Badkobeh[1], Supaporn Chairungsee[1], and Maxime Crochemore[1,2]

[1] King's College London, London, WC2R 2LS, United Kingdom
[2] Université Paris-Est, France
{golnaz.badkobeh,supaporn.chairungsee,maxime.crochemore}@kcl.ac.uk

Abstract. The notion of redundancies in texts, regarded as sequences of symbols, appear under various concepts in the literature of Combinatorics on words and of Algorithms on strings: repetitions, repeats, runs, covers, seeds, and palindromes, for example.

We explore some of the newest aspects of these redundancies.

1 Redundancy: A Versatile Notion

The notion of redundancies in texts, regarded as sequences of symbols, appear under various concepts in the literature of Combinatorics on words and of Algorithms on strings: repetitions, repeats, runs, covers, seeds, and palindromes, for example. Combinatorial and algorithmic aspects of these elements are spread for example in the Lothaire's books [36,37,38].

Squares and cubes (concatenations of 2 or 3 copies of the same non-empty word) are instances of repetitions whose exponent is at least 2 and have been studied for more than a century after the seminal work of Thüe [53] who described infinite words containing none of them.

Repeat often refers to less repetitive segments in text, that is, those having a rational exponent smaller than 2. The frontier between repetitions and repeats is indeed rather blurred in literature. The famous Dejean's conjecture [21] refers mainly to repeats and provides the repetition threshold of each alphabet size: there exist infinite words whose maximal factor exponent is the threshold, and below the threshold there are only finitely many words. After several partial results, including the result of Dejean on the three-letter alphabet, the conjecture has eventually been solved recently (see [47]).

Further works show that maximal exponent repeats, or repetitions on the binary alphabet, occurring in infinite words complying with the threshold can also have a bounded length [51]. Their minimal number is also known for alphabets of size 2 and 3 [4], introducing the finite-repetitions threshold attached to each alphabet size.

The design of methods for computing all the occurrences of repetitions in a string has lead to several optimal algorithms [9,2,41] running in $O(n \log n)$ time. They have been extended to algorithms producing certain repetitions with rational exponent [40], or a unique occurrence for each repetition [28]. They run in $O(n \log a)$ time on an alphabet of size a.

G. Mauri and A. Leporati (Eds.): DLT 2011, LNCS 6795, pp. 1–14, 2011.

What seems to be the right concept on repetitions is that of runs, maximal occurrences of repetitions, whose computation can be done in linear time (see [34], [14]). Their maximal number is an open question ([34]) but tight approximations exist ([48,18]).

Local periodities in strings associated with roots of repetitions are softened by the notion of covers whose consecutive occurrences may overlap or are at least adjacent in the string. A seed instead covers a superstring of the initial string. They reveal redundacies inside strings that are more difficult to discover than runs. Finding the shortest cover of a string can be done in linear time, while its shortest seed is done in $O(n \log n)$ time (see [1]).

Palindromes are yet another kind of redundancies. They can be generalised with palindromes having a bounded-length center. If testing if a string is a palindrome is straightforward, finding palindromic segments is more complex. Under some constraints it can be done in time $O(n \log a)$ [35,7] and even in linear-time using more sophisticated data structures [19].

Beyond theoretical aspect of the problems related to redundancies, they are often the base for string modelling adapted to compression coding. They appear in run-length and Ziv-Lempel compression for example (see [5]). But more is to be done on this to account for all the notions described here.

Repetitions and palindromes receive intensive attentions for the analysis of genetic sequences. Repetitions are called tandem repeats, satellites or SRS and should accept some approximation. Some types of palindromes are crucial for the prediction of the secondary structure of RNA molecules (see [6]).

2 Avoiding Repetitions and Repeats

In this section we show how some infinite string can avoid repeats. The property depends on the alphabet size.

Avoidability in binary words. In 1906 A.Thue established that squares are avoidable on a 3-letter alphabet and cubes are avoidable on a 2-letter alphabet [53]. Iterating the morphism m defined by:

$$\begin{cases} m(\mathtt{0}) = \mathtt{01}, \\ m(\mathtt{1}) = \mathtt{10}. \end{cases}$$

gives Thue-Morse sequence: 0110100110010110100101l0... which is overlap-free. Pansiot [44] observed that the only morphisms generating the Thue-Morse word are powers of m. This was extended by Seebold to:

Theorem 1 ([50]). *Let x be an infinite overlap-free word over the alphabet $\{a, b\}$ that is generated by iterating some morphism h. Then h is a power of m.*

However squares are not avoidable in infinite binary words, but Fraenkel and Simpson [23] proved that some of them can contain only three of them. Indeed all factors of exponent at least 2 occurring in their word should be considered, which adds two cubes to the three squares. Their proof uses a pair of morphisms, one to get an infinite word by iteration, the other to produce the final translation to the binary alphabet. Their result has been proved with different pairs of morphisms by Rampersad et al. [46] (the first morphism is uniform), by Harju and Nowotka [29] (the second morphism accepts any infinite square-free word), and eventually by Badkobeh and Crochemore [4] with the two morphisms f_0:

$$\begin{aligned} f_0(\mathtt{a}) &= \mathtt{abc}, \\ f_0(\mathtt{b}) &= \mathtt{ac}, \\ f_0(\mathtt{c}) &= \mathtt{b}. \end{aligned}$$

and g_0:

$$\begin{aligned} g_0(\mathtt{a}) &= \mathtt{01001110001101}, \\ g_0(\mathtt{b}) &= \mathtt{0011}, \\ g_0(\mathtt{c}) &= \mathtt{000111}. \end{aligned}$$

The infinite binary word $\mathbf{g_0} = g_0({f_0}^{\infty}(\mathtt{a}))$ containing only the three squares 00, 11 and (1010) and the two cubes 000 and (111).

Avoiding large squares in constraint infinite binary words was a noble idea by Shallit [51] who showed extreme cases of infinite binary words under both constrains, maximal exponent and the period length. In his paper he shows that for all t, $t \geq 1$, no infinite binary word simultaneously avoids all squares yy with period $|y| \geq t$ and 7/3-powers. This implies that the number of squares in an infinite binary word is unbounded if it is 7/3-power free. He considers also the period length of the avoided squares when the maximal exponent increases to 5/2 and to 3. Here is the results summary:

Period of avoided squares	Avoidable power	Unavoidable power
2	none	all
3	3^+	3
4, 5, 6	$5/2^+$	5/2
≥ 7	$7/3^+$	7/3

By the way, the avoidability of some period lengths and maximal exponent has been studied by Dekking in [22] and in more details by Ochem in [43].

In [4] the authors show that the two types of constraints for the binary alphabet can be combined: producing an infinite word whose maximal exponent of its factors is the smallest possible while containing the smallest (finite) number of squares. With maximal exponent 7/3 the smallest number of squares is 12 to which is to be added 7/3-powers.

It is known from Karhumäki and Shallit [33] that if an infinite binary word avoids 7/3-powers it contains an infinite number of squares. Proving that it contains more than 12 squares is indeed a matter of simple computation.

Shallit [51] has built an infinite binary word avoiding $7/3^+$-powers and all squares of period at least 7 as mentioned above, However his word contains 18 squares. The infinite binary word in [4] avoids the same powers but contains only 12 squares, the largest having period 8. As before the proof relies on a pair of morphisms satisfying suitable properties. Both morphisms are almost uniform (up to one unit).

The first morphism f_0 is weakly square-free defined from A_6^* to itself by:

$$\begin{array}{ll} f_1(\texttt{a}) = \texttt{abac}, & f_1(\texttt{b}) = \texttt{babd}, \\ f_1(\texttt{c}) = \texttt{eabdf}, & f_1(\texttt{d}) = \texttt{fbace}, \\ f_1(\texttt{e}) = \texttt{bace}, & f_1(\texttt{f}) = \texttt{abdf}. \end{array}$$

And the second morphism g_1 from A_6^* to B^*, does not even corresponds to a uniquely-decipherable code but admits a unique decoding on the words produced by the first morphism. It is defined by:

$$\begin{array}{ll} g_1(\texttt{a}) = \texttt{10011}, & g_1(\texttt{b}) = \texttt{01100}, \\ g_1(\texttt{c}) = \texttt{01001}, & g_1(\texttt{d}) = \texttt{10110}, \\ g_1(\texttt{e}) = \texttt{0110}, & g_1(\texttt{f}) = \texttt{1001}. \end{array}$$

Then the infinite word $\mathbf{g_1} = g_1({f_1}^{\infty}(\texttt{a}))$ has the desired property. It contains the 12 squares $\texttt{0}^2$, $\texttt{1}^2$, $(\texttt{01})^2$, $(\texttt{10})^2$, $(\texttt{001})^2$, $(\texttt{010})^2$, $(\texttt{011})^2$, $(\texttt{100})^2$, $(\texttt{101})^2$, $(\texttt{110})^2$, $(\texttt{01101001})^2$, $(\texttt{10010110})^2$ only. Words 0110110 and 1001001 are the only factors with an exponent larger than 2, that is 7/3.

Looking at repetitions in words on larger alphabets, the subject introduces a new type of threshold, that we call the *finite-repetition threshold.* For the alphabet of a letters, it is defined as the smallest rational number $\mathrm{FRt}(a)$ for which there exists an infinite word avoiding $\mathrm{FRt}(a)^+$-powers and containing a finite number of r-powers, where r is Dejean's repetitive threshold. Karhumäki and Shallit [33] results as well as Badkobeh and Crochemore [4] show that $\mathrm{FRt}(2) = 7/3$, where the associated number of squares is 12.

Fewest repetitions vs maximal-exponent powers in infinite binary words. In this section we provide extra results that deepen the question of avoidable patterns in infinite binary words by introducing another point of view. We analyse the trade-off between the number of (distinct) squares and the number of maximal-exponent repetitions in infinite binary words when the maximal exponent is constant. The interesting results show the behavior of infinite binary words when the maximal exponent varies between 3 to 7/3. The value 7/3 is the Finite-repetition threshold. And the value 3 of the maximal exponent is where the number of squares is the minimum. The next table summarises the results.

Maximal exponent e	Allowed number of e-powers	Minimum number of squares	
7/3	2	12	$\mathbf{g_1} = g_1(f_1^{\infty}(\texttt{a}))$
	1	14	$\mathbf{g_2} = g_2(f_2^{\infty}(\texttt{a}))$
5/2	2	8	$\mathbf{g_3} = g_3(f_2^{\infty}(\texttt{a}))$
	1	11	$\mathbf{g_4} = g_4(f_0^{\infty}(\texttt{a}))$
3	2	3	$\mathbf{g_0} = g_0(f_0^{\infty}(\texttt{a}))$
	1	4	$\mathbf{g_5} = g_5(w)$

The infinite binary word $\mathbf{g_2} = g_2(f_2^{\infty}(\mathtt{a}))$ contains 14 squares, no factor of exponent larger than 7/3 and only one 7/3-power (see Badkobeh [3]). The morphism f_2 is defined from A_5^* to itself by:

$$\begin{aligned}
f_2(\mathtt{a}) &= \mathtt{adcbebc},\\
f_2(\mathtt{b}) &= \mathtt{adcbedc},\\
f_2(\mathtt{c}) &= \mathtt{aebc},\\
f_2(\mathtt{d}) &= \mathtt{aebedc},\\
f_2(\mathtt{e}) &= \mathtt{aebedcbebc}.
\end{aligned}$$

Then we translate $f_2^{\infty}(\mathtt{a})$ to binary using the second morphism g_2 defined by:

$$\begin{aligned}
g_2(\mathtt{a}) &= \mathtt{101001100101},\\
g_2(\mathtt{b}) &= \mathtt{1010011001001},\\
g_2(\mathtt{c}) &= \mathtt{101001011001},\\
g_2(\mathtt{d}) &= \mathtt{101001011001001},\\
g_2(\mathtt{e}) &= \mathtt{101001011001001100101},
\end{aligned}$$

The word $\mathbf{g_2} = g_2({f_2}^{\infty}(\mathtt{a}))$ satisfies the property, contains the 14 squares $\mathtt{0}^2$, $\mathtt{1}^2$, $(\mathtt{01})^2$, $(\mathtt{10})^2$, $(\mathtt{001})^2$, $(\mathtt{010})^2$, $(\mathtt{100})^2$, $(\mathtt{101})^2$, $(\mathtt{0110})^2$, $(\mathtt{1001})^2$, $(\mathtt{100110})^2$, $(\mathtt{0100110})^2$, $(\mathtt{0110010})^2$, $(\mathtt{10010110})^2$, and only one 7/3-power, 1001001.

Proving that it is impossible to have less than 12 squares when avoiding 5/2 powers of binary words needs a simple computation, but there exist infinite $5/2^+$-free binary words with less than 12 squares [3]. Additionally the number of squares varies according to the number of maximal-exponent powers: if there is only one 5/2-power the minimum number of squares is 11, and if there are two 5/2-powers, it becomes 8.

The infinite binary word $\mathbf{g_3} = g_3(f_2^{\infty}(\mathtt{a}))$ contains 8 squares, no factor of exponent larger than 5/2, and only two 5/2-powers [3]. The morphism g_3 is defined by:

$$\begin{aligned}
g_3(\mathtt{a}) &= \mathtt{001100101},\\
g_3(\mathtt{b}) &= \mathtt{0011001011},\\
g_3(\mathtt{c}) &= \mathtt{001101},\\
g_3(\mathtt{d}) &= \mathtt{001101011},\\
g_3(\mathtt{e}) &= \mathtt{00110101100101}.
\end{aligned}$$

The 8 squares $\mathbf{g_3}$ contains are $\mathtt{0}^2$, $\mathtt{1}^2$, $(\mathtt{01})^2$, $(\mathtt{10})^2$, $(\mathtt{0110})^2$, $(\mathtt{1001})^2$, $(\mathtt{011001})^2$, $(\mathtt{100110})^2$, and the two 5/2-powers are 01010, 10101.

The infinite binary word $\mathbf{g_4} = g_4(f_0^{\infty}(\mathtt{a}))$ is a $5/2^+$-free with only one 5/2-power and no more than 11 squares [3], where g_4, from A_3^* to B^*, is defined by:

$$\begin{aligned}
g_4(\mathtt{a}) &= \mathtt{1001001101011001101001011001001101100}\\
&\quad\ \mathtt{10110100110110010011010010110011010 1},\\
g_4(\mathtt{b}) &= \mathtt{100100110100101},\\
g_4(\mathtt{c}) &= \mathtt{1001001101100101101001101}.
\end{aligned}$$

The 11 squares of $\mathbf{g_4}$ are $\mathtt{0}^2$, $\mathtt{1}^2$, $(\mathtt{01})^2$, $(\mathtt{10})^2$, $(\mathtt{001})^2$, $(\mathtt{010})^2$, $(\mathtt{011})^2$, $(\mathtt{100})^2$, $(\mathtt{101})^2$, $(\mathtt{110})^2$, $(\mathtt{0110})^2$, and its 5/2-power is 10101.

Finally, proving that it is impossible to have less than 8 squares when avoiding cubes needs another mere computation, but recalling Fraenkel and Simpson result [23] and the word $\mathbf{g_0} = g_0(f_0^\infty(\mathtt{a}))$ shows the existence of infinite binary word containing only 3 squares and 2 cubes. it has been shown that the number of squares increases to 4 if only one cube is allowed in the infinite binary word. Let us consider the morphism g_5 defined by

$$g_5(\mathtt{a}) = \mathtt{1100010110010100},$$
$$g_5(\mathtt{b}) = \mathtt{1101000110010100},$$
$$g_5(\mathtt{c}) = \mathtt{0110101100010100}.$$

Then the infinite word $\mathbf{g_5} = g_5(w)$, where w is any infinite $7/4^+$-free ternary word, is 3^+-free and contains the 4 squares 00, 11, 0101 and 1010, and the only cube 000 [3].

3 Finding Repetitions

Repetitions and periods in strings constitute one of the most fundamental areas of string combinatorics. They have been studied already in the papers of Axel Thue [53], considered as having founded stringology. While Thue was interested in finding long sequences with few repetitions, in recent times a lot of attention has been devoted to the algorithmic side of the problem.

Detecting repetitions in strings is an important element of several questions: pattern matching, text compression, and computational biology to quote a few. Pattern matching algorithms have to cope with repetitions to be efficient as these are likely to slow down the process; the large family of dictionary-based text compression methods (see [54]) use a weaker notion of repeats (like the software gzip); repetitions in genomes, called satellites or Simple Sequence Repeats, are intensively studied because, for example, some over-repeated short segments are related to genetic diseases [39]; some satellites are also used in forensic crime investigations.

In this section, we recall some of the most significant achievements in the area. We focus on algorithms for finding repetitions and on their analysis that relies on counting various types of repetitions. The main result concerns the linear-time computation of runs in a string as well as combinatorial estimation on their number.

Initially people investigated mostly squares occurrences, but their number can be as high as $\Theta(n \log n)$ [9], hence algorithms computing all of them cannot run in linear time, due to the potential size of the output. Indeed the same result holds for any type of repetition having an integer exponent greater than 1 [11]. The optimal algorithms reporting all positioned squares or just a single square were designed in [9,2,41,10].

Theorem 2 (Crochemore [9], Apostolico-Preparata [2], Main-Lorentz [41]). *There exists an $O(n \log n)$ worst-case time algorithm for computing all the occurrences of primitively-rooted squares in a string of length n.*

Techniques used to design the algorithms are based on partitioning, suffix trees, and naming segments, respectively. A similar result has been obtained by Franek, Smyth, and Tang using suffix arrays [25].

Testing the existence of some repetitions can however be done faster as is the case of squares.

Theorem 3 (Crochemore [10], Main-Lorentz [41]). *Testing if a string of length n is square-free can be done in worst-case time $O(n \log a)$, where a is the alphabet size of the string.*

Looking at (distinct) factors that are repetitions and not their occurrences as discussed above, it is known for example that only $O(n)$ (distinct) squares can appear in a string of length n.

Corollary 1 (Fraenkel and Simpson [24]). *Any string of length n contains at most $2n$ (distinct) squares.*

The proof is a direct consequence of a useful lemma known as the Three-square Lemma (see [20]), but a simple direct proof is due to Ilie [30]. Based on numerical evidence, it has been conjectured that the number of squares is at most n and the best bound to date, $2n - \Theta(\log n)$, is by Ilie [31].

The structure of all squares and of unpositioned runs has been also computed within the running time $O(n \log a)$ in [40] and [28].

Runs. The concept of *maximal occurrence of repetitions*, called runs in [32], has been introduced to represent all repetitions in a succinct manner. The crucial property of runs is that there are only $O(n)$ many of them in a string of length n [34,48,13,45].

Formally, a *run* in a string w is an interval $[i \mathinner{..} j]$ of positions for which both the associated string $w[i \mathinner{..} j]$ is periodic (i.e. has period $p \leq (j-i+1)/2$), and the periodicity cannot be extended to the right nor to the left: $w[i-1] \neq w[x+p-1]$ and $w[j-p+1] \neq w[j+1]$ have higher periods when these words are defined. When the period p of a run is known, we call it a p-run.

As a consequence of the algorithms and of the estimation on the number of squares, the most important result related to repetitions in strings can be formulated as follows.

Theorem 4 (Kolpakov-Kucherov [34], Rytter [48], Crochemore-Ilie [13]).
(i) All runs in a string of length n over an alphabet of size a can be computed in time $O(n \log a)$.
(ii) The number of all runs is linear in the length of the string.

Point (ii) of the statement is very intricate and of purely combinatorial nature. The algorithm for (i) executes in time proportional to the number of runs (on a fixed-size alphabet) which, by (ii), is linear. Indeed, with an reasonable hypothesis on the alphabet, the running time of (i) can be reduced to $O(n)$ as stated in Theorem 5 below.

Let $\rho(n)$ be the maximal number of runs in a string of length n. By item (ii) we have $\rho(n) < cn$ for some constant c. Based on the experiments in [34] Kolpakov and Kucherov conjectured that $c = 1$ for binary alphabets. A stronger conjecture was proposed in [26] where a family of strings is given with the number of runs equal to $\frac{3}{2\Phi} = 0.927\ldots$ (Φ is the golden ratio), thus proving $c \geq 0.927\ldots$. The authors of [26] conjectured that this bound is optimal, but the best known lower bound for c has been shown to be 0.944 by Matsubara et al. [42] and by Simpson [52].

Computing runs. Next, we sketch shortly the basic components of the proof of the point (i) of Theorem 4. The main idea is to use, as for the previous Theorem 3, the *f-factorisation* of the input string (see [10]): a string w is decomposed into factors u_1, u_2, ..., u_k, where u_i is the longest segment which appears before its position in w, i.e. in $u_1u_2\ldots u_iA^{-1}$, possibly overlapping the present occurrence of u_i; if the segment is empty u_i is the letter $u[i]$. The factorisation is similar to Ziv-Lempel factorisation that plays an important role in data compression algorithms.

The runs which fit in a single factor of the f-factorisation are called internal runs, other runs are called here overlapping runs. There are three crucial facts:

- all overlapping runs can be computed in linear time,
- each internal run is a copy of an earlier overlapping run,
- the f-factorisation can be computed in linear time under some hypothesis on the alphabet of the string (see Theorem 5 below).

It follows easily from the definition of the f-factorisation that if a run overlaps two consecutive factors u_{k-1} and u_k then its size is at most twice the total size of these two factors.

The f-factorisation of a string is commonly computed with the suffix tree or the suffix automaton of the string. When the alphabet of the string has a fixed size, thanks to the efficient algorithms for building these data structures, the whole process can be carried on in linear time. Two recent algorithms, in [14] and [8] (see also [17]), use the suffix array of the string to provide linear-time algorithms for integer alphabets (whose sequences of letters can be sorted in linear time). This is done through a useful table called LPF for longest previous factor, that provides for each position on the string the longest factor occurring both at that position and before it. Its computation is done in linear time with the suffix array of the string (see [14,17,16]).

Theorem 5 (Crochemore-Ilie [14], Chen-Puglisi-Smyth [8]). *On an integer alphabet, the f-factorisation of a string and its runs can be computed in linear time.*

Counting runs. The most intriguing question remains the asymptotically tight bound for the maximum number of runs $\rho(n)$ in a string of length n.

The first explicit upper bound, $5\,n$, on $\rho(n)$ for general strings was given by Rytter [48] and improved in a structural and intricate manner in [49] to $3.44\,n$, by introducing a sparse-neighbour technique. Another improvement of the ideas of [48] was done in [45] where the bound $3.48\,n$ is obtained. The neighbours are runs for which both the distance between their starting positions is small and the difference between their periods is also proportionally small according to some fixed coefficient of proportionality. The occurrences of neighbours satisfy certain sparsity properties which imply the linear upper bound. Several variations for the definitions of neighbours and sparsity are possible. Considering runs having close centres (the beginning position of the second period) the bound has been lowered to $1.6\,n$ in [13,15], improved to $1.52\,n$ in [27], and further to $1.029\,n$ as a result of the computation of a tight approximation of the maximal number of 60-runs (see [18]).

It is interesting to note that the approach of [13,15] is somewhat counterintuitive. On the one hand, it is known that there can be only logarithmically many runs starting at the same position and this is how they are counted in [48]. On the other hand, there can be linearly many runs with the same centre, and still counting them this way in [13,15] yields a better bound. This is essentially due to the fact that many runs with the same centre implies strong local periodicities in the string, thus eliminating many other potential runs.

4 Palindromes in DNA Sequence

A palindrome is a word or phrase which reads the same backwards as forwards, for instance "refer", "level" and "stats." Words with palindromic structure are important elements of DNA and RNA sequences, as they reflect the capacity of molecules to form double-stranded stems and loops, which insures a stable state of those molecules with low free energy. Restriction endonucleases usually recognize palindromic sequences of DNA as they are useful for gene isolation and cloning, genetic recombination, examining chromosome structure, and sequencing long DNA fragments. When an RNA is sequenced (digitised) it appears in the form of a mere string on the alphabet of nucleotides `A`, `C`, `G` and `U` that stand for adenine, cytosine, guanine and uracil.

Meaningful palindromic genetic sequences include a gap (spacer) between left and right parts. Those palindromes correspond to hair-pin structures of RNA molecules and they are significant in DNA sequences. These structures are widespread in the natural plasmids, viral and bacterial genomes, eukaryotic chromosomes and cell organelles. The reverse part of a palindrome is to be combined with the complementarity relation on nucleotides, where `A` and `U` are complements as well as `C` and `G` are. For example, the (exact) complimentary strand of the sequence `CGAATGGCTCTT` is `GCTTACCGAGAA`. And `CGAATGGCTCTT`s`AAGAGCCATTCG` is a gapped palindrome with spacer s.

Using LPrF to locate palindromes. In this section, we introduce an efficient technique to find palindrome in a given string. This technique is based on the notion of longest previous reverse factor, LPrF. The LPrF table is a concept close to the LPF table for which the previous occurrence in not reversed (see [16]). This latter table extends the Ziv-Lempel factorisation of a text [55] intensively used for conservative text compression (see [5]). The LPrF table also generalises a factorisation of strings used by Kolpakov and Kucherov [35] to extract certain types of palindromes in molecular sequences. These palindromes play an important role in RNA secondary structure prediction because they signal potential hair-pin loops in RNA folding (see [6]). In addition the reverse complement of a factor has to be considered up to some degree of approximation.

One of the problems is to compute efficiently, for a given string y, the LPrF table that stores at each index i the maximal length of factors that both start at position i on y and occur reversed at a smaller position. Here is the example table for the string $y =$ aababaabab:

position i	0	1	2	3	4	5	6	7	8	9
$y[i]$	a	a	b	a	b	a	a	b	a	b
LPrF$[i]$	0	1	0	1	3	2	4	3	2	1

From the table above, we can find the palindrome of this string y by storing the starting position of the LPrF value as shown in the table below.

position i	0	1	2	3	4	5	6	7	8	9
$y[i]$	a	a	b	a	b	a	a	b	a	b
LPrF$[i]$	0	1	0	1	3	2	4	3	2	1
Pal$[i]$	-	0	-	1	0	0	2	2	1	2

One efficient technique to compute the LPrF table is to use a Suffix Automaton. The Suffix Automaton of a string w, noted $\mathbf{S}(w)$, is the minimal deterministic automaton that accepts the set of suffixes of w (see [12]). Its construction makes use of a table defined on its states, noted F, and known as the failure link of $\mathbf{S}(w)$. We also consider the length function L. They are informally defined as follows. If state q is associated with the nonempty string u, $F[q]$ is the state associated with the longest suffix of u leading from the initial state to a state different from q. And $L[q]$ denotes maximal length of labels of paths from the initial state to q.

Using a Suffix Automaton to compute LPrF. A solution to compute the LPrF table of the input string y in linear time is designed with the Suffix Automaton $\mathcal{S}(y^{\mathrm{R}})$. The structure includes the failure link F and the table L recalled above as well as an additional attribute on states SC described below. Figure 1 displays the Suffix Automaton of babaababaa used for computing the LPrF table of the string aababaabab.

The next table gives the attributes (F, L and SC) of the states of the automaton displayed in Figure 1:

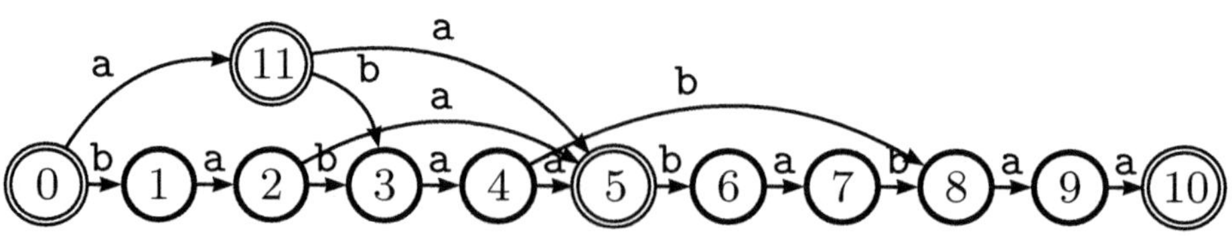

Fig. 1. The Suffix Automaton of babaababaa, reverse of aababaabab

state q	0	1	2	3	4	5	6	7	8	9	10	11
$F[q]$	0	0	11	1	2	11	3	4	3	4	5	0
$L[q]$	0	1	2	3	4	5	6	7	8	9	10	1
$SC[q]$	0	2	1	2	1	0	4	3	2	1	0	0

Here is how the computation is carried on according to the algorithm described below. At a given step, i is a position on y, ℓ is the length of the current match, and q is the current state of the automaton. The principal invariant of the computation is the equality $\delta(\mathit{initial}, y[i-\ell\,..\,i-1]) = q$ where δ denotes the transition function of the automaton and *initial* its initial state.

The condition to extend the match by the letter $a = y[i]$ is that $\delta(q,a)$ is defined and that $y[i-\ell\,..\,i-1]a$ occurs in y^{R} at a position at least as large as $n-i+\ell$. This can be tested efficiently on the automaton if the table SC is available. For a state r, $SC[r]$ is the minimal length of labels of paths from r to a terminal state. The table can be pre-processed via a mere traversal of the automaton. The test becomes $i-\ell \leq \ell+1+SC[\delta(q,a)]$ where the first member is the length of $y[0\,..\,i-\ell-1]$ and the second member the minimal length of suffixes of y^{R} starting with the next match.

When the test is negative, the failure link F is applied to shorten the match whose length is given by L. None of the suffixes of the match of length larger than $L[F[q]]$, which all correspond to the same state q, is able to change the value of the test in line 4. Then, a batch of LPrF values are computed in lines 8–10.

In the code below we assume that $F[\mathit{initial}] = \mathit{initial}$. The value of $F[\mathit{initial}]$ is usually left undefined for Suffix Automata but the assumption simplifies the presentation of the algorithm.

```
LPrF-AUTOMATON(y, n)
 1  (q, ℓ) ← (initial, 0)
 2  i ← 0
 3  repeat a ← y[i]
 4           while (i < n) and (δ(q, a) ≠ NULL)
               and ((i − ℓ) ≥ ℓ + 1 + SC[δ(q, a)]) do
 5                  (q, ℓ) ← (δ(q, a), ℓ + 1)
 6                  i ← i + 1
 7                  a ← y[i]
 8           repeat LPrF[i − ℓ] ← ℓ
 9                  ℓ ← max{0, ℓ − 1}
10           until ℓ = L[F[q]]
11           if q ≠ initial then
12                  q ← F[q]
13           else   i ← i + 1
14  until (i = n) and (ℓ = 0)
15  return LPrF
```

Theorem 6. *The algorithm* LPrF-AUTOMATON *computes the* LPrF *table of a string of length n in time $O(n)$ on a fixed-size alphabet.*

For the computation of genetic palindromes in which the left part in not only reverse but also complemented, a variant of the LPrF table is introduced accordingly (see [7]). The corresponding computation uses the Suffix Automaton $\mathbf{S}(z)$, where z is the reverse complement of the input y.

Note that if the automaton is implemented in linear space on a potentially infinite alphabet, the overall algorithm runs in $O(n \log a)$ time. But the running time can be reduced to $O(n)$ on an integer alphabet using the Suffix Array of the input y and Range Minimum Queries data structures (see [19]).

References

1. Apostolico, A., Breslauer, D.: Of periods, quasiperiods, repetitions and covers, pp. 236–248 (1997)
2. Apostolico, A., Preparata, F.P.: Optimal off-line detection of repetitions in a string. Theoret. Comput. Sci. 22(3), 297–315 (1983)
3. Badkobeh, G.: Fewest repetitions vs maximal-exponent powers in infinite binary words (2011) (submitted)
4. Badkobeh, G., Crochemore, M.: Bounded number of squares in infinite repetition-constrained binary words. In: Holub, J., Zd'árek, J. (eds.) Prague Stringology Conference, pp. 161–166. Czech Technical University in Prague (2010) ISBN 978-80-01-04597-8
5. Bell, T.C., Clearly, J.G., Witten, I.H.: Text Compression. Prentice Hall Inc., New Jersey (1990)
6. Böckenhauer, H.-J., Bongartz, D.: Algorithmic Aspects of Bioinformatics. Springer, Berlin (2007)
7. Chairungsee, S., Crochemore, M.: Efficient computing of longest previous reverse factors. In: Shoukourian, Y. (ed.) Seventh International Conference on Computer Science and Information Technologies (CSIT 2009), pp. 27–30. The National Academy of Sciences of Armenia Publishers, Yerevan (2009)
8. Chen, G., Puglisi, S.J., Smyth, W.F.: Fast and practical algorithms for computing all the runs in a string. In: Ma, B., Zhang, K. (eds.) CPM 2007. LNCS, vol. 4580, pp. 307–315. Springer, Heidelberg (2007)
9. Crochemore, M.: An optimal algorithm for computing the repetitions in a word. Inf. Process. Lett. 12(5), 244–250 (1981)
10. Crochemore, M.: Transducers and repetitions. Theoretical Computer Science 45(1), 63–86 (1986)
11. Crochemore, M., Fazekas, S.Z., Iliopoulos, C., Jayasekera, I.: Number of occurrences of powers in strings. International Journal of Foundations of Computer Science 21(4), 535–547 (2010)
12. Crochemore, M., Hancart, C., Lecroq, T.: Algorithms on Strings. Cambridge University Press, Cambridge (2007)
13. Crochemore, M., Ilie, L.: Analysis of maximal repetitions in strings. In: Kučera, L., Kučera, A. (eds.) MFCS 2007. LNCS, vol. 4708, pp. 465–476. Springer, Heidelberg (2007)

14. Crochemore, M., Ilie, L.: Computing longest previous factors in linear time and applications. Information Processing Letters 106(2), 75–80 (2008), doi:10.1016/j.ipl.2007.10.006
15. Crochemore, M., Ilie, L.: Maximal repetitions in strings. J. Comput. Syst. Sci. 74(5), 796–807 (2008)
16. Crochemore, M., Ilie, L., Iliopoulos, C., Kubica, M., Rytter, W., Waleń, T.: LPF computation revisited. In: Fiala, J., Kratochvíl, J., Miller, M. (eds.) IWOCA 2009. LNCS, vol. 5874, pp. 158–169. Springer, Heidelberg (2009)
17. Crochemore, M., Ilie, L., Smyth, W.F.: A simple algorithm for computing the Lempel-Ziv factorization. In: Storer, J.A., Marcellin, M.W. (eds.) 18th Data Compression Conference, March 25-27, pp. 482–488. IEEE Computer Society, Los Alamitos (2008)
18. Crochemore, M., Ilie, L., Tinta, L.: The "runs" conjecture. In: de Felice, C., Carpi, A. (eds.) Theoretical Computer Science (2010) (in press, corrected proof)
19. Crochemore, M., Iliopoulos, C., Kubica, M., Rytter, W., Waleń, T.: Efficient algorithms for two extensions of LPF table: The power of suffix arrays. In: van Leeuwen, J., Muscholl, A., Peleg, D., Pokorný, J., Rumpe, B. (eds.) SOFSEM 2010. LNCS, vol. 5901, pp. 296–307. Springer, Heidelberg (2010)
20. Crochemore, M., Rytter, W.: Squares, cubes and time-space efficient string-searching. Algorithmica 13(5), 405–425 (1995)
21. Dejean, F.: Sur un théorème de Thue. J. Comb. Theory, Ser. A 13(1), 90–99 (1972)
22. Dekking, F.M.: On repetitions of blocks in binary sequences. J. Comb. Theory, Ser. A 20(3), 292–299 (1976)
23. Fraenkel, A.S., Simpson, J.: How many squares must a binary sequence contain? Electr. J. Comb. 2 (1995)
24. Fraenkel, A.S., Simpson, J.: How many squares can a string contain? J. Comb. Theory, Ser. A 82(1), 112–120 (1998)
25. Franek, F., Smyth, W.F., Tang, Y.: Computing all repeats using suffix arrays. Journal of Automata, Languages and Combinatorics 8(4), 579–591 (2003)
26. Franek, F., Yang, Q.: An asymptotic lower bound for the maximal-number-of-runs function. In: Holub, J., Zdárek, J. (eds.) Proceedings of the Prague Stringology Conference. Department of Computer Science and Engineering, Faculty of Electrical Engineering, pp. 3–8. Czech Technical University (2006)
27. Giraud, M.: Not so many runs in strings. In: Martin-Vide, C. (ed.) 2nd International Conference on Language and Automata Theory and Applications (2008)
28. Gusfield, D., Stoye, J.: Linear time algorithms for finding and representing all the tandem repeats in a string. J. Comput. Syst. Sci. 69(4), 525–546 (2004)
29. Harju, T., Nowotka, D.: Binary words with few squares. Bulletin of the EATCS 89, 164–166 (2006)
30. Ilie, L.: A simple proof that a word of length has at most 2 distinct squares. J. Comb. Theory, Ser. A 112(1), 163–164 (2005)
31. Ilie, L.: A note on the number of squares in a word. Theor. Comput. Sci. 380(3), 373–376 (2007)
32. Iliopoulos, C.S., Moore, D., Smyth, W.F.: A characterization of the squares in a Fibonacci string. Theoret. Comput. Sci. 172(1-2), 281–291 (1997)
33. Karhumäki, J., Shallit, J.: Polynomial versus exponential growth in repetition-free binary words. J. Comb. Theory, Ser. A 105(2), 335–347 (2004)
34. Kolpakov, R., Kucherov, G.: Finding maximal repetitions in a word in linear time. In: Proceedings of the 40th IEEE Annual Symposium on Foundations of Computer Science, pp. 596–604. IEEE Computer Society Press, New York (1999)

35. Kolpakov, R., Kucherov, G.: Searching for gapped palindromes. In: Ferragina, P., Landau, G.M. (eds.) CPM 2008. LNCS, vol. 5029, pp. 18–30. Springer, Heidelberg (2008)
36. Lothaire, M. (ed.): Combinatorics on Words, 2nd edn. Cambridge University Press, Cambridge (1997)
37. Lothaire, M. (ed.): Algebraic Combinatorics on Words. Cambridge University Press, Cambridge (2001)
38. Lothaire, M. (ed.): Appplied Combinatorics on Words. Cambridge University Press, Cambridge (2005)
39. MacDonald, M., Ambrose, C.M.: A novel gene containing a trinucleotide repeat that is expanded and unstable on huntington's disease chromosomes. Cell 72(6), 971–983 (1993)
40. Main, M.G.: Detecting leftmost maximal periodicities. Discret. Appl. Math. 25, 145–153 (1989)
41. Main, M.G., Lorentz, R.J.: An $O(n \log n)$ algorithm for finding all repetitions in a string. J. Algorithms 5(3), 422–432 (1984)
42. Matsubara, W., Kusano, K., Ishino, A., Bannai, H., Shinohara, A.: New lower bounds for the maximum number of runs in a string. In: Holub, J., Zdárek, J. (eds.) Proceedings of the Prague Stringology Conference. Prague Stringology Club, Department of Computer Science and Engineering, Faculty of Electrical Engineering, pp.140–145. Czech Technical University in Prague (2008)
43. Ochem, P.: A generator of morphisms for infinite words. ITA 40(3), 427–441 (2006)
44. Pansiot, J.J.: The morse sequence and iterated morphisms. Inf. Process. Lett. 12(2), 68–70 (1981)
45. Puglisi, S.J., Simpson, J., Smyth, W.F.: How many runs can a string contain? Theor. Comput. Sci. 401(1-3), 165–171 (2008)
46. Rampersad, N., Shallit, J., Wei Wang, M.: Avoiding large squares in infinite binary words. Theor. Comput. Sci. 339(1), 19–34 (2005)
47. Rao, M.: Last cases of Dejean's conjecture. In: Carpi, A., de Felice, C. (eds.) WORDS 2009. University of Salerno, Italy (2009)
48. Rytter, W.: The number of runs in a string: Improved analysis of the linear upper bound. In: Durand, B., Thomas, W. (eds.) STACS 2006. LNCS, vol. 3884, pp. 184–195. Springer, Heidelberg (2006)
49. Rytter, W.: The number of runs in a string. Inf. Comput. 205(9), 1459–1469 (2007)
50. Séébold, P.: Sur les morphismes qui engendrent des mots infinis ayant des facteurs prescrits, pp. 301–311 (1983)
51. Shallit, J.: Simultaneous avoidance of large squares and fractional powers in infinite binary words. Intl. J. Found. Comput. Sci. 15, 317–327 (2004)
52. Simpson, J.: Modified Padovan words and the maximum number of runs in a word. Australasian J. of Comb. 46, 129–145 (2010)
53. Thue: Uber unendliche zeichenreihen. Norske vid. Selsk. Skr. I. Mat. Nat. Kl. Christiana 7, 1–22 (1906)
54. Witten, I.H., Moffat, A., Bell, T.C.: Managing Gigabytes. Van Nostrand Reinhold (1994)
55. Ziv, J., Lempel, A.: A universal algorithm for sequential data compression. IEEE Transactions on Information Theory, 337–343 (1977)

Some Remarks on Automata Minimality

Antonio Restivo and Roberto Vaglica

Dipartimento di Matematica e Informatica,
Università degli Studi di Palermo
Via Archirafi 34, 90123 Palermo, Italy
{restivo,vaglica}@math.unipa.it

1 Introduction

It is well known that the minimization problem of deterministic finite automata (*DFAs*) is related to the indistinguishability notion of states (cf. [HMU00]). Indeed, a well known technique to minimize a DFA, essentially, consists in finding pairs of states that are equivalent (or *indistinguishable*), namely pairs of states (p, q) such that it is impossible to assert the difference between p and q only by starting in each of the two states and asking whether or not a given input string leads to a final state. Since, in the testing states equivalence, the notion of initial state is irrelevant, some of the main techniques for the minimization of automata, such as Moore's algorithm [Moo56] and Hopcroft's algorithm [Hop71], do not care what is the initial state of the automaton, when applied to accessible automata (i.e. such that all states can be reached from the initial state). Therefore a natural question that arises is, for accessible automata, on what does minimality depend? Obviously, it depends on both the automata transitions and the set of final states. In this paper, our main focus is to investigate to what extent minimality depends on the particular subset of final states.

In order to investigate the dependence of minimality of the automaton on the choice of final states, we follow a graph-theoretic approach. For any DFA $\mathcal{A}$, we introduce the *state-pair graph* $G(\mathcal{A})$, having as set of vertices the family $[Q]^2$ of subsets of Q of cardinality 2. We show that, for any choice of the set F of final states, the minimality of $\mathcal{A}$ corresponds to the property that F *separates* all *closed* components of $G(\mathcal{A})$. A set $F \subset Q$ separates $S \subseteq [Q]^2$ if there exists $\{p, q\} \in S$ such that $p \in F$ and $q \notin F$. In this way, in order to check whether $\mathcal{A}$ is minimal with respect to various sets of final states, we need to compute the closed components of $G(\mathcal{A})$ only once, and then test, for each $F \subset Q$, whether F separates such closed components.
In this paper we consider *strongly connected complete* automata $\mathcal{A} = (Q, \Sigma, \delta)$ and, in order to exclude the trivial case, we choose the sets F of final states among the *proper* subsets of Q.

The effectiveness of our approach is shown by analyzing some extremal cases. We introduce the family of *uniformly minimal* automata, i.e. automata which are minimal for any choice of the set of final states. The definitions is consistent: indeed we show that for each positive integer n there exist uniformly minimal automata with n states. We provide a characterization of such a family of automata in terms of closed components the corresponding state-pair graphs, from

G. Mauri and A. Leporati (Eds.): DLT 2011, LNCS 6795, pp. 15–27, 2011.

which one derives a polynomial algorithm to decide whether a given DFA is uniformly minimal. Later we consider the opposite extremal case, i.e. automata that are *never-minimal*, for any choice of the set of final states. Also in this case we prove that there exists an infinite family of never-minimal automata. However, contrary to the previous cases, does not exist a polynomial algorithm for checking whether an automaton is never-minimal, as shown in [RS11]. In the last part of the paper the minimization properties of a DFA are analyzed by looking at its transformation monoid. In particular, we show some connections between the problem to decide whether an automaton is never-minimal and the *syntactic monoid problem* of [GK98]. Synchronization properties of the automaton seem to play a role in such investigation.

2 Basic Definitions and Notation

In this section we recall the basic definitions and fix the notation used in this work. A deterministic finite automaton (DFA) is a triple $\mathcal{A} = (Q, \Sigma, \delta)$ where Q is a finite set of states, Σ is a finite alphabet of input symbols and δ is a map from $Q \times \Sigma$ to Q called the *transition function* of the automaton. The action of the letters in Σ on the states in Q can be extended in a natural way to Σ^*, where Σ^* is the free monoid over the alphabet Σ; this extension is here denoted by δ^*. If the transition function δ is a total function, then we say that the DFA is *complete*. An automaton $\mathcal{A} = (Q, \Sigma, \delta)$ is said *strongly connected* if for every ordered pair of states $q, q' \in Q$ there exists $w \in \Sigma^*$ such that $\delta^*(q, w) = q'$.
A *synchronizing* word for a DFA $\mathcal{A}$ is a sequence of symbols in the input alphabet which sends any state of $\mathcal{A}$ to the same state. A DFA is said to be *synchronizing* if it has a synchronizing word. In this paper, when not specified, we consider only complete and strongly connected automata. If we fix a state $i \in Q$ as *initial* state and a non-empty subset $F \subseteq Q$ as set of *final* states for the automaton $\mathcal{A}$, then we say that the automaton recognizes a language. The language recognized by $\mathcal{A}(i, F)$ is the set $L(\mathcal{A}(i, F)) = \{w \in \Sigma^* : \delta^*(i, w) \in F\}$. The class of languages recognized by DFAs is known as the class of *regular* languages.

Two automata that recognize the same language are called *equivalent*. Finally a DFA is *minimal* if it has the minimum number of states among all its equivalent DFAs. For any finite deterministic automaton $\mathcal{A}(i, F)$ there is a unique (up to labeling of the states) minimal automaton that recognizes the same language as the automaton $\mathcal{A}(i, F)$. As already mentioned in the introduction, the minimal automaton equivalent to a given DFA, can be computed essentially by using the *indistinguishable equivalence I*. More precisely, we say that two states p and q are indistinguishable if, for all input strings w, $\delta^*(p, w) \in F$ iff $\delta^*(q, w) \in F$.

3 Some Graph-Theoretic Tools

In this section we introduce some graph-theoretic tools, whose usefulness in our context will be clear shortly. Given an arbitrary set X, we denote by $[X]^k$ the family of subsets of X of cardinality k:

$$[X]^k = \{Y \subset X \mid |X| = k\}.$$

Definition 1. *Given a deterministic finite automaton $\mathcal{A} = (Q, \Sigma, \delta)$, the state-pair graph of $\mathcal{A}$ is the directed graph $G(\mathcal{A}) = (V_G, E_G)$ defined as follows:*

i. $V_G = [Q]^2$;
ii. $E_G = \{(\{p, q\}, \{p', q'\}) \mid \delta(p, a) = p', \delta(q, a) = q', a \in \Sigma\}$.

An example of state-pair graph of an automaton is given in Fig. 1.

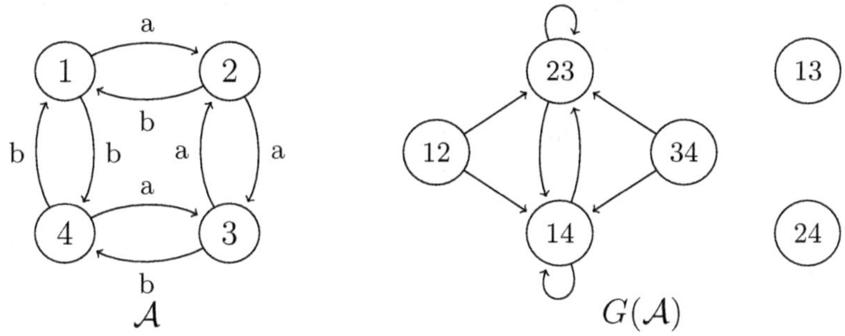

Fig. 1. A DFA $\mathcal{A}$ and the corresponding state-pair graph $G(\mathcal{A})$

As regards the complexity of the state-pair graph $G(\mathcal{A})$, one can verify that:

$$|V_G| = \binom{|Q|}{2}$$

and

$$|E_G| \leq |\Sigma| \cdot |V_G|.$$

The following definition plays an important role in our paper.

Definition 2. *Given a set $F \subseteq Q$ and a set $S \subseteq [Q]^2$, we say that F* separates *S if there exists $\{p, q\} \in S$ such that $p \in F$ and $q \notin F$.*

Another useful notion for our investigation is that of closed component of a directed graph.

Definition 3. *A* closed component *of a directed graph G is a subset S of the set of the vertices of G such that*

- *there exists a path from any element of S to any other element of S (i.e. S is a strongly connected component), and*
- *there is no outgoing edge from one element of S to a vertex of G which is not in S.*

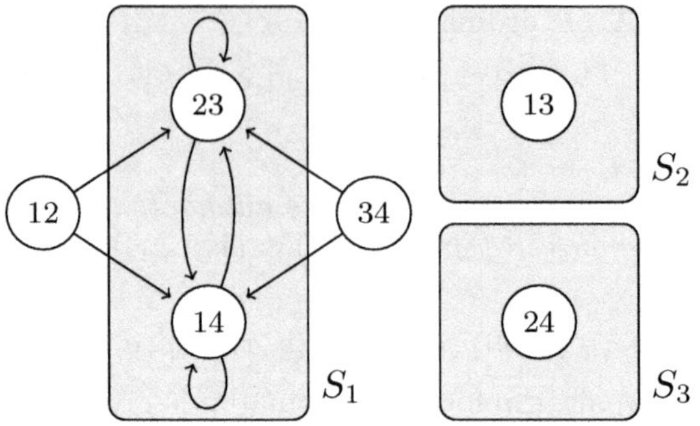

Fig. 2. The state-pair graph $G(\mathcal{A})$ and its closed components marked: $S_1 = \{\{1,4\},\{2,3\}\}, S_2 = \{\{1,3\}\}, S_3 = \{\{2,4\}\}$

The set of all closed components of $G(\mathcal{A})$ is here denoted by $\mathfrak{S}(\mathcal{A})$. Figure 2 shows the state-pair graph $G(\mathcal{A})$ with the closed components marked.

Given an automaton $\mathcal{A}$ such that $\mathfrak{S}(\mathcal{A}) = \{S_1, S_2, ..., S_k\}$ for some integer k, we denote by $\|\mathfrak{S}(\mathcal{A})\|$ the number of vertices of $G(\mathcal{A})$ that are in some closed component:

$$\|\mathfrak{S}(\mathcal{A})\| = \sum_{i=1}^{k} |S_i|.$$

Lemma 1. *If $\mathcal{A} = (Q, \Sigma, \delta)$ is a strongly connected DFA, one has*

$$1 \leq \|\mathfrak{S}(\mathcal{A})\| \leq \binom{|Q|}{2},$$

and the bounds are tight.

The tightness of the upper bound in the previous lemma is proved by the existence of automata having a strongly connected state-pair graph (see for instance the example after Proposition 1). The tightness of the lower bound is obtained by the following infinite family $\{\mathfrak{D}_n\}$ of automata. The automaton $\mathfrak{D}_n$ has $Q = \{1, 2, ..., n\}$ as set of states, $\Sigma = \{a, b\}$ as symbol of input letters and the transition function defined as follows:

$$\delta(i,a) = \begin{cases} i+1, & \text{if } i < n, \\ i, & \text{if } i = n; \end{cases} \qquad \delta(i,b) = 1, \forall i \in Q.$$

The following result follows from the basic properties of minimal DFAs.

Theorem 1. *Let $\mathcal{A}(i, F)$ be an accessible DFA. Then $\mathcal{A}(i, F)$ is minimal if and only if F separates all the closed components of $G(\mathcal{A})$.*

Remark 1. The interest of Theorem 1 is that, in order to check whether $\mathcal{A}$ is minimal with respect to various sets of final states, we need to compute the closed components of $G(\mathcal{A})$ only once, and then test, for each $F \subset Q$, whether F separates such closed component. Moreover, the advantage in considering only the closed components of $G(\mathcal{A})$ relies on the fact that, by Lemma 1, we have that $\|\mathfrak{S}(\mathcal{A})\|$ can be much smaller than the cardinality of $G(\mathcal{A})$.

As an example, with reference to the automaton of Fig. 1, we have that the set of final states $F = \{1, 2\}$ separates all the elements of $\mathfrak{S}(\mathcal{A})$ (see Fig. 2). Thus, since $\mathcal{A}(i, F)$ is accessible for all $i \in \{1, ..., 4\}$ (note that $\mathcal{A}$ is strongly connected), $\mathcal{A}(i, \{1, 2\})$ is minimal for all i. On the contrary, the set $F = \{1\}$ doesn't separate the closed component $S_3 = \{\{2, 4\}\}$, hence $\mathcal{A}(i, \{1\})$ is not a minimal DFA.

Given a set $S \subseteq [Q]^2$, let $Q_S = \{q \in Q \mid \{p, q\} \in S\}$ be the set of states involved in S. To each closed component $S \in \mathfrak{S}(\mathcal{A})$ one can associate the (undirected) graph H_S having Q_S as set of vertices and S as set of edges. Figure 3 illustrates the graphs associated to the closed components S_1, S_2 and S_3 of $\mathfrak{S}(\mathcal{A})$. The graphs H_S play an important role in the characterization of uniformly minimal automata (of Theorem 2).

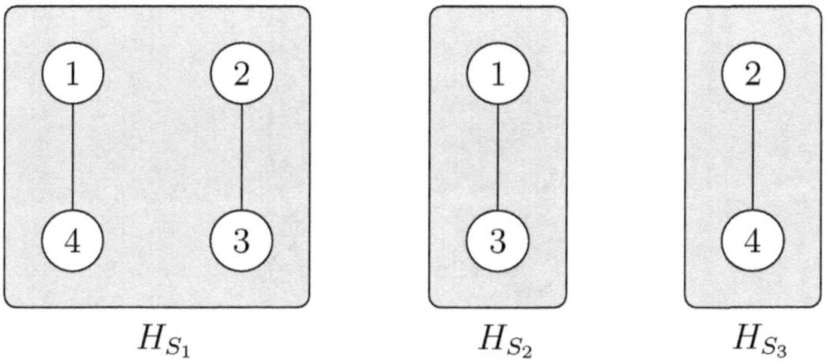

Fig. 3. The graphs H_{S_1}, H_{S_2} and H_{S_3} associated to the elements of $\mathfrak{S}(\mathcal{A})$

4 Uniformly Minimal Automata

In this section we deal with the family of uniformly minimal automata which was introduced firstly in [RV10] (cf. also [RV11]).

Definition 4. *A strongly connected automaton* $\mathcal{A} = (Q, \Sigma, \delta)$ *is called* uniformly minimal *if, for all* proper *subsets* $F \subset Q$ *and* $i \in Q$, $\mathcal{A}(i, F)$ *is minimal.*

Remark 2. We wish to point out that in [RV10] the automata of Definition 4 were called "almost uniformly minimal". The justification for this different terminology is due to the fact that in that work we considered non-necessary complete automata. Thus, in [RV10], we called uniformly minimal the deterministic automata which are minimal for all choices of the set of final states, included the choice $F = Q$.

The following statement is a consequence of the previous definition and of Theorem 1.

Proposition 1. *Given a strongly connected complete automaton $\mathcal{A}$, if the state-pair graph $G(\mathcal{A})$ is strongly connected, then $\mathcal{A}$ is uniformly minimal.*

We observe that a well-known family of uniformly minimal automata is that of Černý automata $\{\mathfrak{C}_n\}$, namely the infinite family of n-state synchronizing automata such that the length of the shortest synchronizing word reaches the bound $(n-1)^2$ of Černý's conjecture (cf. [Č64]). The automaton $\mathfrak{C}_n$ has $Q = \{1, 2, ..., n\}$ as set of states, $\Sigma = \{a, b\}$ as alphabet of input letters and the transition function defined as follows:

$$\delta(q,a) = \begin{cases} 1 & \text{if } q = n, \\ \mathrm{q} & \text{if } q < n; \end{cases} \qquad \delta(q,b) = \begin{cases} 1 & \text{if } q = n, \\ \mathrm{q{+}1} & \text{if } q < n. \end{cases}$$

The automaton $\mathfrak{C}_n$ is illustrates in Fig. 4.

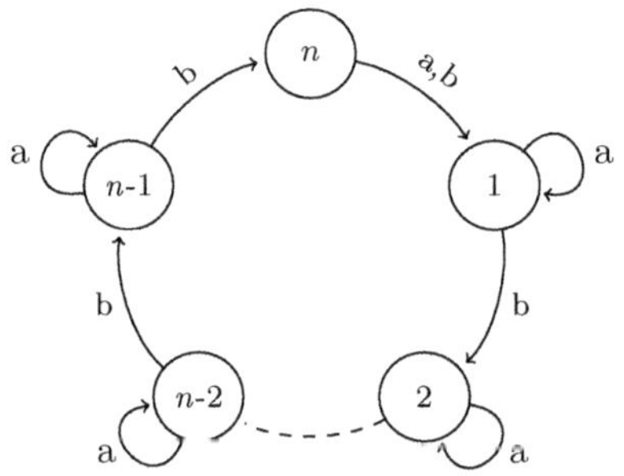

Fig. 4. The Černý automaton $\mathfrak{C}_n$

For each of these automata, it is not hard to see that the corresponding state-pair graph is strongly connected. Thus, by Proposition 1, every Černý automaton is uniformly minimal. Contrary to what we might have expected, synchronization doesn't seem to play a role for the uniform minimality of an automaton. Indeed, the uniformly minimal automata considered in [RV10] belong to an infinite family $\{\mathcal{M}_n\}$ of non synchronizing uniformly minimal automata.

Remark 3. We point out that the condition given by Proposition 1 is not a necessary condition, as shown by the example in Fig. 5.

The following result provides a characterization of uniformly minimal automata in terms of the graphs H_S associated to the closed components $S \in \mathfrak{S}(\mathcal{A})$.

Theorem 2. *A strongly connected automaton $\mathcal{A} = (Q, \Sigma, \delta)$ is uniformly minimal if and only if, for any closed component S of $G(\mathcal{A})$, $Q_S = Q$ and the graph H_S is connected.*

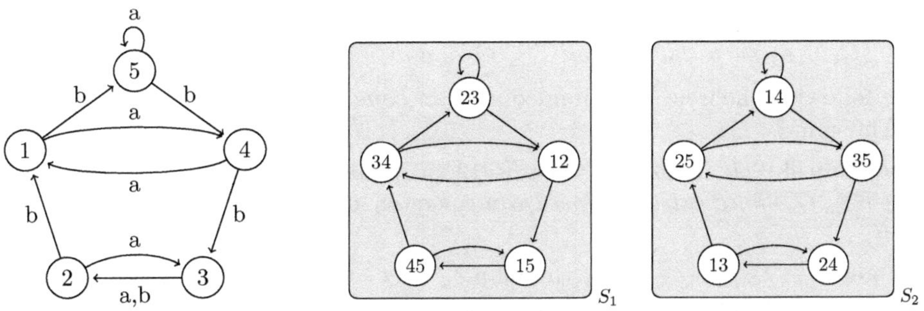

Fig. 5. A uniformly minimal automaton (on the left) and its state-pair graph having the two closed components $S_1 = \{\{1,2\},\{2,3\},\{3,4\},\{4,5\},\{1,5\}\}$ and $S_2 = \{\{1,4\},\{3,5\},\{2,4\},\{1,3\},\{2,5\}\}$

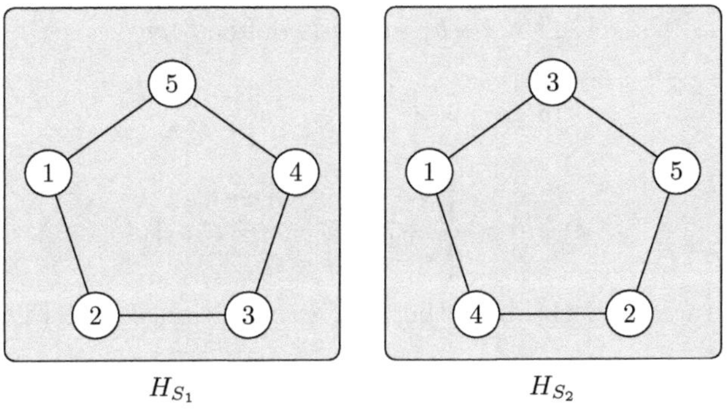

Fig. 6. The graphs associated to the closed components S_1 and S_2 of the state pair-graph of the automaton depicted in Fig. 5

As a consequence of the above result we have the following corollary.

Corollary 1. *There exists a polynomial-time algorithm to test the uniform minimality of an automaton.*

Remark 4. The statement in Corollary 1 is not at all trivial. Indeed, in order to test, by a naive method, the uniform minimality of a DFA $\mathcal{A} = (Q, \Sigma, \delta)$, one has to test its minimality for all $F \subset Q$, i.e. one has to repeat exponentially many times a classical minimization algorithm.

5 Never-Minimal Automata

In this section we consider automata defined by the opposite extremal minimality condition, namely the *never-minimal* automata (cf.[RV10, RV11]).

Definition 5. *A* never-minimal *automaton is a strongly connected automaton* $\mathcal{A} = (Q, \Sigma, \delta)$ *such that, for all* $F \subseteq Q$ *and* $i \in Q$, $\mathcal{A}(i, F)$ *is not minimal.*

The following theorem is an immediate consequence of the above definition and of Theorem 1.

Theorem 3. *The automaton* $\mathcal{A} = (Q, \Sigma, \delta)$ *is never minimal if and only if, for any* $F \subseteq Q$*, there exists a closed component* S *of* $G(\mathcal{A})$ *which is not separated by* F*.*

For any $a \in \Sigma$, denote by δ_a the map $\delta_a : Q \to Q$ defined by $\delta_a(q) = \delta(q, a)$, for all $q \in Q$. The following proposition gives a sufficient condition for a an automaton to be never-minimal.

Proposition 2. *Let* $\mathcal{A}$ *be a DFA. Suppose that there exists a set* $P = \{p, q, r\}$ *of three states such that, for all* $a \in \Sigma$*, the restriction of* δ_a *to* P *is either a constant or the identity function. Then* $\mathcal{A}$ *is never-minimal.*

The above result enables us to construct an infinite family of never-minimal automata. For instance, in [RV10] we have defined the family $\{\mathcal{N}_n = (Q_n, \Sigma, \delta)\}_{n \geq 4}$, where $Q_n = \{1, 2, ..., n\}$, $\Sigma = \{a, b\}$ and δ is defined by

$$\delta(i, a) = \begin{cases} 1, & \text{if } i \leq 3; \\ \text{i-1}, & \text{if } 4 \leq i \leq n; \end{cases}$$

$$\delta(i, b) = \begin{cases} 4, & \text{if } i \leq 3; \\ \text{i+1}, & \text{if } 3 < i \leq n-1; \\ 2, & \text{if } i = n, \end{cases}$$

whose automata satisfy the hypothesis of Proposition 2. See Fig. 7 for the automaton $\mathcal{N}_6$.

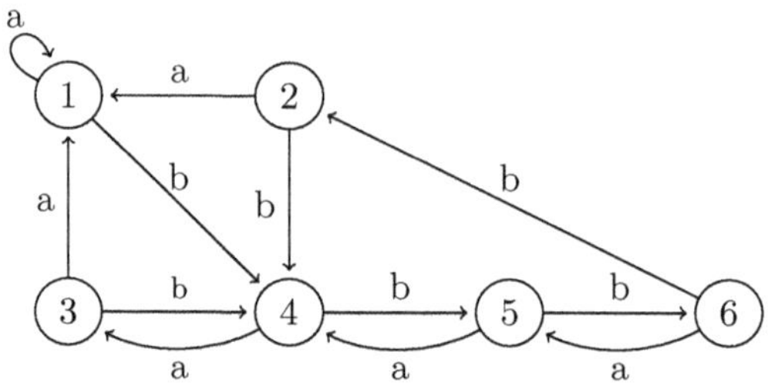

Fig. 7. The automaton $\mathcal{N}_6$

Remark that the condition of Proposition 2 is not necessary, as shown by the never-minimal automaton given in Fig. 8. In [RV10] we raised the question whether $NEVER - MINIMAL$ problem, that is the problem of establish if a given automaton is never-minimal or not, can be solved in polynomial time (as for uniform minimality). A negative answer has been recently given by Rodaro and Silva in [RS11], were they prove the following theorem.

Theorem 4 (Rodaro, Silva). *co-NEVER-MINIMAL is NP-complete.*

6 Automata Minimality and Transformation Monoid

Given a language $L \subseteq \Sigma^*$, the *Nerode equivalence* is defined as follows for two strings $x, y \in \Sigma^*$:

$$x \sim_L y \Leftrightarrow \forall z \in \Sigma^* \ (xz \in L \Leftrightarrow yz \in L) .$$

The Nerode equivalence is *right invariant*, namely $u \sim_L v$ implies $(\forall x)\ ux \sim_L vx$, but it not *left invariant*. So it is not a *congruence*, the latter being defined as an equivalence which is both right and left invariant. In addiction, Nerode equivalence is linked to the *syntactic congruence* σ_L, that is the coarsest congruence on Σ^* saturating L, defined by:

$$x \sigma_L y \ \Leftrightarrow \ \forall u, v \in \Sigma^* \ (uxv \in L \ \Leftrightarrow \ uyv \in L).$$

An important theorem (Myhill–Nerode) says that a language L is regular (i.e., recognized by a DFA) if and only if the number of equivalence classes of σ_L is finite. Moreover, given a regular language L, the quotient Σ^*/σ_L is a (finite) monoid, called the *syntactic monoid* of L.

It is also possible to associate to every automaton $\mathcal{A} = (Q, \Sigma, \delta)$, where the initial and final states are not specified, a finite monoid, called the *transformation monoid* of $\mathcal{A}$. The transformation monoid of $\mathcal{A}$ is the quotient monoid $M(\mathcal{A}) = \Sigma^*/\gamma_\mathcal{A}$ where $\gamma_\mathcal{A}$ is the congruence on Σ^* defined by

$$x \gamma_\mathcal{A} y \Leftrightarrow (\forall q \in Q) \ \delta^*(q, x) = \delta^*(q, y).$$

In addiction, if we choose an initial state i and a set of final states F for the automaton $\mathcal{A}$, also $\gamma_\mathcal{A}$ saturates $L = L(\mathcal{A}(i, F))$. Thus we have that σ_L is coarser than $\gamma_\mathcal{A}$ (namely $x \gamma_\mathcal{A} y$ implies $x \sigma_L y$) and the syntactic monoid of a regular language L coincides with the transformation monoid of the minimal DFA recognizing L. In other words, if $\mathcal{A}(i, F)$ is the minimal DFA recognizing L then $\gamma_\mathcal{A} = \sigma_L$.

Remark that the transformation monoid $M(\mathcal{A})$ only depends on the triple (Q, Σ, δ) that defines the automaton $\mathcal{A}$, and it does not depend on the choice of the initial state $i \in Q$ and on the set $F \subset Q$ of final states. However, if $\mathcal{A}(i, F)$ is minimal for some choice of $i \in Q$ and $F \subset Q$, then $M(\mathcal{A})$ is the syntactic monoid of the language $L(\mathcal{A}(i, F))$. In the case the automaton $\mathcal{A} = (Q, \Sigma, \delta)$ is never minimal, can it happen that the monoid $M(\mathcal{A})$ is the syntactic monoid of some language L? Such a question relates our investigation to the "syntactic monoid problem" introduced in [GK98]. Indeed, it is also possible to introduce the notion of syntacticity of monoids without referring to any language. More precisely, if M is a finite monoid and P a subset of M, there is a largest congruence σ_P saturating P defined by:

$$x \sigma_P y \ \Leftrightarrow \ \forall s, t \in M \ (sxt \in P \ \Leftrightarrow \ syt \in P).$$

The set P is called *disjunctive* if σ_P is the equality in M. A monoid M is *syntactic* if it has a disjunctive subset. In the case of the syntactic monoid of a language L, a disjunctive set is $\{[w] \, | w \in L\} \subseteq \Sigma^*/\sigma_L$.

The *syntactic monoid problem* is to decide whether a finite monoid is syntactic. It is an open problem whether the syntactic monoid problem is polynomial or not. More precisely, in [GK98], Pavel Goralcik and Václav Koubek give a polynomial-time algorithm ($O(|M|^3)$) solving the syntactic monoid problem for a large class of finite monoids and show that a slide generalization of syntactic monoid problem makes it NP-complete. Thus, they conclude that, probably, there is no chance to have a polynomial-time algorithm for the syntactic monoid problem.

The following proposition relates the syntactic monoid problem to the $NEVER - MINIMAL$ problem.

Proposition 3. *Let M be a monoid isomorphic to the transformation monoid of a DFA $\mathcal{A}$. If M is non syntactic then $\mathcal{A}$ is never-minimal.*

However, contrary to what we might have expected, this proposition cannot be reversed.

We show this fact by an example illustrating a never-minimal automaton whose transformation monoid coincides with the the syntactic monoid of some language L.

In order to facilitate the computation of the transformation monoid of an automaton $A = (Q, \Sigma, \delta)$, we observe that each word $w \in \Sigma^*$ defines a transformation $\gamma_w : Q \to Q$ such that, $\forall q \in Q$, $\gamma_w(q) = \delta^*(q, w)$ and each equivalence class of the congruence $\gamma_\mathcal{A}$ corresponds exactly to the set of words that perform the same transformation on Q. Thus, if $Q = \{1, 2, ..., n\}$, we may identify an element $[w] \in \Sigma^*/\gamma_\mathcal{A}$ with the transformation

$$\begin{pmatrix} 1 & 2 & \cdots & n \\ \gamma_w(1) & \gamma_w(2) & \cdots & \gamma_w(n) \end{pmatrix}.$$

Consider the never-minimal automaton $\mathcal{A}$ shown in Fig. 8. The elements of the transformation monoid $M(\mathcal{A})$ are given by

$$[\epsilon] = \begin{pmatrix} 1\,2\,3\,4\,5 \\ 1\,2\,3\,4\,5 \end{pmatrix}, [a] = \begin{pmatrix} 1\,2\,3\,4\,5 \\ 4\,1\,4\,5\,4 \end{pmatrix}, [b] = \begin{pmatrix} 1\,2\,3\,4\,5 \\ 2\,3\,2\,3\,2 \end{pmatrix}, [aa] = \begin{pmatrix} 1\,2\,3\,4\,5 \\ 5\,4\,5\,4\,5 \end{pmatrix},$$

$$[ab] = \begin{pmatrix} 1\,2\,3\,4\,5 \\ 3\,2\,3\,2\,3 \end{pmatrix}, [ba] = \begin{pmatrix} 1\,2\,3\,4\,5 \\ 1\,4\,1\,4\,1 \end{pmatrix}, [aaa] = \begin{pmatrix} 1\,2\,3\,4\,5 \\ 4\,5\,4\,5\,4 \end{pmatrix}, [aba] = \begin{pmatrix} 1\,2\,3\,4\,5 \\ 4\,1\,4\,1\,4 \end{pmatrix}.$$

The Cayley graph of $M(\mathcal{A})$ is illustrated in Fig. 9.

It happens that, if we choose in $\mathcal{A}$ as set of final states the set $F = \{5\}$, the reduction algorithm gives the automaton $\mathcal{B}$ in Fig. 10. Now the reader can verify that $M(\mathcal{B}) = M(\mathcal{A})$, i.e. the transformation monoid of the automaton $\mathcal{B}$ coincides with (is isomorphic to) the transformation monoid of $\mathcal{A}$, given in Fig. 9. Since $\mathcal{B}(\mathrm{A}, \{C\})$ is minimal, $M(\mathcal{B}) = M(\mathcal{A})$ is the syntactic monoid of

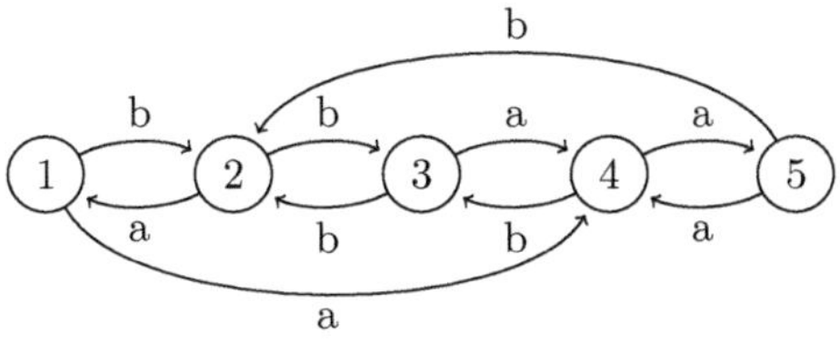

Fig. 8. A never-minimal automaton $\mathcal{A}$

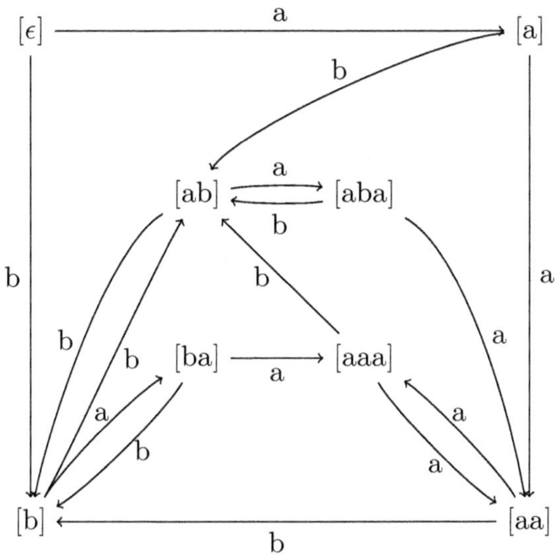

Fig. 9. The Cayley graph of $M(\mathcal{A})$

the language $L(\mathcal{B}(\text{A}, \{C\}))$. So we have shown that the automaton $\mathcal{A}$ in Fig. 8 is never-minimal and its transformation monoid $M(A)$ (given in Fig.9) is syntactic.

In general, given a DFA $\mathcal{A} = (Q, \Sigma, \delta)$ and a set $F \subset Q$ of final states, denote by $\mu_F(\mathcal{A})$ the automaton $\mathcal{B}$ that is obtained by applying the minimization algorithm to the automaton $\mathcal{A}(i, F)$, for some $i \in Q$. It is well known (cf. [How91]) that μ_F induces a morphism φ_F from the transformation monoid $M(\mathcal{A})$ to the transformation monoid $M(\mathcal{B})$, as illustrated by the following diagram:

Previous example shows that it can happen that $\mathcal{A} \neq \mathcal{B}$ and $M(\mathcal{A}) = M(\mathcal{B})$. As a consequence, a proper reduction of the size of an automaton, by merging some indistinguishable states, does not always lead to a reduction of the cardinality of the corresponding transformation monoid. The next lemma shows that the hypothesis that the automaton is *synchronizing* plays a special role in this context.

Lemma 2. *Let $\mathcal{A} = (Q, \Sigma, \delta)$ be a strongly connected synchronizing DFA, and let $\mathcal{B} = \mu_F(\mathcal{A})$ for some $F \subset Q$. If $\mathcal{B} \neq \mathcal{A}$, then $M(\mathcal{B}) \neq M(\mathcal{A})$.*

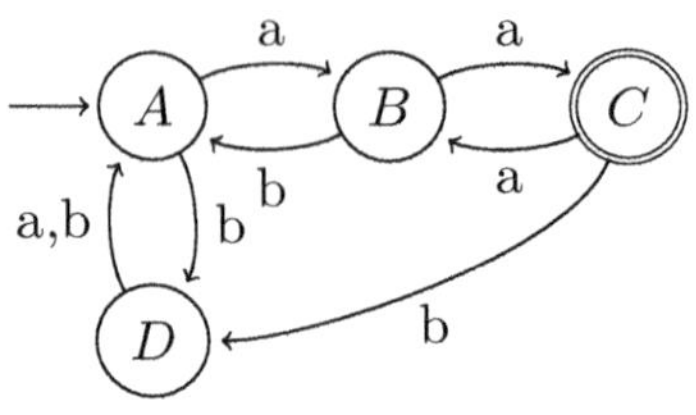

Fig. 10. The automaton $\mathcal{B} = \mu_F(\mathcal{A})$ with $F = \{5\}$

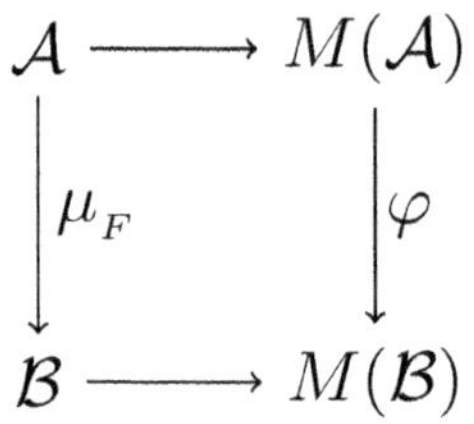

Using the above lemma we can establish a "weak inverse"of Proposition 3 for synchronizing automata. In order to state the next result, we need some notations. If $\mathcal{A} = (Q, \Sigma, \delta)$ is a DFA, for all $i \in Q$ and $F \subseteq Q$ denote $R(i, F) = \{[w] \in M(\mathcal{A}) \mid w \in L(\mathcal{A}(i, F))\}$.

Theorem 5. *Let $\mathcal{A} = (Q, \Sigma, \delta)$ be a strongly connected synchronizing DFA. If $\mathcal{A}$ is never-minimal, then, for all $i \in Q$ and $F \subseteq Q$, $R(i, F)$ is not a disjunctive subsets of $M(\mathcal{A})$.*

We conclude that the relationship between the problem to decide whether an automaton is never minimal and the syntactic monoid problem needs further investigations.

References

[GK98] Goralcik, P., Koubek, V.: On the disjunctive set problem. Theor. Comput. Sci. 204(1-2), 99–118 (1998)

[HMU00] Hopcroft, J.E., Motwani, R., Ullman, J.D.: Introduction to Automata Theory, Languages, and Computation, 2nd edn. Addison Wesley, Reading (2000)

[Hop71] Hopcroft, J.E.: An n log n algorithm for minimizing the states in a finite automaton, pp. 189–196. Academic Press, London (1971)

[How91] Howie, J.M. (ed.): Automata and Languages. Oxford Science Publications. Oxford University Press, Inc., New York (1991)

[Moo56] Moore, E.F.: Gedanken experiments on sequential machines. In: Moore, E.F. (ed.) Automata Studies, pp. 129–153. Princeton U., Princeton (1956)

[RS11] Rodaro, E., Silva, P.: Never minimal automata and the rainbow bipartite subgraph problem. In: Proceedings of the 15th International Conference on Developments in Language Theory. Springer, Heidelberg (2011)

[RV10] Restivo, A., Vaglica, R.: Automata with extremal minimality conditions. In: Gao, Y., Lu, H., Seki, S., Yu, S. (eds.) DLT 2010. LNCS, vol. 6224, pp. 399–410. Springer, Heidelberg (2010)

[RV11] Restivo, A., Vaglica, R.: Extremal minimality conditions on automata (2011) (submitted)

[Č64] Černy, J.: Poznámka k homogénnym experimenton s konečnými automatmi. Mat.-Fyz. Cas. Slovensk. Akad. Vied. 14, 208–215 (1964)

Growth Properties of Power-Free Languages

Arseny M. Shur

Ural State University, Ekaterinburg, Russia

Abstract. The aim of this paper is to give a short survey of the area formed by the intersection of two popular lines of investigation in formal language theory. The first line, originated by Thue in 1906, concerns about *repetition-free* words and languages. The second line is the study of *growth functions* for words and languages; it can be traced back to the classical papers by Morse and Hedlund on symbolic dynamics (1938, 1940). Growth functions of repetition-free languages are investigated since 1980's. Most of the results were obtained for power-free languages, but some ideas can be applied for languages avoiding patterns and Abelian-power-free languages as well.

In this paper, we present key contributions to the area, its state-of-the-art, and conjectures that suggest answers to some natural unsolved problems. Also, we pay attention to the tools and techniques that made possible the progress in the area and suggest some technical results that would be useful to solve open problems.

The growth properties of power-free (more generally, repetition-free) languages constitute a popular area of investigations in formal language theory. For example, one can found about twenty papers devoted to the growth of one particular language, namely, the square-free language over three letters. Since the review by Berstel [1], the general appearance of the area changed significantly. New powerful techniques were developed, and lots of new results were obtained. Also, the boundaries of the area became well-defined after the confirmation of Dejean's conjecture: now we know exactly which power-free languages are infinite (= possess nontrivial growth properties). So, the area certainly needs a new overview and we try our best to provide it.

We start after necessary preliminaries and try to show the area in dynamics, comparing new results and new techniques to the older ones.

1 Preliminaries

We study words over finite alphabets, using standard notions of factors, prefixes and suffixes. The length of a word w is denoted by $|w|$. As usual, we write Σ^* for the set (free monoid) of all words over the alphabet Σ, including the *empty word* λ. A language $L \subseteq \Sigma$ is *factorial* if it is closed under taking factors of its words. A word w is *forbidden* for a language L if it is a factor of no word from L. The set of all minimal (w.r.t. factor order) forbidden words for a language is called the

G. Mauri and A. Leporati (Eds.): DLT 2011, LNCS 6795, pp. 28–43, 2011.

antidictionary of this language. A factorial language is uniquely determined by its antidictionary. Languages with finite antidictionary (*FAD-languages*) form a proper subclass of regular languages. A language is *symmetric* if it is preserved by all automorphisms of Σ^*.

A positive integer p is a *period* of a word w if the ith and the $(i+p)$th letters of w coincide for all possible values of i. For any rational number $\beta \geq 1$ and any word w, we define the β-power and the β^+-power of w as follows:

$$w^{\beta} = \underbrace{w \cdots w}_{\lfloor\beta\rfloor \text{ times}} w', \text{ where } w' \text{ is a prefix of } w, \ \frac{|w^{\beta}|}{|w|} \geq \beta, \text{ and } \frac{|w^{\beta}|-1}{|w|} < \beta,$$

$$w^{\beta^+} = \underbrace{w \cdots w}_{\lfloor\beta\rfloor \text{ times}} w', \text{ where } w' \text{ is a prefix of } w, \ \frac{|w^{\beta}|}{|w|} > \beta, \text{ and } \frac{|w^{\beta}|-1}{|w|} \leq \beta.$$

For convenience, β^+ is considered just as a "number" that covers β in the usual $\leq$ order. Throughout the paper, β is either a rational number, or a "number with plus". The word w is called *β-free* if it contains no β-powers as factors. By *β-free languages* we mean the languages of *all* β-free words over a given alphabet. These languages are obviously factorial and are also called *power-free* languages. 2-free, 3-free, and 2^+-free words are known as *square-free*, *cube-free*, and *overlap-free*, respectively.

Repetition threshold $\mathsf{RT}(k)$ is the infimum of all numbers β such that the k-ary β-free language is infinite. *Dejean's conjecture* [9] states that $\mathsf{RT}(3) = 7/4$, $\mathsf{RT}(4) = 7/5$, and $\mathsf{RT}(k) = k/(k-1)$ for $k = 2$ and $k \geq 5$. This conjecture is now confirmed in all cases due to the efforts of Dejean, Pansiot, Moulin-Ollagnier, Mohammad-Noori, Currie, Carpi, Rampersad, and Rao. The k-ary $\mathsf{RT}(k)^+$-free language is the minimal infinite k-ary power-free language (*threshold language*).

A word w *avoids* a word u if there are no homomorphic images of u among the factors of w. The word u in this case is called a *pattern*. A word $w_1 \cdots w_n$ is an *Abelian n-th power* if $w_1 = \ldots = w_n$ in any commutative monoid. Abelian-n-free words and languages, and also languages avoiding patterns, are defined similar to β-free words and languages.

We consider finite automata with *partial* transition function and view them as digraphs. A *trie* is a deterministic finite automaton that is a tree such that the initial vertex is its root and the set of terminal vertices is the set of all its leaves. A *strong component* of a digraph G is its subgraph G' maximal w.r.t. inclusion such that there is a (directed) walk from any vertex of G' to any other vertex of G'. A digraph is *strongly connected*, if it has only one strong component.

Frobenius root of a nonnegative matrix is its maximal positive eigenvalue. Frobenius root has a nonnegative eigenvector; for some matrices, including the adjacency metrices of strongly connected digraphs, this eigenvector is positive.

For any language L, its growth properties result from the behaviour of its *combinatorial complexity*, which is the function $C_L(n)$ returning the number of words of length n in L. The *growth rate* of L is defined by $\mathsf{Gr}(L) = \limsup_{n\to\infty}(C_L(n))^{1/n}$. By Fekete's lemma, for any factorial language L we have

$$\mathsf{Gr}(L) = \lim_{n\to\infty} (C_L(n))^{1/n} = \inf_{n\in\mathbb{N}} (C_L(n))^{1/n}. \tag{1}$$

The inequality $\mathsf{Gr}(L) > 1$ indicates that L is "big" (has exponential complexity). If $\mathsf{Gr}(L) = 1$, then L is "small" (has subexponential complexity). The complexity $C_L(n)$ is polynomial if it is bounded from above by a polynomial in n. The condition $\mathsf{Gr}(L) = 0$ corresponds to the degenerate case of a finite language. The complexity of a factorial language is either bounded by a constant or strictly increasing.

Concluding the preliminary section, we outline the main directions in the study of the growth properties of power-free languages.

1. Get sharp bounds for the parameters of growth for the most famous power-free languages such as the binary cube-free and overlap-free languages, ternary square-free language.
2. Study the behaviour of the growth rate for the smallset nontrivial (i.e., binary) alphabet.
3. Build universal algorithms for upper and lower bounds on the growth rate.
4. Establish asymptotic laws of behaviour of the growth rate for large alphabets.
5. Extend the line of Dejean's conjecture; in particular, prove the following *Exponential conjecture: threshold languages over three and more letters have exponential complexity*[1].

2 Small Languages: Polynomial Plateau

In this section, all words and languages are binary. It is no surprise that binary power-free languages attract great attention of researchers. The interest was warmed up by the first two results that appears almost simultaneously. Brandenburg [3] proved that the cube-free language CF grows exponentially, while Restivo and Salemi [25] established the polynomial growth of the overlap-free language OF. Kobayashi [15] asked about the "critical exponent", separating subexponential and exponential complexities. Karhumäki and Shallit proved [14] that this exponent equals 7/3. Namely, they showed that the (7/3)-free language has polynomial complexity, while the complexity of the $(7/3)^+$-free language is exponential. So, the complexities of binary power-free languages demonstrate an interesting feature: the β-free languages with $2^+ \le \beta \le 7/3$ form a "polynomial plateau" of complexity. It is very likely that such a plateau is unique, because no arguments is known against Exponential conjecture.

The growth rate of any binary power-free language can be found up to the sixth digit after the dot (or even with better precision), using the results of Sect. 3, 4. So, this section is devoted to the analisys of polynomial complexities. In addition, we point out some other "magic properties" of the constant 7/3.

[1] This folklore conjecture is open in almost all cases. For 3 and 4 letters it is proved by Ochem [21]. Kolpakov and Rao announced lower bounds for the growth rates of threshold languages over $5, \ldots, 10$ letters, but no preprint is currently available.

The proof by Restivo and Salemi is based on the close connection between overlap-free words and the *Thue-Morse morphism* defined by the equalities $\theta(a) = ab$, $\theta(b) = ba$. It is known since Thue [38] that this morphism preserves overlap-freeness. On the other hand, any overlap-free word is "almost" a θ-image of some word, as the following lemma shows.

Lemma 1. *Let w be an overlap-free word. Then w can be obtained from a θ-image of some word by applying one transformation from each of the two lists*

(a) delete the first letter / replace the first letter by different letter / do nothing;
(b) delete the last letter / replace the last letter by different letter / do nothing.

Moreover, if $|w| > 5$, then this pair of transformations is unique.

From this lemma, it is easy to see that any overlap-free word of an odd length $2n-1$ is an "almost θ-image" of an overlap-free word of length n, while each overlap-free word of length n has at most four such "almost θ-images". Hence, $C_{OF}(2n) \leq 8 \cdot C_{OF}(n)$, implying $C_{OF}(n) = O(n^3)$.

Of course, the cubic bound is very rough. It is not very hard to replace it with $n^{1.587}$, see [16]. But the real "bomb" was exploded by Cassaigne [5]: he discovered that *there is no number α such that $C_{OF}(n) = \Theta(n^\alpha)$*. Moreover, he found "fast" and "slow" subsequences of the sequence $\{C_{OF}(n)\}$ and provided upper and lower bounds to the growth of the fastest and the slowest subsequences. The results of [5] were recently refined in [13] with the use of the joint spectral radius of matrices. The following theorem can be considered as the final description of the growth properties of the function $C_{OF}(n)$. Recall that $f(n) = \Theta(n^\alpha)$ if and only if $\lim\limits_{n\to\infty} \frac{\log f(n)}{\log n} = \alpha$.

Theorem 1 ([13]). (1) $\liminf\limits_{n\to\infty} \frac{\log C_{OF}(n)}{\log n} \in [1.2690, 1.2736]$.
(2) $\limsup\limits_{n\to\infty} \frac{\log C_{OF}(n)}{\log n} \in [1.3322, 1.3326]$.
(3) *The ratio $\frac{\log C_{OF}(n)}{\log n}$ has a limit as $n \to \infty$ along some subset $\mathbb{N}' \subset \mathbb{N}$ of density 1, and this limit belongs to the interval* $[1.3005, 1.3098]$.

Now we briefly discuss why we can expect similar results for any language from the polynomial plateau. The reason is that all these languages possess the same restrictive properties as the language OF, while all these restrictions get raised beyond the point $7/3$. We list some of the restrictions.

1. All β-free languages with $2^+ \leq \beta \leq 7/3$ satisfy Lemma 1 [14], and hence their complexities have a cubic upper bound.
2. A morphism preserves β-freeness for some $2^+ \leq \beta \leq 7/3$ if and only if it is equal to some power of θ or to the "negative" of some power of θ (combined [23, 27, 28]).
3. A two-sided infinite word is β-free for some $2^+ \leq \beta \leq 7/3$ if and only if the set of all its factors coincides with the set of all factors of the infinite Thue-Morse word $\theta^\infty(a)$ (combined [10, 28]).

4. Let $|w|_a$ be the number of letters a in the word w, and let $f_a = \liminf_{|w|\to\infty, w\in L} \frac{|w|_a}{|w|}$ be the *minimum density of* a *in the language* L. Then the minimum density of a in the β-free language equals 1/2 if and only if $2^+ \le \beta \le 7/3$ [19].
5. If $2^+ \le \beta \le 7/3$, then any minimal β-power has the length 2^n or $3{\cdot}2^n$ for some n; if $\beta > 7/3$, then the minimal β-power can have any length except for some finite set [35].

The listed properties show that combinatorial complexities of different languages from the polynomial plateau must be quite close to each other. The only interesting question is how fast the (7/3)-free language, i.e., the biggest one, grows. For this language, the following analogs of the bounds given in Theorem 1 (1, 2) were obtained in [2]: the exponent of the slowest growth is in the interval [1.2690, 2.0035], and the one of the fastest growth is in [2.0121, 2.1050].

Overall, we can say that the knowledge about the polynomial plateau is quite satisfactory.

3 Big Languages: Quest for Upper Bounds

The basic idea how to obtain an upper bound for the combinatorial complexity of any factorial language L, is quite straightforward. One can take a "simple" superset of L and estimate its complexity instead of the complexity of L. The role of such "simple" languages is played by FAD-languages. For any factorial language L over a finite alphabet Σ, one can take its antidictionary M and pick up an arbitrary sequence $\{M_i\}$ of finite subsets of M such that

$$M_1 \subseteq M_2 \subseteq \ldots \subseteq M_i \subseteq \ldots \subseteq M, \quad \bigcup_{i=1}^{\infty} M_i = M.$$

For power-free languages, it is usually convenient to take the set of all words from M of period $\le i$ as M_i. The FAD-languages L_i with the antidictionaries M_i are called *regular approximations* of L. We have

$$L \subseteq \ldots \subseteq L_i \subseteq \ldots \subseteq L_1, \quad \bigcap_{i=1}^{\infty} L_i = L.$$

Due to (1), it is easy to check that $\lim_{i\to\infty} \mathsf{Gr}(L_i) = \mathsf{Gr}(L)$. Thus, it is possible in principle to obtain the upper bounds that are arbitrarily close to the growth rate of L. In order to make this possibility real, one needs to solve efficiently the following two problems:

- given a factorial language L and an integer i, calculate the set M_i;
- given a *finite* antidictionary M, calculate the growth rate of the factorial language L having this antidictionary.

Before presenting the solutions to these problems, we mention a simple result that imposes two restrictions on the use of regular languages as approximations of power-free languages. Namely, such approximations are useless for the study of "small" languages and cannot be applied directly for lower bounds.

Proposition 1 ([29]). *If all words in an infinite factorial language L are β-free for some number β, then all regular approximations of L have exponential complexity and all regular subsets of L are finite.*

3.1 Finding Growth Rates of FAD-Languages

There are two approaches to this problem. The first approach uses generating functions for combinatorial complexity. The famous result by Chomsky and Schützenberger [6] says that regular languages (in particular, FAD-languages) have rational generating functions[2]. The least positive pole of such a function is the reciprocal of the growth rate of the considered language. For FAD-languages, this rational function can be derived from the antidictionary by the Goulden-Jackson cluster method, see [11, 20]. After this, the roots of the polynomials can be found with any degree of precision. This method was used for studying the growth rates of some power-free languages, see [20, 6, 24]. Looking at this method and its implementations, we can conclude that it consumes quite a lot of resources. As a result, the capability of the method is restricted to processing the antidictionaries of several hundred words. In some cases, the antidictionaries of this size allows one to obtain reasonable upper bounds, but in many other cases this is insufficient.

The second approach uses finite automata (or digraphs that can be viewed as finite automata, like Rauzy graphs). For the best of our knowledge, this approach was first used by Govorov [12]. The idea of this approach is to calculate walks instead of words, and is based on the following folklore theorem. Recall that the *index* $\mathsf{Ind}(G)$ of a digraph G is the Frobenius root of its adjacency matrix. By the index of a finite automaton we mean the index of the underlying digraph. A finite automaton is *consistent* if any of its vertices belongs to some accepting path.

Theorem 2. *The growth rate of a regular language L equals the index of any consistent deterministic finite automaton recognizing L.*

The same result holds for all *unambiguous* nondeterministic automata, in particular, for Rauzy graphs, see [33]. The original idea by Govorov (see also [15]) was as follows. Suppose that a FAD-language L with some antidictionary M is given, and m is the maximum length of a word from M. Then a word w of length $\geq m$ belongs to L if and only if all its factors of length m are in L. Let us build the *Rauzy graph $\mathcal{R}$ of order m for L*. Its vertices are all words from L of length m, and an edge $u \to v$ exists if and only if the longest proper suffix of u coincides with the longest proper prefix of v. Clearly, any walk of length n in $\mathcal{R}$ uniquely represents a word from L of length $n+m$. So, one can calculate the

---

[2] In simpler terms, this means that for any regular language L there exist polynomials $p(x)$ and $q(x)$ such that the ratio $p(x)/q(x)$ can be expanded into the series $\sum_{n=0}^{\infty} C_L(n)x^n$.

characteristic polynomial of the adjacency matrix of $\mathcal{R}$ and find the Frobenius root by any iterative method[3].

The capability of the straightforward algorithm described above is more or less the same as the capability of the Goulden-Jackson method; one can look, for example, at the bounds obtained in [22, 18]. But the graph approach has significant potential for improvement. Namely, it is possible

- to build from M a much smaller digraph with the same index;
- to calculate the index of a digraph in a much more efficient way.

Consider the calculation of the index of a k-vertex digraph G. Advanced methods of finding the characteristic polynomial or the minimal polynomial of a given matrix, like the methods by Danilevski and Krylov, work in $\Theta(k^3)$ steps using a constant number of additional $k \times k$ matrices. In fact, the situation is a bit worse. As follows from [8, Theorem 1.2], the coefficients of the characteristic polynomial of a graph are likely to grow exponentially with k. So, the straightforward algorithm actually spends $\Theta(k^4)$ time and $\Theta(k^3)$ memory.

Now we roughly describe the algorithm of [31,33], which drastically improves the situation. We split G into strong components and process each component separately, because the index of a digraph equals the maximum index of its strong component. We apply a simple iteration method to find the Frobenius root and the corresponding positive eigenvector; i.e., we calculate the sequence

$$x, Ax, \ldots, A^n x,$$

where x is some initial vector and A is the adjacency matrix of the processed component. The convergence of the iteration method can be guaranteed by some additional argument, and the convergence rate is exponential for the adjacency matrices of all strongly connected digraphs. The approximation error can be verified using the double inequality

$$\min_j \frac{[Ax]_j}{x_j} \leq \alpha(A) \leq \max_j \frac{[Ax]_j}{x_j},$$

where x is an arbitrary positive vector and $\alpha(A)$ is the Frobenius root of A (this inequality is a corollary of the Perron-Frobenius Theorem). Finally, the calculation of the vector Ax from x can be performed without explicit use of the adjacency matrix: instead, we use the representation of the digraphs by adjacency lists or arrays, which are quite short for automata. As a result, we get the following

Theorem 3 ([33]). *Let $G = (V, E)$ be a digraph. There is an algorithm which, given G and a number δ, $0 < \delta < 1$, calculates $\mathsf{Ind}(G)$ with the absolute error at most δ in time $\Theta(\log(1/\delta)\cdot|E|)$ using $\Theta(\log(1/\delta)\cdot|V|)$ additional space.*

[3] Since the Frobenius root is the root of a polynomial equation, in most cases it cannot be found exactly. So, we say "find" or "calculate" it in a sense "approximate with any prescribed absolute error".

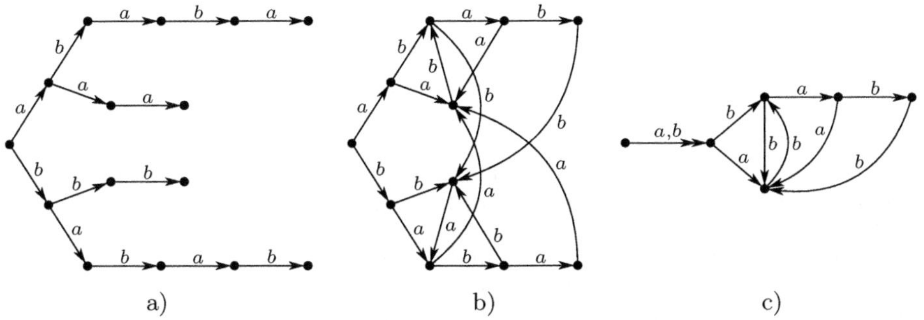

Fig. 1. The trie (a), Aho-Corasick's automaton (b), and the corresponding factor-automaton (c) for the FAD-language with the antidictionary $\{aaa, bbb, ababa, babab\}$

Now we turn to the second improvement: building smaller graphs with the same index $\mathsf{Gr}(L)$ from M. As was shown in [7], a rather small consistent deterministic finite automaton recognizing L can be constructed from M by a modification of the textbook Aho-Corasick algorithm for pattern matching. We illustrate this modification by an example. Obviously, any finite antidictionary is recognized by a trie (see Fig. 1, a). Each vertex of this trie can be identified with the word labeling the path from the root to this vertex. The trie is then modified in order to get the required automaton, see Fig. 1, b. Processing all vertices in the width-first order, one adds missing outgoing edges. The destination vertex of the edge $u \xrightarrow{a} \ldots$ is the destination vertex of the walk from the initial vertex labeled by the longest proper suffix of the word ua. After processing all vertices, the terminal vertices of the trie are deleted. The resulting *Aho-Corasick's automaton* $\mathcal{A}(M)$ recognize exactly the language L (all states are final, and λ is the initial state).

Aho-Corasick's automaton is more compact than the Rauzy graph built from the same antidictionary, but not compact enough. The following idea, described in [33], uses symmetry of power-free languages and their antidictionaries to shrink the size of the automaton $\mathcal{A}(M)$ by a factor of almost $|\Sigma|!$. Partition all vertices of $\mathcal{A}(M)$ into classes such that the class of the vertex u consists exactly of all automorphic images of u. This partition is *equitable* in the sense that, given any two classes C_1 and C_2, each vertex of C_1 has the same number of (forward) adjacent vertices in C_2. This property allows one to build the *factor-graph* of the original graph by the given partition, identifying all vertices in each class. Figure 1, c exhibits the factor-graph of the automaton drawn in Fig. 1, b (the meaning of the labels is described below). By $\mathcal{F}(M)$ we denote the factor-graph of the automaton $\mathcal{A}(M)$. The number of walks of length n from any given vertex of $\mathcal{A}(M)$ obviously equals the number of walks from its class in $\mathcal{F}(M)$. Thus, the indices of $\mathcal{A}(M)$ and $\mathcal{F}(M)$ coincide.

The above "factorization" idea proved really useful, because the factor-graph $\mathcal{F}(M)$ can be built without building the whole automaton $\mathcal{A}(M)$, see [33]. Namely, one can take the set $M' \subset M$ consisting of all words that are lexicographycally smaller than any of their automorphic images (any word with this property is said to be a *lexmin* word). Then one builds the "factor-trie" recognizing M', and constructs the factor-graph directly from this factor-trie by a simple modification of Aho-Corasick's procedure:

- the destination vertex of the edge $u \xrightarrow{a} \ldots$ is the destination vertex of the walk from the initial vertex labeled by the *lexicographycally least automorphic image* of the longest proper suffix of the word ua.

Thus, $\mathcal{F}(M)$ is built as a finite automaton, see Fig. 1, c (but the language it recognizes is not related to L apart from the growth rate).

This modified procedure can be organized in a way that the factor-graph will be built in time $O(N \log N)$, where N is the size of the factor-trie. Hence, the space and time expenses needed to build the factor-graph are also approximately $|\Sigma|!$ times less than the expenses for the Aho-Corasick's automaton.

Overall, the above improvements to Govorov's idea allow one to get the method that can be applied to any FAD-language and has very attractive characteristics:

(∇1) the growth rate of L can be calculated whenever M can be stored in memory;
(∇2) if M is symmetric, then the growth rate of L can be calculated whenever the set M' can be stored in memory.

3.2 Building Finite Antidictionaries

Suppose that L is a β-free language over Σ, M is its antidictionary, and i is a positive integer such that we have to calculate the upper bound $\mathsf{Gr}(L_i)$ for $\mathsf{Gr}(L)$. According to the results discussed above, it is enough to find the set M_i' of lexmin forbidden words. This fact is of critical importance for the languages over big alphabets, because $|M_i'| \approx |M_i|/|\Sigma|!$; actually, the 4-letter alphabet is already big enough.

Due to (∇1) and (∇2), the performance of any algorithm building finite antidictionaries can be evaluated by the following criterion. *Assume that the i is the minimal number such that the factor-trie M_{i+1}' is too big to be stored in the memory of one's computer. Can one build M_i' in a reasonable amount of time?* The following simple algorithm satisfies this criterion in most cases.

The elements of M_i are *minimal* β-powers in the sense that they contain no β-powers as proper factors. In particular, the *root* w of such a power w^β is a β-free word. Thus, we need to check all lexmin β-free words up to the length i as candidates to the roots of minimal β-powers. We perform this check in width-first order, immediately adding each found minimal β-power to the trie. The minimality of each processed β-power is then checked with the use of the current trie. The details can be found in [33].

This algorithm has one bottleneck: if valid roots of minimal β-powers constitute a small share of all β-free words, then storing the queue for the width-first search became too expensive (this is the case of threshold languages over 6 and more letters). A better strategy is the mix of width-first and depth-first search, based on the fact that a non-minimal β-power contains a rather short β-power inside:

Theorem 4. *Let xy be a β-free word, $1 < \beta < 2$. If the β-power $(xy)^\beta$ is not minimal, then the word $(xy)^\beta$ contains a β-power $(zt)^\beta$ such that $|(zt)^\beta| < |xy|$. Moreover, if $\beta \le (4/3)^+$, then $|zt| \le |y|$, and if $\beta \le (5/4)^+$, then $|zt| < |y|$.*

Concluding the section, we can say that the quest for upper bounds is successfully completed. Some numerical results confirming this statement will be given in the next sections.

4 Big Languages: Quest for Lower Bounds

Early attempts to estimate the growth rates of power-free languages from below had a quite restricted goal: just prove that the analysed language has exponential complexity. Morphisms were the main instrument for such estimations. We mention two ways to use morphisms for this purpose.

1. *"Inherited" exponential growth.* If a k-ary β-free language has exponential complexity and some morphism maps k-ary words to k'-ary words such that the image of any β-free word is β'-free, then the k'-ary β'-free language has exponential complexity as well. For example, Karhumäki and Shallit [14] exhibited a morphism that maps quaternary square-free words to binary $(7/3)^+$-free words, thus proving that the binary $(7/3)^+$-free language has exponential complexity. (Note that a morphism with the same property for *ternary* square-free words was presented a few years earlier in [28].) The same idea was used by Ochem [21] to prove that the ternary and quaternary threshold languages grow exponentially.

2. *"Multiple" morphisms* (sometimes referred to as *Brinkhuis tuples*). Suppose we have nonempty sets $W_1, \ldots, W_k$ of β-free words of length $m > 1$ over $\{a_1, \ldots, a_k\}$, $\min_{j\in\{1,\ldots,k\}} |W_j| = l > 1$, and the following property holds: if some word $a_{i_1} \cdots a_{i_n}$ is β-free, then any word $w_1 \cdots w_n$ such that $w_j \in W_{i_j}$ for any j is β-free too. Then for any n, the number of k-ary β-free words of length nm is at least l^n times bigger than the number of such words of length n, yielding the exponential lower bound on the number of these words. This method was first used by Brandenburg [3] and Brinkhuis [4] for ternary square-free words (in [3], the lower bound on the number of binary cube-free words is also given).

Both described methods produce quite weak bounds (the second one works a bit better). What about getting sharp lower bounds?

A method allowing one to get really good lower bounds was proposed by Kolpakov [17, 18] and then modified and extended by the author [32, 34]. The method is based on the convertation of the upper bound obtained from a regular approximation of the target language, to the two-sided bound. In brief, the basic idea is the following. An automaton recognizing some regular approximation L_i of the target language L accepts the words from L and also the words from the set $L_i \backslash L$. This set consists of words containing "long" β-powers; the number of such words is not very big and can be estimated from above, thus giving us the lower bound for the number of words in L. At the moment, we are able to get sharp lower bounds for any β-free language such that $\beta \geq 2$, while no algorithm is known for such bounds in the case of smaller powers. The following theorem is a slightly refined version of [32, Theorem 6].

Theorem 5. *Suppose that $\beta \geq 2$, k and i are positive integers, L is the k-ary β-free language, M_i is the set of all words of period $\leq i$ from the antidictionary of L, L_i is the regular approximation of L with the antidictionary M_i, and the factor-graph $\mathcal{F}(M_i)$ has a unique nonsingleton strong component. Then any number γ such that $\gamma + \frac{1}{\gamma^{i-1}(\gamma-1)} \leq \mathsf{Gr}(L_i)$ satisfies the inequality $\gamma < \mathsf{Gr}(L)$.*

The condition on $\mathcal{F}(M_i)$ in Theorem 5 is not restrictive: we strongly believe that it always holds true (even if $\beta < 2$) and have no counterexamples. But the proof of this fact is probably far from easy. Indeed, this fact is equivalent to the extended version of an old open problem by Restivo: *let L be a k-ary β-free language, containing infinitely many words with some fixed prefix u and infinitely many words with some fixed suffix v. Does L certainly contain a word of the form uwv?*

We mention that Theorem 5 shows high universality of the described method. Indeed, in order to get a lower bound for the growth rate of a power-free language L, we does not need to know k or β; only the parameters i and $\mathsf{Gr}(L_i)$ of the regular approximation are required. In addition, the best possible bound γ can be calculated with any prescribed error in an almost constant time (the condition on $\mathcal{F}(M_i)$ is already checked during the calculation of $\mathsf{Gr}(L_i)$, see Sect. 3).

We conclude this section exhibiting two-sided bounds on the growth rates of some languages. More bounds can be found in [36]. All bounds were obtained using a PC with a 3.0GHz CPU and 2Gb of memory, and are rounded off to 7 digits after the dot. If only one bound is given, then these digits are the same for both lower and upper bounds.

Table 1. *Bounds for the growth rates of β-free languages with $\beta \geq 2$*

k	β	bounds
2	$(7/3)^+$	1.2206318–1.2206448
2	$(5/2)^+$	1.3662971–1.3663011
2	3	1.4575732–1.4575773
2	3^+	1.7951246–1.7951264
2	4	1.8211000
2	4^+	1.9208015
3	2	1.3017597–1.3017619
3	2^+	2.6058789–2.6058791
3	3	2.7015614–2.7015616
3	3^+	2.9119240–2.9119242
3	4	2.9172846
3	4^+	2.9737546

Small alphabets

$k \backslash \beta$	2	2^+	3	3^+
4	2.6215080	3.7284944	3.7789513	3.9487867
5	3.7325386	4.7898507	4.8220672	4.9662411
6	4.7914069	5.8277328	5.8503616	5.9760100
7	5.8284661	6.8537250	6.8705878	6.9820558
8	6.8541173	7.8727609	7.8858522	7.9860649
9	7.8729902	8.8873424	8.8978188	8.9888625
10	8.8874856	9.8988872	9.9074705	9.9908932
11	9.8989813	10.9082635	10.9154294	10.9924142
12	10.9083279	11.9160348	11.9221106	11.9935831
13	11.9160804	12.9225835	12.9278022	12.9945010
14	12.9226167	13.9281788	13.9327109	13.9952350
15	13.9282035	14.9330157	14.9369892	14.9958311

Large alphabets

5 Big Languages: Asymptotic Formulas

Using the results of two previous sections, one can obtain numerical bounds for the growth rates of power-free languages for a wide range of alphabets and powers. As a result, one can consider growth rate as a two variable function $\alpha(k, \beta)$ and derive empirically some laws of behaviour of this function. An example of such a law is the following: *for a fixed $k > 2$, the function $\alpha(k, \beta)$ jumps by more than a unit at $k-1$ points, namely, at the points* $\mathsf{RT}(k), (k-1)/(k-2), \ldots, 3/2, 2$. But how to prove this or other laws? In this section, we sketch the main idea and the most important results and conjectures on the behaviour of $\alpha(k, \beta)$. More details can be found in [34].

The key property that allows one to prove general facts about the function $\alpha(k, \beta)$, is the similarity of the factor-graphs for the regular approximations of different power-free languages. In order to illustrate one type of such similarity, let us take a k-ary β-free language L with the antidictionary M and build the factor-graph $\mathcal{F}(M_i)$ for some i small enough. This graph will have edges of multiplicity 1 and of multiplicities $k-c_j$ for several different constants c_j. If we now replace k by k', retaining the same β and i, then the new factor-graph will be almost the same, with a single distinction: for multiple edges, k will be everywhere replaced by k'. For example, the factor-graph $\mathcal{F}(M_5)$ for the k-ary $(4/3)^+$-free language looks like in Fig. 2, for any $k \geq 5$. From the adjacency matrix, one can derive arbitrarily good asymptotics for the index of $\mathcal{F}(M_5)$. In particular, $\mathsf{Ind}(\mathcal{F}(M_5)) = k-2-3/k+O(1/k^2)$ (compare to Conjecture 1 below).

$$\begin{array}{l} 1231 \\ 1234 \\ 12341 \\ 12345 \\ 123451 \end{array} \quad \begin{bmatrix} 0 & k-3 & 0 & 0 & 0 \\ 1 & 0 & 1 & k-4 & 0 \\ 1 & 0 & 0 & k-4 & 0 \\ 1 & 0 & 1 & k-5 & 1 \\ 1 & 0 & 1 & k-5 & 0 \end{bmatrix}$$

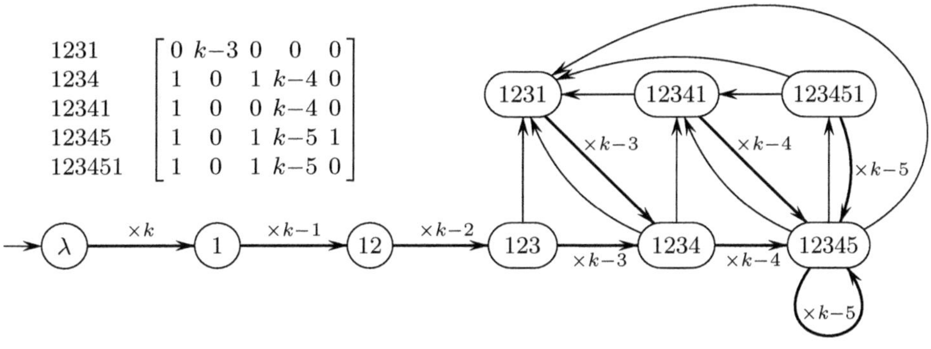

Fig. 2. *The factor-graph $\mathcal{F}(M_5)$ for the $(4/3)^+$-free language over the alphabet $\{1, \ldots, k\}$. Vertices are associated with words. Multiple edges are drawn in boldface. On the left, the adjacency matrix of the nontrivial strong component is given.*

Using similarity of factor-graphs and Theorem 5, one can get uniform asymptotic formulas for $\alpha(k, \beta)$ in the case $\beta \geq 2$. Thus, for $\beta > 2$ we have

Theorem 6. *Let $\beta \in [n^+, n+1]$, where $n \geq 2$ is an integer. Then*

$$\alpha(k, \beta) = \begin{cases} k - \frac{1}{k^{n-1}} + \frac{1}{k^n} - \frac{1}{k^{2n-2}} + O\big(\frac{1}{k^{2n-1}}\big), & \text{if } \beta \in [n^+, n+\frac{1}{2}], \\ k - \frac{1}{k^{n-1}} + \frac{1}{k^n} + O\big(\frac{1}{k^{2n-1}}\big), & \text{if } \beta \in [(n+\frac{1}{2})^+, n+1]. \end{cases}$$

Corollary 1. *For any fixed $\beta \geq 2^+$, the difference $(k - \alpha(k,\beta))$ approaches zero at polynomial rate as $k \to \infty$. For any fixed $k \geq 2$, the same difference approaches zero at exponential rate as $\beta \to \infty$.*

Corollary 2. *For a fixed k, the jumps of the function $\alpha(k,\beta)$ at the endpoints of the interval $[n^+, n{+}1]$ are much bigger than the variation of this function inside this interval. Namely,*

$$\alpha(k,n^+) - \alpha(k,n) = \tfrac{1}{k^{n-2}} + O\big(\tfrac{1}{k^{n-1}}\big),$$
$$\alpha(k,n{+}1) - \alpha(k,n^+) = \tfrac{1}{k^{2n-2}} + O\big(\tfrac{1}{k^{2n-1}}\big).$$

Next we analyze the behaviour of $\alpha(k,\beta)$ at the point $\beta = 2$.

Proposition 2. *The following equalities hold:*

$$\alpha(k{+}1,2) = k - \tfrac{1}{k} - \tfrac{1}{k^3} + O\big(\tfrac{1}{k^5}\big);$$
$$\alpha(k,2^+) = k - \tfrac{1}{k} - \tfrac{1}{k^3} - \tfrac{1}{k^4} + O\big(\tfrac{1}{k^5}\big).$$

Corollary 3. *For any k, the function $\alpha(k,\beta)$ jumps by more than a unit at the point $\beta = 2$. Namely, $\alpha(k,2^+) - \alpha(k,2) = 1 + \frac{1}{k^2} + O(\frac{1}{k^3})$.*

Corollary 4. *At any point $(k,2)$, the increment of k by 1 and the addition of $^+$ to the power almost equally affect the growth rate of the power-free language. Namely, $\alpha(k{+}1,2) - \alpha(k,2^+) = \frac{1}{k^4} + O(\frac{1}{k^5})$.*

All asymptotic formulas given above work perfectly even for small alphabets, predicting the values from Table 1 with a good precision. For $\beta < 2$, the situation is worse, because Theorem 5 is no longer applicable. Consequently, only partial advances were obtained, while the general results have the form of conjectures.

Conjecture 1. The following equalities hold for any fixed integers n,k such that $k > n \geq 3$:

$$\alpha\big(k,\tfrac{n}{n-1}^+\big) = k{+}2{-}n{-}\tfrac{n-1}{k}{+}O\big(\tfrac{1}{k^2}\big),$$
$$\alpha\big(k,\tfrac{n}{n-1}\big) = k{+}1{-}n{-}\tfrac{n-1}{k}{+}O\big(\tfrac{1}{k^2}\big).$$

Conjecture 1 predicts that the properties found above for the point $\beta = 2$ hold at any point $\beta = \frac{n}{n-1}$ such that $2 < n < k$. Indeed, Conjecture 1 implies

Corollary 5. *Let n and k be integers such that $2 < n < k$. Then*

$$\alpha\big(k,\tfrac{n}{n-1}^+\big) - \alpha\big(k,\tfrac{n}{n-1}\big) = 1 + O\big(\tfrac{1}{k^2}\big)$$
$$\alpha\big(k,\tfrac{n}{n-1}\big) - \alpha\big(k,\tfrac{n+1}{n}^+\big) = \tfrac{1}{k} + O\big(\tfrac{1}{k^2}\big)$$
$$\alpha\big(k{+}1,\tfrac{n}{n-1}\big) - \alpha\big(k,\tfrac{n}{n-1}^+\big) = O\big(\tfrac{1}{k^2}\big).$$

If β is not a constant but depends on k, then the most interesting case is when β stays close to repetition threshold (in particular, $\beta = \mathsf{RT}(k)^+$). The similarity between factor-graphs in this case is of different nature than in the case of fixed β. Due to space constraints, we give here only the main conjecture. The case $\beta = \mathsf{RT}(k)^+$ is studied in detail in [37].

Conjecture 2. For any integer $n \geq 0$, the limits

$$\alpha_n = \lim_{k\to\infty} \alpha(k, \tfrac{k-n}{k-n-1}^{+}) \quad \text{and} \quad \alpha'_n = \lim_{k\to\infty} \alpha(k, \tfrac{k-n}{k-n-1})$$

exist. Moreover, $\alpha'_{n+1} = \alpha_n$ and $\alpha_{n+1} - \alpha_n > 1$.

In [37], it is argued that $\alpha_0 \approx 1.242$. This *Growth Rate conjecture* strengthens Exponential conjecture. Thus, the threshold languages are supposed to be big enough in spite of the fact that it was tough to prove that they are even infinite. We also suggest that $\alpha_1 \approx 2.326$, $\alpha_2 \approx 3.376$.

6 Extending the Techniques to Related Classes of Languages

In this section we aim to show two possible applications of the methods described in Sect. 3, 4 beyond the class of power-free languages. Regular approximations (see Sect. 3) can be used to estimate the growth rate of a factorial language L whenever the finite approximations of the antidictionary of L can be calculated. Moreover, if L is symmetric, then factor-graphs can be built to provide better upper bounds. Concerning the lower bounds, the case in which an analog of Theorem 5 is applicable is rather exceptional. Now we exhibit such a case.

Recall that the binary overlap-free language grows polynomially, and that overlap-freeness is equivalent to the avoidance of the set $\{xxx, xyxyx\}$ of patterns. We studied the growth of the binary languages avoiding two quite similar sets of patterns: $\{xxx, xyxxyx\}$ and $\{xxx, xxyxxy\}$. The question was whether any of these two languages has polynomial growth. The answer is no: the first language is finite (the longest word in the antidictionary has length 48), while for the second language the analog of Theorem 5 allowed us to catch the growth rate inside the interval $[1.098814, 1.098891]$.

An interesting, and not easily predictable, effect was observed during the study of the growth of Abelian-power-free languages, see [26]. Namely, the approximating sequences $\{\mathsf{Gr}(L_i)\}$ for the growth rates of Abelian-power-free languages converge very slowly in comparison with such sequences for power-free languages. For example, if L is the binary cube-free language, then the set M'_7 contains only 25 words and $\mathsf{Gr}(L_7) - \mathsf{Gr}(L) < 0.001$. On the other hand, if L is the binary Abelian-4-free language, then M'_{12} contains almost 9 millions of words (M'_{13} cannot be stored in memory), $\mathsf{Gr}(L_{12}) \approx 1.374$, but it is impossible to guess the value of $\mathsf{Gr}(L)$, because $\mathsf{Gr}(L_8) - \mathsf{Gr}(L_9) \approx 0.02$, $\mathsf{Gr}(L_9) - \mathsf{Gr}(L_{10}) \approx 0.037$, $\mathsf{Gr}(L_{10}) - \mathsf{Gr}(L_{11}) \approx 0.011$ and $\mathsf{Gr}(L_{11}) - \mathsf{Gr}(L_{12}) \approx 0.01$. Thus, *the words that contain only long Abelian repetitions constitute a substantial share of all words.* As a result, for any reasonable notion of Abelian *fractional* power, one will get huge but still finite Abelian-power-free languages. Hence, the problem of determining an Abelian analog of repetition threshold looks challenging even for small alphabets.

References

1. Berstel, J.: Growth of repetition-free words – a review. Theor. Comput. Sci. 340(2), 280–290 (2005)
2. Blondel, V.D., Cassaigne, J., Jungers, R.: On the number of α-power-free binary words for $2 < \alpha \leq 7/3$. Theor. Comput. Sci. 410, 2823–2833 (2009)
3. Brandenburg, F.-J.: Uniformly growing k-th power free homomorphisms. Theor. Comput. Sci. 23, 69–82 (1983)
4. Brinkhuis, J.: Non-repetitive sequences on three symbols. Quart. J. Math. Oxford 34, 145–149 (1983)
5. Cassaigne, J.: Counting overlap-free binary words. In: STACS 1993. LNCS, vol. 665, pp. 216–225. Springer, Berlin (1993)
6. Chomsky, N., Schützenberger, M.: The algebraic theory of context-free languages. In: Computer Programming and Formal System, pp. 118–161. North-Holland, Amsterdam (1963)
7. Crochemore, M., Mignosi, F., Restivo, A.: Automata and forbidden words. Inform. Processing Letters 67, 111–117 (1998)
8. Cvetković, D.M., Doob, M., Sachs, H.: Spectra of graphs. In: Theory and Applications, 3rd edn., p. 388. Johann Ambrosius Barth, Heidelberg (1995)
9. Dejean, F.: Sur un Theoreme de Thue. J. Comb. Theory, Ser. A 13, 90–99 (1972)
10. Gottschalk, W.H., Hedlund, G.A.: A characterization of the Morse minimal set. Proc. of Amer. Math. Soc., 15, 70–74 (1964)
11. Goulden, I., Jackson, D.M.: An inversion theorem for cluster decompositions of sequences with distinguished subsequences. J. London Math. Soc. 20, 567–576 (1979)
12. Govorov, V.E.: Graded algebras. Math. Notes 12, 552–556 (1972)
13. Jungers, R.M., Protasov, V.Y., Blondel, V.D.: Overlap-free words and spectra of matrices. Theor. Comput. Sci. 410, 3670–3684 (2009)
14. Karhumäki, J., Shallit, J.: Polynomial versus exponential growth in repetition-free binary words. J. Combin. Theory. Ser. A 104, 335–347 (2004)
15. Kobayashi, Y.: Repetition-free words. Theor. Comput. Sci. 44, 175–197 (1986)
16. Kobayashi, Y.: Enumeration of irreducible binary words. Discr. Appl. Math. 20, 221–232 (1988)
17. Kolpakov, R.M.: On the number of repetition-free words. J. Appl. Ind. Math. 1(4), 453–462 (2007)
18. Kolpakov, R.: Efficient lower bounds on the number of repetition-free words. J. Int. Sequences 10, # 07.3.2 (2007)
19. Kolpakov, R., Kucherov, G., Tarannikov, Y.: On repetition-free binary words of minimal density. Theor. Comput. Sci. 218, 161–175 (1999)
20. Noonan, J., Zeilberger, D.: The Goulden-Jackson Cluster Method: Extensions, Applications, and Implementations. J. Difference Eq. Appl. 5, 355–377 (1999)
21. Ochem, P.: A generator of morphisms for infinite words. RAIRO Theor. Inform. Appl. 40, 427–441 (2006)
22. Ochem, P., Reix, T.: Upper bound on the number of ternary square-free words. In: Proc. Workshop on Words and Automata (WOWA 2006), S.-Petersburg, # 8 (2006) (electronic)
23. Rampersad, N.: Words avoiding (7/3)-powers and the Thue–Morse morphism. Int. J. Foundat. Comput. Sci. 16, 755–766 (2005)
24. Richard, C., Grimm, U.: On the entropy and letter frequencies of ternary square-free words. Electronic J. Combinatorics 11, # R14 (2004)

25. Restivo, A., Salemi, S.: Overlap-free words on two symbols. In: Perrin, D., Nivat, M. (eds.) Automata on Infinite Words. LNCS, vol. 192, pp. 196–206. Springer, Heidelberg (1985)
26. Samsonov, A.V., Shur, A.M.: On Abelian repetition threshold. In: Samsonov, A.V., Shur, A.M. (eds.) Proc. 13th Mons Days of Theoretical Computer Science, pp. 1–11. Univ. de Picardie Jules Verne, Amiens (2010)
27. Séébold, P.: Overlap-free sequences. In: Perrin, D., Nivat, M. (eds.) Automata on Infinite Words. LNCS, vol. 192, pp. 196–206. Springer, Heidelberg (1985)
28. Shur, A.M.: The structure of the set of cube-free Z-words over a two-letter alphabet. Izvestiya Math. 64(4), 847–871 (2000)
29. Shur, A.M.: Factorial languages of low combinatorial complexity. In: Ibarra, O.H., Dang, Z. (eds.) DLT 2006. LNCS, vol. 4036, pp. 397–407. Springer, Heidelberg (2006)
30. Shur, A.M.: Comparing complexity functions of a language and its extendable part. RAIRO Inform. Theor. Appl. 42, 647–655 (2008)
31. Shur, A.M.: Combinatorial complexity of regular languages. In: Hirsch, E.A., Razborov, A.A., Semenov, A., Slissenko, A. (eds.) Computer Science – Theory and Applications. LNCS, vol. 5010, pp. 289–301. Springer, Heidelberg (2008)
32. Shur, A.M.: Two-sided bounds for the growth rates of power-free languages. In: Ibarra, O.H., Dang, Z. (eds.) DLT 2006. LNCS, vol. 4036, pp. 466–477. Springer, Heidelberg (2006)
33. Shur, A.M.: Growth rates of complexity of power-free languages. Theor. Comput. Sci. 411, 3209–3223 (2010)
34. Shur, A.M.: Growth of power-free languages over large alphabets. In: Ablayev, F., Mayr, E.W. (eds.) CSR 2010. LNCS, vol. 6072, pp. 350–361. Springer, Heidelberg (2010)
35. Shur, A.M.: On the existence of minimal β-powers. In: Gao, Y., Lu, H., Seki, S., Yu, S. (eds.) DLT 2010. LNCS, vol. 6224, pp. 411–422. Springer, Heidelberg (2010)
36. Shur, A.M.: Numerical values of the growth rates of power-free languages, arXiv:1009.4415v1 (cs.FL) (2010)
37. Shur, A.M., Gorbunova, I.A.: On the growth rates of complexity of threshold languages. RAIRO Inform. Theor. Appl. 44, 175–192 (2010)
38. Thue, A.: Über die gegenseitige Lage gleicher Teile gewisser Zeichenreihen. Norske Vid. Selsk. Skr. I, Mat. Nat. Kl. 1, 1–67 (1912)

A Functional Program for Regular Expressions Matching

Abstract of Invited Talk

Thomas Wilke

Institut für Informatik, Christian-Albrechts-Universität zu Kiel
wilke@ti.informatik.uni-kiel.de

Regular expressions [4] and tools to handle them, especially tools for regular expression matching—an early one is described in the seminal paper [5] by Ken Thompson—, are one of the major achievements of formal language and automata theory. Google counts 303,000 results for "regular expressions matching" (May 4, 2011); there are numerous command line tools for working with regular expressions such as grep; Google released a regular expression C++ library not long ago [3]; almost every programming language provides support for regular expressions; and even the text editor I am using to produce the source code of this LaTeX document has an extensive regular expression library.

Fig. 1. "Regular Expressions" by xkcd [6]

In the talk, I demonstrate that using ideas from Victor M. Glushkov [2], it is straightforward to write a small Haskell program for regular expression matching that can compete with Perl and Google's regular expression library and, after a slight modification, can check membership in any context-free language. The program is very flexible, because it is based on the theory of weighted automata.

I describe joint work with Sebastian Fischer and Frank Huch [1].

References

1. Fischer, S., Huch, F., Wilke, F.: A play on regular expressions: functional pearl. In: Proceedings of the 15th ACM SIGPLAN International Conference on Functional Programming, ICFP 2010, pp. 357–368. ACM, New York (2010)

G. Mauri and A. Leporati (Eds.): DLT 2011, LNCS 6795, pp. 44–45, 2011.

2. Glushkov, V.M.: On a synthesis algorithm for abstract automata. Ukr. Matem. Zhurnal 12(2), 147–156 (1960)
3. Google: Re2: a principled approach to regular expression matching (March 11, 2010), `http://google-opensource.blogspot.com/2010/03/re2-principled-approach-to-regular.html` (press release)
4. Kleene, S.: Representation of events in nerve nets and finite automata. In: Shannon, C., McCarthy, J. (eds.) Automata Studies, pp. 3–42. Princeton University Press, Princeton (1956)
5. Thompson, K.: Programming techniques: Regular expression search algorithm. Commun. ACM 11, 419–422 (1968)
6. xkcd.com: Regular expressions, `http://xkcd.com/208/`

State Complexity Research and Approximation

Sheng Yu and Yuan Gao

Department of Computer Science
University of Western Ontario
London, Ontario, Canada
{syu,ygao72}@csd.uwo.ca

Abstract. A number of basic questions concerning the state complexity research are discussed, which include why many basic problems weren't studied earlier, whether there is a general algorithm for state complexity, and whether there is a new approach in this area of research. The new concept of state complexity approximation is also discussed. We show that this new concept can be used to obtain good results when the exact state complexities are difficult to find.

1 Introduction

In the past two decades, a large number of research results on state complexity have been published. The state complexities of many individual operations and, recently, the state complexities of a number of combined operations have been studied.

Clearly, the state complexity research is fundamental in automata and formal language theory. All the questions in this research are basic questions. They are also well motivated by automata applications, especially the new applications.

State complexity is a descriptional complexity of regular languages based on the number of states in a minimal finite automaton. It is a descriptional measure of the resulting finite automaton after a transformation or an operation, which are called *transformational state complexity* or *operational state complexity*, correspondingly. Transformations, for example, include a nondeterministic finite automaton (NFA) to an equivalent deterministic finite automaton (DFA) for a subset of regular languages, regular expressions to finite automata, a multi-head finite automaton to a single-head finite automaton, and an alternating finite automaton to a DFA. Operations include catenation, Kleene star, left quotient, intersection combined with reversal, etc.

The finite automata involved in the state complexity research can be either NFAs or DFAs. So, we have deterministic state complexity and also nondeterministic state complexity.

Note that when a class of languages are considered, the state complexity of the class of languages can be the supremum among all the state complexities of the languages in the class, or an average among them (assuming that a certain distribution model is used). Thus, we have worst-case state complexity and also average-case state complexity.

G. Mauri and A. Leporati (Eds.): DLT 2011, LNCS 6795, pp. 46–57, 2011.

Since the early 1990s, most results on state complexity have been on operational, deterministic, and worst-case state complexity.

There are many fundamental questions concerning the state complexity research. We list some of them, which we consider as the most important questions, in the following.

1. Why weren't many of the state complexity questions studied in the 1960s and 1970s? Those questions are very fundamental in automata theory. They also look straightforward and basic.
2. Is there a general algorithm, for a given (individual or combined) operation and a set of regular languages, that can compute the state complexity of the operation on the set of languages? If we could find such an algorithm, the state complexity research would be much easier and we would not have much to do in the future.
3. If the answer to the previous question is no, what would be the best approach for obtaining a state complexity result? We know that for obtaining a worst-case state complexity result, the most difficult part usually is to find a general worst-case example. A proper approach may make this process easier.
4. When the exact state complexity of an operation is extremely difficult to obtain in some cases, is an approximation good enough for theoretical and/or practical purposes? How is an approximation obtained?

We will discuss all these questions and give our thought on them in the following sections, but first we will give the basic definitions and notations in the next section. At the end of the paper, we will consider what would be the next possible steps in state complexity research.

2 Preliminaries

We assume that the reader is familiar with the basics of finite automata and regular languages. Whenever necessary, [20] or [25] should be consulted.

We use the customary notation

$$A = (Q, \Sigma, \delta, q_0, F)$$

for *deterministic finite automata*, DFAs. The five items are, respectively, the state set, the input alphabet, the transition function, the initial state, and the set of final states. A is considered a *complete* automaton if $\delta(q, a)$ is defined for all $q \in Q$ and $a \in \Sigma$.

The (regular) language accepted by the DFA A is denoted by $L(A)$. The *state complexity* of a regular language L is the number of states in the minimal DFA A such that $L = L(A)$.

Formally, the (deterministic) *state complexity* of a regular language L, denoted $sc(L)$, is the number of states in the minimal complete DFA accepting L. The (worst-case) state complexity of a class $\mathcal{L}$ of regular languages, denoted $sc(\mathcal{L})$, is the supremum among all $sc(L)$, $L \in \mathcal{L}$. The state complexity of an operation or a transformation on regular languages is the state complexity of the language resulting from the transformation or operation. The nondeterministic state complexity, as well as the average-case state complexity, can be similarly defined.

3 Why Many State Complexity Questions Were Not Studied Earlier

State complexity questions are more fundamental and basic than time and space complexities in automata theory and applications. It is natural to ask why most of those state complexity questions were not studied earlier in the 1960s and 1970s.

We think that there are at least the following two main reasons:

(1) Those questions were not strongly motivated in applications in e 1960s and 1970s.
(2) Many of the questions were difficult or impossible to solve then due to the lack of the computer and software conditions.

One of the early applications of automata was the lexical analysis in compiler construction. The sizes of the automata used were relatively small (no more than several hundred states) and the main operations, e.g., membership, were simple. Other applications, e.g., circuit design, did not use finite automata of large sizes. Counting the number of states of finite automata was not well motivated at that time.

In the last two decades, there have been many new applications of finite automata that require automata of very large sizes. For example, in natural language and speech processing, the Bell Lab's multilingual TTS system needs 26.6 mbytes for German, 30.0 for French, and 39.0 for Mandarin [14]. Finite automata have also been applied in many other areas, e.g., software testing [5], parallel processing [22], and object-oriented modeling [19]. Finite automata used in applications are becoming very large. State complexity questions are more and more relevant to applications of automata.

Even if state complexity questions were well motivated in the 1960s and 1970s, the research in the area might not go very far then. For example, we now know that the state complexity of $L(A)(L(B) \cap L(C))$, where A, B, and C are DFAs of m, n, and p states, respectively, is $m2^{np} - 2^{np-1}$. If each of the three DFAs is of 5 states, the resulting DFA will have $9 \cdot 2^{24}$ states. General examples, for this combined operation, that can reach the upper bound would be very difficult to construct in the 1960s and 1970s without powerful computers and well-developed softwares. As we know, research in automata and formal language theory in the 1960s and 1970s were done by the researchers with basically only pens and paper. There were some results on state complexity of transformation between models and individual operations on regular languages, e.g., NFA to DFA transformation [15,17] and several basic operations on regular languages [16]. More complicated transformational and operational state complexities, especially for combined operations, would not be able to obtain then.

4 A General Algorithm for State Complexity?

For state complexity of operations, would it be possible to find a general algorithm that, for an arbitrarily given combined operation and a class of regular languages, computes the state complexity of the operation on the class of languages? If such an algorithm would be found, further research on state complexity, especially on combined operations, would become unnecessary. Many efforts in the state complexity research in the past would not be considered very useful. However, the undecidability result proved in [21] implies that such an algorithm does not exist.

The undecidability result in [21] is about the state complexity of the combined operation of intersection and marked catenation. Let $\mathcal{L} = \{L_1, \ldots, L_m\}$ be a set of arbitrary regular languages, $m \geq 2$. A $(\cap, \#)$-composition C is a marked catenation $C_1\#C_2\#\cdots\#C_n$, for some $n > 1$, where each C_i, $1 \leq i \leq n$, is an intersection $L_i^1 \cap L_i^2 \cap \cdots \cap L_i^l$ of l distinct languages from $\mathcal{L}$, for some $l \geq 1$. Let $x_1, \ldots, x_m$ denote the state complexities of $L_1, \ldots, L_m$, respectively. Then clearly, the state complexity C is a polynomial function of x_i, $1 \leq i \leq m$.

The proof of the undecidability is done by a reduction from the problem to a variation of *Hibert's Tenth Problem* [18], which is stated as follows [21]:

Theorem 1. *There is no algorithm of deciding, given a positive integer i, whether or not the inequality*

$$R_i(x_1, \ldots, x_m) + 1 \leq S_i(x_1, \ldots, x_m)$$

holds for all m-tuples $(x_1, \ldots, x_m)$ of nonnegative integers. Here R_i and S_i are effectively constructible polynomials with positive integer coefficients, over a fixed set of variables $\{x_1, \ldots, x_m\}$.

We associate the polynomials R_i with the combined operation of intersection and marked catenation, whereas the polynomials S_i with the proposed state complexity. Thus, the state complexity of such a combined operation on arbitrary regular languages is proved to be undecidable by the reduction. This result implies that there is no general algorithm for state complexity of combined operations.

5 New Approach in State Complexity Research

In the 1960s and 1970s, research in theoretical computer science was done by researchers with pens and paper. The results in automata and formal language theory then were mostly on what could and could not be done, e.g., many decidability results were obtained. This traditional approach in theoretical computer science research alone is no longer appropriate for solving state complexity problems, especially the state complexity of combined operations. Experiments are essential to the research on state complexity and many other topics. Due to the large numbers of states in the resulting automata, even for very small operands of an operation, pens and paper are far from enough for experiments. So, computer and appropriate software are very important to such type of research.

The role of experiments using computer software systems in the research on problems related to state complexity and other topics is described by the diagram in Figure 1.

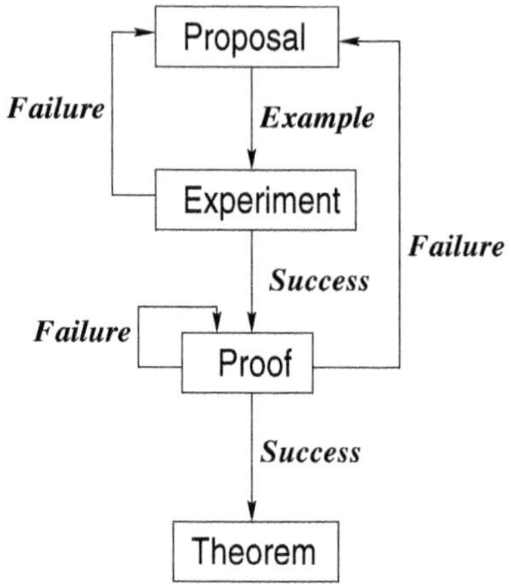

Fig. 1. The role of experiments

For solving a state complexity problem or a problem of a similar nature, we usually first have certain proposed results in mind after studying the problem. Examples are made according to our proposed results and sent to a software system, e.g., Grail, for experiments. If the experiments show that the proposed results are incorrect, we will modify the proposals and try new examples. If the experiments are successful, we will try to prove the results. The results may still be incorrect since we have only done experiments and usually the experiments are limited to only a relatively small number of examples. After several attempts and failures in proving the results, we may gain more insight into the problem and possibly modify the proposal and do experiments again. If we can prove the proposed results successfully, then we have obtained a theorem or theorems. Note that although the experiment part of the process is not the final decisive step, it plays a very important role in the process. Experiments in general cannot prove a general result, but they can, sometime easily, disprove a proposal of a general nature. So, experiments using computer software can greatly speed up the whole process.

6 State Complexity Approximation

There are at least two prominent problems concerning the state complexities of combined operations: (1) the state complexities of many combined operations are extremely difficult to compute, and (2) a large proportion of the results that have been obtained are pretty complex and impossible to comprehend. For example, the state complexity of the catenation for four regular languages accepted by deterministic finite automata (DFAs) of m, n, p, q states, respectively, is

$$9(2m-1)2^{n+p+q-5}-3(m-1)2^{p+q-2}-(2m-1)2^{n+q-2}+(m-1)2^{q}+(2m-1)2^{n-2},$$

which is clearly not intuitive. Close estimations of state complexities are usually good enough in many automata applications. In [24,4], estimations of state complexity of combined operations have been proposed and studied. In this section, we go further in the direction of the study in [24,4] and introduce the concept of state complexity approximation [7].

6.1 Definition of State Complexity Approximation

Briefly speaking, an approximation of a state complexity is an estimate of the state complexity with a ratio bound clearly defined. The ratio bound gives a precise measurement on the quality of the estimate.

The idea of state complexity approximation is from the notion of approximation algorithms which was formalized in early 1970's by David S. Johnson et al. [8,12,13]. Many polynomial-time approximation algorithms have been designed for a quite large number of NP-complete problems, which include the well-known traveling-salesman problem, the set-covering problem, and the subset-sum problem. Obtaining an optimal solution for an NP-complete problem is considered intractable. Near optimal solutions are often good enough in practice. Assuming that the problem is a maximization or a minimization problem, an approximation algorithm is said to have a ratio bound of $\rho(n)$ if for any input of size n, the cost C of the solution produced by the algorithm is within a factor of $\rho(n)$ of the cost C^* of an optimal solution [2]:

$$\max\left(\frac{C}{C^*}, \frac{C^*}{C}\right) \leq \rho(n).$$

The concept of state complexity approximation is in many ways similar to that of approximation algorithms. A state complexity approximation is close to the exact state complexity and normally not equal to it. The ratio bound shows the error range of the approximation. In addition to the property of having a small ratio bound in general, we also consider that a state complexity approximation should be in a simple and intuitive form.

In spite of the similarities, there are fundamental differences between a state complexity approximation and an approximation algorithm. The efforts in the area of approximation algorithms are in finding polynomial algorithms for NP-complete problems such that the results of the algorithms approximate the optimal results. In comparison, the efforts in the state complexity approximation are in searching directly for the estimations of state complexities such that they satisfy certain ratio bounds. The aim of designing an approximation algorithm is to transform an intractable problem into one that is easier to compute and the result is acceptable although not optimal. In comparison, a state complexity approximation result may have two different effects: (1) it gives a reasonable estimation of certain state complexity, with some bound, the exact value of which is difficult or impossible to compute; or (2) it gives a simpler and more comprehensible formula that approximates a known state complexity.

Now we give a formal definition for state complexity approximation. Let ξ be a combined operation on k regular languages. Assume that the state complexity of ξ is θ. We say that α is a state complexity approximation of the operation ξ with the ratio bound ρ if, for any large enough positive integers $n_1, \ldots, n_k$, which are the numbers of states of the DFAs that accept the argument languages of the operation, respectively,

$$\max\left(\frac{\alpha(n_1,\ldots,n_k)}{\theta(n_1,\ldots,n_k)}, \frac{\theta(n_1,\ldots,n_k)}{\alpha(n_1,\ldots,n_k)}\right) \le \rho(n_1,\ldots,n_k).$$

Note that in many cases, ρ is a constant. Since state complexity is a worst-case complexity, an approximation that is not smaller than the actual state complexity is preferred, which is the case for every approximation result in this paper.

6.2 Some Basic Results on State Complexity Approximation

In [24], an estimation method through nondeterministic state complexities was introduced for the (deterministic) state complexities of certain types of combined operations. The method is described in the following.

Assume we are considering the combination of a language operation g_1 with k arguments together with operations g_2^i, $i = 1, \ldots, k$. The *nondeterministic estimation upper bound,* or *NEU-bound* for the deterministic state complexity of the combined operation $g_1(g_2^1, \ldots, g_2^k)$ is calculated as follows:

(i) Let the arguments of the operation g_2^i be DFAs A_j^i with m_j^i states, $i = 1, \ldots, k$, $j = 1, \ldots, r_i$, $r_i \ge 1$.

(ii) The nondeterministic state complexity of the combined operation is at most the composition of the individual state complexities, and hence the language

$$g_1(g_2^1(L(A_1^1), \ldots, L(A_{r_1}^1)), \ldots, g_2^k(L(A_1^k), \ldots, L(A_{r_k}^k)))$$

has an NFA with at most

$$\mathrm{nsc}(g_1)(\mathrm{nsc}(g_2^1)(m_1^1, \ldots, m_{r_1}^1), \ldots, \mathrm{nsc}(g_2^k)(m_1^k, \ldots, m_{r_k}^k))$$

states, where $\mathrm{nsc}(g)$ is the nondeterministic state complexity (as a function) of the language operation g.

(iii) Consequently, the deterministic state complexity of the combined operation $g_1(g_2^1, \ldots, g_2^k)$ is upper bounded by

$$2^{\mathrm{nsc}(g_1)(\mathrm{nsc}(g_2^1)(m_1^1,\ldots,m_{r_1}^1),\ldots,\mathrm{nsc}(g_2^k)(m_1^k,\ldots,m_{r_k}^k))} \qquad (1)$$

The nondeterministic state complexity of the basic individual operations on regular languages has been investigated in [9,10,3].

In the following we show that this estimation method can produce nice approximation results for the state complexities of certain combined operations. The table below shows the actual state complexities and their corresponding NEU-bounds of the four combined operations [24]: (1) star of union, (2) star of intersection, (3) star of catenation, and (4) star of reversal.

Operations	State Complexity	NEU-bound
$(L(A) \cup L(B))^*$	$2^{m+n-1} - 2^{m-1} - 2^{n-1} + 1$	2^{m+n+2}
$(L(A) \cap L(B))^*$	$3/4\ 2^{mn}$	2^{mn+1}
$(L(A)L(B))^*$	$2^{m+n-1} + 2^{m+n-4} - 2^{m-1} - 2^{n-1} + m + 1$	2^{m+n+1}
$(L(B)^R)^*$	2^n	2^{n+2}

The next table shows clearly that each NEU-bound in the previous table gives a very good approximation to its corresponding state complexity.

Operations	Ratio bounds of the approximation
$(L(A) \cup L(B))^*$	≈ 8
$(L(A) \cap L(B))^*$	$8/3$
$(L(A)L(B))^*$	4
$(L(B)^R)^*$	4

In the above cases, although the exact state complexities have been obtained, the approximation results with small ratio bounds are good enough for practical purposes, and they clearly have the advantage of being more intuitive and simpler in formulation.

6.3 Approximation without Knowing Actual State Complexity

In this subsection, we consider two combined operations: (1) star of left quotient and (2) left quotient of star. For each of the combined operations, we do not have the exact state complexity; however, an approximation with a good ratio bound is obtained.

Let R and L be two languages over the alphabet Σ. Then the left quotient of R by L, denoted $L \backslash R$, is the language

$$\{y \mid xy \in R \text{ and } x \in L\}.$$

Let us first investigate a state complexity approximation of star of left quotient. In the following, we assume that all languages are over an alphabet of at least two letters.

Theorem 2. *Let R be a language accepted by an n-state DFA M, $n > 0$, and L be an arbitrary language. Then there exists a DFA of at most 2^n states that accepts $(L \backslash R)^*$.*

Proof. Let $M = (Q, \Sigma, \delta, s, F)$ be a complete DFA of n states and $R = L(M)$. For each $q \in Q$, denote by $L(M_q)$ the set $\{w \in \Sigma^* | \delta(s, w) = q\}$. We construct an NFA M' with multiple initial states to accept $(L\backslash R)^+$ as follows. M' is the same as M except that the initial state s of M is replaced by the set of initial states $S = \{q | L(M_q) \cap L \neq \emptyset\}$ and ε-transitions are added from each final state to the states in S. By using subset construction, we can construct a DFA A' of no more than $2^n - 1$ states that is equivalent to M'. Note that $\emptyset$ is not a state of A'. From the DFA A', we construct a new DFA A by just adding a new initial state that is also a final state and the transitions from this new state that are the same as the transitions from the original initial state of A'. It is easy to see that $L(A) = (L\backslash R)^*$ and A has 2^n states. □

This result gives an upper bound for the state complexity of the combined operation: star of left quotient. It means that for any n-state DFA language R, $n > 0$, and an arbitrary language L, the state complexity of the star of the left quotient of R by L is no more than 2^n.

Theorem 3. *For any integer $n \geq 2$, there exist a DFA M of n states and a language L such that any DFA accepting $(L\backslash L(M))^*$ needs at least $2^{n-1} + 2^{n-2}$ states.*

Proof. For $n = 2$, it is clear that $R = \{w \in \{a, b\}^* | \#_a(w) \text{ is odd}\}$ is accepted by a two-state DFA, and

$$(\{\varepsilon\}\backslash R)^* = R^* = \{\varepsilon\} \cup \{w \in \{a, b\}^* | \#_a(w) \geq 1\}$$

cannot be accepted by a DFA with less than three states.

For $n > 2$, let $M = (Q, \{a, b\}, \delta, 0, \{n-1\})$ where $Q = \{0, 1, \ldots, n-1\}$ and

$$\begin{aligned} \delta(i, a) &= i + 1 \bmod n, \ i = 0, 1, \ldots, n-1, \\ \delta(0, a) &= 0, \\ \delta(j, b) &= j + 1 \bmod n, \ j = 1, \ldots, n-1. \end{aligned}$$

The transition function of M is shown in Figure 2. It has been proved in [26] that the minimal DFA accepting $L(M)^*$ has $2^{n-1} + 2^{n-2}$ states. Let $L = \{\varepsilon\}$. Then $(L\backslash L(M))^* = L(M)^*$. So, any DFA accepting $(L\backslash L(M))^*$ needs at least $2^{n-1} + 2^{n-2}$ states. □

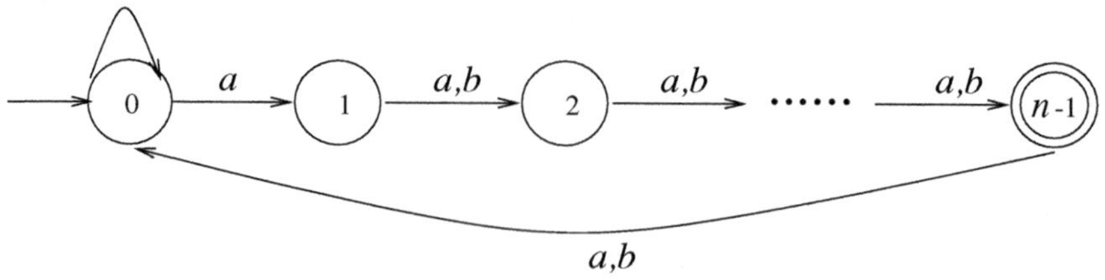

Fig. 2. The transition diagram of DFA M

This result gives a lower bound for the state complexity of star of left quotient. Clearly, the lower bound does not coincide with the upper bound. We still do not know the exact state complexity for this combined operation, yet, which could be difficult to obtain. However, we can easily obtain a good state complexity approximation for the operation. Let 2^n be the approximation. Then the ratio bound would be

$$\frac{2^n}{2^{n-1}+2^{n-2}} = \frac{4}{3} \ .$$

Next, we consider the combined operation: left quotient of star.

Theorem 4. *Let R be a language accepted by an n-state DFA M and L be an arbitrary language. Then there exists a DFA of at most $2^{n+1}-1$ states that accepts $L\backslash R^*$.*

Proof. Let $M = (Q, \Sigma, \delta, s, F)$ be a complete DFA of n states and $R = L(M)$. Then we can easily construct an $(n+1)$-state NFA $M' = (Q \cup \{s'\}, \Sigma, \delta', s', F \cup \{s'\})$ such that $L(M') = R^*$ by adding a new initial state s' and transitions $\delta'(s', \varepsilon) = s$ and $\delta'(f, \varepsilon) = s'$ for each final state $f \in F$. For each $q \in Q \cup \{s'\}$, we denote by $L(M_q)$ the set $\{w \in \Sigma^* | q \in \delta'(s', w)\}$. We construct an NFA N with multiple initial states to accept $L\backslash L(M') = L\backslash R^*$ as follows. N is the same as M' except that the initial state s' of M' is replaced by the set of initial states $S = \{q | L(M_p) \cap L \neq \emptyset\}$. By using subset construction, we can verify that there exists a DFA A of no more than $2^{n+1}-1$ states that is equivalent to N. Note that $\emptyset$ is not a state of A. It is easy to see that

$$L(A) = L(N) = L\backslash L(M') = L\backslash R^*.$$

So, $2^{n+1}-1$ is an upper bound of the state complexity of the left quotient of star operation. □

Theorem 5. *For any integer $n \geq 2$, there exist a DFA M of n states and a language L such that any DFA accepting $L\backslash L(M)^*$ needs at least $2^{n-1}+2^{n-2}$ states.*

Proof. For $n = 2$, we still use $R = \{w \in \{a, b\}^* | \#_a(w) \text{ is odd}\}$ which is accepted by a two-state DFA. $\{\varepsilon\}\backslash R^* = R^*$ cannot be accepted by a DFA with less than three states.

Again we use DFA M shown in Figure 2 for any integer $n > 2$. As stated before, it has been proved that the minimal DFA accepting $L(M)^*$ has $2^{n-1}+2^{n-2}$ states. So any DFA accepting $L\backslash L(M)^*$ needs at least $2^{n-1}+2^{n-2}$ states. □

For this combined operation, we choose 2^{n+1} to be an approximation of its state complexity. Then the ratio bound can be calculated easily as follows:

$$\frac{2^{n+1}}{2^{n-1}+2^{n-2}} = \frac{8}{3} \ .$$

7 Future Directions

In the last two decades, many results on state complexity have been obtained. By now, the state complexities of most individual operations and many combined operation are known. Clearly, there are still many interesting operations, especially combined operations, that should be further studied. We list in the following only two possible new directions in research in state complexity and closely related areas.

(1) In a complete DFA, the number of transitions is linear to the number of states assuming that the size of the alphabet is a constant. However, in practical applications, the size of the alphabet is usually very big (at least over a hundred symbols) and each state has only a few useful transitions, i.e., transitions that are not to the sink state. Therefore, partial automata rather than complete automata are usually constructed. In these cases, the number of transitions rather than the number of states would decide the descriptional size of an automaton. We call the number of transitions of an automaton the *transition complexity* of the automaton. The transition complexity of a regular language we are considering here is the minimum among all the transition complexities of the partial DFAs that accept the language. The transition complexity of NFAs have been studied in a number of papers. However, the transition complexity of partial DFAs basically has not been studied before. For example, the answer to the very basic question "What is the transitional complexity of the catenation of two regular languages, that are accepted by two partial DFAs of m and n transitions, respectively, in the worst case" is not known, yet. Many questions in this direction can be done in the near future.

(2) Average-case state complexities are clearly very important to automata applications. An essential question on average state complexity is the establishment of basic distribution models of automata. Until now only a few results have been obtained. For practical applications, instead of theoretical results on this topic, experimental results are in general satisfactory. The experiments can be based on automata generated by a random generator [1] or pools of examples from specific applications. Those results would be very useful to general or specific applications of automata.

References

1. Almeida, M., Moreira, N., Reis, R.: Enumeration and generation with a string automata representation. Theoretical Computer Science 387(2), 93–102 (2007)
2. Cormen, T.H., Leiserson, C.E., Rivest, R.L.: Introduction to algorithms. The MIT Press and McGraw-Hill, Massachusetts (1990)
3. Ellul, K.: Descriptional complexity measures of regular languages, Master's Thesis, University of Waterloo, Ontario (2002)
4. Ésik, Z., Gao, Y., Liu, G., Yu, S.: Estimation of state complexity of combined operations. Theoretical Computer Science 410(35), 3272–3280 (2009)

5. Fujiwara, S., Bochmann, G.V., Khendek, F., Amalou, M., Ghedamsi, A.: Test development - based on finite state machine (FSM) models. IEEE Transactions on Software Engineering 17(6), 591–603 (1991)
6. Gao, Y., Salomaa, K., Yu, S.: The state complexity of two combined operations: star of catenation and star of reversal. Fundamenta Informaticae 83(1-2), 75–89 (2008)
7. Gao, Y., Yu, S.: State complexity approximation. In: Proceedings of Descriptional Complexity of Formal Systems, pp. 163–174 (2009)
8. Garey, M.R., Graham, R.L., Ullman, J.D.: Worst-case analysis of memory allocation algorithms. In: Proceedings of the 4th Annual ACM Symposium on the Theory of Computing, pp. 143–150 (1972)
9. Holzer, M., Kutrib, M.: Nondeterministic descriptional complexity of regular languages. International Journal of Foundations Computer Science 14, 1087–1102 (2003)
10. Holzer, M., Kutrib, M.: Unary language operations and their nondeterministic state complexity. In: Ito, M., Toyama, M. (eds.) DLT 2002. LNCS, vol. 2450, pp. 162–172. Springer, Heidelberg (2003)
11. Hopcroft, J.E., Ullman, J.D.: Introduction to automata theory, languages, and computation. Addison-Wesley, Reading (1979)
12. Johnson, D.S.: Fast allocation algorithms. In: Proceedings of the 13th Annual IEEE Symposium on Switching and Automata Theory, pp. 144–154 (1972)
13. Johnson, D.S.: Near-optimal bin packing algorithms, PhD Dissertation, Massachusetts Institute of Technology, Cambridge, MA (1993)
14. Kiraz, G.A.: Compressed storage of sparse finite-state transducers. In: Boldt, O., Jürgensen, H. (eds.) Proceedings of CIAA 2001. LNCS, vol. 2214, pp. 109–121. Springer, Heidelberg (2001)
15. Lupanov, O.B.: A comparison of two types of finite automata. Problemy Kibernetiki 9, 321–326 (1963) (in Russian)
16. Maslov, A.N.: Estimates of the number of states of finite automata. Dokl. Akad. Nauk SSSR 194, 1266–1268 (1970) (in Russian); English translation: Soviet Math. Dokl. 11, 1372–1375 (1970)
17. Moore, F.: On the bounds for state-set size in the proofs of equivalence between deterministic, nondeterministic, and two-way finite automata. IEEE Trans. Comput. C-20, 1211–1214 (1971)
18. Rozenberg, G., Salomaa, A.: Cornerstones of Undecidability. Prentice Hall, New York (1994)
19. Rumbaugh, J., Jacobson, I., Booch, G.: The United Modeling Language Reference Manual. Addison-Wesley, Reading (1999)
20. Salomaa, A.: Theory of automata. Pergamon Press, Oxford (1969)
21. Salomaa, A., Salomaa, K., Yu, S.: Undecidability of the state complexity of composed regular operations. In: Proceedings of LATA 2011 (2011) (to be published)
22. Salomaa, K., Yu, S.: Synchronization Expressions with Extended Join Operation. Theoretical Computer Science 207, 73–88 (1998)
23. Salomaa, A., Salomaa, K., Yu, S.: State complexity of combined operations. Theoretical Computer Science 383(2-3), 140–152 (2007)
24. Salomaa, K., Yu, S.: On the state complexity of combined operations and their estimation. International Journal of Foundations of Computer Science 18(4), 683–698 (2007)
25. Yu, S.: Regular languages. In: Rozenberg, G., Salomaa, A. (eds.) Handbook of formal languages, vol. 1, pp. 41–110. Springer, New York (1997)
26. Yu, S., Zhuang, Q., Salomaa, K.: The state complexities of some basic operations on regular languages. Theoretical Computer Science 125(2), 315–328 (1994)

Counting the Orderings for Multisets in Consecutive Ones Property and PQ-Trees*

Giovanni Battaglia[1], Roberto Grossi[1], and Noemi Scutellà[2]

[1] Dipartimento di Informatica, Università di Pisa, Pisa 56127, Italy
gbattag,grossi@di.unipi.it
[2] List SpA, Pisa 56122, Italy
n.scutella@list-group.com

Abstract. A binary matrix satisfies the *consecutive ones property* (C1P) if its columns can be permuted such that the 1s in each row of the resulting matrix are consecutive. Equivalently, a family of *sets* $F = \{Q_1, \ldots, Q_m\}$, where $Q_i \subseteq R$ for some universe R, satisfies the C1P if the symbols in R can be permuted such that the elements of each set $Q_i \in F$ occur consecutively, as a contiguous segment of the permutation of R's symbols. Motivated by combinatorial problems on sequences with repeated symbols, we consider the C1P version on *multisets* and prove that counting the orderings (permutations) thus generated is #P-complete. We prove completeness results also for counting the permutations generated by PQ-trees (which are related to the C1P), thus showing that a polynomial-time algorithm is unlikely to exist when dealing with multisets and sequences with repeated symbols.

1 Introduction

A binary matrix M of size $m \times n$ satisfies the *consecutive ones property* (C1P) if its n columns can be permuted such that the 1s in each row of the resulting matrix are consecutive. An equivalent definition holds for the columns by permuting the rows. The property is often formulated in terms of sets: A family of sets $F = \{Q_1, \ldots, Q_m\}$, where each Q_i is a subset of the universe of symbols $R = \{r_1, \ldots, r_n\}$, satisfies the C1P if the symbols in R can be permuted such that the elements of each set $Q_i \in F$ occur consecutively as a contiguous segment of the permutation of R's symbols.

For example, consider the universe $R = \{\mathtt{a}, \mathtt{b}, \mathtt{c}, \mathtt{d}, \mathtt{e}\}$. The C1P is not satisfied by the family $F = \{\{\mathtt{a}, \mathtt{b}\}, \{\mathtt{b}, \mathtt{c}\}, \{\mathtt{b}, \mathtt{d}\}\}$, since b can have at most two adjacent symbols in any permutation of R. On the other hand, the family $F = \{\{\mathtt{b}, \mathtt{c}\}, \{\mathtt{b}, \mathtt{d}\}\}$ satisfies the C1P: one feasible permutation of R is $x = \mathtt{eacbd}$, but not all permutations of R are feasible (e.g. $y = \mathtt{abcde}$ is not, because the symbols $\{\mathtt{b}, \mathtt{d}\}$ are not consecutive in y).

The C1P on sets can be formulated as a C1P problem on the binary matrix M obtained by associating row i with set $Q_i \in F$, and column j with element $r_j \in R$. Specifically, $M_{ij} = 1$ iff $r_j \in Q_i$, as shown below for our example.

* Research partially supported by MIUR of Italy under project AlgoDEEP prot. 2008TFBWL4. Work done while the third author was at the University of Pisa.

G. Mauri and A. Leporati (Eds.): DLT 2011, LNCS 6795, pp. 58–69, 2011.

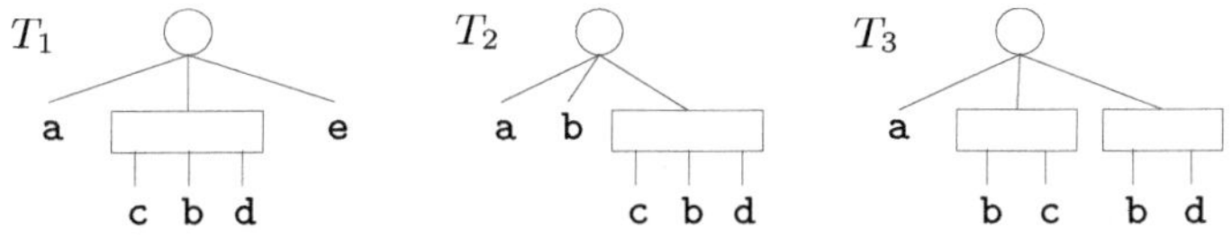

Fig. 1. Some examples of PQ-trees.

	a	b	c	d	e
{b, c}	0	1	1	0	0
{b, d}	0	1	0	1	0

e	a	c	b	d
0	0	1	1	0
0	0	0	1	1

The problem of finding the *orderings*, namely, the permutations of R that are generated by the C1P, arises in several situations. It was first solved efficiently by Fulkerson and Gross [8] in their study on the incidence matrix of interval graphs, using an $O(mn^2)$ time algorithm. Ghosh [10] applied the problem to information retrieval, where R is the set of input records and each Q_i is the set of records satisfying a query: for each Q_i, the C1P guarantees that the corresponding records can be retrieved from consecutive storage locations. Booth and Leuker [3,4] showed how to find any such ordering in linear time, with respect to the number of 1s in M, with applications to some graph problems such as planarity testing. They employed the *PQ-tree* data structure to represent *compactly all the orderings* yielding the C1P for the given matrix M.

The PQ-tree corresponding to our example is denoted by T_1 in Figure 1. The leaves of the PQ-tree contain the symbols of R: when reading these symbols by traversing the leaves in preorder, we obtain a string called the *frontier* of the PQ-tree. As it can be seen, the frontier is one of the orderings yielding the C1P in our example tree T_1. Further orderings can be obtained by rearranging the children of the nodes of the PQ-tree, since they implicitly encode the sets in F. A round node in Figure 1 is called *P-node*, and its children can be rearranged in *any* order. A square node is called *Q-node*, and its children can be only rearranged in *left-to-right* or *right-to-left* order. By conceptually performing all the feasible rearrangements of the nodes in the PQ-tree according to the above rules, we obtain the set of frontiers that are generated by the PQ-tree. These frontiers are in *one-to-one* correspondence with all the orderings yielding the C1P for matrix M, as it can be verified by inspecting our example for T_1: we can represent them as the strings x_1 = acbde, x_2 = adbce, x_3 = aecbd, x_4 = aedbc, x_5 = cbdae, x_6 = dbcae, x_7 = cbdea, x_8 = dbcea, x_9 = ecbda, x_{10} = edbca, x_{11} = eacbd, and x_{12} = eadbc.

Our problems. Since its inception, the C1P has found many applications under several incarnations. Recent fields of application are stringology and bioinformatics, e.g. physical mapping [11,5] and gene analysis [7,12,14], providing the inspiration for the problems in this paper. Motivated by the combinatorial aspects of sequences with repeated symbols, we consider the scenario for the C1P in which the symbols in the input set R are *not* necessarily distinct.

We investigate the problem of how to satisfy the C1P when R and the Q_is are *multisets*. To get the flavor of the problem, consider the universe $R = \{\texttt{a}, \texttt{b}, \texttt{b}, \texttt{c}, \texttt{d}\}$ and the family $F = \{\{\texttt{b}, \texttt{c}\}, \{\texttt{b}, \texttt{d}\}\}$. The situation arises from the fact that the symbol b in both $Q_1 = \{\texttt{b}, \texttt{c}\}$ and $Q_2 = \{\texttt{b}, \texttt{d}\}$ can either match the same occurrence of b in R or not. The former case gives rise to the PQ-tree T_2 in Figure 1, while the latter gives rise to the PQ-tree T_3. The set of frontiers are now strings with repeated symbols: the set of frontiers generated by one PQ-tree is *not* contained in the set of the other PQ-tree. However, the two occurrences of b in R are *indistinguishable.*

In this paper, we consider problems arising from repeated symbols, and show that dealing with the C1P on multisets is hard. Specifically, we study the problem of *counting* the number of orderings. This is "simpler" than listing all the orderings. Note that the counting problem using standard PQ-trees on *sets* takes polynomial time, since we can use the aforementioned one-to-one correspondence between the orderings and the frontiers. The simple algorithm for sets is the following. For a given node u in the PQ-tree, apply a recursive post-order traversal: If u is a leaf, it has just one frontier. Otherwise, let d be the number of children of u, and f_i be the number of frontiers for the ith child in u (where f_i has been recursively computed for $1 \leq i \leq d$). Then, the number f of frontiers for u is $f = d! \times \prod_{i=1}^{d} f_i$ when u is a P-node, and $f = 2 \times \prod_{i=1}^{d} f_i$ when u is a Q-node (e.g. $f = 12 = 3! \times 2$ frontiers for T_1 in Figure 1). It is therefore interesting to study the case of multisets.

Our results. Our first result is to prove that the problem (denoted #FRONT) of counting the frontiers of a PQ-tree whose leaves store the (repeated) symbols of a *multiset* is #$\mathcal{P}$-complete [15]. We refer the reader to Section 3 for a discussion.

As for the original problem (denoted #FMO) of counting the orderings for the C1P, one could hope that a polynomial solution might exist without relying on PQ-trees. This is also unlikely to happen. Our second result is to prove that the problem of counting the orderings for the C1P on multisets is #$\mathcal{P}$-complete. We refer the reader to Section 4 for a discussion.

An interesting implication of our findings is the relation with the well-known counting version #HAM of the Hamiltonian path problem [9]. As previously mentioned, a *direct* mapping of the orderings for the C1P in multisets into the frontiers of PQ-trees has some intrinsic ambiguity. On the other hand, we can prove that both the counting problems #FRONT and #FMO are #$\mathcal{P}$-complete using a reduction from #HAM. By the completeness properties, it follows that there must exist a relation between the latter two problems. However, simply composing the reductions does not immediately produce a single PQ-tree, having the same properties as the input instance, which is an open problem.

Our approach is of independent interest, and the counting nature of the problems emphasizes the combinatorial properties of the strings (orderings) thus generated by the reductions.

2 Definitions and Terminology

We consider a class of strings defined over multisets, where the usual notions of inclusion, equality, and union, take into account the multiplicities of the elements in the multisets. We say that a string $s \equiv s_1 s_2 \cdots s_n$ is drawn from a multiset R of symbols if and only if the multiset $S = \{s_1, s_2, \ldots, s_n\}$ satisfies the condition $S \subseteq R$, where s_i denotes the symbol stored into position i of s, for $1 \leq i \leq n$. We also say that a *multiset* P *occurs* in a *string* s (or equivalently P is *contained* in s), if there is a substring $s_i s_{i+1} \cdots s_j$ of s, where $1 \leq i, j \leq n$, such that $P = \{s_i, s_{i+1}, \ldots, s_j\}$. In the latter case, we say that P occurs at position i in s (and P is called π-pattern [1]). For example, $P = \{\mathtt{a}, \mathtt{c}, \mathtt{a}\}$ occurs at position $i = 1$ in $s = \mathtt{aacb}$, while P is not contained in $s_2 = \mathtt{aabc}$.

Given a decision problem $\mathcal{A}$, we will denote by $\#\mathcal{A}$ its *counting version*, where we are required to count the number of the solutions of $\mathcal{A}$ [15]. We now introduce the #FMO problem, that formalizes the problem of extending the Booth-Leuker approach [4] for the C1P to multisets.

Problem 1 (#FMO = Counting Full Multiset Orderings). Input: an instance $\langle R, F \rangle$, where R is a multiset of symbols, and $F = \{Q_1, \ldots, Q_m\}$ is a family of multisets $Q_i \subset R$. Output: how many strings x can be drawn from *all* symbols in R ($|x| = |R|$), so that each Q_i is contained in x?

For example, given $R = \{\mathtt{a}, \mathtt{b}, \mathtt{b}, \mathtt{c}, \mathtt{d}\}$ and $F = \{\{\mathtt{b}, \mathtt{c}\}, \{\mathtt{b}, \mathtt{d}\}\}$, $x = \mathtt{abcbd}$, is one of the feasible solution of the $\langle R, F \rangle$ #FMO instance. We now introduce our second problem, which requires some additional terminology and is related to Problem 1.

Problem 2 (#FRONT= Counting PQ-trees Frontiers). Input: a PQ-tree T, where its leaves are labeled with symbols that are not necessarily distinct. Output: what is the size of the set of frontiers $Fr(T)$ of T?

We call the *frontier* of a PQ-tree T, denoted by $F(T)$, the permutation of the symbols obtained by reading the labels of the leaves from left to right. Given two PQ-trees T and T', we say that T is *equivalent* to T' (written $T \equiv T'$) if one tree can be obtained from the other by possibly permuting the children of one or more P-nodes, and by possibly reversing the children of some Q-nodes. The set of the frontiers of all the trees that are equivalent to T is denoted by $Fr(T)$.

Since a P-node (or a Q-node) having one child can be removed from T without changing $Fr(T)$, and a P-node with two children can be replaced by a Q-node (it represents the "left to right" and "right to left" permutations only), we define the *canonical form* constraining each Q-node to have at least two children, and each P-node to have at least three children. In the rest of the paper, we assume that each PQ-tree is in canonical form.

We are interested in counting the number of frontiers in Problem 2, namely, the size of $Fr(T)$ for a PQ-tree T. A formal description of the $\#\mathcal{P}$ class is beyond the scope of this paper, and we refer the reader to the textbooks in [2,9,13].

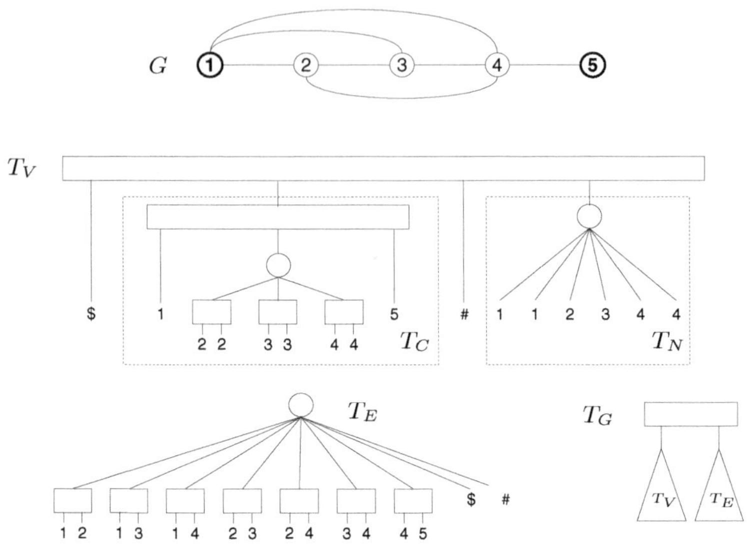

Fig. 2. The PQ-tree T_G associated with the input graph G, where the source and the destination vertices are $w = 1$ and $s = 5$, and T_V and T_E are shown individually

3 Counting the Frontiers of a PQ-tree

We begin by discussing the completeness of the #FRONT problem. We use a reduction from the well-known counting version of Hamiltonian Path (#HAM). We are given an undirected graph G, a source vertex $w \in G$, and a destination vertex $s \in G$. We want to know how many paths H in G start in w and end in s, such that all the vertices in G are traversed exactly once by each H. For example, one such path is $H = \langle 1, 3, 2, 4, 5 \rangle$ in the graph G shown in Figure 2. In the rest of the paper, we assume that G is connected, w and s have degree at least one, and the other vertices have degree at least two (otherwise there is no Hamiltonian path). We also assume that there are no self-loops.

3.1 Construction of the PQ-trees

The main idea is to code the structure of the given graph G in three suitable PQ-trees, T_G, T_V, and T_E, such that each Hamiltonian path H is in one-to-many correspondence with a suitable set of strings from their frontiers. We now describe our reduction from $G = \langle V, E \rangle$ to T_G, T_V, and T_E, using Figure 2 as an illustrative example.

The root of T_G is a Q-node having two PQ-trees T_V and T_E as children.

Tree T_E encodes all the feasible permutations of the edges in E. The root of T_E is a P-node having $|E| + 2$ children. Two of them are special "endmarkers," and are labeled with \$ and #. Each of the remaining children is a Q-node that encodes an edge $e = \{i, j\}$ by two leaves labeled with i and j, respectively, as children. In our example, T_E has $|E| = 7$ Q-nodes with children labeled by $\{1, 2\}$, $\{1, 3\}$, $\{1, 4\}$, $\{2, 3\}$, $\{2, 4\}$, $\{3, 4\}$, and $\{4, 5\}$, plus the endmarkers \$ and #.

Tree T_V enforces a classification of the edges as "coding" a Hamiltonian path, or "non-coding" otherwise. Specifically, the root of T_V is a Q-node with four children: one leaf labeled with \$, a PQ-tree T_C for the coding edges, one more leaf labeled with #, and a PQ-tree T_N for the non-coding edges. The root of T_C is a Q-node with three children. The first child is a leaf labeled with the source w and the last is a leaf labeled with the destination s. The middle child is a P-node with $|V| - 2$ children, each of which is a Q-node with two leaves labeled with the same symbol i, for $i \in V \setminus \{w, s\}$. In our example $w = 1$, $s = 5$, and $|V| = 5$. The root of the non-coding tree T_N is a P-node having $2(|E| - |V| + 1)$ leaves as children. Letting d_i denote the degree of vertex i, there are $d_w - 1$ leaves labeled with w, $d_s - 1$ leaves labeled with s, and $d_i - 2$ leaves labeled with $i \neq w, s$. In our example, the leaves are labeled with $1, 1, 2, 3, 4, 4$, where $2(|E| - |V| + 1) = 6$.

The above construction requires polynomial time, and the rationale will be given in Section 3.2.

Lemma 1. *Given an undirected graph $G = \langle V, E \rangle$, its corresponding PQ-trees T_G, T_V, and T_E can be built in $\mathcal{O}(|V| + |E|)$ time.*

3.2 Properties of the PQ-trees

Consider the Hamiltonian path $H = \langle 1, 3, 2, 4, 5 \rangle$ in our example. (Observe that the reversal of H, namely $\langle 5, 4, 2, 3, 1 \rangle$, is also a Hamiltonian path, but we consider it to be different from H for the counting purposes.) The corresponding strings α^H belonging to the frontiers $Fr(T_G)$ are characterized as follows. First at all, each α^H is a square, namely, the concatenation $\alpha^H = \alpha\,\alpha$ of two equal strings α, where α belongs to both the frontiers $Fr(T_V)$ and $Fr(T_E)$, and is of length $2|E| + 2$. For example, $\alpha =$ \$13322445#121434 is one such feasible string. We can characterize the general structure of the strings α by observing that they match one of the following two patterns. Let π denote an arbitrarily chosen permutation of the pairs in $\{1, 2\}, \{1, 4\}, \{3, 4\}$, which represents the edges *not* traversed by H. (That is, π belongs to the frontiers of the PQ-tree resulting from $\{\{1, 2\}, \{1, 4\}, \{3, 4\}\}$.) The former pattern for α is \$ 13322445 # π, where the initial symbols are fixed and only π may vary; analogously, the latter is π # 13322445 \$. For example, $\alpha =$ 413421 # 13322445 \$ matches the latter pattern.

Having introduced the structure of $\alpha^H = \alpha\,\alpha$ in our example, we show how to make α satisfy the implicit conditions encoded in T_V and T_E. Indeed, T_E guarantees that the two integers in each of the pairs corresponding to the edges in E always occur consecutively in α. Moreover, the subtree T_C in T_V constraints each vertex $i \in V \setminus \{w, s\}$ to appear exactly twice in the chosen subset of edges, while w and s are required to appear just once. Note that the purpose of the subtree T_N is that of "padding" the edges in E that are not traversed by H, since we do not know a priori which ones will be touched by H.

We now generalize the above observations on α. In the following we can restrict our focus on paths of the form $i_1, i_2, \ldots, i_{|V|}$, that are permutations of $\{1, 2, \ldots, |V|\}$ with $i_1 = w$ and $i_{|V|} = s$ (otherwise they cannot be Hamiltonian paths from w to s). We introduce the notation $Perm(Q)$ for a set $Q = \{\{a_1, b_1\}, \{a_2, b_2\}, \ldots, \{a_r, b_r\}\}$ of unordered pairs. It represents the set of all the permutations of $a_1, b_1, a_2, b_2, \ldots, a_r, b_r$ such that a_l and b_l occupy contiguous positions for $1 \le l \le r$. For example, given $Q = \{\{1,2\},\{1,4\},\{3,4\}\}$, we have that 413421 is a valid permutation in $Perm(Q)$, while 413241 is not.

We now show in Lemmas 2–4 that there exists a one-to-many correspondence between the Hamiltonian path H in G and the strings $\alpha \in Fr(T_V) \cap Fr(T_E)$.

Lemma 2. *Let $G = \langle V, E \rangle$ be an undirected graph, and T_G, T_V, and T_E be its corresponding PQ-trees. For any string $\alpha \in Fr(T_V) \cap Fr(T_E)$, there exists a corresponding Hamiltonian path H of G from w to s.*

Lemma 3. *Let $G = \langle V, E \rangle$ be an undirected graph, and T_G, T_V, and T_E be its corresponding PQ-trees. For any Hamiltonian path H of G from w to s, there exists at least one corresponding string $\alpha \in Fr(T_V) \cap Fr(T_E)$.*

Lemma 4. *Let $\Sigma_H \subseteq Fr(T_V) \cap Fr(T_E)$ denote the set of all the strings corresponding to a given Hamiltonian path H, as stated in Lemma 3. Then, for any two Hamiltonian paths $H \neq H'$ of G from vertex w to vertex s, it is $\Sigma_H \cap \Sigma_{H'} = \emptyset$.*

3.3 Reduction from #HAM to #FRONT

We now show how to reduce the problem #HAM of counting the Hamiltonian paths in $G = \langle V, E \rangle$, to the problem #FRONT of counting the frontiers of PQ-trees, namely, T_G, T_V, and T_E. We denote the number of frontiers for a PQ-tree T by $|Fr(T)|$. Here is the polynomial time reduction for the input graph G and its two vertices w and s:

- Build the PQ-trees T_G, T_V, and T_E (see Lemma 1).
- Return the following as the number of Hamiltonian paths from w to s in G:

$$\frac{|Fr(T_V) \cap Fr(T_E)|}{|\Sigma_H|} = \frac{2\,|Fr(T_V)| \times |Fr(T_E)| - |Fr(T_G)|}{2\,(|E| - |V| + 1)! \times 2^{|E|-|V|+1}} \tag{1}$$

Clearly, the formula in (1) can be computed in polynomial time. We now show its correctness.

Lemma 5. *Let $\Sigma_H \subseteq Fr(T_V) \cap Fr(T_E)$ denote the set of strings corresponding to a Hamiltonian path H. Then, for any Hamiltonian path H from w to s, we have $|\Sigma_H| = 2\,(|E| - |V| + 1)! \times 2^{|E|-|V|+1}$.*

Lemma 6. $|Fr(T_G)| = 2\,|Fr(T_V)| \times |Fr(T_E)| - |Fr(T_V) \cap Fr(T_E)|$

We now have all the ingredients to prove the following result.

Theorem 1. *#FRONT is #$\mathcal{P}$-complete.*

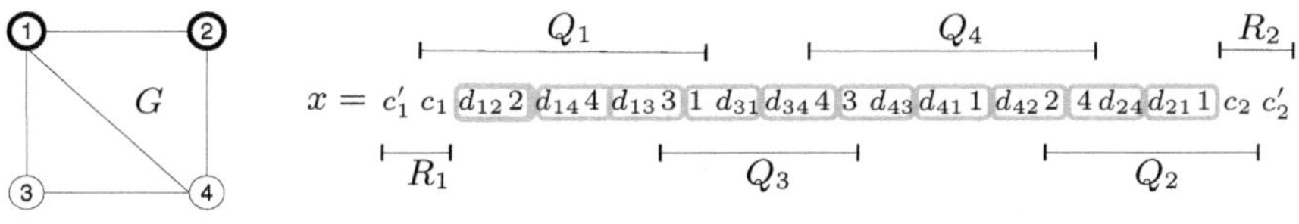

$Q_1 = \{1, d_{12}, d_{13}, d_{14}, 2, 3, 4, c_1\}$
$Q_2 = \{2, d_{21}, d_{24}, 1, 4, c_2\}$
$Q_3 = \{3, 3, d_{31}, d_{34}, 1, 4\}$
$Q_4 = \{4, 4, d_{41}, d_{42}, d_{43}, 1, 2, 3\}$
$R_1 = \{c_1, c_1'\}$
$R_2 = \{c_2, c_2'\}$

$Q_{12} = \{d_{12}, 2\}$
$Q_{13} = \{d_{13}, 3\}$
$Q_{14} = \{d_{14}, 4\}$
$Q_{24} = \{d_{24}, 4\}$
$Q_{34} = \{d_{34}, 4\}$

$Q_{21} = \{d_{21}, 1\}$
$Q_{31} = \{d_{31}, 1\}$
$Q_{41} = \{d_{41}, 1\}$
$Q_{42} = \{d_{42}, 2\}$
$Q_{43} = \{d_{43}, 3\}$

$F = \{ Q_1, Q_2, Q_3, Q_4, R_1, R_2, Q_{12}, Q_{21}, Q_{13}, Q_{31}, Q_{14}, Q_{41}, Q_{24}, Q_{42}, Q_{34}, Q_{43} \}$
$R = \{ d_{12}, d_{21}, d_{13}, d_{31}, d_{14}, d_{41}, d_{24}, d_{42}, d_{34}, d_{43}, 1, 1, 1, 2, 2, 3, 3, 4, 4, 4, c_1, c_1', c_2, c_2' \}$

Fig. 3. Example of reduction from a Hamiltonian Path instance for a graph G, where the source and the destination vertices are $w = 1$ and $s = 2$, into a #FMO instance $\langle R, F\rangle$. Sets Q_{ij} are shown boxed in string x.

4 Hardness Results for #FMO

We now show how to reduce the #HAM problem to the counting version of the Full Multiset Problem (#FMO). For the given undirected graph $G = \langle V, E\rangle$, together with the source and the destination vertices, w and s, we make the same assumptions as in Section 3.

4.1 Instance Construction

Consider the example in Figure 3. On the left we show the input undirected graph G, where the source and the destination vertices $w = 1$ and $s = 2$ are in boldface. The corresponding #FMO instance $\langle R, F\rangle$ is reported on the right, while one of the solution string x, corresponding to the Hamiltonian path $H = \langle 1, 3, 4, 2\rangle$ is represented at the bottom.

We build an instance of #FMO as follows. For each vertex i, we construct the multiset Q_i containing two occurrences of the symbol i (if $i \neq w, s$), or one occurrence of i and one of the special symbol c_i (if $i = w, s$). We also add symbols d_{ij} and j to Q_i, for every incident edge $\{i, j\}$. As a result, each undirected edge $\{i, j\}$ is represented by two different symbols $d_{ij} \in Q_i$ and $d_{ji} \in Q_j$. Formally,

$$Q_i = \begin{cases} \bigcup_{\{i,j\}\in E}\{d_{ij}, j\} \cup \{i, c_i\} & i = w, s \\ \bigcup_{\{i,j\}\in E}\{d_{ij}, j\} \cup \{i, i\} & i \neq w, s \end{cases}$$

To guarantee the condition that w and s are the source and the destination vertices, respectively, we introduce two symbols c'_w and c'_s, and two sets $R_w = \{c_w, c'_w\}$ and $R_s = \{c_s, c'_s\}$, which do not correspond to any vertex of the input graph. They are used to guarantee that Q_w and Q_s will always occur as the first and the last multiset of any solution string x for our #FMO instance.

In general, the intersection between two multisets Q_i and Q_j can contain more symbols than just i and j. For example, the intersection between Q_1 and Q_4 is $I_{14} = \{1, 4, 2, 3\}$ because it contains also 2 and 3, each of them corresponding to the vertex forming a triangle with 1 and 4, respectively. To avoid this situation, $2|E|$ auxiliary multisets $Q_{ij} = \{d_{ij}, j\}$ are used to constraint the intersection between the multisets represented, such that it contains exactly two symbols. Since multisets are represented as substrings, this intersection is represented as a *common substring* inside each solution string x. Observe that each edge $\{i, j\} \in E$ gives rise to two multisets Q_{ij} and Q_{ji}. In the string x shown in Figure 3, the purpose of the multisets Q_{ij} and Q_{ji} is to enforce the common substring for Q_1 and Q_3 inside x to be $1, 3$ or $1, 3$, between Q_3 and Q_4 to be $3, 4$ or $3, 4$, and so on.

We finally choose the multiset $R = Q \setminus R'$ where $Q = \bigcup_i Q_i \cup \{c'_w, c'_s\}$ and $R' = \bigcup_{i \neq w,s} \{i, i\} \cup \{w, s\}$. We also choose $F = \{Q_1, \ldots, Q_{|V|}\} \cup \{R_w, R_s\} \cup \{Q_{ij}, Q_{ji}\}_{\{i,j\} \in E}$. The idea behind the construction of R and F is illustrated in our example. Each Hamiltonian path H from $w = 1$ to $s = 2$ contains only one edge incident to w ($\{1, 3\}$ in our example), one edge incident to s ($\{2, 4\}$), and two edges incident to each of the other vertices in H ($\{1, 3\}$ and $\{3, 4\}$ incident to 3, and $\{3, 4\}$ and $\{2, 4\}$ incident to 4). The path H can always be represented by a string x having size $|R|$. The multisets Q_i occur inside x in the same order as that of the vertices i inside H. The common substring for consecutive Q_i and Q_j is now guaranteed to contain just i, j or j, i in consecutive positions of x. For example, Q_1, Q_3, Q_4, and Q_2 correspond to the vertices in $H = \langle 1, 3, 4, 2 \rangle$, while their intersections correspond to the edges used in H. Here is the role of R': although we do not know a priori which edges will be traversed by H, we known that the multiset of the endpoints of its edges contains one occurrence of w and s, and two occurrences of each other vertex, thus giving rise to R'. Even if we have to remove R' from Q to obtain R, we still guarantee that $\langle R, F \rangle$ is a valid #FMO instance.

Lemma 7. *Each multiset $M \in F$ is contained in R.*

Lemma 8. *Given an undirected graph $G = \langle V, E \rangle$, together with a source and a destination vertex, w and s, the corresponding instance $\langle R, F \rangle$ of #FMO, can be built in $\mathcal{O}(|V| + |E|)$ time.*

4.2 Characterization of the Solutions

We need some technical lemmas, as in Section 3.2. In particular, Lemmas 9–11 follow the same route as that traced in Lemmas 2–4 for #FRONT.

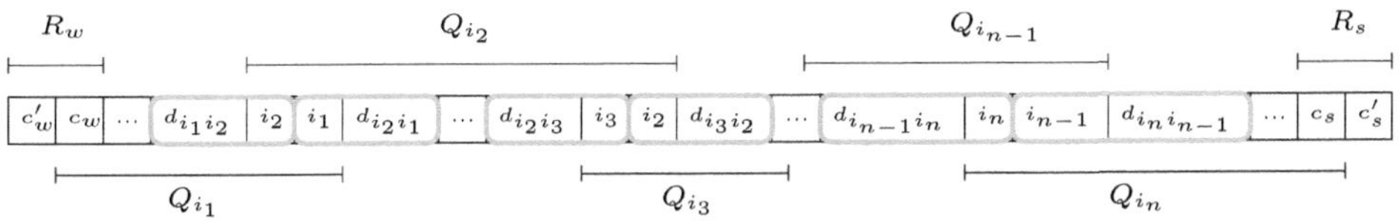

Fig. 4. The string x coding the Hamiltonian path $H = \langle i_1, \ldots, i_n \rangle$ of G. The common substring for Q_i and Q_j has size 2 in x and is constrained to be i, j or j, i.

Lemma 9. *Let $G = \langle V, E \rangle$ be an undirected graph, and $\langle R, F \rangle$ be its corresponding #FMO instance. For any string x that is solution of $\langle R, F \rangle$, there exists a corresponding Hamiltonian path H of G from w to s.*

Lemma 10. *Let $G = \langle V, E \rangle$ be an undirected graph, and $\langle R, F \rangle$ be its corresponding #FMO instance. For any Hamiltonian path H of G from w to s, there exists at least one corresponding solution x of $\langle R, F \rangle$.*

Lemma 11. *Let Σ_H denote the set of all the solutions of $\langle R, F \rangle$ corresponding to a given Hamiltonian path H, as stated in Lemma 10. Then, for any two Hamiltonian paths $H \neq H'$ of G from vertex w to vertex s, it is $\Sigma_H \cap \Sigma_{H'} = \varnothing$.*

We now prove Lemma 9, leaving the proof of Lemmas 10–11 at the end of the section. We consider a solution x of $\langle R, F \rangle$, and make three conceptual steps.

(a) We prove that the multisets Q_i follow a total order $\prec_x$ induced by x.
(b) We show that each Q_i occurs exactly once in x.
(c) For any two consecutive Q_i and Q_j in the total order $\prec_x$, we demonstrate that their intersection in x corresponds to edge $\{i, j\} \in E$.

Observe that steps (a) and (b) select all possible permutations of the vertices in V, while step (c) selects only those permutations (if any) that correspond to paths in G. Putting (a)–(c) together, we can see that the Hamiltonian path corresponding to x is $H = \langle i_1, i_2 \ldots, i_{|V|} \rangle$, where $Q_{i_1} \prec_x Q_{i_2} \prec_x \cdots \prec_x Q_{i_{|V|}}$ is the total order induced by x.

We show a slightly more general property than that stated in (a), using the following lemma.

Lemma 12 (Strict Sperner Property). *The collection of multisets $C = \{R_w, R_s, Q_1, \ldots, Q_{|V|}\}$, is a Strict Sperner collection [6]: no multiset is contained in the union of the others. Hence, there exists a total order $\prec_x$ on the multisets in C.*

We prove the property stated in step (c) by the following lemma.

Lemma 13 (Intersection Size). *Let x be a string of size $|R|$, drawn from all the symbols in R, and containing all the multisets in $C_2 = \{R_w, R_s, Q_i, Q_{ij}\}$. Let $I_{ij} = Q_i \cap Q_j$ denote the intersection between two multisets Q_i and Q_j that occur consecutively in x. Then, (i) $|I_{ij}| = 2$; (ii) $I_{ij} = \{i, j\}$; (iii) $\{i, j\} \in E$.*

Finally, the property stated in step (*b*) is based on the lemma below.

Lemma 14 (Occurrence Uniqueness). *Given a solution x of $\langle R, F\rangle$, each multiset $Q_i \in F$ occurs exactly once inside x.*

We now prove Lemma 10 and Lemma 11.

Let us discuss Lemma 10. Given a Hamiltonian path $H = \langle i_1, i_2, \ldots, i_{|V|}\rangle$ of G, where $i_1 = w$ and $i_{|V|} = s$, in order to construct a solution x of the corresponding #FMO instance $\langle R, F\rangle$, we arrange the multisets Q_i in the same order as the corresponding vertices in H, as shown in Figure 4. The first symbol of x is c'_w and the last one is c'_s. Between them, $Q_{i_1}, Q_{i_1}, \ldots, Q_{i_{|V|}}$ appears in x, where the first symbol of Q_{i_1} is c_w, and the last symbol is i_1, and the first symbol of $Q_{i_{|V|}}$ is $i_{|V|}$ and the last symbol is c_s. For the remaining Q_{i_l}, the first three symbols are i_l,i_{l-1}, and $d_{i_l i_{l-1}}$, and the first two of them overlap with $Q_{i_{l-1}}$ by Lemma 13. Analogously, the last three symbols are $d_{i_l i_{l+1}}$, i_{l+1} and i_l, and the last two of them overlap with $Q_{i_{l+1}}$. The remaining symbols in Q_{i_l} are $d_{i_l j}, j$ for all edges $\{i_l, j\} \in E$, such that $j \neq i_{l-1}, i_{l+1}$.

Each multiset Q_{i_l} intersects $Q_{i_{l+1}}$ in $\{i_l, i_{l+1}\} \in E$. Note that, since H is a Hamiltonian path, the symbols belonging to the union of all the intersections are $R' = \bigcup_{i \neq w,s}\{i, i\} \cup \{w, s\}$. To prove that x is a solution of $\langle R, F\rangle$, note that x contains each multiset Q_i, R_w, R_s by construction. As for each $Q_{ij} = \{d_{ij}, j\}$, we observe that its occurrence is contained in the occurrence of Q_i in x. Moreover, x contains the multiset R and x has size $|R|$, since x is drawn from the multiset $\bigcup_i Q_i \cup \{c'_w, c'_s\} \setminus R'$, that is exactly the way R is defined in $\langle R, F\rangle$. The above discussion proves Lemma 10.

To prove Lemma 11, consider a string $x \in \Sigma_H$, and $x' \in \Sigma_{H'}$ where $H' = \langle i'_1, i'_2, \ldots, i'_{|V|}\rangle$. Since $H \neq H'$, they must differ in at least one position l (i.e. $i_l \neq i'_l$). Assume w.l.o.g. that $|Q_{i_l}| \leq |Q_{i'_l}|$, and select the position k of the leftmost symbol $d_{i_l j} \in Q_{i_l}$ occurring in x for some j. Since the order of the multisets in x is the same as that of the vertices in the Hamiltonian paths, $Q_{i_l} \neq Q_{i'_l}$ (since $i_l \neq i'_l$). By construction of the multisets, we have $d_{ij} \notin Q_{i'_l}$, then the kth symbol in x and x' differs, thus proving the claim.

4.3 Reduction from #HAM to #FMO

The #FMO problem is clearly in $\#\mathcal{P}$, so we focus on its completeness. We are given an undirected graph $G = \langle V, E\rangle$, along with its source w and its destination s. The reduction goes as follows.

- Build an instance $\langle R, F\rangle$ as described in Section 4.1.
- Let z be the number of solutions for the instance $\langle R, F\rangle$.
- Let $a = \prod_{i=1}^{|V|} \alpha_i \neq 0$, where α_i is defined for a vertex i of degree d_i as

$$\alpha_i = \begin{cases} 2^{(d_i-1)}\,(d_i-1)! & i = w, s \\ 2^{(d_i-2)}\,(d_i-2)! & i \neq w, s \end{cases}$$

- Return the integer z/a.

The above reduction takes polynomial time. To see its correctness, it suffices to show that $|\Sigma_H| = a$ for every Hamiltonian path $H = \langle i_1, i_2, \ldots, i_{|V|}\rangle$ in G.

We already proved in Section 4.2 that each solution $x \in \Sigma_H$ has the form reported in Figure 4. Here, the occurrence of each Q_i is a sequence of pairs $Q_{ij} = \{d_{ij}, j\}$ except the first and the last symbol of Q_i. If $i \neq w, s$, the first and the last pairs always stay the same, while the remaining $d_i - 2$ pairs can be permuted in $(d_i - 2)!$ ways. For each such a way, we can permute each pair internally, thus giving an extra factor of $2^{d_i - 2}$. If $i = w, s$, we have $d_1 - 1$ pairs that can be permuted, yielding $2^{(d_i - 1)}\,(d_i - 1)!$ permutations.

Theorem 2. *#FMO is #$\mathcal{P}$-complete.*

As a byproduct of the reduction adopted in this section, we have:

Corollary 1. *Testing the C1P on multisets is $\mathcal{NP}$-complete.*

Acknowledgment. We thank the anonymous Referees for their useful comments.

References

1. Amir, A., Apostolico, A., Landau, G.M., Satta, G.: Efficient text fingerprinting via Parikh mapping. J. Discrete Algorithms 1(5-6), 409–421 (2003)
2. Arora, S., Barak, B.: Computational Complexity A Modern Approach. Cambridge University Press, Cambridge (2009)
3. Booth, K.S.: PQ-tree algorithms. Ph.D. thesis, Univ. of California (December 1975)
4. Booth, K.S., Lueker, G.S.: Testing for the consecutive ones property, interval graphs, and graph planarity using PQ-tree algorithms. JCSS 13(3), 335–379 (1976)
5. Christof, T., Oswald, M., Reinelt, G.: Consecutive ones and a betweenness problem in computational biology. In: Bixby, R.E., Boyd, E.A., Ríos-Mercado, R.Z. (eds.) IPCO 1998. LNCS, vol. 1412, pp. 213–228. Springer, Heidelberg (1998)
6. Engel, K.: Sperner theory. Cambridge University Press, New York (1997)
7. Eres, R., Landau, G.M., Parida, L.: A combinatorial approach to automatic discovery of cluster-patterns. In: Benson, G., Page, R.D.M. (eds.) WABI 2003. LNCS (LNBI), vol. 2812, pp. 139–150. Springer, Heidelberg (2003)
8. Fulkerson, D.R., Gross, D.A.: Incidence matrices and interval graphs. Pacific J. Math. 15(3), 835–855 (1965)
9. Garey, M.R., Johnson, D.S.: Computers and Intractability. Freeman, NY (1979)
10. Ghosh, S.P.: File organization: the consecutive retrieval property. Commun. ACM 15(9), 802–808 (1972)
11. Jain, M., Myers, E.W.: Algorithms for computing and integrating physical maps using unique probes. In: RECOMB 1997, pp. 151–161 (1997)
12. Landau, G.M., Parida, L., Weimann, O.: Gene proximity analysis across whole genomes via PQ-trees. Journal of Computational Biology 12(10), 1289–1306 (2005)
13. Papadimitriou, C.M.: Computational complexity. Addison-Wesley, Reading (1994)
14. Parida, L.: Statistical significance of large gene clusters. Journal of Computational Biology 14(9), 1145–1159 (2007)
15. Valiant, L.G.: The complexity of computing the permanent. Theoretical Computer Science 8(2), 189–201 (1979)

Avoiding Abelian Powers in Partial Words*

Francine Blanchet-Sadri[1] and Sean Simmons[2]

[1] Department of Computer Science, University of North Carolina, P.O. Box 26170, Greensboro, NC 27402–6170, USA
blanchet@uncg.edu

[2] Department of Mathematics, The University of Texas at Austin, 1 University Station C1200, Austin, TX 78712–0233, USA

Abstract. We study abelian repetitions in partial words, or sequences that may contain some unknown positions or holes. First, we look at the avoidance of abelian pth powers in infinite partial words, where $p > 2$, extending recent results regarding the case where $p = 2$. We investigate, for a given p, the smallest alphabet size needed to construct an infinite partial word with finitely or infinitely many holes that avoids abelian pth powers. We construct in particular an infinite binary partial word with infinitely many holes that avoids 6th powers. Then we show, in a number of cases, that the number of abelian p-free partial words of length n with h holes over a given alphabet grows exponentially as n increases. Finally, we prove that we cannot avoid abelian pth powers under arbitrary insertion of holes in an infinite word.

1 Introduction

The study of word structures, or combinatorics on words, plays an important role in theoretical computer science. At the beginning of the twentieth century, Thue discovered in [23] that the consecutive repetitions of non-empty factors, also called squares, can be avoided by an infinite word over a ternary alphabet. Such word is said to be *2-free* or *square-free*. More generally, a word u is *p-free* if it does not contain x^p as a factor, where x is any non-empty factor of u.

The study of abelian repetitions was started by Erdös [14]. He raised the question whether there exist infinite abelian square-free words over a given alphabet (words in which no two consecutive factors are permutations of each other). For example, *abcbac* is an abelian square since *abc* and *bac* are permutations of each

* This material is based upon work supported by the National Science Foundation under Grant No. DMS–0754154. The Department of Defense is also gratefully acknowledged. We thank Dimin Xu for very valuable comments and suggestions. A research assignment from the University of North Carolina at Greensboro for the first author is gratefully acknowledged. Some of this assignment was spent at the LIAFA: Laboratoire d'Informatique Algorithmique: Fondements et Applications of Université Paris 7-Denis Diderot, Paris, France. A World Wide Web server interface has been established at www.uncg.edu/cmp/research/abelianrepetitions2 for automated use of the program.

G. Mauri and A. Leporati (Eds.): DLT 2011, LNCS 6795, pp. 70–81, 2011.

other. It can easily be checked that no infinite abelian square-free word exists over a three-letter alphabet because every word of length eight over three letters contains an abelian square. Evdokimov in [15] showed that abelian squares are avoidable over twenty-five letters. Pleasants [21] showed that there exists an infinite abelian square-free word over five letters using a uniform morphism of size 5×15. Keränen [17] reduced the alphabet size to four by introducing a uniform abelian square-free morphism of size 4×85. Other constructions over four letters appear in [9,18]. Computer-assisted music analysis is an application area of abelian square-free words [19]. Cryptography is another application area [3,22]. For instance, Rivest presented a new way of dithering for iterated hash functions based on such square-free words. Abelian square-free words have also been used in the study of free partially commutative monoids [10,13].

An *abelian pth power* consists of p consecutive factors which are permutations of each other, and a word is *abelian p-free* if it does not contain any abelian pth power. Results for $p > 2$ include: Justin in [16] showed that there exists an abelian 5-free word over a binary alphabet with a uniform morphism of size 2×5. Dekking in [12] proved that over a ternary alphabet, there exists an abelian 3-free infinite word, and using a non-uniform morphism that abelian 4th powers are avoidable over a binary alphabet (abelian cubes cannot be avoided over a binary alphabet). Aberkane and Currie in [1] improved the latter result by introducing a uniform morphism having an abelian 4-free fixed point.

The problem of avoiding abelian squares in partial words was initiated by Blanchet-Sadri et al. in [6]. The idea of partial words was first introduced by Berstel and Boasson in [4] and the one of freeness in partial words by Manea and Mercaş in [20]. Partial words are sequences over a finite alphabet that may have some unknown positions or holes, represented by $\diamond$'s, which are *compatible* with, or match, each letter of the alphabet. For instance, $ab\diamond b\diamond c$ is a partial word with two holes over the ternary alphabet $\{a, b, c\}$. A partial word u over an alphabet A is an *abelian square* if it is possible to substitute letters from A for each hole in such a way that u becomes an abelian square that is a full word (or a partial word without holes). The partial word u is *abelian square-free* if it does not have any full or partial abelian square, except those of the form $\diamond a$ or $a\diamond$, where $a \in A$. For example, $abacaba$ is abelian square-free, while $abcdad\diamond ada$ is not (it contains the abelian square $cdad\diamond a$ compatible with $cdadca$). In particular, in [6], lower and upper bounds were given for the number of letters needed to construct infinite abelian square-free partial words with finitely or infinitely many holes. It was proved that there exists an infinite abelian square-free partial word with one hole over a four-letter alphabet, and none exists with more than one hole. It was also showed that the minimal alphabet size for the existence of an abelian square-free partial word with more than one hole is five. Several of the constructions are based on iterating morphisms.

In this paper, we build on previous work by studying abelian repetitions in partial words. A partial word u is *p-free* if for every non-empty factor $u_0 \cdots u_{p-1}$ of u, there does not exist a full word v compatible with u_i, for all $i \in \{0, \dots, p-1\}$. A partial word u is *abelian p-free* if for every non-empty factor $u_0 \cdots u_{p-1}$ of u, there does not exist a full word v compatible with some permutation of u_i, for all $i \in \{0, \ldots, p-1\}$. We are interested in the following three problems: (1) Study avoidance of abelian pth powers, for $p > 2$. In the context of partial words, which abelian powers can be avoided over a given alphabet? (2) Investigate the number of abelian p-free partial words of a fixed length over an alphabet of a given size. The number of abelian 2-free full words has been studied in [8] and [18] for an alphabet of size four (work has also been done for higher powers [1,2,11]). (3) Can we construct an infinite full word that remains abelian p-free even after arbitrarily many positions are replaced with holes? In [20], a 3-free infinite full word over an optimal alphabet size of four letters was given that remains 3-free after such replacements, while in [7] a 2-free infinite full word over an optimal size of eight letters was given that remains 2-free after such replacements. Here, among other things, we prove that abelian 6th powers can be avoided over a binary alphabet, while such is not the case for abelian 3rd powers. We also show that there is a partial word with infinitely many holes over a three-letter (resp., three-letter, four-letter) alphabet that avoids abelian 5th powers (resp., 4th powers, 3rd powers). Due to the 12 page limit requirement, we have only put sketches for some of the proofs to illustrate a few of our techniques.

We end this section with some background on partial words (we refer the reader to [5] for more information). A *partial word* over an alphabet A is a sequence of symbols from $A_\diamond = A \cup \{\diamond\}$, where $\diamond$ is the symbol representing an unknown position. Any occurrence of $\diamond$ in a partial word is called a *hole* (a *(full) word* is a partial word without holes). The *length* of a partial word u, denoted by $|u|$, represents the number of symbols in u, while $u(i)$ represents the symbol at position i of u, where $0 \leq i < |u|$. The *empty word* is the sequence of length zero and is denoted by ε. The set of distinct letters in u, or the *cardinality* of u, is denoted by $\alpha(u)$. For instance, $u = ab\diamond bba\diamond$, where a, b are distinct letters of the alphabet A, satisfies $\alpha(u) = \{a, b\}$. The set of all words over A is denoted by A^*, while the set of all partial words over A by $A_\diamond^*$. A (right) (resp., two-sided) infinite partial word is a function $u : \mathbb{N} \to A_\diamond$ (resp., $u : \mathbb{Z} \to A_\diamond$).

Let u and v be partial words of equal length. Then u is *contained* in v, denoted by $u \subset v$, if $u(i) = v(i)$, for all i such that $u(i) \in A$. The partial words u and v are *compatible*, denoted by $u \uparrow v$, if there exists a partial word w such that $u \subset w$ and $v \subset w$. For example, $a\diamond bb\diamond \uparrow ab\diamond b\diamond$, since $a\diamond bb\diamond \subset abbb\diamond$ and $ab\diamond b\diamond \subset abbb\diamond$. A partial word u is a *factor* or *subword* of a partial word v if there exist x, y such that $v = xuy$. The factor u is proper if $u \neq \varepsilon$ and $u \neq v$. We say that u is a *prefix* of v if $x = \varepsilon$ and a *suffix* of v if $y = \varepsilon$. The notation $v[i..j)$ represents the factor $v(i) \cdots v(j-1)$, while $v[i..j]$ the factor $v(i) \cdots v(j)$. The powers of u are defined recursively as follows: $u^0 = \varepsilon$, and for any integer $p > 0$, $u^p = uu^{p-1}$.

Given $a \in A$, $|u|_a$ denotes the number of occurrences of a in a partial word u over alphabet A. If we write $A = \{a_0, \ldots, a_{k-1}\}$, where the cardinality of A, $\|A\|$, is k, then the *Parikh vector* of u, $P(u)$, is $P(u) = \langle |u|_{a_0}, \ldots, |u|_{a_{k-1}} \rangle$. For a positive integer p, a non-empty word $u \in A^*$ is an *abelian pth power* if we can write $u = u_0 \cdots u_{p-1}$ so that $P(u_0) = \cdots = P(u_{p-1})$. This is the same as saying that each u_i is a permutation of u_0. A word u is *abelian p-free* if no factor of u is an abelian pth power. A non-empty partial word $u \in A_\diamond^*$ is an *abelian pth power* if it is possible to substitute letters from A for each hole in such a way that u becomes a full word that is an abelian pth power. In other words, u is an abelian pth power if there exists a full word v, compatible with u, that is an abelian pth power. For example, $abca\diamond b$ is an abelian 2nd power, or abelian square, since we can replace the $\diamond$ with letter c and form $abcacb$. Whenever we refer to an abelian pth power $u_0 \cdots u_{p-1}$, it implies that there exists a non-empty word w such that for each i, $0 \leq i < p$, some permutation of u_i is compatible with w. The partial word u is *abelian p-free* if it does not have any full or partial abelian pth power.

Let A and B be alphabets, and 2^{B^*} be the set of all subsets of B^*. A morphism $\phi : A^* \to B^*$ is called *abelian p-free* if $\phi(u)$ is abelian p-free whenever u is abelian p-free. A *multi-valued substitution* is a function $\theta : A^* \to 2^{B^*}$, with the property that if u_1 and u_2 are words in A^*, then $\theta(u_1u_2) = \theta(u_1)\theta(u_2) = \{v_1v_2 \mid v_i \in \theta(u_i)\}$. The multi-valued substitution θ is said to be abelian p-free if, whenever $u \in A^*$ is abelian p-free then so are all of the words in $\theta(u)$. See [11] for more information on multi-valued substitutions.

Finally, inserting a hole in a word u is defined as replacing a position's letter in u with a $\diamond$. An arbitrary insertion of holes is replacing any collection of positions' letters in u with $\diamond$'s, so that every pair of consecutive $\diamond$'s is separated by at least two letters. This restriction is to avoid some trivial occurrences of powers (a pth power $u_0 \cdots u_{p-1}$, $|u_0| = \cdots = |u_{p-1}|$, is *trivial* if $u_i = \diamond$ for some i).

2 Avoiding Abelian Powers Greater Than Two

In [6], it is shown that abelian squares can be avoided by a partial word with infinitely many holes over five letters, where the number of letters is minimal. So we move on to higher abelian powers, that is, we investigate the minimal alphabet size needed to construct partial words with infinitely many holes that avoid abelian pth powers, where $p > 2$.

We begin by proving a result about abelian 4th powers.

Theorem 1. *There exists a partial word with infinitely many holes over a three-letter alphabet that avoids abelian 4th powers.*

Proof. By [12], there exists an infinite word w over $A = \{a, b\}$ which avoids abelian 4th powers. Let c be any letter not in A, and write $B = \{a, b, c\}$. Let $k_i = 5 \times 6^i$. Then we can define an infinite partial word w' over B as follows:

$$w'(j) = \begin{cases} \diamond, & \text{if } j = k_i \text{ for some } i; \\ c, & \text{if } j = k_i + 1 \text{ or } j = k_i - 1 \text{ for some } i; \\ w(j), & \text{otherwise.} \end{cases}$$

Using a proof by contradiction, we can show that w' contains no abelian 4th powers. □

The basic idea of the proof of Theorem 1 is to place the holes so far apart that any abelian 4th power could contain only a few holes. This idea will be used repeatedly in the sequel, and so it is made more precise in the following lemma.

Lemma 1. *Let w be an infinite partial word over a finite alphabet, and let $p > 1$ be an integer. Assume that $u_0 \cdots u_{p-1} = w[i_0..i_1 - 1] \cdots w[i_{p-1}..i_p - 1]$ is an abelian pth power in w. Let $k_0 < k_1 < \cdots$ be a sequence of integers so that $k_0 > 0$, and $k_i > pk_{i-1}$ for all $i \geq 1$. Then the following hold:*

- *If $q > 0$, then there exists at most one j such that $i_q \leq k_j \leq i_{q+1} - 1$.*
- *If $q_1 < q_2$, and if there exist j_1, j_2 such that $i_{q_1} \leq k_{j_1} \leq i_{q_1+1} - 1$ and $i_{q_2} \leq k_{j_2} \leq i_{q_2+1} - 1$, then $q_1 = 0$.*

Consequently, if k_0 is the smallest integer such that $w(k_0) = \diamond$, k_1 the next smallest integer such that $w(k_1) = \diamond$, and so on, then the following hold: (1) if $j > 0$, then u_j contains at most one hole; (2) if $i < j$ and both u_i and u_j contain holes, then $i = 0$.

Then we move on to abelian cubes.

Theorem 2. *There exists a partial word with infinitely many holes over a four-letter alphabet that avoids abelian cubes.*

Proof. Let $A = \{a, b, c\}$ and $B = \{a, b, c, d\}$. Consider the morphism $\phi : A^* \to A^*$ defined as $\phi(a) = aabc$, $\phi(b) = bbc$ and $\phi(c) = acc$. Let w be the fixed point of ϕ. It is known that w is abelian cube-free (see [12]).

Begin by noting that $accaccbbc$ occurs infinitely often in w. We can thus find a sequence $k_0 < k_1 < \cdots$ so that $k_0 > 20$; for each i, $k_{i+1} > 3k_i$; and for each i, $w[k_i - 2..k_i + 6] = accaccbbc$. Then we can define an infinite partial word w' over B as follows:

$$w'(j) = \begin{cases} \diamond, & \text{if } j = k_i \text{ for some } i,\ i \equiv 0 \bmod 19; \\ d, & \text{if } j \in \{k_i - 1, k_i + 2, k_i + 3\} \text{ for some } i,\ i \equiv 0 \bmod 19; \\ d, & \text{if } j = k_i \text{ for some } i,\ i \not\equiv 0 \bmod 19; \\ w(j), & \text{otherwise.} \end{cases}$$

Using Lemma 1, we can show that w' is abelian cube-free. □

To study larger powers, we need the following lemma.

Lemma 2. *Let w be an infinite word over a finite alphabet, and let $p > 1$ be an integer. Assume that there exist infinitely many i's so that, if w' is defined by $w'(j) = \diamond$ if $j = i$, and $w'(j) = w(j)$ otherwise, then w' is abelian p-free. Then we can insert infinitely many holes in w so that the resulting partial word is abelian $(p + 1)$-free.*

Abelian 6th powers can be avoided over a binary alphabet. The construction is not the result of simply inserting holes to a word produced by a morphism.

Theorem 3. *There exists a partial word with infinitely many holes over a two-letter alphabet that avoids abelian 6th powers.*

Proof. Let $A = \{a, b\}$. As mentioned before, there is an infinite binary word w that avoids abelian 4th powers (see [12]). More specifically, it is the fixed point of the morphism $\phi : A^* \to A^*$ defined by $\phi(a) = aaab$ and $\phi(b) = bab$. We also define a morphism $\psi : A^* \to A^*$ by $\psi(a) = b$ and $\psi(b) = a$. Let $v_0 = w[0..119]$ and $v_1 = w[121..240]$. Note that $w(120) = a$. In other words v_0av_1 is the prefix of w where $|v_0av_1| = 241$. We can check with a computer that $\psi(v_0)\diamond v_1$ is abelian 6-free. Similarly, we can check that if u is a subword of w, $30 \leq |u| < 60$, then $|u|_a > |u|_b$. This implies that if $|u| \geq 30$ and u is a subword of w then $|u|_a > |u|_b$, since we can write $u = u_0 \cdots u_n$ for some n, where $30 \leq |u_i| < 60$.

Since v_0av_1 appears once in w, it must appear infinitely often. Therefore define a sequence $k_0 < k_1 < \cdots$ so that $k_0 > 120$, $w[k_i - 120..k_i + 120] = v_0av_1$, and $k_i > 7k_{i-1}$. Let $k_{-1} = -1$. Then define a partial word w' as follows:

$$w'(j) = \begin{cases} \diamond, & \text{if } j = k_i \text{ for some } i; \\ w(j), & \text{if } k_i < j < k_{i+1} \text{ for some } i,\ i \text{ odd}; \\ \psi(w(j)), & \text{otherwise.} \end{cases}$$

Our goal is to show that w' is abelian 6-free. To see this, assume that $u_0u_1u_2u_3u_4u_5$ is an abelian 6th power in w'. From Lemma 1, at most one of u_1, u_2, u_3, u_4, or u_5 contains a hole. This implies that $u_1u_2u_3u_4u_5$ occurs as a factor of w of the form $\psi(w_0v_0)\diamond v_1w_1$ or $w_0v_0\diamond\psi(v_1w_1)$, where w_0 and w_1 are subwords of w, since otherwise $u_1u_2u_3u_4u_5$ would have to contain more than one hole. Assume it occurs in a word of the form $\psi(w_0v_0)\diamond v_1w_1$, the other case being similar. Either u_2, u_3 or u_4 must contain a hole, otherwise we would have that either $u_1u_2u_3u_4$ or $u_2u_3u_4u_5$ is full, so $\diamond$ must occur as a factor in either w or $\psi(w)$. However, both w and $\psi(w)$ are abelian 4-free, so this is impossible. Then consider the possibility that $|u_0| < 30$. However, this implies, since either u_2, u_3 or u_4 contains a hole, that $u_0u_1u_2u_3u_4u_5$ occurs in $\psi(v_0)\diamond v_1$. As mentioned above, from computer testing $\psi(v_0)\diamond v_1$ does not contain any abelian 6th powers. Therefore, assume that $|u_0| \geq 30$. We know that $u_1u_2u_3u_4u_5$ occurs as a subword of $\psi(w_0v_0)\diamond v_1w_1$. Moreover, since the hole is contained in either u_2, u_3 or u_4, we get that u_5 is contained in v_1w_1, while u_1 is contained in $\psi(w_0v_0)$. Since u_1 and u_5 are full and $u_1 \uparrow u_5$, we get that $|u_1|_a = |u_5|_a$ and $|u_1|_b = |u_5|_b$. On the other hand, since u_5 is a subword of w and $|u_5| \geq 30$ we know that $|u_5|_a > |u_5|_b$, while $|u_1|_a < |u_1|_b$ because u_1 is a subword of $\psi(w)$ and $|u_1| \geq 30$. This is a contradiction, so w' does not contain any abelian 6th powers. □

In Section 5, a discussion wraps up what has been shown in this section and what remains to be done.

3 Counting Abelian p-Free Partial Words

We begin by noting that Theorem 3 (resp., Theorem 1) allows us to give a lower bound on the number of binary (resp., ternary) partial words of length n with h holes that are abelian 6-free (resp., 4-free). More specifically, we count, for $p \in \{4, 6\}$, such partial words that avoid abelian pth powers over an alphabet A, where A is the smallest alphabet known to admit partial words with infinitely many holes that avoid abelian pth powers.

Theorem 4. *Let $h > 0$ be an integer. Let $c_{n,h,6}$ (resp., $c_{n,h,4}$) denote the number of partial words over a two-letter (resp., three-letter) alphabet of length n with h holes that avoid abelian 6th (resp., 4th) powers. Then there exist an integer $N > 0$ and real numbers $r > 1, \beta > 0$ such that for all $n > N$, $c_{n,h,6} \geq \beta r^n$ (resp., $c_{n,h,4} \geq \beta r^n$).*

Proof. We prove the result for 6th powers (the proof for 4th powers is similar except that it uses Theorem 1 instead of Theorem 3). Assume that h is even, the other case being similar. Let ϕ, ψ and w be as in Theorem 3. By the proof of Theorem 3, there exists a word $v = w_0 \diamond \psi(w_1) \diamond \cdots \diamond \psi(w_{h-1}) \diamond w_h$, where each w_i is a subword of w, w_h is infinite, and v avoids abelian 6th powers. There exist an infinite subword u of w and a finite word x, $|x| > 5|w_0 \diamond \psi(w_1) \diamond \cdots \diamond \psi(w_{h-1}) \diamond|$, such that if u' is any prefix of u then $x\phi(u')$ is a prefix of w_h.

We can then define the multi-valued substitution $\theta : A^* \to 2^{A^*}$ so that $\theta(a) = \{aaab\}$ and $\theta(b) = \{bab, abb\}$. From [11], θ is abelian 4-free. Therefore let $N = |w_0 \diamond \psi(w_1) \diamond \cdots \diamond \psi(w_{h-1}) \diamond x| + 16$. Consider $n > N$ and denote the prefix of v of length n as v_n. We can write $w_h = x\phi(u')y$ for some prefix u' of u and some word y. Since $|\phi(\alpha)| \leq 4$ for all $\alpha \in A$, we can choose u' so that

$$|u'| \geq \left\lfloor \frac{|v_n| - |w_0 \diamond \psi(w_1) \diamond \cdots \diamond \psi(w_{h-1}) \diamond x|}{4} \right\rfloor = \left\lfloor \frac{n - N + 16}{4} \right\rfloor \geq \frac{n - N + 12}{4}$$

We claim that if $\gamma \in \theta(u')$, then $w_0 \diamond \psi(w_1) \diamond \cdots \diamond \psi(w_{h-1}) \diamond x\gamma y$ is abelian 6-free. To see this, if $w_0 \diamond \psi(w_1) \diamond \cdots \diamond \psi(w_{h-1}) \diamond x\gamma y$ contains an abelian 6th power $u_0u_1u_2u_3u_4u_5$, then one of the u_i's must overlap with γ, since

$$w_0 \diamond \psi(w_1) \diamond \cdots \diamond \psi(w_{h-1}) \diamond x$$

and y are abelian 6-free. However, since $|x| > 5|w_0 \diamond \psi(w_1) \diamond \cdots \diamond \psi(w_{h-1}) \diamond|$, the only u_i that can intersect with $w_0 \diamond \psi(w_1) \diamond \cdots \diamond \psi(w_{h-1}) \diamond$ is u_0, so $u_1u_2u_3u_4$ is contained in $x\gamma y$. By construction $x\phi(u')y$ is a factor of w, so $x\gamma y$ is a factor of some word in $\theta(w)$. However, the fact that w is abelian 4-free implies that all elements in $\theta(w)$ are as well, so $x\gamma y$ is abelian 4-free. It follows that $w_0 \diamond \psi(w_1) \diamond \cdots \diamond \psi(w_{h-1}) \diamond x\gamma y$ is abelian 6-free, which implies that $|\theta(u')| \leq c_{n,h,6}$. Moreover, since u' is a factor of w, it does not have $aaaa$ as a subword. Therefore $|u'|_b \geq \left\lfloor \frac{|u'|}{4} \right\rfloor \geq \frac{|u'| - 3}{4}$. By construction, $|\theta(u')| = 2^{|u'|_b}$. Using the same analysis as in [11], this gives us that $\theta(u')$ contains at least $2^{-\frac{15}{16}} 2^{\frac{1}{16}(n-N+16)}$ words. Therefore, setting $r = 2^{\frac{1}{16}}$ and $\beta = 2^{-\frac{15}{16} - \frac{N}{16} + 1}$, we get that $\beta r^n \leq |\theta(u')| \leq c_{n,h,6}$, so the claim follows. □

4 Inserting Arbitrarily Many Holes

We investigate whether infinite words can be constructed that remain abelian p-free after an arbitrary insertion of holes. We begin by inserting finitely many holes.

Proposition 1. *There exists an infinite word over a two-letter alphabet such that, if we insert a hole anywhere in the word, the resulting partial word is abelian 8-free. Furthermore, there exists an infinite abelian 3-free (resp., 4-free) word over a nine-letter (resp., four-letter) alphabet such that, if we insert a hole anywhere in the word, the resulting partial word is non-trivial abelian 3-free (resp., 4-free).*

Proof. To prove the existence of an abelian 4-free word satisfying the desired properties, consider $w = w_0(0)w_1(0)w_0(1)w_1(1)\cdots$, where w_0 is an abelian 4-free word over $A_0 = \{a_0, b_0\}$ and w_1 is one over $A_1 = \{a_1, b_1\}$. Suppose that $u_0u_1u_2u_3$, where $|u_0| = |u_1| = |u_2| = |u_3| = l$, is a factor of w that becomes an abelian 4th power after inserting a hole. If l is odd, then u_0 and u_2 contain one more letter from one alphabet, while u_1 and u_3 contain one more letter from the other alphabet, which is impossible. Thus l is even, and each u_i contains $\frac{l}{2}$ letters from A_0 and $\frac{l}{2}$ letters from A_1. Since the hole replaces a letter from some alphabet, say A_0, we can remove all the letters from A_0 that are in $u_0u_1u_2u_3$, and the remaining word is an abelian 4th power in w_1, a contradiction. □

Our goal is now to prove that we cannot avoid abelian pth powers under arbitrary insertion of holes. Moreover, we can make the abelian pth powers we get as large as we want. We can also look at the following results as saying that every infinite word has a subword that is almost an abelian pth power. We begin by proving a lemma giving a necessary and sufficient condition for the existence of abelian pth powers.

Lemma 3. *Let $p > 1$ be an integer, and let $v_0 \cdots v_{p-1}$ be a partial word over a k-letter alphabet $A = \{a_0, \ldots, a_{k-1}\}$ such that $|v_i| = |v_0|$, for all i. Let $m_i = \max_j\{|v_j|_{a_i}\}$, for $0 \le i < k$. Then $v_0 \cdots v_{p-1}$ is an abelian pth power if and only if $m_0 + \cdots + m_{k-1} \le |v_0|$.*

Proof. First, assume that $m_0 + \cdots + m_{k-1} \le |v_0| = \cdots = |v_{p-1}|$. Then v_i has $|v_i| - (|v_i|_{a_0} + \cdots + |v_i|_{a_{k-1}}) \ge (m_0 + \cdots + m_{k-1}) - (|v_i|_{a_0} + \cdots + |v_i|_{a_{k-1}}) \ge 0$ holes. We can then produce v_i' from v_i by replacing $m_0 - |v_i|_{a_0}$ holes with the letter a_0, $m_1 - |v_i|_{a_1}$ holes with a_1, and so on. Replace the remaining holes with the letter a_0. The above procedure is possible due to the fact that v_i has at least $m_0 - |v_i|_{a_0} + m_1 - |v_i|_{a_1} + \cdots + m_{k-1} - |v_i|_{a_{k-1}} = m_0 + \cdots + m_{k-1} - (|v_i|_{a_0} + \cdots + |v_i|_{a_{k-1}})$ holes. Then $|v_i'|_{a_j} = m_j = |v_0'|_{a_j}$ for all $j > 0$, while $|v_i'|_{a_0} = |v_0| - m_1 - \cdots - m_{k-1} = |v_0'|_{a_0}$. Therefore each v_i' is a permutation of v_0', so $v_0' \cdots v_{p-1}'$ is an abelian pth power. Therefore $v_0 \cdots v_{p-1}$ is an abelian pth power. On the other hand, assume that $v_0 \cdots v_{p-1}$ is an abelian pth power. This implies that we can fill in the holes to get a full word $v_0' \cdots v_{p-1}'$ that is an abelian pth power. However, $|v_0'|_{a_i} = |v_j'|_{a_i} \ge |v_j|_{a_i}$ for every i, j, so $|v_0'|_{a_i} \ge m_i$. It follows that $|v_0| = |v_0'| = |v_0'|_{a_0} + \cdots + |v_0'|_{a_{k-1}} \ge m_0 + \cdots + m_{k-1}$. □

We now consider a generalization of arbitrary insertion. In our definition of arbitrary insertion, we require two arbitrarily inserted holes to be separated by at least two letters. Below, however, we consider an arbitrary positive integer m

and analyze what happens when we require each pair of consecutive holes to be separated by at least m letters. It actually turns out that this generalization is very useful for our purposes.

The next technical lemma says, in some sense, that if an infinite word contains subwords that are arbitrarily close to being abelian pth powers, then it must contain an abelian pth power under arbitrary insertion.

Lemma 4. *Let w be an infinite word over a finite alphabet, and let $p > 1, m > 0$ be integers. Assume that for every integer $l > 0$ and real number $\epsilon > 0$, there exists a subword $u_0 \cdots u_{p-1}$ of w so that $|u_i| > l$ for all i, and so that, if $a \in A$ then $||u_i|_a - |u_{i'}|_a| < \epsilon |u_j|$ for all i, i', j. Then we can insert holes in w so that each pair of consecutive holes is separated by at least m letters, and so that the resulting partial word contains an abelian pth power $v_0 \cdots v_{p-1}$, where $|v_i| \geq l$ for all $0 \leq i < p$.*

Proof. Let p, m and w be as above, and choose any $l > 0$. Let $A = \{a_0, \ldots, a_{k-1}\}$. Consider $\epsilon > 0$ such that $\epsilon 2k(k(m+1)+1)(2(2p-1)k+1) < 1$. Similarly, consider $L > l$ such that $L > 4m(k+1)$. Then there exists a subword $u_0 \cdots u_{p-1}$ of w so that $|u_i| > L$ for all i and so that, if $a \in A$, then $||u_i|_a - |u_{i'}|_a| < \epsilon |u_j|$ for all i, i' and j. This implies that $||u_i| - |u_{i'}|| < \epsilon k |u_j|$ for all i, i' and j. Let $\mu = \min |u_i| > L$. Then it is clear that $||u_i|_a - |u_{i'}|_a| < \epsilon\mu$, $||u_i| - |u_{i'}|| < \epsilon k \mu$, and $||u_i| - \mu| < \epsilon k \mu$ for all i, i'. Let v_0 be the word consisting of the first μ letters of $u_0 \cdots u_{p-1}$, v_1 the next μ, and so on up to v_{p-1}. Note that the length of the factor of u_0 that does not overlap with v_0 is at most $||u_0| - |v_0|| = ||u_0| - \mu| < k\epsilon\mu$, the length of the factor of v_1 that does not overlap with u_1 plus the length of the factor of u_1 that does not overlap with v_1 is at most $||u_0| - |v_0|| + ||u_0 u_1| - |v_0 v_1|| < 3k\epsilon\mu$, and so on. In particular,

$$||v_0|_a - |u_0|_a| < k\epsilon\mu, ||v_1|_a - |u_1|_a| < 3k\epsilon\mu, \ldots, ||v_{p-1}|_a - |u_{p-1}|_a| < (2p-1)k\epsilon\mu$$

which implies that

$$||v_i|_a - |v_{i'}|_a| \leq ||v_i|_a - |u_i|_a| + ||u_i|_a - |u_{i'}|_a| + ||u_{i'}|_a - |v_{i'}|_a| < (2(2p-1)k+1)\epsilon\mu$$

for all $a \in A$, and all i, i'. Also $|v_i| = \mu > L$ for all i.

Write $v_0 \cdots v_{p-1} = w[i_0..i_1 - 1] \cdots w[i_{p-1}..i_p - 1]$ where $v_j = w[i_j..i_{j+1})$ for all j. We can assume that $|v_i|_{a_0} > \frac{\mu}{k} - (2(2p-1)k+1)\epsilon\mu$ for all i (to see this, by the pigeonhole principle there exists an i such that $|v_0|_{a_i} \geq \frac{\mu}{k}$; we assume that $i = 0$ without loss of generality, and using the fact that $||v_i|_a - |v_{i'}|_a| < (2(2p-1)k+1)\epsilon\mu$ for all $a \in A$ and all i, i', we can get that the assumption is safe to make). Let $\chi = \min |v_j|_{a_0}$, and let $t = \lfloor \frac{\chi - 2m}{m+1} \rfloor$. Note that, since $\mu > L > 4m(k+1)$, we get

$$t > \frac{\frac{\mu}{k} - (2(2p-1)k+1)\epsilon\mu - 2m}{(m+1)} > \frac{\mu}{2k(m+1)} - \frac{2(2p-1)k+1}{m+1}\epsilon\mu$$

Consider $0 \leq j < p$. Then we can define $k_j^{(0)}$ to be the position of the first occurrence of a_0 in v_j, $k_j^{(1)}$ the position of the $(m+2)$nd occurrence, and in general $k_j^{(\alpha)}$ to be the position of the $(\alpha m+\alpha+1)$th occurrence. Assume that β_j is the largest integer where $k_j^{(\beta_j)}$ is defined. Then we produce w' by inserting a hole in w at position $k_j^{(\alpha)}$ for each j, and each $0 \leq \alpha < \beta_j$. Let $v'_j = w'[i_j..i_{j+1})$. Each pair of consecutive holes is separated by at least m letters, and each v'_i contains at least t holes.

Let $m_i = \max|v'_j|_{a_i}$. Then clearly $m_i < |v_0|_{a_i} + (2(2p-1)k+1)\epsilon\mu$ while $m_0 < |v_0|_{a_0} + (2(2p-1)k+1)\epsilon\mu - t$. Then we get

$$\begin{aligned} m_0 + \cdots + m_{k-1} &< \mu + k(2(2p-1)k+1)\epsilon\mu - (\tfrac{\mu}{2k(m+1)} - \tfrac{2(2p-1)k+1}{m+1}\epsilon\mu) \\ &< \mu = |v'_0| \end{aligned}$$

So it follows from Lemma 3 that $v'_0 \cdots v'_{p-1}$ is an abelian pth power, since $|v'_0| > l$, the claim is proved. □

Finally, we look at what happens in general. We can show that, no matter how large p is or how many letters we require between each pair of consecutive inserted holes, it is always possible to insert holes in an infinite word so that the resulting partial word contains arbitrarily long abelian pth powers.

Theorem 5. *Let w be an infinite word over a finite alphabet, and let $p > 1$, $m > 0$, $l > 0$ be integers. Then we can insert holes in w so that each pair of consecutive holes is separated by at least m letters, and so that the resulting partial word contains an abelian pth power $u_0 \cdots u_{p-1}$, where $|u_i| \geq l$ for all $0 \leq i < p$.*

Proof. We use Lemma 4 along with some topological arguments. □

The next corollary relates back to partial words with infinitely many holes that avoid abelian pth powers. In particular, it says that in such a word the holes cannot be too close together.

Corollary 1. *Let w be a partial word with infinitely many holes over a finite alphabet, and let $p > 1, \mu > 0$ be integers. If there are fewer than μ letters between each pair of consecutive holes in w, then w contains an abelian pth power.*

Corollary 1 leads to the question of how close infinitely many holes can be inserted in an abelian p-free word so that the resulting partial word is abelian p-free. We see that they cannot be separated by a constant distance. On the other hand, we know from previous results that there are many cases in which a partial word can be constructed with exponential spacing. We do not know whether it is possible to do this with less separation, perhaps so that the distance between each pair of consecutive holes be bounded by a polynomial.

5 Conclusion

In summary, we investigated the question whether there exist infinite abelian p-free partial words over a given alphabet, that is, words in which we can replace the holes with letters from the alphabet in such a way that no p consecutive factors are permutations of each other. In previous work, infinite abelian 2-free partial words with one hole were constructed over a minimal alphabet size of four, while the minimal size needed for more than one hole was shown to be five [6]. In this paper, we gave lower and upper bounds for the number of letters needed to construct infinite abelian p-free partial words with infinitely many holes, for any $p > 2$. We proved that the minimal alphabet size for 6th or higher powers is two, while for 5th, 4th, and 3rd powers it is at most three, three, and four respectively. We also investigated, in particular, the number of partial words of length n with a fixed number of holes over a binary alphabet that avoid abelian 6th powers and showed that this number grows exponentially with n. In addition, we showed that we cannot avoid abelian pth powers under arbitrary insertion of holes. More specifically, for the problem of avoiding abelian powers in the infinitely many hole case, for any given $p > 2$, we gave lower and upper bounds on the minimal alphabet size so that there exists a word with infinitely many holes over that upper bounded alphabet which avoids abelian pth powers. The following table provides the power p to be avoided, a lower bound on the minimal alphabet size, and an upper bound, all based on results in this paper for $p > 2$ and results in [6] for $p = 2$:

p	LB	UB
2	5	5
3	3	4
4	2	3
5	2	3
≥ 6	2	2

How many letters do we need to construct a partial word with infinitely many holes that avoids abelian 3rd powers (resp., 4th powers, 5th powers)? For instance, we have proved that there exists a partial word with infinitely many holes over an alphabet of size four that avoids abelian 3rd powers, and none exists over an alphabet of size two. Whether or not three is the minimal alphabet size remains open. We also need to investigate whether three is the minimal alphabet size for abelian 4th powers and 5th powers.

References

1. Aberkane, A., Currie, J.: A cyclic binary morphism avoiding abelian fourth powers. Theoretical Computer Science 410, 44–52 (2009)
2. Aberkane, A., Currie, J., Rampersad, N.: The number of ternary words avoiding abelian cubes grows exponentially. Journal of Integer Sequences 7, Article 04.2.7, 13 (2004) (electronic)
3. Andreeva, E., Bouillaguet, C., Fouque, P.A., Hoch, J., Kelsey, J., Shamir, A., Zimmer, S.: Second preimage attacks on dithered hash functions. In: Smart, N. (ed.) EUROCRYPT 2008. LNCS, vol. 4965, pp. 270–288. Springer, Heidelberg (2008)
4. Berstel, J., Boasson, L.: Partial words and a theorem of Fine and Wilf. Theoretical Computer Science 218, 135–141 (1999)

5. Blanchet-Sadri, F.: Algorithmic Combinatorics on Partial Words. Chapman & Hall/CRC Press, Boca Raton, FL (2008)
6. Blanchet-Sadri, F., Kim, J.I., Mercaş, R., Severa, W., Simmons, S.: Abelian square-free partial words. In: Dediu, A.H., Fernau, H., Martín-Vide, C. (eds.) LATA 2010. LNCS, vol. 6031, pp. 94–105. Springer, Heidelberg (2010)
7. Blanchet-Sadri, F., Mercaş, R., Scott, G.: A generalization of Thue freeness for partial words. Theoretical Computer Science 410, 793–800 (2009)
8. Carpi, A.: On the number of abelian square-free words on four letters. Discrete Applied Mathematics 81, 155–167 (1998)
9. Carpi, A.: On abelian squares and substitutions. Theoretical Computer Science 218, 61–81 (1999)
10. Cori, R., Formisano, M.: Partially abelian square-free words. RAIRO-Theoretical Informatics and Applications 24, 509–520 (1990)
11. Currie, J.: The number of binary words avoiding abelian fourth powers grows exponentially. Theoretical Computer Science 319, 441–446 (2004)
12. Dekking, F.M.: Strongly non-repetitive sequences and progression-free sets. Journal of Combinatorial Theory, Series A 27, 181–185 (1979)
13. Diekert, V.: Research topics in the theory of free partially commutative monoids. Bulletin of the European Association for Theoretical Computer Science 40, 479–491 (1990)
14. Erdös, P.: Some unsolved problems. Magyar Tudományos Akadémia Matematikai Kutató Intézete Közl 6, 221–254 (1961)
15. Evdokimov, A.A.: Strongly asymmetric sequences generated by a finite number of symbols. Doklady Mathematics 9, 536–539 (1968)
16. Justin, J.: Characterization of the repetitive commutative semigroups. Journal of Algebra 21, 87–90 (1972)
17. Keränen, V.: Abelian squares are avoidable on 4 letters. In: Kuich, W. (ed.) ICALP 1992. LNCS, vol. 623, pp. 41–52. Springer, Heidelberg (1992)
18. Keränen, V.: A powerful abelian square-free substitution over 4 letters. Theoretical Computer Science 410, 3893–3900 (2009)
19. Laakso, T.: Musical rendering of an infinite repetition-free string. In: Gefwert, C., Orponen, P., Seppänen, J. (eds.) Logic, Mathematics and the Computer, vol. 14, pp. 292–297. Finnish Artificial Intelligence Society, Symposiosarja, Hakapaino, Helsinki (1996)
20. Manea, F., Mercaş, R.: Freeness of partial words. Theoretical Computer Science 389, 265–277 (2007)
21. Pleasants, P.A.B.: Non repetitive sequences. Proceedings of the Cambridge Philosophical Society 68, 267–274 (1970)
22. Rivest, R.L.: Abelian square-free dithering for iterated hash functions. MIT, Cambridge (2005), `http://people.csail.mit.edu/rivest/publications.html`
23. Thue, A.: Über unendliche Zeichenreihen. Norske Vid. Selsk. Skr. I, Mat. Nat. Kl. Christiana 7, 1–22 (1906)

Regular Splicing Languages Must Have a Constant

Paola Bonizzoni[1] and Natasha Jonoska[2]

[1] Dipartimento di Informatica Sistemistica e Comunicazione
Univ. degli Studi di Milano - Bicocca
Viale Sarca 336, 20126 Milano - Italy
bonizzoni@disco.unimib.it

[2] Department of Mathematics and Statistics, University of South Florida, Tampa FL, USA
jonoska@math.usf.edu

Abstract. In spite of wide investigations of finite splicing systems in formal language theory, basic questions, such as their characterization, remain unsolved. In search for understanding the class of finite splicing systems, it has been conjectured that a necessary condition for a regular language L to be a splicing language is that L must have a constant in the Schützenberger's sense. We prove this longstanding conjecture to be true. The result is based on properties of strongly connected components of the minimal deterministic finite state automaton for a regular splicing language.

1 Introduction

A splicing system, originally introduced in [12], is a formal device to generate languages called *splicing languages* by using contextual cross-over operation over words. This operation formalizes the behaviour of basic biomolecular processes involving cut and paste of DNA obtained by restriction enzymes and a ligase.

Restriction enzymes act on double stranded DNA molecules by cleaving certain recognized segments leaving short single stranded overhangs. Molecules with the same overhangs can be joined (in a cross-over fashion) in presence of a ligase enzyme. In the introductory paper, T. Head proved that if the splicing is performed by certain simple rules, then finite splicing can generate the class of strictly locally testable languages [9].

The splicing notion was reformulated by G. Paun at a less restrictive level of generality, giving rise to the splicing operation that is commonly adopted and appears nowadays as a standard [16].

Theoretical results in splicing systems have contributed to new research in formal language theory focused on modeling of biochemical processes [17]. On the other side, the field suggested new ideas in the framework of biomolecular science, for example, the design of automated enzymatic processes.

In this paper, we focus on the original concept of finite splicing systems, called here simply as splicing systems. It is closest to the actual biological process: the

G. Mauri and A. Leporati (Eds.): DLT 2011, LNCS 6795, pp. 82–92, 2011.

splicing operation is meant to have a finite set of rules (modelling enzymes) on a finite set of initial strings (modelling DNA sequences). A splicing system (or H-system) is a triple $H = (A, I, R)$, where A is a finite alphabet, $I \subseteq A^*$ is the initial language and R is the set of rules, (see Section 4 for the definitions). The formal language generated by the splicing system is the smallest language containing I and closed under the splicing operation, which makes the rule intervene.

There have been successes in characterizing certain subclasses of splicing languages, for example those generated by reflexive rules (if $(u_1, u_2), (u_3, u_4)$ is in R then both $(u_1, u_2), (u_1, u_2)$ and $(u_3, u_4), (u_3, u_4)$ are in R) and those generated by symmetric rules (if $(u_1, u_2), (u_3, u_4)$ is in R then $(u_3, u_4), (u_1, u_2)$ is in R) [2].

The general splicing systems have a set of rules R not necessarily symmetric nor reflexive although reflexivity and symmetry are natural properties for splicing systems as originally defined in [12]. Under the formal model, a splicing system is a generative mechanism of a class of languages which turns out to be a proper subclass of the regular languages. This basic result has been firstly proved in [8], and later proved in several other papers by using different approaches (see for example [18], [20]).

In spite of the vast literature on the topic, a characterization of the finite splicing systems is still an open problem.

On the other hand, progress has been recently made towards the characterization of certain sub-clases of splicing systems.

Authors in [11] provide an example of a regular splicing language that is neither reflexive nor symmetric, and prove that it is decidable whether a regular language is a reflexive splicing language. A quite different characterization of reflexive symmetric splicing languages is given in [3] and it has been extended to the general class of reflexive regular languages in [4,5]. This characterization has been given by using the concept of a constant introduced by Schützenberger [19].

Understanding the role of constants for a splicing language seems to be essential to solve the open problem of finding a characterization of the whole class. Indeed, it has been conjectured from the first introduction of splicing systems, and more formally in [10] and in [11], that a necessary condition for a regular language to be splicing is that it must have a constant. In this paper we solve this longstanding open question by proving this conjecture true. This result is proved by investigating structural properties of connected components of the graph given by the minimal finite state automaton for a regular splicing language. More precisely, properties of the factor language of components are related to the notion of synchronizing words [7]. Synchronizing words have been studied in automata theory for a long time and are of interest in both coding theory [1] and symbolic dynamics [15,14]. Our proof uses an old observation that a synchronizing word for an automaton is a constant for the language recognized by the automaton [19].

The paper is organized as follows. In Section 2 we introduce preliminary concepts, including the notion of transitive components of a finite state automaton and the notion of a synchronizing word. In Section 3 we introduce the notion of path-automaton and show results connecting terminal components and synchronizing words. Finally, in Section 4 we present the main result of the paper.

2 Preliminaries

Let A^* be the free monoid over a finite alphabet A and let $A^+ = A^* \setminus 1$, where 1 is the empty word. A *deterministic finite state automaton* (DFA) is a 5-tuple $\mathcal{A} = (Q, A, \delta, q_0, T)$, where Q is a finite set of states, $q_0 \in Q$ is the initial state, $T \subseteq Q$ is the set of terminal (final) states, δ is the transition function $\delta : Q \times A \to Q$. We denote with $L(\mathcal{A})$ the language recognized by $\mathcal{A}$ [13]. As usual, δ is extended to a function on words $Q \times A^* \to Q$. A deterministic finite state automaton is usually depicted as a directed graph with vertices Q and a set of directed edges $\mathcal{E} = \{(q, a, q') \mid \delta(q, a) = q'\}$. For an edge (q, a, q') we say that a is its label. Given a regular language $L \subseteq A^*$ it is well known that there is a unique *minimal* deterministic finite state automaton (mDFA) $\mathcal{A}$ that recognizes L such that all other DFA that recognize L map homomorphically onto $\mathcal{A}$ [13]. This automaton is unique up to a possible renaming of the states, i.e., up to an isomorphism. Given a deterministic finite state automaton $\mathcal{A}$ the transition function δ defines an action of A^* on Q. As usual, we use the standard notation qw to denote $\delta(q, w)$ when δ is understood. Similarly, we write Qw for the image of the set Q under the map $w : Q \to Q$ defined with $w(q) = qw$, which in fact is determined by the action of A^* on Q.

Given a language L, then $F(L)$ is the set of all factors of words in L, where x is a factor of word w iff $w = zxy$ for $z, y \in A^*$.

The *right context of a word* $w \in A^*$ *with respect to a language* L is defined with $\mathcal{R}_L(w) = \{x \in A^* \mid wx \in L\}$.

Similarly, the right context of a word $w \in A^*$ with respect to a DFA $\mathcal{A}$ is defined as follows: $\mathcal{R}_{\mathcal{A}}(w) = \{x \in A^* \mid \exists q \in Q, \delta(q, wx) \in T\}$.

The *right context of a state in* $\mathcal{A}$ is $\mathcal{R}_{\mathcal{A}}(q) = \{x \in A^* \mid qx \in T\}$. It is well known (see for ex. [13]) that given a regular language L, there is a one-to-one correspondence between the right contexts of words with respect to L and the right contexts of the states in the minimal deterministic finite state automaton $\mathcal{A} = (Q, A, \delta, q_0, T)$ for L, i.e.,

$$q_0 w = q \text{ iff } \mathcal{R}_L(w) = \mathcal{R}_{\mathcal{A}}(q).$$

In fact, in the mDFA $\mathcal{A}$ it also holds $\mathcal{R}_L(w) = \mathcal{R}_{\mathcal{A}}(q)$ iff $\mathcal{R}_L(wa) = \mathcal{R}_{\mathcal{A}}(qa)$ for all $a \in A$, and therefore $\mathcal{R}_{\mathcal{A}}(q) = \mathcal{R}_{\mathcal{A}}(q')$ implies $q = q'$.

When the language and the DFA are fixed, we drop the subscripts and write $\mathcal{R}(w)$ and $\mathcal{R}(q)$.

Note that every state in an mDFA is accessible, i.e., for each state $q \in Q$ there is an $x \in A^*$ such that $\delta(q_0, x) = q$. Also, in an mDFA, there is at most one state that is not co-accessible, since for each $q \in Q$, there is $u \in A^*$ such that $qu \in T$ iff $\mathcal{R}(q) \neq \emptyset$. So, q is not co-accessible, iff $\mathcal{R}(q) = \emptyset$. If such a state in $\mathcal{A}$ exists, we call it *zero* and denote it with $\mathbf{z}$. A *trimmed mDFA* for language L is the DFA obtained from the mDFA for L by erasing the state $\mathbf{z}$ and all transitions that terminate in $\mathbf{z}$.

Finally, for a finite set S, by $\#S$, we denote the cardinality of set S.

Recall the definition of a constant, introduced by Schützenberger in [19].

Definition 1. *A word $w \in A^+$ is a* constant *of a language L if w is a factor of some word in L and for all words u_1, u_2, v_1, v_2 in A^* we have:*

$$\begin{array}{l} u_1 w u_2 \in L \\ v_1 w v_2 \in L \end{array} \Rightarrow \begin{array}{l} u_1 w v_2 \in L \\ v_1 w u_2 \in L \end{array}$$

A characterization of constants, which is more or less folklore, is stated below.

Proposition 1. *Let $L \subseteq A^*$ be a regular language and let $\mathcal{A}$ be the mDFA recognizing L. A word $w \in A^+$ is a constant of L if and only if $Qw \setminus \{\mathbf{z}\}$ is a singleton, i.e., there is a unique non-zero state q_w such that $qw \neq \mathbf{z}$ implies $qw = q_w$ for all $q \in Q$.*

In a finite state automaton, if for a word w there is a state q_w such that every path in the automaton with label w terminates in q_w, we say that w is a *synchronizing* word and we say that q_w is a *synchronizing* state, synchronized by w. By Proposition 1 in a trimmed mDFA $\mathcal{A}$ of a regular language L, the set of synchronizing words for $\mathcal{A}$ coincides with the set of constants of L. In general, if w is a synchronizing word for an automaton $\mathcal{A}$ then it is a constant for the language recognized by $\mathcal{A}$.

3 Path-Automata and Synchronizing Words

In this section we provide structural characterizations of a path-automaton that does not have synchronizing words. More precisely, we show that a path-automaton having no synchronizing words has a unique maximal component which is terminal whose language includes all factors of the language accepted by the path-automaton.

Recall the notion of transitive component in a deterministic automaton. A strongly connected component of the directed graph for a deterministic automaton $\mathcal{A}$ is called a *transitive* component for $\mathcal{A}$. If in a transitive component, every edge that starts at a state in this component also ends at the same component, then the transitive component is called *terminal.* For every state in the mDFA for a language L there is a path that leads from that state to a terminal component. For a transitive component $\mathcal{C}$ we say that $\mathcal{C}$ is *induced by* q if q is a state in $\mathcal{C}$. We write $L(\mathcal{C})$ for the set of labels of all paths in $\mathcal{C}$. A transitive component $\mathcal{C}$ is called *trivial* if $L(\mathcal{C}) = \{1\}$.

Two transitive components $\mathcal{C}$ and $\mathcal{C}'$ are called *factor-equivalent* if $L(\mathcal{C}) = L(\mathcal{C}')$. In the following we often use the term component to denote a transitive component.

Definition 2 (path-automaton). *An automaton $\mathcal{A}$ with initial state q_0 is called a* path-automaton *if the following is satisfied:*

(i) There is at most one transition in $\mathcal{A}$ which starts at the component induced by q_0 and terminates in another component.
(ii) There is only one terminal transitive component in $\mathcal{A}$.
(ii) For every transitive component $\mathcal{C}$ which does not contain q_0 there is precisely one transition that starts in a state outside $\mathcal{C}$ but terminates in $\mathcal{C}$, and if $\mathcal{C}$ is not terminal, there is precisely one transition that starts at a state in $\mathcal{C}$ but terminates in a state outside $\mathcal{C}$.

Let $\mathcal{A}$ be a path automaton and $\mathcal{C}_i$ one of its transitive components. The state of $\mathcal{C}_i$ that is the end point of the transition starting outside $\mathcal{C}_i$ but ending at $\mathcal{C}_i$ is called the *entrance* state for $\mathcal{C}_i$ and the state that is the start point of a transition that starts in $\mathcal{C}_i$ but terminates outside $\mathcal{C}_i$ is called the *exit* of $\mathcal{C}_i$. The initial component of $\mathcal{A}$ has no entrance, and the terminal component has no exit.

A path π from an initial state in an automaton $\mathcal{A}$ to a terminal component in $\mathcal{A}$ induces a path-automaton $\mathcal{A}_\pi$ which consists of π and all transitive components in $\mathcal{A}$ induced by states visited by π.

Let $\mathcal{C}$ be a terminal component of the path-automaton $\mathcal{A}_\pi$ and let q be the entrance of $\mathcal{C}$. We define the *language accepted by the component $\mathcal{C}$ induced by the path π*, denoted by $L_\pi(\mathcal{C})$, as the language accepted by the automaton $\mathcal{C}$ with initial state q.

Example 1. Let $\mathcal{A}$ be the mDFA depicted in Figure 1(a). Then $\mathcal{A}$ has two terminal components $\mathcal{C}$ and $\mathcal{C}'$ that are factor-equivalent, that is $L(\mathcal{C}) = L(\mathcal{C}') = a^*$. Moreover if the path π with label cb ends in component denoted $\mathcal{C}$, and path

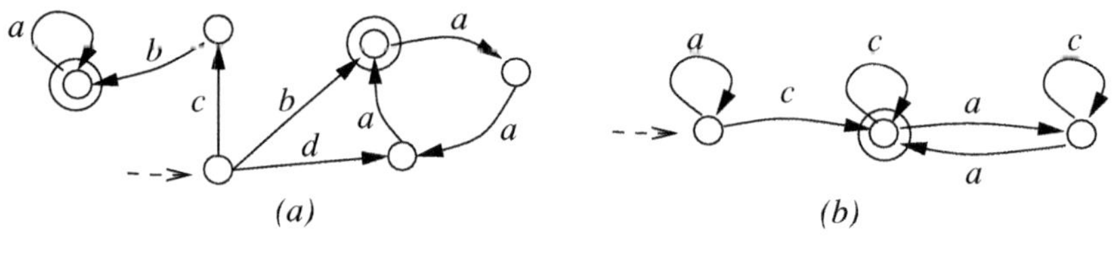

Fig. 1. Two automata, initial states are indicated with an arrow pointing to them and the terminal states are circled; (a) non-reflexive splicing language, (b) path-automaton with no synchronizing words recognizing a non-splicing language

π' labeled ba^3 ends in component denoted $\mathcal{C}'$ of $\mathcal{A}$, then $L_\pi(\mathcal{C}) = (a)^*$ and $L_{\pi'}(\mathcal{C}') = (a^3)^*$.

The following lemma, stated without proof due to space constraints, characterizes a path-automaton with no synchronizing words.

Lemma 1. *Given a deterministic path-automaton $\mathcal{A}$ let $\mathcal{C}_T$ be the terminal component of $\mathcal{A}$. Then one of the following holds:*

(a) $\mathcal{A}$ has a synchronizing word, or,
(b) $F(L(\mathcal{A})) \subseteq L(\mathcal{C}_T)$.

Example 2. The automaton in Figure 1(b) is a path-automaton with no synchronizing words. It has only one terminal component which is maximal and the factors of all words in the language are labels of paths in the terminal component. This illustrates the situation (b) in Lemma 1.

The following two results are used to prove the main result (Proposition 3) of the paper. Although Corollary 1 straight follows Lemma 1, it can also be proved independently.

Corollary 1. *Every deterministic path-automaton with two transitive components whose terminal component is trivial has a synchronizing word.*

Proposition 2. *Let L be a regular language, $x \in F(L)$ and $\mathcal{A}$ be an mDFA for L. At least one of the two cases holds:*

(i) x is a factor of a constant for L,
(ii) there is a path-automaton containing a path labeled x having a non-trivial terminal transitive component with two non-zero states.

Proof. Let $\mathcal{A} = (Q, A, \delta, q_0, T)$ be the mDFA for the language L. Suppose $x \in F(L)$ is not a factor of a constant, i.e. for every $v, v' \in A^*$, vxv' is not a constant for L, and therefore not synchronizing for $\mathcal{A}$. Consider a word w such that $\#Qxw = \min \#\{Qxu \mid u \in A^*\}$ and then pose $P_w = Qxw \setminus \{\mathbf{z}\} \neq \emptyset$. Since xw is not synchronizing, by Proposition 1, $\#P_w > 1$. Then for every word $u \in A^*$ we have that either $Qwu = \emptyset$ or $\#Qwu = \#Qw = \#P_w$. Therefore, we can assume that all states in P_w are in terminal components of $\mathcal{A}$, (if not, we can concatenate w with words that are labels of paths that lead to terminal components). Consequently we have that P_w has cardinality at least 2. If all terminal components are trivial, then because $\mathcal{A}$ is reduced, there is only one terminal transitive component and it is trivial, and hence $\#P_w = 1$ which is a contradiction with our assumption that x does not extend to a constant. Thus there must be at least one terminal transitive component which is not trivial. Assume that each state in P_w is in a transitive component consisting of only one state having a loop at itself. Since $P_w y \neq \emptyset$ implies $P_w y = P_w$, for every $q \in P_w$ we have $qy = q$, i.e., all states in P_w are terminal and their right contexts are equal, hence they cannot be distinct in a reduced automaton, thus again implying that P_w has cardinality 1, a contradiction. Hence, there must be at least two states in P_w that belong to the same terminal transitive component, thus the proposition holds. □

Our investigation produced several additional observation about the properties of transitive components and path-automata with (or without) synchronizing words. These are not necessary for our main result, but might be of interest in general studies of constants and regular languages. They are available at [6].

4 Splicing Languages Must Have a Constant

As mentioned, in this paper we consider the general notion of the splicing operation and the splicing system given by Paun [16], as defined below.

Definition 3. *A* finite splicing system *is a triple* $S = (A, I, R)$*, where* $I \subset A^*$ *is a finite set of strings, called an* initial language*,* R *is a finite set of* rules $r = (u_1, u_2)(u_3, u_4)$*, with* $u_i \in A^*, i = 1, 2, 3, 4$*. Given two words* $x = x_1u_1u_2x_2, y = y_1u_3u_4y_2$*,* $x_1, x_2, y_1, y_2 \in A^*$ *and the rule* $r = (u_1, u_2)(u_3, u_4)$*, the splicing rule produces* $w = x_1u_1u_4y_2$ *denoted* $(x, y)\vdash_r w$*. We also say that* u_1u_2*,* u_3u_4 *are* splice sites *of* r *and* u_1u_4 *is the* paste *site of* r*.*

To simplify the notation, in the following by a splicing system we mean a finite splicing system.

Let $L \subseteq A^*$. We denote $\sigma(L) = \{w \in A^* \mid (x, y)\vdash_r w,\ x, y \in L, r \in R\}$. The (iterated) splicing operation is defined as follows: $\sigma^0(L) = L, \sigma^{i+1}(L) = \sigma^i(L) \cup \sigma(\sigma^i(L))$, $i \geq 0$, $\sigma^*(L) = \bigcup_{i\geq 0} \sigma^i(L)$.

Definition 4 (splicing language). *Given a finite splicing system* $S = (A, I, R)$*, the language* $L(S) = \sigma^*(I)$ *is the language generated by* S*. A language* L *is a* splicing language *if there is a splicing system* S *such that* $L = L(S)$*.*

It is known that every splicing language generated by a finite splicing system is always regular [8,18]. More precisely, regular splicing languages form a proper subclass of the class of regular languages.

Recall that a splicing system S is said to be *reflexive* if for every rule $r = (u_1, u_2)(u_3, u_4)$ in R, both $(u_1, u_2)(u_1, u_2)$ and $(u_3, u_4)(u_3, u_4)$ are rules in R. A language L is said to be *reflexive* splicing language if there is a reflexive splicing system S such that $L = L(S)$. The notion of a constant of a language turned out to be essential in providing a characterization of the subclass of reflexive regular splicing languages [11,3]. Indeed, a fundamental property of a reflexive regular splicing language L is that there exists a splicing system generating L that has rules whose splicing sites consist of constants for the language L.

An example of non-reflexive regular splicing language is the language $L = a^*b^*a^*b^*a^* \cup a^*b^*a^* \cup a^*$ [11]. Considering the importance of constants in characterization of sub-classes of regular splicing languages, it has been conjectured that every splicing language must have a constant [10,11]. Our main result proves this conjecture to be true.

Proposition 3 (Main result). *If* L *is a regular splicing language, then* L *has a constant.*

Example 3. The path-automaton $\mathcal{A}$ of Figure 1(b) has no synchronizing word (see Example 2) and thus the language $L(\mathcal{A}) = a^*c(c^*ac^*a)^*$ has no constant. By Proposition 3, $L(\mathcal{A})$ is not a regular splicing language.

Example 4. The regular language $L = b(a^3)^* + cba^* + da(a^3)^*$ is another example of non-reflexive splicing language, as proved in Lemma 2. Figure 1(a) shows the mDFA graph for language L. Observe that not every path-automaton induced by a path in the mDFA from the initial state q_0 to a terminal component has necessarily a constant. Indeed, the path-automaton recognizing language $b(a^3)^*$ does not have any constant for language L.

Lemma 2. *The regular language* $L = b(a^3)^* + cba^* + da(a^3)^*$ *is a non-reflexive splicing language.*

Proof. First we note that $L \subseteq A^*$, for $A = \{a, b, c, d\}$ is splicing. A splicing system $S = (A, I, R)$ for language L consists of rules $R = \{r_1 = (cba, 1)(cb, a), r_2 = (daa^3, 1)(da, 1), r_3 = (b, a^3)(da, 1)\}$, while the initial language I consists of language $I = \{ba^3, b, cba, cb, daa^3, da\}$. By induction on the number k of iteration steps of splicing rules, we first show that $L(S) \subseteq L$. If $k = 0$, since $I \subseteq L(S)$, the inclusion holds. Assume that $w \in L(S)$ is generated with $k > 0$ iterations by applying a rule r to a pair of words $w_1, w_2 \in L(S)$. By induction $w_1, w_2 \in L$ are obtained with $k-1$ iterations. Checking splice sites in w_1 and w_2 for all of the rules, it is immediate to see that $w \in L$. In order to show that $L \subseteq L(S)$, we observe that language $L_1 = da(a^3)^*$ is generated by rule r_2 applied to words in the same language daa^3. Similarly, we see that language $L_2 = cba^*$ is generated by rule r_1 starting from words from the same language. Language $L_3 = b(a^3)^*$ is generated by rule r_3 applied to words of language $da(a^3)^*$ and of language $b(a^3)^*$. By induction on $i \geq 0$, indeed we can observe that $b(a^3)^i \in L(S)$, $i \geq 0$. If $i = 0$ or $i = 1$, being $b, ba^3 \in I$, the result is immediate. Otherwise, given words $b(a^3)^{i-1} \in L(S)$, for $i > 1$ and word $da(a^3)^i \in L(S)$, by rule r_3 is immediate to generate word $b(a^3)^i \in L(S)$.

Finally, notice that language L is not reflexive, that is, it cannot be generated by a splicing system by reflexive splicing rules. Suppose L is reflexive splicing language generated by a reflexive system S. We obtain a contradiction by considering generation of words in language $b(a^3)^*$. Since in language L the only words that start with a b are those in language $b(a^3)^*$ and there must be splicing rules in S to generate words of the form $b(a^3)^k$ for arbitrarily large k, there must be a rule r with splice site u_1u_2 that is a factor of $b(a^3)^*$. But, because S is reflexive, S must also contain a rule $(u_1, u_2)(u_1, u_2)$. Then when the rule is applied to a word $b(a^3)^k$, for some large $k > 0$ and to word $cb(a)^j$ for some suitable chosen $j > 0$, then word $b(a^2)^l$, for some $l > 0$ can also be generated by such a rule. Therefore the language L cannot be generated by reflexive rules. □

4.1 The Main Result

The proof of Proposition 3 is based on the notions stated below that are used to find special words in a regular splicing language L that must be generated by a splicing rule whose splice site u_3u_4 is a constant of L. For a lack of a better name, we call these words q-canonical and k-special words.

Informally, the q-canonical word of a component $\mathcal{C}$ is a word c such that $qc = q$ and every such path with label c crosses all states in the component, and moreover, the word c is able to identify the language $L(\mathcal{C})$ of the component $\mathcal{C}$.

Definition 5 (q**-canonical).** *Let* $\mathcal{A}$ *be an automaton and let* $\mathcal{C}$ *be a component of* $\mathcal{A}$ *and* q *a state of* $\mathcal{C}$. *Then a word* $c \in A^+$ *such that* $c \in L(\mathcal{C})$, $qc = q$ *is called* q*-canonical for* $\mathcal{C}$, *if* $c \in L(\mathcal{C}')$, *for* $\mathcal{C}'$ *another component of* $\mathcal{A}$ *implies that* $L(\mathcal{C}) \subseteq L(\mathcal{C}')$.

We note that for a deterministic automaton $\mathcal{A}$ and a transitive component $\mathcal{C}$ in $\mathcal{A}$, there is a q-canonical word for each state q in $\mathcal{C}$. We are interested in q-canonical words whose factors include all possible labels of paths of a given length k in the component. Such a word is called *k-special* for $L(\mathcal{C})$ as defined below.

Definition 6 (k-special). *Let $\mathcal{A}$ be an automaton. A word c in $L(\mathcal{A})$ is k-special for the language L if every word of $F(L)$ of length k is a factor of c.*

The following lemma states the existence of a q-canonical word that is a k-special word for every transitive component in $\mathcal{A}$.

Lemma 3. *Given a transitive component $\mathcal{C}$ in a DFA $\mathcal{A}$ let $k = \#Q \cdot \#Q$. Then there is a state q in $\mathcal{C}$ with a q-canonical k-special word $c \in L(\mathcal{C})$.*

Proof. Let $\{x_1, \ldots, x_n\} = L(\mathcal{C}) \cap A^k$. Being $\mathcal{C}$ a transitive component, there are $y_1, \ldots, y_{n-1}$ such that $x_1 y_1 x_2 \cdots y_{n-1} x_n \in L(\mathcal{C})$. Set $c = x_1 y_1 x_2 \cdots y_{n-1} x_n$ and assume that $c \in L(\mathcal{C}')$ for some transitive component $\mathcal{C}'$. Take the shortest word $z \in L(\mathcal{C}) \setminus L(\mathcal{C}')$. Since $L(\mathcal{C}) \setminus L(\mathcal{C}') = L(\mathcal{C}) \cap (L(\mathcal{C}'))^c$, it can be recognized by an automaton with at most $\#Q(\mathcal{C}) \cdot \#Q(\mathcal{C}') \leq k$ states [13], therefore the shortest word in this language has length at most k. Thus $|z| \leq k$ and therefore z must be a factor of c, i.e., z must be in $L(\mathcal{C}')$, contradicting the existence of z. □

The proof of Proposition 3 depends on the existence of a splicing rule $r = (u_1, u_2)(u_3, u_4)$ such that word $u_1 u_4$, ends in a terminal component of the mDFA $\mathcal{A}$.

Definition 7 (paste site). *Let $\mathcal{A}$ be the mDFA for a regular splicing language L. The word $u_1 u_4$ is said to be a* paste site *at a* state *$p \in Q$* for *a* splicing rule *$r = (u_1, u_2)(u_3, u_4)$ if $\mathcal{R}_L(u_3 u_4) \subseteq \mathcal{R}_{\mathcal{A}}(p u_1 u_4)$ and $p u_1 u_2 \neq \mathbf{z}$.*

More precisely, the notion of a paste site at a state q is used to identify states of the automaton where a rule is applied.

In what follows we assume that every splicing system is such that all rules are applied at least once during the generation of the splicing language. The following lemma shows an equivalence between splicing systems with respect to the extension of sites and paste sites of rules.

Lemma 4. *Let $S = (A, I, R)$ be a finite splicing system and $r = (u_1, u_2)(u_3, u_4)$ be a splicing rule in R. Let $c \in \mathcal{R}_{L(S)}(u_1 u_4) \cap \mathcal{R}_{L(S)}(u_3 u_4)$ for some $x, y \in A^*$. Then $L(S)$ is the language generated with the splicing system $S' = (A, I, R')$ where $R' = R \cup \{r'\}$ for $r' = (u_1, u_2)(u_3, u_4 c)$.*

Lemma 5. *Let $S = (A, I, R)$ be a finite splicing system and $\mathcal{A}$ a finite state automaton for $L = L(S)$. If $u_1 u_4$ is a paste site at state p for rule $r = (u_1, u_2)(u_3, u_4) \in R$ then for every $x \in \mathcal{R}_L(u_1 u_4) \cap \mathcal{R}_L(u_3 u_4)$ we have that $u_1 u_4 x$ is a paste site at p for rule $r = (u_1, u_2)(u_3, u_4 x)$.*

We now sketch the main steps of the proof of Proposition 3.

Let L be a regular splicing language and $\mathcal{A}$ the mDFA for L.

- Step 1:
  Assuming that L has no constant, Proposition 2 applies. The mDFA for L, denoted $\mathcal{A}$, has non-trivial terminal components. We consider among all terminal components in the mDFA for L, the component $\mathcal{C}$ and a state q in $\mathcal{C}$ such that the language recognized by the automaton $\mathcal{C}_q$ induced by the component $\mathcal{C}$ having the initial state q is minimal among all terminal components that are factor-equivalent to $\mathcal{C}$. In other words $L(\mathcal{C}_q)$ is minimal among all languages of terminal components that have the same set of labels of paths as $\mathcal{C}$.
- Step 2:
  Given such chosen $\mathcal{C}$, then we consider splicing rules generation of words in L of the form wc^*x, where w is label of a path in the mDFA from the initial state to the component $\mathcal{C}$. The word c is a q-canonical and k-special word for $k = \#Q(\mathcal{A}) \cdot \#Q(\mathcal{A})$.
  We prove the existence of a rule $r = (u_1, u_2)(u_3, u_4)$ such that for arbitrarily large i's, c^i appears in the right-context $\mathcal{R}(u_3u_4)$ of word u_3u_4 (Lemmas 4,5).
- Step 3:
  We show that every state $\bar{q}$ in a terminal component $\bar{\mathcal{C}}$ reached by reading u_3u_4 in automaton $\mathcal{A}$, must be the same state p that is reached by reading vu_1u_4 along the path labeled by wc^*x, for some word $v \in A^*$. This step shows that u_3u_4 is synchronizing for state p, that is u_3u_4 must be constant for L.
  Observe that the proof that state $\bar{q}$ reached by reading u_3u_4 in automaton $\mathcal{A}$ is unique depends on the following fact: the component $\bar{\mathcal{C}}$ having state $\bar{q}$ is terminal (using Corollary 1) and due to the minimality of $\mathcal{C}_q$, $\bar{\mathcal{C}}$ must be the same component as $\mathcal{C}$. This fact also uses the property of splicing rules, more precisely: $\mathcal{R}(\bar{q}) \subseteq \mathcal{R}(u_3u_4) \subseteq \mathcal{R}(qu_1u_4) = \mathcal{R}(p)$ and the minimality of mDFA.

Acknowledgment. P.Bonizzoni is partially supported by MIUR Project *"Mathematical aspects and emerging applications of automata and formal languages"*, N. Jonoska is supported in part by NSF grants CCF-0726396 and DMS-0900671. Parts of this research were done during N. Jonoska's visit to University of Milano-Bicocca in the Spring of 2008.

References

1. Berstel, J., Perrin, D.: Theory of Codes. Academic Press Inc., Orlando (1985)
2. Bonizzoni, P., De Felice, C., Mauri, G., Zizza, R.: Regular Languages Generated by Reflexive Finite Linear Splicing Systems. In: Ésik, Z., Fülöp, Z. (eds.) DLT 2003. LNCS, vol. 2710, pp. 134–145. Springer, Heidelberg (2003)
3. Bonizzoni, P., De Felice, C., Zizza, R.: The structure of reflexive regular splicing languages via Schützenberger constants. Theoretical Computer Science 334(1-3), 71–98 (2005)
4. Bonizzoni, P., Mauri, G.: Regular splicing languages and subclasses. Theoretical Computer Science 340, 349–363 (2005)

5. Bonizzoni, P.: Constants and label-equivalence: A decision procedure for reflexive regular splicing languages. Theoretical Computer Science 411(6), 865–877 (2010)
6. Bonizzoni, P., Jonoska, N.: Splicing languages and constants, manuscript (2011)
7. Černý, J.: Poznámka k homogénnym eksperimentom s konecnými automatami. Matematicko-fyzikalny Časopis Slovenskej Akadémie Vied 14, 208–216 (1964)
8. Culik, K., Harju, T.: Splicing semigroups of dominoes and DNA. Discrete Applied Math. 31, 261–277 (1991)
9. De Luca, A., Restivo, A.: A characterization of strictly locally testable languages and its application to semigroups of free semigroup. Information and Control 44, 300–319 (1980)
10. Goode, E.: Constants and splicing systems, PHD Thesis, Binghamton University (1999)
11. Goode, E., Pixton, D.: Recognizing splicing languages: Syntactic Monoids and Simultaneous Pumping. Discrete Applied Mathematics 155, 989–1006 (2007)
12. Head, T.: Formal Language Theory and DNA: an analysis of the generative capacity of specific recombinant behaviours. Bull. Math. Biol. 49, 737–759 (1987)
13. Hopcroft, J.E., Motwani, R., Ullman, J.D.: Introduction to Automata Theory, Languages, and Computation. Addison-Wesley, Reading (2001)
14. Jonoska, N.: Sofic Systems with Synchronizing Representations. Theoretical Computer Science 158(1-2), 81–115 (1996)
15. Lind, D., Marcus, B.: An Introduction to Symbolic Dynamics. Cambridge University Press, New York (1995)
16. Paun, G.: On the splicing operation. Discrete Applied Math. 70, 57–79 (1996)
17. Paun, G., Rozenberg, G., Salomaa, A.: DNA computing, New Computing Paradigms. Springer, Berlin (1998)
18. Pixton, D.: Regularity of splicing languages. Discrete Applied Math. 69, 101–124 (1996)
19. Schützenberger, M.P.: Sur certaines opérations de fermeture dans le langages rationnels. Symposia Mathematica 15, 245–253 (1975)
20. Verlan, S.: Head systems and applications to bio-informatics. Ph. D. Thesis, University of Metz (2004)

The Average Transition Complexity of Glushkov and Partial Derivative Automata*

Sabine Broda, António Machiavelo, Nelma Moreira, and Rogério Reis

CMUP, Faculdade de Ciências da Universidade do Porto
Rua do Campo Alegre, 4169-007 Porto, Portugal
sbb@dcc.fc.up.pt, ajmachia@fc.up.pt, {nam,rvr}@dcc.fc.up.pt

Abstract. In this paper, the relation between the Glushkov automaton ($\mathcal{A}_{\text{pos}}$) and the partial derivative automaton ($\mathcal{A}_{\text{pd}}$) of a given regular expression, in terms of transition complexity, is studied. The average transition complexity of $\mathcal{A}_{\text{pos}}$ was proved by Nicaud to be linear in the size of the corresponding expression. This result was obtained using an upper bound of the number of transitions of $\mathcal{A}_{\text{pos}}$. Here we present a new quadratic construction of $\mathcal{A}_{\text{pos}}$ that leads to a more elegant and straightforward implementation, and that allows the exact counting of the number of transitions. Based on that, a better estimation of the average size is presented. Asymptotically, and as the alphabet size grows, the number of transitions per state is on average 2.

Broda *et al.* computed an upper bound for the ratio of the number of states of $\mathcal{A}_{\text{pd}}$ to the number of states of $\mathcal{A}_{\text{pos}}$, which is about $\frac{1}{2}$ for large alphabet sizes. Here we show how to obtain an upper bound for the number of transitions in $\mathcal{A}_{\text{pd}}$, which we then use to get an average case approximation. Some experimental results are presented that illustrate the quality of our estimate.

1 Introduction

The conversion methods of regular expressions into equivalent nondeterministic finite automata (NFA) are normally divided in two classes depending on whether ε-transitions are allowed or not in the resulting NFA. Paradigmatic methods of each class are the Thompson's and Glushkov's constructions, respectively. Several optimizations and worst-case descriptional and computational complexity results were obtained for both methods (see Holzer and Kutrib [HK10], and the works cited therein). Given a regular expression with n letters the size of a ε-NFA can be, in the worst-case, $\Theta(n)$. While the size of a Glushkov automaton can be $\Theta(n^2)$, $\Omega(n \log n^2)$ was proved to be a lower bound for the size of a ε-free NFA. In this context, and for practical purposes, it is useful to carry out average-case analysis, both for descriptional and computational complexities, of these methods.

* This work was partially funded by Fundação para a Ciência e Tecnologia (FCT) and Program POSI, and project CANTE (PTDC/EIA-CCO/101904/2008).

G. Mauri and A. Leporati (Eds.): DLT 2011, LNCS 6795, pp. 93–104, 2011.

The framework of analytic combinatorics, by relating the enumeration of combinatorial objects to the algebraic and complex analytic properties of generating functions, provides a powerful tool for asymptotic average-case analysis. Using this framework, Nicaud [Nic09] proved that the average transition complexity of the Glushkov automaton ($\mathcal{A}_{\text{pos}}$) of a regular expression α of size n is $\Theta(n)$. This result was obtained using an upper bound of the number of transitions of $\mathcal{A}_{\text{pos}}$. Here we present a new quadratic construction of the $\mathcal{A}_{\text{pos}}$ that leads to a more elegant and straightforward implementation, and that allows the exact counting of the number of transitions. Based on that, a better estimation of the average size is presented. Asymptotically, and as the alphabet size grows, the number of transitions per state is on average 2.

The partial derivative automaton ($\mathcal{A}_{\text{pd}}$) is a quotient of the $\mathcal{A}_{\text{pos}}$, and thus the states of the former can be seen as mergings of states of the latter. In a previous paper [BMMR11b], we presented a technique for estimating some of those state mergings. This enabled us, in the framework of analytic combinatorics, to compute an upper bound for the ratio of the number of states of $\mathcal{A}_{\text{pd}}$ to the number of states of $\mathcal{A}_{\text{pos}}$, which is about $\frac{1}{2}$ for large alphabet sizes. This upper bound was obtained by estimating the number of regular expressions that have ε as a partial derivative. In this paper, we use an analogous approach to compute an upper bound for the number of transitions in $\mathcal{A}_{\text{pd}}$, and study its asymptotic behaviour. As the alphabet size grows, this upper bound tends to the number of letters of the regular expression, thus it is half the number of transitions in $\mathcal{A}_{\text{pos}}$. The comparison with some experimental results suggests that this upper bound has an error of less than 15%.

2 Regular Expressions and Automata

In this section we briefly review some basic definitions about regular expressions and finite automata. For more details, we refer the reader to Kozen [Koz97] or Sakarovitch [Sak09].

Given an alphabet (set of *letters*) $\Sigma = \{\sigma_1, \ldots, \sigma_k\}$ of size k, the set $\mathcal{R}$ of *regular expressions*, α, over Σ is defined by the following grammar:

$$\alpha := \emptyset \mid \varepsilon \mid \sigma_1 \mid \cdots \mid \sigma_k \mid (\alpha + \alpha) \mid (\alpha \cdot \alpha) \mid \alpha^\star \tag{1}$$

where the operator $\cdot$ (concatenation) is often omitted. The language associated to α is denoted by $\mathcal{L}(\alpha)$ and defined as usual. The *size* $|\alpha|$ of $\alpha \in \mathcal{R}$ is the number of symbols in α (parentheses not counted); the *alphabetic size* $|\alpha|_\Sigma$ is its number of letters. For example, for $\tau = (a+b)(a^\star + ba^\star + b^\star)^\star$ one has $|\tau| = 15$ and $|\tau|_\Sigma = 6$. We define $\varepsilon(\alpha)$ by $\varepsilon(\alpha) = \varepsilon$ if $\varepsilon \in \mathcal{L}(\alpha)$, and $\varepsilon(\alpha) = \emptyset$ otherwise. Also, we denote by α_ε and $\alpha_{\overline{\varepsilon}}$, respectively, the regular expressions such that $\varepsilon(\alpha_\varepsilon) = \varepsilon$ and $\varepsilon(\alpha_{\overline{\varepsilon}}) = \emptyset$.

A *non-deterministic automaton* (NFA) $\mathcal{A}$ is a quintuple $(Q, \Sigma, \delta, q_0, F)$, where Q is a finite set of states, Σ is the alphabet, $\delta \subseteq Q \times \Sigma \times Q$ the transition relation, q_0 the initial state, and $F \subseteq Q$ the set of final states. The *size* of a NFA is $|Q| + |\delta|$. For $q \in Q$ and $\sigma \in \Sigma$, we denote the set $\{p \mid (q, \sigma, p) \in \delta\}$ by $\delta(q, \sigma)$, and we can extend this notation to $w \in \Sigma^\star$, and to $R \subseteq Q$. The *language* accepted by $\mathcal{A}$ is $\mathcal{L}(\mathcal{A}) = \{w \in \Sigma^\star \mid \delta(q_0, w) \cap F \neq \emptyset\}$.

2.1 The Glushkov Automaton

The Glushkov, or position, automaton was independently introduced by Glushkov [Glu61] and McNaughton and Yamada [MY60]. The states in the Glushkov automaton, representing a regular expression α, correspond to the positions of letters in α plus an additional initial state. Let $\overline{\alpha}$ denote the regular expression obtained by marking each letter with its position in α. The marked version of the previous example is $\overline{\tau} = (a_1 + b_2)(a_3^\star + b_4 a_5^\star + b_6^\star)^\star$. The same notation is used to remove the markings, i.e., $\overline{\overline{\alpha}} = \alpha$. Now, let $\mathsf{Pos}(\alpha) = \{1, 2, \ldots, |\alpha|_\Sigma\}$ be the set of positions for $\alpha \in \mathcal{R}$, and let $\mathsf{Pos}_0(\alpha) = \mathsf{Pos}(\alpha) \cup \{0\}$. Then, the construction of the Glushkov automaton is based on the position sets $\mathsf{First}(\overline{\alpha})$, $\mathsf{Last}(\overline{\alpha})$, and $\mathsf{Follow}(\overline{\alpha})$. These sets can be inductively defined as follows:

$$\begin{aligned}
\mathsf{First}(\emptyset) &= \mathsf{First}(\varepsilon) = \emptyset & \mathsf{First}(\alpha + \beta) &= \mathsf{First}(\alpha) \cup \mathsf{First}(\beta) \\
\mathsf{First}(\sigma_i) &= \{i\} & \mathsf{First}(\alpha\beta) &= \begin{cases} \mathsf{First}(\alpha) \cup \mathsf{First}(\beta) & \text{if } \varepsilon(\alpha) = \varepsilon \\ \mathsf{First}(\alpha) & \text{otherwise.} \end{cases} \\
\mathsf{First}(\alpha^\star) &= \mathsf{First}(\alpha)
\end{aligned}$$

The definition of Last is almost identical and differs only for the case of concatenation, which is

$$\mathsf{Last}(\alpha\beta) = \begin{cases} \mathsf{Last}(\alpha) \cup \mathsf{Last}(\beta) & \text{if } \varepsilon(\beta) = \varepsilon \\ \mathsf{Last}(\beta) & \text{otherwise.} \end{cases}$$

The set Follow can be computed as by

$$\begin{aligned}
\mathsf{Follow}(\emptyset) &= \mathsf{Follow}(\varepsilon) = \mathsf{Follow}(\sigma_j) = \emptyset \\
\mathsf{Follow}(\alpha + \beta) &= \mathsf{Follow}(\alpha) \cup \mathsf{Follow}(\beta) \\
\mathsf{Follow}(\alpha\beta) &= \mathsf{Follow}(\alpha) \cup \mathsf{Follow}(\beta) \cup \mathsf{Last}(\alpha) \times \mathsf{First}(\beta) \\
\mathsf{Follow}(\alpha^\star) &= \mathsf{Follow}(\alpha) \cup \mathsf{Last}(\alpha) \times \mathsf{First}(\alpha).
\end{aligned}$$

The *Glushkov automaton* for α is $\mathcal{A}_{\mathrm{pos}}(\alpha) = (\mathsf{Pos}_0(\alpha), \Sigma, \delta_{\mathrm{pos}}, 0, F)$, with $\delta_{\mathrm{pos}} = \{(0, \overline{\sigma_j}, j) \mid j \in \mathsf{First}(\overline{\alpha})\} \cup \{(i, \overline{\sigma_j}, j) \mid (i, j) \in \mathsf{Follow}(\overline{\alpha})\}$ and $F = \mathsf{Last}(\overline{\alpha}) \cup \{0\}$ if $\varepsilon(\alpha) = \varepsilon$, and $F = \mathsf{Last}(\overline{\alpha})$, otherwise. Note that the number of states of $\mathcal{A}_{\mathrm{pos}}(\alpha)$ is exactly $n+1$, where $n = |\alpha|_\Sigma$. On the other hand, the number of transitions in $\mathcal{A}_{\mathrm{pos}}(\alpha)$ is in the worst case $n^2 + n$. Consequently, the time-complexity of any construction algorithm for $\mathcal{A}_{\mathrm{pos}}(\alpha)$ must be at least $O(n^2)$. Considering the simplicity of the recursive definitions of the position sets used for the construction of $\mathcal{A}_{\mathrm{pos}}(\alpha)$, an algorithm of this complexity should not be hard to find. Nevertheless, a naive implementation leads to a $O(n^3)$ algorithm, such as the one proposed by Berry and Sethi [BS86]. This is due to possibly non-disjoint unions of sets in the rule for $\alpha^\star$ in the recursive definition of $\mathsf{Follow}(\alpha)$. To overcome this problem, several techniques for the construction of $\mathcal{A}_{\mathrm{pos}}(\alpha)$ were proposed over the years. The first one, of order $O(m + n^2)$, where $m = |\alpha|$, was proposed by Brüggemann-Klein in 1993 [BK93] and it is primarily based on the prior transformation of α into *star-normal form*. Other quadratic, however sophisticated, algorithms have been introduced in 1996 and 1997, respectively in [PZC97] and [CP97].

Our goal in the next section, is to present an alternative recursive definition of $\mathsf{Follow}(\alpha)$, that only involves disjoint unions of sets, allowing for simple implementations of that construction in time $O(n^2)$. This definition also allows us to define a cost generating function of the exact number of transitions in the Glushkov automaton in Section 4.

3 A New Algorithm for Computing $\mathsf{Follow}(\alpha)$

In this section we define a new function E, such that for every marked regular expression α we have $\mathsf{Follow}(\alpha) = \mathsf{E}(\alpha)$. This function has the advantage that all unions in its definition are clearly disjoint. Our definition of E was inspired by the construction of $\mathcal{A}_{\mathrm{pos}}(\alpha)$ by Leiss [Lei80] and shows some similarities to the transformation algorithm of α into star-normal-form by Brüggemann-Klein. Let E be given by

$$\begin{aligned} \mathsf{E}(\emptyset) &= \mathsf{E}(\varepsilon) = \mathsf{E}(\sigma_i) = \emptyset \\ \mathsf{E}(\alpha+\beta) &= \mathsf{E}(\alpha) \cup \mathsf{E}(\beta) \\ \mathsf{E}(\alpha\beta) &= \mathsf{E}(\alpha) \cup \mathsf{E}(\beta) \cup \mathsf{Last}(\alpha) \times \mathsf{First}(\beta) \\ \mathsf{E}(\alpha^\star) &= \mathsf{E}^\star(\alpha) \end{aligned} \tag{2}$$

$$\begin{aligned} \mathsf{E}^\star(\emptyset) &= \mathsf{E}^\star(\varepsilon) = \emptyset \\ \mathsf{E}^\star(\sigma_i) &= \{(i,i)\} \\ \mathsf{E}^\star(\alpha+\beta) &= \mathsf{E}^\star(\alpha) \cup \mathsf{E}^\star(\beta) \cup \mathsf{Cross}(\alpha,\beta) \\ \mathsf{E}^\star(\alpha\beta) &= \begin{cases} \mathsf{E}^\star(\alpha) \cup \mathsf{E}^\star(\beta) \cup \mathsf{Cross}(\alpha,\beta) & \text{if } \varepsilon(\alpha)=\varepsilon(\beta)=\varepsilon \\ \mathsf{E}^\star(\alpha) \cup \mathsf{E}(\beta) \cup \mathsf{Cross}(\alpha,\beta) & \text{if } \varepsilon(\beta)=\varepsilon \\ \mathsf{E}(\alpha) \cup \mathsf{E}^\star(\beta) \cup \mathsf{Cross}(\alpha,\beta) & \text{if } \varepsilon(\alpha)=\varepsilon \\ \mathsf{E}(\alpha) \cup \mathsf{E}(\beta) \cup \mathsf{Cross}(\alpha,\beta) & \text{otherwise} \end{cases} \\ \mathsf{E}^\star(\alpha^\star) &= \mathsf{E}^\star(\alpha), \end{aligned} \tag{3}$$

with $\mathsf{Cross}(\alpha,\beta) = \mathsf{Last}(\alpha) \times \mathsf{First}(\beta) \cup \mathsf{Last}(\beta) \times \mathsf{First}(\alpha)$.

Proposition 1. *For every regular expression γ we have* $\mathsf{Follow}(\overline{\gamma}) = \mathsf{E}(\overline{\gamma})$.

Proof.
The proof follows by induction on the structure of γ.The result is trivially true for $\overline{\gamma} = \emptyset, \varepsilon, \sigma_i, \alpha+\beta, \alpha\beta$. For $\overline{\gamma} = \delta^\star$, it is sufficient to show that one has $\mathsf{Follow}(\delta) \cup \mathsf{Last}(\delta) \times \mathsf{First}(\delta) = \mathsf{E}^\star(\delta)$.

For $\delta = \emptyset$, $\delta = \varepsilon$ and $\delta = \sigma_i$ this equation evaluates to $\emptyset = \emptyset$, $\emptyset = \emptyset$ and $\{(i,i)\} = \{(i,i)\}$, respectively. For $\delta = \alpha + \beta$ we have

$$\begin{aligned}
&\mathsf{Follow}(\alpha+\beta) \cup \mathsf{Last}(\alpha+\beta) \times \mathsf{First}(\alpha+\beta) = \\
&\quad = (\mathsf{Follow}(\alpha) \cup \mathsf{Last}(\alpha) \times \mathsf{First}(\alpha)) \cup (\mathsf{Follow}(\beta) \cup \mathsf{Last}(\beta) \times \mathsf{First}(\beta)) \cup \\
&\qquad \cup\ \mathsf{Last}(\alpha) \times \mathsf{First}(\beta) \cup \mathsf{Last}(\beta) \times \mathsf{First}(\alpha) = \\
&\quad = \mathsf{E}^\star(\alpha) \cup \mathsf{E}^\star(\beta) \cup \mathsf{Last}(\alpha) \times \mathsf{First}(\beta) \cup \mathsf{Last}(\beta) \times \mathsf{First}(\alpha) = \mathsf{E}^\star(\alpha+\beta).
\end{aligned}$$

We illustrate the proof for $\delta = \alpha\beta$ with the case where $\varepsilon(\alpha) \neq \varepsilon$ and $\varepsilon(\beta) = \varepsilon$:

$$\begin{aligned}
&\mathsf{Follow}(\alpha\beta) \cup \mathsf{Last}(\alpha\beta) \times \mathsf{First}(\alpha\beta) = \\
&\qquad = \mathsf{Follow}(\alpha) \cup \mathsf{Follow}(\beta) \cup \mathsf{Last}(\alpha) \times \mathsf{First}(\beta) \cup \\
&\qquad\quad \cup \mathsf{Last}(\alpha) \times \mathsf{First}(\alpha) \cup \mathsf{Last}(\beta) \times \mathsf{First}(\alpha) \\
&\quad = \mathsf{E}^\star(\alpha) \cup \mathsf{E}(\beta) \cup \mathsf{Last}(\alpha) \times \mathsf{First}(\beta) \cup \mathsf{Last}(\beta) \times \mathsf{First}(\alpha) = \mathsf{E}^\star(\alpha\beta).
\end{aligned}$$

Finally, for $\delta = \alpha^\star$ we have

$$\begin{aligned}
&\mathsf{Follow}(\alpha^\star) \cup \mathsf{Last}(\alpha^\star) \times \mathsf{First}(\alpha^\star) = \\
&\qquad = \mathsf{Follow}(\alpha) \cup \mathsf{Last}(\alpha) \times \mathsf{First}(\alpha) \cup \mathsf{Last}(\alpha) \times \mathsf{First}(\alpha) \\
&\qquad\quad = \mathsf{Follow}(\alpha) \cup \mathsf{Last}(\alpha) \times \mathsf{First}(\alpha) = \mathsf{E}^\star(\alpha) = \mathsf{E}^\star(\alpha^\star). \quad \square
\end{aligned}$$

4 Counting the Number of Transitions in the Glushkov Automaton

Nicaud [Nic09] showed that the average number of transitions in the Glushkov automaton $\mathcal{A}_{pos}(\alpha)$ is $O(|\alpha|)$. However, his computation of the number of transitions was not exact because the definition used for the Follow function did not take into account the possible non-disjoint unions of its results. In this section, based on the algorithm E we compute the exact number of transitions in $\mathcal{A}_{pos}(\alpha)$, $E_k(z)$, as well as its average cardinality, $T_k(z)$. This is done by the use of the standard methods of analytic combinatorics as expounded by Flajolet and Sedgewick [FS08]. These apply to generating functions $A(z) = \sum_n a_n z^n$ for a combinatorial class $\mathcal{A}$ with a_n objects of size n, or cost generating functions $C(z) = \sum_\alpha c(\alpha) z^{|\alpha|}$, where $c(\alpha)$ is some measure of the object $\alpha \in \mathcal{A}$.

In this section we compute and study the cost generating functions $E_k(z)$ and $T_k(z)$, and their asymptotic behaviours. The other functions used herein, as well as details on how to obtain them, can be found in the above cited article and in Broda *et al* [BMMR11b]. A more detailed description of the below computations can be found in a companion technical report of this paper [BMMR11a].

4.1 The Average Number of Transitions in $\mathcal{A}_{\mathrm{pos}}(\alpha)$

For counting purposes, we will consider regular expressions as defined in (1), but without $\emptyset$. Note that this limitation only excludes the empty language.

The functions that count the cardinalities of $\mathsf{First}(\overline{\alpha})$, $\mathsf{Last}(\overline{\alpha})$, and $\mathsf{E}(\overline{\alpha})$, are respectively denoted by $\mathsf{f}(\alpha)$, $\mathsf{l}(\alpha)$, and $e(\alpha)$. Given the definitions of $\mathsf{f}(\alpha)$ and $\mathsf{l}(\alpha)$, $e(\alpha)$ satisfies the following:

$$\begin{aligned} e(\sigma) &= e(\varepsilon) = 0, \\ e(\alpha+\beta) &= e(\alpha) + e(\beta), \\ e(\alpha\beta) &= e(\alpha) + e(\beta) + \mathsf{l}(\alpha)\cdot\mathsf{f}(\beta), \\ e(\alpha^\star) &= e^\star(\alpha), \end{aligned} \tag{4}$$

where $e^\star(\alpha)$ is given by,

$$\begin{aligned} e^\star(\varepsilon) &= 0, \qquad e^\star(\sigma) = 1, \\ e^\star(\alpha+\beta) &= e^\star(\alpha) + e^\star(\beta) + \mathsf{c}(\alpha,\beta), \\ e^\star(\alpha_\varepsilon\beta_\varepsilon) &= e^\star(\alpha_\varepsilon) + e^\star(\beta_\varepsilon) + \mathsf{c}(\alpha_\varepsilon,\beta_\varepsilon), \\ e^\star(\alpha_{\overline{\varepsilon}}\beta_\varepsilon) &= e^\star(\alpha_{\overline{\varepsilon}}) + e(\beta_\varepsilon) + \mathsf{c}(\alpha_{\overline{\varepsilon}},\beta_\varepsilon), \\ e^\star(\alpha_\varepsilon\beta_{\overline{\varepsilon}}) &= e(\alpha_\varepsilon) + e^\star(\beta_{\overline{\varepsilon}}) + \mathsf{c}(\alpha_\varepsilon,\beta_{\overline{\varepsilon}}), \\ e^\star(\alpha_{\overline{\varepsilon}}\beta_{\overline{\varepsilon}}) &= e(\alpha_{\overline{\varepsilon}}) + e(\beta_{\overline{\varepsilon}}) + \mathsf{c}(\alpha_{\overline{\varepsilon}},\beta_{\overline{\varepsilon}}), \\ e^\star(\alpha^\star) &= e^\star(\alpha). \end{aligned} \tag{5}$$

with $\mathsf{c}(\alpha,\beta) = \mathsf{l}(\alpha)\cdot\mathsf{f}(\beta) + \mathsf{l}(\beta)\cdot\mathsf{f}(\alpha)$. Then, the function

$$\mathsf{t}(\alpha) = \mathsf{f}(\alpha) + e(\alpha)$$

computes the number of transitions in the Glushkov automaton of α. The cost generating function associated to t is given by

$$T_k(z) = F_k(z) + E_k(z),$$

where $F_k(z)$ and $E_k(z)$ are the cost generating functions associated to f and e, respectively. By symmetry, the cost generating function $L_k(z)$ associated to l is the same as $F_k(z)$, *i.e.*

$$L_k(z) = F_k(z) = \frac{kz}{1 - z - 3zR_k(z) - zR_{k,\varepsilon}(z)}.$$

In this last expression $R_k(z)$ and $R_{k,\varepsilon}(z)$ denote respectively the generating functions for regular expressions, and for regular expressions whose languages contain ε and are given by

$$R_k(z) = \frac{1 - z - \sqrt{\Delta_k(z)}}{4z} \quad \text{and} \quad R_{k,\varepsilon}(z) = \frac{z + zR_k(z)}{1 - 2zR_k(z)}, \tag{6}$$

where $\Delta_k(z) = 1 - 2z - (7+8k)z^2$. Hence, for the number of regular expressions of size n, one has

$$[z^n]R_k(z) \sim \frac{\sqrt{2(1-\rho_k)}}{8\rho_k\sqrt{\pi}}\rho_k^{-n} n^{-3/2}, \text{ where } \rho_k = \frac{1}{1+\sqrt{8k+8}}. \tag{7}$$

From the equations in (4) one can compute the associated cost generating functions $E_k(z)$ and $E_k^\star(z)$. For instance, the equation for concatenation contributes with the term $2zE_k(z)R_k(z) + zF_k(z)^2$ in the equation for $E_k(z)$. Collecting all terms the following equations must be satisfied

$$E_k(z) = 4zE_k(z)R_k(z) + zF_k(z)^2 + zE_k^\star(z)$$

$$\begin{aligned} E_k^\star(z) = kz + 2zE_k^\star(z)R_k(z) + 2zE_k^\star(z)R_{k,\varepsilon}(z) + \\ 4zF_k(z)^2 + 2zE_k(z)R_{k,\overline{\varepsilon}}(z) + zE_k^\star(z). \end{aligned}$$

After simplification one gets

$$E_k(z) = \frac{kz^2 + zF_k(z)^2\Lambda_k(z) + 4z^2F_k(z)^2}{(1-4zR_k(z))\Lambda_k(z) - 2z^2R_{k,\overline{\varepsilon}}(z)}, \tag{8}$$

where $\Lambda_k(z) = 1 - z - 2zR_{k,\varepsilon}(z) - 2zR_k(z)$. After substituting the functions in (8) by their expressions in terms of z and k, one obtains

$$T_k(z) = \frac{P_k(z)}{Q_k(z)\sqrt{\Delta_k(z)}}, \tag{9}$$

where $P_k(z)$ is a polynomial in $\sqrt{\Delta_k(z)}$ over $\mathbb{Q}[z]$, with $Q_k(z)$ given by

$$\begin{aligned} Q_k(z) = \left(1 - 2\,z - 7\,z^2 + 4\,(1+z)\sqrt{\Delta_k(z)} + 3\,\Delta_k(z)\right)^2 \\ \left(1 - 5\,z^2 + 2\,(1+2z)\sqrt{\Delta_k(z)} + \Delta_k(z)\right). \end{aligned}$$

This function $Q_k(z)$ is positive for all values of z in the real segment $[0, \rho_k]$, because $1 - 2z - 7z^2 = 8kz^2 + \Delta_k(z)$ and $1 - 5z^2 = 2z + 2z^2 + 8kz^2 + \Delta_k(z)$, and $\Delta_k(z)$ is non-negative in that segment. By Pringsheim's Theorem (Theorem IV.6 of [FS08], p. 240) one can conclude that $T_k(z)$ has radius of convergence equal to ρ_k. Moreover, it can be shown that $T_k(z)$ has no singularities on the the boundary of its disc of convergence, $\|z\| = \rho_k$, besides the one at $z = \rho_k$.

Using exactly the same technique employed in the previously referred article, one obtains

$$T_k(z) = \frac{P_k(\rho_k)}{\sqrt{2-2\rho_k}\;Q_k(\rho_k)} \frac{1}{\sqrt{1 - z/\rho_k}} + o\left(\frac{1}{\sqrt{1-z/\rho_k}}\right), \tag{10}$$

from which it follows that

$$[z^n]T_k(z) \sim \frac{P_k(\rho_k)}{\sqrt{\pi}\,\sqrt{2-2\rho_k}\;Q_k(\rho_k)} \rho_k^{-n} n^{-\frac{1}{2}}. \tag{11}$$

Using the actual expression of P_k, which we omit due to lack of space, one can get

$$[z^n]T_k(z) \sim \frac{(1+\rho_k)(2 + 16\rho_k + 10\rho_k^2 - 12\rho_k^3)}{8\rho_k\sqrt{\pi}(1-5\rho_k^2)\sqrt{2-2\rho_k}} \rho_k^{-n} n^{-\frac{1}{2}}. \tag{12}$$

Considering the cost generating function for the number of letters in an element $\alpha \in \mathcal{R}$, computed by Nicaud to be equal to

$$Let_k(z) = \frac{kz}{\sqrt{\Delta_k(z)}},$$

and for which

$$[z^n]Let_k(z) \sim \frac{k\rho_k}{\sqrt{\pi(2-2\rho_k)}}\rho_k{}^{-n}n^{-1/2},$$

one gets an asymptotic expression for the average number of transitions per state:

$$\frac{[z^n]T_k(z)}{[z^n]Let_k(z)} \sim \frac{(1+\rho_k)(2+16\rho_k+10\rho_k^2-12\rho_k^3)}{(1-2\rho_k-7\rho_k^2)(1-5\rho_k^2)}. \tag{13}$$

And finally, one has for the average number of transitions per regular expression the following asymptotic estimation:

$$\frac{[z^n]T_k(z)}{[z^n]R_k(z)} \sim \frac{(1+\rho_k)(1+8\rho_k+5\rho_k^2-6\rho_k^3)}{(1-\rho_k)(1-5\rho_k^2)}n. \tag{14}$$

Since ρ_k tends to 0 as k goes to ∞, it follows that for large k the average number of transitions per state is approximately 2, while the average number of transitions per automaton is approximately the size of the original regular expression.

5 The Average Number of Transitions in $\mathcal{A}_{\text{pd}}$

The partial derivative automaton $\mathcal{A}_{\text{pd}}(\alpha)$ of a regular expression α was defined independently by Mirkin's [Mir66] and Antimirov [Ant96]. Champarnaud and Ziadi stated the equivalence of the two formulations [CZ01], and proved that $\mathcal{A}_{\text{pd}}$ is a quotient of the Glushkov automaton $\mathcal{A}_{\text{pos}}$ [CZ02]. This means that $\mathcal{A}_{\text{pd}}(\alpha)$ can be obtained from $\mathcal{A}_{\text{pos}}(\alpha)$ by the merging of states belonging to the same equivalence class. That, on the other hand, may lead to the merging of transitions. In this section, we estimate the average number of transitions of $\mathcal{A}_{\text{pd}}(\alpha)$ when compared with the ones of $\mathcal{A}_{\text{pos}}(\alpha)$. For this, it is essential to have the exact counting of the number of transitions of $\mathcal{A}_{\text{pos}}(\alpha)$ obtained in Section 4.1.

The $\mathcal{A}_{\text{pd}}(\alpha)$ can be defined using the notion of partial derivative, introduced by Antimirov as a non-deterministic version of Brzozowski's derivative [Brz64].

For a regular expression α and a letter $\sigma \in \Sigma$, the set $\partial_\sigma(\alpha)$ of *partial derivatives* of α w.r.t. σ is defined inductively as follows:

$$\partial_\sigma(\emptyset) = \partial_\sigma(\varepsilon) = \emptyset$$

$$\partial_\sigma(\sigma') = \begin{cases} \{\varepsilon\}, & \text{if } \sigma' = \sigma \\ \emptyset, & \text{otherwise} \end{cases}$$

$$\partial_\sigma(\alpha^\star) = \partial_\sigma(\alpha)\alpha^\star$$

$$\partial_\sigma(\alpha+\beta) = \partial_\sigma(\alpha) \cup \partial_\sigma(\beta)$$

$$\partial_\sigma(\alpha_\varepsilon\beta) = \partial_\sigma(\alpha_\varepsilon)\beta \cup \partial_\sigma(\beta)$$

$$\partial_\sigma(\alpha_{\overline{\varepsilon}}\beta) = \partial_\sigma(\alpha_{\overline{\varepsilon}})\beta$$

where for any $S \subseteq \mathcal{R}$, $S\emptyset = \emptyset S = \emptyset$, and $S\varepsilon = \varepsilon S = S$. This definition can be extended to sets of regular expressions, to words, and to languages in the obvious way. The *set of partial derivatives* of α, $\{\partial_w(\alpha) \mid w \in \Sigma^\star\}$, is denoted by $\mathsf{P}(\alpha)$. The *partial derivative automaton* $\mathcal{A}_{\text{pd}}(\alpha)$ is defined by $\mathcal{A}_{\text{pd}}(\alpha) = (\mathsf{P}(\alpha), \Sigma, \delta_{\text{pd}}, \alpha, \{q \in \mathsf{P}(\alpha) \mid \varepsilon(q) = \varepsilon\})$, where $\delta_{\text{pd}}(q, \sigma) = \partial_\sigma(q)$, for all $q \in \mathsf{P}(\alpha)$ and $\sigma \in \Sigma$. Antimirov proved that $\mathcal{L}(\mathcal{A}_{\text{pd}}(\alpha)) = \mathcal{L}(\alpha)$.

Using Mirkin's formulation one has $\mathsf{P}(\alpha) = \pi(\alpha) \cup \{\alpha\}$, where the set $\pi(\alpha)$ is inductively defined as follows:

$$\begin{aligned} \pi(\emptyset) &= \emptyset & \pi(\alpha + \beta) &= \pi(\alpha) \cup \pi(\beta) \\ \pi(\varepsilon) &= \emptyset & \pi(\alpha\beta) &= \pi(\alpha)\beta \cup \pi(\beta) \\ \pi(\sigma) &= \{\varepsilon\} & \pi(\alpha^\star) &= \pi(\alpha)\alpha^\star. \end{aligned} \tag{15}$$

5.1 Counting the Mergings of Transitions

Broda *et al.* [BMMR11b] gave a lower bound of the number of mergings of states in $\pi(\alpha)$ with respect to $\mathsf{Pos}(\alpha)$, which allowed to obtain an upper bound on the average state complexity of $\mathcal{A}_{\text{pd}}(\alpha)$. There, it was observed that the merging of states is primarily caused by sub-expressions γ of α such that $\varepsilon \in \pi(\gamma)$. In fact, in the presence of sub-expressions with this property, denoted by α_{π_ε}, during the computation of $\pi(\alpha)$ some unions may not be disjoint.

In this section, and using the same technique, we determine a lower bound of the number of mergings of transitions. Considering (2) and in particular, the concatenation case, it is easy to see, that whenever there is a merging of two states in the set $\mathsf{Last}(\alpha)$, there are exactly $\mathsf{f}(\beta) = |\mathsf{First}(\beta)|$ mergings of transitions. Although there can be merging of states of $\mathsf{First}(\beta)$, they will not be considered in the computation of that lower bound. We first compute a lower bound for the number of mergings $i_\ell(\alpha)$ of states i such that $i \in \mathsf{Last}(\alpha)$.

In addition to α_ε and α_{π_ε}, we use the subclass of regular expressions $\alpha_{r,\varepsilon}$ such that $\alpha_{r,\varepsilon} \in \pi(\alpha_{r,\varepsilon})$ and $\varepsilon(\alpha_{r,\varepsilon}) = \varepsilon$.

The grammar for α_{π_ε} and its generating function $R_{k,\pi_\varepsilon}(z)$ are the ones used by Broda *et al.* For $\alpha_{r,\varepsilon}$ one has

$$\alpha_{r,\varepsilon} := \alpha^\star_{\pi_\varepsilon} \mid \alpha_{r,\varepsilon} \cdot \alpha_\varepsilon, \quad \text{and} \quad R_{k,r,\varepsilon}(z) = \frac{zR_{k,\pi_\varepsilon}(z)}{1 - zR_{k,\varepsilon}(z)}.$$

Finally, in order to get a better approximation for the counting of the state mergings, we define $i_\ell(\alpha)$ by the following:

$$\begin{aligned} i_\ell(\emptyset) = i_\ell(\varepsilon) = i_\ell(\sigma) &= 0, \\ i_\ell(\alpha_{\pi_\varepsilon} + \alpha_{\pi_\varepsilon}) &= i_\ell(\alpha_{\pi_\varepsilon}) + i_\ell(\alpha_{\pi_\varepsilon}) + 1, \\ i_\ell(\alpha_{\pi_\varepsilon} + \alpha_{\overline{\pi_\varepsilon}}) &= i_\ell(\alpha_{\pi_\varepsilon}) + i_\ell(\alpha_{\overline{\pi_\varepsilon}}), \\ i_\ell(\alpha_{\overline{\pi_\varepsilon}} + \alpha) &= i_\ell(\alpha_{\overline{\pi_\varepsilon}}) + i_\ell(\alpha), \\ i_\ell(\alpha^\star) &= i_\ell(\alpha), \end{aligned}$$

$$\begin{aligned} i_\ell(\alpha_{\pi_\varepsilon}\alpha_{r,\varepsilon}) &= i_\ell(\alpha_{\pi_\varepsilon}) + i_\ell(\alpha_{r,\varepsilon}) + 1, \\ i_\ell(\alpha_{\pi_\varepsilon}\alpha_{r,\overline{\varepsilon}}) &= i_\ell(\alpha_{r,\overline{\varepsilon}}), \\ i_\ell(\alpha_{\pi_\varepsilon}\alpha_{\overline{r},\varepsilon}) &= i_\ell(\alpha_{\pi_\varepsilon}) + i_\ell(\alpha_{\overline{r},\varepsilon}), \\ i_\ell(\alpha_{\pi_\varepsilon}\alpha_{\overline{r},\overline{\varepsilon}}) &= i_\ell(\alpha_{\overline{r},\overline{\varepsilon}}), \\ i_\ell(\alpha_{\overline{\pi_\varepsilon}}\alpha_\varepsilon) &= i_\ell(\alpha_{\overline{\pi_\varepsilon}}) + i_\ell(\alpha_\varepsilon), \\ i_\ell(\alpha_{\overline{\pi_\varepsilon}}\alpha_{\overline{\varepsilon}}) &= i_\ell(\alpha_{\overline{\varepsilon}}), \end{aligned}$$

where $\gamma_{\overline{x}}$ denotes the complement of γ_x.

To illustrate the previous rules, consider the case of $i_\ell(\alpha_{\pi_\varepsilon}\alpha_{r,\varepsilon})$. Here, one has $\pi(\alpha_{\pi_\varepsilon}\alpha_{r,\varepsilon}) = \pi(\alpha_{\pi_\varepsilon})\alpha_{r,\varepsilon} \cup \pi(\alpha_{r,\varepsilon})$. By definition, $\varepsilon \in \pi(\alpha_{\pi_\varepsilon})$ and $\alpha_{r,\varepsilon} \in \pi(\alpha_{r,\varepsilon})$. Hence $\alpha_{r,\varepsilon}$ belongs to both $\pi(\alpha_{\pi_\varepsilon})\alpha_{r,\varepsilon}$ and $\pi(\alpha_{r,\varepsilon})$, which causes a merging of two states in $\mathsf{Last}(\alpha)$. This merging is accounted for by the 1 in the definition of $i_\ell(\alpha_{\pi_\varepsilon}\alpha_{r,\varepsilon})$. On the other hand, since $\varepsilon(\alpha_{r,\varepsilon}) = \varepsilon$, one also has to count all mergings of states in $\mathsf{Last}(\alpha_{\pi_\varepsilon})$ (counted by $i_\ell(\alpha_{\pi_\varepsilon})$), besides those in $\mathsf{Last}(\alpha_{r,\varepsilon})$.

The generating function, $Il_k(z)$, of i_ℓ satisfies the following:

$$Il_k(z) = \frac{zR_{k,\pi_\varepsilon}(z)^2 + zR_{k,\pi_\varepsilon}(z)R_{k,r,\varepsilon}(z)}{1 - z - 3zR_k(z) - zR_{k,\varepsilon}(z)},$$

where $R_{k,\pi_\varepsilon}(z) = \frac{z^2+3zR_k(z)-1+\sqrt{(z^2+3zR_k(z)-1)^2+4kz^2}}{2z}$.

Using (2) and (3), one can easily define a function that computes a lower bound for the number of transition mergings in the Glushkov automaton:

$$\begin{aligned}
i_t(\varepsilon) &= i_t(\sigma) = 0 & i_t^\star(\alpha+\beta) &= i_t^\star(\alpha) + i_t^\star(\beta) + \mathsf{c}_t(\alpha,\beta)\\
i_t(\alpha+\beta) &= i_t(\alpha) + i_t(\beta) & i_t^\star(\alpha_\varepsilon\beta_\varepsilon) &= i_t^\star(\alpha_\varepsilon) + i_t^\star(\beta_\varepsilon) + \mathsf{c}_t(\alpha_\varepsilon,\beta_\varepsilon)\\
i_t(\alpha\beta) &= i_t(\alpha) + i_t(\beta) + i_\ell(\alpha).\mathsf{f}(\beta) & i_t^\star(\alpha_{\overline{\varepsilon}}\beta_\varepsilon) &= i_t^\star(\alpha_{\overline{\varepsilon}}) + i_t(\beta_\varepsilon) + \mathsf{c}_t(\alpha_{\overline{\varepsilon}},\beta_\varepsilon)\\
i_t(\alpha^\star) &= i_t^\star(\alpha) & i_t^\star(\alpha_\varepsilon\beta_{\overline{\varepsilon}}) &= i_t(\alpha_\varepsilon) + i_t^\star(\beta_{\overline{\varepsilon}}) + \mathsf{c}_t(\alpha_\varepsilon,\beta_{\overline{\varepsilon}})\\
i_t^\star(\varepsilon) &= 0 & i_t^\star(\alpha_{\overline{\varepsilon}}\beta_{\overline{\varepsilon}}) &= i_t(\alpha_{\overline{\varepsilon}}) + i_t(\beta_{\overline{\varepsilon}}) + \mathsf{c}_t(\alpha_{\overline{\varepsilon}},\beta_{\overline{\varepsilon}})\\
i_t^\star(\sigma) &= 1 & i_t^\star(\alpha^\star) &= i_t^\star(\alpha)
\end{aligned}$$

where $\mathsf{c}_t(\alpha,\beta) = i_\ell(\alpha)\cdot\mathsf{f}(\beta) + i_\ell(\beta)\cdot\mathsf{f}(\alpha)$. The corresponding generating function satisfies the following:

$$It_k(z) = \frac{z\Lambda_k(z)Il_k(z)F_k(z) + kz^2 + 4z^2Il_k(z)F_k(z)}{(1-4zR_k(z))\Lambda_k(z) - 2z^2R_{k,\overline{\varepsilon}}(z)},$$

where $\Lambda_k(z) = 1 - z - 2zR_k(z) - 2zR_{k,\varepsilon}(z)$. Analogously to what was done before, one has

$$[z^n]It_k(z) \sim \frac{(1+\rho_k)\,(a(\rho_k)b(\rho_k) + c(\rho_k))}{16\sqrt{\pi}\,\rho_k\,\sqrt{2-2\rho_k}\,(1-5\rho_k^2)d(\rho_k)}\rho_k^{-n}n^{-\frac{1}{2}} \tag{16}$$

where

$$\begin{aligned}
a(z) &= -2 - 23z - 77z^2 - 50z^3 + 92z^4 + 77z^5 - 13z^6 - 4z^7\\
b(z) &= \sqrt{9 - 10z - 55z^2 - 24z^3 + 16z^4}\\
c(z) &= 10 + 89z + 54z^2 - 603z^3 - 1114z^4 - 349z^5 +\\
&\qquad + 130z^6 - 209z^7 - 40z^8 - 16z^9\\
d(z) &= (1 - 2z - 7z^2)(2 + z - 3z^2).
\end{aligned}$$

Therefore, a lower bound for the average number of mergings per transition of the Glushkov automaton is given by

$$[z^n]\frac{It_k(z)}{T_k(z)} \sim \frac{a(\rho_k)b(\rho_k) + c(\rho_k)}{4(1 + 8\rho_k + 5\rho_k^2 - 6\rho_k^3)d(\rho_k)}. \tag{17}$$

This means that, asymptotically with respect to k, the number of transitions in $\mathcal{A}_{\text{pd}}$ is at most half the number of transitions in $\mathcal{A}_{pos}$.

6 Comparison with Experimental Results

We compared the estimates obtained in the previous sections with some experimental results. For each $k \in \{2, 3, 10, 30, 50\}$, the experiment consisted of the comparison of the sizes of $\mathcal{A}_{\text{pos}}$ and $\mathcal{A}_{\text{pd}}$, that were computed for each regular expression in the samples of 1000 uniform random generated regular expressions of size 1000. Table 1 presents the average values obtained, and columns eight and ten give the asymptotic ratios obtained in (13) and in (17), respectively. The quality of the approximation of the asymptotic average number of transitions per state for $\mathcal{A}_{\text{pos}}$, and that the actual values are close to the limit even for relatively small alphabets is evident from the table. The upper bound for the ratio of the transition complexity of $\mathcal{A}_{\text{pd}}$ to the one of $\mathcal{A}_{\text{pos}}$ is within an error less than 15%. The experimental values also suggest that the number of transitions of $\mathcal{A}_{\text{pd}}$ is on average, and as k grows, the alphabetic size of the original regular expression.

Table 1. Experimental results for uniform random generated regular expressions

k	$\lvert\alpha\rvert$	$\lvert\alpha\rvert_\Sigma$	$\lvert\delta_{\text{pos}}\rvert$	$\lvert\mathsf{P}(\alpha)\rvert$	$\lvert\delta_{\text{pd}}\rvert$	$\frac{\lvert\delta_{\text{pos}}\rvert}{\lvert\alpha\rvert_\Sigma+1}$	$\frac{[z^n]T_k(z)}{[z^n]Let_k(z)+1}$	$\frac{\lvert\delta_{\text{pd}}\rvert}{\lvert\delta_{\text{pos}}\rvert}$	$1 - \frac{[z^n]It_k(z)}{[z^n]T_k(z)}$
2	1000	276	3345	187	1806	**12.1**	**12.2**	**0.54**	**0.68**
3	1000	318	2997	206	1564	**9.4**	**9.6**	**0.52**	**0.64**
10	1000	405	2203	236	1079	**5.4**	**5.3**	**0.49**	**0.58**
30	1000	453	1676	247	796	**3.7**	**3.6**	**0.47**	**0.54**
50	1000	466	1516	250	718	**3.3**	**3.2**	**0.47**	**0.53**
100	—	—	—	—	—	—	**2.8**	—	**0.53**
1000	—	—	—	—	—	—	**2.2**	—	**0.49**

7 Conclusions

In this paper we presented a new algorithm for computing Follow, which is quadratic in the alphabetic size of the original regular expression, and that leads to a straightforward and direct implementation. This algorithm allowed us to exactly count the number of transitions of the Glushkov automaton. Using this, we computed the average number of transitions of that automaton, concluding that, for large alphabets, it is approximately the double of the original regular expression alphabetic size.

Considering special sub-classes of regular expressions that are primarily responsible for state mergings, we computed an upper bound for the number of transitions in the partial derivative automaton. We, then, used analytic combinatorial methods to obtain average values and asymptotic limits for that number, concluding that, on average and asymptotically, the partial derivative automaton has at most half the number of transitions of the Glushkov's. Experimental figures corroborate these results.

References

[Ant96] Antimirov, V.M.: Partial derivatives of regular expressions and finite automaton constructions. Theoret. Comput. Sci. 155(2), 291–319 (1996)

[BK93] Brüggemann-Klein, A.: Regular expressions into finite automata. Theoret. Comput. Sci. 48, 197–213 (1993)

[BMMR11a] Broda, S., Machiavelo, A., Moreira, N., Reis, R.: On the average size of Glushkov and partial derivative automata. Technical Report DCC-2011-03, FCUP & CMUP, Universidade do Porto (April 2011)

[BMMR11b] Broda, S., Machiavelo, A., Moreira, N., Reis, R.: On the average state complexity of partial derivative automata. International Journal of Foundations of Computer Science (2011) (accepted to publication)

[Brz64] Brzozowski, J.A.: Derivatives of regular expressions. JACM 11(4), 481–494 (1964)

[BS86] Berry, G., Sethi, R.: From regular expressions to deterministic automata. Theoret. Comput. Sci. 48, 117–126 (1986)

[CP97] Chang, C.-H., Paige, R.: From regular expressions to DFA's using compressed NFA's. Theor. Comput. Sci. 178(1-2), 1–36 (1997)

[CZ01] Champarnaud, J.M., Ziadi, D.: From Mirkin's prebases to Antimirov's word partial derivatives. Fundam. Inform. 45(3), 195–205 (2001)

[CZ02] Champarnaud, J.M., Ziadi, D.: Canonical derivatives, partial derivatives and finite automaton constructions. Theoret. Comput. Sci. 289, 137–163 (2002)

[FS08] Flajolet, P., Sedgewick, R.: Analytic Combinatorics. Cambridge University Press, Cambridge (2008)

[Glu61] Glushkov, V.M.: The abstract theory of automata. Russian Math. Surveys 16, 1–53 (1961)

[HK10] Holzer, M., Kutrib, M.: The complexity of regular(-like) expressions. In: Gao, Y., Lu, H., Seki, S., Yu, S. (eds.) DLT 2010. LNCS, vol. 6224, pp. 16–30. Springer, Heidelberg (2010)

[Koz97] Kozen, D.C.: Automata and Computability. Springer, Heidelberg (1997)

[Lei80] Leiss, E.: Constructing a finite automaton for a given regular expression. SIGACT News 12(3) (September 1980)

[Mir66] Mirkin, B.G.: An algorithm for constructing a base in a language of regular expressions. Engineering Cybernetics 5, 51–57 (1966)

[MY60] McNaughton, R., Yamada, H.: Regular expressions and state graphs for automata. IEEE Transactions on Electronic Computers 9, 39–47 (1960)

[Nic09] Nicaud, C.: On the average size of glushkov's automata. In: Dediu, A., Ionescu, A.M., Martín-Vide, C. (eds.) LATA 2009. LNCS, vol. 5457, pp. 626–637. Springer, Heidelberg (2009)

[PZC97] Ponty, J.-L., Ziadi, D., Champarnaud, J.-M.: A new quadratic algorithm to convert a regular expression into an automaton. In: Raymond, D.R., Yu, S., Wood, D. (eds.) WIA 1996. LNCS, vol. 1260, pp. 109–119. Springer, Heidelberg (1997)

[Sak09] Sakarovitch, J.: Elements of Automata Theory. CUP (2009)

Theory of Átomata*

Janusz Brzozowski[1] and Hellis Tamm[2]

[1] David R. Cheriton School of Computer Science, University of Waterloo, Waterloo, ON, Canada N2L 3G1
brzozo@uwaterloo.ca
[2] Institute of Cybernetics, Tallinn University of Technology, Akadeemia tee 21, 12618 Tallinn, Estonia
hellis@cs.ioc.ee

Abstract. We show that every regular language defines a unique nondeterministic finite automaton (NFA), which we call "átomaton", whose states are the "atoms" of the language, that is, non-empty intersections of complemented or uncomplemented left quotients of the language. We describe methods of constructing the átomaton, and prove that it is isomorphic to the normal automaton of Sengoku, and to an automaton of Matz and Potthoff. We study "atomic" NFA's in which the right language of every state is a union of atoms. We generalize Brzozowski's double-reversal method for minimizing a deterministic finite automaton (DFA), showing that the result of applying the subset construction to an NFA is a minimal DFA if and only if the reverse of the NFA is atomic.

1 Introduction

Nondeterministic finite automata (NFA's) introduced by Rabin and Scott [9] in 1959 play a major role in the theory of automata. For many purposes it is necessary to convert an NFA to a deterministic finite automaton (DFA). In particular, for each NFA there exists a minimal DFA, unique up to isomorphism. This DFA is uniquely defined by every regular language, and uses the left quotients of the language as states. As well, it is possible to associate an NFA with each DFA, and this is the subject of the present paper. Our NFA is also uniquely defined by every regular language, and uses non-empty intersections of complemented and uncomplemented quotients—the "atoms" of the language—as states.

It appears that the NFA most often associated with a regular language is the universal automaton, sometimes appearing under different names. A recent substantial survey by Lombardy and Sakarovitch [7] on the subject of the universal automaton contains its history and a detailed discussion of its properties. We refer the reader to that paper, and mention only that research related to the universal automaton goes back to the 1970's: *e.g.*, in [3] as reported in [1], [4,6].

* This work was supported by the Natural Sciences and Engineering Research Council of Canada under grant No. OGP0000871, by the Estonian Center of Excellence in Computer Science, EXCS, financed by the European Regional Development Fund, and by the Estonian Science Foundation grant 7520.

G. Mauri and A. Leporati (Eds.): DLT 2011, LNCS 6795, pp. 105–116, 2011.

We call our NFA the "átomaton"[1] because it is based on the atoms of a regular language; we add the accent to minimize the possible confusion between "automaton" and "atomaton". Automata isomorphic to our átomaton have previously appeared in 1992 in the little-known master's thesis [10] of Sengoku, and in the 1995 paper [8] by Matz and Potthoff.

We introduce "atomic" automata, in which the right language of any state is a union of some atoms. This generalizes residual automata [5] in which the right language of any state is a left quotient (which we prove to be a union of atoms), and includes also átomata (where the right language of any state is an atom), trim DFA's, and the trim parts of universal automata.

We characterize the class of NFA's for which the subset construction yields a minimal DFA. More specifically, we show that the subset construction applied to a trim NFA produces a minimal DFA if and only if the reverse automaton of that NFA is atomic. This generalizes Brzozowski's method for DFA minimization by double reversal [2].

Section 2 recalls properties of regular languages, finite automata, and systems of language equations. Atoms of a regular language and the átomaton are introduced and studied in Section 3. In Section 4, we examine NFA's in which the right language of every state is a union of atoms. Brzozowski's method of DFA minimization is extended in Section 5, and Section 6 closes the paper.

2 Languages, Automata and Equations

If Σ is a non-empty finite alphabet, then Σ^* is the free monoid generated by Σ. A *word* is any element of Σ^*, and the empty word is ε. The length of a word w is $|w|$. A *language* over Σ is any subset of Σ^*.

The following operations are defined on languages over Σ: *complement* ($\overline{L} = \Sigma^* \setminus L$), *union* ($K \cup L$), *intersection* ($K \cap L$), *product*, usually called *concatenation* or *catenation*, ($KL = \{w \in \Sigma^* \mid w = uv, u \in K, v \in L\}$), *positive closure* ($L^+ = \bigcup_{i\geqslant 1} L^i$), and *star* ($L^* = \bigcup_{i\geqslant 0} L^i$). The *reverse* w^R *of a word* $w \in \Sigma^*$ is defined as follows: $\varepsilon^R = \varepsilon$, and $(wa)^R = aw^R$. The *reverse of a language* L is denoted by L^R and defined as $L^R = \{w^R \mid w \in L\}$.

A *nondeterministic finite automaton* is a quintuple $\mathcal{N} = (Q, \Sigma, \delta, I, F)$, where Q is a finite, non-empty set of *states*, Σ is a finite non-empty *alphabet*, $\delta : Q\times\Sigma \to 2^Q$ is the *transition function*, $I \subseteq Q$ is the set of *initial states*, and $F \subseteq Q$ is the set of *final states*. As usual, we extend the transition function to functions $\delta' : Q \times \Sigma^* \to 2^Q$, and $\delta'' : 2^Q \times \Sigma^* \to 2^Q$. We do not distinguish these functions notationally, but use δ for all three. The *language accepted* by an NFA $\mathcal{N}$ is $L(\mathcal{N}) = \{w \in \Sigma^* \mid \delta(I, w)\cap F \neq \emptyset\}$. Two NFA's are *equivalent* if they accept the same language. The *left language* of a state q of $\mathcal{N}$ is $L_{I,q}(\mathcal{N}) = \{w \in \Sigma^* \mid q \in \delta(I, w)\}$, and the *right language* of q is $L_{q,F}(\mathcal{N}) = \{w \in \Sigma^* \mid \delta(q, w) \cap F \neq \emptyset\}$. So $L(\mathcal{N}) = L_{I,F}(\mathcal{N})$. A state is *empty* if its right language is empty. Two states of an NFA are *equivalent* if their right languages are identical.

[1] The word should be pronounced with the accent on the first a.

A *deterministic finite automaton* is a quintuple $\mathcal{D} = (Q, \Sigma, \delta, q_0, F)$, where Q, Σ, and F are as in an NFA, $\delta : Q \times \Sigma \to Q$ is the transition function, and q_0 is the initial state. A DFA is an NFA in which the set of initial states is $\{q_0\}$ and the range of the transition function is restricted to singletons $\{q\}$, $q \in Q$.

An *incomplete deterministic finite automaton (IDFA)* is a quintuple $\mathcal{I} = (Q, \Sigma, \delta, q_0, F)$, where δ is a partial function such that either $\delta(q, a) = p$ for some $p \in Q$ or $\delta(q, a)$ is undefined. Every DFA is also an IDFA. An IDFA is *minimal* if no two of its states are equivalent.

We use the following operations on automata:

1. The *determinization* operation $\mathbb{D}$ applied to an NFA $\mathcal{N}$ yields a DFA $\mathcal{N}^{\mathbb{D}}$ obtained by the well-known subset construction, where only subsets (including the empty subset) reachable from the initial subset of $\mathcal{N}^{\mathbb{D}}$ are used.

4. The *reversal* operation $\mathbb{R}$ applied to an NFA $\mathcal{N}$ yields an NFA $\mathcal{N}^{\mathbb{R}}$, where initial and final states of $\mathcal{N}$ are interchanged in $\mathcal{N}^{\mathbb{R}}$ and all the transitions between states are reversed.

2. The *trimming* operation $\mathbb{T}$ applied to an NFA $\mathcal{N}$ accepting a non-empty language deletes from $\mathcal{N}$ every state q not reachable from any initial state ($q \notin \delta(I, w)$ for any $w \in \Sigma^*$) and every state q that does not lead to any final state ($\delta(q, w) \cap F = \emptyset$ for all $w \in \Sigma^*$), along with the incident transitions. An NFA that has no such states is said to be *trim*. Note that, if $\mathcal{N}$ is trim, then so is $\mathcal{N}^{\mathbb{R}}$. Also, $\mathcal{N}^{\mathbb{RT}} = \mathcal{N}^{\mathbb{TR}}$ for any NFA $\mathcal{N}$.

If the trimming operation is applied to a DFA $\mathcal{D}$, we obtain the IDFA $\mathcal{D}^{\mathbb{T}}$, which behaves like $\mathcal{D}$, except that it does not have any empty states.

3. The *minimization* operation $\mathbb{M}$ applied to an IDFA (DFA) $\mathcal{D}$ yields the minimal IDFA (DFA) $\mathcal{D}^{\mathbb{M}}$ equivalent to $\mathcal{D}$.

A trim IDFA $\mathcal{I}$ is *bideterministic* if also $\mathcal{I}^{\mathbb{R}}$ is an IDFA. A language is *bideterministic* if its minimal IDFA is bideterministic.

Example 1. Figure 1 (a) shows an NFA $\mathcal{N}$. Its determinized DFA $\mathcal{N}^{\mathbb{D}}$ is in Fig. 1 (b), where braces around sets are omitted. The minimal equivalent $\mathcal{D} = \mathcal{N}^{\mathbb{DM}}$ of $\mathcal{N}^{\mathbb{D}}$ is in Fig. 1 (c), where the equivalent states $\{2\}$, $\{1, 3\}$, and $\{2, 3\}$ are represented by $\{1, 3\}$. The reversed and trimmed version $\mathcal{D}^{\mathbb{RT}}$ of the DFA $\mathcal{D}$ of Fig. 1 (c) is in Fig. 1 (d).

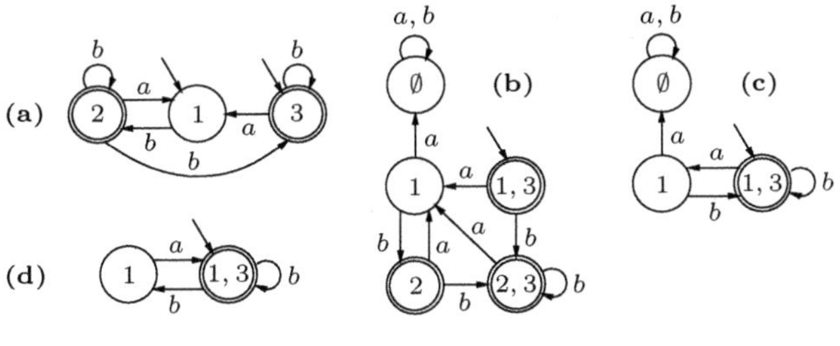

Fig. 1. (a) An NFA $\mathcal{N}$; (b) $\mathcal{N}^{\mathbb{D}}$; (c) $\mathcal{N}^{\mathbb{DM}}$; (d) $\mathcal{N}^{\mathbb{DMRT}}$

The *left quotient*, or simply *quotient*, of a language L by a word w is the language $w^{-1}L = \{x \in \Sigma^* \mid wx \in L\}$. Left quotients are also known as *right residuals*. Dually, the *right quotient* of a language L by a word w is the language $Lw^{-1} = \{x \in \Sigma^* \mid xw \in L\}$. Evidently, if $\mathcal{N}$ is an NFA and x is in $L_{I,q}(\mathcal{N})$, then $L_{q,F}(\mathcal{N}) \subseteq x^{-1}(L(\mathcal{N}))$.

The *quotient DFA* of a regular language L is $\mathcal{D} = (Q, \Sigma, \delta, q_0, F)$, where $Q = \{w^{-1}L \mid w \in \Sigma^*\}$, $\delta(w^{-1}L, a) = a^{-1}(w^{-1}L)$, $q_0 = \varepsilon^{-1}L = L$, and $F = \{w^{-1}L \mid \varepsilon \in w^{-1}L\}$. The *quotient IDFA* of a regular language L is $\mathcal{D}^{\mathbb{T}}$.

The following is from [7]: If $L \subseteq \Sigma^*$, a *subfactorization* of L is a pair (X, Y) of languages over Σ such that $XY \subseteq L$. A *factorization* of L is a subfactorisation (X, Y) such that, if $X \subseteq X'$, $Y \subseteq Y'$, and $X'Y' \subseteq L$ for any pair (X', Y'), then $X = X'$ and $Y = Y'$. The *universal automaton* of L is $\mathcal{U}_L = (Q, \Sigma, \delta, I, F)$ where Q is the set of all factorizations of L, $I = \{(X,Y) \in Q \mid \varepsilon \in X\}$, $F = \{(X,Y) \in Q \mid X \subseteq L\}$, and $(X', Y') \in \delta((X,Y), a)$ if and only if $Xa \subseteq X'$.

For any language L let $L^\varepsilon = \emptyset$ if $\varepsilon \notin L$ and $L^\varepsilon = \{\varepsilon\}$ otherwise. Also, let $n \geqslant 1$ and let $[n] = \{1, \ldots, n\}$. A *nondeterministic system of equations (NSE)* with n variables $L_1, \ldots, L_n$ is a set of language equations

$$L_i = \bigcup_{a \in \Sigma} a\Big(\bigcup_{j \in J_{i,a}} L_j\Big) \cup L_i^\varepsilon \quad i = 1, \ldots, n, \tag{1}$$

where $J_{i,a} \subseteq [n]$, together with an *initial set of variables* $\{L_i \mid i \in I\}$, where $I \subseteq [n]$ is an index set. The equations are assumed to have been simplified by the rules $a\emptyset = \emptyset$ and $K \cup \emptyset = \emptyset \cup K = K$, for any language K. Let $L_{i,a} = \bigcup_{j \in J_{i,a}} L_j$; then $L_{i,a} = a^{-1}L_i$ is the left quotient of L_i by a. The language defined by an NSE is $L = \bigcup_{i \in I} L_i$.

Each NSE defines a unique NFA $\mathcal{N}$ and *vice versa*. States of $\mathcal{N}$ correspond to the variables L_i, there is a transition $L_i \xrightarrow{a} L_j$ in $\mathcal{N}$ if and only if $j \in J_{i,a}$, the set of initial states of $\mathcal{N}$ is $\{L_i \mid i \in I\}$, and the set of final states is $\{L_i \mid L_i^\varepsilon = \{\varepsilon\}\}$.

If each L_i is a left quotient (that is, a right residual) of the language $L = \bigcup_{i \in I} L_i$, then the NSE and the corresponding NFA are called *residual* [5].

A *deterministic system of equations (DSE)* is an NSE

$$L_i = \bigcup_{a \in \Sigma} aL_{i_a} \cup L_i^\varepsilon \quad i = 1, \ldots, n, \tag{2}$$

where $i_a \in [n]$, $I = \{1\}$, and the empty language $\emptyset$ is retained if it appears.

Each DSE defines a unique DFA $\mathcal{D}$ and *vice versa*. Each state of $\mathcal{D}$ corresponds to a variable L_i, there is a transition $L_i \xrightarrow{a} L_j$ in $\mathcal{D}$ if and only if $i_a = j$, the initial state of $\mathcal{D}$ corresponds to L_1, and the set of final states is $\{L_i \mid L_i^\varepsilon = \{\varepsilon\}\}$. In the special case when $\mathcal{D}$ is minimal, its DSE constitutes its *quotient equations*, where every L_i is a quotient of the initial language L_1.

To simplify the notation, we write ε instead of $\{\varepsilon\}$ in equations.

Example 2. For the NFA of Fig. 1 (a), we have the NSE

$$\begin{aligned} L_1 &= bL_2, \\ L_2 &= aL_1 \cup b(L_2 \cup L_3) \cup \varepsilon, \\ L_3 &= aL_1 \cup bL_3 \cup \varepsilon, \end{aligned}$$

with the initial set $\{L_1, L_3\}$. The language $L = L_1 \cup L_3$ accepted by the DFA of Fig. 1 (b) is obtained from this NSE as shown by the equations below on the left. Renaming the unions of variables by new variables corresponding to subsets in the subset construction, we get the equations on the right; for example, $L_1 \cup L_3$ is renamed as $L_{\{1,3\}}$. This is the DSE for the DFA of Fig. 1 (b).

$$\begin{aligned} L_1 \cup L_3 &= aL_1 \cup b(L_2 \cup L_3) \cup \varepsilon, & L_{\{1,3\}} &= aL_{\{1\}} \cup bL_{\{2,3\}} \cup \varepsilon, \\ L_1 &= a\emptyset \cup bL_2, & L_{\{1\}} &= aL_{\emptyset} \cup bL_{\{2\}}, \\ L_2 \cup L_3 &= aL_1 \cup b(L_2 \cup L_3) \cup \varepsilon, & L_{\{2,3\}} &= aL_{\{1\}} \cup bL_{\{2,3\}} \cup \varepsilon, \\ L_2 &= aL_1 \cup b(L_2 \cup L_3) \cup \varepsilon, & L_{\{2\}} &= aL_{\{1\}} \cup bL_{\{2,3\}} \cup \varepsilon, \\ \emptyset &= a\emptyset \cup b\emptyset. & L_{\emptyset} &= aL_{\emptyset} \cup bL_{\emptyset}. \end{aligned}$$

Noting that $L_{\{1,3\}}$, $L_{\{2,3\}}$, and $L_{\{2\}}$ are equivalent, we get the quotient equations for the DFA of Fig. 1 (c), where $L_{\{1\}} = a^{-1}L_{\{1,3\}}$, $L_{\{1,3\}} = b^{-1}L_{\{1,3\}}$, *etc.*

$$\begin{aligned} L_{\{1,3\}} &= aL_{\{1\}} \cup bL_{\{1,3\}} \cup \varepsilon, \\ L_{\{1\}} &= aL_{\emptyset} \cup bL_{\{1,3\}}, \\ L_{\emptyset} &= aL_{\emptyset} \cup bL_{\emptyset}. \end{aligned}$$

3 The Átomaton of a Regular Language

From now on we consider only non-empty regular languages. Let L be a regular language, and let $L_1 = L, L_2, \ldots, L_n$ be its quotients. An *atom* of L is any non-empty language of the form $A = \widetilde{L_1} \cap \widetilde{L_2} \cap \cdots \cap \widetilde{L_n}$, where $\widetilde{L_i}$ is either L_i or $\overline{L_i}$, and at least one of the L_i is not complemented (in other words, $\overline{L_1} \cap \overline{L_2} \cap \cdots \cap \overline{L_n}$ is *not* an atom). A language has at most $2^n - 1$ atoms.

An atom is *initial* if it has L_1 (rather than $\overline{L_1}$) as a term; it is *final* if and only if it contains ε. Since L is non-empty, it has at least one quotient containing ε. Hence it has exactly one final atom, the atom $\widehat{L_1} \cap \widehat{L_1} \cap \cdots \cap \widehat{L_n}$, where $\widehat{L_i} = L_i$ if $\varepsilon \in L_i$, $\widehat{L_i} = \overline{L_i}$ otherwise. The atoms of L will be denoted by $A_1, \ldots, A_m$. By convention, I is the set of initial atoms and A_m is the final atom.

Proposition 1. *The following properties hold for atoms:*

1. *Atoms are pairwise disjoint, that is,* $A_i \cap A_j = \emptyset$ *for all* $i, j \in [m]$, $i \neq j$.
2. *The quotient* $w^{-1}L$ *of* L *by* $w \in \Sigma^*$ *is a (possibly empty) union of atoms.*
3. *The quotient* $w^{-1}A_i$ *of* A_i *by* $w \in \Sigma^*$ *is a (possibly empty) union of atoms.*

Proof. 1. If $A_i \neq A_j$, then there exists $h \in [n]$ such that L_h is a term of A_i and $\overline{L_h}$ is a term of A_j or *vice versa*. Hence $A_i \cap A_j = \emptyset$.

2. The empty quotient, if present, is the empty union of atoms. If $L_i \neq \emptyset$ is a quotient of L, then L_i is the union of all the 2^{n-1} intersections that have L_i as a term. This includes all the atoms that have L_i as a term, and possibly some empty intersections.

3. Consider the quotient equations of L. The quotient of each atom A_i by a letter $a \in \Sigma$ is an intersection X of complemented and uncomplemented quotients of L. If a quotient L_j of L does not appear as a term in X, then we "add

it in" by using the fact that $X = X \cap (L_j \cup \overline{L_j}) = (X \cap L_j) \cup (X \cap \overline{L_j})$. After all the missing quotients are so added, we obtain a union of atoms. Note that the intersection having all quotients complemented does not appear in this construction. It follows that $w^{-1}A_i$ is a union of atoms of L for every $w \in \Sigma^*$. □

Lemma 1. *Let $w, x \in \Sigma^*$. If $wx \in A_i$ and $x \in A_j$ then $wA_j \subseteq A_i$, for $i, j \in [m]$.*

Proof. Assume that $wx \in A_i$ and $x \in A_j$, but suppose $wy \notin A_i$ for some $y \in A_j$. Then $x \in w^{-1}A_i$ and $y \notin w^{-1}A_i$. By Proposition 1, Part 3, $w^{-1}A_i$ is a union of atoms. So, on the one hand, $x \in w^{-1}A_i$ and $x \in A_j$ together imply that $A_j \subseteq w^{-1}A_i$. On the other hand, from $y \notin w^{-1}A_i$ and $y \in A_j$, we get $A_j \not\subseteq w^{-1}A_i$. So if $wy \notin A_i$, we have a contradiction. Hence, $wA_j \subseteq A_i$. □

In the following definition we use a 1-1 correspondence $A_i \leftrightarrow \mathbf{A}_i$ between atoms A_i of a language L and the states $\mathbf{A}_i$ of the NFA $\mathcal{A}$ defined below.

Definition 1. *Let $L = L_1 \subseteq \Sigma^*$ be any regular language with the set of atoms $Q = \{A_1, \ldots, A_m\}$, initial set of atoms $I \subseteq Q$, and final atom A_m. The* átomaton *of L is the NFA $\mathcal{A} = (\mathbf{Q}, \Sigma, \delta, \mathbf{I}, \{\mathbf{A}_m\})$, where $\mathbf{Q} = \{\mathbf{A}_i \mid A_i \in Q\}$, $\mathbf{I} = \{\mathbf{A}_i \mid A_i \in I\}$, and $\mathbf{A}_j \in \delta(\mathbf{A}_i, a)$ if and only if $aA_j \subseteq A_i$, for all $A_i, A_j \in Q$.*

Example 3. Let L be defined by the quotient equations below on the left and accepted by the quotient DFA of Fig. 2 (a).

$$L_1 = aL_2 \cup bL_1, \qquad L_{123} = a(L_{123} \cup L_{\bar{1}23}) \cup b(L_{123} \cup L_{12\bar{3}}),$$
$$L_2 = aL_3 \cup bL_1 \cup \varepsilon, \qquad L_{\bar{1}23} = aL_{\bar{1}\bar{2}3},$$
$$L_3 = aL_3 \cup bL_2. \qquad L_{12\bar{3}} = bL_{1\bar{2}\bar{3}},$$
$$L_{\bar{1}\bar{2}3} = b(L_{\bar{1}23} \cup L_{\bar{1}2\bar{3}}),$$
$$L_{1\bar{2}\bar{3}} = a(L_{12\bar{3}} \cup L_{\bar{1}2\bar{3}}),$$
$$L_{\bar{1}2\bar{3}} = \varepsilon.$$

We find the atoms using the quotient equations. Thus

$$\begin{aligned} L_1 \cap L_2 \cap L_3 &= (aL_2 \cup bL_1) \cap (aL_3 \cup bL_1 \cup \varepsilon) \cap (aL_3 \cup bL_2) \\ &= (aL_2 \cap aL_3 \cap aL_3) \cup (bL_1 \cap bL_1 \cap bL_2) \\ &= a(L_2 \cap L_3) \cup b(L_1 \cap L_2) \\ &= a[(L_1 \cap L_2 \cap L_3) \cup (\overline{L_1} \cap L_2 \cap L_3)] \\ &\cup b[(L_1 \cap L_2 \cap L_3) \cup (L_1 \cap L_2 \cap \overline{L_3})], \textit{etc.} \end{aligned}$$

To simplify the notation, we denote $L_i \cap L_j$ by L_{ij}, $L_i \cap \overline{L_j}$ by $L_{i\bar{j}}$, *etc.* Noting that $L_{1\bar{2}3}$ is empty, we have the equations above on the right, from which we get Fig. 2 (b) for the átomaton of L.

Lemma 2. *For $w \in \Sigma^*$, $\mathbf{A}_j \in \delta(\mathbf{A}_i, w)$ if and only if $wA_j \subseteq A_i$, for $i, j \in [m]$.*

Proof. The proof is by induction on the length of w. If $|w| = 0$ and $\mathbf{A}_j \in \delta(\mathbf{A}_i, \varepsilon)$, then $i = j$ and $\varepsilon A_j \subseteq A_i$. If $|w| = 0$ and $\varepsilon A_j \subseteq A_i$, then $i = j$, since atoms are disjoint; hence $\mathbf{A}_j \in \delta(\mathbf{A}_i, \varepsilon)$. If $|w| = 1$, then the lemma holds by Definition 1.

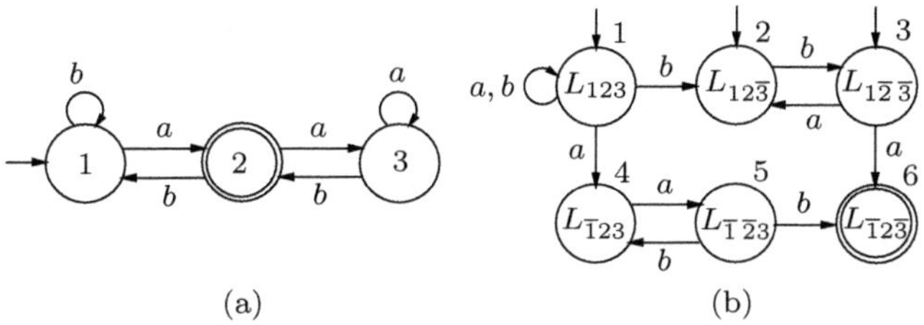

Fig. 2. (a) Quotient automaton; (b) átomaton of L

Now, let $w = av$, where $a \in \Sigma$ and $v \in \Sigma^+$, and assume that lemma holds for v. Suppose that $\mathbf{A}_j \in \delta(\mathbf{A}_i, av)$. Then there exists some state $\mathbf{A}_k$ such that $\mathbf{A}_k \in \delta(\mathbf{A}_i, a)$ and $\mathbf{A}_j \in \delta(\mathbf{A}_k, v)$. Thus, $aA_k \subseteq A_i$ by the definition of átomaton, and $vA_j \subseteq A_k$ by the induction assumption, implying that $avA_j \subseteq A_i$.

Conversely, let $avA_j \subseteq A_i$. Then $vA_j \subseteq a^{-1}A_i$. Let $x \in A_j$. Then $vx \in a^{-1}A_i$. Since by Proposition 1, Part 3, $a^{-1}A_i$ is a union of atoms, there exists some atom A_k such that $vx \in A_k$. Since $x \in A_j$, by Lemma 1 we get $vA_j \subseteq A_k$. Furthermore, because $avA_j \subseteq A_i$ and $x \in A_j$, we have $avx \in A_i$. Since $vx \in A_k$, then $aA_k \subseteq A_i$ by Lemma 1.

As the lemma holds for v and a, $vA_j \subseteq A_k$ implies $\mathbf{A}_j \in \delta(\mathbf{A}_k, v)$, and $aA_k \subseteq A_i$ implies $\mathbf{A}_k \in \delta(\mathbf{A}_i, a)$, showing that $\mathbf{A}_j \in \delta(\mathbf{A}_i, av)$. □

Proposition 2. *The right language of state $\mathbf{A}_i$ of átomaton $\mathcal{A}$ is the atom A_i, that is, $L_{\mathbf{A}_i,\{\mathbf{A}_m\}}(\mathcal{A}) = A_i$, for all $i \in [m]$.*

Proof. Let $w \in L_{\mathbf{A}_i,\{\mathbf{A}_m\}}(\mathcal{A})$; then $\mathbf{A}_m \in \delta(\mathbf{A}_i, w)$. By Lemma 2, we have $wA_m \subseteq A_i$. Since $\varepsilon \in A_m$, we have $w \in A_i$.

Now suppose that $w \in A_i$. Then $w\varepsilon \in A_i$, and since $\varepsilon \in A_m$, by Lemma 1 we get $wA_m \subseteq A_i$. By Lemma 2, $\mathbf{A}_m \in \delta(\mathbf{A}_i, w)$, that is, $w \in L_{\mathbf{A}_i,\{\mathbf{A}_m\}}(\mathcal{A})$. □

Theorem 1. *The language accepted by the átomaton $\mathcal{A}$ of L is L, that is, $L(\mathcal{A}) = L_{I,\{\mathbf{A}_m\}} = L$.*

Proof. We have $L(\mathcal{A}) = \bigcup_{A_i \in I} L_{\mathbf{A}_i,\{\mathbf{A}_m\}}(\mathcal{A}) = \bigcup_{A_i \in I} A_i$, by Proposition 2. Since I is the set of all atoms that have $L = L_1$ as a term, we also have $L = \bigcup_{A_i \in I} A_i$. □

Proposition 3. *The left language of state $\mathbf{A}_i$ of átomaton $\mathcal{A}$ of a language L is $L_{I,\mathbf{A}_i}(\mathcal{A}) = ((x^R)^{-1}L^R)^R$, for every $i \in [m]$ and every word x in A_i.*

Proof. If $w \in L_{I,\mathbf{A}_i}(\mathcal{A})$, then $\mathbf{A}_i \in \delta(\mathbf{A}_{i_0}, w)$ for some $\mathbf{A}_{i_0} \in \mathbf{I}$. Then $wA_i \subseteq A_{i_0}$ by Lemma 2. Since $A_{i_0} \subseteq L$, we also have $wA_i \subseteq L$, that is, $wx \in L$ for every $x \in A_i$. Then $x^R w^R \in L^R$, and $w^R \in (x^R)^{-1}L^R$. Thus, $w \in ((x^R)^{-1}L^R)^R$.

Now, let $w \in ((x^R)^{-1}L^R)^R$, where $x \in A_i$. Then $w^R \in (x^R)^{-1}L^R$, implying that $wx \in L$. By Theorem 1, there is some $\mathbf{A}_{i_0} \in I$ such that $wx \in L_{\mathbf{A}_{i_0},\{\mathbf{A}_m\}}(\mathcal{A})$. By Proposition 2, $wx \in A_{i_0}$. Since $x \in A_i$, by Lemma 1 we have $wA_i \subseteq A_{i_0}$. By Lemma 2, $\mathbf{A}_i \in \delta(\mathbf{A}_{i_0}, w)$, implying that $w \in L_{I,\mathbf{A}_i}(\mathcal{A})$. □

Proposition 4. *The left language of state* $\mathbf{A}_i$ *of átomaton* $\mathcal{A}$ *is non-empty, that is,* $L_{I,\mathbf{A}_i}(\mathcal{A}) \neq \emptyset$*, for every* $i \in [m]$*.*

Proof. Suppose that $L_{I,\mathbf{A}_i}(\mathcal{A}) = \emptyset$ for some $i \in [m]$. Then by Proposition 3, $((x^R)^{-1}L^R)^R = \emptyset$ for any $x \in A_i$. Then also $(x^R)^{-1}L^R = \emptyset$, implying that for any $w \in \Sigma^*$, $wx \notin L$. However, since there is some quotient L_j of L, $j \in [n]$, such that $A_i \subseteq L_j$, and there is an x in A_i, we have $x \in L_j$. Let $v \in \Sigma^*$ be such that $L_j = v^{-1}L$. Then we get $vx \in L$, which is a contradiction. □

Corollary 1. *The átomaton of any regular language is trim.*

Next we recall a (slightly modified version of a) theorem from [2]:

Theorem 2. *For a trim NFA* $\mathcal{N}$*,* $\mathcal{N}^{\mathbb{D}}$ *is minimal if* $\mathcal{N}^{\mathbb{R}}$ *is deterministic.*

We have defined a unique NFA, the átomaton, directly from the quotient equations of a language L, that is, from the minimal DFA recognizing L. In contrast to this, Sengoku [10] defined a unique NFA starting from any NFA accepting L: The *normal automaton* of L is the NFA $\mathcal{N}^{\mathbb{RDMTR}}$. Matz and Potthoff [8] (p. 78) defined an NFA $\overline{\mathcal{E}}$ as the reverse of the trim minimal DFA accepting L^R, that is $\overline{\mathcal{E}} = \mathcal{B}^{\mathbb{TR}}$, where $\mathcal{B}$ is the minimal DFA accepting L^R. We now relate a number of concepts associated with regular languages:

Theorem 3. *Let* L *be any regular language, and let* $\mathcal{A}$ *be its átomaton.*
1. *The reverse* $\mathcal{A}^{\mathbb{R}}$ *of* $\mathcal{A}$ *is an IDFA.*
2. $\mathcal{A}^{\mathbb{R}}$ *is minimal.*
3. *The determinization* $\mathcal{A}^{\mathbb{D}}$ *of* $\mathcal{A}$ *is the minimal DFA of* L*.*
4. *The normal NFA* $\mathcal{N}^{\mathbb{RDMTR}}$ *of any NFA accepting* L *is isomorphic to* $\mathcal{A}$*.*
5. *Matz and Potthoff's NFA* $\overline{\mathcal{E}}$ *is isomorphic to* $\mathcal{A}$*.*
6. $\mathcal{A}$ *is isomorphic to the quotient IDFA of* L *if and only if* L *is bideterministic.*

Proof. Suppose that L has the quotients $L_1, \ldots, L_n$ and atoms $A_1, \ldots, A_m$.

1. Since $\mathcal{A}$ has one accepting state, $\mathcal{A}^{\mathbb{R}}$ has one initial state. Because atoms are disjoint, a word w can belong to at most one atom. If w belongs to A_i, then, by Proposition 2, $w \in L_{\mathbf{A}_i,\{\mathbf{A}_m\}}(\mathcal{A})$ and $w \notin L_{\mathbf{A}_j,\{\mathbf{A}_m\}}(\mathcal{A})$ if $j \neq i$. Hence $w^R \in L_{\{\mathbf{A}_m\},\mathbf{A}_i}(\mathcal{A}^{\mathbb{R}})$ and $w^R \notin L_{\{\mathbf{A}_m\},\mathbf{A}_j}(\mathcal{A}^{\mathbb{R}})$ if $j \neq i$. Thus $\mathcal{A}^{\mathbb{R}}$ is an IDFA.

2. Since $\mathcal{A}$ is trim, so is $\mathcal{A}^{\mathbb{R}}$. Thus, if $\mathcal{A}^{\mathbb{R}}$ is not minimal, there must be states $\mathbf{A}_i, \mathbf{A}_j \in \mathbf{Q}$, $\mathbf{A}_i \neq \mathbf{A}_j$, such that $L_{\mathbf{A}_i,\mathbf{I}}(\mathcal{A}^{\mathbb{R}}) = L_{\mathbf{A}_j,\mathbf{I}}(\mathcal{A}^{\mathbb{R}})$. Let $L_k = u^{-1}L$ be any non-empty quotient of L, where $k \in [n]$ and $u \in \Sigma^*$. Then there are two possibilities: either $u \in L_{\mathbf{I},\mathbf{A}_i}(\mathcal{A})$, or $u \notin L_{\mathbf{I},\mathbf{A}_i}(\mathcal{A})$.

In the first case $u^R \in L_{\mathbf{A}_i,\mathbf{I}}(\mathcal{A}^{\mathbb{R}})$, and, since $L_{\mathbf{A}_i,\mathbf{I}}(\mathcal{A}^{\mathbb{R}}) = L_{\mathbf{A}_j,\mathbf{I}}(\mathcal{A}^{\mathbb{R}})$, we have $u^R \in L_{\mathbf{A}_j,\mathbf{I}}(\mathcal{A}^{\mathbb{R}})$, implying that $u \in L_{\mathbf{I},\mathbf{A}_j}(\mathcal{A})$. Thus, $L_{\mathbf{A}_i,\{\mathbf{A}_m\}}(\mathcal{A}) \subseteq L_k$ and $L_{\mathbf{A}_j,\{\mathbf{A}_m\}}(\mathcal{A}) \subseteq L_k$. In view of Proposition 2, A_i and A_j are both subsets of L_k.

Now, assume that $u \notin L_{\mathbf{I},\mathbf{A}_i}(\mathcal{A})$. Then, as in the first case, we get $u \notin L_{\mathbf{I},\mathbf{A}_j}(\mathcal{A})$. If $L_{\mathbf{A}_i,\{\mathbf{A}_m\}}(\mathcal{A}) \subseteq L_k$, then $ux \in L$ for some $x \in L_{\mathbf{A}_i,\{\mathbf{A}_m\}}(\mathcal{A})$. But then $x^R u^R \in L^R$. Since $x^R \in L_{\{\mathbf{A}_m\},\mathbf{A}_i}(\mathcal{A}^{\mathbb{R}})$ and $\mathcal{A}^{\mathbb{R}}$ is deterministic, we must have $u^R \in L_{\mathbf{A}_i,\mathbf{I}}(\mathcal{A}^{\mathbb{R}})$. This contradicts the assumption that $u \notin L_{\mathbf{I},\mathbf{A}_i}(\mathcal{A})$. Thus $L_{\mathbf{A}_i,\{\mathbf{A}_m\}}(\mathcal{A}) \not\subseteq L_k$, and similarly, $L_{\mathbf{A}_j,\{\mathbf{A}_m\}}(\mathcal{A}) \not\subseteq L_k$, implying that neither A_i nor A_j is a subset of L_k.

So, for every $k \in [n]$, either both atoms A_i and A_j are subsets of L_k or neither of them is. Since A_i and A_j are distinct, there must be an h such that $A_i \subseteq L_h$ and $A_j \subseteq \overline{L_h}$. This contradicts our earlier conclusion.

3. Since $\mathcal{A}^{\mathbb{R}}$ is deterministic, $\mathcal{A}^{\mathbb{D}}$ is minimal by Theorem 2.

4. Let $\mathcal{N}$ be any NFA accepting L. The DFA $\mathcal{N}^{\mathbb{RDM}}$ is the unique minimal DFA accepting the language L^R. By Parts 1 and 2, $\mathcal{A}^{\mathbb{R}}$ is a minimal IDFA, and it accepts $L^{\mathbb{R}}$. Since $\mathcal{N}^{\mathbb{RDMT}}$ is isomorphic to $\mathcal{A}^{\mathbb{R}}$, it follows that the normal automaton $\mathcal{N}^{\mathbb{RDMTR}}$ is isomorphic to $\mathcal{A}$.

5. Since $\mathcal{B}$ is isomorphic to $\mathcal{N}^{\mathbb{RDM}}$ of Part 4, the claim follows.

6. Let $\mathcal{D}$ be the quotient DFA of L, and suppose that $\mathcal{A}$ is isomorphic to $\mathcal{D}^{\mathbb{T}}$. By Part 1, $\mathcal{A}^R$ is an IDFA. Since $\mathcal{A}$ is isomorphic to $\mathcal{D}^{\mathbb{T}}$, $\mathcal{A}$ itself is an IDFA. Hence $\mathcal{A}$, and so also L, are bideterministic.

Conversely, let $\mathcal{B}$ be a trim bideterministic IDFA accepting L.

Since $\mathcal{B}^{\mathbb{R}}$ is deterministic, $\mathcal{B}^{\mathbb{D}}$ is minimal by Theorem 2. Since $\mathcal{B}$ is a trim IDFA, we have $\mathcal{B}^{\mathbb{DT}} = \mathcal{B}$; hence $\mathcal{B}$ is isomorphic to the quotient IDFA of L.

Since $\mathcal{B}$ is deterministic, $\mathcal{B}^{\mathbb{RD}}$ is minimal by Theorem 2, that is $\mathcal{B}^{\mathbb{RDM}} = \mathcal{B}^{\mathbb{RD}}$. Because $\mathcal{B}$ is trim, also $\mathcal{B}^{\mathbb{R}}$ is trim. Since $\mathcal{B}^{\mathbb{R}}$ is deterministic, we get $\mathcal{B}^{\mathbb{RDT}} = \mathcal{B}^{\mathbb{RT}} = \mathcal{B}^{\mathbb{R}}$. Thus $\mathcal{B}^{\mathbb{RDMTR}} = \mathcal{B}^{\mathbb{RDTR}} = \mathcal{B}^{\mathbb{RR}} = \mathcal{B}$ is the átomaton of L by Part 4. Hence $\mathcal{B}$ is both the minimal IDFA of L and its átomaton. □

As noted in [8], for each word w in L there is a unique path in $\mathcal{A}$ accepting w, and deleting any transition from $\mathcal{A}$ results in a smaller accepted language. It is also stated in [8] without proof that the right language $L_{q,F}(\mathcal{N})$ of any state q of an NFA $\mathcal{N}$ accepting L is a subset of a union of atoms. This holds because $L_{q,F}(\mathcal{N})$ is a subset of a (left) quotient of L, and quotients are unions of atoms by Proposition 1, Part 2.

Theorem 3 provides another method of finding the átomaton of L: simply trim the quotient DFA of L^R and reverse it. In view of this we have

Corollary 2. *If $L = L^R$ and L is accepted by quotient DFA $\mathcal{D}$, then $\mathcal{A} = \mathcal{D}^{\mathbb{T}}$.*

Example 4. Let $L = (b \cup ba)^*$; then $L^R = (b \cup ab)^*$, and it is accepted by the minimal DFA $\mathcal{D}$ of Fig. 1 (c). Its trimmed reverse is shown in Fig. 1 (d). Hence the NFA of Fig. 1 (d) is the átomaton of L.

4 Atomic Automata

We now introduce a new class of NFA's and study their properties.

Definition 2. *An NFA $\mathcal{N} = (Q, \Sigma, \delta, I, F)$ is* atomic *if for every state $q \in Q$, the right language $L_{q,F}(\mathcal{N})$ of q is a union of some atoms of $L(\mathcal{N})$.*

Note that, if $L_{q,F}(\mathcal{N}) = \emptyset$, then it is the union of zero atoms.

Recall that an NFA $\mathcal{N}$ is residual, if $L_{q,F}(\mathcal{N})$ is a (left) quotient of $L(\mathcal{N})$ for every $q \in Q$. Since every quotient is a union of atoms (see Proposition 1, Part 2), every residual NFA is atomic. However, the converse is not true: there exist atomic NFA's which are not residual. For example, the átomaton of Fig. 2

is atomic, but not residual. Note also that every trim DFA is a special case of a residual NFA; hence every trim DFA is atomic.

Let us now consider the universal automaton $\mathcal{U}_L = (Q, \Sigma, \delta, I, F)$ of a language L. We state some basic properties of this automaton based on [7]. Let (X, Y) be a factorization of L such that $X \neq \emptyset$ and $Y \neq \emptyset$. Then

(1) $Y = \bigcap_{x \in X} x^{-1}L$ and $X = \bigcap_{y \in Y} Ly^{-1}$.

(2) $L_{I,(X,Y)}(\mathcal{U}_L) = X$ and $L_{(X,Y),F}(\mathcal{U}_L) = Y$.

(3) The universal automaton $\mathcal{U}_L$ accepts L.

To recapitulate what was said above about residual NFA's and DFA's, and also to show that the trim part of the universal automaton is atomic, we have

Theorem 4. *Let L be any regular language. The following automata accepting L are atomic: 1. The átomaton $\mathcal{A}$. 2. Any trim DFA. 3. Any residual NFA. 4. The trim part of the universal automaton $\mathcal{U}_L$.*

Proof. 1. The right language of every state of $\mathcal{A}$ is an atom of L, so $\mathcal{A}$ is atomic.

2. The right language of every state of any trim DFA accepting L is a quotient of L. Since every quotient is a union of atoms, every trim DFA is atomic.

3. The right language of every state of any residual NFA of L is a quotient of L, and hence a union of atoms. Thus, any residual NFA is atomic.

4. Let (X, Y) be a state of $\mathcal{U}_L^{\mathbb{T}}$. Then $X, Y \neq \emptyset$. By (1) and (2), $L_{(X,Y),F}(\mathcal{U}_L) = \bigcap_{x \in X} x^{-1}L$. Let $L_1, \dots, L_n$ be the quotients of L. Then for some $H \subseteq [n]$, $L_{(X,Y),F}(\mathcal{U}_L) = \bigcap_{i \in H} L_i$. Now $\bigcap_{i \in H} L_i = (\bigcap_{i \in H} L_i) \cap (\bigcap_{j \in [n] \setminus H} (L_j \cup \overline{L_j}) = \bigcup (\bigcap_{i \in H} L_i) \cap (\bigcap_{j \in [n] \setminus H} \widetilde{L_j})$, where $\widetilde{L_j}$ is either L_j or $\overline{L_j}$. Thus the right language of (X, Y) is a union of atoms of L. Since $L(\mathcal{U}_L^{\mathbb{T}}) = L(\mathcal{U}_L)$, and $L(\mathcal{U}_L) = L$ by (3), then $\mathcal{U}_L^{\mathbb{T}}$ is atomic. $\square$

5 Extension of Brzozowski's Theorem on Minimal DFA's

Theorem 2 forms the basis for Brzozowski's "double-reversal" minimization algorithm [2]: Given any DFA (or IDFA) $\mathcal{D}$, reverse it to get $\mathcal{D}^{\mathbb{R}}$, determinize $\mathcal{D}^{\mathbb{R}}$ to get $\mathcal{D}^{\mathbb{RD}}$, reverse $\mathcal{D}^{\mathbb{RD}}$ to get $\mathcal{D}^{\mathbb{RDR}}$, and then determinize $\mathcal{D}^{\mathbb{RDR}}$ to get $\mathcal{D}^{\mathbb{RDRD}}$. This last DFA is guaranteed to be minimal by Theorem 2, since $\mathcal{D}^{\mathbb{RD}}$ is deterministic. Hence $\mathcal{D}^{\mathbb{RDRD}}$ is the minimal DFA equivalent to $\mathcal{D}$.

Since this conceptually very simple algorithm carries out two determinizations, its complexity is exponential in the number of states of the original automaton in the worst case. However, its performance is good in practice, often better than Hopcrofts's algorithm [11,12]. Furthermore, this algorithm applied to an NFA still yields an equivalent minimal DFA; see [12], for example.

We now generalize Theorem 2:

Theorem 5. *For a trim NFA $\mathcal{N}$, $\mathcal{N}^{\mathbb{D}}$ is minimal if and only if $\mathcal{N}^{\mathbb{R}}$ is atomic.*

Proof. Let $\mathcal{N} = (Q, \Sigma, \delta, I, F)$ be any trim NFA, and let $\mathcal{N}^{\mathbb{R}} = (Q, \Sigma, \delta_{\mathbb{R}}, F, I)$ be its reverse. Let the atoms of $\mathcal{N}^{\mathbb{R}}$ be $B_1, \dots, B_r$, and let $\mathcal{B}$ be the átomaton of $L(\mathcal{N}^{\mathbb{R}})$.

Assume first that $\mathcal{N}^{\mathbb{D}}$ is minimal. Let q be a state of $\mathcal{N}$, and hence of $\mathcal{N}^{\mathbb{R}}$; since $\mathcal{N}$ is trim, so is $\mathcal{N}^{\mathbb{R}}$, and there are $w, x \in \Sigma^*$ such that $x \in L_{F,q}(\mathcal{N}^{\mathbb{R}})$ and $w \in L_{q,I}(\mathcal{N}^{\mathbb{R}})$. Since $L_{q,I}(\mathcal{N}^{\mathbb{R}}) \subseteq x^{-1}L(\mathcal{N}^R)$, and every quotient of $L(\mathcal{N}^R)$ is a union of atoms, there is some $i \in [r]$ such that $w \in B_i$.

Suppose that $\mathcal{N}^{\mathbb{R}}$ is not atomic; then there must be a state $q' \in Q$ which is not a union of atoms. This means that there is some $i \in [r]$ and words $u, v \in B_i$ such that $u \in L_{q',I}(\mathcal{N}^{\mathbb{R}})$ but $v \notin L_{q',I}(\mathcal{N}^{\mathbb{R}})$. Suppose that z is in $L_{F,q'}(\mathcal{N}^{\mathbb{R}})$. Since $u \in z^{-1}L(\mathcal{N}^{\mathbb{R}})$, $u \in B_i$, and $z^{-1}L(\mathcal{N}^{\mathbb{R}})$ is a union of atoms, we must have $B_i \subseteq z^{-1}L(\mathcal{N}^{\mathbb{R}})$. But now $v \in B_i$ implies that $zv \in L(\mathcal{N}^{\mathbb{R}})$. Hence there must be a state $q'' \in Q$, $q'' \neq q'$, such that $v \in L_{q'',I}(\mathcal{N}^{\mathbb{R}})$. Therefore, we know that $u^R \in L_{I,q'}(\mathcal{N})$, $v^R \notin L_{I,q'}(\mathcal{N})$, and $v^R \in L_{I,q''}(\mathcal{N})$.

Now, since every state of $\mathcal{N}^{\mathbb{D}}$ is a subset of the state set Q of $\mathcal{N}$, there is a state s' of $\mathcal{N}^{\mathbb{D}}$ such that $q' \in s'$ and $u^R \in L_{I,s'}(\mathcal{N}^{\mathbb{D}})$, and there is a state s'' of $\mathcal{N}^{\mathbb{D}}$ such that $q'' \in s''$ and $v^R \in L_{I,s''}(\mathcal{N}^{\mathbb{D}})$. Since $v^R \notin L_{I,q'}(\mathcal{N})$, we have $q' \notin s''$, implying that $s' \neq s''$.

By Theorem 3, Part 4, $(\mathcal{N}^{\mathbb{R}})^{\mathbb{RDMTR}} = \mathcal{N}^{\mathbb{DMTR}}$ is isomorphic to the átomaton $\mathcal{B}$. By the assumption that $\mathcal{N}^{\mathbb{D}}$ is minimal, $\mathcal{N}^{\mathbb{DTR}}$ is isomorphic to $\mathcal{B}$. Thus $L_{s',I}(\mathcal{N}^{\mathbb{DTR}}) = B_k$ and $L_{s'',I}(\mathcal{N}^{\mathbb{DTR}}) = B_l$ for some $k, l \in [r]$. Since $u^R \in L_{I,s'}(\mathcal{N}^{\mathbb{D}})$, we have $u \in L_{s',I}(\mathcal{N}^{\mathbb{DR}}) = L_{s',I}(\mathcal{N}^{\mathbb{DRT}}) = L_{s',I}(\mathcal{N}^{\mathbb{DTR}}) = B_k$. This together with $u \in B_i$, yields $k = i$. Similarly, $v^R \in L_{I,s''}(\mathcal{N}^{\mathbb{D}})$ and $v \in B_i$, implies $l = i$. Thus, $L_{s',I}(\mathcal{N}^{\mathbb{DTR}}) = L_{s'',I}(\mathcal{N}^{\mathbb{DTR}})$. But then $L_{I,s'}(\mathcal{N}^{\mathbb{DT}}) = L_{I,s''}(\mathcal{N}^{\mathbb{DT}})$, which contradicts the inequality $s' \neq s''$. Therefore $\mathcal{N}^{\mathbb{R}}$ is atomic.

To prove the converse, assume that $\mathcal{N}^{\mathbb{R}}$ is atomic; then, for every state q of $\mathcal{N}^{\mathbb{R}}$, there is a set $H_q \subseteq [r]$ such that $L_{q,I}(\mathcal{N}^{\mathbb{R}}) = \bigcup_{i \in H_q} B_i$. This implies that $L_{I,q}(\mathcal{N}) = \bigcup_{i \in H_q} B_i^R$ for every state q of $\mathcal{N}$.

Let $\mathcal{N}^{\mathbb{D}} = (S, \Sigma, \gamma, I, G)$, and suppose that $\mathcal{N}^{\mathbb{D}}$ is not minimal. Then there are at least two states s' and s'' of $\mathcal{N}^{\mathbb{D}}$, $s' \neq s''$, with $L_{s',G}(\mathcal{N}^{\mathbb{D}}) = L_{s'',G}(\mathcal{N}^{\mathbb{D}})$. Let $\mathcal{D}_m = (Q_m, \Sigma, \delta_m, I, F_m)$ be a minimal DFA equivalent to $\mathcal{N}^{\mathbb{D}}$. Then there must be a state s of $\mathcal{D}_m$ such that $L_{s,F_m}(\mathcal{D}_m) = L_{s',G}(\mathcal{N}^{\mathbb{D}}) = L_{s'',G}(\mathcal{N}^{\mathbb{D}})$. Then $L_{I,s'}(\mathcal{N}^{\mathbb{D}}) \subset L_{I,s}(\mathcal{D}_m)$ and $L_{I,s''}(\mathcal{N}^{\mathbb{D}}) \subset L_{I,s}(\mathcal{D}_m)$ must also hold.

Since $\mathcal{N}^{\mathbb{DM}}$ is isomorphic to $\mathcal{D}_m$, and $\mathcal{N}^{\mathbb{DMTR}}$ is isomorphic to the átomaton $\mathcal{B}$ of $L(\mathcal{N}^{\mathbb{R}})$ by Part 4 of Theorem 3, also $(\mathcal{D}_m)^{\mathbb{TR}}$ is isomorphic to $\mathcal{B}$. Thus $L_{s,I}((\mathcal{D}_m)^{\mathbb{TR}}) = B_i$ for some $i \in [r]$. This implies that $L_{I,s}((\mathcal{D}_m)^{\mathbb{T}}) = B_i^R$. Thus $L_{I,s'}(\mathcal{N}^{\mathbb{D}}) \subset B_i^R$.

On the other hand, the left language of state s' of $\mathcal{N}^{\mathbb{D}}$ consists of all words u such that $u \in L_{I,q'}(\mathcal{N})$ for every $q' \in s'$, but $u \notin L_{I,q}(\mathcal{N})$ for any $q \notin s'$. That is, $L_{I,s'}(\mathcal{N}^{\mathbb{D}}) = (\bigcap_{q' \in s'} \bigcup_{i \in H_{q'}} B_i^R) \setminus (\bigcup_{q \notin s'} \bigcup_{i \in H_q} B_i^R)$. Since by Proposition 1, Part 1, $B_i \cap B_j = \emptyset$ for all $i, j \in [r]$, $i \neq j$, then also $B_i^R \cap B_j^R = \emptyset$, and the result of any boolean combination of sets B_i^R where $i \in [r]$, cannot be a proper subset of any B_i^R. Therefore, $L_{I,s'}(\mathcal{N}^{\mathbb{D}}) \subset B_i^R$ cannot hold and thus, $\mathcal{N}^{\mathbb{D}}$ must be minimal. □

Corollary 3. *If $\mathcal{N}$ is a non-minimal DFA, then $\mathcal{N}^{\mathbb{R}}$ is not atomic.*

Theorem 5 can be rephrased as follows: $\mathcal{N}$ is atomic if and only if $\mathcal{N}^{\mathbb{RD}}$ is minimal. Sengoku defines an NFA $\mathcal{N}$ to be *in standard form* [10] if and only if $\mathcal{N}^{\mathbb{RD}}$ is minimal, and also shows that the right language of every state of an NFA in standard form is equal to the union of right languages of some states of the normal automaton (that is, our átomaton).

6 Conclusions

We have introduced a natural set of languages—the atoms—that are defined by every regular language. We then defined a unique NFA, the átomaton, and related it to other known concepts. We introduced atomic automata, and generalized Brzozowski's method of minimization of DFA's by double reversal.

References

1. Arnold, A., Dicky, A., Nivat, M.: A note about minimal non-deterministic automata. Bull. EATCS 47, 166–169 (1992)
2. Brzozowski, J.A.: Canonical regular expressions and minimal state graphs for definite events. In: Proceedings of the Symposium on Mathematical Theory of Automata. MRI Symposia Series, vol. 12, pp. 529–561. Polytechnic Press, Polytechnic Institute of Brooklyn, N.Y (1963)
3. Carrez, C.: On the minimalization of non-deterministic automaton. Technical report, Lille University, Lille, France (1970)
4. Conway, J.: Regular Algebra and Finite Machines. Chapman and Hall, London (1971)
5. Denis, F., Lemay, A., Terlutte, A.: Residual finite state automata. Fund. Inform. 51, 339–368 (2002)
6. Kameda, T., Weiner, P.: On the state minimization of nondeterministic automata. IEEE Trans. Comput. C-19(7), 617–627 (1970)
7. Lombardy, S., Sakarovitch, J.: The universal automaton. In: Flum, J., Grädel, E., Wilke, T. (eds.) Logic and Automata: History and Perspectives. Texts in Logic and Games, vol. 2, pp. 457–504. Amsterdam University Press, Amsterdam (2007)
8. Matz, O., Potthoff, A.: Computing small finite nondeterministic automata. In: Engberg, U.H., Larsen, K.G., Skou, A. (eds.) Proceedings of the Workshop on Tools and Algorithms for Construction and Analysis of Systems. BRICS Note Series, pp. 74–88. BRICS, Aarhus (1995)
9. Rabin, M., Scott, D.: Finite automata and their decision problems. IBM J. Res. and Dev. 3, 114–129 (1959)
10. Sengoku, H.: Minimization of nondeterministic finite automata. Master's thesis, Kyoto University, Department of Information Science, Kyoto University, Kyoto, Japan (1992)
11. Tabakov, D., Vardi, M.: Experimental evaluation of classical automata constructions. In: Sutcliffe, G., Voronkov, A. (eds.) LPAR 2005. LNCS (LNAI), vol. 3835, pp. 396–411. Springer, Heidelberg (2005)
12. Watson, B.W.: Taxonomies and toolkits of regular language algorithms. PhD thesis, Faculty of Mathematics and Computing Science. Eindhoven University of Technology, Eindhoven, The Netherlands (1995)

Syntactic Complexity of Ideal and Closed Languages *

Janusz Brzozowski[1] and Yuli Ye[2]

[1] David R. Cheriton School of Computer Science, University of Waterloo, Waterloo, ON, Canada N2L 3G1
{brzozo@uwaterloo.ca }

[2] Department of Computer Science, University of Toronto
Toronto, ON, Canada M5S 3G4
{y3ye@cs.toronto.edu}

Abstract. The state complexity of a regular language is the number of states in the minimal deterministic automaton accepting the language. The syntactic complexity of a regular language is the cardinality of its syntactic semigroup. The syntactic complexity of a subclass of regular languages is the worst-case syntactic complexity taken as a function of the state complexity n of languages in that class. We prove that n^{n-1} is a tight upper bound on the complexity of right ideals and prefix-closed languages, and that there exist left ideals and suffix-closed languages of syntactic complexity $n^{n-1}+n-1$, and two-sided ideals and factor-closed languages of syntactic complexity $n^{n-2}+(n-2)2^{n-2}+1$.

Keywords: automaton, closed, complexity, ideal, language, monoid, regular, reversal, semigroup, syntactic.

1 Introduction

There are two fundamental congruence relations in the theory of regular languages: the Nerode congruence [16], and the Myhill congruence [15]. In both cases, a language is regular if and only if it is a union of congruence classes of a congruence of finite index. The Nerode congruence leads to the definitions of left quotients of a language and the minimal deterministic finite automaton (DFA) recognizing the language. The Myhill congruence leads to the definitions of the syntactic semigroup and the syntactic monoid of the language.

The *state complexity* of a language is the number of states in the minimal DFA recognizing the language. This concept has been studied quite extensively; for surveys and references see [4,25]. Syntactic complexity of various types of graph languages has been studied by Bozapalidis and Kalampakas [1,2,3,11] in the framework of magmoids. However, in spite of suggestions that syntactic

* This work was supported by the Natural Sciences and Engineering Research Council of Canada under grant No. OGP0000871 and a Postgraduate Scholarship, and by a Graduate Award from the Department of Computer Science, University of Toronto.

G. Mauri and A. Leporati (Eds.): DLT 2011, LNCS 6795, pp. 117–128, 2011.

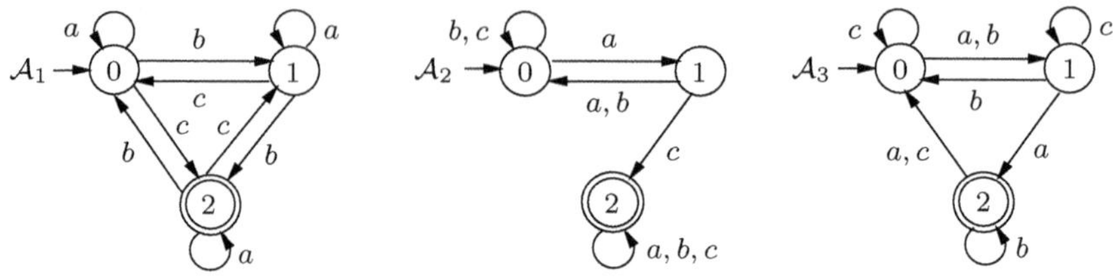

Fig. 1. Automata with various syntactic complexities

semigroups of subsets of free monoids deserve to be studied [9,13], little has been done on the syntactic complexity of a regular language, which we define as the cardinality of its syntactic semigroup. This semigroup is isomorphic to the semigroup of transformations of the set of states of the minimal DFA recognizing the language, where these transformations are performed by non-empty words.

The following example illustrates the significant difference between state complexity and syntactic complexity. The DFA's in Fig. 1 have the same alphabet, are all minimal, and all have the same state complexity. However, the syntactic complexity of $\mathcal{A}_1$ is 3, that of $\mathcal{A}_2$ is 9, and that of $\mathcal{A}_3$ is 27. This shows that syntactic complexity is a much finer measure of complexity than state complexity. The question then arises: Is it possible to find upper bounds to the syntactic complexity of a regular language from its properties or from the properties of its minimal DFA? We shed some light on this question for ideal and closed regular languages.

The set of all n^n transformations of a set Q is a monoid under composition of transformations, with identity as the unit element. The set of all $n!$ permutations of Q is a group, the *symmetric* group of degree n. The fact that two generators are sufficient to form a basis for the set of all permutations of a set of n elements has been known for many years. For example, such a result is stated in the 1895 paper by Hoyer [10], and later in the 1938 paper by Piccard [18]. The fact that three functions suffice to generate all transformations of a set of n elements was proved in 1935 by Piccard [17]. Also in 1935, Eilenberg showed that no two functions suffice, as reported by Sierpiński [24]. Related later work includes that of Salomaa (1960-63) [20,21,22] and Dénes (1968) [8]. In 1970, Maslov [14] dealt with the generators of the semigroup of all transformations in the setting of finite automata. Holzer and König [9], and independently Krawetz, Lawrence and Shallit [12] studied the syntactic complexity of automata with unary and binary alphabets. See also the recent work in [13,23].

The state complexity in prefix-, suffix-, factor-, and subword-closed languages was studied in 2010 by Brzozowski, Jirásková and Zou [6]. A study of state complexity in ideal languages was done in 2010 by Brzozowski, Jirásková and Li [5]. We refer the reader to these papers for a discussion of past work on this topic and additional references. Closed languages are related to ideal languages as follows: A non-empty language is a right (left, two-sided) ideal if and only its complement is a prefix(suffix, factor)-closed language. Since syntactic complexity is preserved under complementation, our proofs use ideals only.

In Section 2 we define our terminology and notation, and some basic properties of syntactic complexity are given in Section 3. The syntactic complexity of right, left, and two-sided ideals is treated in Sections 4–6, some comments about reversal are in Section 7, and Section 8 concludes the paper. For a full version of this paper and omitted proofs see [7].

2 Preliminaries

If Σ is a non-empty finite alphabet, then Σ^* is the free monoid generated by Σ, and Σ^+ is the free semigroup generated by Σ. A *word* is any element of Σ^*, and the empty word is ε. The length of a word $w \in \Sigma^*$ is $|w|$. A *language* over Σ is any subset of Σ^*.

If $w = uxv$ for some $u, v, x \in \Sigma^*$, then u is a *prefix* of w, v is a *suffix* of w, and x is a *factor* of w. A prefix or suffix of w is also a factor of w. A language L is *prefix-closed* if $w \in L$ implies that every prefix of w is also in L. In an analogous way, we define *suffix-closed* and *factor-closed.* A language $L \subseteq \Sigma^*$ is a *right ideal* (respectively, *left ideal, two-sided ideal*) if it is non-empty and satisfies $L = L\Sigma^*$ (respectively, $L = \Sigma^* L$, $L = \Sigma^* L \Sigma^*$). We refer to all three types as *ideal languages* or simply *ideals.*

A *transformation* of a set Q is a mapping of Q *into* itself, whereas a *permutation* of Q is a mapping of Q *onto* itself. In this paper we consider only transformations of finite sets, and we assume without loss of generality that $Q = \{0, 1, \dots, n-1\}$. An arbitrary transformation has the form

$$t = \begin{pmatrix} 0 & 1 & \cdots & n-2 & n-1 \\ i_0 & i_1 & \cdots & i_{n-2} & i_{n-1} \end{pmatrix},$$

where $i_k \in Q$ for $0 \leqslant k \leqslant n-1$. The image of element i under transformation t will be denoted by it. The *identity* transformation maps each element to itself, that is, $it = i$ for $i = 0, \dots, n-1$. A transformation t contains a *cycle* of length k if there exist elements $i_1, \dots, i_k$ such that $i_1 t = i_2, i_2 t = i_3, \dots, i_{k-1} t = i_k, i_k t = i_1$. A cycle is denoted by $(i_1, i_2, \dots, i_k)$. For $i < j$, a *transposition* is the cycle (i, j), and (i, i) is the identity. A *singular* transformation, denoted by $\binom{i}{j}$, has $it = j$, and $ht = h$ for all $h \neq i$, and $\binom{i}{i}$ is the identity. A *constant* transformation, denoted by $\binom{Q}{j}$, has $it = j$ for all i. The following facts are well-known:

Theorem 1. *The complete transformation monoid T_n of size n^n can be generated by any cyclic permutation of n elements together with a transposition and a "returning" transformation $r = \binom{n-1}{0}$. In particular, T_n can be generated by $c = (0, 1, \dots, n-1)$, $t = (0, 1)$ and $r = \binom{n-1}{0}$.*

The *left quotient*, or simply *quotient,* of a language L by a word w is the language $L_w = \{x \in \Sigma^* \mid wx \in L\}$. An equivalence relation $\sim$ on Σ^* is a *left congruence* if, for all $x, y \in \Sigma^*$, $x \sim y \Leftrightarrow ux \sim uy$, for all $u \in \Sigma^*$. It is a *right congruence* if, for all $x, y \in \Sigma^*$, $x \sim y \Leftrightarrow xv \sim yv$, for all $v \in \Sigma^*$.

It is a *congruence* if it is both a left and a right congruence. Equivalently, $\sim$ is a congruence if $x \sim y \Leftrightarrow uxv \sim uyv$, for all $u, v \in \Sigma^*$.

For any language $L \subseteq \Sigma^*$, define the *Nerode congruence* [16] $\rightarrow_L$ of L by

$$x \rightarrow_L y \text{ if and only if } xv \in L \Leftrightarrow yv \in L, \text{ for all } u, v \in \Sigma^*. \quad (1)$$

Evidently, $L_x = L_y$ if and only if $x \rightarrow_L y$. Thus, each equivalence class of this congruence corresponds to a distinct quotient of L.

The *Myhill congruence* [15] $\leftrightarrow_L$ of L is defined by

$$x \leftrightarrow_L y \text{ if and only if } uxv \in L \Leftrightarrow uyv \in L \text{ for all } u, v \in \Sigma^*. \quad (2)$$

This congruence is also known as the *syntactic congruence* of L. The semigroup $\Sigma^+/\leftrightarrow_L$ of equivalence classes of the relation $\leftrightarrow_L$, is the *syntactic semigroup* of L, and $\Sigma^*/\leftrightarrow_L$ is the *syntactic monoid* of L. The *syntactic complexity* $\sigma(L)$ of L is the cardinality of its syntactic semigroup. The *monoid complexity* $\mu(L)$ of L is the cardinality of its syntactic monoid. If the equivalence class of ε is a singleton in the syntactic monoid, then $\sigma(L) = \mu(L) - 1$; otherwise, $\sigma(L) = \mu(L)$.

A *(deterministic) semiautomaton* is a triple, $\mathcal{S} = (Q, \Sigma, \delta)$, where Q is a finite, non-empty set of *states*, Σ is a finite non-empty *alphabet*, and $\delta : Q \times \Sigma \to Q$ is the *transition function*. A *deterministic finite automaton (DFA)* is a quintuple $\mathcal{A} = (Q, \Sigma, \delta, q_0, F)$, where Q, Σ, and δ are as defined in the semiautomaton $\mathcal{S} = (Q, \Sigma, \delta)$, $q_0 \in Q$ is the *initial state*, and $F \subseteq Q$ is the set of *final states*. By the *language of a state* q of $\mathcal{A}$ we mean the language $L(\mathcal{A}_q)$ accepted by the automaton $\mathcal{A}_q = (Q, \Sigma, \delta, q, F)$. States p and q are *equivalent* if $L(\mathcal{A}_p) = L(\mathcal{A}_q)$. A DFA is *minimal* if every state is reachable from the initial state, and no two states are equivalent.

The *ε-function* L^ε of a regular language L is $L^\varepsilon = \emptyset$ if $\varepsilon \notin L$; $L^\varepsilon = \{\varepsilon\}$ if $\varepsilon \in L$. The *quotient automaton* of a regular language L is $\mathcal{A} = (Q, \Sigma, \delta, q_0, F)$, where $Q = \{L_w \mid w \in \Sigma^*\}$, $\delta(L_w, a) = L_{wa}$, $q_0 = L_\varepsilon = L$, $F = \{L_w \mid L_w^\varepsilon = \{\varepsilon\}\}$, and $L_w^\varepsilon = (L_w)^\varepsilon$. The number of states in the quotient automaton of L is the *quotient complexity* $\kappa(L)$ of L. The quotient complexity is the same as the state complexity, but there are advantages to using quotients [4].

In terms of automata, each equivalence class $[w]_{\rightarrow_L}$ is the set of words w that take the automaton to the same state from the initial state. In terms of quotients, it is the set of words w that can all be followed by the same quotient L_w. In terms of automata, each equivalence class $[w]_{\leftrightarrow_L}$ of the syntactic congruence is the set of all words that perform the same transformation on the set of states.

3 Basic Properties of Syntactic Complexity

The *transformation semigroup* of an automaton is the set of transformations performed by words of Σ^+ on the set of states. The transformation semigroup of the quotient automaton of L is isomorphic to the syntactic semigroup of L.

Proposition 1. *For any* $L \subseteq \Sigma^*$ *with* $\kappa(L) = n > 1$, $n - 1 \leqslant \sigma(L) \leqslant n^n$.

Proof. Let $\mathcal{A} = (Q, \Sigma, \delta, q_0, F)$ be the quotient automaton of L. Since every state other than q_0 has to be reachable from the initial state by a non-empty word, there must be at least $n-1$ transformations. If $\Sigma = \{a\}$ and $L = a^{n-1}a^*$, then $\kappa(L) = n$, and $\sigma(L) = n-1$; so the lower bound $n-1$ is achievable. The upper bound is n^n, and by Theorem 1 this upper bound is achievable if $|\Sigma| \geqslant 3$. □

If one of the quotients of L is $\emptyset$ (respectively, $\{\varepsilon\}$, Σ^*, Σ^+), then we say that *L has $\emptyset$* (respectively, $\{\varepsilon\}$, Σ^*, Σ^+). A quotient L_w of a language L is *uniquely reachable (ur)* [4] if $L_x = L_w$ implies that $x = w$. If L_{wa} is uniquely reachable for $a \in \Sigma$, then so is L_w. Thus, if L has a uniquely reachable quotient, then L itself is uniquely reachable by ε, *i.e.*, the minimal automaton of L is *non-returning.*

Theorem 2 (Special Quotients). *Let $L \subseteq \Sigma^*$ and let $\kappa(L) = n \geqslant 1$.*
1. If L has $\emptyset$ or Σ^*, then $\sigma(L) \leqslant n^{n-1}$.
2. If L has $\{\varepsilon\}$ or Σ^+, then $\sigma(L) \leqslant n^{n-2}$.
3. If L is uniquely reachable, then $\sigma(L) \leqslant (n-1)^n$.
4. If L_a is uniquely reachable for some $a \in \Sigma$, then $\sigma(L) \leqslant 1 + (n-2)^n$.
Moreover, all the bounds shown in Table 1 hold.

Table 1. Upper bounds on syntactic complexity for languages with special quotients

$\emptyset$	Σ^*	$\{\varepsilon\}$	Σ^+		L is ur	L_a is ur
√				n^{n-1}	$(n-1)^{n-1}$	$1+(n-3)^{n-2}$
	√			n^{n-1}	$(n-1)^{n-1}$	$1+(n-3)^{n-2}$
√		√		n^{n-2}	$(n-1)^{n-2}$	$1+(n-4)^{n-2}$
	√		√	n^{n-2}	$(n-1)^{n-2}$	$1+(n-4)^{n-2}$
√	√			n^{n-2}	$(n-1)^{n-2}$	$1+(n-4)^{n-2}$
√	√		√	n^{n-3}	$(n-1)^{n-3}$	$1+(n-5)^{n-2}$
√	√	√		n^{n-3}	$(n-1)^{n-3}$	$1+(n-5)^{n-2}$
√	√	√	√	n^{n-4}	$(n-1)^{n-4}$	$1+(n-6)^{n-2}$

Proof. Suppose that $L \subseteq \Sigma^*$, $n \geqslant 1$, and $\kappa(L) = n$.

1. Since $\emptyset_a = \emptyset$ for all $a \in \Sigma$, there are only $n-1$ states in the quotient automaton with which one can distinguish two transformations. Hence there are at most n^{n-1} transformations. If L has Σ^*, then $(\Sigma^*)_a = \Sigma^*$, for all $a \in \Sigma$, and the same argument applies.

2. Since $\{\varepsilon\}_a = \emptyset$ for all $a \in \Sigma$, L has $\emptyset$ if L has $\{\varepsilon\}$. Now there are two states that do not contribute to distinguishing among different transformations. Dually, $(\Sigma^+)_a = \Sigma^*$ for all $a \in \Sigma$, and the same argument applies.

3. If L is uniquely reachable then $L_w = L$ implies $w = \varepsilon$. Thus L does not appear in the image of any transformation by a word in Σ^+, and there remain only $n-1$ choices for each of the n states.

4. If L_a is uniquely reachable, then so is L. Hence L never appears and L_a appears only in one transformation. Therefore there can be at most $(n-2)^n$ other transformations.

The remaining entries in Table 1 are easily verified. □

4 Right Ideals and Prefix-Closed Languages

In this section we characterize the syntactic complexity of right ideals. The DFA of Fig. 2 plays an important role here. For $n \geqslant 4$, let $\mathcal{A}_n = (Q_n, \Sigma, \delta, 0, \{n-1\})$, where $Q_n = \{0, 1, \dots, n-1\}$, $\Sigma = \{a, b, c, d\}$, $a = (0, 1, \dots, n-2)$, $b = (0, 1)$, $c = \binom{n-2}{0}$, and $d = \binom{n-2}{n-1}$. This DFA accepts a right ideal and is minimal.

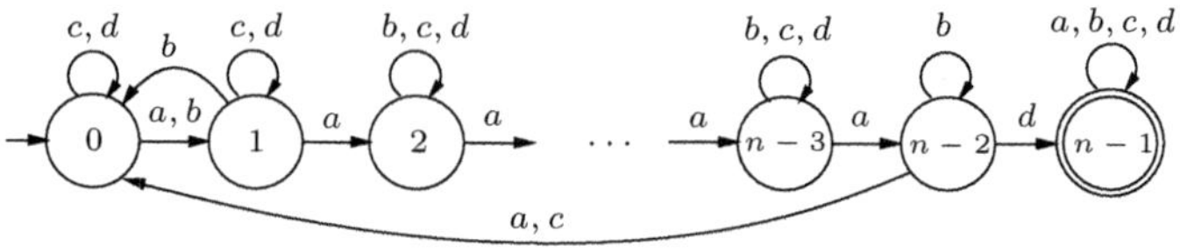

Fig. 2. Automaton $\mathcal{A}_n$ of a right ideal with n^{n-1} transformations

Theorem 3 (Right Ideals and Prefix-Closed Languages). *Suppose that $L \subseteq \Sigma^*$ and $\kappa(L) = n$. If L is a right ideal or a prefix-closed language, then $\sigma(L) \leqslant n^{n-1}$. Moreover, this bound is tight for $n = 1$ if $|\Sigma| \geqslant 1$, for $n = 2$ if $|\Sigma| \geqslant 2$, for $n = 3$ if $|\Sigma| \geqslant 3$, and for $n \geqslant 4$ if $|\Sigma| \geqslant 4$.*

Proof. If L is a right ideal, it has Σ^* as a quotient. By Theorem 2, $\sigma(L) \leqslant n^{n-1}$.

For $n = 1$, $n = 2$, and $n = 3$, the languages a^*, $b^*a\{a, b\}^*$ and $\mathcal{A}_3$ restricted to alphabet $\{a, c, d\}$, respectively, meet the bound.

Next we prove that the language $L(\mathcal{A}_n)$ accepted by the automaton of Fig. 2 meets this bound for $n \geqslant 4$. Consider any transformation t of the form

$$t = \begin{pmatrix} 0 & 1 & 2 & \cdots & n-3 & n-2 & n-1 \\ i_0 & i_1 & i_2 & \cdots & i_{n-3} & i_{n-2} & n-1 \end{pmatrix},$$

where $i_k \in \{0, 1, \dots, n-1\}$ for $0 \leqslant k \leqslant n-2$. There are two cases:

1. Suppose $i_k \neq n-1$ for all k, $0 \leqslant k \leqslant n-2$. By Theorem 1, since all the images of the first $n-1$ states are in $\{0, 1, \dots, n-2\}$, $\mathcal{A}_n$ can do t.
2. If $i_h = n-1$ for some h, $0 \leqslant h \leqslant n-2$, then there exists some j, $0 \leqslant j \leqslant n-2$ such that $i_k \neq j$ for all k, $0 \leqslant k \leqslant n-2$. Define i'_k for all $0 \leqslant k \leqslant n-2$ as follows: $i'_k = j$ if $i_k = n-1$, and $i'_k = i_k$ if $i_k \neq n-1$. Then let

$$s = \begin{pmatrix} 0 & 1 & 2 & \cdots & n-3 & n-2 & n-1 \\ i'_0 & i'_1 & i'_2 & \cdots & i'_{n-3} & i'_{n-2} & n-1 \end{pmatrix}.$$

Also, let $r = (j, n-2)$. Since all the images of the first $n-1$ states in s and r are in $\{0, 1, \dots, n-2\}$, by Theorem 1, s and r can be performed by $\mathcal{A}_n$.

We show now that $t = srdr$, which implies that t can also be performed by $\mathcal{A}_n$. If $kt = n-1$, then $ks = j$, $jr = n-2$, $(n-2)d = n-1$, and $(n-1)r = n-1$. If $kt = n-2$, then $n-2 \neq j$. Now $ks = n-2$, $(n-2)r = j$, $jd = j$, and $jr = n-2$. If $kt = i_k < n-2$, then also $k(srdr) = i_k$. In all cases $t = srdr$.

Since there are n^{n-1} transformations like t, $L(\mathcal{A}_n)$ meets the bound. □

Table 2. Syntactic complexity bounds for right ideals

	$n=1$	$n=2$	$n=3$	$n=4$	$n=5$	...	$n=n$
$\|\Sigma\|=1$	**1**	**1**	**2**	**3**	**4**	...	$\mathbf{n-1}$
$\|\Sigma\|=2$	–	**2**	**7**	**31**	**167**	...	?
$\|\Sigma\|=3$	–	–	**9**	**61**	**545**	...	?
$\|\Sigma\|=4$	–	–	–	**64**	**625**	...	$\mathbf{n^{n-1}}$

Table 2 summarizes our result for right ideals. All the bounds are tight. The bounds for $n \leqslant 5$ have all been verified by a computer program.

5 Left Ideals and Suffix-Closed Languages

We provide strong support for the following conjecture:

Conjecture 1 (Left Ideals and Suffix-Closed Languages). *If L is a left ideal or a suffix-closed language with $\kappa(L) = n \geqslant 1$, then $\sigma(L) \leqslant n^{n-1} + n - 1$.*

We show that the bound can be reached, but first we recall a result from [19]. Consider a semiautomaton $\mathcal{S} = (P \cup \{0\}, \Sigma, \delta)$, where 0 is a sink state, meaning that $\delta(0, a) = 0$ for all $a \in \Sigma$, and P is strongly connected. Such a semiautomaton is *uniformly minimal* if the automaton $\mathcal{A} = (P \cup \{0\}, \Sigma, \delta, q_0, F)$ is minimal for every $q_0 \in P$ and every F such that $\emptyset \neq F \subseteq P$. One can test whether a semiautomaton is uniformly minimal by the method of [19].

Let $n \geqslant 3$, and $\mathcal{S}_n = (Q, \Sigma, \delta)$, where $Q = \{0, \dots, n-1\}$, $\Sigma = \{a, b, c, d, e\}$, $a = (1, 2, \dots, n-1)$, $b = (1, 2)$, $c = \binom{n-1}{1}$, $d = \binom{n-1}{0}$, and $e = \binom{Q}{1}$; see Fig. 3. For $n = 3$, a and b coincide; then we use $\Sigma = \{b, c, d, e\}$. Let $\Sigma' = \Sigma \setminus \{e\}$ and let $\mathcal{R}_n$ be the semiautomaton $\mathcal{R}_n = (Q, \Sigma', \delta')$, where $Q = P \cup \{0\}$, $P = \{1, \dots, n-1\}$, and δ' is the restriction of δ to $Q \times \Sigma'$. Note that 0 is a sink state of $\mathcal{R}_n$.

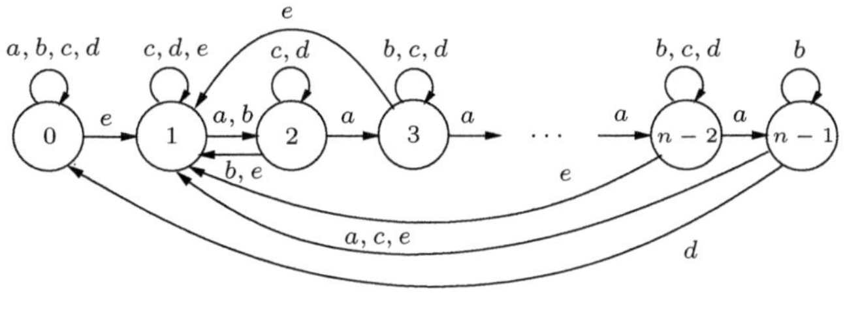

Fig. 3. Semiautomaton $\mathcal{S}_n$ with $n^{n-1} + n - 1$ transformations

Lemma 1. *The set P is strongly connected and $\mathcal{R}_n$ is uniformly minimal.*

Proof. Since a is a cycle of the states in P, $\mathcal{R}_n$ is strongly connected. It can be shown that $\mathcal{R}_n$ is uniformly minimal by using a state-pair graph as in [19]. □

Theorem 4 (Left Ideals and Suffix-Closed Languages). *For $n \geqslant 3$, let $\mathcal{B}_n = (Q, \Sigma, \delta, 0, F)$, where $(Q, \Sigma, \delta) = \mathcal{S}_n$ of Fig. 3, and F is any non-empty subset of $Q \setminus \{0\}$. Then $\mathcal{B}_n$ is minimal, and the language $L = L(\mathcal{B}_n)$ accepted by $\mathcal{B}_n$ is a left ideal and has syntactic complexity $\sigma(L) = n^{n-1} + n - 1$.*

Proof. Since semiautomaton $\mathcal{R}_n$ is uniformly minimal, automaton $\mathcal{B}_n$ is minimal for every choice of F. Hence L has n quotients.

To prove that L is a left ideal it suffices to show that, for any $w \in L$, we also have $hw \in L$ for every $h \in \Sigma$. This is obvious if $h \in \Sigma \setminus \{e\}$, since all transitions from state 0 under h lead to state 0. If $w \in L$, then w has the form $w = uev$, where $\delta(0, u) = 0$, $\delta(0, ue) = 1$, and $v \in L_e$. But $\delta(0, eue) = 1$, since $\delta(i, eue) = 1$ for all $i \in Q$, and $v \in L_e$ gives us $euev = ew \in L$. Thus L is a left ideal.

Consider any transformation t of the form

$$t = \begin{pmatrix} 0 & 1 & 2 & \cdots & n-3 & n-2 & n-1 \\ 0 & i_1 & i_2 & \cdots & i_{n-3} & i_{n-2} & i_{n-1} \end{pmatrix},$$

where $i_k \in \{0, \dots, n-1\}$ for $1 \leqslant k \leqslant n-1$; there are n^{n-1} such transformations. We have two cases:
1. If $i_k \neq 0$ for all k, $1 \leqslant k \leqslant n-1$, then all the images of the last $n-1$ states are in the set $\{1, \dots, n-1\}$. By Theorem 1, t can be performed by $\mathcal{B}_n$.
2. If $i_h = 0$ for some h, $1 \leqslant h \leqslant n-1$, then there exists some j, $1 \leqslant j \leqslant n-1$ such that $i_k \neq j$ for all k, $1 \leqslant k \leqslant n-1$. Define i'_k for all $1 \leqslant k \leqslant n-1$ as follows: $i'_k = j$ if $i_k = 0$, and $i'_k = i_k$, otherwise. Let

$$s = \begin{pmatrix} 0 & 1 & 2 & \cdots & n-3 & n-2 & n-1 \\ 0 & i'_1 & i'_2 & \cdots & i'_{n-3} & i'_{n-2} & i'_{n-1} \end{pmatrix},$$

and $r = (j, n-1)$. By Theorem 1, s and r can be performed by $\mathcal{B}_n$.

Now consider $srdr$. If $kt = 0$, then $ks = j$, $jr = n-1$, $(n-1)d = 0$, and $0r = 0$. If $kt = n-1$, then $ks = n-1$, $(n-1)r = j$, $jd = j$, and $jr = n-1$. Finally, if kt is a state other than 0 or $n-1$, then $srdr$ maps k to that same state. Hence we have $t = srdr$, and t can be performed by $\mathcal{B}_n$ as well.

Now consider any transformation $t = \binom{Q}{j}$ that maps all the states to some state $j \neq 0$; there are $n-1$ such transformations. We have two cases:
1. If $j = 1$, then $t = e$; therefore t can be performed by $\mathcal{B}_n$.
2. Otherwise, let $s = (1, j)$. By Theorem 1, s can be performed by $\mathcal{B}_n$. Since $t = es$, t can also be performed by $\mathcal{B}_n$ as well.

There are no other transformations, since e maps all the states to 1. □

Before considering the cases $n \leqslant 3$, we require some auxiliary results. Let $\mathcal{A} = (Q, \Sigma, \delta, q_0, F)$ be the quotient automaton of a left ideal. For every word $w \in \Sigma^*$, consider the sequence $q_0 = p_0, p_1, p_2 \dots$ of states obtained by applying powers of w to the initial state q_0, that is, let $p_i = \delta(q_0, w^i)$. Since $\mathcal{A}$ has n states, we must eventually have a repeated state in that sequence, that is, we must have some i and $j > i$ such that $p_0, p_1, \dots, p_i, p_{i+1}, \dots p_{j-1}$ are distinct and $p_j = p_i$. The sequence $q_0 = p_0, p_1, \dots, p_i, p_{i+1}, \dots p_{j-1}$ of states with $p_j = p_i$ is called the

Table 3. Syntactic complexity bounds for left ideals

	$n = 1$	$n = 2$	$n = 3$	$n = 4$	$n = 5$		$n = n$
$\|\Sigma\| = 1$	**1**	**1**	**2**	**3**	**4**	...	$\mathbf{n-1}$
$\|\Sigma\| = 2$	–	**2**	**7**	17	34	...	?
$\|\Sigma\| = 3$	–	**3**	**9**	25	65	...	?
$\|\Sigma\| = 4$	–	–	**11**	64	453	...	?
$\|\Sigma\| = 5$	–	–	–	67	629	...	$n^{n-1} + n - 1$

behavior of w *on* $\mathcal{A}$, and the integer $j - i$ is the *period* of that behavior. We will use the notation $\langle p_0, p_1, \ldots, p_i, p_{i+1}, \ldots p_{j-1}; p_j = p_i \rangle$ for such behaviors. If the period of w is 1, then its behavior is *aperiodic*; otherwise, it is *periodic*.

Lemma 2. *If $\mathcal{A}$ is the quotient automaton of a left ideal L, then the behavior of every word $w \in \Sigma^*$ is aperiodic. Moreover, L does not have the empty quotient.*

Proof. Suppose that w has the behavior $\langle q_0 = p_0, p_1, \ldots, p_i, p_{i+1}, \ldots p_{j-1}; p_j = p_i \rangle$, where $j - i \geqslant 2$; then $j - 1 \geqslant i + 1$. Since $\mathcal{A}$ is minimal, states p_i and p_{j-1} must be distinguishable, say by word $x \in \Sigma^*$. If $w^i x \in L$, then $w^{j-1}x = w^i w^{j-i-1} x = w^{j-i-1}(w^i x) \notin L$, contradicting the assumption that L is a left ideal. If $w^{j-1}x \in L$, then $w^j x = w(w^{j-1}x) \notin L$, again a contradiction.

For the second claim, a left ideal is non-empty by definition. If $w \in L$ and L has the empty quotient, say $L_x = \emptyset$, then $xw \notin L$, which is a contradiction. □

Theorem 5 (Left Ideals and Suffix-Closed Languages for n $\leqslant$ 3). *If $1 \leqslant n \leqslant 3$ and L is a left ideal or a suffix-closed language with $\kappa(L) = n$, then $\sigma(L) \leqslant n^{n-1} + n - 1$. Moreover, the bound is tight for $n = 1$ if $|\Sigma| \geqslant 1$, for $n = 2$ if $|\Sigma| \geqslant 3$, and for $n = 3$ if $|\Sigma| \geqslant 4$.*

Table 3 summarizes our results. The figures in bold type are tight bounds, verified by a computer program.

6 Two-Sided Ideals and Factor-Closed Languages

Conjecture 2 (Two-Sided Ideals and Factor-Closed Languages). *If L is a two-sided ideal or a factor-closed language with $\kappa(L) = n \geqslant 2$, then $\sigma(L) \leqslant n^{n-2} + (n-2)2^{n-2} + 1$.*

We show that this complexity can be met. For $n = 2$ and $\Sigma = \{a, b\}$, $\Sigma^* a \Sigma^*$ meets the bound. For $n = 3$ and $\Sigma = \{a, b, c\}$, $(b + c + ac^*b)^* ac^* a \Sigma^*$ meets the bound. Now let $n \geqslant 4$, and let $\mathcal{C}_n = (Q, \Sigma, \delta, 0, \{n-1\})$, where $Q = \{0, \ldots, n-1\}$, $\Sigma = \{a, b, c, d, e, f\}$, $a = (1, 2, \ldots, n-2)$, $b = (1, 2)$, $c = \binom{n-2}{1}$, $d = \binom{n-2}{0}$, $\delta(i, e) = 1$ for $i = 0, \ldots, n-2$ and $\delta(n-1, e) = n-1$, and $f = \binom{1}{n-1}$. For $n = 4$, a and b coincide.

Theorem 6 (Two-Sided Ideals and Factor-Closed Languages). *DFA $\mathcal{C}_n$ is minimal and $L = L(\mathcal{C}_n)$ is a two-sided ideal with $\sigma(L) = n^{n-2} + (n-2)2^{n-2} + 1$.*

Proof. For $i = 1, \dots, n-2$, state i is the only non-final state that accepts $a^{n-1-i}f$; hence all these states are distinguishable. State 0 is distinguishable from these states because it does not accept any words in a^*f. Hence $\mathcal{C}_n$ is minimal. The proof that $\mathcal{C}_n$ is a left ideal is like that in Theorem 4. Since $L_{ef} = \Sigma^*$ is the only accepting quotient, L is a right ideal. Hence it is two-sided.

First consider any transformation t of the form

$$t = \begin{pmatrix} 0 & 1 & 2 & \cdots & n-3 & n-2 & n-1 \\ 0 & i_1 & i_2 & \cdots & i_{n-3} & i_{n-2} & n-1 \end{pmatrix},$$

where $i_k \in \{0, 1, 2, \dots, n-2, n-1\}$ for $1 \leqslant k \leqslant n-2$; there are n^{n-2} such transformations. We have two cases:

1. If $i_k \neq n-1$ for all k, $1 \leqslant k \leqslant n-2$, then all the images of the first $n-2$ states are in the set $\{0, \dots, n-2\}$. Without input f and state $n-1$ we have the semiautomaton of Theorem 4. By that theorem, t can be done by $\mathcal{C}_n$.

2. If $i_h = n-1$ for some h, $1 \leqslant h \leqslant n-2$, then there exists some j, $1 \leqslant j \leqslant n-2$ such that $i_k \neq j$ for all k, $1 \leqslant k \leqslant n-2$. Define i'_k for all $1 \leqslant k \leqslant n-2$ as follows: $i'_k = j$ if $i_k = n-1$, and $i'_k = i_k$ if $i_k \neq n-1$. Let

$$s = \begin{pmatrix} 0 & 1 & 2 & \cdots & n-3 & n-2 & n-1 \\ 0 & i'_1 & i'_2 & \cdots & i'_{n-3} & i'_{n-2} & n-1 \end{pmatrix},$$

and let $r = (1, j)$. By Theorem 4, s and r can be performed by $\mathcal{C}_n$.

Now consider $srfr$. If $kt = n-1$, then $ks = j$, $jr = 1$, $1f = n-1$, and $(n-1)r = n-1$. If $kt = 1$, then $ks = 1$, $1r = j$, $jf = j$, and $jr = 1$. Finally, if t maps k to a state other than 1 or $n-1$, then $srfr$ maps k to the same state. Hence we have $t = srfr$, and t can be performed by $\mathcal{C}_n$ as well.

Next, refer to states in $\{1, \dots, n-2\}$ as the *middle* states. Take any transformation t that maps 0 to $h \in \{1, \dots, n-2\}$, and every middle state to either $\{n-1\}$ or to h. There are $(n-2)2^{n-2}$ such transformations. First consider any middle entry i that is mapped to $n-1$ by t. We can map i to $n-1$ without changing any other states. First, apply a^{n-1-i} to "rotate all the middle states clockwise", so that i is mapped to 1. Then apply f to map i to $n-1$, and then a^i to return all the states other than $n-1$ to their original positions. This is repeated for all the middle states that are mapped to $n-1$ by t. After this is done, apply e to replace all the remaining middle states by 1, and apply a^{h-1} to change 1 to h. Hence t can be done.

Finally, the constant transformation $\binom{Q}{n-1}$ is done by ef.

In summary, if $L = L(\mathcal{C}_n)$ then $\sigma(L) \geqslant n^{n-2} + (n-2)2^{n-2} + 1$.

If 0 is mapped to a middle state i, then the input word must contain an e. But every word of the form xe leaves the automaton in a state in $\{1, n-1\}$. Applying any other word can only result in a state in $\{i, n-1\}$, for some middle state i. Hence no transformations other than the ones we have considered can be done by $\mathcal{C}_n$, and the syntactic complexity of the language accepted by $\mathcal{C}_n$ is precisely $n^{n-2} + (n-2)2^{n-2} + 1$. □

Table 4 summarizes our results for two-sided ideals.

Table 4. Syntactic complexity bounds for two-sided ideals

	$n=1$	$n=2$	$n=3$	$n=4$	$n=5$		$n=n$
$\|\Sigma\|=1$	**1**	**1**	**2**	**3**	**4**	...	$\mathbf{n-1}$
$\|\Sigma\|=2$	–	**2**	5	11	19	...	?
$\|\Sigma\|=3$	–	–	6	16	47	...	?
$\|\Sigma\|=4$	–	–	–	23	90	...	?
$\|\Sigma\|=5$	–	–	–	25	147	...	?
$\|\Sigma\|=6$	–	–	–	–	150	...	$n^{n-2}+(n-2)2^{n-2}+1$

7 Reversal

It is interesting to note that, for our ideals with maximal syntactic complexity, the reverse languages have maximal state complexity. It was shown in [5] that the reverse of a right (left, two-sided) ideal with n quotients has at most 2^{n-1} ($2^{n-1}+1$, $2^{n-2}+1$) quotients, and that these bounds can be met.

Theorem 7. *If $L(\mathcal{A}_n)$, $L(\mathcal{B}_n)$ and $L(\mathcal{C}_n)$ are the languages in Theorems 3, 4, and 6, then their reverses have 2^{n-1}, $2^{n-1}+1$, and $2^{n-2}+1$ quotients, respectively.*

8 Conclusions

Despite the fact that the Myhill congruence has left-right symmetry, there are significant differences between left and right ideals. The major open problem is the upper bound for left ideals. Also, the relation between syntactic complexity and reversal deserves further study.

Acknowledgment. We thank Arto Salomaa for providing some early references.

References

1. Bozapalidis, S., Kalampakas, A.: Recognizability of graph and pattern languages. Acta Inform. 42(8/9), 553–581 (2006)
2. Bozapalidis, S., Kalampakas, A.: On the complexity of the syntax of tree languages. In: Bozapalidis, S., Rahonis, G. (eds.) CAI 2009. LNCS, vol. 5725, pp. 189–203. Springer, Heidelberg (2009)
3. Bozapalidis, S., Kalampakas, A.: On the syntactic complexity of tree series. RAIRO-Theor. Inf. Appl. 44, 257–279 (2010)
4. Brzozowski, J.: Quotient complexity of regular languages. In: Dassow, J., Pighizzini, G., Truthe, B. (eds.) Proceedings of the 11th International Workshop on Descriptional Complexity of Formal Systems, DCFS, pp. 25–42 (2009), Extended abstract at http://arxiv.org/abs/0907.4547
5. Brzozowski, J., Jirásková, G., Li, B.: Quotient complexity of ideal languages. In: López-Ortiz, A. (ed.) LATIN 2010. LNCS, vol. 6034, pp. 208–221. Springer, Heidelberg (2010), Full paper at http://arxiv.org/abs/0908.2083

6. Brzozowski, J., Jirásková, G., Zou, C.: Quotient complexity of closed languages. In: Ablayev, F., Mayr, E.W. (eds.) CSR 2010. LNCS, vol. 6072, pp. 84–95. Springer, Heidelberg (2010)
7. Brzozowski, J., Ye, Y.: Syntactic complexity of ideal and closed languages (October 2010), `http://arxiv.org/abs/1010.3263`
8. Dénes, J.: On transformations, transformation semigroups and graphs. In: Erdös, P., Katona, G. (eds.) Theory of Graphs. Proceedings of the Colloquium on Graph Theory held at Tihany, 1966, pp. 65–75. Akadémiai Kiado (1968)
9. Holzer, M., König, B.: On deterministic finite automata and syntactic monoid size. Theoret. Comput. Sci. 327, 319–347 (2004)
10. Hoyer, M.: Verallgemeinerung zweier sätze aus der theorie der substitutionengruppen. Math. Ann. 46(4), 539–544 (1895)
11. Kalampakas, A.: The syntactic complexity of eulerian graphs. In: Bozapalidis, S., Rahonis, G. (eds.) CAI 2007. LNCS, vol. 4728, pp. 208–217. Springer, Heidelberg (2007)
12. Krawetz, B., Lawrence, J., Shallit, J.: State complexity and the monoid of transformations of a finite set (2003), `http://arxiv.org/abs/math/0306416`
13. Krawetz, B., Lawrence, J., Shallit, J.: State complexity and the monoid of transformations of a finite set. Internat. J. Found. Comput. Sci. 16(3), 547–563 (2005)
14. Maslov, A.N.: Estimates of the number of states of finite automata. Dokl. Akad. Nauk SSSR 194, 1266–1268 (1970) (Russian); English translation: Soviet Math. Dokl. 11 , 1373–1375 (1970)
15. Myhill, J.: Finite automata and representation of events. Wright Air Development Center Technical Report 57–624 (1957)
16. Nerode, A.: Linear automaton transformations. Proc. Amer. Math. Soc. 9, 541–544 (1958)
17. Piccard, S.: Sur les fonctions définies dans les ensembles finis quelconques. Fund. Math. 24, 298–301 (1935)
18. Piccard, S.: Sur les bases du groupe symétrique et du groupe alternant. Commentarii Mathematici Helvetici 11(1), 1–8 (1938)
19. Restivo, A., Vaglica, R.: Automata with extremal minimality conditions. In: Gao, Y., Seki, S., Yu, S. (eds.) DLT 2010. LNCS, vol. 6224, pp. 399–410. Springer, Heidelberg (2010)
20. Salomaa, A.: A theorem concerning the composition of functions of several variables ranging over a finite set. J. Symbolic Logic 25, 203–208 (1960)
21. Salomaa, A.: Some completeness criteria for sets of functions over a finite domain. Ann. Univ. Turkuensis, Ser. AI 53 (1962)
22. Salomaa, A.: On basic groups for the set of functions over a finite domain. Ann. Acad. Scient. Fenn., Ser. A 338 (1963)
23. Salomaa, A.: Composition sequences for functions over a finite domain. Theoret. Comput. Sci. 292, 263–281 (2003)
24. Sierpiński, X.: Sur les suites infinies de fonctions définies dans les ensembles quelconques. Fund. Math. 24, 209–212 (1935)
25. Yu, S.: State complexity of regular languages. J. Autom. Lang. Comb. 6, 221–234 (2001)

Generalized One-Unambiguity

Pascal Caron[1], Yo-Sub Han[2,*], and Ludovic Mignot[1]

[1] LITIS, Université de Rouen, 76801 Saint-Étienne du Rouvray Cedex, France
{pascal.caron,ludovic.mignot}@univ-rouen.fr
[2] Dept. of Computer Science, Yonsei University, Seoul 120-749, Republic of Korea
emmous@cs.yonsei.ac.kr

Abstract. Brüggemann-Klein and Wood have introduced a new family of regular languages, the *one-unambiguous regular languages*, a very important notion in XML DTDs. A regular language L is one-unambiguous if and only if there exists a regular expression E over the operators of sum, catenation and Kleene star such that $L(E) = L$ and the position automaton of E is deterministic. It implies that for a one-unambiguous expression, there exists an equivalent linear-size deterministic recognizer. In this paper, we extend the notion of one-unambiguity to weak one-unambiguity over regular expressions using the complement operator $\neg$. We show that a DFA with at most $(n + 2)$ states can be computed from a weakly one-unambiguous expression and that it is decidable whether or not a given DFA recognizes a weakly one-unambiguous language.

1 Introduction

Regular expressions are basic tools for numerous domains such as pattern-matching or electronic document specification. A regular expression is a compact representation for a set of words. A recurrent question is the membership problem which is to decide whether or not a word belongs to the language denoted by an expression. This problem can be solved by computing a recognizer called automaton from a regular expression. One of the best-known automata construction is the position construction [6,9]. If a regular expression E has n occurrences of symbols, then the corresponding position automaton, which is not necessarily deterministic, has exactly $(n+1)$ states. There always exists a deterministic recognizer equivalent to the position automaton but its size can be exponentially larger.

Brüggemann-Klein and Wood [3] introduced a subfamily of regular languages called *one-unambiguous regular languages*: A regular language is one-unambiguous if and only if there exists an equivalent regular expression the position automaton of which is deterministic. XML DTDs are specified by extended context-free grammars in which the right-hand side of the productions (content models) are one-unambiguous [2]. These content models allow efficient compiling and testing. It turned out that one-unambiguity is very important in XML DTDs. Therefore it

* Han was supported by the Basic Science Research Program through NRF funded by MEST (2010-0009168).

G. Mauri and A. Leporati (Eds.): DLT 2011, LNCS 6795, pp. 129–140, 2011.

is important to investigate the properties of one-unambiguous regular languages. One-unambiguous regular languages are strictly included into regular languages. We want to expand the size of the family of expressions for which there exists a method to compute a deterministic linear-size recognizer. In this purpose, we study generalized expressions since they often enable us to write more compact regular expressions. For instance, Gelade and Neven [4,5] have shown that the size of the smallest regular expression without intersection operators for the intersection of two languages denoted by simple regular expressions can be exponential.

In Section 2, we define some basic notions. We demonstrate in Section 3 that one-unambiguous regular languages are closed neither under intersection nor under complement. Note that Brüggemann-Klein and Wood [3] have shown that one-unambiguous regular languages are closed neither under union, catenation nor under Kleene star. A new family of regular languages closed under complement and containing the one-unambiguous regular languages is then introduced in Section 4. We define weakly one-unambiguous regular expressions and show that a linear-size recognizer can be computed from such an expression. We then structurally characterize the minimal DFA of weakly one-unambiguous regular languages.

2 Preliminaries

A *deterministic finite automaton* (DFA) $A = (\Sigma, Q, i, F, \delta)$ is a 5-tuple defined by Σ a finite set of symbols called the *alphabet*, Q a finite set of *states*, $i \in Q$ the *initial state*, $F \subset Q$ the set of *final states* and $\delta : Q \times \Sigma \to Q$ the *transition function*. The function δ is equivalent to the set defined by: $(q, a, q') \subset \delta$ if and only if $q' = \delta(q, a)$. The function δ is extended to $2^Q \times \Sigma^* \to 2^Q$ as follows: $\forall P \subset Q, \forall a \in \Sigma, \forall w \in \Sigma^*$, $\delta(P, \varepsilon) = P$, $\delta(P, a) = \bigcup_{p \in P} \delta(p, a)$ and $\delta(P, a \cdot w) = \delta(\delta(P, a), w)$. The automaton A *recognizes* the language $L(A) = \{w \in \Sigma^* \mid \delta(i, w) \in F\}$. The *set of recognizable languages* defined by $\{L \mid \exists A \text{ a DFA}, L(A) = L\}$ is written $\mathrm{Rec}(\Sigma^*)$. A DFA is *complete* if $\forall (q, a) \in Q \times \Sigma$, $\mathrm{Card}(\delta(q, a)) = 1$. For every DFA A, there exists an equivalent complete DFA A' such that $L(A') = L(A)$ [13]. The *left language* (resp. *right language*) of a state q of A is the set of words $L_q^{\leftarrow}(A) = \{w \in \Sigma^* \mid \delta(i, w) = q\}$ (resp. $L_q^{\rightarrow}(A) = \{w \in \Sigma^* \mid \delta(q, w) \in F\}$). A state q in Q is *accessible* (resp. *coaccessible*) if and only if $L_q^{\leftarrow}(A) \neq \emptyset$ (resp. $L_q^{\rightarrow}(A) \neq \emptyset$). We say that q is a *sink* state if $L_q^{\rightarrow}(A) = \emptyset$. The automaton A is *trim* if all the states of A are accessible and coaccessible. Two states q and q' are *equivalent* with respect to the Myhill-Nerode congruence [11,12] if and only if $L_q^{\rightarrow}(A) = L_{q'}^{\rightarrow}(A)$. We assume that A has a single sink state since all sink states are equivalent. A DFA A is *minimal* if there exists no DFA recognizing $L(A)$ with less states than A. Notice that for a given language L, all minimal DFAs recognizing $L(A)$ are isomorphic. The minimal DFA of L is computable from any trim DFA recognizing L by merging equivalent states (For computation of the minimal DFA, see [1,7,10]). For a state q in a DFA A, we denote by $[q]$ the equivalent class of q which is a state of the minimal DFA of A. A *regular expression* E over an alphabet Σ is inductively defined by

$E = a$, $E = \emptyset$, $E = \varepsilon$, $E = (F + G)$, $E = (F \cdot G)$, $E = (F^*)$, $E = \neg(F)$ with F and G two regular expressions and a a symbol of Σ. The *alphabetical width* $|E|$ of E is the number of occurrences of symbols of E. Let L_1 and L_2 be two languages. We define $L_1 \cdot L_2 = \{w = w_1 \cdot w_2 \mid w_1 \in L_1 \wedge w_2 \in L_2\}$, $(L_1)^* = \{w = w_1 \cdots w_k \mid k \in \mathbb{N} \wedge \forall j \in \{1, \ldots, k\}, w_j \in L_1\}$ and $\neg L_1 = \{w \in \Sigma^* \mid w \notin L_1\}$. The *language denoted* by a regular expression E over an alphabet Σ is inductively computed by $L(a) = \{a\}$, $L(\emptyset) = \emptyset$, $L(\varepsilon) = \{\varepsilon\}$, $L(F + G) = L(F) \cup L(G)$, $L(F \cdot G) = L(F) \cdot L(G)$, $L(F^*) = (L(F))^*$, and $L(\neg F) = \neg(L(F))$, with F and G two regular expressions and a a symbol in Σ. The regular expression E is said to be a *simple expression* if it only contains sum, catenation and Kleene star operators. As a consequence, the *set of regular languages*, $\text{Rat}(\Sigma^*) = \{L \mid$ there exists a simple expression E such that $L(E) = L\}$, is the smallest family containing $\emptyset$ and $\{a\}$ for all symbol a in Σ and which is closed under catenation, Kleene star and sum. Kleene's Theorem [8] asserts that $\text{Rat}(\Sigma^*) = \text{Rec}(\Sigma^*)$. Moreover, given a complete DFA A, a DFA ${}_cA$ such that $L({}_cA) = \neg L(A)$ can be computed by switching non-final the final states and *vice versa*. The automaton ${}_cA$ is *the complement of* A. Therefore, the set of regular languages is closed under complement.

A subset O of states of a DFA A is an *orbit* if and only if, for every pair of states (q, q') in O^2, there exists a string w in Σ^+ such that $q' = \delta(q, w)$ and if for every state q in $Q \setminus O$, either there exists no word w in Σ^* such that $\delta(q, w) \in O$, or there exists no word w in Σ^* such that $\delta(O, w) \cap \{q\} \neq \emptyset$. Notice that a strongly connected component is not necessarily an orbit, since a singleton without a loop is a strongly connected component but not an orbit. The *gates* of an orbit O are the states belonging to the set gates(O) defined by: $\{o \in O \mid \exists a \in \Sigma, q \in Q \setminus (O \cup \text{sink}(A)), \delta(o, a) = q\} \cup (O \cap F)$. Let j be a state of O. The *automaton of the orbit O for j* is the automaton $A_{O,j} = (\Sigma, O, j, \text{gates}(O), \delta_O)$ where $\delta_O = \delta \cap (O \times \Sigma \times O)$. The language recognized by the automaton $A_{O,j}$ is called *the orbit language of O for j*. The orbit O is *transverse* if and only if the two following conditions are checked:

(1) $\forall q, q' \in \text{gates}(O)$, $\forall a \in \Sigma$: $\delta(q, a) \notin O \setminus \text{sink}(A) \Rightarrow \delta(q, a) = \delta(q', a)$.
(2) $\forall q, q' \in \text{gates}(O)$, q and q' are both either final or non-final.

A symbol a in Σ is *A-consistent* if and only if there exists a state f(a) such that every final state of A has a transition to f(a) labelled by a. A set $\Sigma' \subset \Sigma$ is *A-consistent* if and only if every symbol a of Σ' is A-consistent. If Σ' is an A-consistent set of symbols, the *Σ'-cut of A*, denoted by $A_{\Sigma'}$, is obtained by deleting every transition labelled by a symbol a in Σ' and starting from a final state. The *position automaton* or *Glushkov automaton* of a simple regular expression E is an $(n+1)$-state automaton that recognizes $L(E)$ (see [6,9] for construction rules). A simple expression E is *one-unambiguous* if and only if its Glushkov automaton is deterministic [3]. A regular language is *one-unambiguous* if and only if there exists a one-unambiguous regular expression denoting it. For details on one-unambiguous regular languages and properties, refer to Brüggemann-Klein and Wood [3].

3 One-Unambiguous Languages are Not Closed under Boolean Operators

Regular languages are closed for basic operators such as catenation, union, Kleene star, intersection or complement. On the other hand, for one-unambiguous regular languages, Brüggemann-Klein and Wood [3] noticed that they are closed neither under union, catenation nor under Kleene star. We show that one-unambiguous regular languages are closed neither under intersection (Example 1) nor under complement (Example 2).

Example 1. Let $E_1 = (b(c+\varepsilon))^*$ and $E_2 = b(c(b+\varepsilon))^* + c(b+\varepsilon)(c(b+\varepsilon))^*$ be two one-unambiguous regular expressions. Their minimal DFAs A_1 and A_2 are given in Figure 1. The minimal DFA A_3 of the language $L(A_1) \cap L(A_2)$ is given Figure 2. We can see that $L(A_3)$ is not one-unambiguous, since there does not exist nonempty A-consistent subset of Σ (see Theorem **F** [3]).

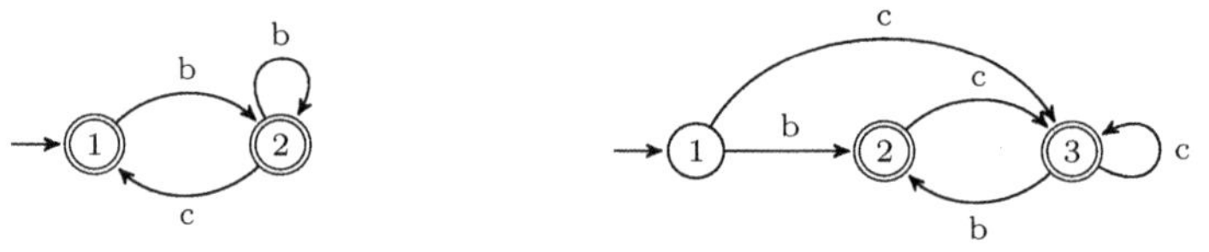

Fig. 1. The Automata A_1 and A_2

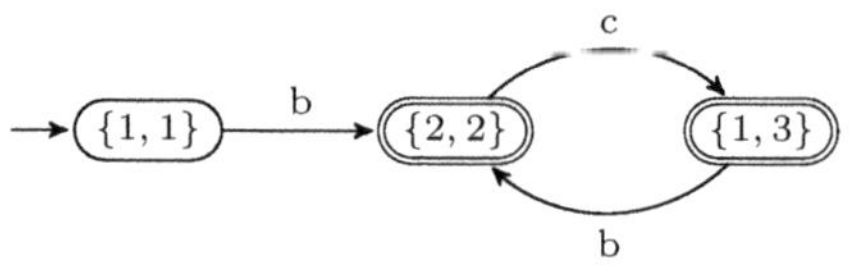

Fig. 2. The Automaton A_3

Example 2. Let A_1 and A_2 be two minimal DFAs given in Figure 3. The automaton A_2 recognizes the language $\neg(L(A_1))$. The language $L(A_1)$ is one-unambiguous, while $L(A_2)$ is not, since there does not exist nonempty A-consistent subset of Σ (see Theorem **F** [3]).

Fig. 3. The Automata A_1 and A_2

From the results (union, catenation and Kleene star) of Brüggemann-Klein and Wood [3] and Examples 1 and 2, we establish the following statement:

Proposition 1. *One-unambiguous regular languages are closed neither under union, catenation, Kleene star, intersection nor under complement.*

We have seen that one-unambiguous languages are not closed under complement. As a consequence, the characterization theorem for one-unambiguous languages has to be rewritten in order to deal with the complement operator. It leads to a new notion, the *weak one-unambiguity*.

4 The Weak One-Unambiguity

We exhibit a new family of languages closed under complement and containing the one-unambiguous languages family. The Kleene-like theorem of Brüggemann-Klein and Wood (Theorem D [3]) is transformed in order to deal with complement closure. This new family of regular languages is called the *weakly one-unambiguous languages*. We first define two particular subsets of symbols in Σ. Let L be a language over an alphabet Σ. The sets $\mathrm{First}(L)$ and $\mathrm{FollowLast}(L)$ are defined as follows: $\mathrm{First}(L) = \{a \in \Sigma \mid \exists w \in \Sigma^*, aw \in L\}$, $\mathrm{FollowLast}(L) = \{a \in \Sigma \mid \exists w, w' \in \Sigma^*, w \neq \varepsilon, w \in L \wedge waw' \in L\}$.

Definition 1. *The family of* weakly one-unambiguous expressions *over an alphabet Σ is the family $\mathcal{E}$ inductively defined as follows:*
(1) $\emptyset$, ε, and a are in $\mathcal{E}$, for each symbol a in Σ,
(2) if $E_1, E_2 \in \mathcal{E}$ and $\mathrm{First}(L(E_1)) \cap \mathrm{First}(L(E_2)) = \emptyset$, then $E_1 + E_2 \in \mathcal{E}$,
(3) if $E_1, E_2 \in \mathcal{E}$, $\mathrm{FollowLast}(L(E_1)) \cap \mathrm{First}(L(E_2)) = \emptyset$ and ($\varepsilon \notin L(E_1) \vee \mathrm{First}(L(E_1)) \cap \mathrm{First}(L(E_2)) = \emptyset$), then $E_1 \cdot E_2 \in \mathcal{E}$,
(4) if $E_1 \in \mathcal{E}$ and $\mathrm{FollowLast}(L(E_1)) \cap \mathrm{First}(L(E_1)) = \emptyset$, then $E_1^* \in \mathcal{E}$,
(5) if $E_1 \in \mathcal{E}$, then $\neg(E_1) \in \mathcal{E}$.

Definition 2. *A language L is* weakly one-unambiguous *if and only if there exists a weakly one-unambiguous expression denoting L.*

The two following definitions are used to characterize the minimal DFA of a weakly one-unambiguous language.

Definition 3. *Let A be a complete DFA and ${}_cA$ be its complement. The automaton A* satisfies the transverse property *if and only if for all orbit O in A, one of the two following propositions is satisfied:*
(1) O is transverse in A and $\forall i \in O$, $A_{O,i}$ satisfies the consistence property;
(2) O is transverse in ${}_cA$ and $\forall i \in O$, $({}_cA)_{O,i}$ satisfies the consistence property.

Definition 4. *Let A be a complete DFA and ${}_cA$ be its complement. The automaton A* satisfies the consistence property *if and only if for all orbit O in A, one of the two following propositions is satisfied:*
(1) there exists a nonempty subset Σ' of Σ such that Σ' is A-consistent and $A_{\Sigma'}$ satisfies the transverse property;
(2) there exists a nonempty subset Σ' of Σ such that Σ' is ${}_cA$-consistent and $({}_cA)_{\Sigma'}$ satisfies the transverse property.

Even if the two previous definitions seem to be dependent, each one checks the other on a smaller automaton (an orbit automaton or a Σ'-cut). Since there are no more orbits at the end, this dependency stops eventually. Note that both properties are satisfied by an acyclic automaton. The transverse property characterizes the minimal DFA of weakly one-unambiguous languages.

Theorem 1. *A regular language is weakly one-unambiguous if and only if its complete minimal DFA satisfies the transverse property. Furthermore, if E is an n symbol weakly one-unambiguous expression, the minimal DFA of $L(E)$ has at most $n+2$ states.*

In order to prove this theorem, we first show that minimization preserves the transverse property (Section 4.1), and then we demonstrate the necessity and the sufficiency of the transverse property to characterize the weak one-unambiguity (Section 4.2 and Section 4.3).

4.1 Minimization Preserves the Transverse Property

Lemma 1. *Let A be a DFA and A' be its minimal DFA. For all orbit O' in A', there exists an orbit O in A such that $\forall p \in O$, $[p] \in O' \wedge (p \in \text{gates}(O) \Leftrightarrow [p] \in \text{gates}(O'))$. The orbit O is said to be a* a lift *of O'.*

Proof. Let $A = (\Sigma, Q, i, F, \delta)$ and $A' = (\Sigma, Q', i', F', \delta')$. Let O' be an orbit in A' and $[p]$ be a state in O'. There exists p in A such that $p \in [p]$. **(1)** For all $[q]$ in O', there exist two words w_1 and w_2 such that $\delta'([p], w_1) = [q]$ and $\delta'([q], w_2) = [p]$. As a consequence, there exists a state $q \in [q]$ such that $\delta(p, w_1) = q$ and $\delta(q, w_2) = p' \in [p]$. This does not imply that p and q are in a same orbit O. However, by repeating these words and since A is a DFA, an orbit O is accessible from p such that for all p' in O, $[p'] \in O'$. **(2a)** Let $[p]$ be in gates(O') and p be a state in O such that $p \in [p]$. If $[p]$ is final, so does p. If there exists a symbol a in Σ such that $\delta'([p], a) \notin O' \cup \text{sink}(A')$, it holds $\delta(p, a) \notin O \cup \text{sink}(A)$ (otherwise contradiction with $p \in [p]$). As a consequence, p is in gates(O). **(2b)** Let O be one of the last accessible orbist in A satisfying **(1)** (considering the partial order of accessible orbits). Let p be a gate of O. Suppose that $[p]$ is not a gate. Consequently, p has to be a non-final state. As a consequence, there exists a transition from p going out of O (but not in the sink state) by a letter a such that $\delta'([p], a) \in O' \cup \text{sink}(A')$. This implies either $\delta(p, a) = \text{sink}(A)$ or there is another orbit O_2 accessible from $\delta(p, a)$ satisfying **(1)**. Contradiction in both cases. ■

Lemma 2. *Let A' be a minimal DFA and O' be a transverse orbit in A'. For every j' in O', $A'_{O',j'}$ is minimal.*

Proof. Let j' be a state in O'. We consider the automata $A' = (\Sigma, Q', i', F', \delta')$ and $A'_{O',j'} = (\Sigma, O', j', \text{gates}(O'), \delta'')$. Let p'_1 and p'_2 be two equivalent states in $(A'_{O',j'})$. For all w in Σ^*, $\delta''(p'_1, w)$ and $\delta''(p'_2, w)$ are equivalent. Since O' is transverse, for every word w such that $\delta'(p'_1, w) \notin O'$, $\delta'(p'_2, w) = \delta'(p'_1, w)$. This equivalence is preserved in A'. Therefore if $p'_1 \neq p'_2$, contradiction with the minimality of A'. ■

Lemma 3. *Let A be a complete DFA and A' be its complete minimal DFA. Let O' be an orbit in A' and O be a lift of O'. For all j in O, if there exists an $A_{O,j}$-consistent symbol a in Σ, then a is $A'_{O',[j]}$-consistent and $(A'_{O',[j]})_{\{a\}}$ is the minimal automaton of $L((A_{O,j})_{\{a\}})$.*

Proof. Let j be a state in O. Let $A'_{O',[j]} = (\Sigma, Q', [j], F', \delta')$ and $(A'_{O',[j]})_{\{a\}} = (\Sigma, Q', [j], F', \delta'')$. **(1)** If there is a symbol a in Σ such that a is not $A'_{O',[j]}$-consistent, there exist two gates $[q'_1]$ and $[q'_2]$ in O' such that $\delta'([q'_1], a) \in O'$ and $\delta'([q'_1], a) \neq \delta'([q'_2], a)$. As a consequence, there exist two gates q_1 and q_2 of O such that $q_1 \in [q'_1]$, $q_2 \in [q'_2]$ and $\delta(q_1, a) \in O$ and $\delta(q_1, a) \neq \delta(q_2, a)$. Finally, a is not $A_{O,j}$-consistent. **(2)** Let us prove that $(A'_{O',[j]})_{\{a\}}$ is minimal and that $L((A_{O,j})_{\{a\}}) = L((A'_{O',[j]})_{\{a\}})$. If $A_{O,j}$ is a-consistent, **(I)** implies that $A'_{O',[j]}$ is a-consistent. **(a)** Let p'_1 and p'_2 be two nonequivalent states in $A'_{O',[j]}$. By definition, there exists a symbol $b \neq a$ in Σ such that $\delta'(p'_1, b) \neq \delta'(p'_2, b)$, and by construction of $(A'_{O',[j]})_{\{a\}}$, $\delta''(p'_1, b) \neq \delta''(p'_2, b)$. By Lemma 2, $A'_{O',[j]}$ is minimal, and so is $(A'_{O',[j]})_{\{a\}}$. **(b)** Let $(A_{O,j})_{\{a\}} = (\Sigma, Q, j, F, \delta)$. The word w is in $L((A_{O,j})_{\{a\}}) \Leftrightarrow \delta(j, w) \in F \Leftrightarrow \delta''([j], w) \in F'' \Leftrightarrow$ The word w is in $L((A'_{O',[j]})_{\{a\}})$. ■

Proposition 2. *Let A be a complete DFA and A' be its complete minimal DFA. If A satisfies the transverse property, so does A'.*

Proof. Assume that A' does not satisfy the transverse property. Then there exists an orbit O' in A' such that one of the four following cases occurs: **(I)** Suppose that O' is not transverse in A'. Let p' and q' be two gates in O'. According to Lemma 1, there exists an orbit O in A containing two nonequivalent states p and q such that p is merged into p' and q is merged into q'; moreover, since p' and q' are two gates of O', so are p and q in O. **(a)** If p' and q' do not have the same finality, neither do p nor q. **(b)** Let a be a symbol in Σ such that $\delta'(p', a) \notin (O' \cup \text{sink}(A'))$ and $\delta'(q', a) \neq \delta'(p', a)$. As a consequence, $\delta(q, a) \neq \delta(p, a)$ and $\delta(p, a) \notin (O \cup \text{sink}(A))$. In both cases, O is not transverse in A. **(II)** If O' is an orbit which is not transverse in ${}_cA'$, the same argument as **(I)** leads to the existence of an orbit O which is not transverse in ${}_cA$. **(III)** Suppose that O' is transverse in A' and there exists $[k]$ in O' such that $A'_{O',[k]}$ does not satisfy the consistence property. Let O be a lift of O'. Let j be a state in O which is in $[k]$. Then $[k] = [j]$. **(a)** Suppose that there exists no nonempty subset Σ' of Σ which is $A'_{O',[j]}$-consistent. As a consequence, every symbol a in Σ is not $A'_{O',[j]}$-consistent. If a in Σ is not $A'_{O',[j]}$-consistent, according to Lemma 3, a is not $A_{O,j}$-consistent. **(b)** Suppose that a is a $A_{O,j}$-consistent symbol. According to Lemma 3, the automaton $(A'_{O',[j]})_{\{a\}}$ is the minimal automaton of $L((A_{O,j})_{\{a\}})$. By recurrence on the number of transitions of the automata, according to **(I)** and **(II)**, if $(A'_{O',[j]})_{\{a\}}$ does not satisfy the transverse property, neither does $(A_{O,j})_{\{a\}}$. **(IV)** If O' is an orbit which is transverse in ${}_cA'$, we let O be a lift of O' and j be a state in O such that $A'_{O',[j]}$ does not satisfy the consistence property. A similar argument as **(III)** leads to the fact that $A_{O,j}$ does not satisfy the consistence property.

Finally, if A' does not satisfy the transverse property, neither does A. ■

4.2 From a Weakly One-Unambiguous Expression to a Linear-Size DFA Satisfying the Transverse Property

We show how to inductively compute a minimal DFA from a weakly one-unambiguous expression E. In a complete minimal DFA, we distinguish, if they exist, two particular states: the sink state and the *c-sink*, the only state q such that $L_q^{\rightarrow} = \Sigma^*$. Notice that the c-sink of A is the sink state of ${}_cA$. For a set Q of states in A, we denote by Q^+ the set $\{q \in Q \mid \exists w \in \Sigma^+, \delta(i,w) = q \wedge q \notin \text{sink}(A) \cup \text{sink}({}_cA)\}\}$. We show that Q^+ has at most $|E|$ elements. As a consequence, since $Q = Q^+ \cup \{i\} \cup \text{sink}({}_cA)$, Q has at most $|E| + 2$ elements.

Lemma 4. *Let $A_1 = (\Sigma, Q_1, i_1, F_1, \delta_1)$ and $A_2 = (\Sigma, Q_2, i_2, F_2, \delta_2)$ be two minimal DFAs satisfying the transverse property such that*

$$\text{First}(L(A_1)) \cap \text{First}(L(A_2)) = \emptyset.$$

Let $A = (\Sigma, Q, i, F, \delta)$ be the minimal DFA of $L(A_1) \cup L(A_2)$. The two following propositions are satisfied:

(1) the automaton A satisfies the transverse property,

(2) $\text{Card}(Q^+) \leq \text{Card}(Q_1^+) + \text{Card}(Q_2^+)$.

Proof. Let us consider the automaton $A' = (\Sigma, Q', i, F', \delta')$ defined by $Q' = Q_1 \cup Q_2 \cup \{i\}$, $F' = F_1 \cup F_2 \cup \{i\}$ if $\varepsilon \in L(A_1) \cup L(A_2)$, $F' = F_1 \cup F_2$ otherwise, and $\delta' = \delta_1 \cup \delta_2 \cup \{(i,a,p) \mid (i_1,a,p) \in \delta_1 \vee (i_2,a,p) \in \delta_2\}$. As $\text{First}(L(A_1)) \cap \text{First}(L(A_2)) = \emptyset$, A' is deterministic.
By construction, since an orbit in A' is an orbit in A_1 or in A_2, A' satisfies the transverse property; moreover, $w \in L(A') \Leftrightarrow \delta'(i,w) \in F' \Leftrightarrow \delta_1(i_1,w) \in F_1 \vee \delta_2(i_2,w) \in F_2 \Leftrightarrow w \in L(A_1) \vee w \in L(A_2)$. Consequently, A' recognizes $L(A_1) \cup L(A_2)$. According to Proposition 2, the complete minimal DFA of $L(E)$ satisfies the transverse property. By construction, $\text{Card}(Q'^+) = \text{Card}(Q_1^+) + \text{Card}(Q_2^+)$; finally, $\text{Card}(Q^+) \leq \text{Card}(Q_1^+) + \text{Card}(Q_2^+)$. ■

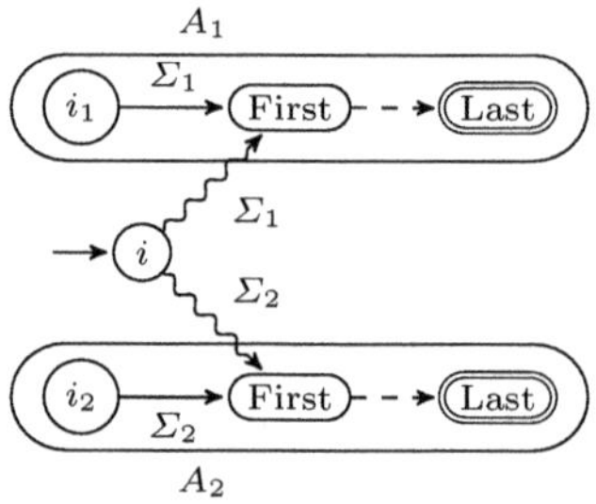

Fig. 4. The Automaton of $L(A_1) \cup L(A_2)$

Lemma 5. *Let $A_1 = (\Sigma, Q_1, i_1, F_1, \delta_1)$ and $A_2 = (\Sigma, Q_2, i_2, F_2, \delta_2)$ be two minimal DFAs satisfying the transverse property such that*

$$(\varepsilon \notin L(A_1) \vee \text{First}(L(A_1)) \cap \text{First}(L(A_2)) = \emptyset)$$
$$\textit{and } \text{FollowLast}(L(A_1)) \cap \text{First}(L(A_2)) = \emptyset.$$

Let $A = (\Sigma, Q, i, F, \delta)$ be the minimal DFA of $L(A_1) \cdot L(A_2)$. The following propositions are satisfied:

(1) the automaton A satisfies the transverse property,
(2) $\mathrm{Card}(Q^+) \leq \mathrm{Card}(Q_1^+) + \mathrm{Card}(Q_2^+)$.

Proof. Let us consider the automaton $A' = (\Sigma, Q', i, F', \delta')$ defined by $Q' = Q_1 \cup Q_2 \cup \{i\}$, $F' = F_1 \cup F_2 \cup \{i\}$ if $\varepsilon \in L(A_1) \wedge \varepsilon \in L(A_2)$, $F' = F_1 \cup F_2$ if $\varepsilon \notin L(A_1) \wedge \varepsilon \in L(A_2)$, $F' = F_2$ otherwise, and $\delta' = \delta_1 \cup \delta_2 \cup \{(i, a, p) \mid (i_1, a, p) \in \delta_1 \vee (i_1 \in F_1 \wedge (i_2, a, p) \in \delta_2\} \cup \{(p, a, p') \mid p \in F_1 \wedge (i_2, a, p') \in \delta_2\}$. Since ($\varepsilon \notin L(A_1) \vee \mathrm{First}(L(A_1)) \cap \mathrm{First}(L(A_2)) = \emptyset$) and $\mathrm{FollowLast}(L(A_1)) \cap \mathrm{First}(L(A_2)) = \emptyset$, A' is deterministic. By construction, an orbit O in A' is an orbit in A_1 or in A_2; if O is in A_2, the transverse property is preserved; if O is in A_1, as the only transitions added are going out of O, the orbit languages are preserved; moreover, the new transitions preserve transversality: if a transition is added, its origin q is a final state, which is a gate in O; if O is transverse in A_1, all the other gates are final and transitions are also added.
If q is a non-final gate in ${}_cA_1$, all these gates are non final in ${}_cA_1$ and final in A_1, consequently transitions out of them are also added. If q is a final state in A_1 which is not a gate in ${}_cA_1$, for all symbol a in Σ, either $\delta'(q, a) = \mathrm{sink}({}_cA_1)$ or $\delta'(q, a) \in O$. As a consequence, every symbol a in Σ is in $\mathrm{FollowLast}(L(E_1))$. Either $L(A_2) = \emptyset$ or $L(A_2) = \{\varepsilon\}$ and there is no transition added, or contradiction with $\mathrm{FollowLast}(L(A_1)) \cap \mathrm{First}(L(A_2)) = \emptyset$.

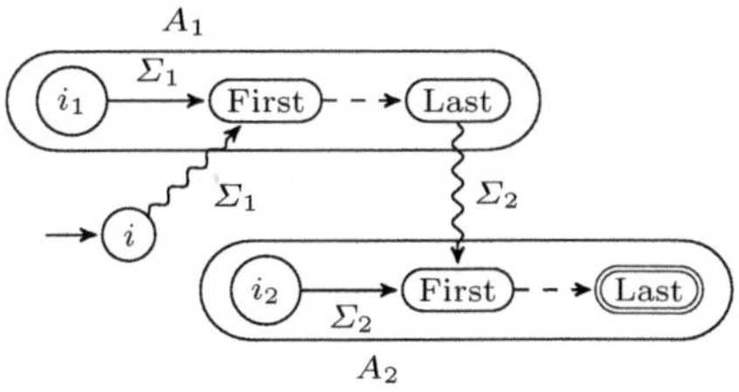

Fig. 5. The Automaton of $L(A_1) \cdot L(A_2)$ in the case where $\varepsilon \notin L(A_1)$ and $\varepsilon \notin L(A_2)$

As a consequence, A' satisfies the transverse property. Moreover, $w \in L(A) \Leftrightarrow \delta'(i, w) \in F' \Leftrightarrow (w = w'aw'' \wedge \delta_2(\delta'(\delta_1(i_1, w'), a), w'') \in F_2) \vee (\delta'(i_2, w) \in F_2 \wedge i_1 \in F_1) \vee (\delta_1(i_1, w) \in F_1 \wedge \varepsilon \in L(A_2)) \Leftrightarrow w \in L(A_1) \cdot L(A_2)$. Consequently, A' recognizes $L(A_1) \cdot L(A_2)$. According to Proposition 2, the complete minimal DFA of $L(E)$ satisfies the transverse property. By construction, $\mathrm{Card}(Q'^+) = \mathrm{Card}(Q_1^+) + \mathrm{Card}(Q_2^+)$; finally, $\mathrm{Card}(Q^+) \leq \mathrm{Card}(Q_1^+) + \mathrm{Card}(Q_2^+)$. ■

Lemma 6. *Let $A_1 = (\Sigma, Q_1, i_1, F_1, \delta_1)$ be a minimal DFA satisfying the transverse property such that*

$$\mathrm{FollowLast}(L(A_1)) \cap \mathrm{First}(L(A_1)) = \emptyset.$$

Let $A = (\Sigma, Q, i, F, \delta)$ be the minimal DFA of $L(A_1)^*$. The two following propositions are satisfied:

(1) the automaton A satisfies the transverse property,
(2) $\mathrm{Card}(Q^+) \leq \mathrm{Card}(Q_1^+)$.

Proof. Let us consider the automaton $A' = (\Sigma, Q', i, F', \delta')$ defined by $Q' = Q_1 \cup \{i\}$, $F' = F_1 \cup \{i\}$, and $\delta' = \delta_1 \cup \{(i, a, p) \mid (i_1, a, p) \in \delta_1\} \cup \{(p, a, p') \mid$

$p \in F_1 \wedge (i_1, a, p') \in \delta_1\}$. Since FollowLast$(L(A_1)) \cap$ First$(L(A_1)) = \emptyset$, A' is deterministic. By construction, there is only one orbit which is transverse. Furthermore, the set First$(L(A_1))$ is A'-consistent, and the First$(L(A_1))$-cut of A' is A_1. As a consequence, A' satisfies the transverse property. Moreover, $w \in L(A) \Leftrightarrow \delta'(i, w) \in F'$ $\Leftrightarrow w = w_1 w_2 \cdots\cdots w_k \wedge \delta'(i, w_1) \in F' \wedge \delta'(i, w_2) \in F' \ldots \delta'(i, w_k) \in F' \Leftrightarrow w_1 \in L(A_1) \wedge w_2 \in L(A_1) \ldots w_k \in L(A_1) \Leftrightarrow w \in L(A_1)^*$. Consequently, A' recognizes $L(A_1)^*$.

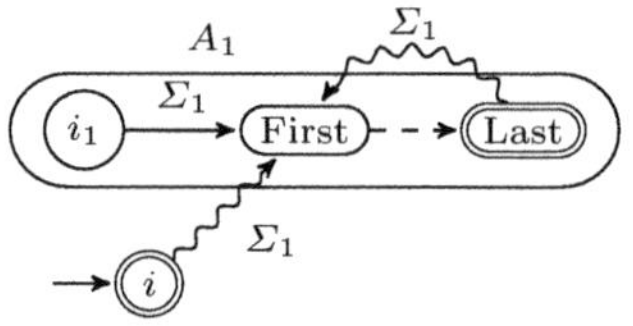

Fig. 6. The Automaton of $L(A_1)^*$

According to Proposition 2, the complete minimal DFA of $L(A')$ satisfies the transverse property. By construction, Card$(Q'^+) =$ Card(Q_1^+); finally, Card$(Q^+) \leq$ Card(Q_1^+). ■

Lemma 7. *Let $A_1 = (\Sigma, Q_1, i_1, F_1, \delta_1)$ be a minimal DFA satisfying the transverse property. Let $A = (\Sigma, Q, i, F, \delta)$ be the minimal DFA of $\neg L(A_1)$. The two following propositions are satisfied:*
(1) the automaton A satisfies the transverse property,
(2) Card$(Q^+) =$ Card(Q_1^+).

Proof. By construction, ${}_cA_1$ is the complete minimal DFA of $\neg(L(A_1))$, which satisfies the transverse property. By construction, Card$(Q'^+) =$ Card(Q_1^+); finally, Card$(Q^+) =$ Card(Q_1^+). ■

Proposition 3. *Let E be a weakly one-unambiguous expression and A' be the minimal DFA of $L(E)$. Then A satisfies the transverse property and A has at most $|E| + 2$ states.*

Proof. According to Lemma 4, Lemma 5, Lemma 6 and Lemma 7.

4.3 From a Minimal Automaton Satisfying the Transverse Property to a Weakly One-Unambiguous Expression

We show that the language denoted by a DFA satisfying the transverse property is weakly one-unambiguous. Let $A = (\Sigma, Q, i, F, \delta)$ be a DFA and q be a state in Q. The *q-starting automaton* A^q of A is the accessible part of the automaton $(\Sigma, Q, q, F, \delta)$. Note that A^q is a subautomaton of A and $L(A^q) = L_q^{\rightarrow}(A)$.

Proposition 4. *Let A be a DFA satisfying the transverse property. Then $L(A)$ is weakly one-unambiguous.*

Proof. We show how to inductively compute $L_q^A = L(A^q)$ from $A = (\Sigma, Q, i, F, \delta)$ and we show that L_q^A is weakly one-unambiguous.
(I) Suppose that Q is not an orbit.

(a) If i is in an orbit O which is transverse in A, let Σ' be the subset $\{a \in \Sigma \mid \forall q \in \text{gates}(O), \delta(q,a) \notin O\}$. Let us consider the language L' defined by $L' = (L_{O,i}) \cdot (\bigcup_{a \in \Sigma'} (\{a\} \cdot L^A_{\mathrm{f}(a)}))$ if $\text{gates}(O) \cap F = \emptyset$ or $L' = (L_{O,i}) \cdot (\{\varepsilon\} \cup \bigcup_{a \in \Sigma'} (\{a\} \cdot L^A_{\mathrm{f}(a)}))$ otherwise, where, $L_{O,i}$ is the orbit language of O beginning in the state i (see **(II)**). A word w is in L_i^A if and only if it can be split into $w = w_1 w_2$ where w_1 is a path from i to a gate g of O and w_2 a path from g to a final state of A which is a nonempty path in A^g if $w_2 \neq \varepsilon$. As a consequence, $L_i^A = L'$. By recurrence on the number of transitions and according to **(II)**, $L_{O,i}$ and $L'' = \bigcup_{a \in \Sigma'} (\{a\} \cdot L^A_{\mathrm{f}(a)})$ are weakly one-unambiguous. If $\varepsilon \in L_{O,i} \wedge \text{First}(L_{O,i}) \cap \text{First}(L'') \neq \emptyset$, then there exist two transitions labelled by the same symbol going out of i, which contradicts the determinism. If $\text{FollowLast}(L_{O,i}) \cap \text{First}(L'') \neq \emptyset$, then there exist two transitions labelled by the same symbol going out of a gate of O, one going in O, and the other in $\mathrm{f}(a)$, which contradicts the determinism. As a consequence, L' is weakly one-unambiguous.
(b) If i is in an orbit O which is transverse in ${}_cA$, then $L_i^A = \neg(L_i^{{}_cA})$ which is weakly one-unambiguous according to **(a)**.
(c) If i is not in an orbit, $L_i^A = \bigcup_{a \in \Sigma} (\{a\} \cdot L^A_{\delta(i,a)})$ if $i \notin F$, $L_i^A = \{\varepsilon\} \cup \bigcup_{a \in \Sigma} (\{a\} \cdot L^A_{\delta(i,a)})$. By recurrence on the number of transitions, $L_i^A = \bigcup_{a \in \Sigma} (\{a\} \cdot L^A_{\delta(i,a)})$ is weakly one-unambiguous.
(II) Suppose that Q is an orbit. Let Σ' be a subset of Σ.
(a) If Σ' is A-consistent, let us consider the language $L' = L_i^{A_{\Sigma'}} \cdot (\bigcup_{a \in \Sigma'} (\{a\} \cdot L^{A_{\Sigma'}}_{\mathrm{f}(a)}))^*$. Since $A_{\Sigma'}$ satisfies the transverse property, according to **(I)**, $L_i^{A_{\Sigma'}}$ is weakly one-unambiguous. By recurrence on the number of transitions, $L'' = \bigcup_{a \in \Sigma'} (\{a\} \cdot L^{A_{\Sigma'}}_{\mathrm{f}(a)})$ is weakly one-unambiguous. Suppose that $\text{FollowLast}(L'') \cap \text{First}(L'') \neq \emptyset$; then there exist two transitions labelled by the same symbol going out of a gate of the orbit, one leading to $\mathrm{f}(a)$ and the other to another state of O. Contradiction with determinism. The same contradiction is implied if $\varepsilon \in L_i^{A_{\Sigma'}} \wedge \text{First}(L_i^{A_{\Sigma'}}) \cap \Sigma' \neq \emptyset$ or $\text{FollowLast}(L_i^{A_{\Sigma'}}) \cap \text{First}(L'') \neq \emptyset$. Finally, since a word w in $L(A)$ can be split up into $w_1 \cdot w_2$ such that w_1 is the label of a path from i to a final state and w_2 the label of a path from a final state to another final state, w is in L'.
(b) If Σ' is ${}_cA$-consistent, then $L_i^A = \neg(L_i^{{}_cA})$, which is weakly one-unambiguous according to **(a)**. ■

Example 3. Consider the automaton A in Figure 7. It is composed by the four orbits $O_1 = \{2,3\}$, $O_2 = \{4,5\}$ and the singleton $\{1\}$ and $\{6\}$. The singletons trivially satisfy the transverse property. The orbit O_1 is transverse in A and O_2

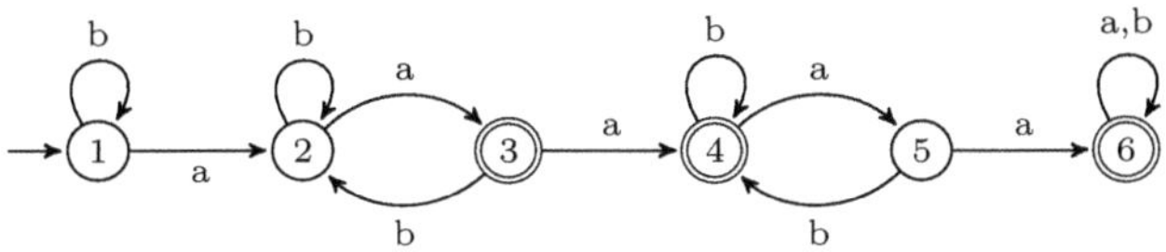

Fig. 7. The automaton A

is transverse in the complement of A. As a consequence, it can be denoted by the weakly one-unambiguous expression $E = b^*ab^*a(bb^*a)^*(\varepsilon + a\neg(b^*a(bb^*a)^*)))$.

5 Conclusion

Content models of XML DTD are one-unambiguous [3]. Usually, the size of a regular expression can be reduced using the complement operator. Thus the closure property of one-unambiguous regular languages is important in XML applications. We have demonstrated that one-unambiguous regular languages are closed neither under intersection nor under complement. We have also considered one-unambiguous regular languages with complement closure property (weak one-unambiguity). We have investigated the closure properties of weakly one-unambiguous regular languages and the state complexity of the languages on some operations. Finally, we have shown that as far as weakly one-unambiguous regular expressions are concerned, a linear-size deterministic recognizer can be computed in order to decide whether or not a word belongs to the language denoted by a non-simple regular expression.

References

1. Blum, N.: An $O(n \log n)$ implementation of the standard method for minimizing n-state finite automata. Inform. Process. Lett. 57(2), 65–69 (1996)
2. Bray, T., Paoli, J., Sperberg-Mc Queen, C.M., Maler, E., Yergeau, F.: Extensible Markup Language (XML) 1.0, 4th edn. (2006), `http://www.w3.org/TR/2006/REC-xml-20060816`
3. Brüggemann-Klein, A., Wood, D.: One-unambiguous regular languages. Inform. Comput. 140, 229–253 (1998)
4. Gelade, W.: Succinctness of regular expressions with interleaving, intersection and counting. Theor. Comput. Sci. 411(31-33), 2987–2998 (2010)
5. Gelade, W., Neven, F.: Succinctness of the complement and intersection of regular expressions. In: Albers, S., Weil, P. (eds.) STACS. Dagstuhl Seminar Proceedings, vol. 08001, pp. 325–336 (2008)
6. Glushkov, V.M.: The abstract theory of automata. Russian Mathematical Surveys 16, 1–53 (1961)
7. Hopcroft, J.E.: An n log n algorithm for minimizing the states in a finite automaton. In: Kohavi, Z. (ed.) The Theory of Machines and Computations, pp. 189–196. Academic Press, New York (1971)
8. Kleene, S.: Representation of events in nerve nets and finite automata. In: Automata Studies, Ann. Math. Studies, vol. 34, pp. 3–41. Princeton U. Press (1956)
9. McNaughton, R.F., Yamada, H.: Regular expressions and state graphs for automata. IEEE Transactions on Electronic Computers 9, 39–57 (1960)
10. Moore, E.F.: Gedanken experiments on sequential machines. In: Automata Studies, pp. 129–153. Princeton Univ. Press, Princeton (1956)
11. Myhill, J.: Finite automata and the representation of events. WADD, TR-57-624, 112–137 (1957)
12. Nerode, A.: Linear automata transformation. In: Proceedings of AMS, vol. 9, pp. 541–544 (1958)
13. Rabin, M.O., Scott, D.: Finite automata and their decision problems. IBM J. Res. 3(2), 115–125 (1959)

Simulations over Two-Dimensional On-Line Tessellation Automata*

Gérard Cécé[1,2] and Alain Giorgetti[1,3]

[1] LIFC - EA 4269 - University of Franche-Comté - France
[2] Centre Numerica, 1 cours Leprince-Ringuet,
BP 21126, 25201 MONTBELIARD Cedex
Gerard.Cece@univ-fcomte.fr
[3] INRIA/CASSIS
16 route de Gray, 25030 BESANCON Cedex
Alain.Giorgetti@univ-fcomte.fr

Abstract. We study the notion of simulation over a class of automata which recognize 2D languages (languages of arrays of letters). This class of two-dimensional On-line Tessellation Automata (2OTA) accepts the same class of languages as the class of tiling systems, considered as the natural extension of classical regular word languages to the 2D case. We prove that simulation over 2OTA implies language inclusion. Even if the existence of a simulation relation between two 2OTA is shown to be a NP-complete problem in time, this is an important result since the inclusion problem is undecidable in general in this class of languages. Then we prove the existence of a unique maximal autosimulation relation in a given 2OTA and the existence of a unique minimal 2OTA which is simulation equivalent to this given 2OTA, both computable in polynomial time.

Keywords: Simulation, 2D words, Tiling systems, Picture languages, Picture automata.

1 Introduction

We are involved in the 'Smart Surface' project [14] whose aim is the realization of an active surface to automatically position and convey micro-items. This active surface is made of an array of smart micromodules. Under the abstraction that a micromodule can evolve only within a small number of states, we can consider those states as letters of a given alphabet. Then, what about representing a set of reachable configurations of the whole system as a recognizable two-dimensional (2D) language and using the regular model-checking (RMC) paradigm [3] on it? Let us recall that the RMC paradigm consists in representing infinite sets

* This work is supported by the ANR, the French national research agency (Agence Nationale de la Recherche), project ANR-06-ROBO-0009-03 and by the INRIA CASSIS project.

G. Mauri and A. Leporati (Eds.): DLT 2011, LNCS 6795, pp. 141–152, 2011.

of configurations of a system by recognizable languages, and developing meta-transitions which can compute infinite sets of successors in one step.

To do this, we first need to clarify what could be recognizable 2D languages. The most accepted class is that recognized by tiling systems [9]. Unfortunately, an important property of classical regular languages is missing in this class, namely decidability of the inclusion problem, which is a necessary property in the RMC paradigm. This led us to seek sufficient conditions to decide inclusion. For words and trees, the existence of a simulation relation (in the sense of [13]) between the underlying automata of two recognizable languages is such a sufficient condition. Moreover, the existence of an autosimulation relation, bigger than the identity, between the states of a finite automaton makes it possible to construct a smaller equivalent automaton by quotient. But tiling systems are not defined in terms of automata with straightforward notions of states and transitions. Fortunately, several kinds of automata recognizing the same class of 2D languages have been defined. Among them, there are two-dimensional On-line Tessellation Automata (2OTA) [10], Wang automata [12] and quadripolic automata [5]. This paper describes our results concerning simulations over 2OTA.[1]

Contributions. We first define simulation relations between two 2OTA. We show that simulation implies language inclusion. From any given autosimulation relation – i.e. a simulation relation between the states of a given automaton – we construct a quotient automaton smaller than the given automaton, simulation equivalent to it (one simulates the other) and which therefore accepts the same language. In a 2OTA A we prove the existence of a unique maximal autosimulation relation. We show that the quotient automaton constructed from this maximal autosimulation relation is the smallest one which is simulation equivalent to A. Then we show how to compute this maximal autosimulation in polynomial time. We also prove that deciding the existence of a simulation relation between two different 2OTA is unfortunately NP-complete in time.

Related work. The study of two-dimensional languages is an active field of research, see [8] for a recent overview. To our knowledge, this is the first work on simulations concerning 2D automata. In the last few years, several works have been done about simulations over tree automata [3,1,2] but mainly to reduce them. For example the complexity of the existence of an upward simulation between two different tree automata has not been investigated. Moreover, the search for a minimal automaton which is simulation equivalent to a given one has not been done for tree automata. Our result concerning this smallest simulation-equivalent automaton in 2OTA is inspired from the one of [6] on Kripke structures. A main difference is that we directly remove what is called "little brothers" in the quotient automata instead of removing them in a second step.

[1] "2OTA" indifferently abbreviates the plural "two-dimensional on-line tessellation automata" and the singular "two-dimensional on-line tessellation automaton".

Outline. The next section introduces pictures (two-dimensional arrays of letters), picture languages and tiling systems. Section 3 is dedicated to 2OTA and some of their properties. Then we define simulation relations over two 2OTA in Section 4, and give the first results of the paper: simulation implies language inclusion, there are a unique maximal autosimulation relation in a given 2OTA and a unique minimal 2OTA which is simulation equivalent to it. We treat the algorithmic and complexity issues in Section 5. Section 6 is about backward simulations between 2OTA. We show that they do not imply language inclusion, contrarily to backward simulations between tree automata. Section 7 finally concludes the paper and suggests some future directions. Detailed proofs for all the results in this paper can be found in [7].

2 Picture Languages and Tiling Systems

A *picture* is a two-dimensional array of letters from a given finite alphabet Σ. The set of all pictures over Σ is noted Σ^{**}. The *size* of a picture p is a couple of integers, $size(p) = (m, n)$, where m is the number of rows and n is the number of columns. By convention, we note ε the empty picture, whose size is $(0, 0)$. There is no picture of size $(0, k)$ or $(k, 0)$ with k positive. For a given picture p, we note $p_{i,j}$ the letter found at the intersection of the ith line and the jth column and we note $\hat{p}$ the picture which consists of p surrounded with a special symbol $\# \notin \Sigma$.

In the following example we show a square picture p of size $(5, 5)$ made of b but the main diagonal which is made of a. The corresponding $\hat{p}$, which size is $(7, 7)$, is also given.

$$p = \begin{array}{ccccc} a & b & b & b & b \\ b & a & b & b & b \\ b & b & a & b & b \\ b & b & b & a & b \\ b & b & b & b & a \end{array} \text{ and } \hat{p} = \begin{array}{ccccccc} \# & \# & \# & \# & \# & \# & \# \\ \# & a & b & b & b & b & \# \\ \# & b & a & b & b & b & \# \\ \# & b & b & a & b & b & \# \\ \# & b & b & b & a & b & \# \\ \# & b & b & b & b & a & \# \\ \# & \# & \# & \# & \# & \# & \# \end{array} \qquad (1)$$

A *picture language* on Σ is a subset of Σ^{**}. For a language $L \subseteq \Sigma^{**}$, we define $\hat{L} = \{\hat{p} \mid p \in L\}$. A *tiling system (TS)* is a tuple $T = (\Sigma, \Gamma, \Theta, \pi)$ such that Σ and Γ are finite alphabets, $\pi : \Gamma \to \Sigma$ is a mapping and Θ, the set of *tiles*, is a finite set of pictures of size $(2, 2)$ on the alphabet Γ. A language $L \subseteq \Sigma^{**}$ is said *recognized* by T if there exists a language $L' \subseteq \Gamma^{**}$ such that $L = \pi(L')$ and all sub-pictures of $\hat{L'}$ of size $(2, 2)$ belong to Θ. If the tile $\begin{array}{|c|c|}\hline \# & \# \\ \hline \# & \# \\ \hline \end{array}$ belongs to Θ we consider by convention that the empty picture ε belongs to L. Given a tiling system T, we note $\mathcal{L}(T)$ the language recognized by T. The family of languages recognized by tiling systems is noted $\mathcal{L}(TS)$ and is called the class of *recognizable picture languages*.

As an example, consider the tiling system $T = (\Sigma, \Gamma, \Theta, \pi)$ with: $\Sigma = \{a, b\}$, $\Gamma = \{0, 1, 2\}$, $\pi(0) = \pi(2) = b$, $\pi(1) = a$, and Θ the set of all the seventeen sub-pictures of size $(2, 2)$ of the following picture:

$$\begin{array}{ccccccc} \# & \# & \# & \# & \# & \# & \# \\ \# & 1 & 0 & 0 & 0 & 0 & \# \\ \# & 2 & 1 & 0 & 0 & 0 & \# \\ \# & 2 & 2 & 1 & 0 & 0 & \# \\ \# & 2 & 2 & 2 & 1 & 0 & \# \\ \# & 2 & 2 & 2 & 2 & 1 & \# \\ \# & \# & \# & \# & \# & \# & \# \end{array}$$

It is easy to show that the picture p in (1) belongs to $\mathcal{L}(T)$ and furthermore that $\mathcal{L}(T)$ is the set of all non empty square pictures whose main diagonal is made of a while the other positions are labeled by b.

3 Two-Dimensional On-Line Tessellation Automata

We consider 2OTA as an extension of classical finite automata from words to pictures. The intuition is as follows. In a finite automaton over words, a transition goes from one state to another state while reading a letter. In the 2D case, two directions have to be taken in account: downward and rightward. A transition in a 2OTA goes from two states to a third state while reading a letter, moving at the same time downward from the first state and rightward from the second state. In [10,9] a 2OTA is considered as a cellular automaton where cells change state in a synchronous way, diagonally across the array. This constraint is not necessary. Therefore, we relax it, and consider a run in a 2OTA like a run in a non-deterministic word automaton or tree automaton: a state is non deterministically associated to each position in the picture and we verify afterwards that this association satisfies the transition relation. Moreover, we do not force the set of initial states to be a singleton.

A (non-deterministic) *two-dimensional on-line tessellation automaton* (2OTA) is a tuple $A = (\Sigma, Q, I, F, \delta)$ where Σ is a finite alphabet, Q is a finite set of states, $I \subseteq Q$ is the set of initial states, $F \subseteq Q$ is the set of final, or accepting, states, and $\delta \subseteq Q^2 \times \Sigma \times Q$ is the transition relation. Given three states q_1, q_2, $q_3 \in Q$ and a letter $a \in \Sigma$, we can note more graphically the transition (q_1, q_2, a, q_3) by $\begin{smallmatrix} & q_1 & \\ q_2 & a & q_3 \end{smallmatrix}$. This emphasizes the fact that q_1 and q_2 are respectively above q_3 and to the left of q_3.

Let $p \in \Sigma^{**}$ be a nonempty picture of size (m, n) over the alphabet Σ. A *run* of the automaton A on the picture p is a sequence of states $q_{i,j}$ for (i, j) in $\{0, \ldots, m\} \times \{0, \ldots, n\} \setminus \{(0, 0)\}$ such that there exists $q_0 \in I$ and for all valid i and j: $q_{i,0} = q_{0,j} = q_0$, $q_{m,n} \in F$ and $\begin{smallmatrix} & q_{i-1,j} & \\ q_{i,j-1} & p_{i,j} & q_{i,j} \end{smallmatrix} \in \delta$. A run of the automaton A on the empty picture ε is a state q in $I \cap F$.

A two-dimensional on-line tessellation automaton A *accepts* a picture p if and only if there exists a run of A on p. The language *recognized by* A is the set $\mathcal{L}(A)$ of pictures accepted by A. The family of picture languages recognized by 2OTA is denoted $\mathcal{L}(2\text{OTA})$.

As an example (inspired from one in [9]), a 2OTA recognizing square pictures with a in the main diagonal and b elsewhere is $A = (\Sigma, Q, I, F, \delta)$ with $\Sigma = \{a, b\}$, $Q = \{0, 1, 2\}$, $I = \{0\}$, $F = \{2\}$ and $\delta = \left\{ {}_{0}\overset{0}{a}_{2}, {}_{2}\overset{0}{b}_{1}, {}_{1}\overset{0}{b}_{1}, {}_{0}\overset{2}{b}_{1}, {}_{1}\overset{1}{a}_{2}, {}_{2}\overset{1}{b}_{1}, {}_{1}\overset{1}{b}_{1}, {}_{0}\overset{1}{b}_{1}, {}_{1}\overset{2}{b}_{1} \right\}$. A visual way to represent a run of A on a picture p is to surround the letters in p with states such that the three surrounding states of a letter form with this letter a transition in δ. For instance:

$$\begin{array}{ccccccccc} & 0 & & 0 & & 0 & & 0 & \\ 0 & a & 2 & b & 1 & b & 1 & b & 1 \\ 0 & b & 1 & a & 2 & b & 1 & b & 1 \\ 0 & b & 1 & b & 1 & a & 2 & b & 1 \\ 0 & b & 1 & b & 1 & b & 1 & a & 2 \end{array} \text{ represents a run of } A \text{ on } p = \begin{array}{cccc} a & b & b & b \\ b & a & b & b \\ b & b & a & b \\ b & b & b & a \end{array}.$$

Let $A = (\Sigma, Q, I, F, \delta)$ be a 2OTA. It is an undecidable problem to know whether a state is useful (whether it appears in a run of a recognized picture) or not, see the next proposition. However, the set of reachable states of A is easily computable as the sets I, Q and δ are finite. Furthermore, given a 2OTA, we can obviously restrict its set of states to its reachable states without changing its recognized picture language. We call the resulting automaton the *restriction* of the given 2OTA.

The following proposition summarizes some of the principal properties concerning 2OTA and tiling systems.

Proposition 1.

1. *The class of recognizable picture languages is closed under union, intersection and projection, but not under complement.*
2. $\mathcal{L}(2OTA) = \mathcal{L}(TS)$.
3. *From a 2OTA, a tiling system recognizing the same language is computable in polynomial time (and vice versa).*
4. *The membership problem for the language of some 2OTA is NP-complete.*
5. *The inclusion problem for recognizable picture languages is undecidable.*
6. *Knowing whether a given state belongs to a run of a given 2OTA is an undecidable problem.*

The proofs are given in [9] for (1), (2), and (3). In [11] it is shown that the membership problem in $\mathcal{L}(TS)$ is NP-complete. With (2) and (3) we therefore have (4). In [9] it is shown that the universality problem (whether a picture language is indeed the set of all pictures) is undecidable in $\mathcal{L}(TS)$, we therefore deduce (5). In [9] it is shown that the emptiness problem (whether a picture language is empty) is undecidable in $\mathcal{L}(TS)$ and thus also in $\mathcal{L}(2OTA)$. Since it is easy to transform a 2OTA such that it has a single accepting state, we therefore deduce (6).

4 Simulations

The first motivation of this paper is to obtain a test of inclusion between the languages accepted by two 2OTA. This is done by the possible existence of a simulation between them. The following definition is therefore an extension of the definition of simulations from the case of classical finite word automata.

Definition 1. *Let $A = (\Sigma, Q, I, F, \delta)$ and $A' = (\Sigma, Q', I', F', \delta')$ be two 2OTA. A relation $S \subseteq Q \times Q'$ is a* simulation *over $A \times A'$, and A' is said to* simulate *A if:*

1. *for all $q \in I$ there exists $r \in I'$ such that $(q, r) \in S$,*
2. *for all $(q_1, r_1), (q_2, r_2) \in S$ and $(q_1, q_2, a, q_3) \in \delta$ there exists $r_3 \in Q'$ such that $(r_1, r_2, a, r_3) \in \delta'$ and $(q_3, r_3) \in S$, and*
3. *$(q, r) \in S$ and $q \in F$ imply $r \in F'$.*

For a simulation S we will occasionally note xSy for $(x, y) \in S$. A and A' are said simulation equivalent *if there exist a simulation over $A \times A'$ and a simulation over $A' \times A$.*

From this definition, we get the following expected theorem.

Theorem 1. *Let A, A' be two 2OTA and S a simulation over $A \times A'$. Then $\mathcal{L}(A) \subseteq \mathcal{L}(A')$.*

Proof. A detailed proof can be found in [7]. □

4.1 Autosimulations

The second motivation of this study on simulations over 2OTA is to reduce them thanks to an autosimulation relation. We obtain more: the existence of a minimal 2OTA which is simulation equivalent to a given 2OTA.

Definition 2. *Let $A = (\Sigma, Q, I, F, \delta)$ be a 2OTA. A relation $S \subseteq Q \times Q$ is a* simulation, *or more precisely an* autosimulation, *over A if:*

1. *S is reflexive,*
2. *for all $(q_1, r_1), (q_2, r_2) \in S$ and $(q_1, q_2, a, q_3) \in \delta$ there exists a state r_3 such that $(r_1, r_2, a, r_3) \in \delta$ and $(q_3, r_3) \in S$, and*
3. *$(q, r) \in S$ and $q \in F$ imply $r \in F$.*

From this definition, autosimulations and simulations over 2OTA are related as follows.

Proposition 2. *Let $A = (\Sigma, Q, I, F, \delta)$ be a 2OTA. A relation $S \subseteq Q \times Q$ is an autosimulation over A iff S is a reflexive simulation over $A \times A$.*

The finite set of autosimulations over a given 2OTA A is partially ordered by inclusion. It consequently admits maximal elements. The following lemma addresses the question of their uniqueness.

Theorem 2. *For any 2OTA A there exists a unique maximal autosimulation, denoted $\preccurlyeq_A$, over A. This maximal autosimulation is furthermore reflexive and transitive.*

Proof. A proof can be found in [7]. □

We will henceforth only consider transitive autosimulations, i.e. autosimulations which are preorders.

4.2 Quotienting 2OTA

From an autosimulation S (that we assume transitive), we can define the equivalence relation (reflexive, symmetric and transitive) $S \cap S^{-1}$. We note $[q]_S$, or simply $[q]$ if S is obvious from the context, the class of the state q by the equivalence relation $S \cap S^{-1}$. We extend S on equivalence classes such that $([q],[r]) \in S$ iff $(q,r) \in S$.

Definition 3. *Let $A = (\Sigma, Q, I, F, \delta)$ be a 2OTA and S a (transitive) autosimulation over A. The* quotient automaton $A_{/S} = (\Sigma, Q_{/S}, I_{/S}, F_{/S}, \delta_{/S})$*, of A by S, is such that:*

1. $Q_{/S} = \{[q] \mid q \in Q\}$ *is the set of equivalence classes of* $S \cap S^{-1}$,
2. $I_{/S} = \{[q] \mid q \in I \wedge \forall r \in I, (qSr \Rightarrow rSq)\}$,
3. $F_{/S} = \{[q] \mid q \in F\}$,
4. $$\delta_{/S} = \left\{ \begin{array}{c|l} \substack{[q_1] \\ [q_2]\ a\ [q]} & \substack{q_1 \\ q_2\ a\ q} \in \delta \wedge \forall q_1', q_2', r \in Q, \\ & \left([q_1'] = [q_1] \wedge [q_2'] = [q_2] \wedge qSr \wedge \substack{q_1' \\ q_2'\ a\ r} \in \delta \right) \Rightarrow rSq \end{array} \right\}.$$

Our definition of the quotient automaton is not the classical one. The idea is to forget unnecessary transitions. In the case of a classical word automaton, this amounts at forgetting a transition $q' \xrightarrow{a} q$ if there already exists a transition $q' \xrightarrow{a} r$ with qSr. In this case, q is said to be a *little brother* of r. We have also adapted the initial set with the same idea: we keep only maximal initial states, maximality being defined with respect to the preorder S. From a 2OTA A and a simulation S over A the quotient $A_{/S}$ can be computed in polynomial time.

Lemma 1. *Let $(\alpha_1, \alpha_2, a, \alpha_3) \in \delta_{/S}$ be a transition in the quotient automaton. Then for all $r_1 \in \alpha_1$, $r_2 \in \alpha_2$ there exists $r_3 \in \alpha_3$ such that $(r_1, r_2, a, r_3) \in \delta$.*

Proof. By definition of $\delta_{/S}$ there exist three states q_1, q_2 and q_3 such that $[q_1] = \alpha_1$, $[q_2] = \alpha_2$, $[q] = \alpha_3$, $(q_1, q_2, a, q_3) \in \delta$ and

$$\forall q_1', q_2', r \in Q, ([q_1'] = [q_1] \wedge [q_2'] = [q_2] \wedge q_3Sr \wedge (q_1', q_2', a, r) \in \delta) \Rightarrow rSq_3. \quad (2)$$

Let $r_1 \in \alpha_1 = [q_1]$ and $r_2 \in \alpha_2 = [q_2]$. Then we have q_1Sr_1 and q_2Sr_2. From $(q_1, q_2, a, q_3) \in \delta$ and the definition of an autosimulation, there exists a state r_3 such that $(r_1, r_2, a, r_3) \in \delta$ and q_3Sr_3. Finally, applying (2) when q_1' is r_1, q_2' is r_2 and r is r_3 leads to r_3Sq_3. Consequently $r_3 \in \alpha_3$ completes the proof. □

Lemma 2. *Let* $(q_1, q_2, a, q_3) \in \delta$ *be a transition in* A. *Then there exists* q *such that* $q_3 S q$ *and* $([q_1], [q_2], a, [q]) \in \delta_{/S}$.

Proof. Let

$$R = \{r \mid q_3 S r \wedge \exists r_1, r_2 \in Q, [r_1] = [q_1] \wedge [r_2] = [q_2] \wedge (r_1, r_2, a, r) \in \delta\}.$$

Let q be such a maximal element of R for the preorder S, i.e. an element of R such that $\forall q' \in R, q S q' \Rightarrow q' S q$. The state q always exists since R is a subset of the finite set Q. By definition of R and q, we have $q_3 S q$. Let r_1 and r_2 be such that $[r_1] = [q_1]$, $[r_2] = [q_2]$ and $(r_1, r_2, a, q) \in \delta$. It remains to prove that $([q_1], [q_2], a, [q]) \in \delta_{/S}$.

Let q_1', q_2' and r be any three states in Q such that qSr and

$$[q_1'] = [r_1] \wedge [q_2'] = [r_2] \wedge (q_1', q_2', a, r) \in \delta. \tag{3}$$

By transitivity of S, $q_3 S r$ holds and means together with (3) that r is an element of R. Consequently rSq. By definition of $\delta_{/S}$ it results from $(r_1, r_2, a, q) \in \delta$ that $([r_1], [r_2], a, [q])$ is in $\delta_{/S}$. The equalities $[r_1] = [q_1]$ and $[r_2] = [q_2]$ complete the proof. □

Theorem 3. *Let* $A = (\Sigma, Q, I, F, \delta)$ *be a 2OTA and* S *be a simulation over* A. *Then* A *and* $A_{/S}$ *are simulation equivalent.*

Proof. Let S' and S'' be the binary relations respectively defined over $A \times A_{/S}$ and $A_{/S} \times A$ by $S' = \{(q, [r]) \mid qSr\}$ and $S'' = \{([q], q) \mid q \in Q\}$. We separately prove that S' and S'' are simulations. A detailed proof can be found in [7]. □

As an immediate consequence, doing such a quotient does not modify the recognized language.

Corollary 1. *Let* $A = (\Sigma, Q, I, F, \delta)$ *be a 2OTA and* S *be a simulation over* A. *Then* A *and* $A_{/S}$ *recognize the same picture language.*

4.3 The Minimal Simulation-Equivalent 2OTA

In the preceding sections, we have proved the existence of a maximal autosimulation. Then we have used this maximal autosimulation to reduce a given 2OTA more than by doing a classical quotient. We are now ready to prove that the restriction of this reduced 2OTA to its reachable part is indeed the minimal one which is still simulation equivalent with the given 2OTA.

Theorem 4. *Let* A *and* A' *be two 2OTA. Let* $\preccurlyeq_A$ *and* $\preccurlyeq_{A'}$ *be their respective maximal simulation relations. Then the restrictions of* $A_{/\preccurlyeq_A}$ *and of* $A'_{/\preccurlyeq_{A'}}$ *are simulation equivalent if and only if they are isomorphic.*

Proof. The reverse implication is straightforward. Two isomorphic 2OTA are obviously simulation equivalent. To prove the direct implication we name S (resp. S') a simulation over $A_{/\preccurlyeq_A} \times A'_{/\preccurlyeq_{A'}}$ (resp. $A'_{/\preccurlyeq_{A'}} \times A_{/\preccurlyeq_A}$). Then we define the binary relation $f = S \cap S'^{-1}$ and we successively prove that f is a partial function, f is total, f is one-to-one and finally that f is an isomorphism between the respective restrictions of $A_{/\preccurlyeq_A}$ and $A'_{/\preccurlyeq_{A'}}$. Due to lack of space, the complete proof is not reproduced here. It can be found in [7]. □

Corollary 2. *Let A be a 2OTA and $\preccurlyeq_A$ its maximal simulation relation. Then the restriction of $A_{/\preccurlyeq_A}$ is the smallest 2OTA which is simulation equivalent to A.*

Proof. A proof can be found in [7]. □

5 How to Compute Simulations

A transition in a 2OTA resembles a transition in a tree automaton. Indeed, the transition (q_1, q_2, a, q) can be viewed as the rule $(q_1, q_2) \xrightarrow{a} q$ in a tree automaton. In [1] a polynomial time algorithm is given to compute what is called the maximal *upward simulation* in a tree automaton. In this section we reduce computation of maximal simulations in 2OTA to computation of maximal upward simulations in tree automata. Before that we shortly recall useful results about upward simulations in binary tree automata. In particular we do not define the semantics of tree automata which is not related to the present subject.

Definition 4. *A* binary Tree Automaton *(bTA) is a tuple $T = (\Sigma, Q, F, \delta)$ where Σ is a finite alphabet, Q is a finite set of states, $F \subseteq Q$ is a set of final states and $\delta \subseteq Q^2 \times \Sigma \times Q$ is a finite set of transitions.*

As usual, we can note $(q_1, q_2) \xrightarrow{a} q$ whenever $(q_1, q_2, a, q) \in \delta$. If we forget that 2OTA recognize pictures and bTA recognize binary trees, their definitions are similar, up to an extra set of initial states for 2OTA. This similarity is used in the remainder of this section. An *upward simulation* over a bTA $T = (\Sigma, Q, F, \delta)$ is a relation $S \subseteq Q \times Q$ such that $(q_1, q_2) \xrightarrow{a} q$ and $q_i S r_i$ for a given $i \in \{1, 2\}$ imply the existence of a state r such that qSr and $(r_1, q_2) \xrightarrow{a} r$ if $i = 1$ and $(q_1, r_2) \xrightarrow{a} r$ if $i = 2$. Note that the set of final states is not present in this definition. So let us call a *simulation without final states* (wfs-simulation) a relation $S \subseteq Q \times Q$ such that $(q_1, q_2) \xrightarrow{a} q$ and $q_i S r_i$ for all $i \in \{1, 2\}$ imply the existence of a state r such that $(r_1, r_2) \xrightarrow{a} r$ and qSr. We immediately get the following lemma.

Lemma 3. *Let T be a bTA and S a reflexive and transitive relation over T. Then S is a wfs-simulation over T iff it is an upward simulation over T.*

Proof. The fact that a reflexive wfs-simulation is also an upward simulation is obvious. Now let us consider a transitive upward simulation S over $T = (\Sigma, Q, F, \delta)$, a transition $(q_1, q_2) \xrightarrow{a} q$ in δ and two states $r_1, r_2 \in Q$ such that $q_i S r_i$ for all $i \in \{1, 2\}$. As S is an upward simulation there exists a state r' and a transition $(r_1, q_2) \xrightarrow{a} r'$ such that qSr'. By the same argument on this new transition there also exists a state r and a transition $(r_1, r_2) \xrightarrow{a} r$ such that $r'Sr$. By transitivity of S we get qSr, which concludes the proof. □

In [3], the existence and the uniqueness of a maximal upward simulation over a bTA T are shown. This maximal upward simulation, noted $\preccurlyeq_T$, is furthermore shown reflexive and transitive. As argued in [1], given a preorder R, it can still be shown that there is a unique maximal upward simulation included in R, noted $\preccurlyeq_T^R$, over a given bTA T. This relation $\preccurlyeq_T^R$ is reflexive and transitive. Still in [1], a polynomial time algorithm is given to compute $\preccurlyeq_T$. But indeed, a straightforward extension of the construction used in their proof (adding R as a constraint on the initial partition-relation pair) leads to the following stronger result.

Theorem 5. *Let $T = (\Sigma, Q, F, \delta)$ be a bTA and $R \subseteq Q \times Q$ be a reflexive and transitive relation (preorder). The maximal upward simulation $\preccurlyeq_T^R$ included in R is reflexive, transitive and computable in polynomial time.*

Corollary 3. *The maximal autosimulation over a 2OTA A is computable in polynomial time.*

Proof. Let $A = (\Sigma, Q, I, F, \delta)$ be a 2OTA. Then $T = (\Sigma, Q, F, \delta)$ is a bTA. Let $R \subseteq Q \times Q$ be the preorder such that qRr iff $(q \in F \Rightarrow r \in F)$. This preorder simply defines a partition of Q in two blocks: F and $Q \setminus F$, with $(Q \setminus F) \times F \subseteq R$. By Condition (3) of Definition 2 any autosimulation over A is included in R. This means that the maximal autosimulation $\preccurlyeq_A$ over A is also the maximal reflexive and transitive wfs-simulation over T included in R. By Lemma 3 it is also the maximal upward simulation $\preccurlyeq_T^R$ included in R, since $\preccurlyeq_T^R$ is reflexive and transitive. By Theorem 5 it is computable in polynomial time. □

Unfortunately, unlike in word automata, deciding the existence of a simulation between two 2OTA is not feasible in polynomial time.

Theorem 6. *Deciding whether a 2OTA is simulated by another 2OTA is a NP-complete problem.*

Proof. A proof can be found in [7]. □

However, in 2OTA the inclusion problem is undecidable. In this perspective, the simulation test is a sufficient condition of inclusion which does not have a worst time complexity than the one of the membership problem.

6 The Case of Backward Simulations

In word automata or tree automata, there are two different simulations: a *forward simulation*, from initial states to final states, and a *backward simulation*, from final states to initial states. They both imply language inclusion. The simulations in 2OTA considered so far in this paper are forward simulations. This section establishes the noticeable fact that what could correspond to backward simulation in the case of 2OTA does not imply language inclusion.

Definition 5. *Let $A = (\Sigma, Q, \{q_0\}, F, \delta)$ and $A' = (\Sigma, Q', \{q_0'\}, F', \delta')$ be two 2OTA. A relation $S \subseteq Q \times Q'$ is a* backward simulation *over $A \times A'$ if:*

1. $(q_0, q_0') \in S$,
2. *for all* $(q_3, r_3) \in S$ *and* $(q_1, q_2, a, q_3) \in \delta$ *there exist* $r_1, r_2 \in Q'$ *such that* $(r_1, r_2, a, r_3) \in \delta'$, $(q_1, r_1) \in S$ *and* $(q_2, r_2) \in S$,
3. *for all* $q \in F$ *there exists* $r \in F'$ *such that* $(q, r) \in S$.

In order to simplify the problem, initial sets are restricted to singletons (this is implicitly done in the case of tree automata where the initial state indeed corresponds to rules with an empty left hand side).

Proposition 3. *In 2OTA, backward simulation does not imply language inclusion.*

Proof. A proof can be found in [7]. □

7 Conclusion and Future Work

In this paper, we have applied the notion of simulation to a class of 2D automata, namely the one of two dimensional on-line tessellation automata. We have obtained many desired results. A first result is a non trivial sufficient condition to decide the inclusion problem between the languages recognized by two 2OTA. This is an important result since this problem is undecidable in general. The fact established here that this sufficient condition is decidable in NP-complete time is the trade-off to pay. We also show how to reduce the size of a 2OTA and obtain the minimal automaton with respect to simulation equivalence. This is also an important result since, although decidable, the membership problem in all comparable classes of 2D automata is NP-complete in time, and reducing the size of a given automaton dramatically reduces the time to obtain a response to the membership question (less states have to be tried). We have shown that this reduction can fortunately be done in polynomial time.

Now that we have a first test for inclusion we can come back to our initial motivation: the extension of the regular model-checking paradigm to two dimensional languages. This will probably require to relax our definition of simulation and take into account the scanning strategy to recognize a picture [4,12] and identify meta-transitions. This will be done in accordance with the examples we will treat.

The structure of 2OTA is similar to the one of binary tree automata. Indeed, both can be viewed as relational structures with two successors relations S_1 and S_2 [15]. The difference lies in the fact that in 2OTA, we necessarily have that the composition of the two relations commutes, i.e. $S_1 \circ S_2 = S_2 \circ S_1$. With this point of view, our work can be seen as the extension of the notion of simulation from trees to tree structures with a constraint. Another noticeable difference between the two models is that in 2OTA what corresponds to downward simulation in trees does not imply language inclusion. The similarity with tree automata has allowed us to use the algorithm of [1] for the computation of the maximal autosimulation in a 2OTA. In the other direction the study of simulations in trees can benefit from our work. For example, applications of the existence of a minimal simulation-equivalent tree automaton have not been investigated. We plan to make this study.

Acknowledgement. The authors are grateful to the anonymous referees for helpful comments and suggestions.

References

1. Abdulla, P.A., Bouajjani, A., Holík, L., Kaati, L., Vojnar, T.: Computing Simulations over Tree Automata. In: Ramakrishnan, C.R., Rehof, J. (eds.) TACAS 2008. LNCS, vol. 4963, pp. 93–108. Springer, Heidelberg (2008)
2. Abdulla, P.A., Chen, Y.F., Holík, L., Mayr, R., Vojnar, T.: When Simulation Meets Antichains. In: Esparza, J., Majumdar, R. (eds.) TACAS 2010. LNCS, vol. 6015, pp. 158–174. Springer, Heidelberg (2010)
3. Abdulla, P.A., Legay, A., d'Orso, J., Rezine, A.: Tree regular model checking: A simulation-based approach. J. Log. Algebr. Program. 69(1-2), 93–121 (2006)
4. Anselmo, M., Giammarresi, D., Madonia, M.: A computational model for tiling recognizable two-dimensional languages. Theor. Comput. Sci. 410(37), 3520–3529 (2009)
5. Bozapalidis, S., Grammatikopoulou, A.: Recognizable Picture Series. Journal of Automata, Languages and Combinatorics 10(2/3), 159–183 (2005)
6. Bustan, D., Grumberg, O.: Simulation-based minimization. ACM Trans. Comput. Logic 4(2), 181–206 (2003)
7. Cécé, G., Giorgetti, A.: Simulations for a Class of Two-Dimensional Automata. Research Report RR-7425, INRIA (October 2010), `http://hal.inria.fr/inria-00527077/en/`
8. Cherubini, A., Pradella, M.: Picture Languages: From Wang Tiles to 2D Grammars. In: Bozapalidis, S., Rahonis, G. (eds.) CAI 2009. LNCS, vol. 5725, pp. 13–46. Springer, Heidelberg (2009)
9. Giammarresi, D., Restivo, A.: Two-Dimensional Languages. In: Salomaa, A., Rozenberg, G. (eds.) Handbook of Formal Languages, Beyond Words, vol. 3, pp. 215–267. Springer, Berlin (1997)
10. Inoue, K., Nakamura, A.: Some properties of two-dimensional on-line tessellation acceptors. Inf. Sci. 13(2), 95–121 (1977)
11. Lindgren, K., Moore, C., Nordahl, M.: Complexity of Two-Dimensional Patterns. Journal of Statistical Physics 91(5-6), 909–951 (1998)
12. Lonati, V., Pradella, M.: Deterministic recognizability of picture languages with Wang automata. Discrete Math. & Theor. Comput. Sci. 12(4), 73–94 (2010)
13. Milner, R.: An Algebraic Definition of Simulation Between Programs. In: IJCAI, pp. 481–489 (1971)
14. The Smart Surface Project (2010), `http://www.smartsurface.cnrs.fr/`
15. Thomas, W.: Uniform and nonuniform recognizability. Theor. Comput. Sci. 292(1), 299–316 (2003)

Δ-Clearing Restarting Automata and CFL*

Peter Černo and František Mráz

Department of Computer Science
Charles University, Faculty of Mathematics and Physics
Malostranské nám. 25, 118 00 PRAHA 1, Czech Republic
petercerno@gmail.com, mraz@ksvi.ms.mff.cuni.cz

Abstract. Δ-clearing restarting automata represent a new restricted model of restarting automata which, based on a limited context, can either delete a substring of the current content of its tape or replace a substring by a special auxiliary symbol Δ, which cannot be overwritten anymore, but it can be deleted later. The main result of this paper consists in proving that besides their limited operations, Δ-clearing restarting automata recognize all context-free languages.

Keywords: analysis by reduction, context-free languages, Δ-clearing restarting automata, formal languages.

1 Introduction

Restarting automata [6] were introduced as a tool for modeling some techniques used for natural language processing. In particular, they are used for analysis by reduction, which is a method for checking (syntactical) correctness or non-correctness of a sentence. Analysis by reduction consists in iterative application of (non)correctness preserving simplifications to the given input sentence until it cannot be simplified anymore. If we obtain a correct simple sentence, we accept, otherwise we reject. While restarting automata are quite general (see [11]), they still lack some properties which could facilitate their wider use. Among others they could benefit from simpler definition.

Recently, Kutrib et al. in [8] and [9] introduced stateless restarting automata. For monotone and/or deterministic version of these automata, if they can use auxiliary symbols in rewriting, then they have the same power as the corresponding versions with states ([8,10]). However, the stateless versions of restarting automata without auxiliary symbols are strictly weaker than the respective versions which can use states.

Černo and Mráz [2] introduced an even more simplified version of restarting automata called clearing restarting automata. While general restarting automata see the whole part of the current sentence (word) to the left (and possibly also

* This work was partially supported by the Grant Agency of Charles University under Grant-No. 272111/MFF/A-INF and by the Czech Science Foundation under Grant-No. P103/10/0783 and Grant-No. P202/10/1333.

G. Mauri and A. Leporati (Eds.): DLT 2011, LNCS 6795, pp. 153–164, 2011.

to the right) of the place they rewrite, the rewriting done by clearing restarting automata depends only on a fixed context around the rewritten subword. Moreover, clearing restarting automata can only "clear" a subword, i.e. completely delete a subword based on limited context around the "cleared" subword. Hence the automata have no states and in one cycle they rewrite (exactly once) at any place according to some of their finitely many instructions. Obviously, such automata are more restricted than the weakest version of the stateless restarting automata (the so-called stateless R-automata). It turned out that clearing restarting automata are rather limited. While they can recognize all regular languages and even some languages that are not context-free, they cannot recognize all context-free languages (see [2]). Hence there were introduced Δ-clearing automata and Δ^*-clearing automata that can use an auxiliary symbol Δ. Besides deleting a subword, Δ-clearing automata can rewrite a subword by the special symbol Δ, which can be deleted in later cycles, too. Δ^*-clearing automata are even stronger, as they can also rewrite a subword by Δ^i, where i is not greater than the length of the rewritten word.

In [1] we have shown that a Δ-clearing automaton can accept the Greibach's "hardest context-free language" and later in [2] there was shown that Δ^*-clearing automata can recognize all context-free languages. [2] conjectured that also Δ-clearing automata can recognize all context-free languages (CFL). In this paper we prove the conjecture.

The paper is divided into several sections. In Section 2 we introduce a general concept called *context rewriting system* which will serve us as a framework for Δ-clearing restarting automata and their extended version Δ^*-clearing restarting automata. The main source for this section is [2]. In Section 3 we describe a special coding used by Δ-clearing restarting automata to encode some information into the tape. In Section 4 we describe the algorithm behind the Δ-clearing restarting automaton recognizing a given context-free language. Because of the page limit we omit most of the proofs in this paper and refer the interested reader to the technical report [3] (`http://popelka.ms.mff.cuni.cz/cerno/files/cerno_mraz_dclra_and_cfl.pdf`).

We use the standard notation from the theory of formal languages and automata as in Hopcroft and Ullman [5].

2 Theoretical Background

In this section we introduce a general concept called *context rewriting system* which will serve us as a framework for Δ-clearing restarting automata and their extended version Δ^*-clearing restarting automata.

Definition 1 ([2]). *Let k be a positive integer. A k-*context rewriting system *(k-CRS for short) is a system $R = (\Sigma, \Gamma, I)$, where Σ is an input alphabet, $\Gamma \supseteq \Sigma$ is a working alphabet not containing the special symbols ¢ and \$, called* sentinels, *and I is a finite set of* instructions *of the form:*

$$(x, z \to t, y) ,$$

where x is called left context, $x \in \Gamma^k \cup ¢ \cdot \Gamma^{\leq k-1}$, *$y$ is called* right context, $y \in \Gamma^k \cup \Gamma^{\leq k-1} \cdot \$$ *and $z \to t$ is called* rule, $z, t \in \Gamma^*$.

A word $w = uzv$ can be rewritten *into utv (denoted as $uzv \to_R utv$) if and only if there exists an instruction $i = (x, z \to t, y) \in I$ such that x is a suffix of $¢ \cdot u$ and y is a prefix of $v \cdot \$$. We often underline the rewritten part of the word w, and if the instruction i is known we use $\to_R^{(i)}$ instead of $\to_R$, i.e. $u\underline{z}v \to_R^{(i)} utv$. The relation $\to_R \subseteq \Gamma^* \times \Gamma^*$ is called* rewriting relation.

The reduction language *(*reduction characteristic language, *respectively) associated with R is defined as $L^-(R) = \{w \in \Sigma^* \mid w \to_R^* \lambda\}$ ($L_C^-(R) = \{w \in \Gamma^* \mid w \to_R^* \lambda\}$, respectively), where $\to_R^*$ is the reflexive and transitive closure of $\to_R$. Note that, by definition, $\lambda \in L^-(R)$ ($\lambda \in L_C^-(R)$, respectively).*

Naturally, if we increase the length of contexts used in instructions of a CRS, we do not decrease their expressiveness ([2]).

It is easy to see that general k-CRS can simulate any type 0 grammar (according to the Chomsky hierarchy [5]). Hence we will study their restricted versions only. First we introduce a *clearing restarting automaton* which is a k-CRS such that $\Sigma = \Gamma$ and all rules in its instructions are of the form $z \to \lambda$, where $z \in \Sigma^+$.

Definition 2 ([2]). *Let k be a positive integer. A k-*clearing restarting automaton *(k-cl-RA for short) is a system $M = (\Sigma, I)$, where $R = (\Sigma, \Sigma, I)$ is a k-CRS such that for each instruction $i = (x, z \to t, y) \in I$ it holds $z \in \Sigma^+$ and $t = \lambda$. Since t is always the empty word, we use the notation $i = (x, z, y)$. The* width *of the instruction $i = (x, z, y)$ is $|i| = |xzy|$.*

The k-cl-RA M recognizes *the language $L(M) = \{w \in \Sigma^* \mid w \vdash_M^* \lambda\} = L^-(M)$, where $\vdash_M$ is the* rewriting relation $\to_R$ *of R.*

Example 1. Let $M = (\Sigma, I)$ be the 1-cl-RA with $\Sigma = \{a, b\}$ and I consisting of the following two instructions:

$$\begin{array}{ll} (1) & (a, ab, b), \\ (2) & (¢, ab, \$). \end{array}$$

Then we have $aaa\underline{ab}bbb \vdash_M^{(1)} aa\underline{ab}bb \vdash_M^{(1)} a\underline{ab}b \vdash_M^{(1)} \underline{ab} \vdash_M^{(2)} \lambda$ which means that $aaaabbbb \vdash_M^* \lambda$. So the word $aaaabbbb$ is accepted by M. It is easy to see that M recognizes the language $L(M) = \{a^n b^n \mid n \geq 0\}$.

In the following, k-cl-RA (cl-RA, respectively) denotes the class of all k-clearing restarting automata (clearing restarting automata, respectively), where cl-RA $= \bigcup_{k=1}^{\infty} k$-cl-RA.

Remark 1. By definition, each cl-RA accepts λ. If we say that a cl-RA M *recognizes* (or *accepts*) a language L, we always mean that $L(M) = L \cup \{\lambda\}$. This implicit acceptance of the empty word can be avoided by a slight modification of the definition of clearing restarting automata, or even context rewriting systems, but in principle, we would not get a more powerful model.

Clearing restarting automata are studied in [2]. We only mention that they can recognize all regular languages, some context-free languages and even some non-context-free languages.

In [2] there were introduced two extended versions of clearing restarting automata – the so-called Δ-clearing restarting automata and Δ^*-clearing restarting automata. Both of them can use a single auxiliary symbol Δ only. Δ-clearing restarting automata can leave a mark – a symbol Δ – at the place of deleting besides rewriting into the empty word λ. Δ^*-clearing restarting automata can rewrite a subword w into Δ^k where k is bounded from above by the length of w.

Definition 3 ([2]). *Let k be a positive integer. A k-Δ-*clearing restarting automaton *(k-Δcl-RA for short) is a system $M = (\Sigma, I)$, where $R = (\Sigma, \Gamma, I)$ is a k-*CRS *such that $\Delta \notin \Sigma$, $\Gamma = \Sigma \cup \{\Delta\}$, and for each instruction $i = (x, z \to t, y) \in I$: $z \in \Gamma^+$ and either $t = \lambda$, or $t = \Delta$.*

*Analogously, a k-Δ^*-*clearing restarting automaton *(k-Δ^*cl-RA for short) is a system $M = (\Sigma, I)$, such that for each instruction $i = (x, z \to t, y) \in I$: $z \in \Gamma^+$ and $t = \Delta^i$, where $0 \le i \le |z|$.*

The k-Δcl-RA (k-Δ^*cl-RA) M recognizes *the language $L(M) = \{w \in \Sigma^* \mid w \vdash_M^* \lambda\} = L^-(M)$, where $\vdash_M$ is the* rewriting relation *$\to_R$ of $R = (\Sigma, \Gamma, I)$.*

The characteristic language *of M is the language $L_C(M) = L_C^-(M)$.*

In what follows we will first repeat a result from [2] showing that 1-Δ^*cl-RA are powerful enough to recognize all context-free languages. The following proof utilizes the same idea as in [2], but a little modified encoding of nonterminals of a context-free grammar using Δ's.

Theorem 1. *For each context-free language L there exists a* 1-Δ^*cl-RA-*automaton M recognizing L.*

Proof. Let L be a context-free language. Then there exists a context-free grammar $G = (V_N, V_T, S, P)$ in Chomsky normal form generating the language $L(G) = L \setminus \{\lambda\}$. Let $V_N = \{N_1, \ldots, N_m\}$, $S = N_1$ and $\Delta \notin V_N \cup V_T$, and let $G' = (V_N, V_T', S, P')$ be the grammar obtained from G by adding a new terminal symbol Δ to V_T ($V_T' = \Sigma \cup \{\Delta\}$), and adding new productions $N_i \to a\Delta^i b$ to P, for all $1 \le i \le m$ and all $a, b \in V_T$. Obviously, $L(G') \cap \Sigma^* = L(G)$. We will show that we can effectively construct a 1-Δ^*cl-RA M such that $L_C(M) = L(G') \cup \{\lambda\}$ and $L(M) = L_C(M) \cap \Sigma^* = (L(G') \cup \{\lambda\}) \cap \Sigma^* = L(G) \cup \{\lambda\}$.

For the automaton M all the words $a\Delta^i b$ for all $a, b \in \Sigma$ represent "codes" for the nonterminal N_i. The letters $a, b \in \Sigma$ serve as separators for distinguishing several consecutive encoded nonterminals.

The automaton M works in a bottom-up manner. If the automaton recognizes that some subword w of the input tape can be derived from some nonterminal N_i, then the automaton can (nondeterministically) replace this subword w by a corresponding code Δ^i. Or to be more precise, the automaton M replaces only the inner part of the subword w by the code Δ^i in order to leave the first and the last letter of w as a separator. If the word on the input tape is short enough and belongs to the language $L(G')$ then the automaton M just erases the whole input word in a single step.

The next proposition ([3]) ensures that for correct recognition of $L(G')$ the automaton M can be restricted to replacing only subwords of length limited by a constant.

Proposition 1 ([3]). *Let* $G' = (V_N, V'_T, S, P')$ *be the grammar constructed above and* w *be a word from* $L(G')$ *of length* $|w| > c = |V_N| + 2$. *Then there exist words* x, y, z *from* $(V'_T)^*$ *such that* $|z| \leq 2c$ *and* $S \Rightarrow^*_{G'} xN_iy \Rightarrow^*_{G'} xzy = w$ *for some nonterminal* $N_i \in V_N$.

Now we construct the 1-Δ^*cl-RA $M = (\Sigma, I)$, where $\Sigma = V_T$, $\Gamma = \Sigma \cup \{\Delta\} = V'_T$. First, we set:

$$I_1 = \{(¢, w \to \lambda, \$) \mid w \in L(G'), |w| \leq c\}.$$

For every $i \in \{1, 2, \ldots, m\}$ let us define:

$$L_i = \{z \in \Gamma^* \mid N_i \Rightarrow^*_{G'} z, c < |z| \leq 2c\}.$$

For every such $z \in L_i$, $z = z_1 z_2 \ldots z_{s-1} z_s$, consider the instruction:

$$(z_1, z_2 \ldots z_{s-1} \to \Delta^i, z_s)\,.$$

This instruction rewrites the inner part of the word z to Δ^i leaving z_1 and z_s as separators. Let I_2 be the set of all such instructions. (Observe that $z_1, z_s \in \Sigma$, and both I_1 and I_2 are finite sets of instructions). Then $M = (\Sigma, I_1 \cup I_2)$ is the required automaton.

This completes the proof of Theorem 1. □

We have shown that 1-Δ^*cl-RA are able to recognize all context-free languages (containing the empty word λ – see Remark 1). This result opens an interesting question whether it is possible to transform each Δ^*cl-RA into an equivalent Δcl-RA. If we are interested only in the problem whether Δcl-RA can recognize all context-free languages, then we do not need to do this transformation to all Δ^*cl-RA. We just need to do this transformation to such Δ^*cl-RA which were obtained from a context-free grammar, as was shown above. Moreover, the aforementioned construction can be generalized, i.e. we can put some extra restrictions on the instructions of the resulting Δ^*cl-RA. We can generalize the construction of the grammar G' and the corresponding Δ^*cl-RA M in the following four ways:

1. We can choose a minimal length $m_0 \geq 1$ of codes for nonterminals, i.e. we code N_i by using at least m_0 consecutive letters Δ.
2. We can choose a minimal length $m_1 \geq 1$ of shortening for each reduction, i.e. for each instruction $(x, u \to \Delta^r, y)$ such that $r \geq 1$, we guarantee that $|u| - |\Delta^r| \geq m_1$.
3. We can choose a number of codes $m_2 \geq 1$ representing one nonterminal, i.e. we code N_i by using $m_0 + m_2(i-1) + j - 1$ consecutive letters Δ, for all $j \in \{1, 2, \ldots m_2\}$.

4. We can choose a length $k \geq 1$ of the separator, i.e. instead of one letter we use k consecutive arbitrary letters from Σ as a separator.

The details of the construction can be found in the technical report [3]. One of the important consequences we obtain is the following Lemma 1.

Lemma 1. *For each $t \geq 1$, we can set the parameters m_1 and k (depending only on t) in such a way, that for each instruction $(x, u \to \Delta^r, y) \in I_2$, there exists a subword $v \in \Sigma^*$ (not containing Δ) in the word u with the length $|v| \geq t$. Moreover, $m_1, k = \Theta(t)$. The parameters m_0 and m_2 can be chosen arbitrarily.*

Now we will outline the basic idea behind the transformation of a k-Δ^*cl-RA M obtained from the generalized construction into an equivalent Δcl-RA N. Clearly, by transforming each instruction $\phi = (x, u \to \Delta^r, y)$ of M, where $r > 1$ and $u = u_1 \ldots u_s$, into the following set of instructions:

$$\begin{aligned}
\phi_1 &= (x, u_1 \to \Delta, u_2 u_3 \ldots u_s y),\\
\phi_2 &= (x\Delta, u_2 \to \Delta, u_3 u_4 \ldots u_s y),\\
\ldots&\\
\phi_{r-1} &= (x\Delta^{r-2}, u_{r-1} \to \Delta, u_r u_{r+1} \ldots u_s y),\\
\phi_r &= (x\Delta^{r-1}, u_r u_{r+1} \ldots u_s \to \Delta, y),
\end{aligned}$$

we get only the inclusion $L(M) \subseteq L(N)$. The opposite inclusion is not guaranteed, as there can exist two different instructions ϕ and ψ in M such that they have different partial instructions ϕ_i and ψ_j applicable in the same context. One possible way how to avoid such situations is to introduce some new special instructions, which will encode some extra information into u by rewriting some letters with auxiliary Δ-symbols. Lemma 1 guarantees a long enough subword $v \in \Sigma^*$ in u, which we can use to encode this information. In the rest of this paper we describe how to accomplish this task by using one specific coding.

3 Coding

We would like to encode information in an arbitrary, sufficiently long word $w \in \Sigma^*$ only by replacing some letters of w by symbols $\Delta \notin \Sigma$. Moreover, we require that it should be possible to recover the original word w at any time. Such encoding is guaranteed by the following theorem which can be easily proven by using Hall's Theorem [4].

Theorem 2 (Coding 1 [3]). *Let Σ be a finite nonempty alphabet and $\Delta \notin \Sigma$. Then there exist a positive integer B and a table T of triples (x, z, y), $xzy \in \Sigma^B$, $z \in \Sigma$, such that:*

1. *$\{xzy \mid (x, z, y) \in T\} = \Sigma^B$,*
2. *For each pair (x, y), $xy \in \Sigma^{B-1}$ there exists exactly one $z \in \Sigma : (x, z, y) \in T$.*

This theorem guarantees that if we take any word $w \in \Sigma^B$ then there exists a factorization $w = xzy$, such that $(x, z, y) \in T$. Now if we replace the letter z by Δ, we do not lose any information, since we are able to recover the letter z from the context (x, y) by using the table T.

Example 2. For $\Sigma = \{a, b, c\}$ we only give the resulting bijection:

$$\begin{array}{lllll} aaa \leftrightarrow \Delta aa, & aab \leftrightarrow \Delta ab, & aac \leftrightarrow aa\Delta, & aba \leftrightarrow \Delta ba, & abb \leftrightarrow a\Delta b, \\ abc \leftrightarrow ab\Delta, & aca \leftrightarrow a\Delta a, & acb \leftrightarrow ac\Delta, & acc \leftrightarrow a\Delta c, & baa \leftrightarrow b\Delta a, \\ bab \leftrightarrow b\Delta b, & bac \leftrightarrow ba\Delta, & bba \leftrightarrow bb\Delta, & bbb \leftrightarrow \Delta bb, & bbc \leftrightarrow \Delta bc, \\ bca \leftrightarrow \Delta ca, & bcb \leftrightarrow bc\Delta, & bcc \leftrightarrow b\Delta c, & caa \leftrightarrow c\Delta a, & cab \leftrightarrow ca\Delta, \\ cac \leftrightarrow \Delta ac, & cba \leftrightarrow cb\Delta, & cbb \leftrightarrow c\Delta b, & cbc \leftrightarrow c\Delta c, & cca \leftrightarrow cc\Delta, \\ ccb \leftrightarrow \Delta cb, & ccc \leftrightarrow \Delta cc. & & & \end{array}$$

Next consider the following sample word $w = accbabccacaabbcabcbcacaa$. We can factorize it into the groups of $B = 3$ letters:

$$w \;=\; acc \mid bab \mid cca \mid caa \mid bbc \mid abc \mid bca \mid caa.$$

We can encode some information, e.g. $i = 11001010$, into the word w without losing any information. Using the above bijection, we *mark* the groups of w that correspond to 1's in the information i by Δ:

$$w' \;=\; a\Delta c \mid b\Delta b \mid cca \mid caa \mid \Delta bc \mid abc \mid \Delta ca \mid caa.$$

It is easy to see that we can recover the original word w.

By applying the encoding from Example 2 we can store an n-bit information in any word w of length at least nB. However, we need to "see" the whole word w in order to correctly define the groups of B letters.

The problem with Coding 1 is that by seeing only a factor w of a whole input word we may not be able to decode Δ symbols. If w does not start with the left sentinel ¢ then we are not able to correctly factorize w into the groups of B letters, and thus we are not able to recover Δ symbols occurring in the word w.

Fortunately, there exists a simple trick how to avoid this problem. In order to correctly factorize the input word w into the groups of B letters we need only some "fixed point", which exactly defines the starting position of the first group of the correct factorization. The left sentinel ¢ is one example of such fixed point. Thus in the first phase we distribute such fixed points throughout the whole input tape starting at the left sentinel ¢. The distances between two consecutive fixed points will be approximately constant. We illustrate this idea on the following simplified example. Suppose that we have the following factor:

$$w = abacc\Delta bacbbacacbcbaacbcbacbacabab$$

The symbol Δ in w represents our fixed point and defines the following factorization $w = abacc\boldsymbol{\Delta} \mid bac \mid bba \mid cac \mid bcb \mid aac \mid bcb \mid acb \mid aca \mid bab$. We place the next fixed point into the 9th group to the right from the highlighted fixed point **Δ** (by using the bijection from Example 2):

$$w' = abacc\Delta \mid bac \mid bba \mid cac \mid bcb \mid aac \mid bcb \mid acb \mid aca \mid b\Delta b$$

As you can see, the number of letters between two consecutive fixed points is either 3×8, or $3 \times 8 + 1$, or $3 \times 8 + 2$. We place another fixed point whenever the factor w is of the form $w \in \cent \cdot \Sigma^{\geq 3\times 9}$ or $w \in \Sigma^* \cdot \Delta \cdot \Sigma^{\geq 3\times 9}$.

4 Idea of the Algorithm

As we already know, every Δ-clearing restarting automaton is defined by some finite set of instructions. However, in this section we prefer the following algorithmic viewpoint. Imagine a Δ-clearing restarting automaton as a nondeterministic machine N which repeatedly executes the following two steps. In the first "choosing" step it nondeterministically chooses a subword w of the input tape $\cent u\$$, such that the length of this subword is limited from above by some constant K. In the second "solving" step it runs a computation on this selected word w which either rejects, or replaces some subword of w by either λ or Δ, while preserving the sentinels ($\cent$ and $\$$). The automaton N accepts u if and only if there exists a computation starting from $\cent u\$$ which clears the whole word u.

If we want to define a Δ-clearing restarting automaton, we only need to define the solving algorithm S, called the *solver*, behind the second step of the above schema, and then show the existence of a suitable limit K. We put no resource limits on the solver S, because due to the constant K there are only finitely many inputs we may ask S.

Now consider a k-Δ^*cl-RA M whose construction was based on a given context-free grammar G as described in Section 2. Our goal is now to construct a solving algorithm S which will somehow imitate the work of the automaton M in such a way, that S will split one instruction of M into several steps. The automaton M works in a bottom-up manner. If the automaton recognizes that some subword w of the input tape can be derived from a nonterminal N_i, then it can replace the inner part of this subword w by the code Δ^r, where $r = m_0 + m_2(i-1) + j - 1$ for some $j \in \{1, 2, \dots, m_2\}$, leaving the first k letters and the last k letters of w as separators. The segment Δ^r together with its separators represents a code for the nonterminal N_i. These separators have a nice and useful property: if one changes some letters in these separators, the acceptance of the whole word on the input tape remains unchanged.

Our solver S is not obliged to preserve the representation used by the automaton M. Moreover, because of some technical reasons, we will not represent the nonterminal N_i by using a continuous segment Δ^r. Instead, we will use the segment $\Delta x \Delta^{r-4} y \Delta$, where $x, y \in \Sigma$ are the so-called "holes". These holes are useful in the sense that they can unambiguously identify the start and the end of the segment $\Delta x \Delta^{r-4} y \Delta$ inside any word marked by other symbols Δ.

As we have already explained, if we want to encode the information into some word $w \in \Sigma^*$, we need to know the factorization of this word w into the groups of B letters (see Section 3). This factorization (once defined) cannot be changed in the course of the algorithm. Otherwise, we could misinterpret the symbols Δ occurring in the word w. In order to correctly factorize the input word we need to find the so-called *fixed point.* We recognize three types of fixed points:

1. *Fixed point* ¢: In the word $¢ \cdot w$, where $w \in \Gamma^*$, the fixed point ¢ defines the following groups:
$$¢ \mid w_1 \mid w_2 \mid w_3 \mid \ldots,$$
where $w = w_1 w_2 w_3 \ldots$, and $|w_1| = |w_2| = |w_3| = \ldots = B$.
2. *Fixed point* Δ^r *with holes*: In the word uw, where u is the segment Δ^r with holes, $r = m_0 + m_2(i-1) + j - 1$ for some $j \in \{1, 2, \ldots, m_2\}$, the fixed point u defines the following groups:
$$w_0 \mid w_1 \mid w_2 \mid w_3 \mid \ldots,$$
where $uw = w_0 w_1 w_2 \ldots$, $|w_0| = m_0 + m_2(i-1)$, and $|w_1| = |w_2| = |w_3| = \ldots = B$.
3. *Fixed point* $\mathbf{u}\Delta\mathbf{v}\Delta$: In the word $\mathbf{u}\Delta\mathbf{v}\Delta w$, where $\mathbf{u} \in \Sigma^{2B}$, $\mathbf{v} \in \Sigma^{\leq 2B-2}$, $w \in \Sigma \cdot \Gamma^*$, the fixed point $\mathbf{u}\Delta\mathbf{v}\Delta$ defines the following groups:
$$\mathbf{u}\Delta\mathbf{v}\Delta \mid w_1 \mid w_2 \mid w_3 \mid \ldots,$$
where $w = w_1 w_2 w_3 \ldots$, and $|w_1| = |w_2| = |w_3| = \ldots = B$.

To prevent the accidental creation of fixed points of the type $\mathbf{u}\Delta\mathbf{v}\Delta$ we introduce the following convention. We do not encode the information straight into the groups of length B, but rather to the so-called *units.* A unit is defined as three consecutive groups of length B. If we want to mark a unit in order to encode one bit of information we always mark the middle group of the unit. The first and the third group of the unit serves as separators. Thanks to these separators any two consecutive symbols Δ in any two neighboring units are separated by at least $2B$ letters from Σ. If we mark two consecutive units, we never get the fixed point of the type $\mathbf{u}\Delta\mathbf{v}\Delta$. However, we can always obtain such a fixed point by marking two consecutive groups, provided that there are at least two unmarked groups preceding the first marked group. We use this observation in the first phase of the algorithm, in which we distribute the fixed points of this type throughout the whole input tape.

In the following we introduce the term *working area.* Consider an input tape with all its fixed points. If we erase these fixed points, the input tape will split into the segments, which we refer to as the working areas (see Figure 1).

Fixed points exactly define the factorization into the groups of length B and thus they also define the corresponding units inside these working areas. The term *working space* refers to the longest subword of the working area containing only the whole units that can be used to encode some information.

Now we are going to sketch the solving algorithm S which imitates the work of the automaton M. Because of the page limit we omit most of the details and refer the reader to the technical report [3]. Suppose that S is given an input word

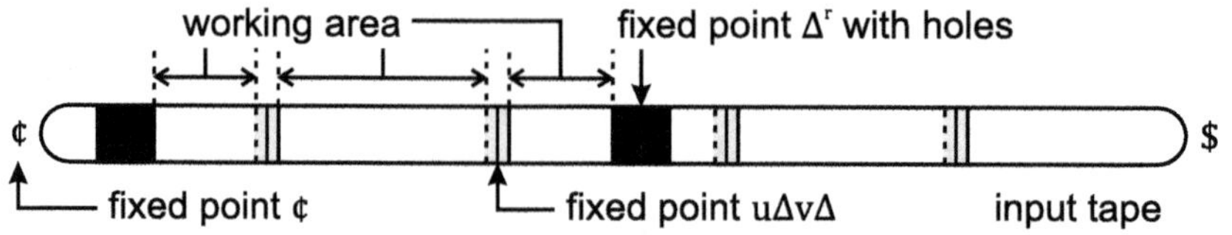

Fig. 1. The segmentation of the input tape into the working areas

w of lenght at most K which represents some bounded part of the current tape content. If w contains both sentinels ¢ and \$, then we either reject or accept, depending on whether w belongs to the target language (there are only finitely many cases to consider). Suppose that w does not contain both sentinels. If w is too short, we reject. Otherwise, we identify all fixed points within w. Suppose that $w = z_0 o_0 z_1 o_1 \dots z_d o_d$, where z_i are the fixed points in w. If it is not possible to factorize w in this way, especially if w does not start with the fixed point, we reject. If $o_d \in \Sigma^{\geq c} \cdot \{\lambda, \$\}$, where c is a suitable constant, we place another fixed point of the type $\mathbf{u}\Delta\mathbf{v}\Delta$ into w. This can be handled in two separate steps – each for one symbol Δ. Otherwise, we recover all symbols Δ in w that can be recovered. Suppose that the instruction $\phi = (x, u \to \Delta^r, y)$ of M can be applied to the recovered w. If there is a working area o_γ in w which already contains some encoded information, then suppose that ϕ is compatible with that information. If there is no such working area, then we choose one such area. The existence of such area is guaranteed by Lemma 1. Note that in our coding this word v can be interrupted by fixed points of the type $\mathbf{u}\Delta\mathbf{v}\Delta$. Now suppose that we want to imitate the instruction $\phi = (x, u \to \Delta^r, y)$, where $x, y \in \Sigma^k$, $r = m_0 + m_2(i-1) + (j-1)$, $1 \leq i \leq m$, $1 \leq j \leq m_2$. The number i is fixed for this instruction. However, we can choose $j \in \{1, 2, \dots, m_2\}$ arbitrarily. We will explain later which value we have to choose for j. Let us suppose that u contains the subword $z_\alpha o_\alpha \dots z_\beta o_\beta$, where z_α is the first and z_β is the last fixed point contained in u. We unfold the process of the realization of the instruction ϕ into several individual steps inside some working area o_γ, $\alpha \leq \gamma \leq \beta - 1$. The interpretation of the corresponding working space p_γ is illustrated in Figure 2.
1. Mark the unit D_2 of p_γ. The corresponding Δ is called the *reference point.* The reference point not only reserves the working area o_γ, but it also exactly defines the relative positions of the first and the last letter of the word u.
Let *left* be the position of the first letter of u relative to the reference point, *right* be the position of the last letter of u relative to the reference point. Compute j and $s = m_0 + m_2(i-1) + (j-1)$. We need to choose the parameter j in

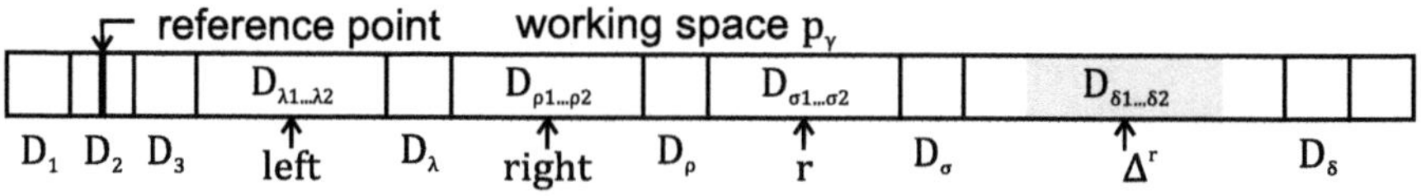

Fig. 2. The interpretation of the working space p_γ

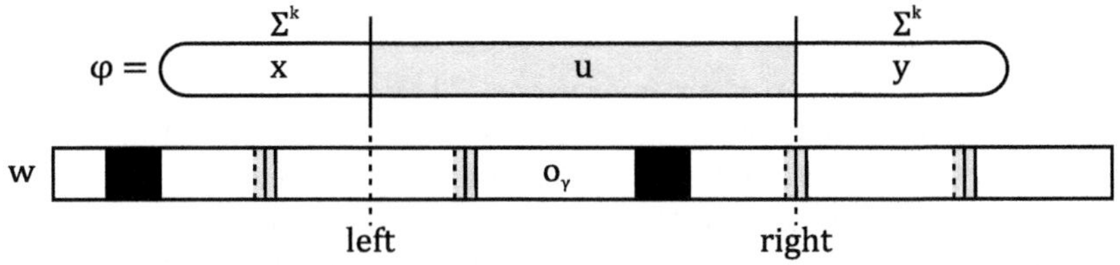

Fig. 3. Problem with the fixed point of the type $\mathbf{u}\Delta\mathbf{v}\Delta$

such a way, that the newly defined groups of length B, defined by the newly created fixed point of the type Δ^r (with holes), are exactly the same groups as the original groups before creating this fixed point.

2. Encode *left* into the units $D_{\lambda_1}, \ldots, D_{\lambda_2}$. Mark the unit D_λ.
3. Encode *right* into the units $D_{\rho_1}, \ldots, D_{\rho_2}$. Mark the unit D_ρ.
4. Encode s into the units $D_{\sigma_1}, \ldots, D_{\sigma_2}$. Mark the unit D_σ.
5. Mark the unit D_δ.
6. Encode the segment Δ^r (with holes) into the units $D_{\delta_1}, \ldots, D_{\delta_2}$.
7. Clear all letters from the end of the segment Δ^r to the position *right*.
8. Clear all letters from the position *left* to the beginning of the segment Δ^r.

The aforementioned algorithm is designed in such a way, that at any time it is possible to determine which steps were already executed and which were not. Each step of the algorithm is consistent with some instruction ϕ of the simulated automaton M, therefore the resulting solver cannot accept more than M. On the other hand, it is possible to define the parameters and constants used in the solver in such a way, that each instruction of M can be simulated by the solver. Moreover, by using the length-reducing version of coding it is possible to obtain a length-reducing solver which shortens the word in each step.

We conclude this section by mentioning one specific problem concerning Step 7 and Step 8. The problem is, that the position *left* (*right*, respectively) can cross the fixed point of the type $\mathbf{u}\Delta\mathbf{v}\Delta$ (see Figure 3).

The positions *left* and *right* are fixed and cannot be changed (since they are defined by the instruction ϕ). Fortunately, it does not matter if we damage the fixed point of the type $\mathbf{u}\Delta\mathbf{v}\Delta$, because the newly created fixed point of the type Δ^r (with holes) defines the groups of length B in exactly the same way as did the damaged fixed point. Moreover, thanks to the holes in the newly created fixed point, we do not lose even the ability to exactly define the borders of this newly created fixed point. We only lose the ability to recover the Δ symbol(s) of the damaged fixed point. Fortunately, these Δ symbols are situated in the separator of the newly created fixed point. Since the letters in the separator can be set arbitrarily, we can use any letters we want to recover these Δ symbols.

Note that the position *left* (*right*, respectively) cannot cross the fixed point of the type Δ^r (with holes), because the word xuy covers always the whole code of the nonterminal (including its separators).

5 Conclusion

Δcl-RA are very limited in their operations. They can in one step either completely delete a subword or they can delete a subword and simultaneously mark its position by a single symbol Δ. Surprisingly, they can accept any context-free language. We have designed a rather complicated coding of information used for encoding nonterminals during a bottom-up analysis. It would be interesting to find a simpler encoding, e.g. without the fixed points.

Another open problem is to characterize exactly the class of languages recognized by Δcl-RA and Δ^*cl-RA. Obviously, $\mathcal{L}(\Delta\text{cl-RA}) \subseteq \mathcal{L}(\Delta^*\text{cl-RA})$, but we do not know whether this inclusion is strict. Of course, Δcl-RA and Δ^*cl-RA should be more precisely related to the stateless restarting automata from [9] and [10]. In contrast to the stateless restarting automata, Δcl-RA- and Δ^*cl-RA can use only single auxiliary symbol Δ, but on the other hand, they need not to be length reducing.

References

1. Černo, P., Mráz, F.: Clearing restarting automata. In: Bordinh, H., Freund, R., Holzer, M., Kutrib, M., Otto, F. (eds.)Workshop on Non-Classical Models for Automata and Applications (NCMA), vol. 256, pp. 77–90. Österreichisches Computer Gesellschaft (2009), books@ocg.at
2. Černo, P., Mráz, F.: Clearing restarting automata. Fundamenta Informaticae 104(1), 17–54 (2010)
3. Černo, P., Mráz, F.: Delta-clearing restarting automata and CFL. Tech. rep., Charles University, Faculty of Mathematics and Physics, Prague (2011), `http://popelka.ms.mff.cuni.cz/cerno/files/cerno_mraz_dclra_and_cfl.pdf`
4. Hall, P.: On Representatives of Subsets. Journal of the London Mathematical Society s1-10(1), 26–30 (1935)
5. Hopcroft, J.E., Ullman, J.D.: Formal Languages and their Relation to Automata. Addison-Wesley, Reading (1969)
6. Jančar, P., Mráz, F., Plátek, M., Vogel, J.: Restarting automata. In: Reichel, H. (ed.) FCT 1995. LNCS, vol. 965, pp. 283–292. Springer, Heidelberg (1995)
7. Kutrib, M., Messerschmidt, H., Otto, F.: On stateless two-pushdown automata and restarting automata. In: Csuhaj-Varjú, E., Ésik, Z. (eds.) Proc. of Automata and Formal Languages, AFL, pp. 257–268. Computer and Automation Research Institute, Hungarian Academy of Sciences, Budapest (2008)
8. Kutrib, M., Messerschmidt, H., Otto, F.: On stateless deterministic restarting automata. In: Nielsen, M., Kučera, A., Miltersen, P.B., Palamidessi, C., Tůma, P., Valencia, F. D (eds.) SOFSEM 2009. LNCS, vol. 5404, pp. 353–364. Springer, Heidelberg (2009)
9. Kutrib, M., Messerschmidt, H., Otto, F.: On stateless deterministic restarting automata. Acta Inf. 47, 391–412 (2010)
10. Kutrib, M., Messerschmidt, H., Otto, F.: On stateless two pushdown automata and restarting automata. International Journal of Foundations of Computer Science 21, 781–798 (2010)
11. Otto, F.: Restarting automata. In: Ésik, Z., Martín-Vide, C., Mitrana, V. (eds.) Recent Advances in Formal Languages and Applications. Studies in Computational Intelligence, vol. 25, pp. 269–303. Springer, Berlin (2006)

Enumeration and Decidable Properties of Automatic Sequences

Émilie Charlier[1], Narad Rampersad[2], and Jeffrey Shallit[1]

[1] University of Waterloo, Waterloo, ON N2L 3G1 Canada
echarlier@uwaterloo.ca, shallit@cs.uwaterloo.ca
[2] Department of Mathematics, University of Liège, Grande Traverse, 12 (Bat. B37), 4000 Liège, Belgium
narad.rampersad@gmail.com

Abstract. We show that various aspects of k-automatic sequences — such as having an unbordered factor of length n — are both decidable and effectively enumerable. As a consequence it follows that many related sequences are either k-automatic or k-regular. These include many sequences previously studied in the literature, such as the recurrence function, the appearance function, and the repetitivity index. We also give a new characterization of the class of k-regular sequences. Many results extend to other sequences defined in terms of Pisot numeration systems.

1 Introduction

Let $\mathbf{x} = (a(n))_{n \geq 0}$ be an infinite sequence over a finite alphabet Δ. We write $\mathbf{x}[i] = a(i)$, and we let $\mathbf{x}[i..i+n-1]$ denote the factor of length n beginning at position i.

An infinite sequence $\mathbf{x}$ is said to be *k-automatic* if it is computable by a finite automaton taking as input the base-k representation of n, and having $a(n)$ as the output associated with the last state encountered [5].

For example, in Figure 1, we see an automaton generating the Thue-Morse sequence $\mathbf{t} = t_0 t_1 t_2 \cdots = \mathtt{011010011001} \cdots$. The input is n, expressed in base 2, and the output is the number contained in the state last reached.

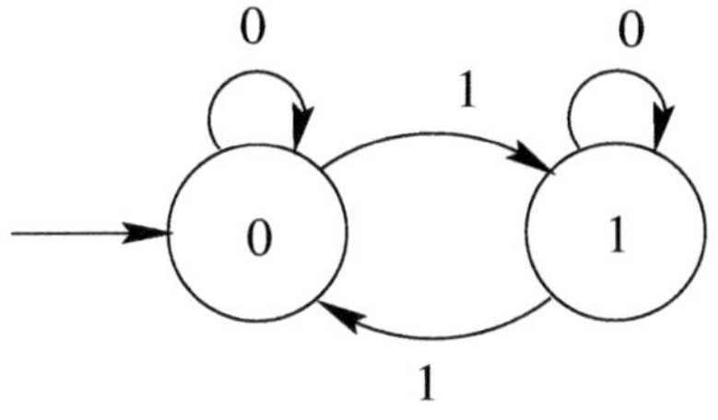

Fig. 1. A finite automaton generating a sequence

G. Mauri and A. Leporati (Eds.): DLT 2011, LNCS 6795, pp. 165–179, 2011.

Honkala [21] showed that, given an automaton, it is decidable if the sequence it generates is ultimately periodic. Later, Leroux [23] gave a polynomial-time algorithm for the problem.

Recently, Allouche, Rampersad, and Shallit [2] found a different proof of Honkala's result using a more general technique. They showed that their technique suffices to show that the following properties (and many more) are decidable for k-automatic sequences:

(a) Given a rational number $r > 1$, whether $\mathbf{x}$ is r-power-free;
(b) Given a rational number $r > 1$, whether $\mathbf{x}$ contains infinitely many occurrences of r-powers;
(c) Given a rational number $r > 1$, whether $\mathbf{x}$ contains infinitely many distinct r-powers;
(d) Given a length l, whether $\mathbf{x}$ avoids palindromes of length $\geq l$.

Related results have recently been given by Halava, Harju, Kärki, and Rigo [20].

In this paper we first show that many additional properties of automatic sequences are decidable using the same general technique. More significantly, we also show that related enumeration questions on automatic sequences (such as counting the number of distinct factors of length n) can be solved using a similar technique, in an entirely effective manner. As a consequence, we recover or improve results due to Mossé [24]; Allouche, Baake, Cassaigne, and Damanik [1]; Currie and Saari [16]; Garel [19]; Fagnot [17]; and Brown, Rampersad, Shallit, and Vasiga [8].

2 Connection with Logic

After the publication of [2], the third author noticed that the technique used there was, at its core, very similar to previous techniques developed by Büchi, Bruyère, Michaux, Villemaire, and others, involving formal logic; see, e.g., [10]. This was later independently observed by the first author, as well as by Véronique Bruyère. As it turns out, the properties (a)–(d) above are decidable because they are expressible as predicates in the first-order structure $\langle \mathbb{N}, +, V_k \rangle$, where $V_k(n)$ is the largest power of k dividing n.

We briefly recall the technique discussed in [2] in the context of a particular example. Suppose we want to decide if an automatic sequence $\mathbf{x}$ is squarefree (contains no nonempty square factor). Given an automaton M generating a k-automatic sequence $\mathbf{x}$, we create, via a series of transformations, a new automaton M' that accepts the base-k representations of integers corresponding to the squares in $\mathbf{x}$. For example, M' could accept those integers corresponding to the starting position of each square, or those integers corresponding to the lengths of the squares. The operations we can use in constructing M' include digit-by-digit addition or subtraction (with carry, if necessary), comparison, and lookup of the corresponding term in $\mathbf{x}$ (which comes from simulation of M). Nondeterminism can be used to implement "$\exists$", and "$\forall$" can be implemented by nondeterminism combined with suitable negations.

Ultimately, then, deciding if $\mathbf{x}$ is squarefree corresponds to verifying that $L(M') = \emptyset$ for the M' we construct. Deciding whether $\mathbf{x}$ contains only finitely many square occurrences corresponds to verifying that $L(M')$ is finite. Both can easily be done by the standard methods for automata.

In this paper, we always assume that numbers are encoded in base k using the digits in $\Sigma_k = \{0, 1, \ldots, k-1\}$, and are expressed with their *least significant digit* first. Thus, 13 can be represented in base 2 by 1011. The *canonical encoding* of n is the one with no leading zeroes (actually, trailing zeroes, since we are working with the reversed representation) and is denoted $(n)_k$.

Sometimes we will need to encode pairs, triples, or r-tuples of integers. We handle these by first padding the reversed representation of the smaller integer with trailing zeroes, and then coding the r-tuple as a word over Σ_k^r. For example, the pair $(20, 13)$ could be represented in base-2 as

$$[0,1][0,0][1,1][0,1][1,0],$$

where the first components spell out 00101 and the second components spell out 10110. Of course, there are other possible representations, such as

$$[0,1][0,0][1,1][0,1][1,0][0,0],$$

which correspond to non-canonical representations having trailing zeroes. In general, we permit these.

Thus, the main idea of [2] can be restated as follows:

Theorem 1. *If we can express a property of a k-automatic sequence $\mathbf{x}$ using quantifiers, logical operations, integer variables, the operations of addition, subtraction, indexing into $\mathbf{x}$, and comparison of integers or elements of $\mathbf{x}$, then this property is decidable.*

We illustrate the idea with the following new result. A word w is *bordered* if it begins and ends with the same word x with $0 < |x| \leq |w|/2$. (An example in English is `ingoing`.) Otherwise it is unbordered.

Theorem 2. *Let $\mathbf{x} = a(0)a(1)a(2)\cdots$ be a k-automatic sequence. Then the associated infinite sequence $\mathbf{b} = b(0)b(1)b(2)\cdots$ defined by*

$$b(n) = \begin{cases} 1, & \text{if } \mathbf{x} \text{ has an unbordered factor of length } n; \\ 0, & \text{otherwise;} \end{cases}$$

is k-automatic.

Proof. The sequence $\mathbf{x}$ has an unbordered factor of length n

iff

$\exists j \geq 0$ such that the factor of length n beginning at position j of $\mathbf{x}$ is unbordered

iff

there exists an integer $j \geq 0$ such that for all possible lengths l with $1 \leq l \leq n/2$, there is an integer i with $0 \leq i < l$ such that the i'th letter in the supposed border of length l beginning and ending the factor of length n beginning at position j of $\mathbf{x}$ actually differs in the i'th position

iff

there exists an integer $j \geq 0$ such that for all integers l with $1 \leq l \leq n/2$ there exists an integer i with $0 \leq i < l$ such that $a(j+i) \neq a(j+n-l+i)$.

To carry out this test, we first create an NFA that given the encoding of (j, l, n) guesses the base-k representation of i, digit-by-digit, checks that $i < l$, computes $j+i$ and $j+n-l+i$ on the fly, and checks that $a(j+i) \neq a(j+n-l+i)$. If such an i is found, it accepts. We then convert this to a DFA, and interchange accepting and nonaccepting states. This DFA M_1 accepts (j, l, n) such that there is no i, $0 \leq i < l$ such that $a(j+i) = a(j+n-l+i)$. We then use M_1 as a subroutine to build an NFA M_2 that on input (j, n) guesses l, checks that $1 \leq l \leq n/2$, and calls M_1 on the result. We convert this to a DFA and interchange accepting and nonaccepting states to get M_3. Finally, this M_3 is used as a subroutine to build an NFA M_4 that on input n guesses j and calls M_3.

The set of such integers n then forms a k-automatic sequence. □

Example 1. Consider the problem of determining for which lengths the Thue-Morse sequence has an unbordered factor. Currie and Saari [16] proved that if $n \not\equiv 1 \pmod 6$, then there is an unbordered factor of length n. (Also see [26, Lemma 4.10 and Problem 4.1].) However, this is not a necessary condition, as

$$\mathbf{t}[39..69] = \mathtt{0011010010110100110010110100101},$$

which is an unbordered factor of length 31. They left it as an open problem to give a complete characterization of the lengths for which $\mathbf{t}$ has an unbordered factor. Our method shows the characteristic sequence of such lengths is 2-automatic.

Further, we conjecture that there is an unbordered factor of length n in $\mathbf{t}$ if and only if the base-2 expansion of n (starting with the most significant digit) is not of the form $1(01^*0)^*10^*1$.

In principle this could be verified, purely mechanically, by our method, but we have not yet done so.

We now turn to deciding if a given automatic sequence has infinite *critical exponent* (e.g., [22]).

Theorem 3. *The following question is decidable: given a k-automatic sequence, does it contain powers of arbitrarily large exponent?*

Proof. $\mathbf{x}$ has powers of arbitrarily high exponent

iff

the set of pairs

$$S := \{(n,j) \ : \ \exists i \geq 0 \text{ such that for all } t \text{ with } 0 \leq t < n \text{ we have } \mathbf{x}[i+t] = \mathbf{x}[i+j+t]\}$$

contains pairs (n,j) with n/j arbitrarily large

iff

for all $i \geq 0$ S contains a pair (n,j) with $n > j \cdot 2^i$

iff

L, the set of base-k encodings of pairs in S, contains, for each i, strings ending in

$$\overbrace{[*,0][*,0]\cdots[*,0]}^{i}[b,0]$$

for some $b \neq 0$, where * means any digit.

But we can easily decide if a regular language contains strings ending in arbitrarily long strings of this form. □

In a similar fashion we can show

Theorem 4. *The following question is decidable: given a k-automatic sequence* $\mathbf{x}$*, does* $\mathbf{x}$ *contain arbitrarily large unbordered factors?*

Now we turn to questions of recurrence.

An infinite word $\mathbf{a} = (a(n))_{n\geq 0}$ is said to be *recurrent* if every factor that occurs at least once in $\mathbf{a}$ occurs infinitely often. Equivalently, a word is recurrent if and only if for each occurrence of a factor of $\mathbf{a}$, there exists a later occurrence of that factor in $\mathbf{a}$. Equivalently, for every $n \geq 0$, $r \geq 1$, there exists $m > n$ such that $a(n+j) = a(m+j)$ for $0 \leq j < r$.

Similarly, an infinite word $\mathbf{a} = (a(n))_{n\geq 0}$ is said to be *uniformly recurrent* if every factor that occurs at least once in $\mathbf{a}$ occurs infinitely often, with bounded gaps between consecutive occurrences. Equivalently, a word $\mathbf{a} = (a(n))_{n\geq 0}$ is uniformly recurrent iff for every $r \geq 1$ there exists $t > 0$ such that for every $n \geq 0$ there exists $m \geq 0$ with $n < m < n+t$ such that $a(n+i) = a(m+i)$ for $0 \leq i < r$.

Thus we recover the following recent result of Nicolas and Pritykin [25]:

Theorem 5. *It is decidable if a k-automatic sequence is recurrent or linearly recurrent.*

We now turn to questions of factors shared by two k-automatic sequences. Fagnot [17] showed that it is decidable whether two such sequences $\mathbf{x} = a(0)a(1)\cdots$ and $\mathbf{y} = b(0)b(1)\cdots$ have exactly the same set of factors. This is also decidable by our methods, as follows:

The sequences $\mathbf{x} = a(0)a(1)\cdots$ and and $\mathbf{y} = b(0)b(1)\cdots$ have the same set of factors

iff

for all $i \geq 0, n \geq 1$ there exists $j \geq 0$ such that $\mathbf{x}[i..i+n-1] = \mathbf{y}[j..j+n-1]$

iff

for all $i \geq 0, n \geq 1$ there exists $j \geq 0$ such that for all $t, 0 \leq t < n$ we have $a(i+t) = b(j+t)$.

In a similar fashion, the question of whether the set of factors of one k-automatic word form a subset of the set of factors of another k-automatic word is decidable.

3 Enumeration

In this section we show that many sequences counting aspects of k-automatic sequences are k-regular. Recall that a sequence $(a(n))_{n\geq 0}$ is *k-regular* if the module generated by the set of all subsequences of the form

$$\{(a(k^e n + c))_{n\geq 0} \;:\; e \geq 0,\; 0 \leq c < k^e\}$$

is finitely generated [3,4,5]. Alternatively, $(a(n))_{n\geq 0}$ is k-regular if $\sum_{n\geq 0} a(n)(n)_k$ is a noncommutative rational series [7], where $(n)_k$ is the canonical base-k encoding of n. The k-regular sequences play the same role for integer-valued sequences as the k-automatic sequences play for sequences over a finite alphabet. Classical examples of k-regular sequences include polynomials in n with non-negative coefficients and $s_k(n)$, the sum of the base-k digits of n.

The idea is based on the following simple lemma:

Lemma 1. *Let Σ be a finite alphabet, and let* $\mathtt{B}$ *be a new symbol not contained in Σ. Let Δ be another finite alphabet, and define $\Sigma' = (\Sigma \cup \{\mathtt{B}\}) \times \Delta$. For $x = [a_1, b_1][a_2, b_2] \cdots [a_n, b_n] \in (\Sigma')^*$ define the projection onto the first coordinate $\pi_1(x) = a_1 a_2 \cdots a_n$.*

Then the formal series $f : \Sigma^* \to \mathbb{N}$ is recognizable if and only if there exists a deterministic finite automaton $M = (Q, \Sigma', \delta, q_0, F)$ such that

$$(f, x) = |\{z \in L(M) \;:\; \pi_1(z) \in x\mathtt{B}^*\}|$$

for all strings x.

Furthermore, given the linear representation of such a formal series, we can compute the corresponding DFA, and vice versa.

Proof. Given M, for each state q define the machine $M_q = (Q, \Sigma', \delta, q, F)$, which is the same as M except that the initial state q_0 is replaced by q.

For each $q \in Q$, define the formal series $f_q : \Sigma^* \to \mathbb{N}$ by

$$(f_q, x) = |\{z \in L(M_q) \;:\; \pi_1(z) \in x\mathtt{B}^*\}|.$$

Now for $a \in \Sigma$ we have

$$(f_q, ax) = \sum_{b\in\Delta} (f_{\delta(q,[a,b])}, x).$$

Hence, using [7, Proposition I.5.1], f is recognizable.

On the other hand, if f is recognizable, then it has a linear representation (λ, μ, γ) satisfying $(f, w) = \lambda\mu(w)\gamma$, where λ is a row vector, $\mu : \Sigma^* \to \mathbb{N}^{m\times m}$ is a matrix-valued morphism, and γ is a column vector, all with entries in $\mathbb{N}$. Furthermore, using [27, Ex. III.3.3, p. 426], we can assume that $\lambda = [1\ 0\ \cdots\ 0]$ and $\gamma = [0\ \cdots\ 0\ 1]^T$.

Now choose Δ to have as many elements as the largest entry of $\mu(a)$ for $a \in \Sigma$. Construct a DFA $M = (Q', \Sigma', \delta, q_0, F)$ where $Q = \{q_0, q_1, \ldots, q_{m-1}\}$ and $Q' = Q \cup \{d\}$ such that, for all $a \in \Sigma$, $p, q \in Q$, there exist exactly $\mu(a)_{p,q}$ distinct letters $b \in \Delta$ with $\delta(p, [a, b]) = q$. Set $F = \{q_{m-1}\}$. By our construction, there are exactly (f, x) paths from q_0 to q_{m-1} labeled with x in the first component. Here d is a dead state and transitions on $\mathtt{B}$ go to this state; the effect is that transitions on $\mathtt{B}$ are not really used. Hence by the construction

$$(f, x) = |\{z \in L(M) \ : \ \pi_1(z) = x\}|. \qquad \square$$

The utility of this lemma is illustrated in the following examples.

Theorem 6. *Let* $\mathbf{x} = a(0)a(1)a(2)\ldots$ *be a* k*-automatic sequence. Let* $b(n)$ *be the number of distinct factors of length* n *in* $\mathbf{x}$*. Then* $(b(n))_{n\geq 0}$ *is a* k*-regular sequence.*

Proof. To count distinct factors of length n, we count the number of the first occurrences of each factor.

The number of distinct factors of length n in $\mathbf{x}$ equals the number of indices i such that there is no index $j < i$ with the factor of length n beginning at position i equal to the factor of length n beginning at position j.

Consider the set

$$S = \{(n, i) \ : \ \text{for all } j \text{ with } 0 \leq j < i \text{ there exists an integer } t \text{ with } 0 \leq t < n \text{ such that } a(i + t) \neq a(j + t)\}.$$

Then, by Theorem 1, the language S' defined to be the base-k encoding of elements of S, forms a regular language. We assume without loss of generality that if one representation of (n, i) appears in S', then they all do, including the ones with leading (actually, trailing zeroes).

We now apply a transducer to S', changing every representation of (n, i) as follows: we change every 0 after the last nonzero digit in the first component to $\mathtt{B}$. This transformation preserves the regularity of S'. Finally, we discard every representation that ends with $[\mathtt{B}, 0]$. The effect of this is to ensure that n in the first component, up to ignoring the $\mathtt{B}$'s, has a single representation, and that each i corresponding to a particular n has a unique representation. Using Lemma 1, we see that $b(n)$ is k-regular. $\square$

Remark 1. Mossé [24] proved, among other things, that a sequence that is the fixed point of a k-uniform morphism has a k-regular subword complexity function. With our technique, we obtain her result for these sequences and also the slightly more general case of k-automatic sequence.

Theorem 7. *The sequence counting the number of palindromic factors of length n is k-regular.*

Proof. The number of distinct palindromes of length n in $\mathbf{x}$

is equal to

the number of indices i such that $\mathbf{x}[i..i+n-1]$ is a palindrome and $\mathbf{x}[i..i+n-1]$ does not appear previously in $\mathbf{x}$

is equal to

the number of indices i such that $\mathbf{x}[i..i+n-1] = \mathbf{x}[i..i+n-1]^R$ and for all j with $0 \leq j < i$, $\mathbf{x}[i..i+n-1]$ is not the same as $\mathbf{x}[j..j+n-1]$

is equal to

the number of indices i such that for all t, $0 \leq t \leq n/2$, $a(i+t) = a(i+n-1-t)$ and for all j with $0 \leq j < i$, there exists u with $0 \leq u < n$ such that $a(i+u) \neq a(j+u)$. Now apply Theorem 1 and Lemma 1. □

Remark 2. Allouche, Baake, Cassaigne, and Damanik [1, Theorem 10] proved that the palindrome complexity of the fixed point of a primitive k-uniform morphism is k-automatic. Our result is more general: it shows that the palindrome complexity of a k-automatic sequence is k-regular, and hence is k-automatic iff it is bounded.

Jean-Paul Allouche kindly informs us that our result has just been obtained independently by Carpi and D'Alonzo [12].

Example 2. Let $f(n)$ denote the number of unbordered factors of length n of the Thue-Morse sequence.

Here is a brief table of the values of $f(n)$:

n	1	2	3	4	5	6	7	8	9	10	11	12	13	14	15	16	17	18	19	20	21	22	23	24	25	26	27	28	29	30	31	32
$f(n)$	2	2	4	2	4	6	0	4	4	4	4	12	0	4	4	8	4	8	0	8	4	4	8	24	0	4	4	8	4	8	4	16

By Lemma 1 we know that f is 2-regular. Conjecturally, f is given by the system of recurrences

$$\begin{aligned}
f(4n+1) &= f(2n+1) \\
f(8n+2) &= f(2n+1) - 8f(4n) + f(4n+3) + 4f(8n) \\
f(8n+3) &= 2f(2n) - f(2n+1) + 5f(4n) + f(4n+2) - 3f(8n) \\
f(8n+4) &= -4f(4n) + 2f(4n+2) + 2f(8n) \\
f(8n+6) &= 2f(2n) - f(2n+1) + f(4n) + f(4n+2) + f(4n+3) - f(8n) \\
f(16n) &= -2f(4n) + 3f(8n) \\
f(16n+7) &= -2f(2n) + f(2n+1) - 5f(4n) + f(4n+2) + 3f(8n) \\
f(16n+8) &= -8f(4n) + 4f(4n+2) + 4f(8n) \\
f(16n+15) &= -8f(4n) + 2f(4n+3) + 4f(8n) + f(8n+7).
\end{aligned}$$

In principle this could be verified by our method, but we have not yet done so.

Theorem 8. *Let* $\mathbf{x} = a(0)a(1)a(2)\cdots$ *be a* k*-automatic sequence. Then the following sequences are also* k*-automatic:*

(a) $b(i) = 1$ *if there is a square beginning at position* i*;* 0 *otherwise*
(b) $c(i) = 1$ *if there is a square centered at position* i*;* 0 *otherwise*
(c) $d(i) = 1$ *if there is an overlap beginning at position* i*;* 0 *otherwise*
(d) $e(i) = 1$ *if there is a palindrome beginning at position* i*;* 0 *otherwise*
(e) $f(i) = 1$ *if there is a palindrome centered at position* i*;* 0 *otherwise*

Remark 3. Brown, Rampersad, Shallit, and Vasiga proved results (a)–(c) for the special case of the Thue-Morse sequence [8].

Theorem 9. *Let* $\mathbf{x}$ *and* $\mathbf{y}$ *be* k*-automatic sequences. Then the following are* k*-regular:*

(a) the number of distinct square factors in $\mathbf{x}$ *of length* n*;*
(b) the number of squares in $\mathbf{x}$ *beginning at (centered at, ending at) position* n*;*
(c) the length of the longest square in $\mathbf{x}$ *beginning at (centered at, ending at) position* n*;*
(d) the number of palindromes in $\mathbf{x}$ *beginning at (centered at, ending at) position* n*;*
(e) the length of the longest palindrome in $\mathbf{x}$ *beginning at (centered at, ending at) position* n*;*
(f) the length of the longest fractional power in $\mathbf{x}$ *beginning at (ending at) position* n*;*
(g) the number of distinct recurrent factors in $\mathbf{x}$ *of length* n*;*
(h) the number of factors of length n *that occur in* $\mathbf{x}$ *but not in* $\mathbf{y}$*.*
(i) the number of factors of length n *that occur in both* $\mathbf{x}$ *and* $\mathbf{y}$*.*

Remark 4. Brown, Rampersad, Shallit, and Vasiga proved results (b)–(c) for the special case of the Thue-Morse sequence [8]. In some cases (e.g., (b)) we may have to define these sequences over $\mathbb{N} \cup \{+\infty\}$.

We now turn to some other measures that have received much attention. If an infinite word $\mathbf{x}$ is recurrent, then its recurrence function $R_{\mathbf{x}}(n) = R(n)$ is the smallest integer t such that every factor of length t of $\mathbf{x}$ contains as a factor every factor of length n. Said otherwise, it is the size of the smallest "window" one can slide along $\mathbf{x}$ and always contain all length-n factors.

Theorem 10. *If* $\mathbf{x}$ *is* k*-automatic, then* $R_{\mathbf{x}}(n)$ *is* k*-regular.*

Proof. We translate the predicate "$R(n) > t$", as follows:
$R(n) > t$
iff
there exists $i \geq 0$, $j \geq 0$ such that $\mathbf{x}[j..j+n-1]$ appears nowhere in $\mathbf{x}[i..i+t-1]$

iff

there exists $i \geq 0$, $j \geq 0$ such that for all integers l with $i \leq l < i+t-1-n$ we have $\mathbf{x}[l..l+n-1] \neq \mathbf{x}[j..j+n-1]$

iff

there exists $i \geq 0$, $j \geq 0$, such that for all integers l with $i \leq l < i+t-1-n$ there exists m, $0 \leq m < n$ such that $\mathbf{x}[l+m] \neq \mathbf{x}[j+m]$.

Now for any fixed n, the number of non-negative integers t for which $R(n) > t$ is equal to $R(n)+1$. Hence $R(n)+1$ is k-regular and hence, by a standard result [5, Thm. 16.2.1], so is $R(n)$. □

Another measure is called "appearance" [5, §10.10]. The appearance function $A_{\mathbf{x}}(n) = A(n)$ is the smallest integer t such that every factor of length n appears in a prefix of length t of $\mathbf{x}$. This can be proved in an analogous manner.

Theorem 11. *If* $\mathbf{x}$ *is* k*-automatic, then* $A_{\mathbf{x}}(n)$ *is* k*-regular.*

Next, we consider a measure due to Garel [19]. The separator length $S_{\mathbf{x}}(n)$ is the length of the smallest factor that begins at position n of $\mathbf{x}$ and does not occur previously.

Theorem 12. *If* $\mathbf{x}$ *is* k*-automatic, then* $S_{\mathbf{x}}(n)$ *is* k*-regular.*

Proof. The predicate "$S_{\mathbf{x}}(n) > t$" is the same as saying that for every $i \leq t$ the word of length i beginning at position n of $\mathbf{x}$ occurs previously in $\mathbf{x}$, which is the same as saying for all $i, 0 \leq i \leq t$, there exists $j, 0 \leq j < n$ such that $\mathbf{x}[n..n+i-1] = \mathbf{x}[j..j+i-1]$. Now look at the pairs (n,t) satisfying this. For each n there are exactly $S_{\mathbf{x}}(n)+1$ different t's that work. □

Remark 5. Garel [19] proved this for the case of a fixed point of a uniform circular morphism; our proof works for the more general case of an arbitrary k-automatic sequence.

Finally, Carpi and D'Alonzo have introduced a measure they called repetitivity index [11]. This measure $I_{\mathbf{x}}(n)$ is the minimum distance between two consecutive occurrences of the same length-n factor in $\mathbf{x}$. But "$I_{\mathbf{x}}(n) > t$" is the same as saying for all $i, j \geq 0$ with $i \neq j$, the equality $\mathbf{x}[i..i+n-1] = \mathbf{x}[j..j+n-1]$ implies that $j - i > t$. Hence we get

Theorem 13. *If* $\mathbf{x}$ *is* k*-automatic, then its repetitivity index is* k*-regular.*

4 A New Characterization of k-Regular Sequences

Carpi and Maggi [13] defined the class of k-synchronized sequences, a class which contains the k-automatic sequences and is properly contained in the class of k-regular sequences. A sequence $(u_n)_{n\geq 0}$ is k-synchronized if the relation $\{((n)_k, (u_n)_k) : n \geq 0\}$ is a right-synchronized rational relation. Roughly speaking, this means that the relation is realized by a length-preserving rational transduction, except that we also permit the presence of "padding" symbols at the end of one or the other component of the input. Here we give a similar transducer-based characterization of the more general class of k-regular sequences.

For us, a *j-uniform transducer* is a nondeterministic finite state machine $T = (Q, \Sigma, \delta, q_0, \tau, \Delta, F)$ where $\delta : Q \times \Sigma \to 2^Q$ and $\tau : \Sigma \to \Delta^j$ is a j-uniform morphism. An accepting path P begins at q_0 and ends at a state of F. The output associated with P is the concatenation of the outputs associated with the transitions. The output of T on an input x is the union of outputs associated with all accepting paths labeled x.

We work with strings over the alphabet $\Sigma' = \Sigma_k \times \Delta$, where $\Sigma_k = \{0, 1, \ldots, k-1\}$. For $x \in (\Sigma')^*$ we let $\pi_i(x)$ denote projection onto the i'th coordinate. For $x \in \Sigma_k^*$, $y \in \Delta^*$ with $|x| = |y|$ we let $x \times y$ denote the element of $(\Sigma')^*$ with $\pi_1(x \times y) = x$ and $\pi_2(x \times y) = y$.

Theorem 14. *Let $(b(n))_{n\geq 0}$ be a sequence taking values in $\mathbb{N} \cup \{+\infty\}$. Let $(n)_k \in \Sigma_k^*$ denote the canonical base-k encoding of n in base k, starting with the least significant digit.*

Then the following are equivalent:

(1) $(b(n))_{n\geq 0}$ is k-regular (more precisely, $(\mathbb{N}, k)$-regular; see [3]);

(2) there exist an integer m and vectors $\lambda \in \mathbb{N}^{1\times m}$, $\gamma \in \mathbb{N}^{m\times 1}$, and a matrix-valued morphism $\mu : \Sigma_k^* \to \mathbb{N}^{m\times m}$ such that $b(n) = \lambda\mu((n)_k)\gamma$;

(3) there exist an alphabet Δ and a DFA $M = (Q, \Sigma_k \times \Delta, \delta, q_0, F)$ such that

$$b(n) = |\{x \in (\Sigma_k \times \Delta)^* \;:\; \pi_1(x) = (n)_k\}|$$

for all $n \geq 1$;

(4) there exist an integer $j \geq 1$ and a j-uniform transducer T with inputs and outputs in Σ_k^* such that $b(n) = |T((n)_k)|$ for all $n \geq 1$. □

Proof. (1) $\iff$ (2): See [3].

(2) $\iff$ (3): Follows from Lemma 1.

(3) $\implies$ (4): Choose j sufficiently large so that $|\Delta| \leq k^j$. We can then identify elements of Δ uniquely with elements of Σ^j, using, say, a morphism σ. On input $x = (n)_k$, our transducer T nondeterministically guesses an element $y \in \Delta^*$ with $|x| = |y|$, simulates M on $x \times y$, and outputs the corresponding string $\sigma(y)$. If the transducer guessed correctly and $x \times y \in L(M)$, then the transducer accepts.

(4) $\implies$ (3): Similar to the previous direction. Here we create Δ such that single symbols correspond to blocks of size j output by the transducer.

Remark 6. In (4), by ensuring that the outputs of the transducer have no 0's, we can also ensure that each transduction of $(n)_k$ corresponds to a unique integer in base k. This handles the "trailing zeroes" problem.

As an application we have:

Theorem 15. *Let E be any finite set of integers, and consider $b(n)$, the sequence that counts the number of base-k representations where the digits are chosen only from E. Then $b(n)$ is k-regular.*

Proof. We construct a transducer obeying characterization (4) above. On input $(n)_k$, it guesses a possible representation of the same length (with trailing zeroes), outputs it, and simultaneously "normalizes" it, and checks that it is equal to the input. □

5 Linear Bounds

Yet another application of our method allows us to obtain linear bounds on many quantities associated with automatic sequences. As a first example, we recover an old result of Cobham [14] on "subword" complexity.

Theorem 16. *The number of distinct factors of length n of an automatic sequence is $O(n)$.*

Proof. Let $\mathbf{x}$ be a k-automatic sequence. By Theorem 1 we know that the base-k encoding S' of

$$S = \{(n, I) \ : \ \text{for all } j < I \text{ the factor of length } n \text{ starting at position } j \text{ is different from the one starting at position } I\}$$

is a regular language.

Suppose that the factor complexity of $\mathbf{x}$ is not $O(n)$. Then for every L there exists some pair $(n, I) \in S$ such that the length of the canonical encoding of I is longer than that of n by at least L digits. So in S' there is some word of the form $(n)_k\mathtt{B}^{\geq L} \times (I)_k$, where $(u)_k$ denotes the canonical encoding of u in base k and $\times$ is how we join separate components to form a word.

Since the length of $(I)_k$ is very much longer than that of $(n)_k$, we can apply the pumping lemma to this word, where we only pump in the portion of $(I)_k$ that is longer than $(n)_k$. Hence when we pump, we only add $\mathtt{B}$'s to the first component, and so its value remains unchanged. In this way by pumping we obtain infinitely many values I' such that $(n, I') \in S$. In other words, there are infinitely many distinct factors of length n, which is clearly absurd. The contradiction proves the result. □

In a similar manner we can prove that all the quantities in Theorem 9 are either linearly bounded, or unbounded.

6 Other Numeration Systems

All our results transfer, *mutatis mutandis*, to the setting of other numeration systems where addition can be performed on numbers using a transducer that processes numbers starting with the least significant digit.

A (generalized) numeration system is given by an increasing sequence of integers $U = (U_i)_{i\geq 0}$ such that $U_0 = 1$ and $C_U := \lim_{i\to+\infty} U_{i+1}/U_i$ exists and is finite. Then the canonical U-representation of n (with least significant digit first), which is denoted by $(n)_U$, is the unique finite word w

over the alphabet $\Sigma_U = \{0, \ldots, C_U - 1\}$ not ending with 0 and satisfying $n = \sum_{i=0}^{|w|-1} w[i]\, U_i$ and $\forall t \in \{0, \ldots, |w|-1\}$, $\sum_{i=0}^{t} w[i]\, U_i < U_{t+1}$. The notion of k-automatic sequence extends naturally to this context: an infinite sequence $\mathbf{x}$ is said to be U-automatic if it is computable by a finite automaton taking as input the U-representation $(n)_U$ of n, and having $\mathbf{x}[n]$ as the output associated with the last state encountered.

A numeration system U is called *linear* if U satisfies a linear recurrence relation over $\mathbb{Z}$. A Pisot system is a linear numeration system U whose characteristic polynomial is the minimal polynomial of a Pisot number. Recall that a Pisot number is an algebraic integer greater than 1, all of whose conjugates have moduli less than 1. For example, all integer base numeration systems and the Fibonacci numeration system are Pisot systems. Frougny and Solomyak [18] proved that addition is U-recognizable within all Pisot systems U, i.e., it can be performed by a finite letter-to-letter transducer reading U-representations with least significant digit first. Bruyère and Hansel [9] then proved the following logical characterization of U-automatic sequences for Pisot systems: a sequence is U-automatic if and only if it is U-definable, i.e., it is expressible as a predicate of $\langle \mathbb{N}, +, V_U \rangle$, where $V_U(n)$ is the smallest U_i occurring in $(n)_U$ with a nonzero coefficient. Therefore, if U is a Pisot system, any combinatorial property of U-automatic words that can be described by a predicate of $\langle \mathbb{N}, +, V_U \rangle$ is decidable.

The notion of k-regular sequences extends to Pisot numeration systems: an infinite sequence $\mathbf{x}$ is said to be *U-regular* if the series $\sum_{n \geq 0} \mathbf{x}[n] (n)_U$ is a non-commutative rational series. Thus we obtain

Theorem 17. *Let U be a Pisot numeration system and let $\mathbf{x}$ be any U-automatic word. The following sequences are U-automatic:*

(a) $a(n) = 1$ if there is a square beginning at (centered at, ending at) position n of $\mathbf{x}$, 0 otherwise;
(b) $b(n) = 1$ if there is a palindrome beginning at (centered at, ending at) position n of $\mathbf{x}$, 0 otherwise;
(c) $c(n) = 1$ if there is an unbordered factor beginning at (centered at, ending at) position n of $\mathbf{x}$, 0 otherwise.

The following sequences are U-regular:

(a) The number of distinct square factors beginning at (centered at, ending at) position n of $\mathbf{x}$;
(b) The number of distinct palindromic factors beginning at (centered at, ending at) position n of $\mathbf{x}$, 0 otherwise;
(c) The number of distinct unbordered factors beginning at (centered at, ending at) position n of $\mathbf{x}$, 0 otherwise.

Berstel showed that the cardinality of the set of unnormalized Fibonacci representations is Fibonacci-regular [6], a result also obtained (but not published) by the third author about the same time. In analogy with Theorem 15 we have

Theorem 18. *The number of unnormalized representations of n in a Pisot numeration system U is U-regular.*

7 Closing Remarks

It may be worth noting that the explicit constructions of automata we have given also imply bounds on the smallest example of (or counterexample to) the properties we consider. The bounds are essentially given by a tower of exponents whose height is related to the number of alternating quantifiers. For example,

Theorem 19. *Suppose* $\mathbf{x}$ *and* $\mathbf{y}$ *are* k*-automatic sequences generated by automata with at most* q *states. If the set of factors of* $\mathbf{x}$ *differs from the set of factors of* $\mathbf{y}$*, then there exists a factor of length at most* $2^{2^{2^{2q^2}}}$ *that occurs in one word but not the other.*

We also note that a question left open in [2], regarding the description of the lexicographically least word in the orbit closure of the Rudin-Shapiro sequence, was recently solved by Currie [15].

Acknowledgments

We thank Jean-Paul Allouche for his helpful comments.

References

1. Allouche, J.-P., Baake, M., Cassaigne, J., Damanik, D.: Palindrome complexity. Theoret. Comput. Sci. 292, 9–31 (2003)
2. Allouche, J.-P., Rampersad, N., Shallit, J.: Periodicity, repetitions, and orbits of an automatic sequence. Theoret. Comput. Sci. 410, 2795–2803 (2009)
3. Allouche, J.-P., Shallit, J.O.: The ring of k-regular sequences. Theoret. Comput. Sci. 98, 163–197 (1992)
4. Allouche, J.-P., Shallit, J.O.: The ring of k-regular sequences, II. Theoret. Comput. Sci. 307, 3–29 (2003)
5. Allouche, J.-P., Shallit, J.: Automatic Sequences: Theory, Applications, Generalizations. Cambridge University Press, Cambridge (2003)
6. Berstel, J.: An exercise on Fibonacci representations. RAIRO Inform. Théor. App. 35, 491–498 (2001)
7. Berstel, J., Reutenauer, C.: Noncommutative Rational Series With Applications. Encyclopedia of Mathematics and Its Applications, vol. 137. Cambridge University Press, Cambridge (2011)
8. Brown, S., Rampersad, N., Shallit, J., Vasiga, T.: Squares and overlaps in the Thue-Morse sequence and some variants. RAIRO Inform. Théor. App. 40, 473–484 (2006)
9. Bruyère, V., Hansel, G.: Bertrand numeration systems and recognizability. Theoret. Comput. Sci. 181, 17–43 (1997)
10. Bruyère, V., Hansel, G., Michaux, C., Villemaire, R.: Logic and p-recognizable sets of integers. Bull. Belgian Math. Soc. 1, 191–238 (1994); Corrigendum, Bull. Belg. Math. Soc. 1, 577 (1994)
11. Carpi, A., D'Alonzo, V.: On the repetitivity index of infinite words. Internat. J. Algebra Comput. 19, 145–158 (2009)

12. Carpi, A., D'Alonzo, V.: On factors of synchronized sequences. Theor. Comput. Sci. (to appear 011)
13. Carpi, A., Maggi, C.: On synchronized sequences and their separators. RAIRO Inform. Théor. App. 35, 513–524 (2001)
14. Cobham, A.: Uniform tag sequences. Math. Systems Theory 6, 164–192 (1972)
15. Currie, J.D.: Lexicographically least words in the orbit closure of the Rudin-Shapiro word (2010), `http://arxiv.org/pdf/0905.4923`
16. Currie, J.D., Saari, K.: Least periods of factors of infinite words. RAIRO Inform. Théor. App. 43, 165–178 (2009)
17. Fagnot, I.: Sur les facteurs des mots automatiques. Theoret. Comput. Sci. 172, 67–89 (1997)
18. Frougny, C., Solomyak, B.: On representation of integers in linear numeration systems. In: Pollicott, M., Schmidt, K. (eds.) Ergodic Theory of $\mathbb{Z}^d$ Actions (Warwick, 1993–1994), London Mathematical Society Lecture Note Series, vol. 228, pp. 345–368. Cambridge University Press, Cambridge (1996)
19. Garel, E.: Séparateurs dans les mots infinis engendrés par morphismes. Theoret. Comput. Sci. 180, 81–113 (1997)
20. Halava, V., Harju, T., Kärki, T., Rigo, M.: On the periodicity of morphic words. In: Gao, Y., Lu, H., Seki, S., Yu, S. (eds.) DLT 2010. LNCS, vol. 6224, pp. 209–217. Springer, Heidelberg (2010)
21. Honkala, J.: A decision method for the recognizability of sets defined by number systems. RAIRO Inform. Théor. App. 20, 395–403 (1986)
22. Krieger, D., Shallit, J.: Every real number greater than 1 is a critical exponent. Theoret. Comput. Sci. 381, 177–182 (2007)
23. Leroux, J.: A polynomial time Presburger criterion and synthesis for number decision diagrams. In: 20th IEEE Symposium on Logic in Computer Science (LICS 2005), pp. 147–156. IEEE Press, Los Alamitos (2005)
24. Mossé, B.: Reconnaissabilité des substitutions et complexité des suites automatiques. Bull. Soc. Math. France 124, 329–346 (1996)
25. Nicolas, F., Pritykin, Y.: On uniformly recurrent morphic sequences. Internat. J. Found. Comp. Sci. 20, 919–940 (2009)
26. Saari, K.: On the Frequency and Periodicity of Infinite Words. PhD thesis, University of Turku, Finland (2008)
27. Sakarovitch, J.: Elements of Automata Theory. Cambridge University Press, Cambridge (2009)

Languages vs. ω-Languages in Regular Infinite Games

Namit Chaturvedi*, Jörg Olschewski**, and Wolfgang Thomas

Lehrstuhl Informatik 7, RWTH Aachen University, Germany
{chaturvedi,olschewski,thomas}@automata.rwth-aachen.de

Abstract. Infinite games are studied in a format where two players, called Player 1 and Player 2, generate a play by building up an ω-word as they choose letters in turn. A game is specified by the ω-language which contains the plays won by Player 2. We analyze ω-languages generated from certain classes $\mathcal{K}$ of regular languages of finite words (called $*$-languages), using natural transformations of $*$-languages into ω-languages. Winning strategies for infinite games can be represented again in terms of $*$-languages. Continuing work of Selivanov (2007) and Rabinovich et al. (2007), we analyze how these "strategy $*$-languages" are related to the original language class $\mathcal{K}$. In contrast to that work, we exhibit classes $\mathcal{K}$ where strategy representations strictly exceed $\mathcal{K}$.

1 Introduction

The theory of regular ω-languages is tied to the theory of regular languages of finite words (regular $*$-languages) in at least two different ways. First, one obtains all regular ω-languages as finite unions of sets $U \cdot V^\omega$ where U, V are regular $*$-languages. This representation is obtained via the model of nondeterministic Büchi automata over infinite words. Second, if one works with deterministic Muller automata, one obtains a representation of all regular ω-languages as Boolean combinations of sets $\lim(U)$ with regular U (cf. [5,17]), where

$$\lim(U) = \{\alpha \in \Sigma^\omega \mid \text{infinitely many finite } \alpha\text{-prefixes are in } U\}.$$

In this paper we focus on the latter approach as we study the connection between $*$-languages and ω-languages in the context of infinite games, where the deterministic model of automata is needed. Another canonical transformation of $*$-languages into ω-languages is the *extension* of words of a $*$-language U:

$$\text{ext}(U) = \{\alpha \in \Sigma^\omega \mid \text{some finite } \alpha\text{-prefix is in } U\}.$$

Boolean combinations of such languages with regular U are recognized by deterministic weak Muller automata (also known as Staiger-Wagner automata [13]).

* Supported by DFG-Graduiertenkolleg AlgoSyn.
** Supported by ESF-Eurocores LogICCC project GASICS.

G. Mauri and A. Leporati (Eds.): DLT 2011, LNCS 6795, pp. 180–191, 2011.

We write BC(lim(REG)), respectively BC(ext(REG)), for the Boolean combinations of sets $\lim(U)$, respectively $\text{ext}(U)$, with regular U. In each case we refer to a fixed alphabet Σ so that complementation is done with respect to Σ^ω. We prefer the notation $\text{ext}(U)$ over the other popular alternative, $U \cdot \Sigma^\omega$, only for the purpose of emphasizing analogies or differences to $\lim(U)$.

The purpose of this paper is to study the connection between $*$-languages and ω-languages in two dimensions of refinement. First, the class REG is replaced by small subclasses, such as the class of piecewise testable languages or levels within the dot-depth hierarchy of star-free languages. In particular, for such a class $\mathcal{K}$ of $*$-languages, we consider the classes $\text{BC}(\lim(\mathcal{K}))$ and $\text{BC}(\text{ext}(\mathcal{K}))$, defined as above for the case REG. Secondly, we study a natural approach for the reverse direction, from ω-languages back to $*$-languages. Here the concept of infinite games is used, in which ω-languages enter as "winning conditions", and $*$-languages arise as representations of "winning strategies". We shall study the question whether games with a winning condition in classes such as $\text{BC}(\text{ext}(\mathcal{K}))$ or $\text{BC}(\lim(\mathcal{K}))$ can be "solved" with winning strategies representable in $\mathcal{K}$.

Let us recall the framework of infinite games in a little more detail. These games are played between two players, namely Player 1 and Player 2. In each round, first Player 1 picks a letter from an alphabet Σ_1 and then Player 2 a letter from an alphabet Σ_2. An infinite play of the game is thus an ω-word over $\Sigma := \Sigma_1 \times \Sigma_2$. One decides the winner of this play by consulting an ω-language $L \subseteq \Sigma^\omega$, also called the *winning condition*: If the play belongs to L, then Player 2 is the winner, otherwise Player 1 is. Games whose winning conditions belong to the class $\text{BC}(\text{ext}(\mathcal{K}))$ are referred to as *weak games* while those whose winning conditions belong to $\text{BC}(\lim(\mathcal{K}))$ are called *strong games*.

A strategy for either player gives the choice of an appropriate letter $a \in \Sigma_1$, resp. $a \in \Sigma_2$, for each possible play prefix where it is Player 1's, resp. Player 2's, turn. We can capture a strategy for a player by collecting, for each letter a, the set K_a of those finite play prefixes that induce the choice of a. For a strategy of Player 1 we have $K_a \subseteq \Sigma^*$, for Player 2 we have $K_a \subseteq \Sigma^*\Sigma_1$. We say that a strategy is in $\mathcal{K}$ if each language K_a is.

The fundamental Büchi-Landweber Theorem [1] (also see [3,17]) says that for a winning condition defined by any regular ω-language $L \in \text{BC}(\lim(\text{REG}))$, one of the two players has a winning strategy, that one can decide who is the winner, and that one can present a *regular* winning strategy (in the sense mentioned above) for the winner. In short, we say that *regular games are determined with regular winning strategies*. An analogous result for the class SF of star-free languages was shown in [9] (see also [10]) and later, by a different method, also in [8]: Star-free games are determined with star-free winning strategies. We shall focus in this paper on subclasses of SF, where – as it will turn out – the situation is more complicated. For instance, we show that for the class DD_1 of languages of dot-depth 1, games with winning conditions in classes $\text{BC}(\text{ext}(\text{DD}_1))$ and $\text{BC}(\lim(\text{DD}_1))$ are in general determined, not with winning strategies in DD_1, but only with those in classes DD_2 and DD_3 respectively. In contrast to this, we show that for games in the more restricted class $\text{BC}(\text{ext}(\text{pos-DD}_1))$, we have determinacy with winning strategies in DD_1. The class pos-DD_1 is the closure of languages $w_0\Sigma^* w_1 \dots \Sigma^* w_n$, where $w_i \in \Sigma^*$, under positive Boolean operations, i.e. union and intersection. The Boolean closure of pos-DD_1 is DD_1.

The paper is structured as follows. In Section 2 we summarize technical preliminaries on infinite games, well known subclasses of the class SF of star-free languages, and the subclasses of infinite languages that we consider in this paper. Subsequently, in Sections 3 to 5 we consider games over these classes of infinite languages and present results pertaining to winning strategies in these games. We conclude with some open questions and perspectives.

2 Technical Preliminaries

2.1 Languages, Automata, Games

We use standard notations [4] regarding languages and automata. As a model of automata with output we use Moore machines, which transform words over an alphabet Σ into words of an alphabet Γ via an output function $\lambda\colon Q \to \Gamma$ over the state set Q.

Over ω-words we use the models of *SW-automata* (Staiger-Wagner automata or weak Muller automata) and *Muller automata.* These are deterministic automata whose acceptance component is a family $\mathcal{F}$ of state sets. An ω-word α is accepted by an SW-automaton if the set of visited states in the unique run over α belongs to $\mathcal{F}$; for Muller automata one refers instead to the set of states visited infinitely often.

For any alphabet $\Sigma = \Sigma_1 \times \Sigma_2$, an ω-language $L \subseteq \Sigma^\omega$ induces a game with winning condition L; we shall just speak of the "game L". In this game, a *strategy for Player 1* is a mapping $\sigma\colon \Sigma^* \to \Sigma_1$ and a *strategy for Player 2* is a mapping $\tau\colon \Sigma^* \to \Theta$, where $\Theta := {\Sigma_2}^{\Sigma_1}$ is the (finite) set of all mappings from Σ_1 to Σ_2. An infinite word $\alpha = (a_i, x_i)_{i \in \mathbb{N}} \in \Sigma^\omega$ is said to be *consistent with* σ, if for all positions $i \in \mathbb{N}$ we have $\sigma(\alpha[0,i)) = a_i$. Analogously, α is *consistent with* τ, if for all $i \in \mathbb{N}$ we have $\tau(\alpha[0,i))(a_i) = x_i$. For two strategies σ and τ there is a (uniquely determined) word $\alpha(\sigma,\tau)$ that is consistent with both σ and τ.

If $\alpha(\sigma,\tau) \in L$ for every Player 1 strategy σ, then τ is called a *winning strategy* for Player 2. The other way around, if $\alpha(\sigma,\tau) \notin L$ for all Player 2 strategies τ, then σ is a winning strategy for Player 1. We say that a strategy σ for Player 1 is in the class $\mathcal{K}$ if for every $a \in \Sigma_1$ the language $K_a = \{w \in \Sigma^* \mid \sigma(w) = a\}$ is in $\mathcal{K}$. A strategy τ for Player 2 belongs to $\mathcal{K}$ if the language $K_{a \mapsto x} = \{w \mid \tau(w)(a) = x\}$ is in $\mathcal{K}$ for every $(a,x) \in \Sigma$. For the language classes we consider, this definition is consistent with the one presented in the introduction.

In this paper we focus on "finite-state strategies" realized by Moore machines. A Moore machine implementing a strategy τ for Player 2, is given by $M_\tau = (Q, \Sigma, \Theta, q_0, \delta, \lambda)$ with $\lambda\colon Q \to \Theta$ such that for all $w \in \Sigma^*$ it holds $\lambda(\delta(q_0, w)) = \tau(w)$. A Moore machine M_σ for Player 1 is obtained analogously by replacing Θ with Σ_1.

For all games of this paper, winning strategies of this kind suffice. In order to obtain this format of a strategy, the given winning condition L has to be cast into automata theoretic form. For ω-languages in a class BC(ext($\mathcal{K}$)), where $\mathcal{K} \subseteq$ REG, it is known that SW-automata (over $\Sigma := \Sigma_1 \times \Sigma_2$) can be used; for ω-languages in a class BC(lim($\mathcal{K}$)) we may take Muller automata.

Usually, classical two-player games are considered in the literature over a graph with two different types of nodes: one type belonging to Player 1, the other to Player 2. Such a game graph can be obtained from the ω-automaton recognizing the winning condition by expanding the state space and splitting the moves by letters of Σ into moves via Σ_1 and Σ_2. However, for our purposes it is more convenient to consider a game graph $G = (Q, \Sigma, q_0, \delta)$ with only one type of nodes, and in every node q we first let Player 1 choose an action from Σ_1 and after that let Player 2 choose from Σ_2. With this "unified" model, the conversions between ω-automata, game graphs, and Moore machines are straightforward.

As is well-known, the nodes (or: states) of the game graph in general do not suffice as states of a Moore machine defining a winning strategy. If each state of the game graph determines the move of the strategy, we speak of a *positional strategy*, which can be represented as a mapping $\sigma\colon Q \to \Sigma_1$ or $\tau\colon Q \to \Theta$, respectively. Positional winning strategies suffice in the case of "reachability games", "Büchi games", and "parity games" (see e.g. [18]). The former two of which correspond to games L of the form ext(K) or lim(K) respectively, for some regular $*$-language K. The latter *parity condition* refers to a coloring of game graph vertices by natural numbers, and a play is won by Player 2 iff the maximal color occurring infinitely often in it is even.

The key results about positional determinacy also hold in our unified model of game graphs. This can easily be shown by splitting the nodes of a unified game graph as mentioned above, and copying the color of the original node to the new ones, thereby transforming it to a game on a classical game graph.

The Boolean combinations as they appear in games with the Staiger-Wagner winning condition or Muller winning condition are handled by converting the winning condition into a parity condition while expanding the game graph by an extra "memory component" [18]. For weak games (with Staiger-Wagner winning conditions), one replaces a state q of the game graph by a pair (q, R) where R is the set of those states visited in the play up to the current point. The set R is called the AR ("appearance record"). For strong games (with Muller winning conditions), one refines this information by listing the visited states in the order of most recent visits, and with a pointer to that place in the sequence where the current state was located in the preceding step. In a normalized presentation over a space $\{q_1, \ldots, q_k\}$, we deal with expanded states (q, R) where R is an LAR ("latest appearance record"): a pair consisting of a permutation of $(q_1, \ldots, q_k)$ and a number $h \in \{1, \ldots, k\}$. Over the expanded state-space it suffices to satisfy the mentioned parity condition.

2.2 Classes of Regular Languages

In the subsequent definitions we recall some basic subclasses of the star-free languages; for more background see [6,15].

A $*$-language $K \subseteq \Sigma^*$ is *piecewise testable* if it is a Boolean combination of *basic* PT-*sets* $\Sigma^* a_1 \Sigma^* a_2 \cdots \Sigma^* a_n \Sigma^*$ where $a_1, a_2, \ldots, a_n \in \Sigma$. Denote the class of piecewise testable languages by PT. The class of *positive* Boolean combinations of basic PT-sets (in which only $\cup$ and $\cap$ are used) is denoted by pos-PT.

A $*$-language $K \subseteq \Sigma^*$ is *generalized definite* if it is a Boolean combination of sets $w\Sigma^*$ and $\Sigma^* w$ with $w \in \Sigma^*$. We denote the class of generalized definite languages by GDEF.

The dot-depth hierarchy, introduced by Cohen and Brzozowski [2], is a sequence of language classes $\mathrm{DD}_0, \mathrm{DD}_1, \ldots$ where $\mathrm{DD}_0 = \mathrm{GDEF}$ and DD_{n+1} can be obtained as the class of Boolean combinations of languages $K_1 \cdot \mathcal{K}_2 \cdot \ldots \cdot K_\ell$ (over a given alphabet) with $K_1, \ldots, K_\ell \in \mathrm{DD}_n$. As a special case let us mention the languages of *dot-depth 1*; they are the Boolean combinations of *basic* DD_1-*sets* $w_0 \Sigma^* w_1 \Sigma^* \cdots w_{n-1} \Sigma^* w_n$ where $w_0, w_1, \ldots, w_n \in \Sigma^*$. Also note that $\mathrm{PT} \subsetneq \mathrm{DD}_1$. In analogy to the class pos-PT we define pos-DD_1 as the class of positive Boolean combinations of basic DD_1-sets.

The dot-depth hierarchy is strict, and it exhausts the class SF of star-free languages. For later use we also recall a logical characterization of DD_n (see [16]): L (not containing the empty word) is in DD_n iff it can be defined by a first-order sentence that is a Boolean combination of Σ_n sentences (first-order sentences in prenex normal form with n alternating quantifier blocks starting with an existential block) where the signature has symbols for the minimal and the maximal position of a word, the predecessor and successor functions, the ordering $<$ of positions, and the predicates Q_a giving the positions with letter a.

The study of the language classes above is based on corresponding congruences over a given alphabet. We recall these congruences for the case of languages of dot-depth 1.

For $k, m \in \mathbb{N}$ and an m-tuple $\nu = (w_1, \ldots, w_m)$ of words of length $k+1$, we say that ν *appears* in a word u if for $1 \le i \le m$, u can be written as $u = u_i w_i v_i$ with suitable words u_i, v_i such that $1 \le i < j \le m$ implies $|u_i| < |u_j|$. We say that ν *appears* in an ω-word α if for $1 \le i \le m$, α can be written as $u = u_i w_i \alpha_i$ with suitable words u_i, α_i such that $1 \le i < j \le m$ implies $|u_i| < |u_j|$. With $\mu_{m,k}(w)$ (resp. $\mu_{m,k}(\alpha)$) we denote the set of all m-tuples of words of length $k+1$ that appear in w (resp. in α).

Two words $u, v \in \Sigma^*$ are (m,k)-equivalent ($u \sim_{m,k} v$) if

1. u and v have the same k first letters,
2. the same m-tuples of words of length $k+1$ appear in u and v, and
3. u and v have the same k last letters.

Then we have: $(*)$ *A $*$-language $K \subseteq \Sigma^*$ is of dot-depth one iff it is a union of $\Sigma^*/\sim_{m,k}$ equivalence classes for some $m, k \in \mathbb{N}$* [11]. (In the definition of $\sim_{m,k}$ we refer to possibly overlapping infixes; this does not affect statement $(*)$.)

Let us proceed to ω-languages. For two ω-words α, β we write $\alpha \sim_{m,k} \beta$ if α and β have the same k first letters, and the same m-tuples of words of length $k+1$ appear in α and β. Then we clearly have: $(**)$ *An ω-language $L \subseteq \Sigma^\omega$ is in* $\mathrm{BC}(\mathrm{ext}(\text{pos-}\mathrm{DD}_1))$ *iff it can be written as a union of $\Sigma^\omega/\sim_{m,k}$ equivalence classes for some $m, k \in \mathbb{N}$.*

3 Winning Strategies in Restricted Weak Games

We start with games in BC(ext(pos-DD$_1$)) which coincides with the class of Boolean combinations of sets ext(K) where K is a basic DD$_1$-set, or in other words: Boolean combinations of sets $w_0\Sigma^*w_1\Sigma^* \cdots w_{n-1}\Sigma^*w_n\Sigma^\omega$.

Theorem 1. *Games in* BC(ext(pos-DD$_1$)) *are determined with winning strategies in* DD$_1$.

Proof. By the characterization (**) at the end of the preceding section, we can write an ω-language L in BC(ext(pos-DD$_1$)) as a union of $\Sigma^\omega/{\sim_{m,k}}$ equivalence classes $L = \bigcup_{i=1}^n [\alpha_i]$ where each $\alpha_i \in \Sigma^\omega$. We show how to obtain a game graph with a parity winning condition that captures the game with winning condition L.

In the graph, the play prefix w will lead to the $\sim_{m,k}$-class $[w]$ of w. The game graph consists of the set of nodes $Q = \Sigma^*/{\sim_{m,k}}$. For every $(a, x) \in \Sigma$, we have edges from $[w]$ to $[w(a, x)]$. Note that this relation is well-defined, as from the set of m-tuples of length $k+1$ occurring in w, the suffix of length k of w, and the new letter (a, x), one can determine the set of m-tuples of length $k+1$ occurring in $w(a, x)$. We designate $q_0 = [\epsilon]$ as the start node of a play. For the winning condition, we assign a color $\chi(q)$ to every node q, namely $\chi([w]) = 2\cdot|\mu_{m,k}(w)|$ if there is an $\alpha \in L$ such that the prefix of α of length k equals the length k prefix of w and $\mu_{m,k}(\alpha) = \mu_{m,k}(w)$; $\chi([w]) = 2 \cdot |\mu_{m,k}(w)| - 1$ otherwise. Note that χ is increasing since for $w \in \Sigma^*$, and $(a, x) \in \Sigma$ we have $\chi([w]) \le \chi([w(a, x)])$. A play is won by Player 2 in the game for L iff the corresponding play in the graph game reaches ultimately an even color (and stays there).

By a well-known result on parity games, the parity game is determined, and the winning player has a uniform positional winning strategy. This means in particular, that in the parity game the winning player has a positional winning strategy from q_0. We show that she also has a DD$_1$ winning strategy in the original game.

Let $\lambda\colon Q \to \Sigma_1$ be a positional winning strategy of Player 1 in the parity game. Define $\sigma\colon \Sigma^* \to \Sigma_1$ to be $\sigma(w) = \lambda([w])$. The strategy σ is in DD$_1$, because for each $a \in \Sigma_1$ we know that $\sigma^{-1}(a) = \bigcup_{\lambda(w)=a}[w]$ is in DD$_1$. We still have to show that σ is winning for Player 1 in the game with winning condition L. For this purpose, let $\alpha = (a_0, x_0)(a_1, x_1)(a_2, x_2)\cdots \in \Sigma^\omega$ be consistent with σ. We have to show that $\alpha \notin L$. Then

$$\rho = [\epsilon], [(a_0, x_0)], [(a_0, x_0)(a_1, x_1)], \ldots$$

is a play in the parity game that is consistent with λ. So Player 1 wins ρ and thus the maximal color p that occurs infinitely often in ρ is odd. Let $i \in \mathbb{N}$ such that $\chi(\rho(i)) = p$. Then all following positions must have the same priority $p = \chi(\rho(i)) = \chi(\rho(i+1)) = \ldots$, because χ is increasing. This means the set $\mu_{m,k}(w)$ of m-tuples appearing in a word w from $\rho(i)$ does not change from i onwards. So the set of m-tuples of α is $\mu_{m,k}(\alpha) = \mu_{m,k}(w)$ for any $w \in \rho(i)$. Furthermore the prefix of α of length k is equal to the length k prefix of w for any $w \in \rho(i)$.

Since p is odd, and by the definition of χ there does not exist such a word $\alpha \in L$, so $\alpha \notin L$. This proves that σ is winning for Player 1.

In the analogous way it is shown that if Player 2 has a positional winning strategy in the parity game from q_0, then Player 2 has a DD_1 winning strategy in the game L. □

Next we turn to pos-PT, the class of positive combinations of basic piecewise testable languages; basic piecewise testable languages are languages of the form $\Sigma^* a_1 \Sigma^* a_2 \ldots a_n \Sigma^*$. We show that in this case we can proceed with a much simpler approach that avoids the formation of equivalence classes.

As a preparation we recall a result of I. Simon [12] about the transition structure of automata that accept piecewise testable languages. For a DFA $\mathcal{A} = (Q, \Sigma, q_0, \delta, F)$ and any $\Gamma \subseteq \Sigma$, let $G(\mathcal{A}, \Gamma)$ denote the directed graph underlying the automaton $\mathcal{A}$, such that it only retain edges of $\mathcal{A}$ that are labeled with elements of Γ.

Proposition 2 ([12]). *Let $\mathcal{A}$ be the minimal DFA accepting the $*$-language K. Then K is piecewise testable iff*

1. *$G(\mathcal{A}, \Sigma)$ is acyclic, and*
2. *for every $\Gamma \subseteq \Sigma$, each component of $G(\mathcal{A}, \Gamma)$ has a unique maximal state.*

Theorem 3. *Games in* BC(ext(pos-PT)) *are determined with winning strategies in* PT.

Proof. For every ω-language $L \in$ BC(ext(pos-PT)), there exists a regular language $K \in$ PT such that $L = \lim(K)$. This is shown easily by induction over Boolean combinations (cf. [7]). The minimal DFA $\mathcal{A} = (Q, \Sigma, q_0, \delta, F)$ accepting K can be considered as an ω-automaton accepting L. We thus obtain a game graph with a Büchi winning condition. Since Büchi games are determined with positional winning strategies (see e.g. [3]), the strategy of the winning player only depends on the current state in the play. Assume, without loss of generality, that Player 2 has a winning strategy. Then for every mapping $\theta\colon \Sigma_1 \to \Sigma_2$, let $F_\theta \subseteq Q$ be the set of states that induce a choice of θ. Consider the automaton $\mathcal{A}_\theta = (Q, \Sigma, q_0, \delta, F_\theta)$. Since $G(\mathcal{A}_\theta, \Sigma) = G(\mathcal{A}, \Sigma)$, we conclude from the proposition above that the language accepted by $\mathcal{A}_\theta$ is a piecewise testable language. □

It is worth noting that the game graphs for games in BC(ext(pos-PT)) are obtained directly from the finite automata that recognize the piecewise testable languages in question, and that piecewise testable winning strategies are obtained by observing a certain form of the associated transition graphs. For games in BC(ext(pos-DD_1)) we had to resort to the domain of congruences or, in algebraic terms, to the concept of syntactic monoids. This results in exponentially larger game graphs. In order to avoid this blow-up one might try to apply a result of Stern [14] that gives a property of transition graphs of (minimal) automata which characterizes the languages in DD_1; however, it seems that the necessary step towards obtaining parity games (as done in Theorem 1) spoils this property – which prevents a direct approach as that for pos-PT.

4 Winning Strategies in Weak Games

Theorem 4. *There are games in* BC(ext(PT)), *and therefore in* BC(ext(DD_1)), *that do not admit* DD_1 *winning strategies.*

Proof. Let $\Sigma := \{a, b, c, d\} \times \{0, 1\}$, and define $*$-languages $K_1 := (a, 0)^*(b, 0)$, $K_2 := \Sigma^*(d, 1)\Sigma^*$, and $K_d := \Sigma^*(d, 0)\Sigma^* \cup \Sigma^*(d, 1)\Sigma^*$, and for every letter $x \in \{a, b, c\}$ define $K_x := \Sigma^*(x, 1)\Sigma^*$. Let L be the ω-language over Σ, that contains all ω-words α such that

$$\left[\alpha \in \text{ext}(K_d) \Rightarrow \left(\alpha \in \text{ext}(K_1) \Leftrightarrow \alpha \in \text{ext}(K_2)\right)\right] \wedge \bigwedge_{x \in \{a,b,c\}} \alpha \in \overline{\text{ext}(K_x)}$$

All the $*$-languages above are in PT, so L is in BC(ext(PT)) and a fortiori in BC(ext(DD_1)). We see that L is won by Player 2: she remembers whether a prefix in K_1 has occurred; if so, then she responds to a later occurrence of d with 1, otherwise with 0. We claim there is no DD_1 winning strategy (and a fortiori no PT winning strategy) for Player 2. Assume there is such a winning strategy $\tau\colon \Sigma^* \to \Theta$, which is implemented by a DD_1 Moore machine. Then there are $*$-languages $K_{\theta_1}, \ldots, K_{\theta_n}$ of dot-depth one, implementing this strategy. In particular, $K_{d1} := \{w \mid \tau(w)(d) = 1\}$ is a dot-depth one language as a finite union of K_{θ_i} languages. So it is a finite union of equivalence classes $[w_i]_{\sim_{m,k}}$.

Note that in the word $w := (a^k b^k c^k a^k c^k b^k a^k)^m$ all possible m-tuples of length $k + 1$ over the alphabet $\{a, b, c\}$ appear. Let Player 1 play a strategy σ_1 that chooses $w \cdot d \cdot a^\omega$. Consider the unique word α_1 that is consistent with both σ_1 and τ. Since τ is a winning strategy for Player 2 and σ_1 plays $a^k b$ in the beginning, we have $(d, 1)$ occurring in α. So the word $w \times \{0\}$ is in K_{d1}.

Now let Player 1 play the strategy σ_2 which chooses $a^k c^k \cdot w \cdot d \cdot a^\omega$. The word $w_2 := a^k c^k \cdot w$ contains all possible m-tuples of length $k + 1$ over the alphabet $\{a, b, c\}$, as well. Then we have $w_1 \sim_{m,k} w_2$ and thus $w_2 \times \{0\} \in K_{d1}$. Then the unique word α_2 that is consistent with both σ_2 and τ contains $(d, 1)$ as an infix. This contradicts that τ is a winning strategy for Player 2. □

Theorem 5. *For each* $i \in \mathbb{N}$, *games in* BC(ext(DD_i)) *are determined with winning strategies in* DD_{i+1}.

Proof. Given the language L as a Boolean combination of ω-languages ext(K) with $K \in DD_i$, we first proceed to a game graph where every node is a DD_i equivalence class (and hence a $*$-language in DD_i). The game graph is equipped with a Staiger-Wagner winning condition.

As explained in Section 2.1, we transform the game graph via the AR construction to a new game graph with the parity winning condition. A state in the new game graph is a pair (q, R) consisting of a state $q \in \Sigma^*/{\sim_{DD_i}}$ of the original graph and an AR R. Over this game graph (with the parity winning condition), the winner has a positional winning strategy. We have to show that each node (q, R) corresponds to a DD_{i+1} language in the sense that the play prefixes leading to (q, R) form a language $K_{(q,R)}$ in DD_{i+1}. (Then the play prefixes that cause the winner to choose a fixed letter a are obtained as a union of languages $K_{(q,R)}$ and hence in DD_{i+1} as desired.)

For this purpose, it is convenient to apply the logical characterization of DD_i-languages as mentioned in Section 2.2. Each vertex q corresponds to the language K_q consisting of play prefixes leading to q. Each K_q is a language in DD_i, defined by a Boolean combination ψ_q of Σ_i-sentences. For clarity we write $\psi_q(\max)$ emphasizing the last position max of the current play prefix; so $\psi_q(x)$ expresses that the play prefix up to position x belongs to q. We have to express the restriction that such a play prefix leads to the AR $R \subseteq Q$ for the state space Q. This is formalized by the sentence

$$\varphi_{(q,R)} = \psi_q(\max) \wedge \bigwedge_{r \in R} \exists x \psi_r(x) \wedge \bigwedge_{r \notin R} \neg \exists x \psi_r(x).$$

Since ψ_r is a Boolean combination of Σ_i-sentences, we obtain (in prenex form) a Boolean combination of Σ_{i+1}-sentences. In this way we obtain the membership of $K_{(q,R)}$ in DD_{i+1}. □

5 Winning Strategies in Strong Games

Theorem 6. *Games in* BC(lim(PT)) *are determined with winning strategies in* PT.

Proof. For every ω-language $L \in \mathrm{BC}(\lim(\mathrm{PT}))$, there exists a regular language $K \in \mathrm{PT}$ such that $L = \lim(K)$ (see [7]). The remainder of this proof is identical to that of Theorem 3. □

Theorem 7. *There are games in the class* $\mathrm{BC}(\lim(\mathrm{DD}_1))$ *that do not admit* DD_1 *winning strategies.*

Proof. Let L be the ω-language over $\{a, b\} \times \{0, 1\}$ where $(a, 1)$ does not occur and where $(b, 0)$ occurs infinitely often iff $(b, 1)$ occurs infinitely often. The language L is in $\mathrm{BC}(\lim(\mathrm{DD}_1))$. We can easily see that L is won by Player 2. We claim there is no DD_1 winning strategy for Player 2. Assume there is such a winning strategy $\tau \colon \Sigma^* \to \Theta$, which is implemented by a DD_1 Moore machine. Then there are $*$-languages K_{a0}, K_{a1}, K_{b0}, and K_{b1} of dot-depth one, implementing this strategy (cf. the proof of Theorem 4). In particular K_{b0} and K_{b1} are dot-depth one languages, and we have that K_{b0} is a finite union of equivalence classes $[w]_{\sim_{m,k}}$:

$$K_{b0} = \bigcup_{i=1}^{n} [w_i]_{\sim_{m,k}}$$

Let Player 1 play a strategy σ which chooses $(ba^k)^\omega$. Consider the unique word α that is consistent with both σ and τ. Since τ is a winning strategy for Player 2 and σ plays infinitely many letters b, we have both $(b, 0)$ and $(b, 1)$ occurring infinitely often in α.

For any prefix $u_i := \alpha[0,i)$ of α, the set of m-tuples of words of length $k+1$ that appear in u_i is a subset of the set of m-tuples that appear in u_{i+1}. This means that there exists a $j \in \mathbb{N}$ such that the set of m-tuples is the same for all u_{j+i}, $i \in \mathbb{N}$. We choose $\hat{\jmath} \geq j$ such that $k+1$ divides $\hat{\jmath}$, which yields that the suffix of $u_{\hat{\jmath}}$ of length k is equal to $(a,0)^k$. But then we have $[u_{\hat{\jmath}}]_{\sim_{m,k}} = [u_{\hat{\jmath}+i(k+1)}]_{\sim_{m,k}}$, $i \in \mathbb{N}$, since $u_{\hat{\jmath}}$ and $u_{\hat{\jmath}+i(k+1)}$ have the same m-tuples, the same prefix $(b,0)(a,0)^{k-1}$ resp. $(b,1)(a,0)^{k-1}$, and the same suffix $(a,0)^k$ of length k.

We conclude that either $u_{\hat{\jmath}+i(k+1)} \in K_{b0}$ for all $i \in \mathbb{N}$, or $u_{\hat{\jmath}+i(k+1)} \in K_{b1}$ for all $i \in \mathbb{N}$. In the first case, we have only finitely many occurrences of $(b,1)$ in α, whereas in the second case we have only finitely many letters $(b,0)$ in α. This contradicts that τ is a winning strategy for Player 2. □

While staying at the same level of the dot-depth hierarchy does not yield winning strategies for strong games, the final result shows that there are winning strategies at most two levels higher in the hierarchy. Whether winning strategies are also located in the level between remains open.

Theorem 8. *For each $i \in \mathbb{N}$, games in* $\mathrm{BC}(\lim(\mathrm{DD}_i))$ *are determined with winning strategies in* DD_{i+2}.

Proof. We proceed as in the proof of Theorem 5. We first construct a graph where every node is a DD_i equivalence class – a $*$-language in DD_i. Now, for languages $K \in \mathrm{DD}_i$, we are given a game over the ω-language $L \in \mathrm{BC}(\lim(K))$. We obtain the game graph for L when we equip the graph constructed above with a Muller winning condition. As explained in Section 2.1, we transform this game graph via the LAR construction into a new game graph with a parity winning condition. A state in the new game graph is a pair (q,R) consisting of a state $q \in \Sigma^*/{\sim_{\mathrm{DD}_i}}$ and an LAR R. Over this parity game graph, the winner has a positional winning strategy.

We know that each vertex q collects the play prefixes that belong to a language $K_q \in \mathrm{DD}_i$, defined by a Boolean combination ψ_q of Σ_i-sentences (cf. Section 2.2). Let $\psi_q(x)$ express the fact that the play prefix up to position x belongs to q, with $\psi_q(\max)$ qualifying the last position of the current play prefix. With the help of these formulae, we now express the fact that each play prefix leading to any state (q,R) in the parity game graph forms a language $K_{(q,R)} \in \mathrm{DD}_{i+2}$.

Given a permutation perm of the state space of the original Muller game, and an index h, an LAR can be defined as $R = (\mathrm{perm}, h)$. Let the sentence φ_R express the fact that a play prefix has led to R, then it is evident that $K_{(q,R)} = \psi_q(\max) \wedge \varphi_R$. In order to avoid overloaded notation, we only provide a description for an example: the most recent prefix types in perm are q, r, s, in that order; the index value is $h = 3$. With φ_R, we express that the most recent prefix types are q, r, s in this order and that for the previous prefix this sequence is r, s, q: (1) the current play prefix (at position max) is q, at the previous position is r, and any preceding position that is not occupied by r is occupied by s, and (2) for the play prefix at position $\max - 1$ the most recent play prefixes are in r, s, q in this order. This can be formally described as:

$$\begin{aligned}\varphi_R = {} & \psi_q(\max) \wedge \psi_r(\max - 1) \\ & \wedge \exists x, y, z\big(\max > x > y > z \wedge \psi_r(x) \wedge \psi_s(y) \wedge \psi_q(z) \\ & \qquad \wedge \forall x'(\max > x' > x \rightarrow \psi_r(x')) \\ & \qquad \wedge \forall y'(x > y' > y \rightarrow \psi_s(y'))\big)\end{aligned}$$

Since ψ_q, ψ_r, and ψ_s are Boolean combinations of Σ_i-sentences, we obtain (in prenex form) a Σ_{i+2}-sentence. Thus, we obtain languages $K_{(q,R)} \in \mathrm{DD}_{i+2}$. □

6 Conclusion

The present paper continues the study of a question that was raised already by Büchi and Landweber in their pioneering paper [1, Sect.3]: to analyze "how simple winning strategies do exist" for a given class of games. Complementing the results of [1,8,9] where solvability of regular and star-free infinite games was established with corresponding winning strategies (again regular and star-free strategies, respectively), we showed in this paper that for games of lower complexity three levels need to be distinguished.

1. When we take the basic (pattern-) languages K underlying the piecewise testable languages and the languages of dot-depth 1 and work with Boolean combinations of sets $\mathrm{ext}(K)$, then determinacy with piecewise testable winning strategies, respectively of dot-depth 1, holds.
2. Games with winning conditions in $\mathrm{BC}(\mathrm{ext}(\mathcal{K}))$, where $\mathcal{K}$ is now the full class of piecewise testable languages or languages of dot-depth 1, are determined only with winning strategies beyond $\mathcal{K}$, namely in DD_2.
3. This situation is no better when games in $\mathrm{BC}(\lim(\mathrm{DD}_1))$ are considered; there are winning strategies in DD_3 but not in DD_1. For $\mathrm{BC}(\lim(\mathrm{PT}))$ we fall back to case 1, and obtain winning strategies again in PT.

Finally, there remain some open problems. First, it is left open here whether the bound $i + 2$ of Theorem 8 can be improved to $i + 1$. We also did not exclude the possibility that from some level onward the classes of dot-depth languages become rich enough to describe the winning strategies for weak or strong games. A more general problem is to study complexity issues, e.g. how the sizes of automata for game presentations and strategy presentations can diverge. Finally, the results of this paper motivate setting up an abstract framework of passing from $*$-language classes to corresponding ω-language classes (as winning conditions of games) and back (by considering winning strategies), so that classes beyond the special cases of the present paper are covered as well.

References

1. Büchi, J.R., Landweber, L.H.: Solving sequential conditions by finite-state strategies. Transactions of the American Mathematical Society 138, 295–311 (1969)
2. Cohen, R.S., Brzozowski, J.A.: Dot-depth of star-free events. Journal of Computer and System Sciences 5, 1–16 (1971)

3. Grädel, E., Thomas, W., Wilke, T. (eds.): Automata, Logics, and Infinite Games. LNCS, vol. 2500. Springer, Heidelberg (2002)
4. Hopcroft, J.E., Motwani, R., Ullman, J.D.: Introduction to automata theory, languages, and computation, 2nd edn. Addison-Wesley series in computer science. Addison-Wesley-Longman (2001)
5. Perrin, D., Pin, J.-É.: Infinite Words. Elsevier, Amsterdam (2004)
6. Pin, J.-É.: Varieties of Formal Languages. In: North Oxford, London (1986)
7. Pin, J.-É.: Positive varieties and infinite words. In: Lucchesi, C.L., Moura, A.V. (eds.) LATIN 1998. LNCS, vol. 1380, pp. 76–87. Springer, Heidelberg (1998)
8. Rabinovich, A., Thomas, W.: Logical refinements of church's problem. In: Duparc, J., Henzinger, T. (eds.) CSL 2007. LNCS, vol. 4646, pp. 69–83. Springer, Heidelberg (2007)
9. Selivanov, V.L.: Fine hierarchy of regular aperiodic ω-languages. In: Harju, T., Karhumäki, J., Lepistö, A. (eds.) DLT 2007. LNCS, vol. 4588, pp. 399–410. Springer, Heidelberg (2007)
10. Selivanov, V.L.: Fine hierarchy of regular aperiodic ω-languages. International Journal of Foundations of Computer Science 19(3), 649–675 (2008)
11. Simon, I.: Hierarchies of events with dot-depth one. PhD thesis, University of Waterloo (1972)
12. Simon, I.: Piecewise testable events. In: Brakhage, H. (ed.) Automata Theory and Formal Languages. LNCS, vol. 33, pp. 214–222. Springer, Heidelberg (1975)
13. Staiger, L., Wagner, K.W.: Automatentheoretische und automatenfreie Charakterisierungen topologischer Klassen regulärer Folgenmengen. Elektronische Informationsverarbeitung und Kybernetik 10(7), 379–392 (1974)
14. Stern, J.: Characterizations of some classes of regular events. Theoretical Computer Science 35, 17–42 (1985)
15. Straubing, H.: Finite Automata, Formal Logic, and Circuit Complexity. Birkhäuser Verlag, Basel (1994)
16. Thomas, W.: Classifying regular events in symbolic logic. Journal of Computer and System Sciences 25(3), 360–376 (1982)
17. Thomas, W.: Languages, automata, and logic. In: Rozenberg, G., Salomaa, A. (eds.) Handbook of Formal Languages, vol. 3, pp. 389–455. Springer, New York (1997)
18. Thomas, W.: Church's problem and a tour through automata theory. In: Avron, A., Dershowitz, N., Rabinovich, A. (eds.) Pillars of Computer Science. LNCS, vol. 4800, pp. 635–655. Springer, Heidelberg (2008)

Solving Word Problems in Group Extensions over Infinite Words

Volker Diekert[1] and Alexei G. Myasnikov[2]

[1] FMI, Universität Stuttgart, Universitätsstr. 38, D-70569 Stuttgart, Germany
[2] Department of Mathematics, Stevens Institute of Technology, Hoboken, NJ 07030, USA

Abstract. Non-Archimedean words have been introduced as a new type of infinite words which can be investigated through classical methods in combinatorics on words due to a length function. The length function, however, takes values in the additive group of polynomials $\mathbb{Z}[t]$ (and not, as traditionally, in $\mathbb{N}$), which yields various new properties. Non-Archimedean words allow to solve a number of algorithmic problems in geometric and algorithmic group theory. There is a connection to the first-order theory in free groups (Tarski Problems), too.

In the present paper we provide a general method to use infinite words over a discretely ordered abelian group as a tool to investigate certain group extensions for an arbitrary group G. The central object is a group $\mathrm{E}(A, G)$ which is defined in terms of a non-terminating, but confluent rewriting system. The group G as well as some natural HNN-extensions of G embed into $\mathrm{E}(A, G)$ (and still "behave like" G), which makes it interesting to study its algorithmic properties. The main result characterizes when the Word Problem (**WP**) is decidable in all finitely generated subgroups of $\mathrm{E}(A, G)$. We show that this property holds if and only if the Cyclic Membership Problem "$u \in \langle v \rangle$?" is decidable for all $v \in G$. Our methods combine combinatorics on words, string rewriting, and group theory.

Introduction

Combinatorics on words is a basic tool for various theories in theoretical computer science and mathematics, [7]. It has a long history dating back to the classic work of Thue at the beginning of the 20th century, [3]. Traditionally one studies finite words or words over linear orders in connection with logic and formal languages, [1]. Other generalizations use ordinals or (algebraic) linear orders, see e.g. [4,2,11].

The theory of non-Archimedean words appeared first in [17] in connection with group actions on trees. Group actions split naturally in two cases. The Archimedean case concerns with group actions on $\mathbb{R}$-trees. A complete description of finitely generated groups acting freely on $\mathbb{R}$-trees was obtained in a series of papers. It is known now as Rips' Theorem, see [6] for a detailed discussion. For non-Archimedean actions the focus is on groups acting freely on $\mathbb{Z}^n$-trees. This gives the link to non-Archimedean words. Elements in these groups can be represented as words where the length takes values in the ring of integer polynomials $\mathbb{Z}[t]$. It turned out that the representation of group elements as infinite words over $\mathbb{Z}[t]$ is quite intuitive and leads to the solution of various algorithmic problems for $F^{\mathbb{Z}[t]}$ using the standard Nielsen cancellation argument for the length function, [17]. The importance of Lyndon's group $F^{\mathbb{Z}[t]}$ became

G. Mauri and A. Leporati (Eds.): DLT 2011, LNCS 6795, pp. 192–203, 2011.

prominent due to its relation to algebraic geometry over groups and the solution of the Tarski Problems [12].

There are several contributions here. We start with any group G, and we use infinite words over a discretely ordered abelian group as a tool to investigate certain extensions of G. All precedent papers on this topic were restricted to the consideration of free groups. However, even for free groups we provide an important new result.

Studying free groups in the context of words means to work over an alphabet Σ where first, every letter a has an inverse $a^{-1} \in \Sigma^{-1}$, and second, aa^{-1} and $a^{-1}a$ are forbidden patterns. This generalizes immediately to infinite words and yields the notion of being *freely-reduced*. However, inside infinite words infinitely many cancellations may appear (see Fig. 1), and it is not clear how to cope with them. The only solution so far has been to define the concatenation as a partially defined operation. Our first main result shows that we can extend the concatenation to a totally defined function without collapsing the free group $F(\Sigma)$. Thereby we obtain a natural group extension $\mathrm{E}(A, F(\Sigma))$ of a free group $F(\Sigma)$. The proof that $F(\Sigma)$ embeds into its extension $\mathrm{E}(A, F(\Sigma))$ is rather tedious. The good news is that it generalizes to all groups, see Cor. 1. The result is obtained by the proof that some (non-terminating) rewriting system is strongly confluent, thus confluent. Non-terminating and strongly confluent string rewriting systems were successfully applied in [8], too.

The second part of the paper addresses algorithmic questions. Back in 1910 Max Dehn formulated the *Word Problem* (**WP**): "Given a presentation of a group G. Decide on an input word w whether w is the trivial element 1_G in G." It was only in the 1950s when Boone and Novikov independently showed that there is a fixed finitely presented group with an undecidable **WP**. Since then **WP** in groups became an active research area and various aspects are studied today in geometric and algorithmic group theory [10].

Our main result characterizes when the **WP** is decidable in all finitely generated subgroups of $\mathrm{E}(A, G)$. We show in Thm. 2 that the Word Problem is decidable in all finitely generated subgroups of $\mathrm{E}(A, G)$ if and only if the Cyclic Membership Problem "$u \in \langle v \rangle$?" is decidable for all $v \in G$. Thus, it is a transfer result with a non-trivial condition. There are known examples where G has a soluble Word Problem, but Cyclic Membership Problem is not decidable for some specific v. On the other hand, the Cyclic Membership Problem is uniformly decidable in many natural classes (which encompasses classes of groups with decidable Membership Problem w.r.t. subgroups) like hyperbolic groups, one-relator groups or effective HNN-extensions, see Rem. 1. In the final section we show that some interesting HNN extensions can be embedded into $\mathrm{E}(A, G)$, see Prop. 2. Thus, natural extensions of G still "behave like" G and the theory of non-Archimedean words applies to them.

Our methods combine combinatorics on words ([13]), string rewriting ([5]) and group theory ([14]). The proof techniques are non-trivial but elementary. No particular knowledge about non-Archimedean words or any deep results from group theory is required. Missing proofs will appear in the journal version of this abstract. A preliminary version with full proofs is accessible on-line on the ArXiv [9], too.

1.1 Preliminaries

Rewriting techniques provide a convenient tool to prove that certain constructions have the expected properties. Typically we extend a given group by new generators and defining equations and we want that the original group embeds in the resulting quotient structure. We assume that the reader is familiar with the basic concepts about rewriting systems (over monoids) such as (strong) confluence, termination or the Church-Rosser property. The material can be found in the text books like [5]. If S is a rewriting system, then $\mathrm{IRR}(S)$ means the set of (*irreducible*) elements where no rule can be applied. If a quotient monoid is given by a finite, confluent and terminating (i.e., finite and convergent) string rewriting system $S \subseteq \Gamma^* \times \Gamma^*$ over some alphabet Γ, then the monoid has a decidable **WP**, which yields a major interest in these systems.

Example 1. Let Σ be a set and Σ^{-1} be a disjoint copy. Then the set of rules

$$\{aa^{-1} \longrightarrow 1 \mid a \in \Sigma\} \cup \{a^{-1}a \longrightarrow 1 \mid a \in \Sigma\}$$

defines a strongly confluent and terminating system over $(\Sigma \cup \Sigma^{-1})^*$. The quotient monoid yields the *free group* $F(\Sigma)$ with basis Σ.

Discretely ordered abelian groups. An *ordered abelian group* is an abelian group A together with a linear order $\leq$ such that $x \leq y$ implies $x + z \leq y + z$. It is *discretely ordered*, if, in addition, there is a least positive element 1_A. Here, as usual, an element x is *positive*, if $0 < x$. An ordered abelian group is *Archimedean*, if for all $0 < a \leq b$ there is some $n \in \mathbb{N}$ such that $b < na$, otherwise it is *non-Archimedean*.

If B is any ordered abelian group, then the group $A = \mathbb{Z} \times B$ is discretely ordered with $1_A = (1, 0)$ and the lexicographical ordering:

$$(a, b) \leq (c, d) \text{ if } b < d \text{ or } b = d \text{ and } a < c.$$

In particular, $\mathbb{Z} \times \mathbb{Z}$ is a non-Archimedean discretely ordered abelian group. It serves as our main example. Iterating the process all finitely generated free abelian $\mathbb{Z}^k$ are viewed as being discretely ordered; and by a transfinite iteration we can consider arbitrary direct sums of $\mathbb{Z}$. In the paper we deal with $A \subseteq \mathbb{Z}[t]$, only. Thus, A is a subgroup of the additive group of the polynomial ring over $\mathbb{Z}$ in one variable t. Elements of A are finite sums $\alpha = \sum_i n_i t^i$ with $i \in \mathbb{N}$ and $n_i \in \mathbb{Z}$. Since the sum is finite, either $\alpha = 0$ or there is a greatest $d = \deg(\alpha) \in \mathbb{N}$ (its *degree*) with $n_d \neq 0$. By convention, $\deg(0) = -\infty$. The additive group of integers $\mathbb{Z}$ is viewed as a subgroup of $\mathbb{Z}[t]$ via the embedding $n \mapsto nt^0$. An element $\alpha = \sum_i n_i t^i \in A$ is called *positive*, if $n_{\deg(\alpha)} > 0$. We let $\alpha \leq \beta$, if $\alpha = \beta$ or $\beta - \alpha$ is positive. Thus, $1 \in \mathbb{N}$ is also the smallest positive element in A. If, for example, $A = \mathbb{Z} \times \mathbb{Z}$, then we have identified $1 \in \mathbb{N}$ with the pair $1_A = (1, 0)$. Moreover, for $\alpha, \beta \in A$ we define the *closed interval* $[\alpha, \beta] = \{\gamma \in A \mid \alpha \leq \gamma \leq \beta\}$. Its *length* is defined to be $\beta - \alpha + 1$. A typical interval contains infinitely many elements. For $\mathbb{Z} \times \mathbb{Z}$ the interval $[(-3, 0), (2, 1)]$ has length $(6, 1)$.

Sometimes we simply illustrate intervals of length $(m, 1)$ as $[\cdots)(\cdots]$ and intervals of length $(m, 2)$ as $[\cdots)(\ \cdots\ )(\cdots]$. This will become clearer later.

Non-Archimedean words over a group G. A *monoid with involution* is a monoid M with a bijection $x \mapsto \overline{x}$ such that $\overline{\overline{x}} = x$ and $\overline{xy} = \overline{y}\,\overline{x}$ for all $x, y \in M$ and, as a consequence, $\overline{1} = 1$. Every group is a monoid with involution $x \mapsto x^{-1}$. If M is a monoid with involution $x \mapsto \overline{x}$, then the quotient $M/\{x\overline{x} = 1 \mid x \in M\}$ is a group.

Let $a \mapsto \overline{a}$ denote a bijection between sets Σ and $\overline{\Sigma}$, hence $\overline{\Sigma} = \{\overline{a} \mid a \in \Sigma\}$. The map $a \mapsto \overline{a}, \overline{a} \mapsto a$ is an involution on $\Sigma \cup \overline{\Sigma}$ with $\overline{\overline{a}} = a$. It extends to an involution $x \mapsto \overline{x}$ on the free monoid $(\Sigma \cup \overline{\Sigma})^*$ with basis $\Sigma \cup \overline{\Sigma}$ by $\overline{a_1 \cdots a_n} = \overline{a_n} \cdots \overline{a_1}$.

Throughout G denotes a finitely generated group with group generators in a finite alphabet Σ. We let $\Gamma = \Sigma \cup \overline{\Sigma}$, where $\overline{\Sigma} = \Sigma^{-1} \subseteq G$ and $\overline{a} = a^{-1}$ for $a \in \Gamma$. The inclusion $\Gamma \subseteq G$ induces the canonical homomorphism (presentation) $\pi : \Gamma^* \to G$ from Γ^* onto the group G. Clearly, for every word $w \in \Gamma^*$ we have $\pi(\overline{w}) = \pi(w)^{-1}$.

A *partial A-map* is a map $p : D \to \Gamma$ with domain $D \subseteq A$. Two partial maps $p : D \to \Gamma$ and $p' : D' \to \Gamma$ are termed *equivalent*, if p' is an *α-shift* of p for some $\alpha \in A$, i.e., $D' = \{\alpha + \beta \mid \beta \in D\}$ and $p'(\alpha + \beta) = p(\beta)$ for all $\beta \in D$. This is an equivalence relation on partial A-maps, and an equivalence class of partial A-maps is called a *partial A-word*. If the domain $D = [\alpha, \beta] = \{\gamma \in A \mid \alpha \leq \gamma \leq \beta\}$ is an interval, then the equivalence class of $p : [\alpha, \beta] \to \Gamma$ is called a *closed A-word*. Its *length* $|p|$ is defined by $\beta - \alpha + 1 \in A$. For simplicity, a closed (resp. partial) A-word is called a *word* (resp. *partial word*), too. A word $p : [\alpha, \beta] \to \Gamma$ is *finite* if the set $[\alpha, \beta]$ is finite, otherwise it is *infinite*. Usually, we identify finite words with the corresponding elements in Γ^*. A (one-sided) infinite ω-word is a special case of a partial A-map, but its domain $\mathbb{N}$ is not an interval, hence it is a partial word, which is not closed.

If $p : [\alpha, \beta] \to \Gamma$ and $q : [\gamma, \delta] \to \Gamma$ are closed A-words, then we define their concatenation as follows. (Due to shift equivalence we may assume that $\gamma = \beta + 1$.)

$$
\begin{aligned}
p \cdot q : [\alpha, \delta] &\to \Gamma \\
x &\mapsto p(x) \ \text{ if } x \leq \beta \\
x &\mapsto q(x) \ \text{ otherwise.}
\end{aligned}
$$

It is clear that this operation is associative. Hence, the set of closed A-words forms a monoid, which we denote by $W(A, \Gamma)$. It is the natural analog of the free monoid Γ^* over non-Archimedean words. The empty word is denoted by 1. The standard representation of an A-word p is a mapping $p : [1, \alpha] \to \Gamma$, where $0 \leq \alpha$. In particular, $|p| = \alpha$. The *degree* of p is the degree of α; we also write $\deg(p) = \deg(\alpha)$. For a partial word $p : D \to \Gamma$ and $[\alpha, \beta] \subseteq D$ we denote by $p[\alpha, \beta]$ the restriction of p to the interval $[\alpha, \beta]$. Hence $p[\alpha, \beta]$ is a closed word. Sometimes we write $p[\alpha]$ instead of $p[\alpha, \alpha]$. Thus, $p[\alpha] = p(\alpha)$. The monoid $W(A, \Gamma)$ is a monoid with involution $p \mapsto \overline{p}$ where for $p : [1, \alpha] \to \Gamma$ we define $\overline{p} \in W(A, \Gamma)$ by $\overline{p} : [-\alpha, -1] \to \Gamma, -\beta \mapsto \overline{p(\beta)}$.

If $x \in W(A, \Gamma)$ and $x = pfq$ for some $p, f, q \in W(A, \Gamma)$ then p is called a *prefix*, q is called a *suffix*, and f is called a *factor* of x. If $1 \neq f \neq x$ then f is called a *proper factor*. As usual, a factor is finite, if $|f| \in \mathbb{N}$. Thus, a finite factor can be written as $x[\alpha, \beta]$ where $\beta = \alpha + n$, $n \in \mathbb{N}$.

A closed word $x : [1, \alpha] \to \Gamma$ is called *freely reduced* if $x(\beta) \neq \overline{x(\beta + 1)}$ for all $1 \leq \beta < \alpha$. It is called *cyclically reduced* if x^2 is freely reduced. As a matter of fact we need a stronger condition than being freely reduced. A closed word word x is called *G-reduced* , if no finite factor $x[\beta, \beta + n]$ with $n \in \mathbb{N}$, $n \geq 1$, becomes the identity 1 in the group G. Note that all G-reduced words are freely reduced by definition. By

$R(A, G)$ we denote the set of all G-reduced words in $W(A, \Gamma)$, and by $R^*(A, G)$ we mean the submonoid of $W(A, \Gamma)$ which is generated by $R(A, G)$.

Fig. 2 shows a word w with a sloppy notation $[aaa \cdots)(\cdots abab \cdots)(\cdots bbb]$. For the same word w we have $aw \neq wb$ (because $aw[(0, 1)] = a$ and $wb[(0, 1)] = b$), but we have $aaw = wbb$ in the monoid $W(A, \Gamma)$. Recall, that two elements x, y in a monoid M are called *conjugated*, if $xw = wy$ for some $w \in M$. Note that all finite words $x, y \in \Gamma^*$ are conjugated in $W(A, \Gamma)$ provided they have the same length $|x| = |y|$ and A contains $\mathbb{Z} \times \mathbb{Z}$. Indeed $w = [xxx \cdots)(\cdots yyy]$ does the job $xw = wy$. This phenomenon is in sharp contrast to the behavior of free monoids. The fact that all non-Archimedean words of the same length are conjugated is a main feature in applications.

1.2 The Group Extension $\mathrm{E}(A, G)$ of G by Infinite Words over A

We realize $\mathrm{E}(A, G)$ inside the following quotient monoid of $W(A, \Gamma)$:

$$\mathcal{M}(A, G) = W(A, \Gamma)/ \left\{ u\ell\overline{r}\,\overline{u} = 1 \mid u \text{ is } G\text{-reduced},\ \ell, r \in \Gamma^*, \pi(\ell) = \pi(r) \right\}.$$

We define $\mathrm{E}(A, G)$ as the image of $R^*(A, G)$ in $\mathcal{M}(A, G)$ under the canonical epimorphism $W(A, \Gamma) \to \mathcal{M}(A, G)$.

Our goal here is to construct a confluent rewriting system S over the monoid $W(A, \Gamma)$ such that $\mathcal{M}(A, G) = W(A, \Gamma)/S$ and S has the following form:

$$S = S_0 \cup \{ u\overline{u} \to 1 \mid u \in R(A, G) \text{ and } u \text{ is infinite} \} \tag{1}$$

where $S_0 \subseteq \Gamma^* \times \Gamma^*$ is a rewriting system for G satisfying the following conditions:

1. $\Gamma^*/S_0 = G$ and S_0 is confluent. Moreover, $1 \in \Gamma^*$ is S_0-irreducible.
2. For all $a \in \Gamma$ we have $(a\overline{a},\ 1) \in S_0$; and if $(\ell, r) \in S_0$, then $(\overline{\ell}, \overline{r}) \in S_0$, too.

It is easy to see that for every finitely generated group G there is a rewriting system $S_0 \subseteq \Gamma^* \times \Gamma^*$ satisfying the conditions above. In general S_0 is neither finite nor terminating. However, if G is finitely presented, then one can choose S_0 to be finite.

Due to Eq. 1 we have $\mathcal{M}(A, G) = W(A, \Gamma)/S$. The following lemma is crucial.

Lemma 1. *Let $x \in R(A, G)$ be a non-empty G-reduced word. Then $x \overset{*}{\underset{S}{\Longrightarrow}} y$ implies both $x \overset{*}{\underset{S_0}{\Longrightarrow}} y$ and y is a non-empty word.*

Theorem 1. *The system $S \subseteq \Gamma^* \times \Gamma^* \cup R(A, G) \times R(A, G)$ defined in Equation 1 is confluent on $W(A, \Gamma)$.*

The theorem above yields our first main result:

Corollary 1. *The canonical homomorphism $G \to \mathrm{E}(A, G)$ is an embedding.*

Proof. Let $x, y \in \Gamma^*$ be finite words such that $x = y$ in $\mathrm{E}(A, G)$. Then we have $x \overset{*}{\underset{S}{\Longrightarrow}} w \overset{*}{\underset{S}{\Longleftarrow}} y$ for some $w \in \Gamma^*$. But this implies $x \overset{*}{\underset{S_0}{\Longrightarrow}} w \overset{*}{\underset{S_0}{\Longleftarrow}} y$. Hence $x = y$ in G.

Corollary 2. *The canonical mapping $\mathrm{IRR}(S_0) \cap R(A, G) \to \mathrm{E}(A, G)$ is injective.*

A main idea in the proof of Thm. 1 is to replace the rewrite system $\underset{S}{\Longrightarrow}$ by a new system $\underset{\text{Big}}{\Longrightarrow} = \overset{*}{\underset{S_0}{\Longrightarrow}} \circ \underset{S}{\Longrightarrow} \circ \overset{*}{\underset{S_0}{\Longrightarrow}}$. The complicated part in the proof of Thm. 1 is to show that the (non-terminating!) rewriting system $\underset{\text{Big}}{\Longrightarrow}$ is strongly confluent on $W(A, \Gamma)$. Once, strong confluence is established, the assertion of Thm. 1 is immediate.

Example 2. Let S_0 be the system of Ex. 1 defining the free group $F(\Sigma)$. Let $a \in \Sigma$ and $u, v \in F(\Sigma)$ be represented by non-empty cyclically reduced words in Γ^*. (For example u, v are themselves letters.) Consider the following two infinite closed words:

$$w = [uuu \cdots)(\cdots vvv]$$
$$z = [uuu \cdots)(\cdots aaa][\bar{a}\bar{a}\bar{a} \cdots)(\cdots \bar{v}\bar{v}\bar{v}]$$

The words uw and wv are freely reduced and hence irreducible. By Cor. 2 we have $uw = wv$ in $\mathrm{E}(A, G)$ if and only if $|u| = |v|$.

The word z is not freely reduced, and if $z = z' \in \mathrm{E}(A, G)$, then S_0 is not terminating on z'. No rules from $S \setminus S_0$ apply to z and $w \neq z \in \mathrm{E}(A, G)$. Although the word z has no well-defined length one can infer the same conclusion as for w. First let $|u| = |v|$, then $z = uz\bar{v}$ in $\mathrm{E}(A, G)$ and hence $uz = zv$. For the other direction write $z = z'\bar{v}$ as words and let $uz = zv = z'$ in $\mathrm{E}(A, G)$. Then $uz \overset{*}{\underset{S}{\Longrightarrow}} \widetilde{z} \overset{*}{\underset{S}{\Longleftarrow}} z'$ for some word $\widetilde{z}$. After cancellation of factors $a^m \bar{a}^m$ inside $(\cdots aaa] \cdot [\bar{a}\bar{a}\bar{a} \cdots)$ the borderline between a's and $\bar{a}$'s must match inside $\widetilde{z}$. So exactly $|u|$ more cancellations of type $a\bar{a} \longrightarrow 1$ inside uz took place than in z'. Hence $|u| = |v|$.

1.3 Group Extensions over $A = \mathbb{Z}[t]$

Proper periods. Let $w \in W(A, \Gamma)$ be a word of length $\alpha \in A$, given as a mapping $w : [1, \alpha] \to \Gamma$. An element $\pi \in A$ is called a *period* of w, if for all $\beta \in A$ such that $1 \leq \beta$, $\beta + \pi \leq \alpha$ we have $w(\beta) = w(\beta + \pi)$. A period π is called a *proper period* of w, if $\deg(\pi) < \deg(w)$. In the following we are interested in proper periods, only. We have the following basic fact: Let $w \in W(A, \Gamma)$ of degree $\deg(w) = d$ with $0 \leq d$, then the set $\Pi(w)$ of proper periods forms a subgroup of $A^{\deg < d} = \{\alpha \in A \mid \deg(\alpha) < d\}$. This leads to the following proposition:

Proposition 1. *Let $w_0, w_1, w_2, w_3, \ldots$ be an infinite sequence of elements of $W(A, \Gamma)$ such that w_{i+1} is always a non-empty factor of w_i. Let*

$$\Pi_0, \Pi_1, \Pi_2, \Pi_3, \ldots$$

be the corresponding sequence of proper periods in A. Then this sequence of groups becomes stationary, i.e., there exists m such that $\Pi_m = \Pi_n$ for all $m \leq n$.

Deciding the WP in $\mathbf{E}(A, G)$. Recall that for a finitely generated group the decidability of the **WP** is a property of the group and does not depend on its presentation. The main difficulty for deciding the **WP** in $\mathrm{E}(A, G)$ is due to periodicity.

Let S be the system defined in Equation 1 which is confluent by Thm. 1. If we have $x \xRightarrow[S]{*} y$ then we have $\deg(x) \geq \deg(y)$. Thus, we can define the *reduced degree* by

$$\text{red-deg}(x) = \min\left\{\deg(y) \;\middle|\; x \xRightarrow[S]{*} y\right\}.$$

Note that red-deg(x) is well-defined for group elements $x \in \mathrm{E}(A, G)$ due the confluence of S. The following lemma is an easy consequence of Lem. 1:

Lemma 2. *Let $u \in R(A, G)$ be a non-empty G-reduced word. Then we have $0 \leq \deg(u) = \text{red-deg}(u)$.*

Our goal is to solve the **WP** in $\mathrm{E}(A, G)$ via the following strategy. We compute on input $w \in W(A, \Gamma)$ some $w' \in W(A, \Gamma)$ such that both $w \xLeftrightarrow[S]{*} w'$ and $\deg(w') = \text{red-deg}(w)$. If $\deg(w') > 0$, then $w \neq 1$ in $\mathrm{E}(A, G)$. Otherwise w' is a finite word over Γ and we can use the algorithm for G which decides whether or not $w' = 1$ in $G \subseteq \mathrm{E}(A, G)$. Our main result is the following theorem, which gives the precise answer in terms of the group G whether or not the **WP** in finitely generated subgroups of $\mathrm{E}(A, G)$ is decidable.

Theorem 2. *The following assertions are equivalent:*

i) For each $v \in \Gamma^*$ there exists an algorithm $\mathcal{C}(v)$ which decides on input $u \in \Gamma^*$ the Cyclic Membership Problem "$u \in \langle v \rangle$?"
*ii) The **WP** is decidable for each finitely generated subgroup of $\mathrm{E}(A, G)$.*

Observe that ii) implies that G itself has a decidable **WP**. But this is a weaker condition than i) Indeed, there is a finitely presented group G with a decidable **WP**, for which one can construct a specific word v such that the Cyclic Membership Problem "$u \in \langle v \rangle$?" is undecidable, see [18]. Thus, the assertion of Thm. 2 constitutes a non-trivial transfer result.

Proof of Thm. 2: For lack of space we restrict ourselves to show the more difficult and more important direction that $i)$ implies $ii)$. It is enough to show that for each finite subset $\Delta \subseteq R(A, G)$ of G-reduced words, there is an algorithm $\mathcal{A}(\Delta)$, which computes on input $w \in \Delta^*$ its reduced degree and some $w' \in W(A, \Gamma)$ such that both $w \xLeftrightarrow[S]{*} w'$ and $\deg(w') = \text{red-deg}(w)$.

PART I: Preprocessing: It is clear that we may replace Δ by any other finite set $\widehat{\Delta}$ such that $\Delta \subseteq \widehat{\Delta}^*$. We apply the following transformation rules in any order as long as possible, and we stop if no rule changes Δ anymore. The result is $\widehat{\Delta}$ which is, as we will see, still a set of G-reduced words. (This will follow from the fact that every factor of a G-reduced word is G-reduced).

1.) Replace Δ by $(\Delta \cup \Gamma) \setminus \{1\}$. (Recall that Γ is finite.)
2.) If we have $g \in \Delta$, but $\overline{g} \notin \Delta$, then insert $\overline{g}$ to Δ.
3.) If we have $g \in \Delta$ with $g = fh$ in $W(A, \Gamma)$ and $\deg(g) = \deg(f) = \deg(h)$, then remove g and $\overline{g}$ from Δ and insert f and h to Δ.

Define $g \sim h$, if for some $x, y, z, t, u \in W(A, \Gamma)$ with $\deg(xyzt) < \deg(u)$:

$$g = xuy \quad \text{and} \quad h = zut.$$

Note that the condition implies $\deg(g) = \deg(u) = \deg(h)$. The effect of the next rule is that for each equivalence class there is at most one group generator in Δ.

4.) If we have $g, h \in \Delta$ with $g \notin \{h, \overline{h}\}$, but $g = xuy$ and $h = zut$ for some x, y, z, t, and u with $\deg(xyzt) < \deg(u)$, then remove $g, h, \overline{g}, \overline{h}$ from Δ and insert x, y, z, t (those which are non-empty) and u to Δ.

5.) If we have $g \in \Delta$ with $g \neq \overline{g}$, but $g = xuy = z\overline{u}t$ for some x, y, z, t, and u with $\deg(xyzt) < \deg(u)$, then write $u = pq$ (or write $\overline{u} = pq$) with $\deg(p) < \deg(q) = \deg(u)$ and $q = \overline{q}$. Remove g and $\overline{g}$ from Δ and insert $x, y, z, t, p, \overline{p}$ (those which are non-empty) and q to Δ. (Note that $g \sim q$.)

The next rules deal with periods.

6.) If we have $g \in \Delta$ and $g = xuy$ for some x, y, and u with $\deg(xy) < \deg(u)$ such that u has a proper period which is not a period of g, then remove $g, \overline{g}$ from Δ and insert x, y (those which are non-empty) and u to Δ.

The following final rule below makes Δ larger again, and the rule adds additional information to each generator. For each $g \in \Delta$ let $\Pi(g) \subseteq A$ the group of proper periods. Let $B(g)$ be a set of generators of $\Pi(g)$. We may assume that for each possible degree d there is at most one element $\beta \in B(g)$ of degree d. Moreover, we may assume $0 \leq \beta$ and for each g the set $B(g)$ is fixed. In particular, for $\pi \in \Pi(g)$ with $\deg(\pi) = d \geq 0$ there is exactly one $\beta \in B(g)$ such that $\deg(\beta) = d$ and $\pi = m\beta + \ell$ for some unique $m \in \mathbb{Z}$ and $\ell \in \Pi(g)$ with $\deg(\ell) < d$. For each $\beta \in B(g)$ let $r(\beta)$ be the prefix and $s(\beta)$ be the suffix of length β of g. (In particular, $r(\beta)\, g = g\, s(\beta)$ in $W(A, \Gamma)$.) Note that the number of $r(\beta)$, $s(\beta)$ is bounded by $2\deg(g)$.

7.) If we have $g \in \Delta$, then let $B(g)$ be a set of generators for the set of proper periods $\Pi(g)$ as above. If necessary, enlarge Δ by finitely many elements of degree less than $\deg(g)$ (and which are factors of elements of Δ) such that $r(\beta)$, $s(\beta) \in \Delta^*$ for all $\beta \in B(g)$.

Note that the rules 1.) to 7.) can be applied only a finite number of times (König's Lemma and Prop. 1). The preprocessing is not asked to be effective.

PART II: An algorithm to compute the reduced degree: We may assume that Δ has passed the preprocessing, i.e., $\Delta = \widehat{\Delta}$ and no rule above changes Δ anymore. The input w (to the algorithm we are looking for) is given as a word $g_1 \cdots g_n$ with $g_i \in \Delta$. Let

$$d = \max\{\deg(g_i) \mid 1 \leq i \leq n\}.$$

We may assume that $d > 0$. Either $\deg(w) = \text{red-deg}(w)$ (and we are done) or $\deg(w) > \text{red-deg}(w)$ and $w \in W(A, \Gamma)$ contains a factor $uv\overline{u}$ such that:

1.) The word u is G-reduced and has length $|u| = t^d + \ell$ with $\deg(\ell) < d$,
2.) $\deg(v) < d$ and $v \overset{*}{\underset{S}{\Longrightarrow}} 1$.

We may assume that the factor $uv\overline{u}$ starts in some g_i and ends in some g_j with $i < j$, because the leading coefficient of each length $|g_i| \in \mathbb{Z}[t]$ is 1. Moreover, by making u smaller and thereby v larger, we may in fact assume that u is a factor of g_i and $\overline{u}$ is a factor of g_j. Thus, $\deg(g_i) = \deg(u) = \deg(g_j) = d$ and we can write $g_i = xuy$ and $g_j = z\overline{u}z'$. By preprocessing on Δ (Rule 4), we must have $g_i \in \{g_j, \overline{g_j}\}$. Assume $g_i = g_j$, then we have $g_i = xuy = z\overline{u}z'$ and, by preprocessing on Δ (Rule 5), we may conclude $g_i = \overline{g_i}$. Thus in any case we know $g_i = \overline{g_j}$.

Thus, henceforth we can assume that for some $1 \leq i < j \leq n$ we have in addition to the above:

3.) $g_i = xuy, \quad v = yg_{i+1} \cdots g_{j-1}z,$ and $g_j = z\overline{u}z' = \overline{g_i}$.

Since $g_j = z\overline{u}z' = \overline{g_i}$ we have $xuy = \overline{z'}u\overline{z}$, and by symmetry (in i and j) we may assume that $|y| \geq |z|$. This implies $y = q\overline{z}$ for some $q \in W(A, \Gamma)$ with $\deg(q) < d$ and $uq = q'u$ for $\overline{z'} = xq'$.

Therefore $|q|$ is a proper period of u, and hence, by preprocessing on Δ (Rule 6), we see that $|q|$ is a proper period of g_i. Thus there are $p', p \in \Delta^*$ with $|p'| = |p| = |q|$ such that $p'g_i = g_i p$. But $\overline{z}$ and y are suffixes of g_i, hence $y = \overline{z}p$. Therefore:

4.) $pg_{i+1} \cdots g_{j-1} \overset{*}{\underset{S}{\Longrightarrow}} 1$, where p is a suffix of g_i and $|p|$ is a proper period of g_i.

We know $\deg(g_{i+1} \cdots g_{j-1}) < d$. Hence by induction on d we can compute $h \in \Delta^*$ such that both $g_{i+1} \cdots g_{j-1} \overset{*}{\underset{S}{\Longleftrightarrow}} h$ and $\deg(h) = \text{red-deg}(g_{i+1} \cdots g_{j-1})$. This implies $\deg(h) = \text{red-deg}(p)$, too. But p is a factor of a G-reduced word, hence actually $\deg(h) = \deg(p)$ by Lem. 2.

We distinguish two cases. Assume first that $\deg(h) \leq 0$. Then $h, p \in \Gamma^*$ are finite words. If $h = 1$ in G, then we can replace the input word w by

$$g_1 \cdots g_{i-1} g_{j+1} \cdots g_n$$

since $g_i g_{i+1} \cdots g_{j-1} \overline{g_i} \overset{*}{\underset{S}{\Longrightarrow}} 1$, and we are done by induction on n.

If $h \in \Gamma^*$ is a finite word, but $h \neq 1$ in G, then $p = h^{-1} \neq 1$ in G, too. Consider the smallest element $\rho \in B(g_i)$ and let $r \in \Gamma^*$ be the suffix of g_i with $|r| = \rho$. It follows that p is a positive power of r because $|p|$ is a period of g_i. This means that h is in the subgroup of G generated by r. For this test we have the algorithm $\mathcal{C}(r)$ at our disposal by our hypothesis on G. (Note that r belongs to a finite list depending on Δ and not on the specific input word to the **WP**.) According to our assumption $\deg(w) > \text{red-deg}(w)$ (about the input word w), the answer of the algorithm $\mathcal{C}(r)$ is "*Yes*": h is in the subgroup generated by r. This allows to find $m \in \mathbb{Z}$ with $\overline{h} = r^m$ in the group G. We find some finite word s of length $|s| = |r^m|$ such that $sg_i = g_i r^m$; and we can replace the input word w by $g_1 \cdots g_{i-1} \overline{s} g_{j+1} \cdots g_n$. We are done by induction on the number of generators of degree d (because g_i is missing and $\deg(g_i) = d > 0$).

The final case is $\deg(h) > 0$. We write $|h| = m't^e + \ell$ with $\deg(\ell) < e = \deg(h)$. According to our preprocessing on Δ (Rule 7) there are words $r, s \in \Delta^*$ such that

$\deg(r) = \deg(h)$, r is a suffix of g_i with $sg_i = g_i r$. Note that we effectively find such r and s (or we detect $w \neq 1$), and we know m', too.

For some m with $m \leq m'$ we must have red-deg$(r^m h) < e$. By induction we can compute some word f with $\deg(f) =$ red-deg$(r^m h)$ and $f \overset{*}{\underset{S}{\Longleftrightarrow}} r^m h$. Like above we can replace the input word w by

$$g_1 \cdots g_{i-1} \overline{s}^m g_i f g_j \cdots g_n.$$

We are done by induction on the degree e which is the reduced degree of the factor $r^m g_{i+1} \cdots g_{j-1}$. We can apply this induction since $g_i r^m g_{i+1} \cdots g_{j-1} g_j$ now has a factor $uv'\overline{u}$ such that the following conditions hold (with $e < d$):

1.) The word u is G-reduced and $\deg(u) = d > 0$, 2.) $\deg(v') < e$ and $v' \overset{*}{\underset{S}{\Longrightarrow}} 1$.

□

Remark 1. The assertions in Thm. 2 hold, if G has a decidable *Generalized Word Problem,* i.e., the Membership Problem w.r.t. finitely generated subgroups is decidable. Examples of groups G where the Generalized Word Problem is decidable include metabelian, nilpotent or, more general, abelian by nilpotent groups, see [20]. However, there are also large classes of groups, where the Membership Problem is undecidable, but the Cyclic Membership Problem is easy. For example, the Cyclic Membership Problem is decidable in linear time in a direct product of free groups, but as soon as G contains a direct product of free groups of rank 2, the Generalized Word Problem becomes undecidable by [16]. For hyperbolic groups a construction of Rips shows that the Generalized Word Problem is undecidable ([19]), but the Cyclic Membership Problem "$u \in \langle v \rangle$?" is decidable by [15]. Decidability of the Cyclic Membership Problem is also preserved e.g. by effective HNN extensions. This means, if H is an HNN-extension of G by a stable letter t such that we can effectively compute Britton reduced forms c.f. [14]. In particular the Cyclic Membership Problem is decidable in one-relator groups, but it is a long standing open problem, whether the Generalized Word Problem is decidable for this class.

1.4 Realization of Some HNN-Extensions

The purpose of this section is to show that the group $\mathrm{E}(A, G)$ contains some important HNN-extensions of G which therefore can be studied within the framework of infinite words. Moreover, we show that $\mathrm{E}(A, G)$ realizes more HNN-extensions than it is possible in the approach of [17]. The reason is that [17] is working with cyclically reduced decompositions, only. We need a few more technical terms which have not been introduced so far. A non-empty word $w \in W(A, G)$ is called *primitive*, if first, w does not appear as a factor of ww other than as its prefix or as its suffix, and second, $\overline{w}$ is not a factor of ww. In particular, a primitive word does not have any non-trivial proper period. We say $x \in W(A, G)$ is *cyclically G-reduced* , if every finite power x^k with $k \in \mathbb{N}$ is G-reduced. Over a free group G with basis Σ a word is G-reduced if and only if it is freely-reduced; and it is cyclically G-reduced if and only if x^2 is freely-reduced. As usual the *centralizer* of an element u in a group H is the subgroup $\{v \in H \mid uv = vu\}$.

Proposition 2. *Let H be a finitely generated subgroup of* $\mathrm{E}(A, G)$ *and let* $u, v, w \in H$ *be (not necessarily different) cyclically G-reduced elements such that* $|u| = |v| = |w|$ *and such that w is primitive. Let u and v have cyclic centralizers in H. Then the HNN extension of H with stable letter s* $H' = \langle H, s \mid s^{-1}us = v \rangle$ *embeds into* $\mathrm{E}(A, G)$.

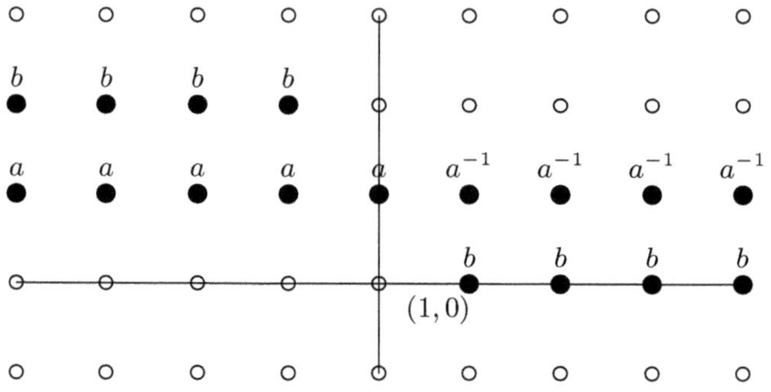

Fig. 1. Infinitely many cancellations aa^{-1} in the middle line

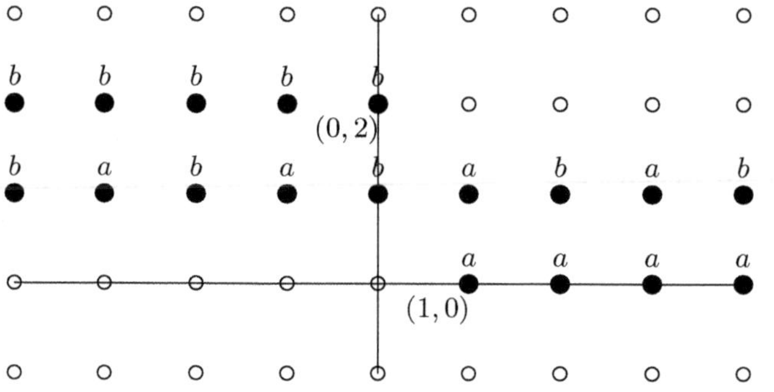

Fig. 2. A word $w = [aaa \cdots)(\cdots abab \cdots)(\cdots bbb]$ where $aw \neq wb$ but $aaw = wbb$

References

1. Bedon, N., Bès, A., Carton, O., Rispal, C.: Logic and rational languages of words indexed by linear orderings. Theory Comput. Syst. 46(4), 737–760 (2010)
2. Bedon, N., Rispal, C.: Schützenberger and eilenberg theorems for words on linear orderings. In: De Felice, C., Restivo, A. (eds.) DLT 2005. LNCS, vol. 3572, pp. 134–145. Springer, Heidelberg (2005)
3. Berstel, J., Perrin, D.: The origins of combinatorics on words. Eur. J. Comb. 28(3), 996–1022 (2007)
4. Bogopolski, O., Zastrow, A.: The word problem for some uncountable groups given by countable words. ArXiv e-prints (March 2011)

5. Book, R., Otto, F.: Confluent String Rewriting. Springer, Heidelberg (1993)
6. Chiswell, I.: Introduction to Λ-trees. World Scientific, Singapore (2001)
7. Choffrut, C., Karhumäki, J.: Combinatorics of words. In: Rozenberg, G., Salomaa, A. (eds.) Handbook of Formal Languages, vol. 1, pp. 329–438. Springer, Heidelberg (1997)
8. Diekert, V., Duncan, A.J., Myasnikov, A.G.: Geodesic rewriting systems and pregroups. In: Bogopolski, O., Bumagin, I., Kharlampovich, O., Ventura, E. (eds.) Combinatorial and Geometric Group Theory. Trends in Mathematics, pp. 55–91. Birkhäuser, Basel (2010)
9. Diekert, V., Myasnikov, A.G.: Group extensions over infinite words. ArXiv e-prints (November 2010)
10. Epstein, D.B.A., Cannon, J.W., Holt, D.F., Levy, S.V.F., Paterson, M.S., Thurston, W.P.: Word Processing in Groups. Jones and Bartlett, Boston (1992)
11. Ésik, Z., Bloom, S.L.: Algebraic linear orderings. International Journal of Foundations of Computer Science (2011)
12. Kharlampovich, O., Myasnikov, A.: Elementary theory of free non-abelian groups. J. of Algebra 302, 451–552 (2006)
13. Lothaire, M.: Combinatorics on Words. Encyclopedia of Mathematics and its Applications, vol. 17. Addison-Wesley, Reading (1983) (reprinted by Cambridge University Press, 1997)
14. Lyndon, R., Schupp, P.: Combinatorial Group Theory. Classics in Mathematics. Springer, Heidelberg (2001)
15. Lysenok, I.: On some algorithmic problems of hyperbolic groups. Math. USSR Izvestiya 35, 145–163 (1990)
16. Mihailova, K.A.: The occurrence problem for direct products of groups. Dokl. Akad. Nauk SSSR 119, 1103–1105 (1958); english translation in: Math. USSR Sbornik, 70, 241–251 (1966)
17. Myasnikov, A., Remeslennikov, V., Serbin, D.: Regular free length functions on Lyndon's free $\mathbb{Z}[t]$-group $F^{\mathbb{Z}[t]}$. Contemp. Math. Amer. Math. Soc. 378, 37–77 (2005)
18. Olshanskii, A.Y., Sapir, M.V.: Length functions on subgroups in finitely presented groups. In: Groups — Korea 1998 (Pusan), de Gruyter, Berlin (2000)
19. Rips, E.: Subgroups of small cancellation groups. Bull. London Math. Soc. 14, 45–47 (1982)
20. Romanovskii, N.S.: The occurrence problem for extensions of abelian by nilpotent groups. Sib. Math. J. 21, 170–174 (1980)

Abelian Primitive Words*

Michael Domaratzki[1] and Narad Rampersad[2]

[1] Department of Computer Science, University of Manitoba, Winnipeg, MB, R3T 2N2, Canada
mdomarat@cs.umanitoba.ca
[2] Department of Mathematics, University of Liège, 4000 Liège, Belgium
narad.rampersad@gmail.com

Abstract. We investigate Abelian primitive words, which are words that are not Abelian powers. We show the set of Abelian primitive words is not context-free. We can determine whether a word is Abelian primitive in linear time. Also different from classical primitive words, we find that a word may have more than one Abelian root. We also consider enumeration of Abelian primitive words.

1 Introduction

Repetition in words is a well-studied topic, and many of the results in this area can be classified into two distinct research areas: the theory of formal languages and the study of combinatorics on words. In these two areas, the focus on repetition is slightly different: in formal language theory, research focuses on the properties of languages containing words with different types of repetition, while in combinatorics on words, research typically concentrates on the existence or non-existence of individual words which avoid certain repetitions, and combinatorial enumeration of words with or without repetitions.

An example of a long-standing area of research relating to repetition in both the theory of formal languages and combinatorics on words are primitive words: a word x is primitive if it cannot be expressed as a repetition of some shorter word y. In combinatorics on words, an elegant proof of the number of primitive words of a given length is given using Möbius inversions (see, e.g., Lothaire [11]). However, in formal language theory, it is unknown whether the set of primitive words is a context-free language or not (see, e.g., Dömösi *et al.* [8]). However, it is known that a closely related set, the set of Lyndon words, is not context-free [4].

In combinatorics on words, a parallel notion to standard repetition is Abelian repetition. A word x is an Abelian power if it can be divided into blocks $x = x_1x_2\cdots x_n$ where every block x_i is a permutation of every other block.

In this paper, we consider the application of Abelian repetition to the concept of primitivity. Despite the naturalness of this application, the concept does not

* A full version of this paper, including proofs omitted for space reasons, is available at arXiv:1006.4104v2.

G. Mauri and A. Leporati (Eds.): DLT 2011, LNCS 6795, pp. 204–215, 2011.

appear to have attracted much attention before[1]. In a related concept, Czeizler *et al.* [6] study repetitions with only limited rearrangement. We study the language of Abelian primitive words, a formal language theoretic question, and find that the set of Abelian primitive words is not context-free. We give a linear time algorithm for determining if a word is Abelian primitive or not. We also consider the problem of counting the number of Abelian primitive words of a given length, a problem from combinatorics on words.

2 Definitions

For additional background in formal languages and automata theory, see Rozenberg and A. Salomaa [14]. Let Σ be a finite set of letters, called an alphabet. A word over Σ is any finite sequence of letters from Σ. The word containing no symbols, the empty word, is denoted ϵ. The set Σ^* is the set of all words over Σ. A language L is any subset of Σ^*. If $x = a_1 a_2 \cdots a_n$ is a word, with $a_i \in \Sigma$, then the length of x, denoted by $|x|$, is n. For $a \in \Sigma$ and $w \in \Sigma$, $|w|_a$ is the number of occurrences of a in w. Given words x, y, x is a subword of y (also called a factor) if there exist words u, v such that $y = uxv$.

For languages $L_1, L_2 \subseteq \Sigma^*$ the left quotient of L_1 by L_2, denoted $L_2^{-1}L_1$, is defined by

$$L_2^{-1}L_1 = \{x \in \Sigma^* \ : \ \exists y \in L_2 \text{ such that } yx \in L_1\}.$$

Given an (ordered) alphabet $\Sigma = \{a_1, \ldots, a_n\}$, the Parikh vector of a word $w \in \Sigma^*$ is $\Psi(w) = (|w|_{a_1}, |w|_{a_2}, \ldots, |w|_{a_n})$. For the alphabet $\Sigma = \{a, b\}$, we assume $a < b$. Thus, for example $\Psi(abbab) = (2, 3)$.

We first recall the standard notion of primitive words. A word w is primitive if w cannot be written as z^k for $z \in \Sigma^*$ and $k \geq 2$. If w is not primitive, then there is a unique primitive word u such that $w = u^k$ for some $k \geq 2$. For an alphabet Σ, the set of all primitive words $w \in \Sigma^*$ is denoted $Q(\Sigma)$ or simply Q if Σ is understood.

We now turn to the generalization of these notions to Abelian repetitions. A word w is a n-th Abelian power if $w = u_1 u_2 \cdots u_n$ for some $u_1, u_2, \ldots, u_n$ such that for all $1 \leq i, j \leq n$, $\Psi(u_i) = \Psi(u_j)$. That is, each u_j with $j \geq 2$ is a permutation of u_1.

We say that a word w is Abelian primitive (or A-primitive, for short) if w fails to be a k-th Abelian power for every $k \geq 2$. For an alphabet Σ, the set of all A-primitive words $w \in \Sigma^*$ is denoted by $AQ(\Sigma)$ or simply AQ if Σ is understood.

Example 1. The word $w = aabbab$ is A-primitive, while $u = aabbabab$ is not, as $u = xy$ where $\Psi(x) = \Psi(y) = (2, 2)$.

---

[1] We have found a reference to a research project studying Abelian primitive words on the web at `http://bit.ly/9NWqSI`, but have been unable to obtain a copy of any associated works.

Let w be an Abelian power. Then we say that a word u is an Abelian root (or A-root) of w if $w = uu_1u_2 \cdots u_n$ for some $u_1, \ldots, u_n \in \Sigma^*$ with $\Psi(u) = \Psi(u_i)$ for all $1 \leq i \leq n$. If w has an A-root u which is also A-primitive, then we say that u is an A-primitive root of w. Two A-primitive roots u, v of a word w are distinct if $|u|$ does not divide $|v|$ or vice versa. On the other hand, we note the following simple but useful fact:

Observation 1. *If a word x of length n has an A-root of length k, then x also has an A-root of length k' for all k' where k divides k' and k' divides n.*

We recall some notation from number theory. Recall that if r, z are integers, $r \mid z$ denotes that r divides z, i.e., $z = rk$ for some $k \geq 0$. We say that a set of integers S is division-free if $x \nmid y$ for all distinct elements $x, y \in S$. For all $n \geq 2$, let $\omega(n)$ denote the number of prime divisors of n, while $\omega'(n)$ is the number of prime divisors of n with multiplicity[2]. Thus, if $n \geq 2$ and $n = p_1^{\alpha_1} p_2^{\alpha_2} \cdots p_k^{\alpha_k}$ is its prime factorization, then $\omega(n) = k$ and $\omega'(n) = \sum_{i=1}^{k} \alpha_i$. We also let $d(n)$ be the number of divisors of n, i.e., $d(n) = \prod_{i=1}^{k}(1 + \alpha_i)$.

3 Non-context-Freeness of *AQ*

We now show that the set AQ of all A-primitive words is not context-free. This is in contrast to the set of ordinary primitive words Q, for which it is unknown whether they are a context-free language or not. We begin with two preliminary lemmas.

Lemma 1. *Let $p \geq 2$ be a prime and $x = aabb(ab)^{p-2}$. Then x is A-primitive.*

Proof. Note that $|x| = 2p$. If x is not A-primitive, then one of three cases occurs:

(a) $x = u^{2p}$ for some letter u,
(b) $x = u_1u_2 \cdots u_p$ for words $u_1, \ldots, u_p$ of length two, or
(c) $x = v_1v_2$ for words v_1, v_2 of length p.

The first of these possibilities cannot occur, as x contains occurrences of both a and b. The second case is also not possible, since if so, we would have $u_1 = aa$ and $u_2 = bb$, which do not have matching Parikh vectors. Thus, we must have that $x = v_1v_2$ for $|v_1| = |v_2| = p$. We have three subcases:

(a) if $p = 2$, then we are in the previous case, i.e., $v_1 = aa$.
(b) if $p = 3$, then $v_1 = aab$ and $v_2 = bab$.
(c) otherwise $p > 3$ and $v_1 = aabb(ab)^{(p-5)/2}a$ which has Parikh vector $((p-5)/2 + 3, (p-5)/2 + 2)$, and $v_2 = b(ab)^{(p-1)/2}$ which has Parikh vector $((p-1)/2, (p-1)/2 + 1)$. We can see that the number of occurrences of a in v_1 is even, while in v_2 it is odd or vice versa.

□

[2] The notation $\Omega(n)$ is also used for what we call $\omega'(n)$, but we reserve Ω for denoting asymptotic function growth. Our notation is from Bach and Shallit [3].

Lemma 2. *Let $M = AQ \cap aabb(ab)^*$. Then $M = \{aabb(ab)^{p-2} : p \text{ is prime.} \}$.*

Proof. The right-to-left inclusion is immediate from Lemma 1.

For the reverse inclusion, let $x \in M$. Then $|x| = 2n$ for some $n \geq 2$. Suppose, contrary to what we want to prove, that x is not of the form $aabb(ab)^{p-2}$ for some prime p. Then we must have that n is not prime. Let q be a prime factor of n and note that

$$x = (aabb(ab)^{q-2}) \cdot ((ab)^q)^{n/q-1}$$

and that all factors of length $2q$ have q occurrences of a and q occurrences of b. Further, $aabb(ab)^{q-2}$ is an A-primitive root by Lemma 1. Thus, x is not A-primitive, a contradiction to our choice of $x \in M$. □

We can now show that the set of all A-primitive words is not context-free.

Theorem 1. *The set AQ is not context-free.*

Proof. We prove that M is not context-free. Let $M' = h^{-1}(\{aabb\}^{-1}M)$ where $h : \{a\}^* \to \{a,b\}^*$ is the morphism $h(a) = ab$. Then $M' = \{a^{p-2} : p \text{ is prime} \}$. As the context-free languages are closed under quotient by regular sets and inverse homomorphism, M' is context-free if M is. But as M' is unary, if it is a context-free language then it is also regular. But by the pumping lemma, we can see that M' is not regular. Thus, neither M nor AQ are context-free.

The set of all non-trivial Abelian powers, $\overline{AQ}$, is also non-context-free, as can be seen through, e.g., the intersection $\overline{AQ} \cap a^*ba^*ba^*b = \{a^nba^nba^nb : n \geq 0\}$. Using the interchange lemma, Gabarró [9] has proven that the language $\{uw_1w_2v : u, w_1, w_2, v \in \Sigma^*, \Psi(w_1) = \Psi(w_2)\}$ of words containing an Abelian square is not context-free.

4 Complexity of *AQ*

Through an elegant pattern matching algorithm [12, Thm. 13], it is known that we can determine whether a word is primitive in linear time. We now consider this problem for A-primitive words. Throughout this section, we consider the size of the alphabet to be a fixed constant. In order to illustrate the basic principles of the algorithm, we begin with an $O(n \log n / \log \log n)$ algorithm:

```
def isAbelPrim(w):
  n = len(w)
  if n==1:
    return True
  PF = { p : p is prime, p | n }
  D = { n/p : p in PF }
  for d in D:
    if w has an A-root of length d:
      return False
  return True
```

Suppose that $w \in AQ$. Then w certainly does not have an A-root whose length is any of the periods in D, thus `isAbelPrim` returns true. On the other hand, if $w \notin AQ$ with $|w| > 1$, then w has an A-root of length r for some $r \mid n$ with $r < n$. There exists $d_r \in D$ such that $r \mid d_r$ (r may also divide other $d \in D$, but it is enough to know it divides some d_r). By Observation 1, on the loop of `isAbelPrim` with $d = d_r$, the algorithm will return false.

One iteration of the loop in `isAbelPrim` will take time $O(n)$, by walking across w and computing the Parikh vectors for each block of length $d \in D$. Thus, the runtime of the algorithm is $O(\rho(n) + n\omega(n))$ where $\rho(n)$ is the time required to calculate the set `PF`.

We claim that even using trial division, we have $\rho(n) \in O(\sqrt{n}\log n)$. In particular, we repeatedly find the least prime p dividing n and factor out the largest power of p^α dividing n. This process is then repeated on n/p^α. In this way, we can compute the prime factors of n. If $n = \prod_{i=1}^{k} p_i^{\alpha_i}$ is the prime factorization of n, we find the smallest prime dividing n $\omega(n)$ times, once for each p_i dividing n. Upon finding such a p_i, calculating α_i takes $O(\sqrt{n}+\alpha_i)$ time. Thus, the total runtime is $O(\sum_{p_i|n}(\sqrt{n} + \alpha_i)) = O(\sqrt{n}\omega(n) + \omega'(n))$. As $\omega(n) \in O(\log n/\log\log n)$ [3, Thm. 8.8.10] and $\omega'(n) \in O(\log n)$ [10, Sect. 22.10], this gives the claimed worst case running time for calculating `PF`.

Thus, the running time of `isAbelPrim` is $O(n\omega(n))$. Using the same estimate on the worst-case growth of $\omega(n)$, we obtain the following result:

Theorem 2. *Given x, there is an algorithm to determine if $x \in AQ$ which runs in $O(n\frac{\log n}{\log\log n})$ time in the worst case.*

4.1 A Linear Time Algorithm for Recognizing AQ

We can improve the algorithm `isAbelPrim` from the previous section by caching commonly used Parikh vectors, and obtain a linear time algorithm. Let $\mathrm{gpf}(n)$ be the greatest prime factor of n. Then we note that if $\mathrm{gpf}(n)^2 \mid n$, every $d \in D$ is divisible by $\mathrm{gpf}(n)$, while if $\mathrm{gpf}(n)^2 \nmid n$, then every $d \in D$ is divisible by $\mathrm{gpf}(n)$ except $d = n/\mathrm{gpf}(n)$. In both cases, we will precompute the Parikh vectors of length $\mathrm{gpf}(n)$ in order to compute the Parikh vectors of length d for all $d \in D$ which are divisible by $\mathrm{gpf}(n)$.

Let w be our input word of length n and write $w = w_1 w_2 \cdots w_{n/\mathrm{gpf}(n)}$ where each block has length $\mathrm{gpf}(n)$. Let $\mathbf{u}_i = \Psi(w_i)$ for $1 \le i \le n/\mathrm{gpf}(n)$. Note then that if $\mathrm{gpf}(n) \mid d$, then the blocks of w of length d have Parikh vectors of the form

$$\sum_{j=1}^{d/\mathrm{gpf}(n)} \mathbf{u}_{kd/\mathrm{gpf}(n)+j}$$

for some $1 \le k < \mathrm{gpf}(n)$. Thus, we can compute these Parikh vectors quickly by summing the precomputed $\mathbf{u}_i$.

```
def isAbelPrimLin(w):
  n = len(w)
```

```
PF = { p : p is prime, p|n }
gpf = max(PF)
D = { n/p : p in PF }
if ( n % (gpf**2) != 0):
  D.remove(n/gpf)
  calculate Parikh vectors of length n/gpf.
  if w has an A-root of length n/gpf:
     return False
for i in range(0,n/gpf):
  u[i] = Parikh(w,i,gpf)
for d in D:
  calculate Parikh vectors of length d (using u[i])
  if w has an A-root of length d:
    return False
return True
```

Here, we let `Parikh(w,i,j)` be a method which computes the `i`-th Parikh vector of length `j` in the word `w`.

This modified implementation has the same correctness as the previous implementation, as the same tests are performed. We now show the claimed $O(n)$ run time. Computing `D` and `PF` is the same as in **`isAbelPrim`** and can be done in linear time. While computing `PF`, we can also keep track of $\text{gpf}(n)$. In the case where $\text{gpf}(n)^2 \nmid n$, the time to execute the additional statements is $O(n)$ time. Similarly, the computation of the Parikh vectors `u[i]` takes time $O(n)$.

Consider the execution of the final for loop. For $d \in D$, we need $O(d/\text{gpf}(n))$ time to compute one Parikh vector of a subword of w of length d, so to compute all n/d such vectors requires time $O(n/\text{gpf}(n))$. To test the equalities of all these n/d vectors (implied by the if statement) requires time $O(n/d) = O(p)$ where $d = n/p$. Thus, the worst case running time of the loop is

$$\sum_{p|n} \left(O\left(\frac{n}{\text{gpf}(n)}\right) + O(p) \right) = O\left(n \frac{\omega(n)}{\text{gpf}(n)}\right) + O\left(\sum_{p|n} p\right).$$

We now give an estimate of the first quantity. The following lemma is easily established:

Lemma 3. *For all integers n, $\omega(n)/gpf(n) \leq 2/3$.*

Finally, we have that $\sum_{p|n} p \leq n$. Thus, the total running time of the loop is $O(n)$.

Theorem 3. *Given x, there is an algorithm to determine if $x \in AQ$ which runs in time $O(n)$ time in the worst case.*

5 Number of Abelian Primitive Roots

We now turn to the number of A-primitive roots a word may have, as a function of its length. We show that unlike classical primitive words, a word may have multiple distinct A-primitive roots. This fact was essentially noted by Constantinescu and Ilie [5] who constructed an infinite word w with two distinct Abelian periods. We generalize this to show a tight bound on the number of A-primitive roots a word may have.

We begin with the following proposition which is of independent interest:

Proposition 1. *If w has two distinct A-primitive roots u, v where $|u| = \ell_1$, $|v| = \ell_2$, then* $\gcd(\ell_1, \ell_2) \geq 2$.

Proof. Assume that w has two distinct A-primitive roots: $w = u_1 u_2 \cdots u_m$ and $w = v_1 v_2 \cdots v_n$ where $|u_i| = \ell_1$ for all $1 \leq i \leq m$ and $|v_j| = \ell_2$ for all $1 \leq j \leq n$. Assume, contrary to what we want to prove that $\gcd(\ell_1, \ell_2) = 1$.

First note that $m \geq \ell_2$. To see this, note that $|w| = m\ell_1 = n\ell_2$ and so we have that $\ell_2 \mid \ell_1 m$. Since we assume that ℓ_1 and ℓ_2 are coprime, then we must have that $\ell_2 \mid m$, which implies that $m \geq \ell_2$.

Thus $m \geq \ell_2$ and $n \geq \ell_1$ as well. As $\gcd(\ell_1, \ell_2) = 1$, there exist $r, s \geq 0$ such that $r\ell_1 = s\ell_2 - 1$ (or $r\ell_1 = s\ell_2 + 1$, which is proven similarly). As $m \geq \ell_2$ and $n \geq \ell_1$, we can assume that $s \leq n$ and $r \leq m$.

Thus, the prefix $v' = v_1 v_2 \cdots v_s$ of w of length $s\ell_2$ is one letter longer than the prefix $u' = u_1 u_2 \cdots u_r$. Without loss of generality, let a be the last symbol of v_s, which is also the first symbol of u_{r+1}. Let $\alpha = |u_1|_a$ and $\beta = |v_1|_a$. Counting the occurrences of a in u' and v', we get

$$r\alpha = s\beta - 1. \tag{1}$$

Now consider that the prefix of w of length $\ell_1\ell_2$ is $u_1 \cdots u_{\ell_2} = v_1 \cdots v_{\ell_1}$. Considering $v'' = v_{s+1} \cdots v_{\ell_1}$ and $u'' = u_{r+1} \cdots u_{\ell_2}$, and again counting the occurrences of a, we also have

$$(\ell_2 - r)\alpha = (\ell_1 - s)\beta + 1. \tag{2}$$

Equating both (1) and (2) in terms of α, we get

$$r((\ell_1 - s)\beta + 1) = (\ell_2 - r)(s\beta - 1).$$

Solving for β gives $\beta = \ell_2$. Thus, we have that $v_1 \in a^+$ and thus w only has A-primitive root a, a contradiction. Thus, $\gcd(\ell_1, \ell_2) \geq 2$. □

5.1 Upper Bound

We next give an upper bound on the number of A-primitive roots a word may have. We first need the estimate $d(n) \in O(2^{\log n / \log \log n})$. We will also use a result by de Bruijn *et al.* [7] (see also Anderson [2]):

Theorem 4. *Let $n = p_1^{\alpha_1} p_2^{\alpha_2} \cdots p_k^{\alpha_k}$ be the prime factorization of $n \geq 2$. Let $D(n)$ be the set of integers defined by*

$$D(n) = \{p_1^{\beta_1} p_2^{\beta_2} \cdots p_k^{\beta_k} \ : \ \forall i(\beta_i \leq \alpha_i) \text{ and } \sum_{i=1}^{k} \beta_i = \lfloor \omega'(n)/2 \rfloor\}.$$

Then $D(n)$ is a maximal anti-chain in the divisor lattice of n.

In other words, $D(n)$ is the largest division-free set of divisors of n. Anderson [2] gives the following estimate on the size of $D(n)$, which we denote $s(n)$:

Theorem 5. *Let $n = p_1^{\alpha_1} p_2^{\alpha_2} \cdots p_k^{\alpha_k}$ be the prime factorization of $n \geq 2$. Let $A(n) = \frac{1}{3}\sum_{i=1}^{k} \alpha_i(\alpha_i + 2)$. Then the maximal anti-chain in the divisor lattice of n has size $s(n) = \Theta(d(n)/\sqrt{A(n)})$.*

Now a word w of length n has at most $|D(n)|$ A-primitive roots: if r is an A-primitive root, then $|r|$ divides $|w|$ and $|r|$ is not divisible by the length of any other A-primitive root. Thus, we can obtain the following result:

Theorem 6. *If w is a word of length n, the number of distinct A-primitive roots is $s(n) \in o(2^{\log n/\log\log n})$.*

Proof. By Theorem 4, if w is a word of length n, then w has at most $s(n)$ distinct A-primitive roots. By Theorem 5, $s(n) \in o(2^{\log n/\log\log(n)})$. □

We can use a result of Anderson [1] which gives the average order of $s(n)$:

Theorem 7. *As $\omega'(n) \to \infty$, we have*

$$s(n) \leq \left(\sqrt{\frac{2}{\pi}} + o(1)\right) \frac{d(n)}{\sqrt{\omega'(n)}}.$$

As $n \to \infty$,

$$\sum_{m \leq n} \frac{d(m)}{\sqrt{\omega'(m)}} \sim \frac{n \log n}{\sqrt{2 \log\log n}}.$$

5.2 Lower Bound

For a lower bound on the number of A-primitive roots a word may have, we give an explicit construction. For any $n \geq 2$, let $T(n) = \{kd \ : \ k \in \mathbb{N}, d \in D(n), kd \leq n\}$. Let $t_1 < t_2 < \cdots < t_{m_n} = n$ be the m_n elements of $T(n)$ in sorted order. Define

$$z_n = a^{t_1} b^{t_1} \prod_{i=2}^{m_n} a^{t_i - t_{i-1}} b^{t_i - t_{i-1}}.$$

Note that z_n is a word of length $2n$ with $\Psi(z_n) = (n, n)$.

Example 2. If $n = 30$, then $D(30) = \{2, 3, 5\}$. In this case

$$T(30) = \{2, 3, 4, 5, 6, 8, 9, 10, 12, 14, 15, 16, 18, 20, 21, 22, 24, 25, 26, 27, 28, 30\}.$$

With this, we have

$$z_{30} = aabbababababaabbababaabbaabbababaabbaabbababaabbababababaabb.$$

Proposition 2. *Let $n \geq 2$ and $t \in D(n)$. Then z_n has an A-primitive root of length $2t$.*

Proof. Let $1 \leq j \leq m_n$ be the index such that $t = t_j$. As $t \in D(n) \subset T(n)$, we have that the prefix of z_n of length $2t$ is

$$w_n = a^{t_1} b^{t_1} a^{t_2 - t_1} b^{t_2 - t_1} \cdots a^{t - t_{j-1}} b^{t - t_{j-1}}.$$

Note that $\Psi(w_n) = (t, t)$. Now, each additional block of length $2t$ from has the form

$$a^{t_\alpha} b^{t_\alpha} \cdots a^{t_\beta} b^{t_\beta}$$

for some α, β which are differences of successive t_i. To see this, note that these factors of z_n begin and end at positions which are multiples of $t \in D(n)$, so each of the breakpoints are elements of $T(n)$. By telescoping, each of these factors has Parikh vector (t, t). Thus, z_n is a n/t-th A-power.

Further, w_n must be an A-primitive root of z_n. Otherwise, there is some $z \in T(n)$ such that $z \mid t$, but in this case, z is divisible by some element in $D(n)$, by definition of $T(n)$. But this gives a contradiction, since $t \in D(n)$ and $D(n)$ is an anti-chain of divisors. □

Corollary 1. *For all $n \geq 2$, there exists a word of length $2n$ with $s(n)$ distinct A-primitive roots.*

6 Counting Abelian Primitive Words

Let $\psi_k(n)$ be the number of primitive words of length n over a k-letter alphabet, $\psi_k^A(n)$ be the number of A-primitive words of length n over a k-letter alphabet and $\Delta_k(n) = \psi_k(n) - \psi_k^A(n)$. Note that $\Delta_k(n) \geq 0$ for all n, but we can observe, e.g., that $\Delta_k(p) = 0$ for all primes p. Small values of $\psi_k^A(n)$ are given in Figure 1.

↓ k n →	1	2	3	4	5	6	7	8	9	10
2	2	2	6	10	30	36	126	186	456	740
3	3	6	24	66	240	612	2184	5922	19302	54300
4	4	12	60	228	1020	3792	16380	62820	260952	1016880
5	5	20	120	580	3120	15000	78120	382740	1950420	9637400

Fig. 1. Number of A-primitive words $\psi_k^A(n)$ by length (n) and alphabet size (k)

The function $\psi_k(n)$ is well-known (see, e.g., Lothaire [11]). The formula $\psi_k(n) = \sum_{d|n} \mu(d) k^{n/d}$ expresses ψ_k in terms of the Möbius function μ defined by $\mu(1) = 1$, $\mu(n) = (-1)^k$ if n is a product of k distinct primes and $\mu(n) = 0$ if $p^2 \mid n$ for some prime p.

We can characterize Δ_k for prime powers exactly:

Lemma 4. *For all primes p and all $r \geq 2$,*

$$\Delta_k(p^r) = \sum_{n_1 + n_2 + \cdots + n_k = p^{r-1}} \binom{p^{r-1}}{n_1\ n_2 \ldots n_k} \left(\binom{p^{r-1}}{n_1\ n_2 \ldots n_k}^{p-1} - 1 \right).$$

Here, the sum is taken over all partitions $n_1 + n_2 + \cdots + n_k$ of p^{r-1}.

Proof. Let $x \in Q - AQ$ of length p^r. As x is not A-primitive, it has an A-primitive root of length p^i for some $1 \leq i < r$. But then x can also be written as $x = x_1 x_2 \cdots x_p$ where $|x_i| = p^{r-1}$ and $\Psi(x_i) = \Psi(x_j)$ for all $1 \leq i, j \leq p$. Thus, it suffices to count only those x of this form.

Consider that there are $\binom{p^{r-1}}{n_1\ n_2 \ldots n_k}$ different words x_1 of length p^{r-1} such that $\Psi(x_1) = (n_1, n_2, \ldots, n_k)$ for each partition $n_1 + n_2 + \cdots + n_k = n$. As recently noted by Richmond and Shallit [13], for a fixed choice of x_1, the remainder of the words $x_2, \ldots, x_p$ must satisfy $\Psi(x_j) = \Psi(x_1)$, which can be done in $\binom{p^{r-1}}{n_1\ n_2 \ldots n_k}$ ways for each $2 \leq j \leq p$. Thus, we get a total of $\binom{p^{r-1}}{n_1\ n_2 \ldots n_k}^{p-1}$ possibilities, and we must exclude the choice $x_1 = x_2 = x_3 = \cdots = x_p$, as this word is not primitive.

Thus, multiplying the number of choices of the word x_1 and the words $x_2, \ldots, x_p$ and summing over all possible Parikh vectors, we get the result. □

The problem of giving a closed form of $\Delta_k(n)$ or $\psi_k^A(n)$ for all values of n is still open.

7 Equivalence Relations on A-Primitive Words

In this section, we consider classical results such as the Lyndon-Schützenberger Theorem for classical words in the context of Abelian primitivity. To do so, we define appropriate equivalence relations to replace equality.

We first note that the A-primitive words are not closed under conjugation. For example, note that *bbababaa* $\in AQ$ but *aabbabab* $\notin AQ$. For all $n \geq 1$, let $\sim_n$ be the binary relation defined on words by $u \sim_n x$ if we can write $u = \alpha_1 \alpha_2 \cdots \alpha_m$ and $x = \beta_1 \beta_2 \cdots \beta_m$ where

(a) for all $1 \leq i \leq m$, $|\alpha_i| = |\beta_i| = n$.
(b) for all $1 \leq i, j \leq m$, $\Psi(\alpha_i) = \Psi(\beta_j)$.

Thus, $\sim_n$ represents that two words can be broken into blocks of length n, all of which have the same image under Ψ.

Example 3. Let $n = 3$. Then *abc acb abc* $\sim_3$ *cba bca bca* as each block β of length three in both words satisfies $\Psi(\beta) = (1, 1, 1)$.

We use $\sim_n$ to investigate relationships with the theory of codes in the context of commutation.

Theorem 8. *For all words $u, x \in \Sigma^*$, $ux \sim_n xu$ if and only if there exists $r \geq 1$, $\alpha_1, \ldots, \alpha_r, \beta_1, \ldots, \beta_r \in \Sigma^*$ such that*

(a) for all $1 \leq i \leq r$, $|\alpha_i \beta_i| = n$.
(b) for all $1 \leq i, j \leq r$, $\Psi(\alpha_i) = \Psi(\alpha_j)$ and $\Psi(\beta_i) = \Psi(\beta_j)$.
(c) there exists $1 \leq s < r$ such that

$$u = \alpha_1 \beta_1 \cdots \alpha_{s-1} \beta_{s-1} \alpha_s \text{ and } x = \beta_s \alpha_{s+1} \beta_{s+1} \cdots \alpha_r \beta_r.$$

Example 4. If $x = abca$ and $u = cbabc$ then $xu \sim_3 ux$ (which was shown in Example 3). Note that x and u have different lengths and thus cannot share an A-primitive root.

The case where both x and u have A-primitive roots of length n is of particular interest:

Corollary 2. *Let $u, x \in \Sigma^*$ with $ux \sim_n xu$. If u has an A-primitive root of length n, then x does as well, and these A-roots are the same.*

Corollary 2 is analogous to the second Lyndon-Schützenberger theorem (see e.g., Lothaire [11]) which can be interpreted (in part) as $ux = xu$ if and only if x and u both have the same primitive root.

8 Conclusions

We have studied the formal language theoretic and combinatorial properties of Abelian primitive words. Unlike classical primitive words, the number of Abelian primitive words is a nontrivial combinatorial problem. On the other hand, we show that the set of Abelian primitive words are not context-free, unlike the long-standing open problem for primitive words. Future research problems include an exact enumeration of the number of Abelian primitive words of length n.

Acknowledgments. The authors would like to thank Jeff Shallit for making several helpful comments and corrections on an earlier version of the paper. We are also grateful to Terry Visentin for helpful discussions.

References

1. Anderson, I.: On primitive sequences. Journal London Math. Soc. 42, 137–148 (1967)
2. Anderson, I.: Combinatorics of finite sets. Dover Publications, Mineola (2002)
3. Bach, E., Shallit, J.: Algorithmic Number Theory, vol. 1. MIT Press, Cambridge (1997)
4. Berstel, J., Boasson, L.: The set of Lyndon words is not context-free. Bull. EATCS 63, 139–140 (1997)
5. Constantinescu, S., Ilie, L.: Fine and Wilf's theorem for Abelian periods. Bull. EATCS 89, 167–170 (2006)
6. Czeizler, E., Kari, L., Seki, S.: On a special class of primitive words. Theoretical Computer Science 411, 617–630 (2010)
7. de Bruijn, N., van Ebbenhorst Tengbergen, C., Kruyswijk, D.: On the set of divisors of a number. Nieuw Arch. Wiskunde 23, 191–193 (1951)
8. Dömösi, P., Horváth, S., Ito, M., Kászonyi, L., Katsura, M.: Formal languages consisting of primitive words. In: Ésik, Z. (ed.) FCT 1993. LNCS, vol. 710, pp. 194–203. Springer, Heidelberg (1993)
9. Gabarró, J.: Some applications of the interchange lemma. Bull. EATCS 25, 19–21 (1985)

10. Hardy, G., Wright, E.: An Introduction to the Theory of Numbers, 5th edn. Oxford Science Publications, Oxford (2000)
11. Lothaire, M.: Combinatorics on words. Cambridge University Press, Cambridge (1997)
12. Petersen, H.: The ambiguity of primitive words. In: Enjalbert, P., Mayr, E., Wagner, K. (eds.) STACS 1994. LNCS, vol. 775, pp. 679–690. Springer, Heidelberg (1994)
13. Richmond, L.B., Shallit, J.: Counting Abelian squares. Elec. J. Combinatorics 16, R72 (2009)
14. Rozenberg, G., Salomaa, A.: Handbook of Formal Languages, vol. 1. Springer, Heidelberg (1997)

Scattered Context-Free Linear Orderings

Zoltán Ésik*

Department of Informatics
University of Szeged
Szeged, Hungary

Abstract. We show that it is decidable in exponential time whether the lexicographic ordering of a context-free language is scattered, or a well-ordering.

1 Introduction

When the alphabet A of a language $L \subseteq A^*$ is linearly ordered, L may be equipped with the lexicographic order turning L into a linearly ordered set. Every countable linear ordering may be represented as the lexicographic ordering of a language (over the two-letter alphabet). A (deterministic) context-free linear order is a linear ordering that can be represented as the lexicographic ordering of a (deterministic) context-free language. The study of context-free linear orderings has been initiated in [3]. In [4], it was shown that a well-ordering is deterministic context-free (or equivalently, definable by an algebraic recursion scheme) iff its order type is less than ω^{ω^ω}. Then, in [5] it was shown that the Hausdorff rank of any scattered deterministic context-free linear ordering is less than ω^ω. For an extension of these results to linear orderings definable by higher order recursion schemes we refer to [2].

Any monadic second-order definable property is decidable for deterministic context-free linear orders (given by $LR(1)$ grammars, say). This fact follows form a general decidability result for graphs in the pushdown hierarchy [6], more exactly from the "uniform version" of this result. In particular, it is decidable whether a deterministic context-free linear ordering is dense, or scattered, or a well-ordering. The results of [4,5] implicitly give rise to practical algorithms for deterministic context-free languages and algebraic recursion schemes. In contrast, as shown in [9], it is undecidable for a context-free linear ordering whether it is dense. The main results of this paper show that on the contrary, there is an exponential time algorithm to decide whether a context-free linear ordering is scattered, or a well-ordering. The fact that these properties are decidable for context-free linear orderings was first announced in [7].

* Partially supported by the project TÁMOP-4.2.1/B-09/1/KONV-2010-0005 "Creating the Center of Excellence at the University of Szeged", supported by the European Union and co-financed by the European Regional Fund, the TÁMOP-4.2.2/08/1/2008-0008 program of the Hungarian National Development Agency, and by the National Foundation of Hungary for Scientific Research, grant no. K 75249.

G. Mauri and A. Leporati (Eds.): DLT 2011, LNCS 6795, pp. 216–227, 2011.

2 Linear Orderings

In this paper, by a linear ordering $L = (L, <)$ we shall mean a countable linear ordering. We will use standard terminology as in [11]. The isomorphism class of a linear ordering is its *order-type*.

A linear ordering L is *dense* if it has at least two elements and for all $x, y \in L$, if $x < y$ then there is some z with $x < z < y$. Up to isomorphism there are four (countable) dense linear orderings, the ordering of the rationals whose order-type is denoted η, possibly endowed with a least or greatest element, or both. A *scattered* linear order is a linear ordering that has no dense sub-order. A *well-ordering* is a linear ordering that has no sub-ordering isomorphic to the ordered set of the negative integers. Every well-ordering is scattered.

A linear ordering is *quasi-dense* if it is not scattered. It is well-known that any scattered sum or finite union of scattered linear orderings is scattered. Thus, if I is a scattered linear ordering and for each $i \in I$, L_i is a scattered linear ordering, then so is $\sum_{i \in I} L_i$. Moreover, if a linear ordering L is the finite union of sub-orderings L_i, $i = 1, \ldots, n$, then L is quasi-dense iff at least one the L_i is quasi-dense.

Suppose that A is an alphabet whose letters are ordered by $a_1 < \ldots < a_k$. Then we define the *strict order* $<_s$ on the set of words A^* by $u <_s v$ iff $u = xa_iy$ and $v = xa_jy$ for some $x, y, y' \in A^*$ and letters a_i and a_j with $a_i < a_j$. The *prefix order* is defined by $u <_p v$ iff u is a proper prefix of v. The strict order and the prefix order are partial orders. The *lexicographic order* $<_\ell$ is the union of the two, so that $x <_\ell y$ iff $x <_s y$ or $x <_p y$. Clearly, $(A^*, <_\ell)$ is a linear ordering.

If $L \subseteq A^*$ then $(L, <_\ell)$ is a linear ordering, called the *lexicographic ordering of* L. We call L dense, scattered or well-ordered if $(L, <_\ell)$ has the appropriate property. When L is a (deterministic) context-free language, we call $(L, <_\ell)$, and sometimes any linear ordering isomorphic to $(L, <_\ell)$ a *(deterministic) context-free linear ordering*. Every (deterministic) context-free linear ordering is isomorphic to the lexicographic ordering of a (deterministic) context-free language over the alphabet $\{0, 1\}$, ordered by $0 < 1$. Indeed, when $L \subseteq A^*$ and A has k letters $a_1 < \ldots < a_k$, say, then we may encode each letter a_i with a binary word $h(a_i)$ of length $\lceil \log k \rceil$ over $\{0, 1\}$ so that $h(a_i) <_\ell h(a_j)$ whenever $a_i < a_j$, then $(L, <_\ell)$ is isomorphic to $(h(L), <_\ell)$.

When $L \subseteq \{0, 1\}^*$ then $T(L)$ is the binary tree whose vertices are the words in the prefix closure of L. $T(L)$ is nonempty if L is nonempty. A vertex y is a descendant of vertex x if x is a prefix of y. The following fact is quite standard:

Proposition 1. *Suppose that* $L \subseteq \{0, 1\}^*$ *and consider the corresponding tree* $T(L)$. *Then* L *is quasi-dense iff the full binary tree has an embedding in* $T(L)$.

For completeness, a proof of Proposition 1 is given in the Appendix.

Proposition 2. *Suppose that* $L, L' \subseteq \{0, 1\}^*$. *If* $(L, <_\ell)$ *and* $(L', <_\ell)$ *are both scattered, then so is* $(LL', <_\ell)$.

Proof. We will prove that if $(LL', <_\ell)$ is quasi-dense, then one of $(L, <_\ell)$ and $(L', <_\ell)$ is quasi-dense. Assuming that LL' is quasi-dense, $T(LL')$ has an embedded copy T_0 of the full binary tree. Let us color a vertex u of $T(LL')$ blue if $uv \in L$ for some $v \in \{0,1\}^*$, i.e., when u has a descendant in L. There are two cases to consider, either each subtree of T_0 contains a blue vertex, or there is a subtree of T_0 having no blue vertex.

Case 1. Suppose that each subtree of T_0 contains a blue vertex. Then each vertex of T_0 is colored blue, so that L is quasi-dense.

Case 2. Suppose that T_0 contains a subtree having no blue vertex. Let T_1 denote such a subtree and let u denote the root of T_1. Let $u_0, \ldots, u_k$ be all the (proper) prefixes of u that are in L. Now let us color each vertex x of T_1 with the set of all integers i, $0 \le i \le k$, such that x has a descendant in $T(LL')$ which is a word in $u_i L'$. Then each vertex x of T_1 is labeled by a nonempty subset of the set $\{0, \ldots, k\}$, and if x' is a descendant of x in T_1, then the label of x' is included in the label of x. Let H be a minimal set that appears as the label of a vertex v of T_1. Then all descendants of v in T_1 are labeled H. Thus, if $i \in H$, then the full binary tree embeds in $T(u_i L')$ and thus in $T(L')$, so that L' is quasi-dense. □

3 Scattered Context-Free Linear Orderings

In this section, we assume that $G = (N, \{0,1\}, P, S)$ is a context-free grammar with nonterminal alphabet N, terminal alphabet $\{0,1\}$, rules P and start symbol S that contains no useless nonterminals or ϵ-rules. Moreover, we assume that G is left-recursion free and that $L(G)$ is not empty. These can be assumed for the results of the paper, since there is an easy polynomial time transformation of a context-free grammar to a grammar over the alphabet $\{0,1\}$ that generates an isomorphic language (with respect to the lexicographic order) not containing ϵ, and each grammar not generating the empty word can be transformed in polynomial time into an equivalent grammar that contains no useless nonterminals or ϵ-rules or any left-recursive nonterminal. See [1,8].

We let X, Y, Z (sometimes decorated) denote nonterminals, u, v, w, x, y terminal words in $\{0,1\}^*$, and we let p, q, r denote words in $(N \cup \{0,1\})^*$. For every word p, we denote by $L(p)$ the set of all words $w \in \{0,1\}^*$ with $p \Rightarrow^* w$. Thus, the language $L(G)$ generated by G is $L(S)$. The length of p is denoted $|p|$.

For nonterminals X and Y we define $Y \preceq X$ iff there exist p, q with $X \Rightarrow^* pYq$, and we define $X \approx Y$ if both $X \preceq Y$ and $Y \preceq X$ hold. When $X \approx Y$, we say that X and Y belong to the same *strong component*. When $Y \preceq X$ but $X \not\approx Y$, we also write $Y \prec X$. The *height* of a nonterminal X is the length k of one of the longest sequences $Y_1 \prec \ldots \prec Y_k = X$. When $\mathcal{C}$ is a strong component and $X \in \mathcal{C}$ has height k, we also say that $\mathcal{C}$ has height k.

A *primitive word* is a nonempty word that is not a proper power. For elementary properties of primitive words and conjugacy of words we refer to [10].

Theorem 1. *The following conditions are equivalent for a context-free grammar* $G = (N, \{0,1\}, P, S)$*:*

1. $(L(G), <_\ell)$ *is a scattered linear ordering.*
2. *There exist no nonterminal* X *and words* $u, v \in \{0,1\}^*$ *such that neither* u *is a prefix of* v *nor* v *is a prefix of* u*, moreover,* $X \Rightarrow^* uXp$ *and* $X \Rightarrow^* vXq$ *hold for some* p, q*.*
3. *For each recursive nonterminal* X *there is a primitive word* $u_0 = u_0^X$ *such that whenever* $X \Rightarrow^+ wXp$ *then* $w \in u_0^+$*.*
4. *For each strong component* $\mathcal{C}$ *containing a recursive nonterminal there is a primitive word* $u_0 = u_0^{\mathcal{C}}$*, unique up to conjugacy, such that for all* $X, Y \in \mathcal{C}$ *there is a (necessarily unique) conjugate* v_0 *of* u_0 *and a proper prefix* v_1 *of* v_0 *such that if* $X \Rightarrow^+ wYp$ *for some* $w \in \{0,1\}^*$ *and* $p \in (N \cup \{0,1\})^*$ *then* $w \in v_0^* v_1$*.*

Proof. It is easy to prove that the first condition implies the second. Suppose that $L(G)$ is scattered and let $X \Rightarrow^* uXp$ and $X \Rightarrow^* vXq$. If neither u is a prefix of v nor v is a prefix of u, then u and v are nonempty and comparable with respect to the strict order, say $u <_s v$. Suppose that $S \Rightarrow^* wXp$. The vertices $w(u+v)^*$ determine an embedding of the full binary tree in $T(L)$. Thus, by Proposition 1, L is quasi-dense, a contradiction. Thus, either u is a prefix of v or vice versa.

Suppose now that the second condition holds. We prove that the third condition also holds. Let X be a recursive nonterminal and suppose that $X \Rightarrow^+ uXp$. Then u is nonempty (since G is left recursion free) and thus has a primitive root u_0. We claim that whenever $X \Rightarrow^+ wXq$ then w is a power of u_0. Indeed, if $X \Rightarrow^+ wXq$ then w is also nonempty and thus there exist $m, n > 0$ with $|u^n| = |w^m|$. Since $X \Rightarrow^+ u^n X p^n$ and $X \Rightarrow^+ w^m X q^m$, it follows that $u^n = w^m$, so that u_0 is also the primitive root of w.

Next we prove that the third condition implies the fourth. So assume that the third condition holds. Note that if a strong component contains a recursive nonterminal, then all nonterminals in that strong component are recursive.

Lemma 1. *Suppose that* X, Y *are different recursive nonterminals that belong to the same strong component. Then* u_0^X *and* u_0^Y *are conjugate.*

Proof. Since X, Y belong to the same strong component, there exist x, y and p, q with $X \Rightarrow^+ xYp$ and $Y \Rightarrow^+ yXq$. Thus, $X \Rightarrow^+ xyXqp$ and $Y \Rightarrow^+ yxYpq$. Thus, xy is a power of u_0^X and yx is a power of u_0^Y. Since xy and yx are conjugate and u_0^X and u_0^Y are primitive, this is possible only if u_0^X and u_0^Y are conjugate. □

Using the lemma, we now complete the proof of the fact that the third condition implies the fourth.

Suppose that the strong component $\mathcal{C}$ contains a recursive nonterminal and $X_0 \in \mathcal{C}$. Let $u_0^{\mathcal{C}} = u_0^{X_0}$. For the sake of simplicity, below we will just write u_0 for this word. Let $X, Y \in \mathcal{C}$ with $X \Rightarrow^+ wYp$ and $Y \Rightarrow^* xXq$, where w, x, p, q are appropriate words, so that $X \Rightarrow^+ wxXqp$. By Lemma 1 we have that wx is a power of a primitive word v_0 which is a conjugate of u_0. It is clear that v_0 is unique. Also, $w = v_0^n v_1$ for some $n \geq 0$ and some proper prefix v_1 of v_0.

We still need to show that if $X \Rightarrow^+ w'Yp'$ for some w' and p', then w' can be written as $v_0^m v_1$ for some m. But in this case $X \Rightarrow^+ w'xXqp'$ and $w'x$ is a power of v_0. Since the length of w' is congruent to the length of w modulo the length of v_0, it follows that $w' = v_0^m v_1$ for some $m \geq 0$. This ends the proof of the fact that the third condition implies the fourth.

Suppose finally that the fourth condition holds. Then clearly, the third condition also holds. We want to prove that $L(G)$ is scattered. To this end, we establish several preliminary facts.

Definition 1. *Suppose that X is a recursive nonterminal and let $u_0 = u_0^X$. For each $n \geq 0$ and prefix ui of u_0, where $i = 0, 1$, let $L(X, n, ui)$ denote the set of all words of the form $u_0^n u\bar{i}w$ in $L(X)$, where $w \in \{0,1\}^*$ and $\bar{i} = 1$ iff $i = 0$.*

Let X be a recursive nonterminal. The following facts are clear. (Below we continue writing u_0 for u_0^X.)

Proposition 3. *Each word in $L(X)$ is either in $L(X, n, ui)$ for some $n \geq 0$ and prefix ui of u_0, or is a word of the form $u_0^n u$ where $n \geq 0$ and u is a proper prefix of u_0.*

Proposition 4. *For each $n \geq 0$ and prefix ui of u_0 there is only a finite number of left derivations*

$$X \Rightarrow_\ell^* wYp \Rightarrow_\ell u_0^n u\bar{i}q \tag{1}$$

such that $u_0^n u\bar{i}$ is not a prefix of w.

Let us denote by $F(X, n, ui)$ the finite set of all words q that occur in derivations (1).

Proposition 5. *If Y is a nonterminal that occurs in a word $q \in F(X, n, ui)$ for some $n \geq 0$ and prefix ui of u_0, then $Y \prec X$.*

Proof. Suppose that (1) is a left derivation and Y occurs in q, so that $q = q_1Yq_2$ for some q_1, q_2. If $X \approx Y$ then there exist some r_1, r_2 with $Y \Rightarrow^* r_1Xr_2$. Thus, $q = q_1Yq_2 \Rightarrow^* q_1r_1Xr_2q_2$. Let v denote a terminal word with $q_1r_1 \Rightarrow^* v$. Then we have

$$X \Rightarrow^* u_0^n u\bar{i}q \Rightarrow^* u_0^n u\bar{i}vXr_2q_2.$$

Since $u_0^n u\bar{i}v$ is not a power of u_0, this contradicts the third condition. □

We now complete the proof of Theorem 1.

Let X be a nonterminal. We prove the following fact: If $(L(Y), <_\ell)$ is scattered for all nonterminals Y whose height is less than the height of X, then $(L(X), <_\ell)$ is scattered.

If X is not a recursive nonterminal, then the height of each nonterminal appearing on the right side of a rule $X \to p$ is less than the height of X.

Thus $L(X)$ is the finite union of all languages $L(p)$ where $X \to p$ is in P. By the induction hypothesis and Proposition 2, each linear ordering $(L(p), <_\ell)$ is scattered. Since any finite union of scattered linear orderings is scattered, $(L(X), <_\ell)$ is also scattered.

Suppose now that X is recursive. Then by Proposition 3,

$$L(X) = L_0 \cup \bigcup_{n \geq 0,\ ui \in \mathrm{Pref}(u_0)} L(X, n, ui)$$

where ui ranges over the prefixes of $u_0 = u_0^X$ and each word of L_0 is of the form $u_0^n v$ for some $n \geq 0$ and some proper prefix v of u_0. It is clear that L_0 is scattered (in fact, either a finite linear ordering or an ω-chain). Thus, it suffices to show that

$$\Big(\bigcup_{n \geq 0,\ ui \in \mathrm{Pref}(u_0)} L(X, n, ui),\ <_\ell \Big)$$

is scattered. But this linear ordering is isomorphic to the ordered sum

$$\sum_{n \geq 0,\ u1 \in \mathrm{Pref}(u_0)} (L(X, n, u1), <_\ell) + \sum_{n \leq 0,\ u0 \in \mathrm{Pref}(u_0)} (L(X, -n, u0, <_\ell)$$

where integers are ordered as usual. Since a scattered sum of scattered linear orderings is scattered, it remains to show that each $L(X, n, ui)$ is scattered. But by Proposition 4 and Proposition 5, for each n and ui, $L(X, n, ui)$ is a finite union of languages of the form $u_0^n u\bar{i} L(q)$ where q contains only nonterminals of height strictly less than the height of X. Thus, by the induction hypothesis and Proposition 2, each such language is scattered. Since any finite union of scattered linear orderings is scattered, it follows that $L(X, n, ui)$ is scattered. This ends the proof of the fact that the fourth condition implies the first. The proof of Theorem 1 is complete. □

Theorem 2. *$L(G)$ is well-ordered iff $L(G)$ is scattered and there is no recursive nonterminal X such that $L(X)$ contains a word w such that $u_0^n <_s w$ for some n, where $u_0 = u_0^X$.*

Proof. Note that the extra condition is equivalent to that for all recursive nonterminals X and for any prefix $u0$ of $u_0 = u_0^X$, the language $L(X, n, u0)$ is empty. Now by repeating the last part of the proof of Theorem 1, it follows that under this condition, if $L(G)$ is scattered, then $L(X)$ is well-ordered for all X. One uses the well-known fact that if a linear order is a finite union of well-orderings, then it is also a well-ordering, and that a well-ordered sum of well-orderings is well-ordered.

On the other hand, if the extra condition is not satisfied for the recursive nonterminal X, then $L(X)$ is not well-ordered. For suppose that $L(X, n, u0)$ contains the word $u_0^n u1x$. We know that there is some $m \geq 1$ and some w with $X \Rightarrow^+ u_0^m X w$. Thus, the words $u_0^{km} u_0^n u1xv^{km}$ for $k = 0, 1, \ldots$ form a strictly decreasing sequence in $L(X)$. We conclude by noting that if $L(X)$ is not well-ordered for some X, then $L(G)$ is not well-ordered either, since G contains no useless nonterminals. □

At this point, we are already able to show that it is decidable whether $L(G)$ is scattered, or well-ordered.

Corollary 1. *There exists an algorithm to decide whether $L(G)$ is scattered.*

Proof. As before, we may assume that $G = (N, \{0, 1\}, P, S)$ contains no useless nonterminals or ϵ-rules. Moreover, we may assume that G is left-recursion free and that $L(G)$ is not empty. By Theorem 1 we know that $L(G)$ is scattered iff for each recursive nonterminal X there is a primitive word u_0 such that whenever $X \Rightarrow^+ wXp$ then $w \in u_0^+$. We are going to test this condition. Given a recursive nonterminal X in the strong component $\mathcal{C}$, we find a word u such that $X \Rightarrow^+ uXp$ for some p. Clearly, $u \neq \epsilon$. Let u_0 denote the primitive root of u. Then consider the following grammar G_X. The nonterminals are the nonterminals of G together with the nonterminals $\overline{Y}$, where $Y \in \mathcal{C}$. The rules are those of G together with the rules

$$\overline{Y} \to p\overline{Z}$$

such that $Y, Z \in \mathcal{C}$ and there is some q with $Y \to pZq \in P$. There is one more rule, $\overline{X} \to \epsilon$. Let $\overline{X}$ be the start symbol. Then $L(G_X) \subseteq u_0^*$ iff for all w such that $X \Rightarrow^+ wXp$ for some p in G, it holds that $w \in u_0^+$. Now $L(G_X) \subseteq u_0^*$ iff the intersection of $L(G_X)$ with the complement of u_0^* is empty, which is decidable. □

Corollary 2. *There exists an algorithm to decide whether $L(G)$ is well-ordered.*

Proof. The extra condition introduced in Theorem 2 can be effectively tested, since it says that for each recursive nonterminal X, the intersection of $L(X)$ with the regular language of all words of the form $u_0^n u1x$, where $n \geq 0$ and $u0$ is a prefix of u_0, is empty. □

4 Decidability in Exponential Time

In this section, we give somewhat more efficient algorithms. First we need some preparation.

Suppose that $u_0 \in \{0, 1\}^*$ is a fixed primitive word, and consider the set $\mathcal{S}$ of all pairs (x_1, x_2), where x_1 is a proper suffix of u_0 and x_2 is a proper prefix of u_0. In particular, $(\epsilon, \epsilon) \in \mathcal{S}$. With each $(x_1, x_2) \in \mathcal{S}$ we associate the language $L(x_1, x_2) = x_1 u_0^* x_2$, if $|x_1 x_2| < |u|$, and $L(x_1, x_2) = x_1 u_0^* x_2 + z$ where z is the suffix of $x_1 x_2$ obtained by removing its prefix of length $|u_0|$, if $|x_1 x_2| \geq |u_0|$. (Note that the prefix of length $|u_0|$ of $x_1 x_2$ is a primitive word which is a conjugate of u_0.) We call a word w *legitimate* if it belongs to $L(x_1, x_2)$ for some $(x_1, x_2) \in \mathcal{S}$. Clearly, a word is legitimate iff it is a factor of some power of u_0 iff it is in $v_0^* z$ for some conjugate v_0 of u_0 and some necessarily unique proper prefix z of v_0. Moreover, for each $(x_1, x_2) \in \mathcal{S}$ there is a unique conjugate v_0 of u_0 and a unique proper prefix z of v_0 with $L(x_1, x_2) = v_0^* z$. It follows from this fact that any two languages $L(x_1, x_2)$ and $L(y_1, y_2)$ for $(x_1, x_2) \neq (y_1, y_2)$ in $\mathcal{S}$ are either disjoint or have a single common element which is a proper factor of u_0. In particular, for any legitimate word u with $|u| \geq |u_0|$ there is a unique $(x_1, x_2) \in \mathcal{S}$ with $u \in L(x_1, x_2)$.

It is also clear that any factor of a legitimate word is legitimate, and if $u \in L(x_1, x_2)$, say, and v is obtained from u by removing a factor of length $|u_0|$, then v is legitimate with $v \in L(x_1, x_2)$. Also, if v is obtained by duplicating a factor of u of length $|u_0|$ then v is legitimate with $v \in L(x_1, x_2)$. Moreover, when $u, v \in L(x_1, x_2)$, then $|u|$ is congruent to $|v|$ modulo $|u_0|$.

Let $(x_1, x_2), (y_1, y_2) \in \mathcal{S}$. Then $L(x_1, x_2)L(y_1, y_2)$ contains only legitimate words iff $x_2 y_1 \in \{u_0, \epsilon\}$, in which case $L(x_1, x_2)L(y_1, y_2) \subseteq L(x_1, y_2)$. This motivates the following definition. For any (x_1, x_2) and (y_1, y_2) in $\mathcal{S}$, let

$$(x_1, x_2) \otimes (y_1, y_2) = \begin{cases} (x_1, y_2) & \text{if } x_2 y_1 \in \{u_0, \epsilon\} \\ \text{undefined otherwise,} & \end{cases}$$

so that $\otimes$ is a partial operation on S. Thus, if $(x_1, x_2) \otimes (y_1, y_2) = (x_1, y_2)$, then $L(x_1, x_2)L(y_1, y_2) \subseteq L(x_1, y_2)$, moreover, (x_1, y_2) is the only element (z_1, z_2) of $\mathcal{S}$ with $L(x_1, x_2)L(y_1, y_2) \subseteq L(z_1, z_2)$.

Now let $(x_1, x_2) \in \mathcal{S}$ and consider a word y. Then $L(x_1, x_2)y$ contains only legitimate words iff $y \in L(y_1, y_2)$ for some $(y_1, y_2) \in \mathcal{S}$ such that $x_2 y_1 \in \{u_0, \epsilon\}$, in which case (x_1, y_2) is the unique element of $\mathcal{S}$ with $L(x_1, x_2)y \subseteq L(x_1, y_2)$. Thus we define $(x_1, x_2) \otimes y = (x_1, y_2)$ if this holds, otherwise $(x_1, x_2) \otimes y$ is not defined. We define $y \otimes (x_1, x_2)$ symmetrically. The partial operation $\otimes$ is associative in a strong sense.

Using the above notions, the fourth condition of Theorem 1 can be rephrased as follows. For each strong component $\mathcal{C}$ containing a recursive nonterminal there is a primitive word $u_0 = u_0^{\mathcal{C}}$ (unique up to conjugacy) such that for all $X, Y \in \mathcal{C}$ there is (a necessarily unique) $(x_1, x_2) \in \mathcal{S}$ such that whenever $X \Rightarrow^+ wYp$ then $w \in L(x_1, x_2)$.

As before, let us assume that $G = (N, \{0, 1\}, P, S)$ is a context-free grammar that contains no useless nonterminals or ϵ-rules. Moreover, we assume that G is left-recursion free and that $L(G)$ is not empty.

Lemma 2. *Suppose that the fourth condition of Theorem 1 holds and let $\mathcal{C}$ be a strong component containing a recursive nonterminal. Let $u_0 = u_0^{\mathcal{C}}$, and suppose that each nonterminal generates at least two terminal words. Then for each X such that $X_0 \Rightarrow^* pXqYr$ for some $X_0, Y \in \mathcal{C}$ and words p, q, r there is a unique $(x_1, x_2) \in \mathcal{S}$ with $L(X) \subseteq L(x_1, x_2)$.*

Proof. Let (y_1, y_2) denote the unique element of $\mathcal{S}$ such that $w \in L(y_1, y_2)$ whenever $X_0 \Rightarrow^+ wYs$ for some s. Then $L(pXq) \subseteq L(y_1, y_2)$, so that $uL(X)v \subseteq L(y_1, y_2)$ for any fixed $u \in L(p)$ and $v \in L(q)$. This is possible only if $L(X) \subseteq L(x_1, x_2)$ for some $(x_1, x_2) \in \mathcal{S}$. Since $L(X)$ contains at least two words, (x_1, x_2) is unique. □

Theorem 3. *Suppose that each nonterminal generates a language of* at least two *words. Then* $(L(G), <_\ell)$ *is a scattered linear ordering iff the following holds for each strong component* $\mathcal{C}$ *containing a recursive nonterminal: There exists a primitive word* u_0 *such that for any two not necessarily different nonterminals* X *and* Y *in* $\mathcal{C}$ *there is some* $\varphi(X,Y) \in \mathcal{S}$ *and for each nonterminal* Z *there is some* $\psi(Z) \in \mathcal{S}$ *such that*

$$\varphi(X,Y) \otimes \varphi(Y,Z) = \varphi(X,Z) \tag{2}$$

for all $X, Y, Z \in \mathcal{C}$, *and such that the following hold for all productions* $X \to w_0Y_1 \dots Y_kw_k$:

1. *If* $X \in \mathcal{C}$ *and* $Y_i \in \mathcal{C}$ *for some* i, *then*

$$\varphi(X, Y_i) = w_0 \otimes \psi(Y_1) \otimes w_1 \otimes \dots \otimes \psi(Y_{i-1}) \otimes w_{i-1}. \tag{3}$$

2. *If there is derivation* $X_0 \Rightarrow^* pXqYr$ *for some* $X_0, Y \in \mathcal{C}$, *then*

$$\psi(X) = w_0 \otimes \psi(Y_1) \otimes \dots \otimes \psi(Y_k) \otimes w_k. \tag{4}$$

(In the degenerate case when $k = 0$ in the last equation, we mean that w_0 belongs to the language $L(\psi(X))$.

Proof. Suppose that the conditions of the Theorem hold. Consider a strong component $\mathcal{C}$ containing a recursive nonterminal and the corresponding primitive word u_0. Then for any X such that there is derivation $X_0 \Rightarrow^* pXqYr$ for some $X_0, Y \in \mathcal{C}$ we have that $L(X) \subseteq L(\psi(X))$:

Claim 1. Suppose that (4) holds for all appropriate rules. Then for each X such that there is a derivation $X_0 \Rightarrow^* pXqYr$ for some $X_0, Y \in \mathcal{C}$ it holds that $L(X) \subseteq L(\psi(X))$.

Indeed, suppose that $X \Rightarrow^* w$. We prove that $w \in L(\psi(X))$ by induction on the length of the derivation. When the length of the derivation is 1, the claim is clear by (4). Suppose that the length is greater than 1. Then there exist some rule $X \to w_0Y_1w_1 \dots Y_kw_k$ and words $z_1, \dots, z_k$ with $w = w_0z_1 \dots z_kw_k$ and $Y_i \Rightarrow^* z_i$ for all i. By the induction hypothesis we have that each z_i is in $L(\psi(Y_i))$. Since (4) holds, we conclude that $w = w_0z_1 \dots z_kw_k \in L(\psi(X))$. This ends the proof of Claim 1.

Also, for any $X, Y \in \mathcal{C}$ and words w and p with $X \Rightarrow^+ wYp$ we have that $w \in L(\varphi(X,Y))$ as shown by the following claim:

Claim 2. Suppose that (2), (3) and (4) hold. Then if $X \Rightarrow^+ wYp$, where $X, Y \in \mathcal{C}$, then $w \in L(\varphi(X,Y))$.

To see this, consider a derivation tree whose root is labeled X and whose frontier is wYp. Let $Y_1 = X, Y_2, \dots, Y_\ell, Y_{\ell+1} = Y$ be all the nonterminal labels along the path from the root to the leaf labeled Y. Moreover, let $Y_i \to p_iY_{i+1}q_i$ denote the rule used to rewrite Y_i, for $i = 1, \dots, \ell$. By (3) we have that

$$\varphi(Y_1, Y_2) = \psi(p_1), \dots, \varphi(Y_\ell, Y_{\ell+1}) = \psi(p_\ell)$$

where if $p_i = z_0 Z_1 \dots Z_k z_k$, say, then $\psi(p_i) = z_0 \otimes \psi(Z_1) \otimes \dots \otimes \psi(Z_k) \otimes z_{k+1}$. Now let us write $w = w_1 \dots w_\ell$ with $p_i \Rightarrow^* w_i$ for all i. Using Claim 1 and the equality $\varphi(Y_i, Y_{i+1}) = \psi(p_i)$, we obtain $w_i \in L(\varphi(Y_i, Y_{i+1}))$. Since this holds for all i, we obtain by (2) that $w \in L(\varphi(X, Y))$.

We conclude that the fourth condition of Theorem 1 holds, so that $L(G)$ is scattered.

Suppose now that $L(G)$ is scattered. Then the fourth condition of Theorem 1 holds. Suppose that $\mathcal{C}$ is a strong component containing a recursive nonterminal. Let $u_0 = u_0^{\mathcal{C}}$. By assumption, for each $X, Y \in \mathcal{C}$ there exists a unique $(x_1, x_2) \in \mathcal{S}$ such that whenever $X \Rightarrow^+ wYp$ then $w \in L(x_1, x_2)$. Define $\varphi(X, Y) = (x_1, x_2)$. By Lemma 2, for each X such that there is derivation $X_0 \Rightarrow^* pXqYr$ for some words p, q, r and nonterminals $X_0, Y \in \mathcal{C}$, there is a unique $(x_1, x_2) \in \mathcal{S}$ with $L(X) \subseteq L(x_1, x_2)$. Define $\psi(X) = (x_1, x_2)$. The pairs so defined solve the system of equations in the Theorem. □

Theorem 4. *It is decidable in exponential time whether a context-free grammar G generated a scattered language.*

Proof. Without loss of generality we may assume that the terminal alphabet is $\{0, 1\}$ and that the grammar G contains no useless nonterminals or ϵ-rules. Moreover, we may assume that G is left-recursion free and each nonterminal generates at least two terminal words.

First, for each $\mathcal{C}$ containing a recursive nonterminal, one can compute in exponential time a primitive word u_0 which is the only candidate for $u_0^{\mathcal{C}}$. This is done by finding in exponential time a left derivation $X \Rightarrow^+ wXp$, with $X \in \mathcal{C}$, then u_0 is the primitive root of w. Second, in the same way, for any $X, Y \in \mathcal{C}$, we can determine in exponential time the only candidate for $\varphi(X, Y)$ by computing a left derivation $X \Rightarrow^+ wYp$, where the length of w is between $|u_0|$ and $2|u_0|$. Also, we can compute in exponential time the only candidate for $\psi(X)$, for all appropriate X. Then it remains to check that the equations of Theorem 3 hold. But there are a polynomial number of them, and the validity of each can be checked in exponential time. □

The same result holds for deciding whether a context-free language is well-ordered.

Theorem 5. *It is decidable in exponential time whether a context-free grammar generates a well-ordered language.*

Proof. Again, we may restrict the grammars as in the previous proof. The extra condition introduced in Theorem 2 can be tested in exponential time. Hint: if $X \Rightarrow_\ell^+ u_0^n u_1 Y p \Rightarrow u_0^n u 1 q$ is a left derivation, where $u0$ is a prefix of u_0, then the length of the derivation can be bounded by a exponential. □

Acknowledgement

In a previous version of this paper, the algorithm presented in the last section was claimed to run in polynomial time. The author would like to thank Jean

Berstel, Luc Boasson, Olivier Carton and Isabelle Fagnot for pointing out this error. As communicated by them, recently they have also found a proof of the fact that it is decidable for a context-free grammar whether it generates a scattered language.

References

1. Blum, N., Koch, R.: Greibach normal form transformation. Information and Computation 150, 112–118 (1999) (revisited)
2. Braud, L., Carayol, A.: Linear orders in the pushdown hierarchy. In: Abramsky, S., Gavoille, C., Kirchner, C., Meyer auf der Heide, F., Spirakis, P.G. (eds.) ICALP 2010. LNCS, vol. 6199, pp. 88–99. Springer, Heidelberg (2010)
3. Bloom, S.L., Ésik, Z.: Regular and Algebraic Words and Ordinals. In: Mossakowski, T., Montanari, U., Haveraaen, M. (eds.) CALCO 2007. LNCS, vol. 4624, pp. 1–15. Springer, Heidelberg (2007)
4. Bloom, S.L., Ésik, Z.: Algebraic ordinals. Fundamenta Informaticæ 99, 383–407 (2010)
5. Bloom, S.L., Ésik, Z.: Algebraic linear orderings. In: Int. J. Foundations of Computer Science (to appear)
6. Caucal, D.: On infinite graphs having a decidable monadic theory. Theoretical Computer Science 290, 79–115 (2003)
7. Ésik, Z.: Algebraic and context-free linear orderings. Slides Presented at, Workshop on Higher-Order Recursion Schemes & Pushdown Automata, Paris, March 10–12 (2010), `http://www.liafa.jussieu.fr/~serre/WorkshopSchemes/`
8. Ésik, Z., Iván, S.: Büchi context-free languages. Theoretical Computer Science 412, 805–821 (2011)
9. Ésik, Z.: An undecidable property of context-free linear orders. Information Processing Letters 111, 107–109 (2001)
10. Lothaire, M.: Combinatorics on Words. Cambridge University Press, Cambridge (1997)
11. Rosenstein, J.G.: Linear Orderings. Pure and Applied Mathematics, vol. 98. Academic Press, New York (1982)

Appendix

Proof of Proposition 1. Let L_0 be the regular prefix language $(00+11)^*01$ whose lexicographic ordering has order type η, and consider the tree $T(L_0)$. If the full binary tree embeds in $T(L)$, then so does $T(L_0)$. Consider an embedding of $T(L_0)$ in $T(L)$ which maps each vertex x of $T(L_0)$ to a vertex $h(x)$ of $T(L)$. For each leaf x of $T(L_0)$ select a leaf v_x of $T(L)$ which is a descendant of $h(x)$. The words v_x form a dense subset of L with respect to the lexicographic order. (Note also that any two words v_x are actually related by the strict order.)

For the reverse direction, suppose that L is quasi-dense. Let us color a vertex x of $T(L)$ blue if $x \in L$. Call a vertex x of $T(L)$ *appropriate* if the blue vertices of the subtree T_x rooted at x form a quasi-dense linear ordering with respect to the lexicographic order. If x is appropriate, then it has at least two proper descendants y and z which are appropriate vertices with $y <_s z$. Indeed, x has a proper descendant x' such that both the set of blue vertices y' of T_x with $y' <_s x'$ and the set of blue vertices z' of T_x with $x' <_\ell z'$ form quasi-dense linear orderings with respect to the lexicographic order. Suppose that $x' = xu$, where u is a nonempty word. Then one of the vertices $xv0$ where $v1$ is a prefix of u is appropriate, as is one of the vertices $xv1$ and x', where $xv0$ is a prefix of x. Let y and z be these vertices.

Thus, starting from the root of $T(L)$, we can construct a set V of appropriate vertices such that each $x \in V$ has two (proper) descendants y and z in V with $y <_s z$. The vertices in V determine an embedding of the full binary tree in $T(L)$. □

On Prefix Normal Words

Gabriele Fici[1] and Zsuzsanna Lipták[2]

[1] I3S, CNRS & Université de Nice-Sophia Antipolis, France
fici@i3s.unice.fr
[2] AG Genominformatik, Technische Fakultät, Bielefeld University, Germany
zsuzsa@cebitec.uni-bielefeld.de

Abstract. We present a new class of binary words: the prefix normal words. They are defined by the property that for any given length k, no factor of length k has more a's than the prefix of the same length. These words arise in the context of indexing for jumbled pattern matching (a.k.a. permutation matching or Parikh vector matching), where the aim is to decide whether a string has a factor with a given multiplicity of characters, i.e., with a given Parikh vector. Using prefix normal words, we give the first non-trivial characterization of binary words having the same set of Parikh vectors of factors. We prove that the language of prefix normal words is not context-free and is strictly contained in the language of pre-necklaces, which are prefixes of powers of Lyndon words. We discuss further properties and state open problems.

Keywords: Parikh vectors, pre-necklaces, Lyndon words, context-free languages, jumbled pattern matching, permutation matching, non-standard pattern matching, indexing.

1 Introduction

Given a finite word w over a finite ordered alphabet Σ, the Parikh vector of w is defined as the vector of multiplicities of the characters in w. In recent years, Parikh vectors have been increasingly studied, in particular Parikh vectors of factors (substrings) of words, motivated by applications in computational biology, e.g. mass spectrometry [1, 5, 9, 10]. Among the new problems introduced in this context is that of *jumbled pattern matching* (a.k.a. permutation matching or Parikh vector matching), whose decision variant is the task of deciding whether a given word w (the text) has a factor with a given Parikh vector (the pattern). In [8], Cicalese *et al.* showed that in order to answer decision queries for binary words, it suffices to know, for each k, the maximum and minimum number of a's in a factor of length k. Thus it is possible to create an index of size $O(n)$ of a text of length n, which contains, for every k, the maximum and minimum number of a's in a factor of length k, and which allows answering decision queries in constant time.

In this paper, we introduce a new class of binary words, *prefix normal words*. They are defined by the property that for any given length k, no factor of length k appearing in the word has more a's than the prefix of the word of the same

G. Mauri and A. Leporati (Eds.): DLT 2011, LNCS 6795, pp. 228–238, 2011.

length. For example, the word *aabbaaba* is not a prefix normal word, because the factor *aaba* has more *a*'s then the prefix of the same length, *aabb*.

We show that for every binary word w, there is a prefix normal word w' such that, for every $0 \leq k \leq |w|$, the maximum number of a's in a factor of length k coincide for w and w' (Theorem 1). We refer to w' as the *prefix normal form* of w (with respect to a).

Given a word w, a *factor Parikh vector* of w is the Parikh vector of a factor of w. An interesting characterization of words with the same multi-set of factor Parikh vectors was given recently by Acharya *et al.* [1]. In this paper, we give the first non-trivial characterization of the *set* of factor Parikh vectors, by showing that two words have the same set of factor Parikh vectors if and only if their prefix normal forms, with respect to a and b, both coincide (Theorem 2).

We explore the language of prefix normal words and its connection to other known languages. Among other things, we show that this language is not context-free (Theorem 3) by adapting a proof of Berstel and Boasson [3] for Lyndon words, and that it is properly included in the language of pre-necklaces, the prefixes of powers of Lyndon words (Theorem 4). We close with a number of open problems.

Connection to Indexed Jumbled Pattern Matching. The current fastest algorithms for computing an index for the binary jumbled pattern matching problem were concurrently and independently developed by Burcsi *et al.* [6] and Moosa and Rahman [14]. In order to compute an index of a text of length n, both used a reduction to min-plus convolution, for which the current best algorithms have a runtime of $O(n^2/\log n)$. Very recently, Moosa and Rahman [13] introduced an algorithm with runtime $O(n^2/\log^2 n)$ which uses word-RAM operations. Our characterization of the set of factor Parikh vectors in terms of prefix normal forms yields a new approach to the problem of indexed jumbled pattern matching: Given the prefix normal forms of a word w, the index for the jumbled pattern matching problem can be computed in linear time $O(n)$. This implies that any algorithm for computing the prefix normal form with runtime $o(n^2/\log^2 n)$ will result in an improvement for the indexing problem for binary jumbled pattern matching. Since on the other hand, the prefix normal forms can be computed from the index in $O(n)$ time, we also have that a lower bound for the computation of the prefix normal form would yield a lower bound for the binary jumbled pattern matching problem.

2 The Prefix Normal Form

We fix the ordered alphabet $\Sigma = \{a, b\}$, with $a < b$. A word $w = w_1 \cdots w_n$ over Σ is a finite sequence of elements from Σ. Its length n is denoted by $|w|$. We denote the empty word by ε. For any $1 \leq i \leq |w|$, the i-th symbol of a word w is denoted by w_i. As is standard, we denote by Σ^n the words over Σ of length n, and by $\Sigma^* = \cup_{n \geq 0} \Sigma^n$ the set of finite words over Σ. Let $w \in \Sigma^*$. If $w = uv$ for some $u, v \in \Sigma^*$, we say that u is a *prefix* of w and v is a *suffix* of w. A *factor* of w is a prefix of a suffix of w (or, equivalently, a suffix of a prefix). We denote by *Pref*(w), *Suff*(w), *Fact*(w) the set of prefixes, suffixes, and factors of the word w, respectively.

For a letter $a \in \Sigma$, we denote by $|w|_a$ the number of occurrences of a in the word w. The *Parikh vector* of a word w over Σ is defined as $p(w) = (|w|_a, |w|_b)$. The *Parikh set* of w is $\Pi(w) = \{p(v) \mid v \in Fact(w)\}$, the set of Parikh vectors of the factors of w.

Finally, given a word w over Σ and a letter $a \in \Sigma$, we denote by $P_a(w,i) = |w_1 \cdots w_i|_a$, the number of a's in the prefix of length i of w, and by $pos_a(w,i)$ the position of the i'th a in w, i.e., $pos_a(w,i) = \min\{k : |w_1 \cdots w_k|_a = i\}$. When w is clear from the context, we also write $P_a(i)$ and $pos_a(i)$. Note that in the context of succint indexing, these functions are frequently called *rank* and *select*, cf. [15]: We have $P_a(w,i) = rank_a(w,i)$ and $pos_a(w,i) = select_a(w,i)$.

Definition 1. *Let $w \in \Sigma^*$. We define, for each $0 \leq k \leq |w|$,*

$$F_a(w,k) = \max\{|v|_a \mid v \in Fact(w) \cap \Sigma^k\},$$

the maximum number of a's in a factor of w of length k. When no confusion can arise, we also write $F_a(k)$ for $F_a(w,k)$. The function $F_b(w)$ is defined analogously by taking b in place of a.

Example 1. Take $w = ababbaabaabbbaaabbab$. In Table 1, we give the values of F_a and F_b for w.

Table 1. The sequences F_a and F_b for the word $w = ababbaabaabbbaaabbab$

k	0	1	2	3	4	5	6	7	8	9	10	11	12	13	14	15	16	17	18	19	20
F_a	0	1	2	3	3	4	4	4	5	5	6	7	7	7	8	8	9	9	9	10	10
F_b	0	1	2	3	3	3	4	4	5	5	6	6	7	7	7	8	8	9	9	10	10

Lemma 1. *Let $w \in \Sigma^*$. The function $F_a(\cdot) = F_a(w,\cdot)$ has the following property:*

$$F_a(j) - F_a(i) \leq F_a(j-i) \qquad \text{for all } 0 \leq i \leq j \leq |w|.$$

Proof. Assume otherwise. Then there are indices $i \leq j$ such that $F_a(j) - F_a(i) > F_a(j-i)$. Let $v \in Fact(w)$ be a word that realizes $F_a(j)$, i.e., $|v| = j$ and $|v|_a = F_a(j)$. Let us write $v = v_1 v_2 \cdots v_j$. Then for the word $u = v_{i+1} \cdots v_j$, we have $|u|_a = |v|_a - |v_1 \cdots v_i|_a \geq F_a(j) - F_a(i) > F_a(j-i)$, in contradiction to the definition of F_a, since $|u| = j - i$. □

We are now ready to show that for every word w there is a word w' which realizes the function $F_a(w)$ as its prefix function $P_a(w')$.

Theorem 1. *Let $w \in \Sigma^*$. Then there exists a unique word w' s.t. for all $0 \leq k \leq |w|$, $F_a(w,k) = F_a(w',k) = P_a(w',k)$. We call this word w' the* prefix normal form *of w (with respect to a), and denote it $\mathrm{PNF}_a(w)$. Analogously, there exists a unique word w'', such that for all $0 \leq k \leq |w|$, $F_b(w,k) = F_b(w'',k) = P_b(w'',k)$, the prefix normal form of w with respect to b, denoted $\mathrm{PNF}_b(w)$.*

Proof. We only give the proof for w'. The construction of w'' is analogous. It is easy to see that for $1 \leq k \leq |w|$, one has either $F_a(w,k) = F_a(w,k-1)$ or $F_a(w,k) = 1 + F_a(w,k-1)$. Now define the word w' by

$$w'_k = \begin{cases} a & \text{if } F_a(w,k) = 1 + F_a(w,k-1) \\ b & \text{if } F_a(w,k) = F_a(w,k-1) \end{cases}$$

for every $1 \leq k \leq |w|$.

By construction, we have $P_a(w',k) = F_a(w,k)$ for every $1 \leq k \leq |w|$. We still need to show that $P_a(w',k) = F_a(w',k)$ for all k, i.e., that w' is in prefix normal form. By definition, $P_a(w',k) \leq F_a(w',k)$ for all k. Now let $v \in Fact(w')$, $|v| = k$, and $v = w_{i+1} \cdots w_j$. Then $|v|_a = P_a(w',j) - P_a(w',i) = F_a(w,j) - F_a(w,i) \leq F_a(w,j-i) = P_a(w',j-i) = P_a(w',k)$, where the inequality holds by Lemma 1. We have thus proved that $F_a(w',k) \leq P_a(w',k)$, and we are done. □

Example 2. Let $w = ababbaabaabbbaaabbab$. The prefix normal forms of w are the words

$$\mathrm{PNF}_a(w) = aaababbabaabbababbab,$$

and

$$\mathrm{PNF}_b(w) = bbbaababababaabababa.$$

The operators PNF_a and PNF_b are idempotent operators, that is, if $u = \mathrm{PNF}_x(w)$ then $\mathrm{PNF}_x(u) = u$, for any $x \in \Sigma$. Also, for any $w \in \Sigma^*$ and $x \in \Sigma$, it holds that $\mathrm{PNF}_x(w) = \mathrm{PNF}_x(\tilde{w})$, where $\tilde{w} = w_n w_{n-1} \cdots w_1$ is the reversal of w.

The prefix normal forms of a word allow one to determine the Parikh vectors of the factors of the word, as we will show in Theorem 2. We first recall the following lemma from [8], where we say that a Parikh vector q *occurs* in a word w if w has a factor v with $p(v) = q$.

Lemma 2 (Interval Lemma, Cicalese et al. [8]). *Let $w \in \Sigma^*$. Fix $1 \leq k \leq |w|$. If the Parikh vectors $(x_1, k - x_1)$ and $(x_2, k - x_2)$ both occur in w, then so does $(y, k - y)$ for any $x_1 \leq y \leq x_2$.*

The lemma can be proved with a simple sliding window argument.

Theorem 2. *Let w, w' be words over Σ. Then $\Pi(w) = \Pi(w')$ if and only if $\mathrm{PNF}_a(w) = \mathrm{PNF}_a(w')$ and $\mathrm{PNF}_b(w) = \mathrm{PNF}_b(w')$.*

Proof. Let $f_a(w,k)$ denote the minimum number of a's in a factor of w of length k. As a direct consequence of Lemma 2, we have that for a Parikh vector $q = (x,y)$, $q \in \Pi(w)$ if and only if $f_a(w,x+y) \leq x \leq F_a(w,x+y)$. Thus for two words

w, w', we have $\Pi(w) = \Pi(w')$ if and only if $F_a(w) = F_a(w')$ and $f_a(w) = f_a(w')$. It is easy to see that for all k, $f_a(w,k) = k - F_b(w,k)$, thus the last statement is equivalent to $F_a(w) = F_a(w')$ and $F_b(w) = F_b(w')$. This holds if and only if $\mathrm{PNF}_a(w) = \mathrm{PNF}_a(w')$ and $\mathrm{PNF}_b(w) = \mathrm{PNF}_b(w')$, and the claim is proved. □

There is a simple geometrical construction for computing the prefix normal forms of a word w, and hence, by Theorem 2, the set $\Pi(w)$ of Parikh vectors occurring in w. An example of this construction is given in Fig. 1.

Draw in the Euclidean plane the word w by linking, for every $0 \le i \le |w|$, the points (i, j), where j is the difference between the number of a's and the number of b's in the prefix of w of length i. That is, draw w by representing each letter a by an upper unit diagonal and each letter b by a lower unit diagonal, starting from the origin $(0, 0)$.

Then draw all the suffixes of w in the same way, always starting from the origin. The region of the plane so delineated is in fact the region of points (x, y) such that there exists a factor v of w such that $x = |v| = |v|_a + |v|_b$ and $y = |v|_a - |v|_b$. Hence $(|v|_a, |v|_b) = (\frac{x+y}{2}, \frac{x-y}{2}) = p(v)$ belongs to $\Pi(w)$.

The region is connected by Lemma 2, in the sense that all internal points belong to $\Pi(w)$. The prefix normal forms $\mathrm{PNF}_a(w)$ and $\mathrm{PNF}_b(w)$ are obtained by connecting the upper and the lower points of the region, respectively.

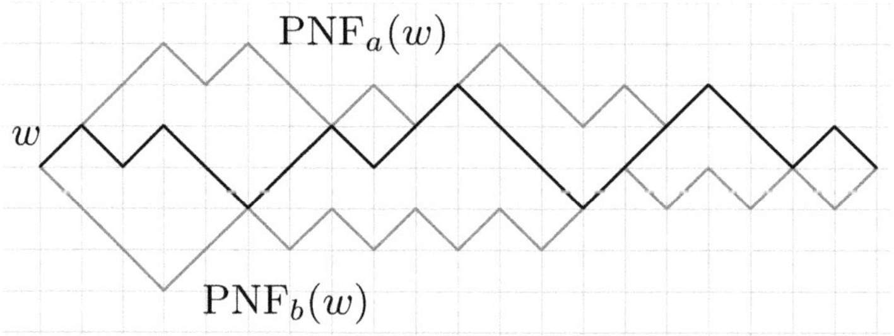

Fig. 1. The word $w = ababbaabaabbbaaabbab$, its prefix normal forms $\mathrm{PNF}_a(w) = aaababbabaabbababbab$ and $\mathrm{PNF}_b(w) = bbbaababababaababababa$, and the region delineated by $\Pi(w)$, the Parikh set of w

3 The Language of Prefix Normal Words

In this section, we take a closer look at those words which are in prefix normal form, which we refer to as *prefix normal words*. For simplicity of exposition, from now on we only refer to prefix normality with respect to the letter a.

Definition 2. *A* prefix normal word *is a word* $w \in \Sigma^*$ *such that for every* $0 \le k \le |w|$, $F_a(w,k) = P_a(w,k)$. *That is, a word such that* $w = \mathrm{PNF}_a(w)$. *We denote by* $L_a \subset \Sigma^*$ *the language of prefix normal words.*

The following proposition gives some characterizations of prefix normal words. Recall that $P_a(w,i) = |w_1 \cdots w_i|_a$ is the number of a's in the prefix of length i, and $pos_a(w, i) = \min\{k : |w_1 \cdots w_k|_a = i\}$ is the position of the i'th a. When no confusion can arise, we write simply $P_a(i)$ and $pos_a(i)$. In particular, we have $P_a(pos_a(i)) = i$ and $pos_a(P_a(i)) \leq i$.

Proposition 1. *Let $w \in \Sigma^*$. The following properties are equivalent:*

1. *w is a prefix normal word;*
2. *$\forall i, j$ where $0 \leq i \leq j \leq |w|$, we have $P_a(j) - P_a(i) \leq P_a(j - i)$;*
3. *$\forall v \in Fact(w)$ such that $|v|_a = i$, we have $|v| \geq pos_a(i)$;*
4. *$\forall i, j$ such that $i+j-1 \leq |w|_a$, we have $pos_a(i)+pos_a(j)-1 \leq pos_a(i+j-1)$.*

Proof. (1) $\Rightarrow$ (2). Follows from Lemma 1, since $P_a(w) = F_a(w)$.

(2) $\Rightarrow$ (3). Assume otherwise. Then there exists $v \in Fact(w)$ s.t. $|v| < pos_a(k)$, where $k = |v|_a$. Let $v = w_{i+1} \cdots w_j$, thus $j - i = k$. Then $P_a(j) - P_a(i) = k$. But $P_a(j - i) = P_a(|v|) \leq k - 1 < k = P_a(j) - P_a(i)$, a contradiction.

(3) $\Rightarrow$ (4). Again assume that the claim does not hold. Then there are i, j s.t. $pos_a(i + j - 1) < pos_a(i) + pos_a(j) - 1$. Let $k = pos_a(j)$ and $l = pos_a(i + j - 1)$ and define $v = w_k \cdots w_l$. Then v has i many a's. But $|v| = pos_a(i + j - 1) - pos_a(j) + 1 < pos_a(i) + pos_a(j) - 1 - pos_a(j) + 1 = pos_a(i)$, in contradiction to (3).

(4) $\Rightarrow$ (1). Let $v \in Fact(w)$, $|v|_a = i$. We have to show that $P_a(|v|) \geq i$. This is equivalent to showing that $pos_a(i) \leq |v|$. Let $v = w_{l+1} \cdots w_r$, thus $P_a(r) - P_a(l) = i$. Let $j = P_a(l) + 1$, thus the first a in v is the j'th a of w. Note that we have $l < pos_a(j)$ and $r \geq pos_a(i + j - 1)$. By the assumption, we have $pos_a(i) \leq pos_a(i + j - 1) - pos_a(j) + 1 \leq r - l = |v|$. □

We now give some simple facts about the language L_a.

Proposition 2. *Let L_a be the language of prefix normal words.*

1. *L_a is prefix-closed, that is, any prefix of a word in L_a is a word in L_a.*
2. *If $w \in L_a$, then any word of the form $a^k w$ or wb^k, $k \geq 0$, also belongs to L_a.*
3. *Let $|w|_a < 3$. Then $w \in L_a$ iff either $w = b^n$ for some $n \geq 0$ or the first letter of w is a.*
4. *Let $w \in \Sigma^*$. Then there exist infinitely many $v \in \Sigma^*$ such that $vw \in L_a$.*

Proof. The claims *1.*, *2.*, *3.* follow easily from the definition. For *4.*, note that for any $n \geq |w|$, the word $a^n w$ belongs to L_a. □

We now deal with the question of how a prefix normal word can be extended to the right into another prefix normal word.

Lemma 3. *Let $w \in L_a$. Then $wa \in L_a$ if and only if for every $0 \leq k < |w|$ the suffix of w of length k has less a's than the prefix of w of length $k + 1$.*

Proof. Suppose $wa \in L_a$. Fix k and let va be the suffix of wa of length $k + 1$. By definition of L_a one has $|va|_a \leq P_a(k + 1)$, and therefore $|v|_a < P_a(k + 1)$.

Conversely, let v be the suffix of w of length k. Since $w \in L_a$ one has $|v|_a \leq P_a(k)$. We cannot have $|v|_a = P_a(k)$ and $w_{k+1} = b$ since by hypothesis we must have $|v|_a < P_a(k + 1)$. Thus either $|v|_a < P_a(k)$ or $w_{k+1} = b$. In both cases we have then $|va|_a \leq P_a(k + 1)$. Since no suffix of wa has more a's than the prefix of wa of the same length, and since $w \in L_a$, it follows that $wa \in L_a$. □

We close this section by proving that L_a is not context-free. Our proof is an easy modification of the proof that Berstel and Boasson gave for the fact that the language of binary Lyndon words is not context-free [3].

Theorem 3. *L_a is not context-free.*

Proof. Recall that Ogden's iteration lemma (see e.g. [2]) states that, for every context-free language L there exists an integer N such that, for any word $w \in L$ and for any choice of at least N distinguished positions in w, there exists a factorization $w = xuyvz$ such that

1. either x, u, y each contain at least one distinguished position, or y, v, z each contain at least one distinguished position;
2. the word uyv contains at most N distinguished positions;
3. for any $n \geq 0$, the word $xu^n yv^n z$ is in L.

Now, assume that the language L_a is context-free, and consider the word $w = a^{N+1}ba^N ba^{N+1}$ where N is the constant of Ogden's Lemma. It is easy to see that $w \in L_a$. Distinguish the central run of N letters a. We claim that for every factorization $w = xuyvz$ of w, pumping u and v eventually results in a word in which the first run of a's is not the longest one. Such a word cannot belong to L_a.

If x, u, y each contain at least one distinguished position, then u is non-empty and it is contained in the central run of a's. Now observe that every word obtained by pumping u and v is prefixed by $a^{N+1}b$. Pumping u and v, we then get a word $a^{N+1}ba^m s$, for some word s, where $m > N+1$. This word is not in L_a.

Suppose now that y, v, z each contain at least one distinguished position. Then v is non-empty and it is contained in the central run of a's.

If u is contained in the first run of a's and it is non-empty, then, pumping *down*, one gets a word of the form $a^k ba^m ba^{N+1}$ with $k \leq N$. This word is not in L_a. In all other cases (u is contained in the second run of a's and it is non-empty, or $u = \varepsilon$, or u contains the first b of w), every word obtained by pumping u and v is prefixed by $a^{N+1}b$. Again, pumping u and v, we obtain a word in which the first run of a's is not the longest one. □

4 Prefix Normal Words vs. Lyndon Words

In this section, we explore the relationship between the language L_a of prefix normal words and some known classes of words defined by means of lexicographic properties.

A *Lyndon word* is a word which is lexicographically (strictly) smaller than any of its proper non-empty suffixes. Equivalently, w is a Lyndon word if it is the (strictly) smallest, in the lexicographic order, among its conjugates, i.e., for any factorization $w = uv$, with u, v non-empty words, one has that the word vu is lexicographically greater than w [12]. Note that, by definition, a Lyndon word is primitive, i.e., it cannot be written as $w = u^k$ for a $u \in \Sigma^*$ and $k > 1$. Let us denote by *Lyn* the set of Lyndon words over Σ. One has that $Lyn \not\subseteq L_a$ and

$L_a \not\subseteq Lyn$. For example, the word $w = abab$ belongs to L_a but is not a Lyndon word since it is not primitive. An example of Lyndon word which is not in prefix normal form is $w = aabbabaabbb$.

A power of a Lyndon word is called a *prime word* [11] or *necklace* (see [4] for more details and references on this definition).

Let us denote by *PL* the set of prefixes of powers of Lyndon words, also called sesquipowers (or fractional powers) of Lyndon words [7], or *preprime words* [11], or also *pre-necklaces* [16]. It is easy to see that *PL* is in fact the set of prefixes of Lyndon words plus the powers of the letter b.

The next proposition shows that any prefix normal word different form a power of the letter b is a prefix of a Lyndon word.

Proposition 3. *Let $w \in L_a$ with $|w|_a > 0$. Then the word $wb^{|w|}$ is a Lyndon word.*

Proof. We have to prove that any non-empty suffix of $wb^{|w|}$ is greater than $wb^{|w|}$. Suppose by contradiction that there exists a non-empty suffix v of $wb^{|w|}$ that is smaller than $wb^{|w|}$, and let u be the longest common prefix between v and $wb^{|w|}$. This implies that u is followed by different letters when it appears as prefix of v and as prefix of $wb^{|w|}$. Since we supposed that v is smaller than $wb^{|w|}$, we conclude that ub is prefix of $wb^{|w|}$ and ua is prefix of v. Since ua is a factor of $wb^{|w|}$ ending with a, ua must be a factor of w and therefore ub is a prefix of w. Thus the factor ua of w has one more a than the prefix ub of w, contradicting the fact that w is a prefix normal word. □

We can now state the following result.

Theorem 4. *Every prefix normal word is a pre-necklace. That is, $L_a \subset PL$.*

Proof. If w is of the form b^n, $n \geq 1$, then w is a power of the Lyndon word b. Otherwise, w contains at least one a and the claim follows by Proposition 3. □

The languages L_a and *PL*, however, do not coincide. The shortest word in *PL* that does not belong to L_a is $w = aabbabaa$. Below we give the table of the number of words in L_a of each length, up to 16, compared with that of pre-necklaces. This latter sequence is listed in Neil Sloane's On-Line Encyclopedia of Integer Sequences [17].

Table 2. The number of words in L_a and in *PL* for each length up to 16

n	1	2	3	4	5	6	7	8	9	10	11	12	13	14	15	16
$L_a \cap \Sigma^n$	2	3	5	8	14	23	41	70	125	218	395	697	1273	2279	4185	7568
$PL \cap \Sigma^n$	2	3	5	8	14	23	41	71	127	226	412	747	1377	2538	4720	8800

5 The Prefix Normal Equivalence

The prefix normal form PNF_a induces an equivalence relation on Σ^*, namely $u \equiv_{\text{PNF}_a} v$ if and only if $\text{PNF}_a(u) = \text{PNF}_a(v)$. In Table 3, we give all prefix normal words of length 4, and their equivalence classes.

An interesting question is how to characterize two words that have the same prefix normal form. The classes of this equivalence do not seem to follow regular patterns. For example, the words *aabababa*, *aabbaaba*, *abaababa*, *abaabbaa*,

Table 3. The classes of words of length 4 having the same prefix normal form

PNF_a	class	card.
aaaa	{*aaaa*}	1
aaab	{*aaab*, *baaa*}	2
aaba	{*aaba*, *abaa*}	2
aabb	{*aabb*, *baab*, *bbaa*}	3
abab	{*abab*, *baba*}	2
abba	{*abba*}	1
abbb	{*abbb*, *babb*, *bbab*, *bbba*}	4
bbbb	{*bbbb*}	1

Table 4. The cardinalities of the 70 classes of words of length 8 having the same prefix normal form. There are 7 classes of length 1, 24 classes of length 2, 5 classes of length 3, 16 classes of length 4, 2 classes of length 5, 9 classes of length 6, 1 class of length 7, 4 classes of length 8, 1 class of length 9 and 1 class of length 10.

PNF_a	card.	PNF_a	card.	PNF_a	card.	PNF_a	card.
aaaaaaaa	1	*aaabaabb*	6	*aabababa*	6	*abababba*	2
aaaaaaab	2	*aaababaa*	2	*aabababb*	9	*abababbb*	4
aaaaaaba	2	*aaababab*	6	*aababbaa*	2	*ababbaba*	1
aaaaaabb	3	*aaababba*	4	*aababbab*	8	*ababbabb*	6
aaaaabaa	2	*aaababbb*	8	*aababbba*	4	*ababbbab*	4
aaaaabab	4	*aaabbaaa*	1	*aababbbb*	10	*ababbbba*	2
aaaaabba	2	*aaabbaab*	4	*aabbaabb*	3	*ababbbbb*	6
aaaaabbb	4	*aaabbaba*	2	*aabbabab*	4	*abbabbab*	2
aaaabaaa	2	*aaabbabb*	6	*aabbabba*	3	*abbabbba*	2
aaaabaab	4	*aaabbbaa*	2	*aabbabbb*	8	*abbabbbb*	5
aaaababa	3	*aaabbbab*	4	*aabbbaab*	2	*abbbabbb*	4
aaaababb	6	*aaabbbba*	2	*aabbbaba*	2	*abbbbabb*	3
aaaabbaa	2	*aaabbbbb*	6	*aabbbabb*	6	*abbbbbab*	2
aaaabbab	4	*aabaabaa*	1	*aabbbbaa*	1	*abbbbbba*	1
aaaabbba	2	*aabaabab*	4	*aabbbbab*	4	*abbbbbbb*	8
aaaabbbb	5	*aabaabba*	2	*aabbbbba*	2	*bbbbbbbb*	1
aaabaaab	2	*aabaabbb*	4	*aabbbbbb*	7		
aaabaaba	4	*aababaab*	2	*abababab*	2		

ababaaba, *abababaa* all have the same prefix normal form *aabababa*, so that no simple statement about the lengths of the runs of the two letters seems to provide a characterization of the classes. This example also shows that the prefix normal form of a word w is in general more complicated that just a rotation of w or of its reversal (which is in fact the case for small lengths).

Recall that for word length $n \leq 16$, we listed the number of equivalence classes in Table 2. The sizes of the equivalence classes seem to exhibit an irregular behaviour. We report in Table 4, for each of the 70 equivalence classes for words of length 8, the prefix normal form and the number of words in the class. Furthermore, we report the cardinality of the largest class of words for each length up to 16 (Table 5).

Table 5. The maximum cardinality of a class of words having the same prefix normal form

n	1	2	3	4	5	6	7	8	9	10	11	12	13	14	15	16
$\max \lvert [w] \rvert$	1	2	3	4	5	6	8	10	12	18	24	30	40	60	80	111

6 Conclusion and Open Problems

In this paper, we introduced the prefix normal form of a binary word. This construction arises in the context of indexing for jumbled pattern matching and provides a characterization of the set of factor Parikh vectors of a binary word. We then investigated the language L_a of words which are in prefix normal form (w.r.t. the letter a).

Many open problems remain. Among these, the questions regarding the prefix normal equivalence were explored in Section 5: How can we characterize two words that have the same prefix normal form? Can we say anything about the number or the size of equivalence classes for a given word length?

Although we showed that the language L_a is strictly contained in the language of pre-necklaces (prefixes of powers of Lyndon words), we were not able to find a formula for enumerating prefix normal words. A possible direction for attacking this problem would be finding a characterization of those pre-necklaces which are not prefix normal. Indeed, an enumerative formula for the pre-necklaces is known [17].

Another open problem is to find an algorithm for testing whether a word is in prefix normal form. The best offline algorithms at the moment are the ones for computing an index for the jumbled pattern matching problem, and thus the prefix normal form [6, 14, 13]; these have running time $O(n^2/\log n)$ for a word of length n, or $O(n^2/\log^2 n)$ in the word-RAM model. However, testing may be easier than actually computing the PNF. Note also that Lemma 3 gives us an online testing algorithm, with time complexity $O(n^2)$. Another, similar, online testing algorithm is provided by condition *4.* of Proposition 1, with running time $O(|w|_a^2)$, which is, of course, again $O(n^2)$ in general.

Acknowledgements. We would like to thank Bill Smyth for interesting discussions on prefix normal words. We are also grateful to an anonymous referee who helped us improve the proof of Theorem 3.

References

1. Acharya, J., Das, H., Milenkovic, O., Orlitsky, A., Pan, S.: Reconstructing a string from its substring compositions. In: Proceedings of IEEE International Symposium on Information Theory, ISIT 2010. pp. 1238–1242 (2010)
2. Berstel, J., Boasson, L.: Context-free languages. In: Handbook of Theoretical Computer Science, Volume B: Formal Models and Sematics (B), pp. 59–102. Elsevier, Amsterdam (1990)
3. Berstel, J., Boasson, L.: The set of Lyndon words is not context-free. Bull. Eur. Assoc. Theor. Comput. Sci. EATCS 63, 139–140 (1997)
4. Berstel, J., Perrin, D.: The origins of combinatorics on words. Eur. J. Comb. 28, 996–1022 (2007)
5. Böcker, S.: Simulating multiplexed SNP discovery rates using base-specific cleavage and mass spectrometry. Bioinformatics 23(2), 5–12 (2007)
6. Burcsi, P., Cicalese, F., Fici, G., Lipták, Zs.: On table arrangements, scrabble freaks, and jumbled pattern matching. In: Boldi, P., Gargano, L. (eds.) FUN 2010. LNCS, vol. 6099, pp. 89–101. Springer, Heidelberg (2010)
7. Champarnaud, J., Hansel, G., Perrin, D.: Unavoidable sets of constant length. Internat. J. Algebra Comput. 14, 241–251 (2004)
8. Cicalese, F., Fici, G., Lipták, Zs.: Searching for Jumbled Patterns in Strings. In: Holub, J., Zdárek, J. (eds.) Prague Stringology Conference, PSC 2009. Proceedings, pp. 105–117. Czech Tech. Univ. in Prague (2009)
9. Cieliebak, M., Erlebach, T., Lipták, Zs., Stoye, J., Welzl, E.: Algorithmic complexity of protein identification: combinatorics of weighted strings. Discrete Appl. Math. 137(1), 27–46 (2004)
10. Eres, R., Landau, G.M., Parida, L.: Permutation pattern discovery in biosequences. J. Comput. Biol. 11(6), 1050–1060 (2004)
11. Knuth, D.E.: Generating All Tuples and Permutations. The Art of Computer Programming, Vol. 4, Fascicle 2. Addison-Wesley, Reading (2005)
12. Lothaire, M.: Algebraic Combinatorics on Words. Encyclopedia of Mathematics and its Applications. Cambridge Univ. Press, Cambridge (2002)
13. Moosa, T.M., Rahman, M.S.: Sub-quadratic time and linear size data structures for permutation matching in binary strings . J. Discrete Algorithms (to appear)
14. Moosa, T.M., Rahman, M.S.: Indexing permutations for binary strings. Inf. Process. Lett. 110, 795–798 (2010)
15. Navarro, G., Mäkinen, V.: Compressed full-text indexes. ACM Comput. Surv. 39(1) (2007)
16. Ruskey, F., Savage, C., Wang, T.M.Y.: Generating necklaces. J. Algorithms 13(3), 414–430 (1992)
17. Sloane, N.J.A.: The On-Line Encyclopedia of Integer Sequences, Sequence A062692, available electronically at `http://oeis.org`

On Non-complete Sets and Restivo's Conjecture

Vladimir V. Gusev and Elena V. Pribavkina*

Ural State University, Lenina st. 51, 620083, Ekaterinburg, Russia
vl.gusev@gmail.com, elena.pribavkina@usu.ru

Abstract. A finite set S of words over the alphabet Σ is called non-complete if $Fact(S^*) \neq \Sigma^*$. A word $w \in \Sigma^* \setminus Fact(S^*)$ is said to be uncompletable. We present a series of non-complete sets S_k whose minimal uncompletable words have length $5k^2 - 17k + 13$, where $k \geq 4$ is the maximal length of words in S_k. This is an infinite series of counterexamples to Restivo's conjecture, which states that any non-complete set possesses an uncompletable word of length at most $2k^2$.

1 Introduction

Let Σ be a finite alphabet. A finite set S of words over the alphabet Σ is called *complete* if $Fact(S^*) = \Sigma^*$, i.e. every word over the alphabet Σ is a factor of a word of S^*. If S is not complete, $\Sigma^* \setminus Fact(S^*)$ is not empty and a word of minimal length in this set is called a *minimal uncompletable word* (with respect to the non-complete set S). Its length will be denoted by $uwl(S)$.

The problem of finding minimal uncompletable words and their length was introduced in 1981 by Restivo. In his paper [6] he conjectured that a non-complete set S always possesses an uncompletable word w of length at most $2k^2$, where k is the maximal length of words in S, and w is of the form $w = uv_1uv_2 \cdots uv_{k-1}u$, where $u \notin S$, $|u| = k$ and $|v_i| \leq k$ for all $i = 1, 2, \ldots, k-1$. An example giving a lower bound $k^2 + k - 1$ for the length of minimal uncompletable words was presented in [4]. However Restivo's conjecture appeared to be false by means of a counterexample found in [2]. Namely, let $k > 6$ and let $R_k = \Sigma^k \setminus \{a^{k-2}bb\} \cup \Sigma ba^{k-4}\Sigma \cup \Sigma ba \cup b^4 \cup J_k$, where $J_k = \bigcup_{i=1}^{k-3}(ba^i\Sigma \cup a^ib)$. In [2] the authors computed for $7 \leq k \leq 12$ that the length of a minimal uncompletable word for R_k is equal to $3k^2 - 9k + 1$ but were unable to prove it in general.

In this paper we present a new series of non-complete sets S_k whose minimal uncompletable words have length $5k^2 - 17k + 13$ for $k \geq 4$.

As far as the upper bound is concerned, only trivial exponential one is known. More precisely, the length of a minimal uncompletable word is at most $2^{\|S\|-m+1}$, where m is the number of elements in S and $\|S\|$ is the sum of lengths of all elements in S. It comes from the connection between non-complete sets and

* The authors acknowledge support from the Russian Foundation for Basic Research, grant 10-01-00524, and from the Federal Education Agency of Russia, grant 2.1.1/13995.

G. Mauri and A. Leporati (Eds.): DLT 2011, LNCS 6795, pp. 239–250, 2011.

synchronizing automata studied in [4]. However this bound is not likely to be precise.

An interesting related question is to decide whether a given regular language L satisfies one of the properties $\Sigma^* = \mathit{Fact}(L)$, $\Sigma^* = \mathit{Pref}(L)$, $\Sigma^* = \mathit{Suff}(L)$. Computational complexity of the aforementioned problems has been recently considered by Rampersad et al. in [5]. They gave complete solution in case L is represented by a deterministic or non-deterministic finite automaton. In particular case $L = S^*$ for S being a finite set of words the authors mention that the complexity of deciding whether or not $\Sigma^* = \mathit{Fact}(S^*)$ is still an open problem.

2 The Set S_k

We assume the reader's acquaintance with the basics of combinatorics on words. Additional information on the subject can be found in [3]. To fix the notation, let us recall some basic definitions. By $|w|$ we denote the *length* of a word w. The length of the *empty word* ε is equal to zero. By Σ^+ we denote the set of all non-empty words over the alphabet Σ; by Σ^k – the set of all words of length exactly k over Σ and by $\Sigma^{\leq k}$ – the set of all words of length at most k over Σ. A word $u \in \Sigma^+$ is a *factor* of w (*prefix* or *suffix* respectively) if w can be decomposed as $w = xuy$ ($w = uy$ or $w = xu$ respectively) for some $x, y \in \Sigma^*$. A factor (prefix, suffix) u of w is called *proper* if $u \neq w$. Given a word $u = a_1a_2 \cdots a_n$ with $a_i \in \Sigma$ by $u[i \dots j]$ with $1 \leq i, j \leq n$ we denote the factor $a_ia_{i+1} \cdots a_j$ if $i \leq j$, and the empty word if $i > j$. Moreover, we put $u[0] = \varepsilon$.

Let $\Sigma = \{a, b\}$. Consider the set

$$S_k = \left(\Sigma^k \setminus \{ba^{k-1}, b^{k-1}a\}\right) \cup \left(\Sigma^{k-1} \setminus \{a^{k-1}, b^{k-1}\}\right).$$

In section 3 we show that this set is not complete for $k \geq 4$ and possesses an uncompletable word of length $5k^2 - 17k + 13$. In section 4 we show that this upper bound is precise. Our results considerably rely upon the notion of a *forbidden position* in a word. This notion was introduced in [4]. Let S be any non-complete set and let w be an uncompletable word for the set S. We say that $0 \leq j \leq |w| - 1$ is a forbidden position in w with respect to S, if $w[j + 1, \dots, |w|] \notin \mathit{Pref}(S^*)$, i.e. the suffix of the word w starting from position j is not a prefix of any word in S^*. If the set S is clear from context we will omit reference to S. Note that, if $S \subseteq \Sigma^{\leq k}$ and positions $0, 1, \dots, k - 1$ are forbidden in some word w, then w is uncompletable for S. So to prove that a set $S \subseteq \Sigma^{\leq k}$ is not complete, it is enough to find a word with first k forbidden positions.

Lemma 1. *Let $j \geq k$ be a forbidden position in a word w with respect to S_k. The position $j - k$ is forbidden in w with respect to S_k if and only if one of the following conditions holds true:*

(i) $j - 1$ is forbidden with respect to S_k;
(ii) $w[j - k + 1, \dots, j - 1] \in \{a^{k-1}, b^{k-1}\}$.

Proof. Let $j-k$ be forbidden in w. Then by definition $w[j-k+1,\ldots,|w|] \notin \mathit{Pref}(S_k^*)$. Suppose $j-1$ is not forbidden in w, i.e. $w[j,\ldots,|w|] = x \in \mathit{Pref}(S_k^*)$. If the factor $y = w[j-k+1,\ldots,j-1]$ of length $k-1$ is in S_k, then $w[j-k+1,\ldots,|w|] = yx \in \mathit{Pref}(S_k^*)$, which is a contradiction. Hence $y \in \Sigma^{k-1} \setminus S_k = \{a^{k-1}, b^{k-1}\}$.

Conversely, arguing by contradiction suppose $j-k$ is not forbidden. Then since the length of the suffix $w[j-k+1,\ldots,|w|]$ is at least k, it can be factorized as xy where $x \in S_k$ and $y \in \mathit{Pref}(S_k^*)$. The case $|x| = k$ contradicts the condition that j is forbidden, since we get $w[j+1,\ldots,|w|] = y \in \mathit{Pref}(S_k^*)$. Hence $|x| = k-1$ and since $x \in S_k$ we have that x is different both from a^{k-1} and b^{k-1}. But then position $j-1$ is not forbidden, which is a contradiction.

In the rest of the paper we strictly fix the following notation: $u = ba^{k-1}$ and $v = b^{k-1}a$. We will consider forbidden positions only in occurrences of u and v in w. In each such occurrence for convenience we will enumerate forbidden positions locally from 0 to $k-1$.

Example 1. Consider the following word:

$$w = \underbrace{'b'a'a}_{\{0,1,2\}}\; a'b\, \underbrace{'b\,\overbrace{b'a\,a}^{\{1\}}}_{\{0,2\}}\, a\; \underbrace{'bba}_{\{0\}}\,.$$

Using the definition and lemma 1 it is easy to calculate the set of its forbidden positions with respect to S_3: $\{0,1,2,4,5,7,10\}$. There are two occurrences of u and two occurrences of v in w (the first occurrence of v overlaps with the second occurrence of u). Locally enumerated sets of forbidden positions are: $\{0,1,2\}$ in the first occurrence of u, $\{1\}$ in the second occurrence of u, $\{0,2\}$ in the first occurrence of v, and $\{0\}$ in the second occurrence of v. Note that, since first three positions are forbidden in w, this word is uncompletable for S_3.

Position 0 may be forbidden in an occurrence of u or v. The following statement gives necessary and sufficient conditions for this to happen. It is an easy consequence of lemma 1.

Lemma 2. *Position* 0 *is forbidden in an occurrence of* u *in a word* w *iff position* $k-1$ *is forbidden in the same occurrence of* u. *Position* 0 *is forbidden in any occurrence of* v *in* w.

Two occurrences $p, q \in \{u, v\}$ in a word w are said to be *consecutive* if they either overlap or are the only occurrences from $\{u, v\}$ in the factor pxq of w.

Lemma 3. *Let* $p, q \in \{u, v\}$ *be two consecutive occurrences without overlap in a word* w, *and let* pxq *be a factor of* w *with* $x \in \Sigma^*$. *Let* F_p *and* F_q *be the sets of forbidden positions in* p *and* q *respectively. Then*

$$F_p \subseteq \{j + |x| \mod k \mid j \in F_q\} \cup \{0\}.$$

Proof. Consider a forbidden position i in p such that $i \neq 0$ and consider the factor y of length multiple of k in w from position i in p to some position j in q $(0 \leq j < k)$. This factor is in S_k^+, since p and q are consecutive occurrences of words from $\Sigma^k \setminus S_k$. Thus, if position $j \notin F_q$, then neither position i is forbidden. On the one hand, we have $|y| \equiv 0 \mod k$, on the other hand $|y| = k - i + |x| + j$, hence $i \equiv j + |x| \mod k$.

Note that, two words from $\Sigma^k \setminus S_k$ overlap only in case of v and u. More precisely, two last letters of v overlap with first two letters of u leading to the word $b^{k-1}a^{k-1}$. The following statement can be easily proved using the same argument as in the previous lemma.

Lemma 4. *Let v and u be two consecutive overlapping occurrences in a word w, and let F_v and F_u be the corresponding sets of forbidden positions. Then*

$$F_v \subseteq \{j - 2 \mod k \mid j \in F_u\} \cup \{0\}.$$

Previous lemmas allow us to make the following observation. Let $p, q \in \{u, v\}$ be two consecutive occurrences in w. Then forbidden positions in p except 0 are inherited from forbidden positions in q, and position 0 may appear in F_p according to lemma 2. In our proofs we will trace backwards forbidden positions only in occurrences of words from $\Sigma^k \setminus S_k$ starting from the last one. Besides, the number of forbidden positions in consecutive occurrences increases by at most 1.

Example 2. Let $k = 5$ and let $paaaq$ be a factor of some word w such that $p = q = u = baaaa$. Let F_p and F_q be the sets of forbidden positions in p and q respectively. Let $F_q = \{1, 3\}$ (see Fig. 1). Since $|x| = |aaa| = 3$ by lemma 3 we have $F_p \subseteq \{0, 1, 4\}$. The position 4 is forbidden in p. Indeed, the only word in S_k which can be read from this position is a^4b of length 5 leading to the forbidden position 1 in q. This means that the condition *(ii)* of lemma 1 holds. By lemma 2 the position 0 is in F_p. The position 1 is not forbidden in p, since from this position we can read the word $aaaaa{\cdot}aaba \in S_k^+$ and get to the position 2 in F_q, which is not forbidden. Informally speaking, the forbidden position 3 in q is lost as we step backwards by 5 letters.

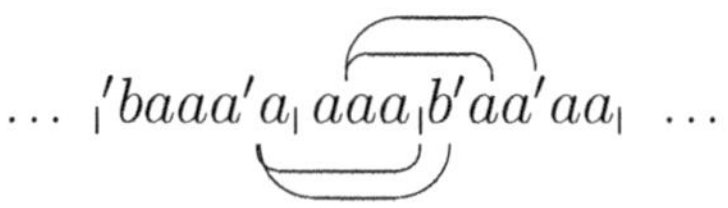

Fig. 1. Inheritance of forbidden positions

3 Upper Bound for $uwl(S_k)$

In this section we prove that the set S_k is non-complete by presenting an uncompletable word w of length $5k^2 - 17k + 13$ for $k \geq 4$.

Theorem 1. *For $k \geq 4$ the set S_k is not complete and there exists an uncompletable word of length $5k^2 - 17k + 13$.*

Proof. For clarity by r we denote overlapping occurrences of v and u, i.e. $r = b^{k-1}a^{k-1}$. Consider the word $\omega = u \centerdot \prod_{i=1}^{k-3} (ra^i \centerdot b^{k-2-i}r \centerdot)v$, where $\centerdot$ can be replaced by any letter of the alphabet Σ. Let us enumerate occurrences of v counting backwards from the last one (the first occurrence of v in this order will have number 0), and let $F_v^i \subseteq \{0, \ldots, k-1\}$ be the set of forbidden positions in the ith occurrence of v. Occurrences of u are counted in the same way (but starting from 1 instead of 0) and F_u^i are defined analogously. Note that, $F_v^0 = \{0\}$ by lemma 2. By lemma 1 applied to the position 0 in F_v^0 we get $F_u^1 = \{1\}$. Note that, the position 1 in F_u^1 has number $k-1$ in F_v^1. By lemma 2 we have $0 \in F_v^1$. Thus, $F_v^1 = \{0, k-1\}$. Lemma 3 implies $F_u^2 \subseteq \{0, k-2, k-1\}$. After applying lemma 1 consecutively to positions $k-1$ and 0 in F_v^1 we obtain that positions $k-2$ and $k-1$ are in F_u^2. Finally, by lemma 2 we have $0 \in F_u^2$. So, $F_u^2 = \{0, k-2, k-1\}$. Now we aim to show that $F_v^2 = \{0, k-3, k-2\}$. Lemma 4 implies $F_v^2 \subseteq \{0, k-4, k-3, k-2\}$. By lemma 2 we get $0 \in F_v^2$. Note that, the position 0 in F_u^2 is the same position as $k-2$ in F_v^2. Since the condition (i) of lemma 1 applied to the position $k-1 \in F_u^2$ is satisfied, the corresponding position $k-3$ is in F_v^2. When we apply the same lemma to the position $k-2$ in F_u^2 we see that none of the conditions $(i), (ii)$ holds true. Thus, $k-4 \notin F_v^2$. Finally, we have $F_v^2 = \{0, k-3, k-2\}$.

Now assume by induction $F_v^{2i} = \{0, k-i-2, \ldots, k-2\}$ for some $1 \leq i \leq k-4$, and let us show that $F_v^{2(i+1)} = \{0, k-i-3, \ldots, k-2\}$. Applying step by step lemmas 1-4 we obtain the following sets of forbidden positions:

$$\begin{aligned} F_u^{2i+1} &= \{0, 1, k-i, \ldots, k-1\}, \\ F_v^{2i+1} &= \{0, k-i-1, \ldots, k-1\}, \\ F_u^{2i+2} &= \{0, k-i-2, \ldots, k-1\}, \\ F_v^{2i+2} &= \{0, k-i-3, \ldots, k-2\}. \end{aligned}$$

Thus, for $i = k-3$ we have $F_v^{2(k-3)} = \{0, 1, \ldots, k-2\}$. It is not hard to see that the set of forbidden positions in the last occurrence of u is $F_u^{2(k-3)+1} = \{0, 1, \ldots, k-1\}$, which means that the word ω is uncompletable. Its length equals $5k^2 - 17k + 13$.

4 Lower Bound for $uwl(S_k)$

First we prove some nice properties of a minimal uncompletable word in S_k.

Theorem 2. *Consider a minimal uncompletable word w. Then u is a prefix of w and v is its suffix.*

Proof. The word w has either u or v as a factor, otherwise $w \in Pref((S_k \cap \Sigma^k)^*)$. Let $w = w'x$, where $|x| = k$, and suppose $x \neq v$. Let $x = x'z$, where $z \in \Sigma$. Since

w is minimal, we conclude that $w'x' \in Fact(S_k^*)$, which means $rw'x' = qy$, for some $r \in \Sigma^*$, $q \in S_k^*$ and $|y| \le k-1$. If $|y| < k-1$, then $yz \in Pref(S_k)$, because all the words of length at most $k-1$ are prefixes of some words in S_k. Thus, $rw = qyz \in Pref(S_k^*)$ and $w \in Fact(S_k^*)$, which is impossible. If $|y| = k-1$, then $yz = x$. If $x \neq u$, then $yz \in S_k$, and $qyz = rw \in S_k^*$. If $x = u$, then $y = ba^{k-2}$, $z = a$ and we have for instance $rwa^{k-1} = rw'(ba^{k-2})(a^k) \in S_k^*$. In any case we get a contradiction with the fact that w is uncompletable. Thus, w has v as a suffix.

Now we are going to investigate one particular symmetry property of uncompletable words for S_k. It is trivial that the mirror image $\overleftarrow{S}$ of a non-complete set S is again non-complete. Moreover, mirror images $\overleftarrow{w}$ of uncompletable words w for S are uncompletable for $\overleftarrow{S}$. The same property holds true for renaming morphism: $\varphi(a) = b$ and $\varphi(b) = a$. Applying these statements to our set we get $T_k = \overleftarrow{S_k} = \varphi(S_k)$, where

$$T_k = \left(\Sigma^k \setminus \{a^{k-1}b, ab^{k-1}\}\right) \cup \left(\Sigma^{k-1} \setminus \{a^{k-1}, b^{k-1}\}\right).$$

So, if w is an uncompletable word for S_k, then $\varphi(\overleftarrow{w})$ is also uncompletable for S_k. As we have already shown, every minimal uncompletable word has v as a suffix. From the symmetry property it follows that every such word has u as a prefix.

The suffix v of a minimal uncompletable word w has only one forbidden position, namely 0, and in the prefix u of w all the positions from 0 to $k-1$ are forbidden. Thus, we have to analyze how forbidden positions change from one occurrence of a word from $\Sigma^k \setminus S_k$ to the next one.

Consider an arbitrary occurrence of a word from $\Sigma^k \setminus S_k$ in w. Let F be the set of its forbidden positions. We will make use of the following representation of F:

$$F = [f_{1,1}, f_{1,2}, \ldots, f_{1,m_1}; f_{2,1}, \ldots, f_{2,m_2}; \ldots; f_{n,1}, \ldots, f_{n,m_n}],$$

where $f_{i,j+1} \equiv f_{i,j} + 1 \mod k$, $f_{i+1,1} > f_{i,m_i}$ and $n \ge 1$. Simply speaking, we partition the set F into blocks of consecutive (with respect to cyclic order) forbidden positions.

Example 3. Consider the word ba^{10}, and let $F = \{0, 1, 2, 5, 6, 8, 10\}$ be the set of its forbidden positions with respect to S_{11}. Then according to our representation $F = [5, 6; 8; 10, 0, 1, 2]$.

Theorem 3. *Let p and q be two consecutive occurrences of words from $\{u, v\}$ in a minimal uncompletable word w. Let F_p and F_q be the sets of forbidden positions in p and q respectively. If $|F_p| > |F_q|$, then one of the following holds true:*

(i) $p = q = u$, $F_p = \{0, k-j, \ldots, k-1\}$ *and* $F_q = \{1, \ldots, j\}$, *where* $1 \le j \le k-2$;
(ii) $p = v$, $q = u$, *these occurrences overlap,* $F_p = \{0, k-1\}$ *and* $F_q = \{1\}$;
(iii) $p = u$, $q = v$, $F_p = \{0, j-i-1, \ldots, k-1\}$, $F_q = \{0, \ldots, i, j, \ldots, k-1\}$, *where* $j \not\equiv i+1 \mod k$ *and there are* $k-1-i \mod k$ *letters between these occurrences;*

Proof. Let $F_q = [f_{1,1}, \ldots, f_{1,m_1}; \ldots; f_{n,1}, \ldots, f_{n,m_n}]$. First assume that p and q do not overlap, so let pxq be the corresponding factor of w.

Case 1. Let $p = q = u$. Consider the case $n \geq 2$. Then there exists $1 \leq i \leq n$ such that $f_{i,1} \geq 2$. If $f_{i,1} > 2$, then by lemma 1 applied to $f_{i,1}$ we obtain that $f_{i,1} + |x| \mod k \notin F_p$. Indeed, let ℓ be the number of position $f_{i,1}$ according to the global enumeration. The position $f_{i,1} - 1$ is not forbidden by our block representation. Thus, the condition (i) of lemma 1 is not satisfied. Since $w[\ell-k+1, \ldots, \ell-1]$ contains ba as a factor we conclude that the codition (ii) of lemma 1 is not satisfied. Thus, position $\ell - k$ is not forbidden. Note, that the factor from position $f_{i,1}+|x| \mod k$ in p to position $\ell-k$ in w belongs to S_k^*. Since the length of this factor is multiple of k, and p, q are consecutive occurrences. Therefore, position $f_{i,1} + |x| \mod k \notin F_p$. In what follows we will apply lemma 1 without repeating the reasoning as above. If $f_{i,1} = 2$, then applying lemma 1 to position $f_{i,1}$ in q we see that, $f_{i,1} + |x| \mod k \in F_p$ implies $x = x'b^{k-2}$. But then v is a factor of pxq which contradicts the fact that p and q are two consecutive occurrences from $\{u, v\}$. Thus, $f_{i,1} + |x| \mod k$ is not forbidden in p. From lemma 3 it follows that $F_p \subseteq \{j+|x| \mod k \mid j \in F_q\} \cup \{0\}$, whence $|F_p| \leq |F_q|$, a contradiction. Consequently $n = 1$ and $F_q = [f_{1,1}, \ldots, f_{1,m}]$. If $f_{1,1} \geq 2$, then following the same argument as above, we conclude that $|F_p| \leq |F_q|$, hence $f_{1,1} \in \{0, 1\}$. Note that, if $f_{1,1} = 0$, by lemma 2 we have $k - 1 \in F_q$. It means that $F_q = \{0, 1, \ldots, k-1\}$ which contradicts minimality of w. Thus $f_{1,1} = 1$. If $f_{1,m} = k - 1$, then by lemma 2 we obtain that $0 \in F_q$ and in this case $f_{1,1}$ cannot be equal to 1. So $f_{1,m} \leq k - 2$. Now it remains to prove that F_p has form as stated in (i). Since $|F_p| > |F_q|$, by lemma 3 we have $0 \in F_p$. Then by lemma 2 we obtain that $k - 1$ is also in F_p. So there exists a position $i \in F_q$ satisfying $i + |x| \equiv k - 1 \mod k$. If $i < f_{1,m}$, then $i + 1 + |x| \equiv 0 \mod k$, hence $|F_p| \leq |F_q|$. Thus $i = f_{1,m}$, $|x| \equiv k - f_{1,m} - 1 \mod k$ and by lemma 3 we deduce $F_p = \{0, k - f_{1,m}, \ldots, k-1\}$.

Case 2. Let $p = v$, $q = u$. We are to show that either $f_{1,1} + |x| \mod k$ is not forbidden in p or $f_{1,1} + |x| \equiv 0 \mod k$, so in both cases by lemma 3 we have $|F_p| \leq |F_q|$. It holds for $f_{1,1} \geq 2$ by the argument as in the previous case. Note that, $f_{1,1} \neq 0$, otherwise by lemma 2 position $k - 1 \in F_q$, which contradicts our block representation. Therefore, we have $f_{1,1} = 1$. Suppose $1 + |x| \equiv i \mod k$. If $0 < i < k - 1$, then by lemma 1 applied to position $f_{1,1}$ in q we conclude that i is not forbidden in p. If $i = 0$, then lemma 3 implies $|F_p| \leq |F_q|$. If $i = k - 1$, then by lemma 1 we have $i \in F_p$ only if a^{k-2} is a prefix of x, but then p and q are not consecutive occurrences from $\{u, v\}$.

Case 3. Let $p = u$, $q = v$. Suppose $n \geq 2$, and consider arbitrary positions $f_{i,1} < f_{j,1} \in F_q$. We show that either $f_{i,1} + |x| \mod k \notin F_p$ or $f_{j,1} + |x| \mod k \notin F_p$. Arguing by contradiction, suppose both positions are forbidden in p. Then by lemma 1 applied to position $f_{j,1}$ in q the word x must have suffix $b^{k-f_{j,1}}$ (if $|x| < k - f_{j,1}$, then $f_{j,1} + |x| \mod k \notin F_p$ by lemma 1). Analogously $b^{k-f_{i,1}}$ must be the suffix of x, and if $|x| < k - f_{i,1}$, then $f_{i,1} + |x| \mod k \notin F_p$ by lemma 1. But then by the same lemma $k + (k - f_{j,1})$ last letters of x are b's. Continuing this argument we get that $\ell k - f_{j,1}$ last letters of x are b's for any positive integer ℓ. It means that there exists no finite word x such that both positions $f_{i,1}+|x| \mod k$ and $f_{j,1}+|x| \mod k$ are forbidden in p. Hence $n=1$, and $F_q=[f_{1,1},\ldots,f_{1,m}]$. Since $|F_p|>|F_q|$, by lemmas 3 and 2 there is no j such that $f_{1,j}+|x|\equiv 0 \mod k$. Besides $f_{1,i}+|x|\equiv k-1 \mod k$ for some i. If $i<m$, then $f_{1,i+1}+|x|\equiv 0 \mod k$, which is impossible. Thus, $i=m$. Moreover, $|x|\equiv k-1-f_{1,m} \mod k$. Since 0 is always forbidden in v, we can represent F_q as $\{0,\ldots i,j,\ldots,k-1\}$ for some $0\le i<j\le k-1$. Then by lemma 3 we have $F_p=\{0,j-i-1,\ldots,k-1\}$.

Case 4. Let $p=v$, $q=v$. Let $i\equiv f_{1,1}+|x| \mod k$. Suppose $0<i<k-1$. Let ℓ be the number of the position i in the global enumeration. Note that $y=w[\ell+1,\ldots,\ell+k-1]$ contains ba as a factor, thus $y\in S_k$. It is not hard to see that the length of the factor of w from the position $\ell+k-1$ to the position $f_{1,1}-1 \mod k$ in q is a multiple of k, and this factor belongs to S_k^*. Since the position $f_{1,1}-1 \mod k$ is not forbidden in q due to our block representation, the position i is not forbidden in p. But then $|F_p|\le|F_q|$. Thus, we have either $f_{1,1}+|x|\equiv 0 \mod k$ or $f_{1,1}+|x|\equiv k-1 \mod k$. In the first case by lemma 3 we get $|F_p|\le|F_q|$. In the latter case by lemma 1 the word a^{k-2} have to be the prefix of x, which contradicts the fact that p and q are consecutive occurrences.

Now assume that $p=v$, $q=u$, and they overlap. If there exists i such that $f_{i,1}>2$ then by lemma 1 applied to position $f_{i,1}$ in q we have $f_{i,1}-2\notin F_p$. Therefore, lemma 4 implies $|F_p|\le|F_q|$. Note that, $f_{1,1}\ne 0$, otherwise by lemma 2 position $k-1\in F_q$, which contradicts our block representation. Hence, for all i we have $f_{i,1}\in\{1,2\}$. It immediately implies that $n=1$. If $f_{1,j}=2$ for some j then by lemma 4 we get that $f_{1,j}-2$ is equal to 0 and $|F_p|\le|F_q|$. Thus $F_q=\{1\}$ and $F_p=\{0,k-1\}$.

Lemma 5. *Let p and q be two consecutive occurrences of u in a word w, F_p and F_q be the corresponding sets of forbidden positions. If $F_q=\{0,i,\ldots,k-1\}$ and $|F_p|=|F_q|$, then $F_p=F_q$.*

Proof. If $i=1$, then the statement of lemma obviously holds true. So we may assume $i>1$. Let pxq be the factor of w. Then by lemma 3 we have $F_p\subseteq\{0,|x| \mod k, i+|x| \mod k, i+1+|x| \mod k,\ldots,k-1+|x| \mod k\}$. We aim to show that the position $i+|x| \mod k\notin F_p$. Suppose $i=2$, and $i+|x| \mod k\in F_p$. Then by lemma 1 we must have b^{k-2} as a suffix of x. But then we have overlapping occurrences of v and u, which contradicts the fact that p and q are consecutive occurrences. So we may assume $i>2$. We see that neither condition (i) nor (ii) of lemma 1 applied to the position i is satisfied. Therefore, in this case we also have $i+|x| \mod k\notin F_p$. Thus in order to have $|F_p|=|F_q|$, it is necessary that $0\in F_p$ and $j+|x|\not\equiv 0 \mod k$ for all $i+1\le j\le k$. Lemma 2 implies $k-1\in F_p$. It means that $j+|x|\equiv k-1 \mod k$ for some $i+1\le j\le k$. It is straightforward that $j=k$, $|x|\equiv k-1 \mod k$ and $F_p=\{0,i,i+1,\ldots,k-1\}=F_q$.

Lemma 6. *Let p and q be consecutive occurrences of v and u respectively in a minimal uncompletable word w. Let F_p and F_q be the corresponding sets of* *forbidden positions. If $F_q = \{0, i, \ldots, k-1\}$ and $|F_p| = |F_q|$, then these occurrences overlap and $F_p = \{0, i-1, \ldots, k-2\}$.*

Proof. If $i = 1$, then trivially w is not a minimal uncompletable word, so $i > 1$. Suppose p and q do not overlap, so there exists x such that pxq is a factor of w. By lemma 3 we have $F_p \subseteq \{0, |x| \mod k, i+|x| \mod k, i+1+|x| \mod k, \ldots, k-1+|x| \mod k\}$ and from lemma 1 it follows that either $i+|x| \mod k \notin F_p$ or $i+|x| \equiv 0 \mod k$. Moreover, if $j+|x| \equiv 0 \mod k$ for some $i+1 \le j \le k$, then we immediately get $|F_p| < |F_q|$. So we may assume that $j+|x| \not\equiv 0 \mod k$ for all $i+1 \le j \le k$. First let $|x| > k-(i+1)$. Then by lemma 1 if position $i+1+|x| \mod k$ is forbidden in p, then either $i+1+|x| \equiv 0 \mod k$ or $i+1+|x| \equiv k-1 \mod k$. The first case contradicts our assumption. In the latter case $x = a^{k-2}x'$, but this contradicts the fact that p and q are consecutive occurrences. Thus, both cases are impossible. So $0 \le |x| \le k-(i+1)$, but then we have $i+1 \le j = k-|x| \le k$. It means that $j \mod k$ is a forbidden position in q and $j+|x| \equiv 0 \mod k$, which again contradicts our assumption that $j+|x| \not\equiv 0 \mod k$ for all $i+1 \le j \le k$. Therefore occurrences p and q overlap. By lemmas 1 and 4 we get $F_p = \{0, i-1, \ldots, k-2\}$.

Lemma 7. *Let p and q be two consecutive occurrences of v in a minimal uncompletable word w, and let F_p and F_q be the corresponding sets of forbidden positions. If $F_q = \{0, i, \ldots, k-2\}$ then $|F_p| < |F_q|$.*

Proof. Consider the factor pxq of w. By lemma 3 we get $F_p \subseteq \{0, |x| \mod k, i+|x| \mod k, i+1+|x| \mod k, \ldots, k-2+|x| \mod k\}$. We are going to show that $|x| \mod k$ is not forbidden in p. From lemma 1 it trivially follows that the position $|x| \mod k \in F_p$ if either $|x| \equiv 0 \mod k$ or $|x| \equiv k-1 \mod k$. The first case contradicts minimality of w, for we would have $F_p = F_q$. In the latter case by the same lemma we conclude $x = a^{k-2}x'$, which contradicts the fact that p and q are consecutive occurrences. Similar arguments can be applied to $i+|x| \mod k$. Namely, if $i+|x| \equiv k-1 \mod k$, then again $x = a^{k-2}x'$, a contradiction. So either $i+|x| \equiv 0 \mod k$ or $i+|x| \mod k \notin F_p$. In both cases we have $|F_p| < |F_q|$.

Lemma 8. *Let p and q be consecutive occurrences of u and v respectively in a word w. Let F_p and F_q be the corresponding sets of forbidden positions, and let $|F_p| = |F_q|$. If $F_q = \{0, i, \ldots, k-2\}$ with $i > 1$ and $F_p = [f_{1,1}, \ldots, f_{1,m}]$, then either $F_p = \{0, 1, i+2, i+3, \ldots, k-1\}$ or $F_p = \{0, i+1, i+2, \ldots, k-1\}$.*

Proof. Consider the factor pxq of w. By lemma 3 we get $F_p \subseteq \{0, |x| \mod k, i+|x| \mod k, i+1+|x| \mod k, \ldots, k-2+|x| \mod k\}$. Arguing as in case 3 of the theorem 3 we easily obtain that either $|x| \mod k \notin F_p$ or $i+|x| \mod k \notin F_p$. Suppose that position $|x| \mod k$ is not forbidden in F_p. Since $|F_p| = |F_q|$ it is necessary that $0 \in F_p$, therefore $k-1$ is also in F_p by lemma 2. Every $i \le j \le k-3$ satisfies $j+|x| \not\equiv k-1 \mod k$, since otherwise we would have $j+1+|x| \equiv 0 \mod k$, which would imply $|F_p| < |F_q|$. Therefore, $k-2+|x| \equiv k-1 \mod k$, whence $|x| \equiv 1 \mod k$ and $F_p = \{0, i+1, \ldots, k-1\}$. Assume now that position

$i + |x| \mod k \notin F_p$. Following the same argument as in the previous case, we get $k-1 \in F_p$ and either $|x| \equiv k-1 \mod k$ or $k-2+|x| \equiv k-1 \mod k$. In the first case $F_p = \{0, i, \ldots, k-3, k-1\}$, and it has more than one block of consecutive forbidden positions, which contradicts the statement of lemma. In the latter case $|x| \equiv 1 \mod k$ and $F_p = \{0, 1, i+2, \ldots, k-1\}$.

Theorem 4. *The length of a minimal uncomletable word for S_k is at least $5k^2 - 17k + 13$ for $k \geq 4$.*

Proof. Let w be an arbitrary minimal uncomletable word for S_k. By theorem 2 the word u is a prefix of w and v is its suffix. Note that, we have $F_u = \{0, 1, \ldots, k-1\}$ and $F_v = \{0\}$ in the aforementioned occurrences of u and v.

If s and t are two consecutive occurrences from $\{u, v\}$ such that $|F_s| > |F_t|$, then we say that s is an *increasing occurrence.* Recall that by lemma 3 it means that $|F_s| = |F_t| + 1$. Since there is only one forbidden position in the last occurrence of v in w and k forbidden positions in the first occurrence of u, there must be at least $k-1$ increasing occurrences in w. Now we are going to estimate the length of a factor between two consecutive such occurrences. Consider a factor pxq of w such that p and q are the only increasing occurrences inside this factor. Note that, for an occurrence $r \in \{u, v\}$ in pxq, different from p and q (if any), $|F_r| = |F_q|$. Otherwise p and q are not consecutive increasing occurrences.

Let $3 \leq |F_q| < |F_p| \leq k$. Since q is an increasing occurrence and $|F_q| \geq 3$ by theorem 3 we must have $q = u$ and $F_q = \{0, i, \ldots, k-1\}$. Analogously, $p = u$ and $F_p = \{0, i-1, \ldots, k-1\}$. Moreover, p and q are not the only occurrences from $\{u, v\}$ in pxq. Assume first $i > 2$, i.e. $|F_p| < k$.

Suppose that there is only one occurrence $r \in \{u, v\}$ in pxq different from p and q. Then from lemma 5 it follows that $r = v$ and since $|F_r| = |F_q|$, applying lemma 6 we obtain $F_r = \{0, i-1, \ldots, k-2\}$. But then since $i > 2$ the set F_r does not have the form required by theorem 3 for the condition $|F_p| > |F_r|$ to hold true.

Assume now that $r_1, r_2 \in \{u, v\}$ are the two occurrences in pxq different from p and q. By the same argument as above $r_2 = v$ and $F_{r_2} = \{0, i-1, \ldots, k-2\}$. If $r_1 = u$, then on the one hand by theorem 3 it should be $F_{r_1} = \{1, 2, \ldots, k-i+1\}$. On the other hand by lemma 8 position 0 have to be forbidden in F_{r_1}, a contradiction. If $r_2 = v$, then by lemma 7 we have $|F_{r_1}| < |F_{r_2}|$, which is impossible. So there are at least three occurrences of words from $\{u, v\}$ in pxq except p and q.

Suppose that $r_1, r_2, r_3 \in \{u, v\}$. As we have already seen before $r_3 = v$ and $F_{r_3} = \{0, i-1, \ldots, k-2\}$. It immediately follows from lemma 7 that $r_2 = u$. Then F_{r_2} is either $\{0, 1, i+1, i+2, \ldots, k-1\}$ or $\{0, i, i+1, \ldots, k-1\}$ by lemma 8. The latter case contradicts minimality of w, since we would have $|F_{r_2}| = |F_q|$. Assume $r_1 = u$, then by theorem 3 we have $F_{r_1} = \{1, 2, \ldots, k-i+1\}$. Let $r_1 y r_2$ be a factor of pxq. Note that, $0 \notin F_{r_1}$ and by lemma 1 position $i + 1 + |y| \mod k \notin F_{r_1}$. Thus $|F_{r_1}| < |F_{r_2}|$, a contradiction. Therefore $r_1 = v$. This case is possible and exactly this situation takes place in the word presented in theorem 1.

Now we are going to estimate the length of x in this case. By lemma 6 the factor r_3 overlaps with q, so we can factorize x in the following way: either $x = y_1 r_1 y_2 r_2 y_3 b^{k-2}$ or, if r_1 and r_2 overlap, $x = z_1 b^{k-1} a^{k-1} z_2 b^{k-2}$. By the proof of lemma 8, since $F_{r_2} = \{0, 1, i+1, i+2, \ldots, k-1\}$, we have $|y_3| \equiv |z_2| \equiv 1 \mod k$. Let us first estimate the length of z_1. Since r_1 and r_2 overlap, lemma 4 implies $F_{r_1} = \{0, i, i+1 \ldots, k-1\}$. Then by theorem 3 there are $k-1 \mod k$ letters between p and r_1, thus $|z_1| \equiv k-1 \mod k$. So in this case we have $|x| \geq 4k-4$. Now let us assume that r_1 and r_2 are not overlapping. By theorem 3 the factor r_1 must have only one block of consecutive forbidden positions. By lemma 3 we have $F_{r_1} \subseteq \{0, i+1+|y_2| \mod k, i+2+|y_2| \mod k, \ldots, k-1+|y_2| \mod k, |y_2| \mod k, |y_2|+1 \mod k\}$. Note that, by lemma 1 either $i+1+|y_2| \mod k \notin F_{r_1}$ or $i+1+|y_2| \equiv 0 \mod k$. So for the set F_{r_1} to have only one block of consecutive forbidden positions and the same cardinality as F_{r_2} we must have either $1+|y_2| \equiv k-1 \mod k$ or $i+2+|y_2| \equiv 1 \mod k$. In the first case we have $|y_2| \equiv k-2 \mod k$ and $F_{r_1} = \{0, i, i+1, \ldots, k-1\}$, but this is the same as in the case of overlapping occurrences r_1 and r_2, so this is impossible in a minimal uncompletable word. In the second case we have $|y_2| \equiv k-i-1 \mod k$ and $F_{r_1} = \{0, 1, 2, \ldots, k-i\}$. Then by theorem 3 we conclude that $|y_1| \equiv i-1 \mod k$. Therefore in this case we have $|x| \geq 4k-3$.

It is not hard to see that, if there are more than 3 occurrences from $\{u, v\}$ different from p and q in pxq, then even if some of them overlap, the total length of the word x is at least $4k-4$. So we conclude that $|x| \geq 4k-4$. Note that, $|x| = 4k-4$ for the word from theorem 1.

Now let $i = 2$, i.e. $F_p = \{0, 1, \ldots, k-1\}$ and $F_q = \{0, 2, 3, \ldots, k-1\}$. Arguing as above, by lemmas 5 and 6 there is an occurrence $r = v$ just before q, overlapping with q and $F_r = \{0, 1, \ldots, k-2\}$. If this is the only occurrence and $pxq = pyb^{k-2}q$, then by theorem 3 we have $|y| \equiv 1 \mod k$ and $F_p = \{0, 1, \ldots, k-1\}$, so here we have $|x| \geq k+1$. More occurrences from $\{u, v\}$ inside pxq will obviously give a longer factor x.

The previous argument implies that any minimal uncompletable word has prefix $pyb^{k-2}q$ with $p = q = u$ and $|y| \equiv 1 \mod k$. From the symmetry property observed in theorem2 we deduce that any minimal uncompletable word has suffix $q'a^{k-2}\hat{y}p'$ with $p' = q' = v$ and $|\hat{y}| \equiv 1 \mod k$. Clearly $F_{p'} = \{0\}$. To calculate $F_{q'}$ note that, there is an occurrence $r = u$ overlapping with q' and $F_r = \{1\}$. From theorem 3 we deduce that q' is an increasing occurrence and $F_{q'} = \{0, k-1\}$. Let p be the next increasing occurrence. If $|F_p| = 3$, then the same theorem implies that $p = u$, $F_p = \{0, k-2, k-1\}$ and there are at least $k-1$ letters between p and q'. If $|F_p| \leq 2$, then we would obviously require at least k letters between q' and an increasing occurrence with three forbidden positions.

Thus to increase the number of forbidden positions from 1 in the suffix v of w to 2 we need at least $k-1$ letters; from 2 to 3 – at least $k-1$; from ℓ to $\ell+1$ for $3 \leq \ell \leq k-2$ we need at least $4k-4$ letters, and finally, from $k-1$ to k

we need $k-1$ letters. Besides we have at least $k-1$ increasing occurrences and the suffix v of w with only one forbidden position. Thus the length of a minimal uncompletable word is at least $3(k-1)+(4k-4)(k-4)+k(k-1)+k = 5k^2-17k+13$.

5 Conclusion

The series of sets S_k was found during exhaustive computational experiment. We searched for maximal with respect to inclusion non-complete sets among all the subsets of $\Sigma^{\leq 3}$; we were interested in such sets having longest possible minimal uncompletable word. We have found two extreme sets up to renaming letters and taking mirror image. Namely, $S_3 = \big(\Sigma^3 \setminus \{baa, bba\}\big) \cup \big(\Sigma^2 \setminus \{aa, bb\}\big)$ and $\big(\Sigma^3 \setminus \{baa, bba\}\big) \cup \big(\Sigma^2 \setminus \{ab, ba\}\big)$. Computation was based on representation of a set S^* as a flower automaton and on the fact that S is non-complete if and only if the corresponding non-deterministic automaton is synchronizing. Moreover, the set of uncompletable words coincides with the language of synchronizing words, see [1] and [4] for more details. The same task for $k=4$ was unfeasible for a typical laptop, so the search was performed with restriction $|\Sigma^4 \cap S| \geq 11$. There is only one extreme non-complete set up to renaming letters and taking mirror image in this class. Namely, $\big(\Sigma^4 \setminus \{aabb, abaa, abbb\}\big) \cup \big(\Sigma^3 \setminus \{aba, bba, bbb\}\big)$. The length of a minimal uncompletable word for this example is 31 compared to 25 for the set S_k. So S_k is not optimal even for $k=4$. Thus, the lower bound $5k^2-17k+13$ is likely to be improved. Nevertheless, the most interesting question whether the tight bound is quadratic remains open.

References

1. Berstel, J., Perrin, D., Reutenauer, C.: Codes and automata. Cambridge University Press, Cambridge (2009)
2. Fici, G., Pribavkina, E., Sakarovitch, J.: On the Minimal Uncompletable Word Problem. In: CoRR (2010), `http://arxiv.org/abs/1002.1928`
3. Lothaire, M.: Combinatorics on Words. Cambridge University Press, Cambridge (1997)
4. Pribavkina, E.V.: Slowly synchronizing automata with zero and incomplete sets. In: CoRR (2009), `http://arxiv.org/abs/0907.4576`
5. Rampersad, N., Shallit, J., Xu, Z.: The computational complexity of universality problems for prefxes, suffixes, factors, and subwords of regular languages. In: CoRR (2009), `http://arxiv.org/abs/0907.0159`
6. Restivo, A.: Some remarks on complete subsets of a free monoid. Quaderni de "La ricerca Scientifica", CNR Roma 109, 19–25 (1981)

Self-organization in Cellular Automata: A Particle-Based Approach

Benjamin Hellouin de Menibus[1] and Mathieu Sablik[2]

[1] Laboratoire d'Informatique Fondamentale,
Université de Provence, Marseille, France
benjamin.hellouin-de-menibus@ens-lyon.fr
[2] Laboratoire d'Informatique Fondamentale,
Université de Provence, Marseille, France
sablik@latp.univ-mrs.fr

Abstract. For some classes of cellular automata, we observe empirically a phenomenon of self-organization: starting from a random configuration, regular strips separated by defects appear in the space-time diagram. When there is no creation of defects, all defects have the same direction after some time. In this article, we propose to formalise this phenomenon. Starting from the notion of propagation of defect by a cellular automaton formalized in [Piv07b, Piv07a], we show that, when iterating the automaton on a random configuration, defects in one direction only remain asymptotically.

1 Introduction

Cellular automata (CA) were introduced by J. von Neumann and S. Ulam as simplified models of biological systems which can exhibit self-reproduction and universal computation. A cellular automaton is a complex system defined by a local rule which acts synchronously and uniformly on the configuration space. These simple models have a wide variety of dynamical behaviours.

A first empirical classification was suggested by S. Wolfram [Wol84]. He introduced four behaviour classes and we are interested in the fourth one: *"The fourth class of cellular automata exhibits still more complicated behaviour [...]. Even starting from disordered or random initial configurations, cellular automata evolve to generate characteristic patterns. Such self-organizing behaviour occurs by virtue of the irreversibility of cellular automaton evolution."* Indeed, for some cellular automata, starting from a random configuration we observe the emergence and the persistence of homogeneous regions separated by particles which propagate and sometimes collide over time (see Fig. 1).

The persistence of these regions under the action of a CA was studied empirically [Wol84, BNR91, HC97] and theoretically [Elo94]. M. Pivato proposed a general formalism to describe this phenomenon: regions are characterized by a subshift Σ and particles are defects in a configuration of Σ. In particular, he develops some invariants to characterize the persistence of a defect [Piv07b] and he describes the different dynamics of propagation of a defect [Piv07a].

G. Mauri and A. Leporati (Eds.): DLT 2011, LNCS 6795, pp. 251–263, 2011.

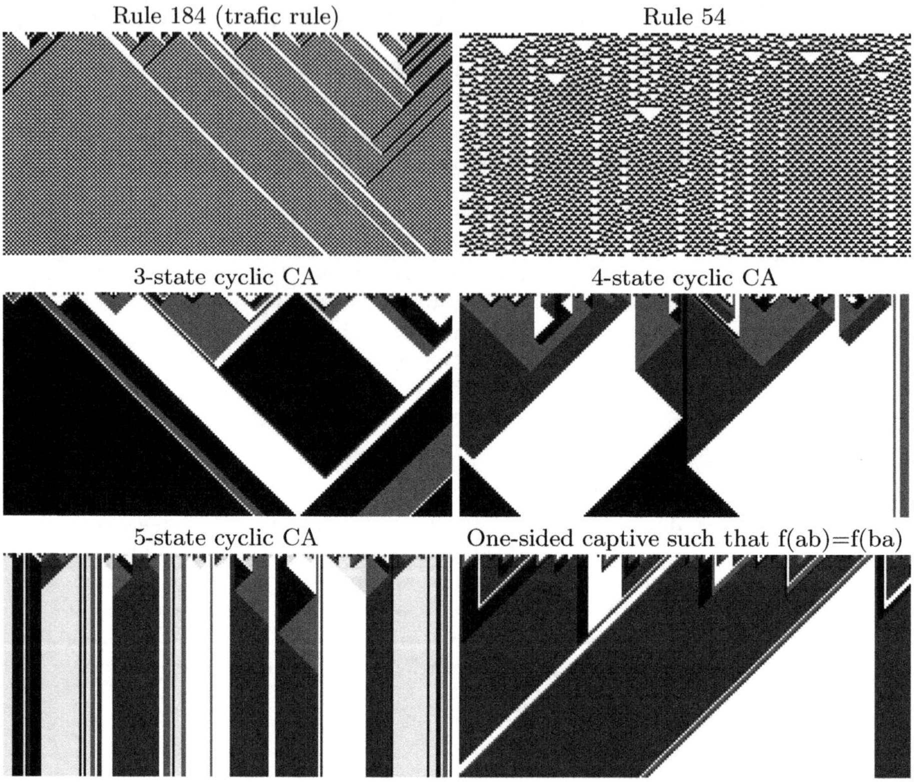

Fig. 1. Space-time diagrams of some cellular automata starting from a random configuration

To explain the emergence of a particular subshift when starting from a configuration chosen randomly according to a measure μ, P. Kůrka and A. Maass introduced the μ-limit set, which is the subshift whose forbidden patterns are exactly those for which the probability to appear tends to zero as time tends to infinity [KM00]. This set corresponds to the configurations observed when a random configuration is iterated. The μ-limit set of an arbitrary CA is difficult to compute: for example, it is undecidable to determine if it contains only one configuration [BPT06]. In order to compute it in some given cases, P. Kůrka suggests an approach based on particle weight function which assigns weights to certain words [Kůr03]. However, this method does not explain why some defects remain in the μ-limit set.

In this article we combine the notions of defect of a subshift Σ and μ-limit set to explain how structures can emerge from interactions of defects. More precisely, we show that for a subshift Σ such that defects have good collision properties, only defects in one particular direction can remain in the μ-limit set for a given σ-ergodic measure μ. In the last section, we show that this theorem can be applied to different cellular automata, thus explaining the behaviours observed in the examples shown in Fig. 1.

2 Definitions

2.1 Configurations and Cellular Automata

Let $\mathcal{A}$ be a finite alphabet. We consider the spaces $\mathcal{A}^* = \cup_{n\in\mathbb{N}}\mathcal{A}^{[0,n]}$ of finite *words* and $\mathcal{A}^{\mathbb{Z}}$ of bi-infinite *configurations*. For $n \in \mathbb{N}, u \in \mathcal{A}^*, a \in \mathcal{A}^{\mathbb{Z}}$, we write $u \sqsubset_n a$ for $a_{[n,n+|u|-1]} = u$ and $u \sqsubset a$ for $\exists n, u \sqsubset_n a$. The product topology on $\mathcal{A}^{\mathbb{Z}}$ is metrizable with the *Cantor metric* on $\mathcal{A}^{\mathbb{Z}}$ defined by $d(a,b) = 2^{-\Delta(a,b)}$, where $\Delta(a,b) = \min\{|z| : z \in \mathbb{Z} \text{ and } a_z \neq b_z\}$. For $u \in \mathcal{A}^*$ and $m \in \mathbb{Z}$, we define the *cylinder* $[u]_m = \{a \in \mathcal{A}^{\mathbb{Z}} : u \sqsubset_m a\}$, if $m = 0$ we denote $[u] = [u]_0$. Cylinders are clopen sets and a base for the Cantor topology. If $U \subseteq \mathcal{A}^*$, we also note $[U]_m = \bigcup_{u\in U}[u]_m \subseteq \mathcal{A}^{\mathbb{Z}}$ which is a borelian and $[U] = [U]_0$.

The *shift* function $\sigma : \mathcal{A}^{\mathbb{Z}} \to \mathcal{A}^{\mathbb{Z}}$ is defined by $(\sigma(a))_v = a_{v+1}$ for all $v \in \mathbb{Z}$. The *language* of a set $\Sigma \subseteq \mathcal{A}^{\mathbb{Z}}$ is $\mathcal{L}(\Sigma) = \{u \in \mathcal{A}^* : \exists x \in \Sigma, u \sqsubset x\}$. Also, we note $\mathcal{L}_r(\Sigma) = \mathcal{L}(\Sigma) \cap \mathcal{A}^{[0,r-1]}$. A subset $\Sigma \subseteq \mathcal{A}^{\mathbb{Z}}$ is a *subshift* if it is closed for Cantor topology and σ-invariant (i.e. $\sigma(\Sigma) \subseteq \Sigma$). In particular, Σ is a *subshift of finite type (SFT)* if there is an order $r > 0$ such that Σ is entirely defined by $\mathcal{L}_r(\Sigma)$, in the sense that $\Sigma = \{a \in \mathcal{A}^{\mathbb{Z}} : \forall z \in \mathbb{Z}, a_{[z,z+r-1]} \in \mathcal{L}_r(\Sigma)\}$. A subshift is *transitive* if there exists an $a \in \Sigma$ such that the orbit $\{\sigma^z(a)\}_{z\in\mathbb{N}}$ is dense in Σ. For a word $u \in \mathcal{A}^*$, we note ${}^{\infty}u^{\infty}$ the σ-periodic configuration of period $|u|$ such that $({}^{\infty}u^{\infty})_{[0,|u|-1]} = u$.

A *cellular automaton* is a continuous function $F : \mathcal{A}^{\mathbb{Z}} \to \mathcal{A}^{\mathbb{Z}}$ which commutes with σ. Equivalently [Hed69], F is defined by a *local rule* $f : \mathcal{A}^{\mathbb{B}_r} \to \mathcal{A}$ such that $F(a)_z = f(a_{z+\mathbb{B}_r})$ (r is the *radius* of the automaton) where $\mathbb{B}_r = [-r, r]$. We study the action of F on $\mathcal{A}^{\mathbb{Z}}$, and especially the values of $(F^n(x))_{n\in\mathbb{N}}$ for some initial configuration x, which we represent as a *space-time diagram.*

2.2 Measures and Density of Configuration

We consider probability measures on the borelians of $\mathcal{A}^{\mathbb{Z}}$, noted $\mathcal{M}(\mathcal{A}^{\mathbb{Z}})$. For some property P, if $\mu(\{x \in \mathcal{A}^{\mathbb{Z}} : P(x)\}) = 1$, we say that P is true for μ-almost all x.

A probability measure μ is *σ-invariant* if for any borelian set X, we have $\mu(\sigma(X)) = \mu(X)$. A probability measure μ is *σ-ergodic* if it is σ-invariant and if for any σ-invariant borelian $X \subseteq \mathcal{A}^{\mathbb{Z}}$ (i.e. $\sigma(X) \subseteq X$) one has $\mu(X) = 0$ or 1. We call $\mathcal{M}_\sigma(\mathcal{A}^{\mathbb{Z}})$ and $\mathcal{M}_\sigma^{erg}(\mathcal{A}^{\mathbb{Z}})$, respectively, the set of σ-invariant measures and the set of σ-ergodic measures. Of course $\mathcal{M}_\sigma^{erg}(\mathcal{A}^{\mathbb{Z}}) \subset \mathcal{M}_\sigma(\mathcal{A}^{\mathbb{Z}}) \subset \mathcal{M}(\mathcal{A}^{\mathbb{Z}})$.

For a configuration $a \in \mathcal{A}^{\mathbb{Z}}$, we define the *Dirac measure* as $\delta_a(U) = 1$ if $a \in U$ and 0 if not for any borelian U. We also define the *Bernoulli measure* μ associated at a sequence $(p_a)_{a\in\mathcal{A}}$ (which verifies $\sum_{a\in\mathcal{A}} p_a = 1$) as $\mu([u]_0) = p_{u_0}p_{u_1}\cdots p_{u_{|u|-1}}$. The action of F on a probability measure μ is $F\mu(X) = \mu(F^{-1}(X))$ for any borelian X. Thus we obtain a function $F : \mathcal{M}_\sigma(\mathcal{A}^{\mathbb{Z}}) \to \mathcal{M}_\sigma(\mathcal{A}^{\mathbb{Z}})$, with $F(\mathcal{M}_\sigma^{erg}(\mathcal{A}^{\mathbb{Z}})) \subseteq \mathcal{M}_\sigma^{erg}(\mathcal{A}^{\mathbb{Z}})$.

Define the *density* of $\mathbb{U} \subseteq \mathbb{Z}$ as $d_{\mathbb{U}} = \limsup \frac{1}{2n+1} |\mathbb{U} \cap \mathbb{B}_n|$. For a set $U \subseteq \mathcal{A}^*$ and $a \in \mathcal{A}^{\mathbb{Z}}$ denote $\mathbb{U}(a) = \{n \in \mathbb{Z} : \exists u \in U, u \sqsubset_n a\} \subseteq \mathbb{Z}$ the *set of positions of* U *in* a. For $\mu \in \mathcal{M}_\sigma^{erg}(\mathcal{A}^{\mathbb{Z}})$ and if U is a finite set of words, the Birkhoff ergodic theorem [Wal00] applied to characteristic functions of cylinders can be restated in terms of density:

$$\text{For } \mu\text{-almost all } a \in \mathcal{A}^{\mathbb{Z}}, d_U(a) = d_{\mathbb{U}(a)} = \limsup_{n\to\infty} \frac{1}{2n+1} |\mathbb{U}(a) \cap \mathbb{B}_n| = \mu([U]).$$

Moreover, we also have for any two open sets A and B:

$$\frac{1}{N} \sum_{k=0}^{N} \mu(A \cap \sigma^{-k}(B)) \underset{N\to\infty}{\longrightarrow} \mu(A) \cdot \mu(B)$$

2.3 Limit and μ-Limit sets

The study of self-organization leads to an interest in the behaviour of the cellular automaton when time tends to infinity. The set of configurations which appear infinitely often is the *limit set* of F defined by $\Omega(F) = \bigcap_{n=0}^{\infty} F^n(\mathcal{A}^{\mathbb{Z}})$. This set can be viewed as the largest attractor: a closed set A is an *attractor* if there exists an open set X such that $F(\overline{X}) \subset X$ and $A = \bigcap_{n=0}^{\infty} F^n(X)$ [Hur90a].

However, these topological notions do not capture the empirical point of view where the initial configuration is randomly chosen according to a measure μ. That is why the notion of μ-attractor is introduced by [Hur90b]: for $\mu \in \mathcal{M}_\sigma(\mathcal{A}^{\mathbb{Z}})$, a closed set A is a μ-*attractor* if A is an attractor of X and $\mu(X) > 0$. As discussed in [KM00] with many examples, this notion is not satisfactory empirically and the authors introduced the notion of μ-*limit set*. It corresponds to the configurations whose subwords are actually observed in simulations:

$$\Lambda_\mu(F) = \left\{ x \in \mathcal{A}^{\mathbb{Z}} : \forall u \sqsubset x, F^n\mu([u]_0) \underset{n\to+\infty}{\not\to} 0 \right\}.$$

3 Defects

In this section, we recall the formalism introduced in [Piv07b, Piv07a] to describe defects with respect to a subshift Σ, and we introduce a formalism to study their dynamics under the action of a cellular automaton. More precisely, we focus our study on interfaces and dislocations.

3.1 General Definitions

The *defect field* of $a \in \mathcal{A}^{\mathbb{Z}}$ with respect to a subshift Σ is defined for all $z \in \mathbb{Z}$ by $\mathcal{F}_a^{\Sigma}(z) = \max\left\{ r \in \mathbb{N} : a_{z+[-\lfloor\frac{r-1}{2}\rfloor, \lfloor\frac{r}{2}\rfloor]} \in \mathcal{L}_r(\Sigma) \right\}$ where the result is possibly 0 or ∞ if the set is empty or infinite. Intuitively, this function returns the size of the largest admissible word centered on a cell. The set of *defects* $\mathbb{D}^{\Sigma}(a)$ is the set of local minima of $\mathcal{F}_a^{\Sigma}$. The *successor* of $d \in \mathbb{D}^{\Sigma}(a)$ is $s_{\mathbb{D}^{\Sigma}(a)}(d) = \min\{z \in \mathbb{D}^{\Sigma}(a) : z > d\}$, and the interval $[d+1, s_{\mathbb{D}^{\Sigma}(a)}(d)]$ is a *homogeneous region* in the sense that $a_{[d+1,s_{\mathbb{D}^{\Sigma}(a)}(d)]} \in \mathcal{L}(\Sigma)$. If there is no ambiguity, we write $\mathbb{D}$ and $s(d)$.

If Σ is a SFT of order r, any defect d of a satisfies $\mathcal{F}_a(d) \leq r$, thus this notion can be extended to finite words of size $\geq r$ except for the first $\lfloor \frac{r-1}{2} \rfloor$ and last $\lfloor \frac{r}{2} \rfloor$ cells. Thus for a word $u \in \mathcal{A}^{[0,n-1]}$, we have $\mathbb{D}(u) \subseteq [\lfloor \frac{r-1}{2} \rfloor; n-1-\lfloor \frac{r}{2} \rfloor]$.

The examples given in Fig. 1 suggest that, in each case, defects can be classified according to their behaviour in two ways:

- Regions correspond to different subshifts and defects behave according to their surrounding regions (*interfaces* - e.g. cyclic automaton);
- Regions correspond to the same periodic subshift and defects correspond to a "phase transition" (*dislocations* - e.g. rule 184 automaton).

3.2 Interfaces

We now assume that the subshift Σ can be decomposed as a disjoint union $\Sigma_1 \sqcup \cdots \sqcup \Sigma_n$ of σ-transitive SFT. Since the different *domains* $(\Sigma_k)_{k\in[1,n]}$ are disjoint SFTs, Σ is also an SFT, and there is some $\alpha > 0$ such that $u \in \mathcal{L}_\alpha(\Sigma) \Leftrightarrow \exists! k, u \in \mathcal{L}(\Sigma_k)$: we say that u *belongs to the domain* k. Thus, for a configuration a, we can associate a domain with each homogeneous region $[d+1; s(d)]$, and only one choice is possible if $s(d) > d + \alpha$.

A *domain signature* is a continuous σ-invariant function $\kappa_d : \mathcal{A}^{\mathbb{Z}} \to \{1 \ldots n\}^{\mathbb{Z}}$ that satisfies the following conditions:

- $\kappa_d(a)_z \neq \kappa_d(a)_{z+1}$ only if $z \in \mathbb{D}(a)$;
- if $\forall z \in [d+1, s(d)], \kappa_d(a)_z = k$, then $a_{[d+1,s(d)]} \in \mathcal{L}(\Sigma_k)$.

We can classify interfaces according to the domain signatures of the surrounding regions: we write $\mathbb{D}_{i,j}^{\kappa_d}(a) = \{d \in \mathbb{D}(a) : \kappa_d(a)_d = i, \kappa_d(a)_{d+1} = j\}$. It is possible to define those sets for finite words except for the first $\alpha - 1$ and last α cells.

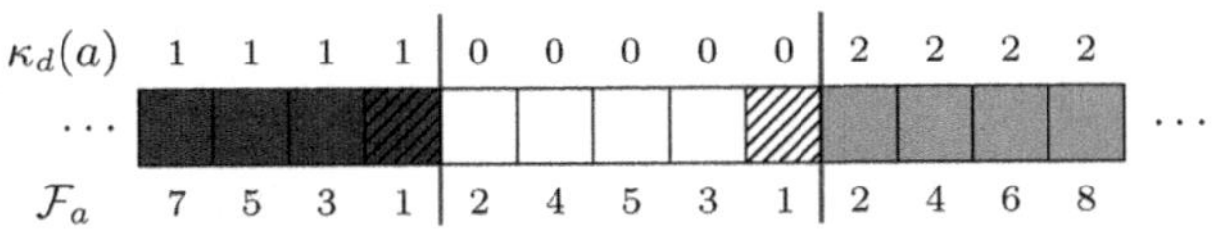

Fig. 2. Interfaces between monochromatic domains

3.3 Dislocations

Let Σ be a σ-transitive SFT of order $r > 1$. We say that Σ is *P-periodic* if there exists a partition $V_1, \ldots, V_P$ of $\mathcal{L}_{r-1}(\Sigma)$ such that $a_1, \ldots, a_r \in \mathcal{L}_r(\Sigma)$ if and only if there exists $i \in \mathbb{Z}/P\mathbb{Z}$ such that $a_1, \ldots, a_{r-1} \in V_i$ and $a_2, \ldots, a_r \in V_{i+1}$. The *period* of Σ is the maximal $P \in \mathbb{N}$ such that Σ is P-periodic.

We thus associate to each $a \in \Sigma$ its *phase* $\varphi(a) \in \mathbb{Z}/P\mathbb{Z}$ such that $a_{[0,r-2]} \in V_{\varphi(a)}$. Obviously, $\varphi(\sigma^k(a)) = \varphi(a)+k$. For $a \in \mathcal{A}^{\mathbb{Z}}$, we say that the homogeneous region $[d+1, s(d)]$ *is in phase* k if $\exists b \in \Sigma, \varphi(b) = k, a \sqsubset_{d+1} b$. If $s(d) > d+r-2$, the phase is unique and corresponds to $a_{[d+1,d+r-1]} \in V_{k+d+1}$.

A *phase signature* $\kappa_\varphi : \mathcal{A}^{\mathbb{Z}} \to (\mathbb{Z}/P\mathbb{Z})^{\mathbb{Z}}$ is a continuous function that satisfies:

- $\kappa_\varphi(a)_z \neq \kappa_\varphi(a)_{z+1}$ only if $z \in \mathbb{D}(a)$;
- if $\forall z \in [d+1, s(d)], \kappa_\varphi(a)_z = k$, then $\exists b \in \Sigma, \varphi(b) = k$ and $a_{[d+1,s(d)]} \sqsubset_{d+1} b$
- $\kappa_\varphi(\sigma(a))_z = \kappa_\varphi(a)_z + 1$

When $s(d) > d + r - 2$, the second condition is equivalent to $\kappa_\varphi(a)_z = \varphi(a_{[d+1,d+r-1]})+d+1$ and shows that the phase signature is defined locally. Since we want our classification of defects to be σ-invariant, and considering the last condition, we define $\mathbb{D}_{i,j}^{\kappa_\varphi}(a) = \{d \in \mathbb{D}(a) : \kappa_\varphi(a)_d + d = i, \kappa_\varphi(a)_{d+1} + d+1 = j\}$. These sets can be extended to defects in finite words except for the first $r-2$ and last $r-1$ cells.

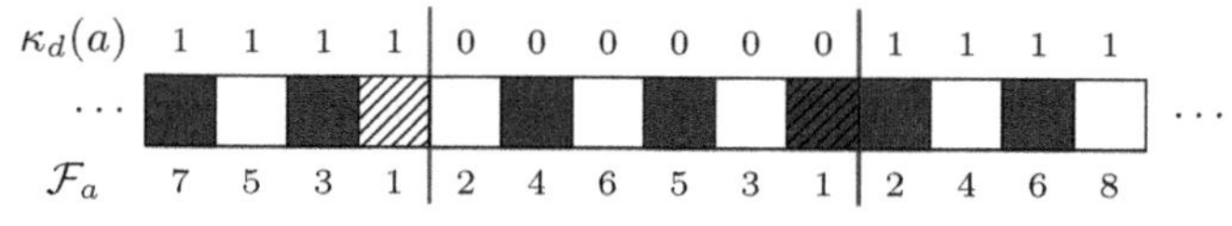

Fig. 3. Dislocations in the checkerboard subshift

3.4 Dynamics

We now consider the general case: assume that Σ can be decomposed into disjoint transitive SFTs $\Sigma_1 \sqcup \cdots \sqcup \Sigma_n$ of respective periods $P_1 \ldots P_n$ (possibly 1). Moreover, we suppose that the Σ_i are F-invariant to give sense to the notion of dynamics of defects. $\kappa = (\kappa_d, \kappa_\varphi) : \mathcal{A}^{\mathbb{Z}} \to S^{\mathbb{Z}}$, where $S = \{(i, x) : i \in [1; n], x \in \mathbb{Z}/P_i\mathbb{Z}\}$ is a generalized *signature* if κ_d and κ_φ respect the conditions of sections 3.2 and 3.3, respectively. We classify the defects according to the signature of the surrounding phases $\mathbb{D} = \bigcup_{s_1,s_2 \in S} \mathbb{D}_{s_1,s_2}$ and there is some α such that we can extend this classification to finite words except for the first and last α cells.

To describe the dynamics of defects, we study the evolution from $\mathbb{D}(a)$ to $\mathbb{D}(F(a))$ for $a \in \mathcal{A}^{\mathbb{Z}}$. An *interpretation of the action of* F *on a configuration* $a \in \mathcal{A}^{\mathbb{Z}}$ is a function $\psi_a : \mathbb{D}(a) \to \mathcal{I}(\mathbb{D}(F(a)))$, where $\mathcal{I}(\mathbb{U})$ is the set of intervals of $\mathbb{U}$, that satisfies:

Locality $\forall d \in \mathbb{D}(a), \forall d' \in \psi_a(d), |d' - d| \leq r$;

Growth If $d_1 < d_2 \in \mathbb{D}(a)$, then $\psi_a(d_1) = \psi_a(d_2)$ or $\max(\psi_a(d_1)) < \min(\psi_a(d_2))$ (where $\max \emptyset = -\infty$ and $\min \emptyset = \infty$).

Surjectivity $\mathbb{D}(F(a)) = \bigcup_{d \in \mathbb{D}(a)} \psi_a(d)$.

Equivalently, an interpretation is a decomposition of $\mathbb{D}(a)$ and $\mathbb{D}(F(a))$ into disjoint "increasing" intervals I_k and I'_k ($k \in \mathbb{Z}$) of size $\leq 2r+1$, satisfying locality (we have $\psi_a(I_k) = I'_k$). We distinguish different situations:

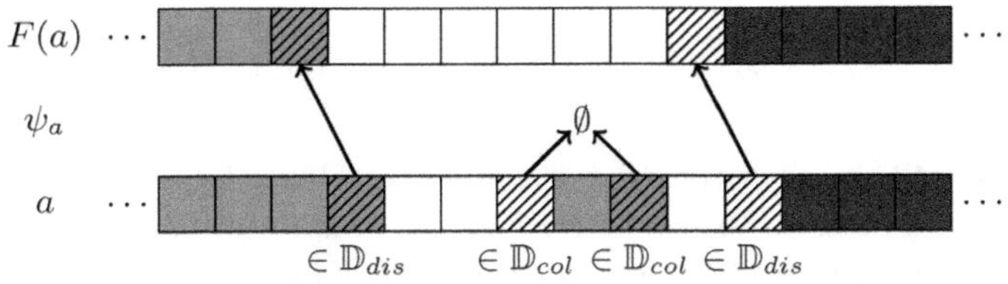

Fig. 4. An interpretation for the 3-state cyclic automaton

$\lvert I_k\rvert$ \ $\lvert I'_k\rvert$	1	>1	0
1	displacement	explosion	(destructive)
>1	collision	collision	collision

Thus, we decompose $\mathbb{D}(a)$ into $\mathbb{D}_{dis}(a) \sqcup \mathbb{D}_{col}(a) \sqcup \mathbb{D}_{expl}(a)$. If we have a set of interpretations $(\psi_a)_{a\in\mathcal{A}^{\mathbb{Z}}}$, we consider the iterated interpretation $\psi_a^2 : d \mapsto \bigcup_{d'\in\psi_a(d)} \psi_{F(a)}(d')$, which is an interpretation for F^2, and it extends to $n > 2$.

An interpretation ψ_a is *coalescent* if it contains only displacements ($|I_k| = |I'_k| = 1$) and decreasing collisions ($|I_k| > |I'_k|$). In this case, a defect $d \in \mathbb{D}_{s_1,s_2}(a)$ has *speed* $(p, q) \in \mathbb{Z} \times \mathbb{N}^*$ if $\forall k < q$, $\psi_a^k(d) \subseteq \mathbb{D}_{dis}(F^k(a))$ and $\psi_a^q(d) = \{d + p\} \subseteq \mathbb{D}_{s_1,s_2}(F^k(a))$. An interpretation ψ_a respects a *velocity function* $V : S^2 \to \mathbb{Z}\times\mathbb{N}^*$ if for any $s_1, s_2 \in S$ and any $d \in \mathbb{D}_{s_1,s_2}(a)$, writing $V(s_1, s_2) = \frac{p}{q}$, either $\psi_a^k(d) \subseteq \mathbb{D}_{col}(F^k(a))$ for some $k < q$, or d has speed $V(s_1, s_2)$. The *order* of a velocity function is the least common multiple of the q appearing in the image of V.

4 A Step towards Self-organization

Proposition 1. *Let F be a CA, $\Sigma = \Sigma_1 \sqcup \cdots \sqcup \Sigma_n$ a decomposition into F-invariant SFTs, κ a signature, $a \in \mathcal{A}^{\mathbb{Z}}$ and ψ_a a coalescent interpretation. Then,*

1. $d_{\mathbb{D}}(F(a)) \leq d_{\mathbb{D}}(a) - \frac{1}{2r+1} d_{\mathbb{D}_{col}}(a)$;
2. *if ψ_a respects V of order q,* $d_{\mathbb{D}_{s_1,s_2}}(F^q(a)) \leq d_{\mathbb{D}_{s_1,s_2}}(a) + \sum_{k=0}^{q} d_{\mathbb{D}_{col}}(F^k(a))$.

Proof (of 1). By surjectivity and locality,

$$\forall n \in \mathbb{N}, \mathbb{D}(F(a)) \cap \mathbb{B}_n \subseteq \psi_a(\mathbb{D}(a) \cap \mathbb{B}_{n+r}).$$

Besides, $|\psi_a(\mathbb{D}_{dis}(a) \cap \mathbb{B}_{n+r})| = |\mathbb{D}_{dis}(a) \cap \mathbb{B}_{n+r}|$, and $|\psi_a(\mathbb{D}_{col}(a) \cap \mathbb{B}_{n+r})| \leq \frac{2r}{2r+1} |\psi_a^{-1}(\psi_a(\mathbb{D}_{col}(a) \cap \mathbb{B}_n))| \leq \frac{2r}{2r+1} |\mathbb{D}_{col}(a) \cap \mathbb{B}_{n+2r}|$.

Since the automaton is coalescent, there is no defect in explosion. Therefore, we have $\forall n \in \mathbb{N}, |\mathbb{D}(F(a)) \cap \mathbb{B}_{n+r}| \leq |\mathbb{D}_{dis}(a) \cap \mathbb{B}_{n+r}| + \frac{2r}{2r+1} |\mathbb{D}_{col}(a) \cap \mathbb{B}_{n+2r}|$, and we conclude by passing to the upper limit. □

Proof (of 2). We only prove the case q=1. Again,

$$\forall n \in \mathbb{N}, \mathbb{D}_{s_1,s_2}(F(a)) \cap \mathbb{B}_n \subseteq \psi_a(\mathbb{D}(a) \cap \mathbb{B}_{n+r}).$$

If $d \in \mathbb{D}_{dis}(a)$ and $\psi_a(d) \in \mathbb{D}_{s_1,s_2}(F(a))$, we have $d \in \mathbb{D}_{s_1,s_2}(a)$ since ψ_a respects V. Since $\mathbb{D}(a) = \mathbb{D}_{dis} \sqcup \mathbb{D}_{col}$, we conclude by passing to the upper limit. □

Theorem 1 (Main result). *Let F be a CA, $\Sigma = \Sigma_1 \sqcup \cdots \sqcup \Sigma_n$ a decomposition into σ-transitive F-invariant SFTs, κ a signature and $(\psi_a)_{a \in \mathcal{A}^{\mathbb{Z}}}$ a set of coalescent interpretations that respect a velocity function V.*

Then for all $\mu \in \mathcal{M}_\sigma^{erg}(\mathcal{A}^{\mathbb{Z}})$, there is a speed $v \in \mathbb{Q}$ such that

$$\forall s_1, s_2 \in S, \forall a \in \Lambda_\mu(F), \mathbb{D}_{s_1,s_2}(a) \neq \emptyset \Rightarrow V(s_1,s_2) = (p,q) \text{ with } \frac{p}{q} = v.$$

We only prove the result for velocity functions of order 1; the result can be easily extended by considering F^q. We note $V(s_1,s_2) = p$ instead of $(p,1)$. Let α be the order of Σ and κ a signature, we introduce the following sets of words:

$$\begin{aligned} I_{s_1,s_2} &= \{u \in \mathcal{A}^{2\alpha+1} : \alpha+1 \in \mathbb{D}_{s_1,s_2}(u)\} \\ J_{s_1,s_2,s_3,s_4}(n) &= \{u \in \mathcal{A}^{n+2\alpha+1} : \alpha+1 \in \mathbb{D}_{s_1,s_2}(u), n+\alpha+1 \in \mathbb{D}_{s_3,s_4}(u)\} \\ I &= \bigcup_{s_1,s_2 \in S} I_{s_1,s_2} \end{aligned}$$

It is obvious that for $a \in \mathcal{A}^{\mathbb{Z}}$, for all $s_1, s_2 \in S, d_{\mathbb{I}_{s_1,s_2}(a)} = d_{\mathbb{D}_{s_1,s_2}(a)}$ (recall that $\mathbb{U}$ is the set of positions of U).

Lemma 1. *Under the previous assumptions, let $s_i \in S$ such that $V(s_1,s_2) > V(s_3,s_4)$. Then:*

$$\forall a \in \mathcal{A}^{\mathbb{Z}}, \forall n \in \mathbb{N}, d_{\mathbb{D}}(F^n(a)) \leq d_{\mathbb{D}}(a) - \frac{1}{4r+2} d_{\mathbb{J}_{s_1,s_2,s_3,s_4}(n)}(a).$$

Proof. For $a \in \mathcal{A}^{\mathbb{Z}}$, we proceed by induction on n.

- *Initialization ($n=1$):* let $x \in \mathbb{J}_{s_1,s_2,s_3,s_4}(1)$, that is $z = x+\alpha+1 \in \mathbb{D}_{s_1,s_2}(a)$ and $z+1 \in \mathbb{D}_{s_3,s_4}(a)$.

Assume that $z, z+1 \in \mathbb{D}_{dis}(a)$: then $\psi_a(z) = z + V(s_1,s_2) \geq \psi_a(z+1) = z+1+V(s_3,s_4)$, which is a contradiction with the growth property. Therefore, either z or $z+1$ is in collision, and $d_{\mathbb{D}_{col}}(a) \geq \frac{1}{2} d_{\mathbb{J}_{s_1,s_2,s_3,s_4}(1)}(a)$. We conclude by Proposition 1 (1).

- *Heredity ($n>1$):* we assume the lemma is true for all $k<n$, and we consider $x \in \mathbb{J}_{s_1,s_2,s_3,s_4}(n)$, that is $z = x+\alpha+1 \in \mathbb{D}_{s_1,s_2}(a)$ and $z+n \in \mathbb{D}_{s_3,s_4}(a)$.

Assume that $z, z+n \in \mathbb{D}_{dis}(a)$: then $\psi_a(z) = z + V(s_1,s_2)$ and $\psi_a(z+n) = z+n+V(s_3,s_4)$, and so $x + V(s_1,s_2) \in \mathbb{J}_{s_1,s_2,s_3,s_4}(k)(F(a))$ where $k = n - V(s_1,s_2) + V(s_3,s_4) < n$. We conclude that $z \in \mathbb{D}_{col}(a)$ or $z+n \in \mathbb{D}_{col}(a)$ or $z, z+n \in \mathbb{D}_{dis}(a)$ and $x + V(s_1,s_2) \in \mathbb{J}_{s_1,s_2,s_3,s_4}(k)(F(a))$.

Therefore, we have $d_{\mathbb{J}_{s_1,s_2,s_3,s_4}(n)}(a) \leq 2 d_{\mathbb{D}_{col}}(a) + d_{\mathbb{J}_{s_1,s_2,s_3,s_4}(k)}(F(a))$.

If we apply the induction hypothesis,

$$\begin{aligned} d_{\mathbb{D}}(F^{k+1}(a)) &\leq d_{\mathbb{D}}(F(a)) - \frac{1}{4r+2} d_{\mathbb{J}_{s_1,s_2,s_3,s_4}(k)}(F(a)) \\ &\leq d_{\mathbb{D}}(a) - \frac{1}{2} d_{\mathbb{D}_{col}}(a) - \frac{1}{4r+2} d_{\mathbb{J}_{s_1,s_2,s_3,s_4}(k)}(F(a)) \\ &\leq d_{\mathbb{D}}(a) - \frac{1}{4r+2} d_{\mathbb{J}_{s_1,s_2,s_3,s_4}(k)}(a) \end{aligned}$$

Since $d_{\mathbb{D}}(F^n(a)) \leq d_{\mathbb{D}}(F^{k+1}(a))$ (Proposition 1 (1)), we conclude. □

Proof (of Theorem 1). By Birkhoff's theorem, we have for almost all $a \in \mathcal{A}^{\mathbb{Z}}, F^n\mu([I]) = d_{\mathbb{D}}(F^n(a))$. By prop. 1 (1), $(F^n\mu([I]))_{n\in\mathbb{N}}$ is decreasing and has a limit d_∞.

First step. Assume $\exists s_i \in S, V(s_1, s_2) > V(s_3, s_4)$ and $F^n\mu([I_{s_1,s_2}]) \cdot F^n\mu([I_{s_3,s_4}]) \not\to 0$; let $\varepsilon > 0$ such that $\forall n_0, \exists n \geq n_0, F^n\mu([I_{s_1,s_2}]) \cdot F^n\mu([I_{s_3,s_4}]) > \varepsilon$.

Consider n large enough so that $(F^n\mu([I])) - d_\infty < \frac{\varepsilon}{8r+4}$ and $F^n\mu([I_{s_1,s_2}]) \cdot F^n\mu([I_{s_3,s_4}]) > \varepsilon$. Since $F^n\mu \in \mathcal{M}_\sigma^{erg}(\mathcal{A}^{\mathbb{Z}})$, we have:

$$\frac{1}{N}\sum_{k=0}^{N} F^n\mu([I_{s_1,s_2}] \cap \sigma^k([I_{s_3,s_4}])) \underset{N\to\infty}{\longrightarrow} F^n\mu([I_{s_1,s_2}]) \cdot F^n\mu([I_{s_3,s_4}])$$

$$\frac{1}{N}\sum_{k=0}^{N} F^n\mu([J_{s_1,s_2,s_3,s_4}(k)]) \underset{N\to\infty}{\longrightarrow} F^n\mu([I_{s_1,s_2}]) \cdot F^n\mu([I_{s_3,s_4}])$$

We can choose N large enough so that $\frac{1}{N}\sum_{k=0}^{N} F^n\mu([J_{s_1,s_2,s_3,s_4}(k)]) > \frac{\varepsilon}{2}$, so we have $F^n\mu([J_{s_1,s_2,s_3,s_4}(k_0)]) > \frac{\varepsilon}{2}$ for some k_0.

By the preliminary lemma, we have:

$$\forall a \in \mathcal{A}^{\mathbb{Z}}, \qquad d_{\mathbb{D}}(F^{n+k_0}(a)) \leq d_{\mathbb{D}}(F^n(a)) - \frac{1}{4r+2} d_{\mathbb{J}_{s_1,s_2,s_3,s_4}(k_0)}(F^n(a))$$

$$F^{n+k_0}\mu([I]) \leq F^n\mu([I]) - \frac{\varepsilon}{8r+4}$$

Which is in contradiction with $F^n\mu([I]) - d_\infty < \frac{\varepsilon}{8r+4}$.

Second step. Assume $\exists s_i \in S$ such that $V(s_1, s_2) > V(s_3, s_4)$ and $F^n\mu([I_{s_1,s_2}]) \not\to 0, F^n\mu([I_{s_3,s_4}]) \not\to 0$. Since $F^n\mu([I_{s_1,s_2}]) \cdot F^n\mu([I_{s_3,s_4}]) \to 0$, 0 is an accumulation point of at least one of the sequences. Let $\varepsilon > 0$: w.l.o.g, we can choose n large enough so that $F^n\mu([I]) - d_\infty < \frac{\varepsilon}{4r+2}$ and $F^n\mu([I_{s_1,s_2}]) \leq \frac{\varepsilon}{2}$.

By applying iteratively proposition 1 (1) and (2), we have

$$\forall a \in \mathcal{A}^{\mathbb{Z}}, \forall k \in \mathbb{N}, \qquad d_{\mathbb{D}_{s_1,s_2}}(F^k(a)) - d_{\mathbb{D}_{s_1,s_2}}(a) \leq (2r+1) \cdot \left(d_{\mathbb{D}}(a) - d_{\mathbb{D}}(F^k(a))\right)$$

$$\forall k \in \mathbb{N}, \qquad F^k\mu([I_{s_1,s_2}]) - \mu([I_{s_1,s_2}]) \leq (2r+1) \cdot \left(\mu([I]) - F^k\mu([I])\right)$$

By applying this to the measure $F^n\mu$, we have:

$$\forall k \in \mathbb{N}, F^{n+k}\mu([I_{s_1,s_2}]) \leq \frac{\varepsilon}{2} + \frac{\varepsilon}{2}$$

From which we deduce $F^n\mu([I_{s_1,s_2}]) \underset{n\to\infty}{\longrightarrow} 0$, a contradiction.

Summary : For all s_1, s_2, s_3, s_4 such that $V(s_1,s_2) \neq V(s_3,s_4)$, $F^n\mu([I_{s_1,s_2}]) \to 0$ or $F^n\mu([I_{s_3,s_4}]) \to 0$. □

5 Applications

5.1 *n*-state Cyclic Automaton

The n-state cyclic automaton is a particular captive cellular automaton defined on the alphabet $\mathcal{A} = \mathbb{Z}/n\mathbb{Z}$ by the local rule

$$f(a_{i-1}, a_i, a_{i+1}) = \begin{cases} a_i + 1 & \text{if } a_{i-1} = a_i + 1 \text{ or } a_{i+1} = a_i + 1 \\ a_i & \text{otherwise} \end{cases}$$

This automaton was introduced by [Fis90]. In this paper, the author shows that for all Bernoulli measure μ, the set $[i]_0$ (for $i \in \mathcal{A}$) is a μ-attractor iff $n \geq 5$. Simulations starting from a random configuration suggest the following: for $n = 3$ or 4, monochromatic regions keep increasing in size; for $n \geq 5$, we observe the convergence to a fixed point where small regions are delimited by vertical lines. We are going to apply the main result to explain this observation.

We consider the decomposition $\Sigma = \bigsqcup_{i\in\mathcal{A}} \Sigma_i$ where $\Sigma_i = \{{}^\infty i^\infty\}$ of periods $P_i = 1$ (no dislocations). Here, $\kappa_d(a,z) = a_z$ and since $\kappa_\varphi = 1$, we write $\kappa(a,z) = i$ for $(i,1)$. Defects are exactly transitions between colors, and we define the velocity function as $V(i+1,i) = (1,1)$, $V(i,i+1) = (-1,1)$ and $V(i,j) = (0,1)$ for $i,j \in \mathcal{A}$ with $i+1 \neq j \neq i-1$.

For any $a \in \mathcal{A}^{\mathbb{Z}}$, we define $\mathbb{D}_k(a) = \bigcup_{V(i,j)=(k,1)} \mathbb{D}_{i,j}(a)$. We also define for any $a \in \mathcal{A}^{\mathbb{Z}}$ the function ψ_a by:

$$\forall d \in \mathbb{D}_k, \psi_a(d) = \begin{cases} \emptyset & \text{if } \exists d' \in \mathbb{D}_{k'}, sign(d-d') \neq sign(d+k-(d'+k')) \\ \{d+k\} & \text{otherwise} \end{cases}$$

This interpretation corresponds to the behaviour of defects as observed in simulations. It is straightforward to prove that it is well-defined (that is, it maps a defect to a interval of defects) and that it satisfies the properties of locality, growth and surjectivity.

Since no image interval has size bigger than 1, it is coalescent and respects the velocity function V. By applying the main result, we show that for all $\mu \in \mathcal{M}_\sigma^{erg}(\mathcal{A}^{\mathbb{Z}})$, defects in only one direction remain in the μ-limit set, that is $\exists k \in \{-1,0,1\}, \forall a \in \Lambda_\mu(F), \mathbb{D}(a) = \mathbb{D}_k(a)$.

In particular, for any Bernoulli measure μ, if we consider the "mirror" application $\gamma((a_i)) = (a_{-i})$, we have $\mu(\gamma([u])) = \mu([u^{-1}]) = \mu([u])$, where $u_1 \ldots u_n^{-1} = u_n \ldots u_1$. But $d \in \mathbb{D}_1(a) \Leftrightarrow -d \in \mathbb{D}_{-1}(\gamma(a))$, and conversely; since this is true for any $F^k\mu$, one has $\mathbb{D}_1(a) = \mathbb{D}_{-1}(a) = \emptyset$ for all $a \in \Lambda_\mu(F)$. We deduce the following properties of Λ_μ for each n-cyclic cellular automaton:

- If $n = 3$, there is no defect of speed 0. Therefore, one has $\mathbb{D}(a) = \emptyset$ for all $a \in \Lambda_\mu(F)$, which means that $\Lambda_\mu(F) = \Sigma$ is a set of monochromatic configurations.

- If $n = 4$, the result of [Fis90] shows that $[i]_0$ cannot be a μ-attractor for all i. Thus one has $\mathbb{D}_0(a) = \emptyset$ for all $a \in \Lambda_\mu(F)$, and $\Lambda_\mu(F) = \Sigma$ is a set of monochromatic configurations.
- If $n \geq 5$, the result of [Fis90] shows that $[i]_0$ is a μ-attractor for all i. Thus for some $a \in \Lambda_\mu(F)$ one has $\mathbb{D}(a) = \mathbb{D}_0(a) \neq \emptyset$. This means that they contain homogeneous regions separated by vertical lines.

5.2 Automaton #184

On the #184 "traffic" cellular automaton, we consider the defects according to $\Sigma = \{^\infty(01)^\infty, {}^\infty(10)^\infty\}$ (checkerboard pattern) with $P = 2$ (no interfaces). Since $\kappa_d = 1$, we write $\kappa(a, z) = i$ for $(1, i)$. If we define the phases $\varphi(^\infty(01)^\infty) = 0$ and $\varphi(^\infty(10)^\infty) = 1$, we can see that κ is unambiguous and that $\kappa(a, z) = 0$ if $a_z = z \mod 2$, 1 otherwise. We define the velocity function as $V(0, 0) = (1, 1)$ and $V(1, 1) = (-1, 1)$ (this corresponds to {□□} and {■■}, respectively).

For any $a \in \mathcal{A}^{\mathbb{Z}}$, we define ψ_a by

$$\forall d \in \mathbb{D}_{0,0}, \psi_a(d) = \begin{cases} \emptyset & \text{if } d + 2 \in \mathbb{D}_{1,1} \\ \{d + 1\} & \text{otherwise} \end{cases}$$

And symetrically for $d \in \mathbb{D}_{1,1}(a)$. Similarly, we can check that this interpretation is well-defined, respects the properties of locality, surjectivity and growth, is coalescent and respects the velocity function V. By applying the previous theorem, we have for all $\mu \in \mathcal{M}_\sigma^{erg}(\mathcal{A}^{\mathbb{Z}})$ either $\mathbb{D}_{0,0}(a) = \emptyset$ (checkerboard and monochromatic black patterns) or $\mathbb{D}_{1,1}(a) = \emptyset$ (checkerboard and monochromatic white patterns).

In particular, for the uniform Bernoulli measure μ, we consider the application $\gamma'((a_i)) = (1 - a_{-i})$, and we can see that $\mu(\gamma'([u])) = \mu([\overline{u^{-1}}]) = \mu([u])$, where $\overline{u_1 \dots u_n} = (1 - u_1) \dots (1 - u_n)$. But $d \in \mathbb{D}_{0,0}(a) \Leftrightarrow -d \in \mathbb{D}_{1,1}(\gamma'(a))$, and conversely; therefore, for all $a \in \Lambda_\mu(F)$, $\mathbb{D}_{0,0}(a) = \mathbb{D}_{1,1}(a) = \emptyset$. We deduce that $\Lambda_\mu(F)$ is the checkerboard subshift and by σ-invariance, $(F^n\mu)$ admits a single accumulation point $\frac{1}{2}\delta_{^\infty 01^\infty} + \frac{1}{2}\delta_{^\infty 10^\infty}$.

5.3 Captive One Sided Cellular Automata

Now consider a captive cellular automaton $F : \mathcal{A}^{\mathbb{Z}} \to \mathcal{A}^{\mathbb{Z}}$ of neighborhood $[0; 1]$, which means that the local rule $f : \mathcal{A}^{[0;1]} \to \mathcal{A}$ verifies $f(a_0a_1) \in \{a_0, a_1\}$. Captive cellular automata were introduced in [The04] and have some interesting algebraic propeties.

We consider the decomposition $\Sigma = \bigsqcup_{i \in \mathcal{A}} \Sigma_i$ where $\Sigma_i = \{^\infty i^\infty\}$ of periods $P_i = 1$ (no dislocations). We define the velocity function as $V(i, j) = (-1, 1)$ if $f(ij) = j$ and $V(i, j) = (0, 1)$ if $f(ij) = i$, and we define $\mathbb{D}_{-1}$ and $\mathbb{D}_0$ as in 5.1. For all $a \in \mathcal{A}^{\mathbb{Z}}$, we define:

$$\forall d \in \mathbb{D}_{-1}(a), \psi_a(d) = \begin{cases} \emptyset & \text{if } d - 1 \in \mathbb{D}_0 \\ \{d - 1\} & \text{otherwise} \end{cases}$$

and symetrically if $d \in \mathbb{D}_0$. As in the two previous examples, we can check that this is well-defined and respects the properties of locality, growth, surjectivity, coalescence and the velocity function.

Thus, for any σ-ergodic measure μ, $\Lambda_\mu(F)$ contains defects in one direction only. If moreover, for all $a, b \in \mathcal{A}$, the local rule verifies $f(ab) = f(ba)$ and μ verifies $\mu([ab]) = \mu([ba])$ (e.g. Bernoulli measures), we have $\Lambda_\mu(F) = \Sigma$.

6 Conclusion

In this article we have presented a formalism to link the notion of defect with respect to a subshift Σ introduced by M. Pivato and the emergence of homogeneous regions separated by defects when we iterate a random configuration. Under some assumptions on the collisions of defects, we proved the only defects that possibly remain in the μ-limit set all have the same direction. This explains the behaviour observed in simulations for large classes of cellular automata.

Acknowledgments

The authors are grateful to Guillaume Theyssier for pointing out the problematic around the captive cellular automata class. We also thank the project ANR EMC: ANR-09-BLAN-0164 for financial support.

References

[BNR91] Boccara, N., Nasser, J., Roger, M.: Particlelike structures and their interactions in spatiotemporal patterns generated by one-dimensional deterministic cellular-automaton rules. Phys. Rev. A 44(2), 866–875 (1991)

[BPT06] Boyer, L., Poupet, V., Theyssier, G.: On the Complexity of Limit Sets of Cellular Automata Associated with Probability Measures. In: Královič, R., Urzyczyn, P. (eds.) MFCS 2006. LNCS, vol. 4162, pp. 190–201. Springer, Heidelberg (2006)

[Elo94] Eloranta, K.: The dynamics of defect ensembles in one-dimensional cellular automata. Journal of Statistical Physics 76, 1377–1398 (1994), doi:10.1007/BF02187067

[Fis90] Fisch, R.: The one-dimensional cyclic cellular automaton: A system with deterministic dynamics that emulates an interacting particle system with stochastic dynamics. Journal of Theoretical Probability 3, 311–338 (1990), doi:10.1007/BF01045164

[HC97] Hanson, J.E., Crutchfield, J.P.: Computational mechanics of cellular automata: An example. Physica D: Nonlinear Phenomena 103(1-4), 169–189 (1997)

[Hed69] Hedlund, G.A.: Endomorphisms and automorphisms of the shift dynamical system. Theory of Computing Systems 3(4), 320–375 (1969)

[Hur90a] Hurley, M.: Attractors in cellular automata. Ergodic Theory Dynam. Systems 10(1), 131–140 (1990)

[Hur90b] Hurley, M.: Ergodic aspects of cellular automata. Ergodic Theory Dynam. Systems 10(4), 671–685 (1990)

[KM00] Kůrka, P., Maass, A.: Limit sets of cellular automata associated to probability measures. Journal of Statistical Physics 100(5), 1031–1047 (2000)

[Kůr03] Kůrka, P.: Cellular automata with vanishing particules. Fundamenta Informaticae 58, 1–19 (2003)

[Piv07a] Pivato, M.: Defect particle kinematics in one-dimensional cellular automata. Theoretical Computer Science 377(1-3), 205–228 (2007)

[Piv07b] Pivato, M.: Spectral domain boundaries in cellular automata. Fundamenta Informaticae 78(3), 417–447 (2007)

[The04] Theyssier, G.: Captive cellular automata. In: Fiala, J., Koubek, V., Kratochvíl, J. (eds.) MFCS 2004. LNCS, vol. 3153, pp. 427–438. Springer, Heidelberg (2004)

[Wal00] Walters, P.: An introduction to ergodic theory. Springer, Heidelberg (2000)

[Wol84] Wolfram, S.: Computation theory of cellular automata. Communications in Mathematical Physics 96(1), 15–57 (1984)

Chop Operations and Expressions: Descriptional Complexity Considerations

Markus Holzer and Sebastian Jakobi

Institut für Informatik, Universität Giessen,
Arndtstr. 2, 35392 Giessen, Germany
{holzer,jakobi}@informatik.uni-giessen.de

Abstract. The chop or fusion operation was recently introduced in [S. A. Babu, P. K. Pandya: Chop Expressions and Discrete Duration Calculus. *Modern Applications of Automata Theory*, World Scientific, 2010], where a characterization of regular languages in terms of chop expressions was shown. Simply speaking, the chop or fusion of two words is a concatenation were the touching letters are coalesced, if both letters are equal; otherwise the operation is undefined. We investigate the descriptional complexity of the chop operation and its iteration for deterministic and nondeterministic finite automata as well as for regular expressions. In most cases tight bounds are shown. Moreover, we also consider the conversion problem between finite automata, regular expressions, and chop expressions. Again, for most conversions we get tight bounds in order of magnitude. It is worth mentioning that regular expressions can be transformed into equivalent chop expressions of polynomial size, but chop expressions can be exponentially more succinct than regular expressions.

1 Introduction

Regular languages obey a legion of lovely characterizations. The most prominent ones are deterministic (DFAs) and nondeterministic finite automata (NFAs), and regular expressions (REs). Although these models are computationally equivalent, their descriptional complexity may vary significantly. It is well known that NFAs can offer exponential saving in space compared with DFAs. More precisely, if A is an n-state NFA, then 2^n states are sufficient and necessary in the worst case for a DFA to accept the language $L(A)$ [21,23,24,25]. For the other conversions, from regular expressions to finite automata a tight bound of $n+1$ states is shown in [10], and for converting NFAs to regular expressions an asymptotic bound of $2^{\Theta(n)}$ expression size (alphabetic width) is known [8,9,11]. The study of the descriptional complexity of certain devices is not limited to conversion problems. From the very beginning of the field of descriptional complexity operation problems were also investigated. That is, determine how the size of the description changes when applying an (regularity preserving) operation to the accepted language, staying with the same device type for accepting the new language. For instance, consider the Kleene star operation. There a tight bound of $3 \cdot 2^{n-2}$ states is known for DFAs [29] to accept the Kleene star of a language accepted

G. Mauri and A. Leporati (Eds.): DLT 2011, LNCS 6795, pp. 264–275, 2011.

Table 1. Descriptional complexity result for the chop $\odot$, chop-star $\otimes$, and chop-plus $\oplus$ operation on deterministic finite automata (DFA), nondeterministic finite automata (NFA), and regular expressions (RE); all bounds are tight. Here t is the number of accepting states of the "left" automaton.

Operation	DFA	NFA	RE
$\odot$	$m \cdot 2^n - t \cdot 2^{n-\min(\|\Sigma\|,n)} + 1$	$m+n$	$\|\Sigma\| \cdot O(n^2)$
$\otimes$	$2^n - 1 + \min(\|\Sigma\|, n)$	$n+2$	$2^{\Theta(n)}$, for $\|\Sigma\| = n$
$\oplus$	2^n	$n+1$	$2^{O(\|\Sigma\|)} \cdot O(n^4)$

by an n-state DFA. For NFAs this bound is $n+1$ on the number of states [17], and thus linear, and for regular expressions the operation under consideration is trivial. Some of these results emphasize even more the difference in descriptional complexity of certain devices, even if they are computationally equivalent.

Recently, an alternative characterization of regular languages in terms of so called *chop expressions* (CEs) was introduced in [1]. These expressions are inspired by logic, to be more precise by the duration calculus, which is an interval logic for real time systems. It was shown that the model checking problem for the discrete duration calculus can be reduced to so-called (extended) chop expressions (with tests). Simply speaking a CE is nothing but a RE, where the operations of concatenation and Kleene star are replaced by the chop operation and its iterated version. The *chop*[1] or *fusion* of two words is built by coalescing the last and first letter of the first and second word, respectively, if these letters coincide; otherwise the operation is not defined. Thus, the chop operation can be seen as a generalization of concatenation that allows one to verify whether the end of a the first word overlaps by a single letter with the beginning of the second word. The chop operation is well known in the logic community. Various other operations that are more or less closely related to the herein studied chop operation can be found in [4,6,16,19,20,22]. The focus in [1] was more on the connection to the duration calculus and on the computational complexity of CEs than on descriptional complexity issues. Nevertheless, a few upper bounds for conversion problems were shown, namely a linear upper bound for CEs to automata and an exponential upper bound for the conversion to REs. On the other hand, it turned out that polynomial size CEs can be build from REs. Whether these bounds are tight, was left open in [1]. We solve most of these open problems on the descriptional complexity of chop operations and expressions. Many upper bounds from that paper turn out to be optimal.

We concentrate on two topics: (i) the operation problem for the chop operation and its iteration, and (ii) on conversion problems between automata, REs, and CEs and *vice versa*. For the operation problem we obtain tight bounds in the exact number of states for DFAs and NFAs and tight asymptotic bounds for

[1] One should not confuse our operation with the chop operation discussed in [26, Example 8.21, page 181], which deletes the middle letter of a word if its length is odd, and is not regularity preserving.

Table 2. Descriptional complexity results for conversions between (non)deterministic finite automata (NFA), regular expressions (RE), and chop expressions (CE) and *vice versa*. For automata the size is the number of states, while for expressions the measure of alphabetic width is used (appropriately generalized to chop expressions). For comparison we have listed also the known conversions between NFAs and REs.

<table>
<tr><th>Convert ... to ...</th><th>NFA</th><th>RE</th><th>CE</th></tr>
<tr><td>NFA</td><td>–</td><td>$2^{\Theta(n)}$, for $|\Sigma| \geq 2$</td><td>$2^{O(n)}$, for $|\Sigma| = n^{O(1)}$</td></tr>
<tr><td>RE</td><td rowspan="2">$\Theta(n)$</td><td>–</td><td>$|\Sigma| \cdot O(n^2)$</td></tr>
<tr><td>CE</td><td>$2^{\Theta(n)}$, for $|\Sigma| = n$</td><td>–</td></tr>
</table>

REs. The former bounds on finite automata are similar to those bounds for concatenation, Kleene star, and Kleene plus. The main difference is that for some chop operations the size of the alphabet is an explicit parameter in some of our tight bounds. For instance, the chop of two languages over the alphabet Σ accepted by m- and n-state DFAs, respectively, requires at most $m \cdot 2^n - t \cdot 2^{n-\min(|\Sigma|,n)} + 1$ states, where t is the number of accepting states of the "left" automaton, while for NFAs $m+n$ states are sufficient. In both cases these bounds are tight in the exact number of states. For REs an upper bound of $|\Sigma| \cdot O(n^2)$ for the chop of two REs is determined. We summarize our results on the operation problem for chop operations in Table 1. More on the descriptional complexity of chop operations, in particular on unary and finite languages, can be found in [15]. For the conversion problems, we also obtain tight bounds. Here it turns out that chop expressions are exponentially more succinct than regular expressions, while still having a linear descriptional complexity bound for the conversion to finite automata. It is worth mentioning again, that regular expressions can be converted to equivalent chop expressions by only a polynomial increase in size. Our results on conversion problems are summarized in Table 2. Most proofs are omitted due to space constraints.

2 Definitions

We introduce some basic notions in formal language and automata theory—for a thorough treatment, the reader might want to consult a textbook such as [18]. In particular, let Σ be a finite alphabet and Σ^* the set of all words over Σ, including the *empty word* λ. The length of a word w is denoted by $|w|$, where $|\lambda| = 0$. A *formal language* over the alphabet Σ is a subset of Σ^*.

A *nondeterministic finite automaton* (NFA) is a quintuple $A = (Q, \Sigma, \delta, q_0, F)$, where Q is the finite set of *states*, Σ is the finite set of *input symbols*, $q_0 \in Q$ is the *initial state*, $F \subseteq Q$ is the set of *accepting states*, and $\delta : Q \times \Sigma \to 2^Q$ is the *transition function*, where 2^Q refers to the powerset of Q. If $p \in \delta(q, a)$, for $q \in Q$ and $a \in \Sigma$, then we say that (q, a, p) is a *transition* of A. As usual the transition function is extended to $\delta : Q \times \Sigma^* \to 2^Q$ reflecting sequences of inputs: $\delta(q, \lambda) = \{q\}$ and $\delta(q, aw) = \bigcup_{p \in \delta(q,a)} \delta(p, w)$, for $q \in Q$, $a \in \Sigma$, and

$w \in \Sigma^*$. A word $w \in \Sigma^*$ is *accepted* by A if $\delta(q_0, w) \cap F \neq \emptyset$. The *language accepted* by A is $L(A) = \{ w \in \Sigma^* \mid w \text{ is accepted by } A \}$. A finite automaton is *deterministic* (DFA) if and only if $|\delta(q, a)| = 1$, for all $q \in Q$ and $a \in \Sigma$. In this case we simply write $\delta(q, a) = p$ for $\delta(q, a) = \{p\}$ assuming that the transition function is a mapping $\delta : Q \times \Sigma \to Q$. So, any DFA is complete, that is, the transition function is total, whereas for NFAs it is possible that δ maps to the empty set.

It is well known that finite automata and *regular expressions* (REs) are equally powerful, i.e., for every finite automaton one can construct an equivalent regular expression and *vice versa*. Here we allow both $\emptyset$ and λ as regular expression primitives. The *size* or *alphabetic width* of a RE r over the alphabet Σ, denoted by $\text{alph}(r)$, is defined as the total number of occurrences of letters of Σ in r. For a regular language L, we define its alphabetic width, $\text{alph}(L)$, as the minimum alphabetic width among all REs describing L.

We investigate the descriptional complexity of the chop operation, which was recently used in [1] to describe chop expressions—here we use the notation from [7]: the *chop* or *fusion* of two words u and w in Σ^* is defined as $u \odot v = u'av'$ if $u = u'a$ and $v = av'$, for $u', v' \in \Sigma^*$, and $a \in \Sigma$, otherwise $u \odot v$ is undefined. This is extended to languages by $L_1 \odot L_2 = \{ u \odot v \mid u \in L_1 \text{ and } v \in L_2 \}$, for languages $L_1, L_2 \subseteq \Sigma^*$. We like to point out that this operation differs from the *latin product* defined in [22] only if the words have no overlap. In that case, the latin product is just the concatenation of both words. For the chop iteration, we set the ith iteration, for $i \geq 0$, to $L^{\otimes_0} = \Sigma$, and $L^{\otimes_i} = L \odot L^{\otimes_{i-1}}$, for $i \geq 1$. Then the *iterated chop* or *chop-star* of a language L is defined as $L^{\otimes} = \bigcup_{i \geq 0} L^{\otimes_i}$. Moreover the chop-plus is denoted by $L^{\oplus} = \bigcup_{i \geq 1} L^{\otimes_i}$. It is easy to see that the chop operation $\odot$ is associative, and by definition the set Σ acts as the neutral element on all languages L from Σ^+. This is compatible with the definition of chop-star, because $\emptyset^{\otimes} = \Sigma$. In general, an application of the chop operation with Σ will cancel λ from the language L. Therefore, for every language $L \subseteq \Sigma^*$, we have $\Sigma \odot L = L \odot \Sigma = L \setminus \{\lambda\}$.

Finally we introduce chop expressions [1] in a similar vein as regular expressions. More precisely, the *chop expressions* (CEs) over an alphabet Σ and the languages they describe are defined inductively as follows:[2] $\emptyset$, λ, every letter a with $a \in \Sigma$, and Δ are CEs, and when s and t are CEs, then $(s + t)$, $(s \odot t)$, and $(s)^{\otimes}$ are also CEs. The language defined by a CE r, denoted by $L(r)$, is defined as follows: $L(\emptyset) = \emptyset$, $L(\lambda) = \{\lambda\}$, $L(a) = \{a\}$, $L(\Delta) = \Sigma^2$, $L(s+t) = L(s) \cup L(t)$, $L(s \odot t) = L(s) \odot L(t)$, and $L(s^{\otimes}) = L(s)^{\otimes}$. In accordance with REs we define the *size* of a CE r over the alphabet Σ to be the total number of occurrences of letters of Σ plus the number of occurrences of Δ symbols in r. In our definition we have slightly deviated from that given in [1] also allowing λ as a primitive element. Otherwise one could describe λ-free regular languages only. Hence our CEs are equally powerful to ordinary REs.

---

2 As for regular expression, for convenience, parentheses in chop expressions are sometimes omitted. The priority of operators is specified in the usual fashion: chop is performed before union, and chop-star before both chop and union.

3 State Complexity of Chop Operations

In this section we investigate the state complexity of the operations $\odot$, $\otimes$, and $\oplus$. Since $\odot$ is some sort of concatenation, the results in this section are very similar to those for concatenation and its iteration. Let us briefly recall the known result from the literature for comparison reasons. For DFAs with m and n states, respectively, tight bounds of $m \cdot 2^n - t \cdot 2^{n-1}$ for concatenation, where t is the number of accepting states of the "left" automaton, and $3 \cdot 2^{n-2}$ for Kleene star and Kleene plus in the exact number of states for DFAs were shown in [29]. For NFAs the situation is slightly different as proven in [17], where tight bounds of $m + n$ for concatenation, $n + 1$ for Kleene star, and n for Kleene plus can be found for NFAs. In fact, we obtain a similar scenario since the chop operation variants will be cheap for NFAs but costly for DFAs. On the other hand, some of the tight bounds for the considered operations will depend on the size of the alphabet. We first give constructions for the operation under consideration for NFAs and show that they are best possible. Then, in the next subsection, the constructed NFAs will be determinized to obtain (tight) upper bounds for DFAs.

3.1 Nondeterministic State Complexity

In [1] it is shown how to construct an NFA for the chop of two languages. We prove this construction to be optimal in general.

Theorem 1. *Let A be an m-state and B be an n-state nondeterministic finite automaton for any integers $m, n \geq 1$. Then $m + n$ states are sufficient and necessary in the worst case for any nondeterministic finite automaton to accept the language $L(A) \odot L(B)$.* □

Next we show how to construct an NFA for the chop-star language of any given NFA. Our construction is a slight modification of the one given in [1], which works for non-returning NFAs only—an automaton is non-returning if the initial state doesn't have any ingoing transitions.

Theorem 2. *Let A be an n-state nondeterministic finite automaton for $n \geq 4$. Then $n + 2$ states are sufficient and necessary in the worst case for any nondeterministic finite automaton to accept $L(A)^{\otimes}$.*

Proof. We start by giving a construction that provides the upper bound. Given an NFA $A = (Q, \Sigma, \delta, q_0, F)$ with $|Q| = n$, we define $A^{\otimes} = (Q \cup \{s, f\}, \Sigma, \delta^{\otimes}, s, F^{\otimes})$, with $F^{\otimes} = F \cup \{f\}$, and transition function $\delta^{\otimes}$ such, that for all states $p, q \in Q$ and $a \in \Sigma$ we have $\delta(p, a) \subseteq \delta^{\otimes}(p, a)$, and additionally (i) $q \in \delta^{\otimes}(s, a)$ if $q \in \delta(q_0, a)$, (ii) $f \in \delta^{\otimes}(s, a)$ if $\delta(q_0, a) \cap F = \emptyset$, (iii) $q \in \delta^{\otimes}(p, a)$ if $\delta(p, a) \cap F \neq \emptyset$ and $q \in \delta(q_0, a)$. The proof of $L(A^{\otimes}) = L(A)^{\otimes}$ is left to the reader.

For the lower bound consider the language $(a^n)^*$, for $n > 3$, accepted by a minimal finite automaton having a circle of states $0, 1, \ldots, n-1$ with the transitions $(q, a, (q + 1) \bmod n)$. State 0 is the initial and sole accepting state. Constructing an NFA for the corresponding chop-star language L as described above, results in two additional states s and f, where s is the new initial and f is a new accepting state, and additionally we get the transitions $(n-1, a, 1)$, $(s, a, 1)$, and (s, a, f). One can show the minimality of this automaton, using the fooling set technique [2]. It remains to prove that

$$S = \{ (a^i, a^{n-i}) \mid 2 \leq i \leq n-2 \} \\ \cup \{(\lambda, a), (a, \lambda), (a^{n-1}, a^{n+1}), (a^{n+1}, a^{n-1}), (a^n, a^n)\}$$

is a fooling set for L. The tedious details are left to the reader. □

Since $\oplus$ does not add Σ to the language, there is no need for the additional accepting state f as in the construction for $\otimes$. So we obtain the following.

Theorem 3. *Let A be an n-state nondeterministic finite automaton for $n \geq 1$. Then $n+1$ states are sufficient and necessary in the worst case for any nondeterministic finite automaton to accept $L(A)^{\oplus}$.*

3.2 Deterministic State Complexity

The foregoing constructions will be adapted *via* subset construction to DFAs. As for (iterated) concatenation, we get an exponential blow-up for all three operations. But unlike (iterated) concatenation on DFAs, for the chop and chop-star operations on DFAs, the obtained bounds will also depend on the size of the alphabet.

Theorem 4. *Let A be an m-state and B be an n-state deterministic finite automaton, both over alphabet Σ with $m \geq 2$, and $n, |\Sigma| \geq 3$. Let F_1 be the set of final states of A. Then $m \cdot 2^n - |F_1| \cdot 2^{n-\min(|\Sigma|,n)} + 1$ states are sufficient and necessary in the worst case for any deterministic finite automaton to accept $L(A) \odot L(B)$.*

Proof. The upper bound can be seen as follows: Let $A_i = (Q_i, \Sigma, \delta_i, s_i, F_i)$, for $i \in \{1, 2\}$, be two DFAs and $A = (Q, \Sigma, \delta, q_0, F)$ be a DFA with $Q = Q_1 \times 2^{Q_2}$, $q_0 = \langle s_1, \emptyset \rangle$, $F = \{ \langle q, P \rangle \mid P \cap F_2 \neq \emptyset \}$, and

$$\delta(\langle q, P \rangle, a) = \begin{cases} \langle \delta_1(q, a), \bigcup_{p \in P} \delta_2(p, a) \rangle & \text{if } \delta_1(q, a) \notin F_1, \\ \langle \delta_1(q, a), \bigcup_{p \in P} \delta_2(p, a) \cup \delta_2(s_2, a) \rangle & \text{if } \delta_1(q, a) \in F_1. \end{cases}$$

It should be clear that $L(A) = L(A_1) \odot L(A_2)$ and $|Q| = |Q_1| \cdot 2^{|Q_2|}$. This bound can be further improved, since some states may not be reachable. In the following let $k = \min(|\Sigma|, n)$. Note that for all reachable states $\langle q, P \rangle \in Q_1 \times 2^{Q_2}$ different from the initial state $\langle s_1, \emptyset \rangle$, we have $q \notin F_1$ or $P \cap \{ \delta_2(s_2, a) \mid a \in \Sigma \} \neq \emptyset$. So, out of all states in $Q_1 \times 2^{Q_2}$, besides the initial state, we can reach at most $(m - |F_1|) \cdot 2^n$ states where the first component is not in F_1, and at most $|F_1| \cdot (2^k - 1) \cdot 2^{n-k}$ states with first component an element from the set F_1 and the second component contains a nonempty subset of states that are a successor of the initial state from A_2. Together with the initial state we get at most $m \cdot 2^n - |F_1| \cdot 2^{n-k} + 1$ states.

We now prove that this bound is tight, where it suffices to consider the alphabet $\Sigma = \{a, b, c\} \cup \Sigma_d$ for some possibly empty $\Sigma_d \subseteq \{d_4, d_5, \ldots, d_k\}$. Define DFAs $A_i = (Q_i, \Sigma, \delta_i, 1, F_i)$ for $i \in \{1, 2\}$ with state sets $Q_1 = \{1, 2, \ldots, m\}$ and $Q_2 = \{1, 2, \ldots, n\}$, final states $F_1 = Q_1 \setminus \{2, \ldots, t\}$ with $2 \leq t \leq m$ and $F_2 = \{1\}$ and transitions as follows. For any state $q \in Q_1$, A_1 goes to $(q \bmod m) + 1$ on input a, and stays in q on all other inputs. In any state $q \in Q_2$, A_2 stays in q on input a, and goes to $(q \bmod n) + 1$ on input b; input c takes A_2 from state q to state $(q \bmod n) + 1$ if $4 \leq q \leq n$ and to state 3 if $q \leq 3$; on inputs d_i, automaton A_2 stays in state q for any $q \in Q_2 \setminus \{1\}$, while state 1 leads A_2 to state i.

Let A be the DFA for $L(A_1) \odot L(A_2)$ with transition function δ constructed as above. We first prove that any distinct states $\langle q, P\rangle$ and $\langle q', P'\rangle$ are pairwise inequivalent, starting with the case $P \neq P'$. In the forthcoming $\triangle$ refers to the operation of the symmetric difference of two sets. If $1 \in P \triangle P'$, both states are inequivalent. Otherwise, for $r = \max(P \triangle P')$, the states can be distinguished by reading the word b^{n-r+1}. Consider now states $\langle q, P\rangle$ and $\langle q', P\rangle$ with $q \neq q'$. By reading enough a symbols, we may assume $q = 2$ and $1 \in P$, and after reading enough c symbols, we reach the states $\langle 2, \{3\}\rangle$ and $\langle q', \{3\}\rangle$. If $q' \in F_1$, then input b leads to $\langle 2, \{4\}\rangle$ and $\langle q', \{2, 4\}\rangle$, which are already proven to be inequivalent. If $q' \notin F_1$, then $2 < q' \leq t$, and the states can be distinguished by reading $a^{t+1-q'}$.

Finally we prove that exactly $m \cdot 2^n - |F_1| \cdot 2^{n-k} + 1$ states are reachable in A. In detail, we show that all $(m - |F_1|) \cdot 2^n$ states of the form $\langle q, P\rangle$ with $q \notin F_1$ and all $|F_1| \cdot (2^k - 1) \cdot 2^{n-k}$ states $\langle q, P\rangle$ with $P \cap \{\, \delta_2(1, e) \mid e \in \Sigma \,\} \neq \emptyset$ and $q \in F_1$ are reachable from the initial state $\langle 1, \emptyset\rangle$, which proves the claim. Let $q \notin F_1$, then $\langle q, \emptyset\rangle$ is reachable with the word a^{q-1}. Continuing by induction, assume all states $\langle q, P\rangle$ with $q \notin F_1$ and $|P| \leq i$ to be reachable, and consider $q' \in Q_1 \setminus F_1$ and $P' = \{p_0, p_1, \ldots, p_i\} \subseteq Q_2$ with $1 \leq p_0 < p_1 < \cdots < p_i \leq n$. Then $\langle q', P'\rangle$ is reachable from $\langle 2, \{p_1 - (p_0 - 1), \ldots, p_i - (p_0 - 1)\}\rangle$ by reading $a^{m+q'-2}b^{p_0-1}$. So all states $\langle q, P\rangle$ with $q \notin F_1$ and $P \subseteq Q_2$ are reachable. Now let $q \in F_1$ and $S = P \cap \{\, \delta_2(1, e) \mid e \in \Sigma \,\} \neq \emptyset$. If $1 \in P$, then $\langle q, P\rangle = \delta(\langle 2, P\rangle, a^{m+q-2})$. If $\delta_2(q, d_i) \in P$ for some $i \geq 4$, then $\langle q, P\rangle$ is reachable from $\langle 2, P\rangle$ with the word $a^{m+q-2}d_i$. Otherwise it must be $S \subseteq \{2, 3\}$. Let $P_{-1} = \{p - 1 \mid p \in P\}$, then $\langle q, P\rangle$ is reachable from $\langle 2, P_{-1}\rangle$ with the word $a^{m+q-2}b$ if $2 \in S$, and with $a^{m+q-2}c$, if $S = \{3\}$. This concludes our proof. □

We now investigate deterministic state complexity of $\otimes$, where we get an exponential blow-up of states. We prove the upper bound to be tight already for a binary alphabet, which is best possible in the sense that for unary languages, the upper bound is strictly less than in the general case [15].

Theorem 5. *Let A be an n-state deterministic finite automaton with $n \geq 3$ over an alphabet Σ with $|\Sigma| \geq 2$. Then $2^n - 1 + \min(|\Sigma|, n)$ states are sufficient and necessary in the worst case for a deterministic finite automaton to accept the language $L(A)^{\otimes}$.* □

For $\oplus$ there is no dependency on the alphabet size since, unlike $\otimes$, the operation $\oplus$ does not add the elements of the alphabet to the resulting language.

Theorem 6. *Let A be an n-state deterministic finite automaton for $n \geq 3$. Then 2^n states are sufficient and necessary in the worst case for a deterministic finite automaton to accept $L(A)^{\oplus}$.* □

4 Complexity of Chop Expressions

We consider the descriptional complexity of conversion results for CEs to equivalent finite automata or REs, and *vice versa.* Briefly recall what is known for REs and finite automata. The simplest way to transform REs into equivalent finite automata is Thompson's construction [28], which yields an NFA with λ-transitions with a linear number of states, which in turn is transformed into an ordinary NFA without increasing the number of states. Simple examples, such as the singleton set $\{a^n\}$ show that this bound is asymptotically tight. For the backward conversion, i.e., from finite automata to equivalent REs, most classical approaches are reformulations [27] of the same underlying algorithmic idea of *state elimination* [3], which produces REs of alphabetic width at most $2^{O(n)}$, where n is the number of states of the finite automaton over an alphabet polynomial in n. While this exponential bound was known to be tight for growing size alphabets [8], recent research effort [9,11] resulted in a tight lower bound of $2^{\Omega(n)}$, even for constant alphabets with at least two letters.

Now let us turn to the descriptional complexity of CEs. Due to the similarities in the definition of chop and regular expressions it is obvious that Thompson's construction applies to CEs as well, if appropriately adapted, leading to an equivalent NFA with a linear number of states—this conversion approach is also undertaken in [1]. Since the singleton set $\{a^n\}$ can be written as $a \odot (\Delta \odot a)^{\otimes_{n-1}}$ in the form of a CE, this linear bound is asymptotically tight.

Theorem 7. *Let r be a chop expression of size n. Then $O(n)$ states are sufficient and necessary in the worst case for a nondeterministic finite automaton to accept the language $L(r)$.* □

What about the other conversion direction from finite automata to CEs? Before we discuss this direction in more detail, we take a closer look at conversions between expressions. The conversion of REs to equivalent CEs results in expressions of polynomial size, where alphabet size appears as a parameter. This was shown in [1]. For the convenience of the reader we recall this theorem.

Theorem 8. *Let r be a regular expression over the alphabet Σ of alphabetic width n. Then size $|\Sigma| \cdot O(n^2)$ is sufficient for a chop expression describing the language $L(r)$.*

This result can be used for the conversion of finite automata to equivalent chop expressions by converting to regular expressions first, and then applying Theorem 8. This naive approach results in an upper bound of $2^{O(n)}$, where n is the number of states of the finite automaton.

Theorem 9. *Let A be an n-state finite automaton over an alphabet polynomial in n. Then size $2^{O(n)}$ is sufficient for a chop expression to describe the language $L(A)$.* □

We come back to conversion problems between expressions. For the backward conversion from CEs of size n to equivalent REs an exponential upper bound of $2^{n^{O(1)}}$ was shown in [1]. Here the naive approach to construct a RE by converting the given CE into an NFA, and in turn into a RE is doing better because it gives an exponential upper bound of $2^{O(n)}$. Thus we can state the following:

Theorem 10. *Let r be a chop expression of size n over an alphabet of size polynomial in n. Then alphabetic width $2^{O(n)}$ is sufficient for a regular expression to describe the language $L(r)$.* □

Now the question arises whether this bound is tight in order of magnitude. Our arguments on lower bounds for the alphabetic width of regular languages is based on a recent result that utilizes the star height of regular languages [11]. The star height of a regular language L, denoted by $h(L)$, is then defined as the minimum star height among all REs describing L.

Theorem 11. *Let $L \subseteq \Sigma^*$ be a regular language. Then* $\mathrm{alph}(L) \geq 2^{\frac{1}{3}(h(L)-1)}-1$.

The star height of a regular language appears to be more difficult to determine than its alphabetic width, see, e.g., [13]. Fortunately, the star height is known for some languages. For instance, the next theorem is due to [5].

Theorem 12. *Let J_n be the complete digraph on n vertices with self-loops, where each edge (i, j) carries a unique label a_{ij}. Let W_n denote the set of all walks $a_{i_0 i_1} a_{i_1 i_2} \cdots a_{i_{r-2} i_{r-1}} a_{i_{r-1} i_r}$ in J_n, including the empty walk λ. Then the star height of language W_n equals n.*

This means that any RE describing W_n requires at least $2^{\Omega(n)}$ alphabetic width. This exponential lower bound holds even if we apply a star height preserving morphism σ to W_n in order to decrease the size of the alphabet—for star height preserving morphisms we refer to [14]. We utilize these facts in the next theorem, proving an exponential lower bound for the conversion of CEs into equivalent REs.

Theorem 13. *There exists an infinite family of languages L_n over an alphabet of size n having chop expressions of size $O(n)$, such that the alphabetic width of any regular expression describing L_n is at least $2^{\Omega(n)}$.*

Proof. We show that a star height preserving encoding of the set of walks W_n in a complete n-vertex digraph as defined in Theorem 12 allows a compact representation using CEs. Let $E = \{ a_{ij} \mid 1 \leq i, j \leq n \}$ denote the set of edges in the digraph. Then we define the alphabet $\Sigma = \{ a_i \mid 1 \leq i \leq n \}$ and the star height preserving morphism $\sigma : E^* \to \Sigma^*$ by $\sigma(a_{ij}) = a_i a_j$, for $a_{ij} \in E$. Now we show how to describe $\sigma(W_n)$ in terms of a linear size CE. Then consider the finite set $M = \{ abb \mid a, b \in \Sigma \}$ and observe that

$$\sigma(W_n) = M^{\otimes} \odot \Delta + \lambda = M^{\oplus} \odot \Delta + \Delta + \lambda, \tag{1}$$

since the iterated chop forces to fuse the same letters in a possible extension of a valid walk described by a word with an element of M. Finally, we have to describe M by a CE. This expression reads as $\sum_{a\in\Sigma} \Delta \odot a \odot \Delta \odot a$ and has size $4n$. Since by Theorems 11 and 12 the set $\sigma(W_n)$ has alphabetic width at least $2^{\Omega(n)}$ w.r.t. REs, but CE size at most $O(n)$, the proof is complete. □

Obviously, the RE $(\sum_{a\in\Sigma} a)(\sum_{b\in\Sigma} bb)$ of alphabetic width $3n$ describes the set M from the previous proof. Because of Equation (1) both languages $M^{\otimes}$ and $M^{\oplus}$ must still have alphabetic width at least $2^{\Omega(n)}$, because otherwise we would get a contradiction on the alphabetic width for $\sigma(W_n)$ and W_n, respectively. Thus, we can state the following corollary.

Corollary 14. *There exists an infinite family of languages L_n over an alphabet of size n with $\text{alph}(L_n) = O(n)$, such that $\text{alph}(L_n^{\otimes}) = 2^{\Omega(n)}$. A similar result holds when considering the alphabetic width of $L_n^{\oplus}$ instead.* □

The corresponding upper bound of alphabetic width $2^{O(n)}$ for a RE describing $L(r)^{\otimes}$ or $L(r)^{\oplus}$ for any RE r of alphabetic width n is seen as follows. First convert r into an $O(n)$-state NFA A for the language $L(r)$, then apply the $\otimes$ or $\oplus$ construction described earlier, which results in an $O(n)$-state NFA accepting $L(r)^{\otimes}$ or $L(r)^{\oplus}$, respectively, which in turn can be converted to a RE of alphabetic width of at most $2^{O(n)}$. Together with Corollary 14 we obtain:

Corollary 15. *Let r be regular expression of alphabetic width n over an alphabet of size n. Then alphabetic width $2^{\Theta(n)}$ is sufficient and necessary in the worst case for a regular expression to describe the language $L(r)^{\otimes}$. A similar tight bound holds when describing the language $L(r)^{\oplus}$ instead.* □

An immediate question is whether the size of the alphabet can be further improved in the previous corollary to, e.g., binary alphabet. Unfortunately it seems that the lower bound proof breaks down, because $\otimes$ can only fuse letters and not small length subwords. For the slightly easier setting, the chop-star problem for REs, as described in Corollary 14 we can prove that this is in fact the case. To this end we need some notation. The (left) derivative of a language L with respect to a letter a, written as $a^{-1}L$, is defined as $\{ x \mid ax \in L \}$. The right derivative of L by a, written as La^{-1}, is similarly defined. The next theorem is from [12].

Theorem 16. *Let r be a regular expression of alphabetic width n denoting the language $L \subseteq \Sigma^*$, and let $a \in \Sigma$. Then there is a regular expression of alphabetic width $O(n^2)$ denoting $a^{-1}L$. A similar statement is valid for regular expressions describing La^{-1}.*

Consider the chop operation problem for REs. Let $r = s \odot t$ for two REs s and t. Then there are two REs s_a and ${}_at$ describing the languages $L(s)a^{-1}$ and $a^{-1}L(t)$, respectively, and we have $L(r) = \sum_{a\in\Sigma}(L(s_a) \cdot a) \cdot L({}_at)$. Since by Theorem 16 both REs s_a and ${}_at$ are of polynomial size the chop operation problem for REs results in an upper bound of $|\Sigma| \cdot O(n^2)$. The idea for the simulation of the chop operation can be generalized to the chop-star problem for REs. We do so in the proof of the next theorem.

Theorem 17. *Let r be a regular expression over the alphabet Σ of alphabetic width n. Then alphabetic width $2^{O(|\Sigma|)} \cdot O(n^4)$ is sufficient for a regular expression to describe the language $L(r^{\otimes})$.* □

Therefore the iterated chop of a regular language over a constant size alphabet transforms into an equivalent ordinary RE of polynomial size. This is in sharp contrast to the lower bound for growing size alphabet as given in Corollary 14. We believe that a similar statement is valid for the conversion of CEs in general to equivalent REs, too, but we have no proof yet. It is worth mentioning that a positive answer to this would induce a tight exponential lower bound of $2^{\Omega(n)}$ for the conversion of finite automata to equivalent CEs—compare with Theorem 9. This is seen as follows: in [11] it was shown, that there exists an infinite family of n-state finite automata A_n over a binary alphabet, such that $\text{alph}(L(A_n)) = 2^{\Omega(n)}$. If the conversion of CEs over a *constant* alphabet into an equivalent REs would be polynomial, the conversion of a finite automata into an equivalent REs *via* CEs must induce that the first conversion, namely that from the automaton to the CE, is at least $2^{\Omega(n)}$. Otherwise, this would contradict the aforementioned result of [11]. We have to leave open, whether this is in fact the case.

References

1. Babu, S.A., Pandya, P.K.: Chop expressions and discrete duration calculus. In: D'Souza, D., Shankar, P. (eds.) Modern Applications of Automata Theory. IISc research Monographs Series, vol. 2. World Scientific, Singapore (2010)
2. Birget, J.-C.: Intersection and union of regular languages and state complexity. Inform. Process. Lett. 43, 185–190 (1992)
3. Brzozowski, J.A., McCluskey, E.J.: Signal flow graph techniques for sequential circuit state diagrams. IEEE Trans. Comput. C-12(2), 67–76 (1963)
4. Cărăuşu, A., Păun, G.: String intersection and short concatenation. Rev. Roumaine Math. Pures Appl. 26, 713–726 (1981)
5. Cohen, R.S.: Star height of certain families of regular events. J. Comput. System Sci. 4(3), 281–297 (1970)
6. Domaratzki, M.: Minimality in Template-Guided Recombination. Inform. and Comput. 207, 1209–1220 (2009)
7. Harel, D., Peleg, D.: Process logic with regular formulas. Theoret. Comput. Sci. 38, 307–322 (1985)
8. Ehrenfeucht, A., Zeiger, H.P.: Complexity measures for regular expressions. J. Comput. System Sci. 12(2), 134–146 (1976)

9. Gelade, W., Neven, F.: Succinctness of complement and intersection of regular expressions. In: STACS, Bordeaux, France, pp. 325–336. Schloss Dagstuhl–Leibniz-Zentrum für Informatik, Dagstuhl (2008)
10. Glushkov, V.M.: The abstract theory of automata. Russian Math. Surveys 16, 1–53 (1961)
11. Gruber, H., Holzer, M.: Finite automata, digraph connectivity, and regular expression size. In: Aceto, L., Damgård, I., Goldberg, L.A., Halldórsson, M.M., Ingólfsdóttir, A., Walukiewicz, I. (eds.) ICALP 2008, Part II. LNCS, vol. 5126, pp. 39–50. Springer, Heidelberg (2008)
12. Gruber, H., Holzer, M.: Language operations with regular expressions of polynomial size. Theoret. Comput. Sci. 410(35), 3281–3289 (2009)
13. Hashiguchi, K.: Algorithms for determining the relative star height and star height. Inform. Comput. 78(2), 124–169 (1988)
14. Hashiguchi, K., Honda, N.: Homomorphisms that preserve star height. Inform. Comput. 30(3), 247–266 (1976)
15. Holzer, M., Jakobi, S.: State complexity of chop operations on unary and finite languages (2011) (in preparation)
16. Holzer, M., Jakobi, S., Kutrib, M.: The chop of languages. In: AFL, Debrecen, Hungary (to appear, 2011)
17. Holzer, M., Kutrib, M.: Nondeterministic descriptional complexity of regular languages. Internat. J. Found. Comput. Sci. 14(6), 1087–1102 (2003)
18. Hopcroft, J.E., Ullman, J.D.: Introduction to Automata Theory, Languages and Computation. Addison-Wesley, Reading (1979)
19. Ito, M., Lischke, G.: Generalized periodicity and primitivity. Math Logic Q 53, 91–106 (2007)
20. Kallas, J., Kufleitner, M., Lauser, A.: First-order fragments with successor over infinite words. In: STACS, Dortmund, Germany, pp. 356–367. Schloss Dagstuhl–Leibniz-Zentrum für Informatik, Dagstuhl (2011)
21. Maslov, A.N.: Estimates of the number of states of finite automata. Soviet Math. Dokl. 11, 1373–1375 (1970)
22. Mateescu, A., Salomaa, A.: Parallel composition of words with re-entrant symbols. An. Univ. Bucuresti, Mat.-Inform. 45, 71–80 (1996)
23. Meyer, A.R., Fischer, M.J.: Economy of description by automata, grammars, and formal systems. In: SWAT, pp. 188–191. IEEE Computer Society Press, Los Alamitos (1971)
24. Moore, F.R.: On the bounds for state-set size in the proofs of equivalence between deterministic, nondeterministic, and two-way finite automata. IEEE Trans. Comput. C-20, 1211–1219 (1971)
25. Rabin, M.O., Scott, D.: Finite automata and their decision problems. IBM J. Res. Dev. 3, 114–125 (1959)
26. Rich, E.A.: Automata, Computability, and Complexity: Theory and Applications. Prentice Hall, Englewood Cliffs (2007)
27. Sakarovitch, J.: The language, the expression, and the (small) automaton. In: Farré, J., Litovsky, I., Schmitz, S. (eds.) CIAA 2005. LNCS, vol. 3845, pp. 15–30. Springer, Heidelberg (2006)
28. Thompson, K.: Regular expression search algorithm. Com. ACM 11(6), 419–422 (1968)
29. Yu, S.: State complexity of regular languages. J. Autom. Lang. Comb. 6, 221–234 (2001)

Nodes Connected by Path Languages

Markus Holzer, Martin Kutrib, and Ursula Leiter

Institut für Informatik, Universität Giessen,
Arndtstraße 2, 35392 Giessen, Germany
{holzer,kutrib}@informatik.uni-giessen.de

Abstract. We investigate reachability problems on different types of labeled graphs constrained to formal languages from a family $\mathscr{L}$. If every language in $\mathscr{L}$ is accepted by a one-way nondeterministic storage automaton, then we give an appealing characterization of the computational complexity of the labeled graph reachability problem in terms of two-way nondeterministic storage automata with auxiliary worktape that is logarithmic-space bounded. Moreover, we also consider acyclic graphs in the underlying reachability instance, obtaining a lower bound result for auxiliary storage automata that are simultaneously space and time restricted.

1 Introduction

Many problems in computer science can be reformulated as (labeled) graph reachability or accessibility problems, which are defined as follows. Given a graph G with designated nodes s and t and with a single label attached to each edge, and a language L from the family $\mathscr{L}$, does there exist a path from node s to node t such that the concatenation of the labels along the path belongs to L. For instance, in [7] the complexity of the solution concept of iterated (weak) dominance on self-anonymous strategic games with two actions is shown to be closely related to a labeled grid graph reachability problem (GGR), where the solution path must fulfill a certain matching condition described by means of a regular expression. Another example where labeled graph reachability problems play an important role is the field of model checking (see for example [3]), which has gained a lot of interest in the research community during the last two decades. It is not surprising that the complexity of (labeled) graph reachability problems vary for different path forms such as, paths in general, shortest paths, simple paths, etc., and for different formal language families. From the literature it is known that the labeled graph reachability problem (GR) constrained to context-free languages is P-complete in general, and LOGCFL-complete if G is acyclic [11,19]. Restricted to shortest paths the problem remains efficiently solvable in polynomial time, while it becomes NP-complete for simple paths [4], even when considering regular instead of context-free languages. As seen from this examples, there are a lot of parameters such as, graph topology, formal language family, path property, which significantly influence the computational complexity of (labeled) graph reachability.

G. Mauri and A. Leporati (Eds.): DLT 2011, LNCS 6795, pp. 276–287, 2011.

The complexity of *unlabeled* graph reachability is partially solved and related to complexity classes in the hierarchy $\mathsf{AC}^0 \subset \mathsf{NC}^1 \subseteq \mathsf{L} = \mathsf{SL} \subseteq \mathsf{NL} \subseteq \mathsf{P}$. All the relationships depicted in the inclusion chain are known for a quarter of a century, except for $\mathsf{L} = \mathsf{SL}$, shown in [18]. Alternatively the class NL can be characterized by the sets of problems logarithmic-space reducible to the directed graph reachability (GR) problem. Similarly, the class SL is characterized by the sets of problems logarithmic-space reducible to the undirected graph reachability (UGR) problem. Besides general graphs, reachability on grid graphs is of particular interest. A grid graph is an $n \times n$ grid of nodes such that an edge only connects immediate vertical or horizontal neighbors. The undirected grid graph reachability problem (UGGR) is solvable in deterministic logarithmic-space [6], which has been known long before is was shown that the general undirected graph reachability problem (UGR) is in L [18]. In contrast to UGR, which is actually L-complete [18], the problem UGGR is only known to be NC^1-hard [2] and, thus, seems to be of lower complexity, because even the general grid graph reachability problem (GGR) is not known to be hard for L under AC^0 reductions.

In this paper our focus is on the influence of graph topology and formal language family to the complexity of labeled graph reachability. We study this question in the general setting of abstract storage automata, which were introduced in [10]. Simply speaking, an abstract storage automata is a device with a finite control and an additional storage such as, for example, a pushdown, that can be manipulated by a finite number of operations. Abstract storage automata obey the attractive closure property of a (full) TRIO [13], namely closure under intersection with regular sets, inverse homomorphism, and homomorphism. In a first step we prove that the graph topology is irrelevant when the formal language family $\mathscr{L}$ is an effectively closed TRIO, that is, the labeled versions of GR, UGR, GGR, and UGGR are computationally equivalent. Secondly we give a precise characterization of the complexity of labeled graph reachability constrained to a language from family $\mathscr{L}$. If $\mathscr{L}$ is accepted by a one-way nondeterministic abstract storage automaton, then GR for languages from $\mathscr{L}$ is logarithmic-space many-one complete for two-way nondeterministic abstract storage automata with an auxiliary worktape that is logarithmic-space bounded. This is a very general result that applies to a variety of formal language families, such as, for example, regular languages, linear context-free languages, context-free languages, and stack automata languages and variants. Hence in turn, one adds complete problems to the complexity classes NL, P, PSPACE, and $\mathsf{DTIME}(2^{n^{O(1)}})$. Finally, we also discuss the change in complexity if acyclic graphs are used in the reachability instances. There it turns out that in this case our approach leads to a lower bound proof for simultaneously space and time restricted auxiliary storage automata.

The paper is organized as follows: in the next section we introduce the necessary notation on labeled graphs and labeled graph reachability. Then Section 3 is two-folded. In the first subsection we show that labeled GR, UGR, GGR, and UGGR are all computationally equivalent with respect to logarithmic-space many-one reducibilities under weak closure properties of the formal language family, while in the second subsection we prove our main result on the complexity of labeled GR in terms of automata with abstract storage. Finally we summarize our results and discuss some directions for future research.

2 Preliminaries

We assume the reader to be familiar with some basic notions of formal language theory as presented in [13], and with the basic concepts of complexity theory [17]. Concerning our notations, we denote the powerset of a set S by 2^S. We write Σ^* for the set of all words over the finite alphabet Σ. The empty word is denoted by λ, and $\Sigma^+ = \Sigma^* \setminus \{\lambda\}$. The reversal of a word w is denoted by w^R and for the length of w we write $|w|$. We use $\subseteq$ for inclusions and $\subset$ for strict inclusions.

Next we turn to the problem in question, that is, graph reachability on graphs with labeled edges. A *labeled (un)directed graph* is a triple (G, Σ, μ), where G is an (un)directed graph with nodes V and edges $E \subseteq V \times V$, alphabet Σ, and a finite labeling function $\mu : E \to 2^{\Sigma^*}$, such that $\mu(e)$ is a regular language, for every $e \in E$. The regular languages are represented by the state transition graphs of nondeterministic finite automata. The labeled graph is said to be *normalized*, if all of its edges e are labeled by letters from Σ, that is, $\mu(e) = \{a\}$, for some a in Σ. In order to define the reachability problems we need the concept of a paths and its value. A *path* of length k from node s to node t in G is a sequence of edges $(e_1, e_2, \ldots, e_k)$, such that $e_1 = (s, v_1)$, $e_k = (v_k, t)$, and $e_i = (v_{i-1}, v_i)$, for $1 < i < k$. The *value* of a path $p = (e_1, e_2, \ldots, e_k)$ is defined as $\mu(p) = \mu(e_1)\mu(e_2)\cdots\mu(e_k)$, that is, a formal language which is obtained by concatenating the labels of the edges of the path in sequence. Then the *labeled graph reachability problem* for a family of formal languages $\mathscr{L}$ is defined as follows:

- Given an (un)directed labeled graph G, with designated nodes s and t, and a formal language L from $\mathscr{L}$. Is there a path p from node s to node t, such that there exists a word w in $\mu(p)$ belongs to L, that is, such that $\mu(p) \cap L \neq \emptyset$?

The reachability problems on undirected grid graphs, directed grid graphs, undirected graphs, and directed graphs constrained to languages from the formal language family $\mathscr{L}$ are referred to as UGGR[$\mathscr{L}$], GGR[$\mathscr{L}$], UGR[$\mathscr{L}$], and GR[$\mathscr{L}$].

In order to classify the computational complexity of the reachability problems constrained to a formal language, we consider formal language families with effective closure properties. We say that a family of languages $\mathscr{L}$ is *effectively closed* under a binary operation $\circ$, if there is an effective *deterministic logarithmic-space bounded* construction that, given some pair of descriptors A_1 and A_2 such that both languages $L(A_1)$ and $L(A_2)$ belong to $\mathscr{L}$, yields a descriptor for $L(A_1) \circ L(A_2)$, which is again in $\mathscr{L}$. Effective closure under other types of language operations is similarly defined. A formal language family is called an *effective TRIO* [13], if it is effectively closed under λ-free homomorphism, inverse homomorphism, and intersection with regular languages. Well-known examples for effective TRIOs are the families of regular, context-free, and context-sensitive languages. Every TRIO is also effectively closed under concatenation with regular languages. Although we do not need the closure under λ-free homomorphism in the sequel, we will require language families to be TRIOs, since these properties are well established and widely known.

3 Reachability Problems on Labeled Graphs

First we prove some basic computational equivalences between different reachability problems on labeled graphs by varying the topology and the properties of the underlying graph structure. For these equivalences we mostly use logarithmic-space many-one reductions denoted by $\leq_m^{\log}$.

3.1 Relations between Reachability Problems on Labeled Graphs

We start with the two obvious relations, which also hold in the unlabeled case: UGGR[$\mathscr{L}$] is logarithmic-space many-one reducible to UGR[$\mathscr{L}$] and GGR[$\mathscr{L}$] is logarithmic-space many-one reducible to GR[$\mathscr{L}$], where $\mathscr{L}$ is an arbitrary language family. Second, an undirected graph can be turned into a directed graph without changing the reachability problem by replacing one undirected edge by two directed edges with the same edge label. Therefore, UGGR[$\mathscr{L}$] is logarithmic-space many-one reducible to GGR[$\mathscr{L}$] and UGR[$\mathscr{L}$] is logarithmic-space many-one reducible to GR[$\mathscr{L}$].

For our investigations on labeled reachability problems we utilize a special normal form. The next theorem shows that is suffices to consider normalized graphs. Due to space constraints we omit the proof.

Theorem 1. *Let $\mathscr{L}$ be an effective TRIO. Then both problems GGR[$\mathscr{L}$] and GR[$\mathscr{L}$] are logarithmic-space many-one reducible to normalized GR[$\mathscr{L}$].* □

The reduction of the previous theorem does not preserve the property of being a grid graph. However, we next show that any normalized GR instance can be reduced to a normalized grid graph.

Theorem 2. *Let $\mathscr{L}$ be an effective TRIO. Then normalized GR[$\mathscr{L}$] is logarithmic-space many-one reducible to normalized GGR[$\mathscr{L}$].*

Proof. Let (G, Σ, μ) be a labeled graph, $G = (V, E)$, and $L \subseteq \Sigma^*$ be a formal language from $\mathscr{L}$ forming a normalized GR instance. We assume $V = \{1, 2, \ldots, n\}$ without loss of generality.

Now, a normalized GGR instance (G', Σ', μ') with language L' is constructed. A schematic drawing of the construction is depicted in Figure 1. Grid graph G' is of size $((n+1)\cdot n) \times ((n+1)\cdot n)$. We refer to the nodes by pairs (i, j), for $0 \leq i \leq n$, $1 \leq j \leq n$. Consider $n+1$ copies of the nodes of G and arrange them in the order $(0,1), (0,2), \ldots, (0,n), (1,1), \ldots, (1,n), \ldots, (n,1), \ldots, (n,n)$ from top left to bottom right in a diagonal fashion in G'. In order to simplify our presentation we say that the nodes $(i,1), (i,2), \ldots, (i,n)$ form the ith block, $0 \leq i \leq n$, where

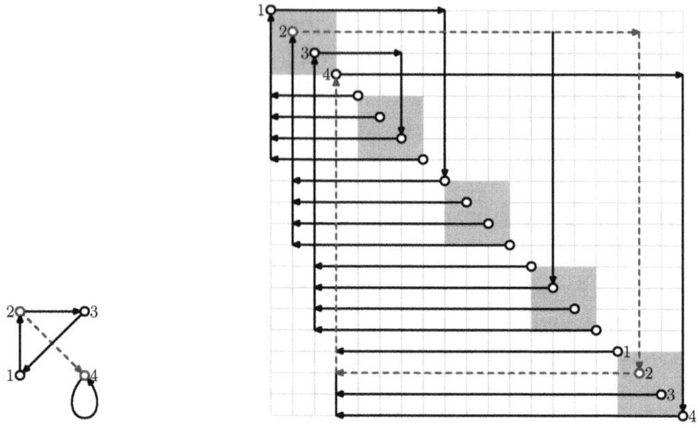

Fig. 1. Example of the normalized GR[$\mathscr{L}$] to normalized GGR[$\mathscr{L}$] reduction. Labels on edges are not depicted. The dashed drawn edge (2, 4) with label $\mu(2,4)$ of the GR instance on the left is represented by the dashed path of the GGR instance on the right. The value of the path is $2r^{15}d^{14}\mu(2,4)2$ concatenated with $4\ell^{13}u^{13}4$.

we can refer to the jth node within the block, for $1 \leq j \leq n$. The alphabet of the labeled grid graph instance has to be chosen in such a way that we can encode some path information of the given graph G. To this end, we set $\Sigma' = \Sigma \cup V \cup \{\ell, r, u, d\}$ (the unions being disjoint). Now each edge from a node j to a node k of G is represented by a path from node $(0, j)$ to (k, j) to $(0, k)$ of G' as follows.

1. Starting from the jth node of the zeroth block $(0, j)$ the path goes horizontally to the right, and then vertically down to meet the jth node of the kth block (k, j). The value of the path is a subset of $jr^*d^*\mu(j,k)j$, where r labels edges to the right and d edges down.
2. From node (k, j) the path goes horizontally to the left and then vertically up to meet the kth node of the zeroth block $(0, k)$. The value of the path is of the form $k\ell^*u^*k$, where ℓ labels edges to the left and u edges up.

Observe, that these paths may intersection each other. The labeling function μ' can be easily deduced from the above given description. Finally, we choose the start node as sth node of the zeroth block $(0, s)$ and the target node as the tth node of the zeroth block $(0, t)$. This completes the construction of the labeled grid graph instance (G', Σ', μ').

Next, we modify the language L to $L' = h^{-1}(L) \cap R$, where h is the homomorphism mapping Σ'^* to Σ^* by $h(a) = a$, for $a \in \Sigma$, $h(i) = \lambda$, for $1 \leq i \leq n$, and $h(\ell) = h(r) = h(u) = h(d) = \lambda$. The regular language R is defined to be

$$R = \left(\bigcup_{(j,k)\in E} jr^*d^*\mu(j,k)jk\ell^*u^*k \right)^* .$$

It remains to be shown that the construction is in fact a logarithmic-space many-one reduction from the given normalized GR instance to the normalized GGR instance constructed as (G', Σ', μ') with language L'.

By construction we have inserted control information into the language L' which allows us to show that one cannot deviate from a given path. So, assume that (G, Σ, μ) with language $L \subseteq \Sigma^*$, start node s, and target node t has a solution. That is, there is a path p linking s with t such that there is a word w in $\mu(p)$ which also belongs to L. Let $p = (e_1, e_2, \ldots, e_z)$ be this solution. Then, by construction, for each edge $e_i = (j, k)$, for $1 \leq i \leq z$, there is a path $(0, j)$ to (k, j) to $(0, k)$ of G' whose value is of the form $jr^*d^*\mu(j,k)jk\ell^*u^*k$. This implies that (G', Σ', μ') with language L', start node $(0, s)$, and target node $(0, t)$ has a solution.

Conversely, language L' forces any path of G' starting in some node $(0, j)$, $1 \leq j \leq n$, to move to the right and then down until it meets a node on the diagonal. In order to be a subpath of a solution the value of this path which starts with j has to end with j. This implies that the path meets a node on the diagonal that is the jth node in some block k, for $1 \leq k \leq n$. The only possibility to continue the path according to L' is to move left and then up until it meets a node in the zeroth block. Since this second part of the subpath starts with an edge labeled by k it has to end with an edge labeled by k. Therefore, it ends at node $(0, k)$. Altogether the subpath has a value of the form $jr^*d^*\mu(j,k)jk\ell^*u^*k$. Its existence ensures that there is a path from node j to node k in G labeled by $\mu(j, k)$. This implies that (G, Σ, μ) with language L, start node s, and target node t has a solution.

Concerning the space complexity of the reduction we recall the remark at the end of the proof of Theorem 1. □

Hence, up to this point we have obtained the following chain of reductions:

$$\begin{array}{ccccc}
\mathrm{GGR}[\mathscr{L}] & \leq_m^{\log} & \mathrm{GR}[\mathscr{L}] & \leq_m^{\log} \text{ norm. } \mathrm{GR}[\mathscr{L}] \leq_m^{\log} \text{ norm. } \mathrm{GGR}[\mathscr{L}] \\
\rotatebox{90}{$\leq_m^{\log}$} & & \rotatebox{90}{$\leq_m^{\log}$} & \\
\mathrm{UGGR}[\mathscr{L}] & \leq_m^{\log} & \mathrm{UGR}[\mathscr{L}] &
\end{array}$$

Finally, we show that, in fact, all of these previous problems are computationally equivalent, that is, they are logarithmic-space many-one reducible to each other. To this end, it suffices to close the previous chain of reductions by proving that normalized GGR is logarithmic-space many-one reducible to normalized UGGR. In addition, this shows that directed edges may be replaced by undirected ones.

Theorem 3. *Let $\mathscr{L}$ be an effective TRIO. Then normalized GGR[$\mathscr{L}$] is logarithmic-space many-one reducible to normalized UGGR[$\mathscr{L}$].* □

Altogether we obtain the following corollary.

Corollary 4. *Let $\mathscr{L}$ be an effective TRIO. Then the reachability problems on labeled graphs (normalized) UGGR[$\mathscr{L}$], (normalized) UGR[$\mathscr{L}$], (normalized) GGR[$\mathscr{L}$], and (normalized) GR[$\mathscr{L}$] are computationally equivalent with respect to logarithmic-space many-one reductions.* □

This shows that there is a significant difference between ordinary reachability problems and labeled reachability problems on graphs (unless $\mathsf{NC}^1 = \mathsf{NL}$). In fact, the technique to encode information on paths and, thus, on the structure of the given graph into the formal language can be used to obtain further relations, even for simpler instances than undirected grid graphs.

Theorem 5. *Let $\mathscr{L}$ be an effective TRIO. Then normalized acyclic GR[$\mathscr{L}$] and normalized acyclic GGR[$\mathscr{L}$] are computationally equivalent with respect to logarithmic-space many-one reductions.* □

3.2 Reachability of Labeled Graphs and Word Problems of Language Families

In this subsection we show that the reachability problem for labeled graphs has close relations to general word problems. Here the language families are represented by certain types of automata. For example, it is well known that the reachability problem for directed graphs is NL-complete [16]. The reachability problem for directed graphs over the regular languages (represented by nondeterministic finite automata) is still in NL and, thus, NL-complete. On the other hand, the word problem of a nondeterministic automaton which is equipped with a two-way input head and a logarithmic-space bounded working tape is clearly NL-complete. In fact, there is a general relation of this type which is shown next. To this end, we have to introduce the concept of (two-way) abstract storage automata. Here we follow the lines of [10] with the slight modifications introduced in [9].

A *storage type* is a quintuple $S = (C, P, I, C_0, C_f)$, where C is a set of *storage configurations*, P is the finite set of *predicates* over C, I is the finite set of *storage modification instructions*, and $C_0 \subseteq C$ and $C_f \subseteq C$ are the sets of *initial* and *final* configurations. In order to clarify our notation we give a small example (cf. [10]).

Example 6. The storage type *pushdown* (*PD*) is defined by (C, P, I, C_0, C_f), where $C = \Gamma^*$, for some fixed alphabet Γ of pushdown symbols, the set of predicates $P = \{ \mathrm{top}_a \mid a \in \Gamma \} \cup \{\mathrm{bottom}\}$, the set of instructions $I = \{ \mathrm{push}_a \mid a \in \Gamma \} \cup \{\mathrm{pop}\}$, the set of initial configurations $C_0 = \{\lambda\}$, and the set of final configurations $C_f = C$. The predicates are defined to be

$$\mathrm{top}_a(bx) = \begin{cases} \mathit{true} & \text{if } a = b \\ \mathit{false} & \text{otherwise} \end{cases} \quad \text{and} \quad \mathrm{bottom}(x) = \begin{cases} \mathit{true} & \text{if } x = \lambda \\ \mathit{false} & \text{otherwise} \end{cases}$$

for $x \in \Gamma^*$ and $a, b \in \Gamma$. Finally, let $\mathrm{push}_a(x) = ax$ and $\mathrm{pop}(ax) = x$, for $x \in \Gamma^*$ and $a \in \Gamma$, and $\mathrm{pop}(\lambda)$ is undefined. □

Based on a storage type $S = (C, P, I, C_0, C_f)$, we define a *two-way (nondeterministic) S-automaton* as a 6-tuple $A = (Q, \Sigma, \delta, w_0, c_0, F)$, where Q is a finite set of *states*, Σ is the *input alphabet*, $q_0 \in Q$ is the *initial state*, $c_0 \in C_0$ is the *initial storage configuration*, $F \subseteq Q$ is the set of *accepting states*, and the *transition relation* δ is a finite subset of $Q \times \Sigma \times \mathcal{B}(P) \times Q \times \{L, N, R\} \times I^*$, where L, N, and R are interpreted as moves of the input head to the left, no move, and right, and $\mathcal{B}(P)$ denotes the set of Boolean expressions over P.

The computation of A is naturally defined on configurations. Here a *configuration* is a quintuple (q, u, v, c), where q is a state in Q, both u and v are words of input symbols, and c is a storage configuration. A two-way S-automaton A is said to be in configuration (q, u, v, c), if A is in state q, word uv is the inscription of the input tape, the input head is scanning the leftmost symbol of v, and the storage configuration is c. A transition (p, a, B, p', d, g) is applicable to a configuration (q, u, v, c) if and only if $p = q$, the leftmost symbol of v is a, and the Boolean expression B over predicates evaluates to true on c. Then the successor configuration is determined by the new state p', an input head movement according to $d \in \{L, N, R\}$, and a new storage configuration that is obtained by applying the instructions g to c. As usual the transition relation between configurations is denoted by $\vdash_A$ and its reflexive and transitive closure by $\vdash_A^*$.

The *language accepted* by A is defined as

$$L(A) = \{\, w \in \Sigma^* \mid (q_0, \lambda, w, c_0) \vdash_A^* (q, u, v, c) \text{ for some } q \in F \text{ and } c \in C_f \,\}.$$

An S-automaton is a *one-way* S-automaton, if the head cannot move to the left. The family of languages accepted by nondeterministic two-way (one-way, respectively) S-automata is denoted by 2N-S-A (1N-S-A, respectively). Let S be a family of storage types, then 2N-S-A $= \bigcup_{S \in \mathsf{S}}$ 2N-S-A. Similarly we define the family of languages 1N-S-A $= \cup_{S \in \mathsf{S}}$1N-$S$-A.

The next theorem shown in [9] reveals the relation between S-automata and effective TRIOs.

Theorem 7. *Let S be a storage type. Then the language family* 1N-S-A *is an effective TRIO.*

As mentioned before, we are interested in S-automata that are equipped with an additional working tape, so-called *auxiliary S-automata*. One can see these devices as Turing machines having access to a type S storage in addition to its working tape. Note that the available type-S storage may be unlimited and is, in fact, not subject to possible restrictions of the Turing machine such as space or time bounds. We denote the family of languages accepted by auxiliary S-automata with $s(n)$-space bounded working tape by the infix Aux and the suffix $(s(n))$, that is, 2AuxN-S-A$(s(n))$ and 1AuxN-S-A$(s(n))$, and adapt the notation for families of storage types.

Now we are prepared for the main theorem of this subsection. Its proof resembles an idea used in the proof to show the relation between 2AuxN-S-A$(\log n)$ and the non-emptiness problem for 1N-S-A in [9].

Theorem 8. *Let S be a storage type. Then the (normalized) (un)directed graph reachability problem constrained to languages from the family* 1N-S-A *and the general membership problem for* 2AuxN-S-A$(\log n)$ *are computationally equivalent with respect to logarithmic-space many-one reductions.*

Proof (Sketch). By Corollary 4 we may assume without loss of generality that the graph reachability instance is directed and normalized. So, let (G, Σ, μ) be a labeled graph, $G = (V, E)$, and $L \subseteq \Sigma^*$ be a formal language from 1N-S-A forming a normalized GR instance with starting node s and target node t. Then there is an S-automaton $A_1 = (Q, \Sigma, \delta, q_0, c_0, F)$ with storage type $S = (C, P, I, C_0, C_f)$ accepting language L. We construct a two-way S-automaton $A_2 = (Q', \Sigma, \delta', q_0', c_0, F')$ with $Q' = Q \times (V \cup E)$, $q_0' = (q_0, s)$, and $F' = F \times (\{t\} \cup (V \times \{t\}))$. The transition relation δ' is specified as follows. We fix a letter $a \in \Sigma$, and set for all $p, q \in Q$, $i, j \in V$, $e \in E$, $B \in \mathcal{B}(P)$, and $g \in I^*$,

1. $((p, i), a, B, (q, j), N, g) \in \delta'$, if $(p, \mu(e), B, q, R, g) \in \delta$ and $e = (i, j)$,
2. $((p, i), a, B, (q, e), N, g) \in \delta'$, if $(p, \mu(e), B, q, N, g) \in \delta$ and $e = (i, j)$,
3. $((p, e), a, B, (q, j), N, g) \in \delta'$, if $(p, \mu(e), B, q, R, g) \in \delta$ and $e = (i, j)$,
4. $((p, e), a, B, (q, e), N, g) \in \delta'$, if $(p, \mu(e), B, q, N, g) \in \delta$.

Basically, the idea is that transitions from (1) to (4) cause A_2 to simulate A_1 step-by-step while simultaneously following a path from s to t in G. In the simulation we have to distinguish two cases, namely whether A_1 will make a right move of the input head or will stay stationary. The former case is covered by the transitions in (1) and (3), while the latter is simulated by the transitions in (2) and (4). We have to deal with the latter case separately, because A_1 may read an edge label without moving the head, and then read the same edge label again. So, A_2 cannot follow this edge while simulating the first reading of A_1. Instead, A_2 has to remember the edge for a correct simulation of the next step of A_1. The only word that may belong to language $L(A_2)$ is a. So, it is easy to see that there is a solution to the normalized GR instance if and only if the two-way auxiliary S-automaton A_2 accepts the word a.

Conversely, let A_2 be a two-way auxiliary S-automaton whose working tape is logarithmic-space bounded, and w the input word. In order to construct a normalized GR instance with respect to a language accepted by a 1N-S-A, we first sketch the idea of the 1N-S-A A_1. Since content w of the input tape is a parameter of the construction, and we consider logarithmic-space many-one reductions, the entire input word w and any possible content of the working tape can be encoded into the states of the 1N-S-A A_1 to be constructed. So, the two-way movement can be simulated in the finite control of A_1 while A_1 does not move its input head at all. Automaton A_1 reads the same input symbol, say a, in every transition. As above the only word that may belong to language $L(A_1)$ is a. Now, the normalized GR instance (G, Σ, μ) with respect to language $L(A_1)$ can be chosen as $G = (\{1\}, \{(1, 1)\})$, $\Sigma = \{a\}$, and $\mu((1, 1)) = a$. The starting and target node is 1. By construction, automaton A_2 accepts the input w if and only if A_1 accepts the input a, and the normalized GR instance has a solution if and only if automaton A_1 accepts the input a. Therefore, the claim stated follows. □

Since in [9] it has been shown that the general membership problem for 2AuxN-S-A($\log n$) and the non-emptiness problem for 1N-S-A are computationally equivalent with respect to logarithmic-space many-one reductions, we obtain the next corollary.

Corollary 9. *Let S be a storage type. Then the (normalized) (un)directed graph reachability problem constrained to languages from the family* $1N\text{-}S\text{-}A$ *and the non-emptiness problem for* $1N\text{-}S\text{-}A$ *are computationally equivalent with respect to logarithmic-space many-one reductions.* □

In the remainder of this subsection we give some applications utilizing known complexity results for the word problem of language families characterized by different types of two-way auxiliary S-automata. We start with some language families from the Chomsky hierarchy, and denote the family of storage types that are pushdown storages by PD, and the empty family of storage types by F. Obviously, $2AuxNFA(s(n)) = NSPACE(s(n))$ and Cook's [8] seminal result reads as $2AuxNPDA(s(n)) = \bigcup DTIME(2^{c \cdot s(n)})$, where unions are over c and $s(n) \geq \log n$. Restricting the pushdown storage to act as counter (C) we obtain $2AuxNCA(s(n)) = NSPACE(s(n))$, for space bounds $s(n) \geq \log n$. A similar characterization is true for one-turn pushdown storages. So, applying Theorem 8 and specializing the characterizations of auxiliary finite automata, pushdown automata, one-turn pushdown, and counter automata to $s(n) = \log n$ we obtain:

Corollary 10. *The directed graph reachability problem constrained to (i) regular languages, counter languages, or linear context-free languages is* NL*-complete and (ii) to context-free languages is* P*-complete.* □

What about the context-sensitive languages? It is clear that the directed graph reachability problem constrained to context-sensitive languages is already undecidable, which follows by a direct construction, since the emptiness problem for context-sensitive languages is undecidable. This shows that Theorem 8 has its limitations.

Next we consider variants of stack automata. A *stack* (S) is a pushdown storage allowing its interior content (that is, symbols other than the topmost symbol) to be *read* at any time, a *nonerasing stack* (NES) is a stack which cannot be popped, and a *checking stack* (CS) is a nonerasing stack which forbids any push operation once an interior stack symbol has been read. The previous characterizations are in sharp contrast with Ibarra's [14] results, who proved that $2AuxNSA(s(n)) = \bigcup DTIME(2^{2^{c \cdot s(n)}})$, $2AuxNNESA(s(n)) = \bigcup DSPACE(2^{c \cdot s(n)})$, and $2AuxNCSA(s(n)) = \bigcup NSPACE(2^{c \cdot s(n)})$, where again unions are over c and $s(n) \geq \log n$. These results turn into the following corollary, where we write DEXPTIME short for $DTIME(2^{n^{O(1)}})$.

Corollary 11. *The directed graph reachability problem constrained to (i) nonerasing stack or checking stack languages is* PSPACE*-complete and (ii) to stack languages is* DEXPTIME*-complete.* □

It is worth mentioning that the indexed languages introduced in [1] as a generalization of context-free can be characterized by an automaton model that is a variant of stack automata. These so-called *nested stack automata* (NestS) also have an S-automata characterization. Auxiliary nested stack automata were considered in the literature [5], and it was shown that the family of languages accepted by two-way nondeterministic nested stack automata with $s(n)$-space bounded auxiliary working tape is equal to the family of languages accepted by two-way nondeterministic *stack* automata with the same space bound on the auxiliary working tape. The latter family in turn equals $\bigcup \mathsf{DTIME}(2^{2^{c \cdot s(n)}})$. So, from our previous discussion on stack automata we obtain the following completeness result.

Corollary 12. *The directed graph reachability problem constrained to indexed languages is* DEXPTIME-*complete.* □

Finally, Theorem 8 also applies to certain Lindenmayer language families. Here one has to be careful because, for instance, the automaton model of pushdown array of counters cannot be described in terms of S-automata.

In the remainder of this subsection we discuss the complexity of the labeled *acyclic* GR problem. For the next theorem we have to adapt our notation on two-way nondeterministic auxiliary S-automata to cope with simultaneously space- and time-bounded automata. If an $s(n)$-space bounded 2AuxN-S-A is also $t(n)$ time bounded we address the language family by 2AuxN-S-A$(s(n), t(n))$.

Theorem 13. *Let S be a storage type. Then the general membership problem for* 2AuxN-S-A$(\log n, n^{O(1)})$ *logarithmic-space many-one reduces to the normalized directed* acyclic *graph reachability problem constrained to languages from the family* 1N-S-A. □

It has to be left open whether the converse relation of Theorem 13 and, thus, computational equivalence holds. A problem is that a 1N-S-A may perform λ-moves on the input while operating on the storage structure for more than a polynomial number of steps. In order to handle this situation we need explicit knowledge about the storage structure.

An immediate consequence of Theorem 13 is, for instance, that normalized acyclic GR constrained to regular languages is NL-hard, and for context-free languages it becomes hard for LOGCFL. The latter follows from the equality LOGCFL $=$ 2AuxNPDA$(\log n, n^{O(1)})$ [20]. Further NP-hardness results follow from normalized acyclic GR instances, where the constrained languages are from one of the various aforementioned stack language families, since NP $=$ 2AuxNSA$(\log n, n^{O(1)})$ $=$ 2AuxNNESA$(\log n, n^{O(1)})$ $=$ 2AuxNCSA$(\log n, n^{O(1)})$ by [12,15]. A closer look reveals that in all these cases we, in fact, obtain completeness results with respect to logarithmic-space many-one reductions by direct constructions of upper bounds.

4 Conclusions

We have investigated the computational complexity of labeled graph reachability problems constrained to language families in the setting of abstract storage automata. The advantage of our approach is proven in a series of corollaries applying well-known characterizations of complexity classes in terms of nondeterministic abstract storage automata. Finally, let us stress that some of the results obtained can be generalized to deterministic language families such as, for example, deterministic context-free languages or deterministic stack languages.

References

1. Aho, A.V.: Indexed grammars–An extension of context-free grammars. J. ACM 15, 647–671 (1968)
2. Allender, E., Barrington, D.A.M., Chakraborty, T., Datta, S., Roy, S.: Planar and grid graph reachability problems. Theoret. Comput. Sci. 45, 675–723 (2009)
3. Baier, C., Katoen, J.P.: Principles of Model Checking. MIT Press, Cambridge (2008)
4. Barrett, C.L., Jacob, R., Marathe, M.V.: Formal-language-constrained path problems. SIAM J. Comput. 30, 809–837 (2000)
5. Beeri, C.: Two-way nested stack automata are equivalent to two-way stack automata. J. Comput. System Sci. 10 (1975)
6. Blum, M., Kozen, D.: On the power of the compass (or, why mazes are easier to search than graphs). In: Symposium on Foundations of Computer Science (FOCS 1978), pp. 132–142. IEEE, Los Alamitos (1978)
7. Brandt, F., Fischer, F.A., Holzer, M.: On iterated dominance, matrix elimination, and matched paths. In: Symposium on Theoretical Aspects of Computer Science (STACS 2010). LIPIcs, vol. 5, pp. 107–118. Schloss Dagstuhl - Leibniz-Zentrum für Informatik (2010)
8. Cook, S.A.: Characterizations of pushdown machines in terms of time-bounded computers. J. ACM 18, 4–18 (1971)
9. Dassow, J., Lange, K.J.: Computational calculus and hardest languages of automata with abstract storages. In: Budach, L. (ed.) FCT 1991. LNCS, vol. 529, pp. 200–209. Springer, Heidelberg (1991)
10. Engelfriet, J., Hoogeboom, H.J.: Automata with storage on infinite words. In: Ronchi Della Rocca, S., Ausiello, G., Dezani-Ciancaglini, M. (eds.) ICALP 1989. LNCS, vol. 372, pp. 289–303. Springer, Heidelberg (1989)
11. Greenlaw, R., Hoover, H.J., Ruzzo, W.L.: Limits to Parallel Computation: P-Completeness Theory. Oxford University Press, Oxford (1995)
12. Holzer, M., McKenzie, P.: Alternating and empty alternating auxiliary stack automata. Theoret. Comput. Sci. (1-3), 307–326 (2003)
13. Hopcroft, J.E., Ullman, J.D.: Introduction to Automata Theory, Languages, and Computation. Addison-Wesley, Reading (1979)
14. Ibarra, O.H.: Characterizations of some tape and time complexity classes of turing machines in terms of multihead and auxiliary stack automata. J. Comput. System Sci. 5, 88–117 (1971)
15. Jenner, B., Kirsig, B.: Characterizing the polynomial hierarchy by alternating auxiliary pushdown automata. RAIRO Inform. Theor. 23, 87–99 (1989)
16. Jones, N.D., Lien, Y.E., Laaser, W.T.: New problems complete for nondeterministic log space. Math. Systems Theory 10, 1–17 (1976)
17. Papadimitriou, C.H.: Computational Complexity. Addison-Wesley, Reading (1994)
18. Reingold, O.: Undirected connectivity in log-space. J. ACM 55, 1–24 (2008)
19. Ruzzo, W.L.: Complete pushdown languages (1979), `http://www.cs.washington.edu/homes/ruzzo/papers/complete-pushdown-languages.pdf`
20. Sudborough, I.H.: On the Tape Complexity of Deterministic Context-Free Languages. J. ACM 25, 405–414 (1978)

Characterizing the Regular Languages by Nonforgetting Restarting Automata

Norbert Hundeshagen and Friedrich Otto

Fachbereich Elektrotechnik/Informatik
Universität Kassel, 34109 Kassel, Germany
{hundeshagen,otto}@theory.informatik.uni-kassel.de

Abstract. We study nonforgetting R- and nonforgetting deterministic RR-automata of window size one, that is, nf-R(1)- and det-nf-RR(1)-automata. Our main result shows that the monotone variants of these two types of restarting automata characterize the regular languages. On the other hand, we prove that already the non-monotone deterministic nonforgetting R(1)-automata accept a class of languages that is incomparable to the class of semi-linear languages with respect to inclusion, but that properly includes the class of languages that are accepted by globally deterministic cooperating distributed systems of stateless deterministic R(1)-automata.

1 Introduction

Restarting automata can be seen as an extension of finite-state acceptors by certain restricted rewrite and restart operations (see, e.g., [2]). Therefore it is not surprizing that each type of restarting automaton accepts a superclass of the regular languages. In fact, also characterizations of the regular languages in terms of certain restricted types of restarting automata have been obtained: in [11] Mráz has shown that (deterministic) R-automata with window size one (R(1)-automata) accept only regular languages, and Reimann has carried this result over to deterministic RR(1)-automata in [14]. Here we extend these results even further by showing that the monotone variants of nonforgetting R(1)- and nonforgetting deterministic RR(1)-automata only accept regular languages.

Nonforgetting restarting automata were introduced by Messerschmidt and Stamer in [10]. In contrast to the situation for standard restarting automata, the restart operations of a nonforgetting restarting automaton are combined with a change of its internal state just like the other operations, that is, it is not reset to the initial state when it executes a restart operation. As shown in [7,8] (deterministic) nonforgetting R- and RR-automata correspond to (globally deterministic) cooperating distributed systems (CD-systems) of R- and RR-automata working in mode = 1. Therefore, we also compare the expressive power of non-monotone deterministic nonforgetting R-automata to that of the CD-systems of stateless deterministic R(1)-automata studied in [12].

This paper is structured as follows. In Section 2 we repeat in short the definitions of the various types of restarting automata we are interested in in this

G. Mauri and A. Leporati (Eds.): DLT 2011, LNCS 6795, pp. 288–299, 2011.

paper. Then in Section 3 we prove our main result characterizing the regular languages in terms of nonforgetting R(1)- and deterministic nonforgetting RR(1)-automata, and in Section 4 we study the non-monotone det-nf-R(1)-automaton. In the concluding section we address the question of the descriptional complexity for monotone nf-R(1)- and det-nf-RR(1)-automata in short.

2 Nonforgetting Restarting Automata

A *nonforgetting* RR-*automaton* (nf-RR-automaton, for short) is a one-tape machine with a finite-state control and a read/write window. It is described by a 7-tuple $M = (Q, \Sigma, ¢, \$, q_0, k, \delta)$. Here Q is a finite set of internal states, Σ is a finite (tape) alphabet, the symbols $¢, \$ \notin \Sigma$ serve as markers for the left and right border of the workspace, respectively, $q_0 \in Q$ is the initial state, the size of the *read/write window* is $k \geq 1$, and δ is the *transition relation* that associates a finite set of transition steps to pairs of the form (q, u), where $q \in Q$ is a state and u is a possible contents of the read/write window. There are four types of transition steps: *move-right steps* of the form (q', MVR), which shift the window one step to the right and change the internal state to q', *rewrite steps* of the form (q', v), where v is a proper scattered subword of u, that replace the contents u of the read/write window by the word v and change the internal state to q', *restart steps* of the form $(q', \mathsf{Restart})$ that place the read/write window over the left end of the tape and change the internal state to q', and *accept steps* (Accept), which cause the automaton to halt and accept. Some additional restrictions apply in that the sentinels ¢ and \$ must not be deleted, and that the window must not move right across the \$-symbol. Further, M is *deterministic* if δ is a partial function. We use the prefix det- to denote deterministic types of RR-automata.

A *configuration* of M is described by a word $\alpha q \beta$, where $q \in Q$ and either $\alpha = \varepsilon$ (the empty word) and $\beta \in \{¢\} \cdot \Sigma^* \cdot \{\$\}$ or $\alpha \in \{¢\} \cdot \Sigma^*$ and $\beta \in \Sigma^* \cdot \{\$\}$; here q is the current internal state, $\alpha\beta$ is the current content of the tape, and it is understood that the head scans the first k symbols of β (or all of β if $|\beta| \leq k$). A *restarting configuration* is of the form $q¢w\$$, and an *initial configuration* is of the form $q_0¢w\$$. By $\vdash_M$ we denote the single-step computation relation of M, and $\vdash_M^*$ denotes the reflexive transitive closure of $\vdash_M$.

The automaton M proceeds as follows. Starting from a restarting configuration $q¢w\$$, the window is shifted to the right by a sequence of move-right steps until a configuration of the form $¢xquy\$$ is reached such that $(q', v) \in \delta(q, u)$, where $w = xuy$. Now the latter configuration can be transformed into the configuration $¢xvq'y\$$, and the computation proceeds with further move-right steps until eventually a restart operation is executed, which yields a restarting configuration of the form $p¢xvy\$$. This sequence of computational steps, which is called a *cycle*, is expressed as $q¢w\$ \vdash_M^c p¢xvy\$$. A computation of M consists of a finite sequence of cycles that is followed by a tail computation, which consists of a sequence of move-right operations (and possibly a single application of a rewrite operation) that is possibly followed by an accept step. An input word $w \in \Sigma^*$ is *accepted* by M, if there exists a computation of M which starts with the initial configuration $q_0¢w\$$ and finishes by executing an accept step. By $L(M)$ we denote the language consisting of all words accepted by M.

Each cycle C of a restarting automaton M contains a unique configuration $\alpha q \beta$ in which a rewrite step is applied. Then $|\beta|$ is called the *right distance* of C, denoted as $D_r(C)$. A sequence of cycles $(C_1, C_2, \ldots, C_n)$ of M is called *monotone* if $D_r(C_1) \geq D_r(C_2) \geq \cdots \geq D_r(C_n)$. A computation of M is called *monotone* if the corresponding sequence of cycles is monotone. Finally, M itself is called *monotone* if all its computations that start from an initial configuration are monotone. We use the prefix mon- to denote monotone types of restarting automata.

M is called a *nonforgetting* R*-automaton* (nf-R-automaton, for short) if it is a nonforgetting RR-automaton for which each rewrite operation is immediately followed by a restart operation. To simplify the description we combine each rewrite operation of an R-automaton with the subsequent restart operation. Of course, the notions of determinism and monotonicity also apply to R-automata.

Observe that nonforgetting R- and RR-automata differ from the standard R- and RR-automata as defined in [2] in that they are not necessarily reset to their initial states when executing a restart operation. This feature increases the expressive power of nonforgetting R- and RR-automata considerably. In fact, it is shown in [7,8] that (deterministic) nonforgetting R- and RR-automata correspond to (globally deterministic) cooperating distributed systems (CD-systems) of R- and RR-automata working in mode = 1. In Section 4 we will consider certain restricted variants of these CD-systems.

3 Characterizing the Regular Languages

The following results are known on the expressive power of deterministic (monotone) restarting automata. Here $\mathcal{L}(\mathsf{X})$ denotes the class of languages that are accepted by restarting automata of type X, DCFL denotes the class of deterministic context-free languages, and CRL is the class of Church-Rosser languages [6].

Proposition 1. [4,9,10]

(a) $\mathcal{L}(\text{det-mon-RR}) = \mathcal{L}(\text{det-mon-R}) = \mathcal{L}(\text{det-mon-nf-R}) = \text{DCFL}$
$\subsetneq \mathcal{L}(\text{det-mon-nf-RR}) \subsetneq \text{CRL}.$

(b) $\mathcal{L}(\text{det-nf-R})$ *and* $\mathcal{L}(\text{det-nf-RR})$ *are incomparable to* CRL *under inclusion.*

Of course, $\mathcal{L}(\text{det-nf-R}) \subseteq \mathcal{L}(\text{det-nf-RR})$, but it is still open whether this inclusion is strict. For each $k \geq 1$, let $\mathsf{X}(k)$ denote the restarting automata of type X that have a read/write window of size k. Concerning the expressive power of restarting automata with window size one, the following results have been obtained in [3,11,14], where REG denotes the class of regular languages. Observe that each rewrite operation of a nf-RR(1)-automaton simply erases a single symbol from the tape.

Proposition 2. (a) $\mathcal{L}(\text{R}(1)) = \mathcal{L}(\text{mon-R}(1)) = \mathcal{L}(\text{det-R}(1)) = \text{REG}$.
(b) $\mathcal{L}(\text{det-RR}(1)) = \text{REG} \subsetneq \mathcal{L}(\text{RR}(1))$.
(c) $\mathcal{L}(\text{det-R}(5))$ *contains a language that is not context-free.*

It is easily verified that deterministic R(1)- and RR(1)-automata are necessarily monotone. Here we extend the above characterizations of REG to nonforgetting restarting automata.

Theorem 1. $\mathcal{L}(\text{det-mon-nf-R}(1)) = \text{REG}$.

Proof. It remains to prove the inclusion from left to right. Let $M = (Q, \Sigma, ¢, \$, q_0, 1, \delta)$ be a det-mon-nf-R(1)-automaton for $L \subseteq \Sigma^*$. W.l.o.g. we may assume that $Q = \{q_0, q_1, \ldots, q_{n-1}\}$, and that M executes Accept-instructions only on reading the \$-symbol. Below we describe a deterministic finite-state acceptor (DFA) $A = (Q_A, \Sigma \cup \{\$\}, q_0^{(A)}, F, \delta_A)$ for $L \cdot \$$. In its finite-state control A stores the following information on the computation of M that it tries to simulate:

- The *current restart state* CRS that contains the state $q \in Q$ with which the cycle of M starts that is currently active. Initially CRS is set to q_0.
- A *state table* T that initially contains the list of all pairs (q_i, q_i') $(i = 0, 1, \ldots, n-1)$, where q_i' is the state that M enters from state q_i on seeing the ¢-symbol. If $\delta(q_i, ¢)$ is undefined, then T contains the item $(q_i, -)$.
- A *buffer* B of length n that is initially empty. It will be used to store information on possible rewrite (that is, delete) operations encountered during the current simulation.

The intended simulation of M by the DFA A crucially depends on the property of M of being monotone. Consider an accepting computation of M on input w, and assume that this computation consists of a sequence of at least two cycles that is followed by a tail computation, that is, it has the following form:

$$\begin{aligned} q_0¢w\$ &= q_0¢uav\$ \vdash^+_{\mathsf{MVR}} ¢up_0av\$ \vdash_M q_{i_1}¢uv\$ = q_{i_1}¢xby\$ \vdash^+_{\mathsf{MVR}} ¢xp_{i_1}by\$ \\ &\vdash_M q_{i_2}¢xy\$ \vdash^{c^*}_M q_{i_m}¢z\$ \vdash^+_{\mathsf{MVR}} ¢zp_{i_m}\$ \vdash_M \mathsf{Accept}. \end{aligned}$$

The right distance of the first cycle is $d_1 = |v| + 2$, and the right distance of the second cycle is $d_2 = |y| + 2$. As M is monotone, we have $d_1 \geq d_2$, that is, $|v| \geq |y|$. Since $uv = xby$, this means that y is a suffix of v. If y is a proper suffix of v, then $v = x_2by$ for a suffix x_2 of x. In this case $w = uav = uax_2by$, which implies that the second delete operation is executed at a place that is strictly to the right of the place where the first delete operation was executed. In this situation A first encounters the letter a deleted in the first cycle and later it encounters the letter b deleted in the second cycle.

If, however, $v = y$, then $uv = xby = xbv$ implies that $u = xb$, and so $w = uav = xbav$. Thus, in this situation the second delete operation is executed immediately to the *left* of the place where the first delete operation was executed. Hence, in this case the b deleted in the second cycle is encountered by A before the a that is deleted in the first cycle. However, this can happen only if M completes the first cycle by restarting in the correct state q_{i_1}. In addition, as M is deterministic, we have $q_{i_1} \neq q_0$.

Obviously, the second case above can occur more than once in a row. However, as all the corresponding restarting states must differ from one another, the length of such a sequence is bounded from above by the number n of states of M.

The DFA A will use its buffer B to record information on sequences of delete operations of M of the form described above. Because of the restriction on the length of such sequences, the size of B is sufficient for this task. In the situation above A will store the following information:

- CRS still contains the initial state q_0;
- the table T contains the pairs (q_i, p_i) $(i = 0, 1, \ldots, n-1)$, where p_i is the state that M reaches from the restarting configuration $q_i ¢w\$ = q_i ¢xbav\$$ by moving right across the prefix $¢x$;
- on realizing that M can execute the rewrite operation $\delta(p_{i_1}, b) = (q_{i_2}, \varepsilon)$, A stores the pair $(b, (q_{i_1}, p_{i_1}, q_{i_2}, T_b))$ in B, and moves right to the next letter. Here T_b is the table that is obtained from T by replacing each pair (q_i, p_i) by (q_i, p_i'), if $\delta(p_i, b) = (p_i', \mathsf{MVR})$, by leaving it as it is, if $\delta(p_i, b)$ is a rewrite operation, and by replacing it by $(q_i, -)$ if $\delta(p_i, b)$ is undefined. In fact, the second component of the above pair stored in B contains a 4-tuple of the form (q_i, p_i, p_i', T_b) for each index i such that $\delta(p_i, b)$ is a rewrite operation.

Next A realizes that M can execute the rewrite operation $\delta(p_0, a) = (q_{i_1}, \varepsilon)$. Hence, it has detected that the computation of M on input $w = xbav$ begins with a sequence of two cycles that deletes the factor ba. In this case A sets CRS to q_{i_2}, it leaves T unchanged, and it empties the buffer B. Thus, A now simulates the computation of M that begins with the restarting configuration $q_{i_2} ¢xv\$$, and as M is deterministic, A can do so by starting with the first letter of v and by considering the state transitions of M that are induced by moving right across the prefix $¢x$ which are already recorded in table T.

In general the situation can be more complicated than in the above example. In fact, there are three cases that we need to deal with.

Case 1. The word w has a factorization of the form $w = xa_m a_{m-1} \cdots a_1 v$ such that the computation of M on input w begins with a sequence of m cycles such that, in the j-th cycle $(1 \leq j \leq m)$, the letter a_j is deleted. Thus, while the first cycle has the form

$$q_0 ¢w\$ = q_0 ¢xa_m \cdots a_2 a_1 v\$ \vdash^{+}_{\mathsf{MVR}} ¢xa_m \cdots a_2 p_0 a_1 v\$ \vdash q_{i_1} ¢xa_m \cdots a_2 v\$,$$

the j-th cycle $(2 \leq j \leq m)$ has the form

$$q_{i_{j-1}} ¢xa_m \cdots a_{j+1} a_j v\$ \vdash^{+}_{\mathsf{MVR}} ¢xa_m \cdots a_{j+1} p_{i_{j-1}} a_j v\$ \vdash q_{i_j} ¢xa_m \cdots a_{j+1} v\$.$$

This case is dealt with in analogy to the example above.

Case 2. The word w has a factorization of the form $w = xa_m \cdots a_{s+1} a_s v$ such that, for each $j = s+1, \ldots, m$, there is a possible cycle of M of the form

$$q_{i_{j-1}} ¢xa_m \cdots a_{j+1} a_j a_s v\$ \vdash^{+}_{\mathsf{MVR}} ¢xa_m \cdots a_{j+1} p_{i_{j-1}} a_j a_s v\$ \vdash q_{i_j} ¢xa_m \cdots a_{j+1} a_s v\$,$$

but from the restarting configuration $q_i ¢xa_m \cdots a_{s+1} a_s v\$$, M does not apply a rewrite operation to the letter a_s for any state $q_i \in Q$. While reading the prefix $xa_m \cdots a_{s+1}$, A has collected the following information in its finite-state control:

- CRS is q_0;
- the table T contains the pairs of the form (q_i, q'_i) $(i = 0, 1, \ldots, n-1)$, where q'_i is the state that M reaches from the restarting configuration $q_i¢w\$ = q_i¢xa_m \cdots a_s v\$$ by moving right across the prefix $¢x$;
- the buffer B contains the sequence of pairs

$$\left((a_m, (q_{i_{m-1}}, p_{i_{m-1}}, q_{i_m}, T_m)), \ldots, (a_{s+1}, (q_{i_s}, p_{i_s}, q_{i_{s+1}}, T_{s+1}))\right),$$

where T_j is obtained from T by replacing each pair (q_i, q'_i) by the pair $(q_i, r_i^{(j)})$. Here $r_i^{(j)}$ is the state that M reaches from state q'_i by a sequence of MVR-steps that reads across the word $a_m \cdots a_j$ if that is possible. Otherwise, the pair (q_i, q'_i) is replaced by $(q_i, -)$.

On encoutering the letter a_s, A realizes that M cannot execute a rewrite step that completes the partial computation on input w consisting of the sequence of cycles above in the sense of Case 1. In fact, as M cannot execute any rewrite step on a_s, no matter in which state it starts the current cycle, monotonicity of M implies that the partial computation above cannot be part of the computation of M on input w. Therefore, A empties the buffer B completely, and it replaces each pair (q_i, q'_i) of T by the pair $(q_i, \hat{q}_i)$, if starting from the restarting configuration $q_i¢xa_m \cdots a_{s+1}a_s v\$$, M reaches state $\hat{q}_i$ by moving right across the prefix $¢xa_m \cdots a_s$.

Case 3. This case is a combination of the two cases above. The word w has a factorization of the form $w = xa_m \cdots a_{s+1}a_s \cdots a_1 v$ such that, for each $j = s+1, \ldots, m$, there is a possible cycle of M of the form

$$q_{i_{j-1}}¢x \cdots a_j a_s \cdots a_1 v\$ \vdash^+_{\mathsf{MVR}} ¢x \cdots p_{i_{j-1}} a_j a_s \cdots a_1 v\$ \vdash q_{i_j}¢x \cdots a_{j+1} a_s \cdots a_1 v\$,$$

and from a restarting configuration $q'_{i_{s-1}}¢xa_m \cdots a_{s+1}a_s \cdots a_1 v\$$, M executes a rewrite operation of the form $\delta(p'_{i_{s-1}}, a_s) = (q'_{i_s}, \varepsilon)$ for some state $q'_{i_s} \neq q_{i_s}$. In addition, for each $\mu = 1, \ldots, s-1$, there is a possible cycle of M of the form

$$q'_{i_{\mu-1}}¢xa_m \cdots a_{\mu+1}a_\mu v\$ \vdash^+_{\mathsf{MVR}} ¢xa_m \cdots a_{\mu+1}p'_{i_{\mu-1}}a_\mu v\$ \vdash q'_{i_\mu}¢xa_m \cdots a_{\mu+1} v\$$$

such that $q'_{i_0} = q_0$. While reading the prefix $xa_m \cdots a_{s+1}$, A has collected the same information as in Case 2.

On encoutering the letter a_s, A realizes that M cannot execute a rewrite step that extends the partial computation on input w consisting of the sequence of cycles above in the sense of Case 1. As, however, there is a rewrite operation that M may apply to the letter a_s, A continues as follows:

- CRS remains unchanged;
- the table T remains unchanged;
- the buffer B is extended by the pair $(a_s, (q'_{i_{s-1}}, p'_{i_{s-1}}, q'_{i_s}, T_s))$, where T_s is obtained from T by replacing each pair $(q_i, r_i^{(s+1)})$ by the pair $(q_i, r_i^{(s)})$. Here $r_i^{(s)}$ is the state that M reaches from state $r_i^{(s+1)}$ by a MVR-step that reads across the letter a_s if that is possible. Otherwise, the pair $(q_i, r_i^{(s+1)})$ is replaced by $(q_i, -)$.

For each letter $a_{s-1}, \ldots, a_2$, A extends the buffer B in the same way as before. Observe that $m \leq n$ must still hold due to the fact that M is deterministic. On encoutering the letter a_1, A realizes that it has found a sequence of s cycles that is the first part of the computation of M on input w. Accordingly, it acts as in Case 1. Of course, the above situation may appear repeatedly, but the overall length of the longest sequence of possible rewrite operations that is stored in B is always bounded in length from above by the number n of states of M. This completes the description of Case 3.

Now the overall computation of A on input w proceeds as follows. Assume that A reads the letter a.

- If, for no pair (q_i, q_i') stored in the table T, $\delta(q_i', a)$ is a rewrite operation, and if the buffer B is empty, then the table T is updated by replacing (q_i, q_i'), for all i, by $(q_i, \hat{q}_i)$, if $\delta(q_i', a) = (\hat{q}_i, \mathsf{MVR})$, and by $(q_i, -)$, if $\delta(q_i', a)$ is undefined. In this situation, CRS remains unchanged.
- If, for no pair (q_i, q_i') stored in the table T, $\delta(q_i', a)$ is a rewrite operation, but the buffer is non-empty, then we are in the situation described by Case 2 above.
- If CRS is q_i, and for the pair (q_i, q_i') stored in T, $\delta(q_i', a) = (q_j, \varepsilon)$, then the rewrite operation of the current cycle of M has been detected. If B is empty, then CRS is set to q_j, and T and B remain unchanged. Observe that this is a special case of Case 1 above. If B is non-empty, then A proceeds as detailed in Cases 1 and 3 above.
- If $\delta(q_i', a) = (q_j, \varepsilon)$ for a pair (q_i, q_i') stored in T, where q_i is different from the state stored in CRS, then the corresponding information is pushed onto the buffer B as in Cases 1 and 3.

The finite-state acceptor A keeps on reading the word w letter by letter from left to right until it encounters the right sentinel \$. Now the actual state of A is accepting, if the pair (q_i, q_i') in the current table T that corresponds to the current state q_i of M stored in CRS satisfies the condition that $\delta(q_i', \$) = \mathsf{Accept}$. It follows that $L(A) = L \cdot \$$. Since the class of regular languages is closed under right quotients, this implies that the language L is regular. This completes the proof of Theorem 1. □

Actually, Theorem 1 even extends to nondeterministic nf-R-automata.

Theorem 2. $\mathcal{L}(\mathsf{mon\text{-}nf\text{-}R(1)}) = \mathsf{REG}$.

Proof. Let $M = (Q, \Sigma, \text{¢}, \$, q_0, 1, \delta)$ be a mon-nf-R(1)-automaton for $L \subseteq \Sigma^*$. Then M can be simulated by a nondeterministic finite-state acceptor (NFA) $A = (Q_A, \Sigma \cup \{\$\}, q_0^{(A)}, F, \delta_A)$ for the language $L \cdot \$$ by using exactly the same strategy as in the proof of Theorem 1. It only remains to verify the following claim.

Claim. Let $w = x a_m a_{m-1} \cdots a_1 v$ be an input word such that M has a computation on input w that begins with a sequence of m cycles such that, in the j-th cycle ($1 \leq j \leq m$), the letter a_j is deleted. Then $m \leq n$, where n is the number of internal states of M.

Thus, whenever in some simulation of a computation of M by A a sequence of cycles is detected that contains a repetition of a restarting state, then this sequence of cycles cannot possibly be part of a computation of M that begins with a proper initial configuration. Accordingly, the actual computation of A can be terminated in a non-accepting state. It follows that $L(A) = L \cdot \$$, which in turn implies that the language L itself is regular. □

Finally we want to extend Theorem 1 to nonforgetting RR-automata. For doing so we need the following technical result on deterministic two-way finite-state acceptors (2DFA) from [1] (pages 212–213).

Lemma 1. *Let B be a DFA. For each word x and each integer i, $1 \leq i \leq |x|$, let $q_B(x,i)$ be the internal state of B after processing the prefix of length i of x. Then there exists a 2DFA B' such that, for each input x and each $i \in \{2, 3, \ldots, |x|\}$, if B' starts its computation on x in a state corresponding to $q_B(x,i)$ with its head on the i-th symbol of x, then B' finishes its computation in a state that corresponds to $q_B(x, i-1)$ with its head on the $(i-1)$-th symbol of x. During this computation B' only visits (a part of) the prefix of length i of x.*

Theorem 3. $\mathcal{L}(\text{det-mon-nf-RR}(1)) = \text{REG}$.

Proof. Obviously it remains to prove the inclusion from left to right. So let $M = (Q, \Sigma, ¢, \$, q_0, 1, \delta)$ be a det-mon-nf-RR(1)-automaton for $L \subseteq \Sigma^*$. W.l.o.g. we may assume that $Q = \{q_0, q_1, \ldots, q_{n-1}\}$, and that M executes Restart- and Accept-instructions only on reading the \$-symbol. Below we describe a 2DFA $A = (Q_A, \Sigma \cup \{¢, \$\}, q_0^{(A)}, F, \delta_A)$ for $¢ \cdot L \cdot \$$. Essentially, A works in the very same way as the DFA in the proof of Theorem 1, but there is a technical problem that we must overcome:

Whenever M executes a rewrite operation, then we need to know whether this operation is within an accepting or a rejecting tail computation, or whether it is part of a cycle of M. In the latter case, we also need to know which state of M is entered by the restart operation of this cycle.

To solve this problem A will execute a preprocessing stage given an input of the form $¢w\$$ ($w \in \Sigma^*$).

Preprocessing Stage: Let $w \in \Sigma^*$, and let $¢w\$$ be the input for A.

Step 1. Starting from the initial configuration $q_0^{(A)}¢w\$$, A scans its input from left to right until it encounters the \$-symbol, that is, $q_0^{(A)}¢w\$ \vdash_A^+ ¢wq_1^{(A)}\$$.

Step 2. For each suffix v of w, and for each state $q \in Q$, let

$$P_{v\$}(q) = \{\, p \in Q \mid \exists p' \in Q : ¢pv\$ \vdash_{\mathsf{MVR}}^{|v|} ¢vp'\$ \vdash_{\mathsf{Restart}} q¢v\$ \,\},$$

and let

$$P_{v\$}(+) = \{\, p \in Q \mid \exists p' \in Q : ¢pv\$ \vdash_{\mathsf{MVR}}^{|v|} ¢vp'\$ \vdash \mathsf{Accept} \,\}.$$

The initial sets $P_{\$}(q)$ and $P_{\$}(+)$, which are easily obtained from M, are stored in A's finite-state control.

Step 3. Now A reads its tape from right to left, letter by letter. Assume that $w = xav$, where $x, v \in \Sigma^*$ and $a \in \Sigma$, and that A has moved left across the suffix v thereby computing the sets $P_{v\$}(q)$ $(q \in Q)$ and $P_{v\$}(+)$. Now A moves left reading the symbol a, and while doing so it updates the set $P_{v\$}(q)$ to

$$P_{av\$}(q) = \{\, p \in Q \mid \exists p' \in P_{v\$}(q) : \delta(p, a) = (p', \mathsf{MVR}) \,\}$$

for each $q \in Q$, and it updates the set $P_{v\$}(+)$ to

$$P_{av\$}(+) = \{\, p \in Q \mid \exists p' \in P_{v\$}(+) : \delta(p, a) = (p', \mathsf{MVR}) \,\}.$$

This process continues until A reaches the ¢-symbol. At that moment it has stored the sets $P_{w\$}(q)$ $(q \in Q)$ and $P_{w\$}(+)$ in its finite-state control.

Unfortunately, A can only store one collection of sets in its finite-state control, that is, when storing the sets $P_{av\$}(q)$ $(q \in Q)$ and $P_{av\$}(+)$, it forgets the sets $P_{v\$}(q)$ $(q \in Q)$ and $P_{v\$}(+)$. Fortunately, we can now apply Lemma 1 to the DFA realizing Steps 2 and 3 of the above preprocessing stage. Just observe that this acceptor works from right to left, and so we need the symmetric version of the above lemma.

We combine this 2DFA with the DFA from the proof of Theorem 1. For each rewrite operation of the form $\delta(q', a) = (p, \varepsilon)$ encountered in the simulation of M, we check whether $p \in P_{v\$}(+)$ or whether $p \in P_{v\$}(q)$ for some $q \in Q$, where v is the corresponding suffix of the input word w. Based on this information the simulation then continues as in the proof of Theorem 1. □

As the non-regular language $L = \{\, a^n b^n, a^n b^{n+1} \mid n \geq 0 \}$ is accepted by a monotone RR(1)-automaton, we see that Theorem 3 does not extend to nondeterministic nonforgetting RR(1)-automata.

4 Deterministic Nonforgetting R(1)-Automata

In [5] the *stateless* variants of R-automata were introduced, and in [12] *cooperating distributed systems* (CD-systems) of stateless deterministic R(1)-automata were studied. Here an R(1)-automaton $M = (Q, \Sigma, ¢, \$, q_0, 1, \delta)$ is called *stateless* if $Q = \{q_0\}$ holds. Thus, in this case M can simply be described by the 5-tuple $M = (\Sigma, ¢, \$, 1, \delta)$. A *CD-system of* stateless deterministic R(1)-*automata* (or a stl-det-local-CD-R(1)-system, for short) consists of a finite collection $\mathcal{M} = ((M_i, \sigma_i)_{i \in I}, I_0)$ of stateless deterministic R(1)-automata $M_i = (\Sigma, ¢, \$, 1, \delta_i)$ $(i \in I)$, *successor relations* $\sigma_i \subseteq I$ $(i \in I)$, and a subset $I_0 \subseteq I$ of *initial indices.* It is required that $I_0 \neq \emptyset$ and that $\sigma_i \neq \emptyset$ for all $i \in I$.

The computation of $\mathcal{M}$ (in mode $= 1$) on an input word w proceeds as follows. First an index $i_0 \in I_0$ is chosen nondeterministically. Then the R-automaton M_{i_0} starts the computation with the initial configuration that corresponds to input w

and executes one cycle. Thereafter an index $i_1 \in \sigma_{i_0}$ is chosen nondeterministically, and M_{i_1} continues the computation by executing one cycle. This continues until, for some $l \geq 0$, the automaton M_{i_l} accepts. Should at some stage M_{i_l} be unable to execute a cycle or to accept, then the computation fails.

By $L(\mathcal{M})$ we denote the language that the CD-R-system $\mathcal{M}$ accepts. It consists of all words $w \in \Sigma^*$ that are accepted by $\mathcal{M}$ as described above. By $\mathcal{L}$(stl-det-local-CD-R(1)) we denote the class of languages that are accepted by stl-det-local-CD-R(1)-systems.

A stl-det-local-CD-R(1)-system $\mathcal{M} = ((M_i, \sigma_i)_{i \in I}, I_0)$ is called *globally deterministic* if $|I_0| = 1$, and if each restart operation of M_i $(i \in I)$ is associated with a successor $j \in \sigma_i$ of M_i. Thus, while computations of locally deterministic CD-R(1)-systems are in general still nondeterministic, those of globally deterministic CD-R(1)-systems are completely deterministic. By $\mathcal{L}$(stl-det-global-CD-R(1)) we denote the class of languages that are accepted by stl-det-global-CD-R(1)-systems. It is known that $\mathcal{L}$(stl-det-local-CD-R(1)) only contains languages with a semi-linear Parikh image, and that it contains all rational trace languages [12]. On the other hand, it has been shown that the language class $\mathcal{L}$(stl-det-global-CD-R(1)) is incomparable to the class of rational trace languages with respect to inclusion [13].

As globally deterministic CD-systems can be simulated by deterministic nonforgetting restarting automata [8], we have the following inclusion result.

Proposition 3. $\mathcal{L}$(stl-det-global-CD-R(1)) $\subsetneq$ $\mathcal{L}$(det-nf-R(1)).

This inclusion is strict, as witnessed by the language $L_{pr} = \{\, wc \in \{a,b\}^* \mid |w|_a \geq |w|_b \geq 0 \,\} \in \mathcal{L}$(det-nf-R(1)), which is not accepted by any stl-det-global-CD-R(1))-system [13]. In fact, the language class $\mathcal{L}$(det-nf-R(1)) even contains languages that are not semi-linear.

Lemma 2. $L_{ex2} = \{\, (ab)^{2^n}(cd)^n e^2 \mid n \geq 0 \,\} \in \mathcal{L}$(det-nf-R(1)).

Proof. It can be shown that L_{ex2} is accepted by the det-nf-R(1)-automaton M_{ex2} that is given through the following meta-instructions (see, e.g., [9]):

(1) $(q_0, ¢ \cdot (ab)^+ \cdot (cd)^* \cdot e, e \to \varepsilon, q_1)$,
(2) $(q_1, ¢ae\$, \mathsf{Accept})$,
(3) $(q_1, ¢ \cdot (aab)^* \cdot a, b \to \varepsilon, q_1)$,
(4) $(q_1, ¢ \cdot (aab)^+, c \to \varepsilon, q_2)$,
(5) $(q_2, ¢a, a \to \varepsilon, q_2)$,
(6) $(q_2, ¢ \cdot (ab)^+ \cdot a, a \to \varepsilon, q_2)$,
(7) $(q_2, ¢ \cdot (ab)^+, d \to \varepsilon, q_1)$. □

As the Parikh image of L_{ex2} is not a semi-linear subset of $\mathbb{N}^5$, it follows that $\mathcal{L}$(det-nf-R(1)) is not contained in the class of semi-linear languages.

The language $L_\vee = \{\, w \in \{a,b\}^* \mid \exists n \geq 0 : |w|_a = n \text{ and } |w|_b \in \{n, 2n\} \,\}$ is a rational trace language that is not accepted by any stl-det-global-CD-R(1)-system [13]. Actually, the following stronger result can be shown.

Lemma 3. $L_\vee \notin \mathcal{L}$(det-nf-R(1)).

Together with the above result on L_{ex2} this lemma yields the following incomparability results.

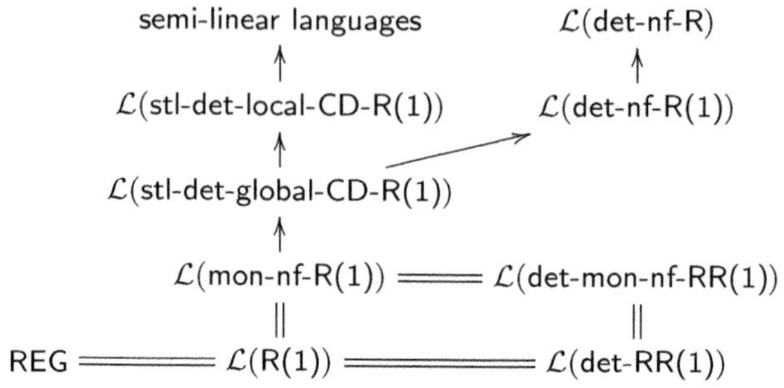

Fig. 1. Hierarchy of language classes accepted by various types of nonforgetting R-automata with window size 1. Each arrow represents a proper inclusion.

Corollary 1. $\mathcal{L}$(det-nf-R(1)) *is incomparable to* $\mathcal{L}$(stl-det-local-CD-R(1)) *and to the class of semi-linear languages with respect to inclusion.*

In fact, we think that Lemma 3 even extends to $\mathcal{L}$(det-nf-R). One can easily design a det-nf-R(2)-automaton that accepts the language $L_{\text{pal}} = \{ w \in \{a,b\}^* \mid w = w^R \}$. On the other hand, it can be shown that $L_{\text{pal}} \notin \mathcal{L}$(det-nf-R(1)) holds. Thus, we have the following proper inclusion.

Corollary 2. $\mathcal{L}$(det-nf-R(1)) $\subsetneq$ $\mathcal{L}$(det-nf-R).

Figure 1 summarizes the inclusion relations on $\mathcal{L}$(det-nf-R(1)).

5 Concluding Remarks

As monotone nf-R(1)- and det-nf-RR(1)-automata accept just the regular languages, the question for their descriptional complexity arises: Do these more involved types of automata offer more succinct representations for regular languages than (standard) R(1)- and det-RR(1)-automata? Here we have the following first preliminary result.

Proposition 4. *For each $n \geq 2$, there exists a language $L_n \subseteq \{a,b\}^*$ that is accepted by a* det-mon-nf-RR(1)*-automaton with $O(n)$ states, but every (standard)* det-RR(1)*-automaton accepting L_n has at least $O(2^n)$ many states.*

Proof. For $n \geq 2$, let $L_n = \{ w \in \{a,b\}^m \mid m > n,\ w_n = a, \text{ and } w_{m+1-n} = b \}$, where w_i $(1 \leq i \leq |w|)$ denotes the i-th symbol of w. A det-mon-nf-RR(1)-automaton M for L_n can be described by the following meta-instructions, where $x \in \{a,b\}$:

$$(1)\ (q_0, ¢, x \to \varepsilon, \{a,b\}^{n-2} \cdot a \cdot \{a,b\}^+ \cdot \$, q_1),$$
$$(2)\ (q_1, ¢ \cdot a^*, b \to \varepsilon, \{a,b\}^{n-1} \cdot \{a,b\}^+ \cdot \$, q_1),$$
$$(3)\ (q_1, ¢ \cdot a^*, b \to \varepsilon, \{a,b\}^{n-1} \cdot \$, \mathsf{Accept}).$$

For realizing these meta-instructions $O(n)$ states suffice, as M must be able to count from 1 to n. On the other hand, a deterministic RR(1)-automaton M' for

L_n must accept each word $w \in L_n$ satisfying $|w| = m \leq 2n$ in a tail computation because of the correctness preserving property (see, e.g., [4]). Hence, it behaves essentially just like a DFA, which implies that it needs $O(2^n)$ states to check the condition $w_{m+1-n} = b$. □

In [14] Reimann studied the descriptional complexity of R(1)-automata and of deterministic RR(1)-automata. It remains to carry his studies over to (deterministic) mon-nf-R(1)- and det-mon-nf-RR(1)-automata.

References

1. Aho, A., Hopcroft, J., Ullman, J.: A general theory of translation. Math. Systems Theory 3, 193–221 (1969)
2. Jančar, P., Mráz, F., Plátek, M., Vogel, J.: Restarting automata. In: Reichel, H. (ed.) FCT 1995. LNCS, vol. 965, pp. 283–292. Springer, Heidelberg (1995)
3. Jančar, P., Mráz, F., Plátek, M., Vogel, J.: On restarting automata with rewriting. In: Păun, G., Salomaa, A. (eds.) New Trends in Formal Languages. LNCS, vol. 1218, pp. 119–136. Springer, Heidelberg (1997)
4. Jančar, P., Mráz, F., Plátek, M., Vogel, J.: On monotonic automata with a restart operation. J. Autom. Lang. Comb. 4, 287–311 (1999)
5. Kutrib, M., Messerschmidt, H., Otto, F.: On stateless two-pushdown automata and restarting automata. In: Csuhaj-Varju, E., Esik, Z. (eds.) Proc. of AFL 2008, pp. 257–268. Computer and Automation Research Institute, Hungarian Academy of Sciences (2008)
6. McNaughton, R., Narendran, P., Otto, F.: Church-Rosser Thue systems and formal languages. J. ACM 35, 324–344 (1988)
7. Messerschmidt, H., Otto, F.: Cooperating distributed systems of restarting automata. Intern. J. Found. Comp. Sci. 18, 1333–1342 (2007)
8. Messerschmidt, H., Otto, F.: On deterministic CD-systems of restarting automata. Intern. J. Found. Comp. Sci. 20, 185–209 (2009)
9. Messerschmidt, H., Otto, F.: A hierarchy of monotone deterministic nonforgetting restarting automata. Theory Comput. Syst. 48, 343–373 (2011)
10. Messerschmidt, H., Stamer, H.: Restart-Automaten mit mehreren Restart-Zuständen. In: Bordihn, H. (ed.) Proc. of Workshop 'Formale Sprachen in der Linguistik' und 14. Theorietag 'Automaten und Formale Sprachen', pp. 111–116. Institut für Informatik, Universität Potsdam (2004)
11. Mráz, F.: Lookahead hierarchies of restarting automata. J. Autom. Lang. Comb. 6, 493–506 (2001)
12. Nagy, B., Otto, F.: CD-systems of stateless deterministic R(1)-automata accept all rational trace languages. In: Dediu, A., Fernau, H., Martín-Vide, C. (eds.) LATA 2010. LNCS, vol. 6031, pp. 463–474. Springer, Heidelberg (2010)
13. Nagy, B., Otto, F.: Globally deterministic CD-systems of stateless R(1)-automata. In: Dediu, A.-H., Inenaga, S., Martin-Vide, C. (eds.) LATA 2011. LNCS, vol. 6638, pp. 390–401. Springer, Heidelberg (2011)
14. Reimann, J.: Beschreibungskomplexität von Restart-Automaten. PhD thesis, Naturwissenschaftliche Fachbereiche, Justus-Liebig-Universität Giessen (2007)

On Two-Way Transducers

Oscar H. Ibarra[1,*] and Hsu-Chun Yen[2,**]

[1] Dept. of Computer Science, Univ. of California, Santa Barbara, CA 93106, USA
ibarra@cs.ucsb.edu

[2] Dept. of Electrical Engineering, National Taiwan Univ., Taipei, Taiwan 106, ROC
yen@cc.ee.ntu.edu.tw

Abstract. We look at some classes of two-way transducers with auxiliary memory and investigate their containment and equivalence problems. We believe that our results are the strongest known to date concerning two-way transducers.

Keywords: two-way transducer, containment problem, equivalence problem.

1 Introduction

It is known that the equivalence problem for two-way deterministic finite transducers is decidable [4]. We generalize this result for some models of two-way transducers with auxiliary memory.

We consider two-way transducers, i.e., two-way finite automata (with input end markers # and \$) augmented with reversal-bounded counters and a one-way output tape. Call the nondeterministic (resp., deterministic) version 2NCMT (resp., 2DCMT). The relation defined by such a transducer A is $R(A) = \{(x, y) \mid A$, when started in its initial state on the left end marker of $\#x\$$, outputs y and falls off the right end marker in an accepting state $\}$. The transducer is *finite-crossing* if there is some fixed k such that in every accepting computation on any input $\#x\$$, the number of times the input head crosses the boundary between any two adjacent symbols of $\#x\$$ is at most k. Note that the number of turns (i.e., changes in direction from left-to-right and right-to-left and vice-versa) the input head makes on the input may be unbounded. Also note that the requirement is only for accepting computations. So if $R(A) = \varnothing$, then A is finite-crossing and, in fact, k-crossing for any k. We assume that when we are given a finite-crossing machine, the integer k for which the machine is k-crossing is also specified. Unfortunately, as we will see, it is undecidable to determine if a 2DCMT is finite-crossing, or k-crossing for a given k.

We show that the following problems are decidable:

1. Given a finite-crossing 2NCMT A_1 and a finite-crossing 2DCMT A_2, is $R(A_1) \subseteq R(A_2)$? Hence, equivalence of finite-crossing 2DCMTs is decidable.

* Supported in part by NSF Grant CCF-0524136.

** Supported in part by NSC Grant NSC-97-2221-E-002-095-MY3, Taiwan.

G. Mauri and A. Leporati (Eds.): DLT 2011, LNCS 6795, pp. 300–311, 2011.

2. Given a one-way nondeterministic pushdown transducer with reversal -bounded counters (1NPCMT) A_1 and a finite-crossing 2DCMT A_2, is $R(A_1) \subseteq R(A_2)$?
3. Given a finite-crossing 2NCMT A_1 and a 1DPCMT A_2 (the deterministic version of 1NPCMT), is $R(A_1) \subseteq R(A_2)$?
4. Given a finite-crossing 2DCMT A_1 and a 1DPCMT A_2, is $R(A_1) = R(A_2)$?

We believe that these results are the strongest known to date concerning containment and equivalence of transducers. We note that in the above results, the "finite-crossing" assumption is necessary, since when the two-way input is unrestricted, the equivalence problem becomes undecidable, as we shall see. We also note that the 1NPCMT and 1DPCMT in (2), (3) and (4) above cannot be generalized to be two-way, since we can show that it is undecidable to determine, given a 2DPCMT A, whether $R(A) = \varnothing$, even when A makes only two turns on the input. However, we show:

5. It is decidable to determine, given two finite-crossing 2DPCMTs whose inputs come from a bounded language (i.e., from $w_1^* \cdots w_k^*$ for some non-null strings $w_1, \ldots, w_k$) A_1 and A_2, whether $R(A_1) \subseteq R(A_2)$. (Hence, equivalence is also decidable.)

The proofs for the results above use the decidability of emptiness for a large class of acceptors, called 3-phase finite-crossing 2NPCMs, which we introduce in Section 3. We show that it is decidable to determine, given a 3-phase finite-crossing 2NPCM M, whether the language it accepts is empty.

It should also be noted that, as we shall see later, the assumption of A_2 being "deterministic" in (1) - (4) above is required. In fact, the equivalence problem is known to be undecidable even for one-way nondeterministic finite transducers (1NFTs) [3,9].

As "*single-valuedness*" is a natural extension of the notion of determinism and equivalence of single-valued 1NFTs (i.e., 1NFTs that output at most one value for every input) is decidable [1,12], we also study containment and equivalence between single-valued finite-crossing two-way nondeterministic finite transducers (2NFTs) and various finite-crossing two-way transducers with auxiliary memory. We show the following to be decidable:

6. Given a finite-crossing 2NCMT (or a 1NPCMT) A_1 and a single-valued finite-crossing 2NFT A_2, is $R(A_1) \subseteq R(A_2)$?
7. Given a single-valued finite-crossing 2NFT A_1 and a finite-crossing 2DCMT (or a 1DPCMT) A_2, is $R(A_1) \subseteq R(A_2)$?
8. Given a single-valued finite-crossing 2NFT A_1 and a finite-crossing 2DCMT (or a 1DPCMT) A_2, is $R(A_1) = R(A_2)$?

To put our results into proper perspective, main decidability and undecidability results of the containment problem for a variety of transducers considered in this paper are summarized in Table 1. The undecidability results follow from the fact that the emptiness and universality problems for 2-way deterministic reversal-bounded counter machines are undecidable [8] and that the containment problem for deterministic pushdown automata is undecidable [2].

Table 1. Decidability/undecidability of the question "$R(A_1) \subseteq R(A_2)$?". (D: decidable; U: undecidable; *fc-2DCMT*: finite-crossing 2DCMT; *fc-2NCMT*: finite-crossing 2NCMT; *sv-fc-2NFT*: single-valued finite-crossing 2NFT.)

$R(A_1) \subseteq R(A_2)$?	A_2			
A_1	*2DCMT*	*fc-2DCMT*	*1DPCMT*	*sv-fc-2NFT*
2DCMT	U	U	U	U
fc-2NCMT	U	D	D	D
1NPCMT	U	D	U	D
sv-fc-2NFT	U	D	D	D

2 Preliminaries

A one-way nondeterministic reversal-bounded multicounter machine (1NCM) M is a one-way NFA augmented with multiple 1-reversal counters which are initially set to zero. At each step, every counter can be incremented by 1, decremented by 1, or left unchanged, and can be tested for zero. A zero counter cannot be decremented. M is 1-reversal in that it has the property that once a counter is decremented, it can no longer be incremented. A 1NCM augmented with a pushdown stack is called a 1NPCM. The deterministic versions are called 1DCM and 1DPCM, respectively. A machine has reversal-bounded counters if there is a given r such that each counter makes at most r reversals during the computation. Clearly, a counter that makes r reversals can be simulated by $\lceil \frac{r+1}{2} \rceil$ counters each of which makes 1 reversal. Hence, in this paper, when the number of reversal-bounded counters is not a parameter in the problem being investigated, we may assume that the counters are 1-reversal. The following result is known [8]:

Theorem 1. *The emptiness problem (given M, is $L(M) = \varnothing$?) for 1NPCMs (hence, also for 1NCMs) is decidable.*

A 2NCM (2DCM) is a two-way NCM (DCM) with input left and right end markers # and \$. A 2NCMT (2DCMT) A is a 2NCM (2DCM) with outputs. The relation it defines is $R(A) = \{(x, y) \mid A$, when started in its initial state on the left end marker of $\#x\$$, outputs y and falls off the right end marker in an accepting state $\}$. Let k be a positive integer. Transducer A is said to be *k-valued* if for every input x, the cardinality of the set $\{y \mid (x, y) \in R(A)\}$ is $\leq k$. A is *finite-valued* if it is k-valued for some k. A 2NFT (2DFT) is a 2NCMT (2DCMT) with no reversal-bounded counters. The one-way versions (with no input end markers) are denoted by 1NCMT, 1DCMT, 1NFT, 1DFT. For machines with reversal-bounded counters, we write the suffix "(k)" to denote the fact that there are k counters. So, e.g., *2DCM*(k) means a 2DCM with k reversal-bounded counters.

It is known that the emptiness problem for 2DCM(2)s is undecidable [8]. In fact, this result holds for machines operating on letter-bounded languages (i.e., subsets of $a_1^* \cdots a_k^*$ for some k and distinct symbols $a_1, \ldots, a_k$). Since a 2DCM can trivially be made a 2DCMT by having it output ε at each step, it follows that it is undecidable to determine, given a 2DCMT(2) A, whether $R(A) = \varnothing$. Hence,

Proposition 1. *The containment and equivalence problems for 2DCMT(2)s (even on letter-bounded language inputs) are undecidable.*

A *finite-crossing* 2NCM (2DCM, 2NCMT, 2DCMT) is a machine with the property that for some specified k, in every accepting computation on any input $\#x\$$, the number of times the input head crosses the boundary between any two adjacent symbols of $\#x\$$ is at most k. Note that the number of turns the input head makes on the input may be unbounded. We assume that when we are given a finite-crossing machine, we are also given the integer k for which the machine is k-crossing. This is because of the following undecidability result:

Proposition 2. *It is undecidable to determine, given a 2DCMT(2) A, whether it is finite-crossing (or whether it is k-crossing for a given k).*

Proof. Let A be a 2DCMT(2) with input alphabet Σ. Let d be a new symbol not in Σ. We construct a 2DCMT(2) A', which when given xd^n, where $x \in \Sigma^*$ and $n \geq 1$, simulates the computation of A on x. When A accepts, then A' makes n left-to-right and right-to-left turns on x and accepts. Clearly, A' is k-crossing for any k if and only if $R(A) = \varnothing$, which is undecidable by Proposition 1. It should be noted that the finite crossing condition involves only accepting computations. □

We will need the following result which was shown in [5].

Theorem 2. *We can effectively construct, given a finite-crossing 2NCM M, a 1NCM M' such that $L(M') = L(M)$. Hence, the emptiness problem for finite-crossing 2NCMs is decidable.*

Consider the language $L = \{x \mid x \text{ in } \{a, b, c, d\}^+$, the sum of the lengths of all runs of c's occurring between symbols a and b in this order (i.e., in substring ayb for some y in $\{c, d\}^*$) equals the number of d's $\}$. L can be accepted by a 5-crossing 2DCM with only one 1-reversal counter M. However, as noted in [5], L cannot be accepted by any 1DCM or by a 2DCM which makes a fixed number of turns on the input. Clearly, we can construct a 5-crossing 2DCMT A from the 5-crossing 2DCM which outputs ε on every move. The relation defined by A is then $R(A) = L \times \{\varepsilon\}$, which cannot be defined by any 1DCMT. Hence, finite-crossing 2DCMTs can define more relations than 1DCMTs. Note that L can easily be accepted by a 1NCM (with one 1-reversal counter) by just guessing and verifying, as it moves left-to-right on the input x, the locations of the substrings of the form ayb in x.

From Theorems 1 and 2, we get the following corollary:

Corollary 1. *It is decidable to determine, given a finite-crossing 2NCM M_1 and a 1NPCM M_2, whether $L(M_1) \cap L(M_2) = \varnothing$.*

Proof. From Theorem 2, we construct a 1NCM M_1' from finite-crossing 2NCM M_1 such that $L(M_1') = L(M_1)$. Then we construct a 1NPCM M which, on a given input, simulates M_1' and M_2 in parallel. Then $L(M_1) \cap L(M_2) = \varnothing$ if and only if $L(M) = \varnothing$, which is decidable by Theorem 1. □

3 3-Phase Finite-Crossing 2NPCMs

We can generalize Theorems 1 and 2. Define a 3-phase finite-crossing 2NPCM M which operates in three phases: In the first phase, M operates as a finite-crossing 2NCM without using the stack. In the second phase, with the configuration (state, input head position, and counter values) the first phase left off, M operates as a 1NPCM where the head can only move right on the input. Finally, in the third phase with the configuration (state, head position, counter values but not the stack) the second phase left off, M operates again as a finite-crossing 2NCM without using the stack. Note that not all phases may be present. So, e.g., M can accept with only Phase 1, or with only Phases 1 and 2.

We will need the following result for the proofs in the next section.

Theorem 3. *It is decidable to determine, given a 3-phase finite-crossing 2NPCM M, whether $L(M) = \varnothing$.*

Proof. Assume that M has disjoint state set and input alphabet. Suppose M has n counters. Let $0, 1$ be new symbols. Define the following languages:

1. $L_1 = \{q1^i01^{j_1}0 \cdots 01^{j_n}xp1^s01^{k_1}0 \cdots 01^{k_n} \mid$ in Phase 1, M on input x ends the phase in state q with the input head on position i of x and the 1-reversal counters with values $j_1, \ldots, j_n\}$.
2. $L_2 = \{q1^i01^{j_1}0 \cdots 01^{j_n}xp1^s01^{k_1}0 \cdots 01^{k_n} \mid$ in Phase 2, M when started in state q with the input head on position i of x and the counters with values $j_1, \ldots, j_n$, ends the phase in state p with the input head on position s of x and the 1-reversal counters with values $k_1, \ldots, k_n\}$.
3. $L_3 = \{q1^i01^{j_1}0 \cdots 01^{j_n}xp1^s01^{k_1}0 \cdots 01^{k_n} \mid$ in Phase 3, M when started in state p with the input head on position s of x and the counters with values $k_1, \ldots, k_n$, accepts$\}$.

We construct a finite-crossing 2NCM M_1 which, when given input
$q1^i01^{j_1}0 \cdots 01^{j_n}xp1^s01^{k_1}0 \cdots 01^{k_n}$
simulates Phase 1 of M on x and at the end of Phase 1, checks that the state is q and, using additional counters, checks that the head is on position i and the counters have values $j_1, \ldots, j_n$.

Similarly, we can construct a 1 NPCM M_2 and a finite-crossing 2NCM M_3 accepting L_2 and L_3, respectively. One can easily verify that $L(M) = \varnothing$ if and only if $L(M_1) \cap L(M_2) \cap L(M_3) = \varnothing$. We then construct a finite-crossing 2NCM M_4 such that $L(M_4) = L(M_1) \cap L(M_3)$. Then $L(M) = \varnothing$ if and only if $L(M_4) \cap L(M_2) = \varnothing$, which is decidable by Corollary 1. □

Theorem 3 seems to be the strongest result that we can prove in the sense that we cannot generalize the 1NPCM in the second phase to be a finite-crossing 2NPCM. In fact, as we will see in the next section, a 2DPDA which makes only 3 reversals on the stack and 2-turns on the input has an undecidable emptiness problem. However, for machines accepting bounded languages, we have the following:

Theorem 4. *It is decidable to determine, given a finite-crossing 2DPCM M accepting a bounded language (i.e., $L(M) \subseteq w_1^* \cdots w_k^*$ for some given $k \geq 1$ and strings $w_1, \ldots, w_k$), whether $L(M) = \varnothing$.*

Proof. It was recently shown in [11] that the set $Q = \{(i_1, \ldots, i_k) \mid w_1^{i_1} \cdots w_k^{i_k}$ is in $L(M)\}$ is a semilinear set effectively constructable from M. The theorem then follows since emptiness of semilinear sets is decidable. □

4 Proofs of the Main Results

We begin with the following lemma.

Lemma 1. *We can effectively convert a finite-crossing 2DCM M_1 to an equivalent finite-crossing 2DCM M_2 such that on every input, M_2 eventually falls off the right end marker in either an accepting state or a rejecting state.*

Proof. Let M_1 be a k-crossing 2DCM with s transition rules and d 1-reversal counters. Suppose $x = a_1 \cdots a_n$ is an input (where a_1 and a_2 denote the end markers) accepted by M_1 (hence M_1 halts on x). Denote by $v(i)$, the number of times the head of M_1 moves to the right or to the left of symbol a_i during the computation. We do not count (i.e., $v(i)$ does not include) the times when the head remains on symbol a_i , which can be unbounded. Clearly, $v(1)+\cdots+v(n) \leq s^k n$.

We construct a halting finite-crossing 2DCM M_2 simulating M_1 as follows. M_2 will have a new counter C, in addition to the d counters of M_1. M_2 starts by moving its head on the input incrementing C by s^k for each symbol (thus, M_2 remains on each symbol s^k times, since it can only increment the counter by 1 at each step). When $s^k n$ has been stored in C, M_2 increments C by 1 (hence C will have value $s^k n + 1$) and moves the input head back to the left end marker. Then M_2 simulates M_1 keeping track of the following situations:

1. If the input head remains on a symbol a_i more than s^k steps without a counter being decremented, then M_1 is in an infinite loop. M_2 moves the input head to the right and falls off the right end marker in a rejecting state.
2. If the input head remains on a symbol a_i more than s^k steps but with at least one counter getting decremented during this time period, the simulation continues.
3. Every time the input head moves left or right of a symbol, M_2 decrements C.

4. If during the simulation C becomes zero, then M_1 is in an infinite loop. M_2 moves the input head to the right and falls off the right end marker in a rejecting state.
5. If M_1 enters a configuration where there is no next move, M_2 moves the input head to the right and falls off the right end marker in a rejecting state.

Since the counters are 1-reversal, when a counter becomes zero, it remains zero. It follows that only the 5 situations above can happen. □

Notation: If R is a relation, $domain(R) = \{x \mid (x, y) \text{ is in } R \text{ for some } y\}$. If A s a transducer, the underlying acceptor M of A is the transducer with its outputs suppressed. Thus, $L(M) = domain(R(A))$.

Theorem 5. *The following problems are decidable:*

1. *Given a finite-crossing 2NCMT A_1 and a finite-crossing 2DCMT A_2, is $R(A_1) \subseteq R(A_2)$?*
2. *Given two finite-crossing 2DCMTs A_1 and A_2, is $R(A_1) = R(A_2)$?*

Proof. Clearly, we only need to prove (1). Let M_1 be the underlying finite-crossing 2NCM of A_1 and M_2 be the underlying finite-crossing 2DCM of A_2. Thus, $L(M_1) = domain(R(A_1))$ and $L(M_2) = domain(R(A_2))$. By Lemma 1, we may assume that M_2 always accepts or rejects a given input.

Clearly, $R(A_1) \not\subseteq R(A_2)$ if and only if there exists an x such that the following two conditions are satisfied:

(a) For some y, (x, y) is in $R(A_1)$.
(b) x is not in $domain(R(A_2))$, or x is in $domain(R(A_2))$ and the only z such that (x, z) is in $R(A_2)$ is different from y.

Given A_1, M_1, A_2, M_2, we construct a finite-crossing 2NCM M such that $L(M) \neq \varnothing$ if and only if the conditions above are satisfied. The details of the operation of M are as follows:

1. M stores a nondeterministically chosen number r in two special counters C_1 and D_1. (Thus, M increments these counters simultaneously r times.) Then M simulates an accepting computation of A_1 (suppressing the outputs) and determines the r-th symbol, say a, of output y of the computation. M decrements counter D_1 to locate the symbol. Note that $a = \varepsilon$ if $r > |y|$.
2. M then simulates M_2 on x. If M_2 rejects x (thus x is in $domain(R(A_1))$ but not in $domain(R(A_2))$), then M accepts. If M_2 accepts x (thus, x is in $domain(R(A_1))$ and in $domain(R(A_2))$), then, as in (1), M stores a nondeterministically chosen number s in another set of counters C_2 and D_2. M then simulates A_2 and determines the s-th symbol, say b, of z. M uses counter D_2 for this purpose. Again, $b = \varepsilon$ if $s > |z|$. Then M verifies that $a \neq b$ and $C_1 = C_2$ (by decrementing these counters to check that they become zero at the same time), and accepts.

The result follows since we can decide if $L(M) = \varnothing$ by Theorem 2. □

We note that the condition that A_2 in Theorem 5 is deterministic is necessary, as the following shows.

Proposition 3. *There is a* fixed *single-valued real-time 1-reversal 1NCMT(1) A (thus the transducer has one 1-reversal counter and the input head moves right at every step and only outputs one value for every input string) over some input alphabet Σ such that it is undecidable to determine, given a positive integer d, whether $R(A_d) = \Sigma^* \times \{\varepsilon\}$, where A_d denotes A with initial counter value d. (Note that d is the only parameter in the decision problem.)*

Proof. Theorem 1 in a recent paper [10] showed that there is a *fixed* nondeterministic real-time machine M with one 1-reversal counter over some input alphabet Σ such that it is undecidable to determine, given an arbitrary positive integer d, whether $L(M_d) = \Sigma^*$ (where M_d denotes M with initial counter value d.)

We construct from M a fixed real-time 1-reversal 1NCMT(1) that outputs ε at every step. Hence, $R(A_d) = L(M_d) \times \{\varepsilon\}$, and $R(A_d) = \Sigma^* \times \{\varepsilon\}$ if and only if $L(M_d) = \Sigma^*$, which would be undecidable. □

Proposition 3 contrasts the decidability of equivalence for finite-valued 1NFTs (i.e., when transducers have no counter) [6,1,12]. However, the following result (which we will need later) shows that we can decide if a finite-crossing 2NCMT is k-valued for a given k.

Theorem 6. *It is decidable to determine, given a finite-crossing 2NCMT A and a positive integer k, whether A is k-valued.*

Proof. First consider $k = 1$. Given a finite-crossing 2NCMT A, we construct a finite-crossing 2NCM M, which on input $\#x\$$, simulates A and uses two additional counters C_1 and C_2. M simulates A suppressing the output and storing in counter C_1 a position (which is nondeterministically selected) on the output in this accepting computation where another accepting computation may differ. It also remembers the output symbol a in this location. When A accepts, M resets all the counters (except C_1) to zero and simulates another accepting computation of A and stores in counter C_2 a position (again nondeterministically selected) of a symbol b in this second accepting computation and makes sure that $a \neq b$. When A accepts, M checks that counters C_1 and C_2 have the same values by decrementing C_1 and C_2 simultaneously.

Clearly the above idea generalizes for any k. In this case, M nondeterministically simulates $k+1$ accepting computations, say $T_1, \ldots, T_{k+1}$ of A on input $\#x\$$ and verifies that these computations produce pair-wise distinct outputs (without actually generating the outputs). M uses two 1-reversal counters to check that two computations give different outputs (i.e., disagree on the output symbols in some position). Thus, M uses a total of $k(k+1)$ 1-reversal counters. Then A is not k-valued if and only if $L(M) \neq \varnothing$, which is decidable by Theorem 2. □

We also note that Theorem 5 cannot be generalized to the case when A_1 or A_2 has an unrestricted counter, even when the input head is constrained to make only 2 turns, as the following result shows:

Proposition 4. *There is a fixed 2DCAT (i.e., a 2DFA transducer with one unrestricted counter) A which makes only 2 input head turns such that it is undecidable to determine, given an arbitrary positive integer d, whether* $R(A_d) = \varnothing$, *where* A_d *is A with initial counter value d. (Thus, d is the only parameter to the decision problem.)*

Proof. It was shown in [10] (Theorem 9) that there is a *fixed* deterministic real-time counter machine M and a *fixed* deterministic real-time counter machine M' such that it is undecidable to determine, given an arbitrary positive integer d, whether $L(M_d) \cap L(M') = \varnothing$, where M_d is M with initial counter value d. (M' always starts with counter value 0.) The counters of the machines are unrestricted.

Clearly, we can construct a fixed 2DCAT A that outputs ε at every step which, with initial counter value d, simulates A and when it accepts, zeros out the counter, returns the head to the left end marker and simulates M'. Then $R(A_d) = (L(M_d) \cap L(M')) \times \{\varepsilon\}$. It follows that it is undecidable to determine, given A_d, whether $R(A_d) = \varnothing$. □

We can generalize Theorem 5. A finite-crossing 2UDCMT A is a finite union of finite-crossing 2DCMTs. Thus, $R(A) = R(A^1) \cup \cdots \cup R(A^n)$ for some n, where A^i is a finite-crossing 2DCMT. We assume that the state sets of $A^1, \ldots, A^n$ are disjoint. Note that A is a special case of a 2NCMT in that the only nondeterministic move of A is at the start of the computation: With its input head on the left end marker, A outputs ε, remains on the end marker, and has a choice of entering the initial state q_0^i of A^i, $i = 1, \ldots, n$.

Theorem 7. *The following problems are decidable:*

1. *Given a finite-crossing 2NCMT* A_1 *and a finite-crossing 2UDCMT* A_2, *is* $R(A_1) \subseteq R(A_2)$?
2. *Given two finite-crossing 2UDCMTs* A_1 *and* A_2, *is* $R(A_1) = R(A_2)$?

Proof. Again, we only need to prove (1). Let M_1 be the underlying finite-crossing 2NCM of A_1 (hence $L(M_1) = domain(R(A_1))$) and let n finite-crossing 2DCMTs $A_2^1, \ldots, A_2^n$ (for some n) be such that $R(A_2) = R(A_2^1) \cup \ldots \cup R(A_2^n)$. Let M_2^i be the underlying finite-crossing 2DCM of A_2^i (hence $L(M_2^i) = domain(R(A_2^i))$) for $1 \le i \le n$. Then $R(A_1) \not\subseteq R(A_2)$ if and only if there exists an x such that the following two conditions are satisfied:

(a) For some y, (x, y) is in $R(A_1)$.
(b) For $1 \le i \le n$: x is not in $domain(R(A_2^i))$, or if x is in $domain(R(A_2^i))$, then there is a z_i such that (x, z_i) is in $R(A_2^i)$ and $y \neq z_i$. (Note that z_i is unique.)

We construct from finite-crossing 2NCMT A_1 and its underlying finite-crossing NCM M_1, finite-crossing 2DCMTs $A_2^1, \ldots, A_2^n$ and their underlying finite-crossing 2DCMs $M_2^1, \ldots, M_2^n$, a finite crossing 2NCM M, which when given an input x, accepts if conditions (a) and (b) of the Claim above are satisfied. Hence $L(M) \neq \varnothing$ if and only if $R(A_1) \not\subseteq R(A_2)$.

The construction of M generalizes the idea in the construction in the proof of Theorem 5. Since x may be in the domain of all the A_2^i's (i.e., accepted by all the M_2^i's), and M needs to check that y is different from all the z_i's, M may need to use $4n$ 1-reversal counters for this purpose. Note that M_2^i is used to determine if there exists a z_i such that (x, z_i) is in $R(A_2^i)$. We omit the details. □

Lemma 2. *We can effectively convert a 1DPCM M_1 to an equivalent 1DPCM M_2 that always accepts or rejects a given input.*

Proof. (Sketch) The proof is similar to that of showing the class of languages accepted by deterministic pushdown automata (DPDA) to be closed under complementation (see pp. 235-239 in [7] for a proof, see also [2]). The key in the conversion is to be able to identify the situation when M_1 falls into an infinite loop comprising of only ε-moves. Due to space limitation, the details are omitted here. □

Next we show that Theorem 7 part 1 also holds when A_1 is a 1NPCMT.

Theorem 8. *The following problems are decidable:*

1. *Given a 1NPCMT A_1 and a finite-crossing 2UDCMT A_2, is $R(A_1) \subseteq R(A_2)$?*
2. *Given a finite-crossing 2NCMT A_1 and a 1DPCMT A_2, is $R(A_1) \subseteq R(A_2)$?*
3. *Given a finite-crossing 2UDCMT A_1 and a 1DPCMT A_2, is $R(A_1) = R(A_2)$?*

Proof. Again, (3) follows from (1) and (2). The proof of (1) is similar to the proof of part 1 of Theorem 7, but now M would be a 3-phase finite-crossing 2NPCM operating only with phases 2 and 3, and noting that emptiness of 3-phase finite-crossing 2NPCM is decidable (Theorem 3).

A similar construction applies to (2), noting that, according to Lemma 2, the underlying 1DPCM of a 1DPCMT can always be converted to one that always accepts or rejects a given input. The finite-crossing M will be a 3-phase finite-crossing 2NPCM operating only with phases 1 and 2. We omit the details. □

Parts (2) and (3) of Theorem 8 cannot be strengthened by generalizing the 1DPCMT to be a finite-crossing 2DPCMT as the following shows.

Proposition 5. *There is a* fixed *3-reversal 2DPDT (i.e., a 2DPDA transducer A which makes only 3 reversals on its stack) which makes only 2 input head turns such that it is undecidable to determine, given an arbitrary non-null string z, whether $R(A_z) = \varnothing$, where A_z is A with initial stack content z. (Thus, the only parameter to the decision problem is the initial stack content z.)*

Proof. Theorem 5 in [10] showed that there is a *fixed* deterministic real-time 1-reversal 1DPDA M (i.e., the stack makes 1 reversal) and a *fixed* deterministic real-time 1-reversal 1DPDA M' such that it is undecidable to determine, given an arbitrary non-null string z, whether $L(M_z) \cap L(M') = \varnothing$, where M_z is M with initial stack content z. (M' always starts with a fixed stack symbol.)

As in Proposition 4. we can construct a fixed 3-reversal 2DPDA which with initial stack content z, simulates M and when it accepts returns the head to the left end marker and simulates M'. Then $R(A_z) = (L(M_z) \cap L(M')) \times \{\varepsilon\}$. It follows that is undecidable to determine, given A_z, whether $R(A_z) = \varnothing$. □

Special Case: Define a 3-phase finite-crossing 2NPCMT to be a 3-phase finite-crossing 2NPCM with outputs. The deterministic version is called a 3-phase finite-crossing 2DPCMT. Then Theorem 8 can be generalized so that the 1NPCMT A_1 in part (1) is a 3-phase finite-crossing 2NPCMT and the 1DPCMT A_2 in parts (2) and (3) is a 3-phase 2DPCMT. (We omit the proof.)

A finite-crossing 2UDPCMT A is a finite union of finite-crossing 2DPCMTs. Thus, $R(A) = R(A^1) \cup \cdots \cup R(A^n)$ for some n, where A^i is a finite-crossing 2DPCMT.

Theorem 9. *Over bounded inputs (i.e., the inputs to the transducers come from $w_1^* \cdots w_k^*$ for some non-null strings $w_1, \ldots, w_k$), the following problems are decidable:*

1. *Given a finite-crossing 2NCMT A_1 and a finite-crossing 2UDPCMT A_2, is $R(A_1) \subseteq R(A_2)$?*
2. *Given two finite-crossing 2UDPCMTs A_1 and A_2, is $R(A_1) = R(A_2)$?*

Proof. (Idea) By a rather intricate construction (which we do not give here), it can be shown that over bounded inputs, any finite-crossing 2UDPCMT A can effectively be converted to an equivalent finite-crossing 2UDCMT (i.e., the stack can be removed). The result then follows from Theorem 7. □

5 Finite-Crossing 2NFTs

In this section, we look at the special case of finite-crossing 2NFTs (i.e., there are no counters). Equivalence of 1NFTs is undecidable [3,9], although equivalence of single-valued (in fact, even for finite-valued) 1NFTs is decidable [1,12].

Recall from Theorem 6 that it is decidable to determine, given a finite-crossing 2NCMT A and a positive integer k, whether A is k-valued; hence this result applies to the special case when A is a finite-crossing 2NFT.

Here we show that containment (hence, also equivalence) of single-valued finite-crossing 2NFTs is decidable.

Theorem 10. *The following problems are decidable:*

1. *Given a finite-crossing 2NCMT A_1 and a single-valued finite-crossing 2NFT A_2, is $R(A_1) \subseteq R(A_2)$?*
2. *Given a single-valued finite-crossing 2NFT A_1 and a finite-crossing 2DCMT A_2, is $R(A_1) \subseteq R(A_2)$?*
3. *Given a single-valued finite-crossing 2NFT A_1 and a finite-crossing 2DCMT A_2, is $R(A_1) = R(A_2)$?*

Proof. Part (3) follows from (1) and (2). Part (2) is a special case of (1) of Theorem 5. Part (1) is similar to the proof of (1) of Theorem 5 by noting the fact that we can effectively construct from the underlying 2NFA M_2 of the 2NFT A_2 a 1DFA M_2' accepting the (regular) language $L(M_2)$ and, therefore, M_2' always accepts or rejects any given input. Hence, we can decide if $L(M_1) \not\subseteq L(M_2)$, where M_1 is the underlying finite-crossing 2NCM of A_1, as described in the proof of Theorem 5. □

We also have the following theorem whose proof is omitted.

Theorem 11. *The following problems are decidable:*

1. *Given a 3-phase finite-crossing 2NPCMT A_1 and a single-valued finite-crossing 2NFT A_2, is $R(A_1) \subseteq R(A_2)$?*
2. *Given a single-valued finite-crossing 2NFT A_1 and a 3-phase finite-crossing 2DPCMT A_2, is $R(A_1) \subseteq R(A_2)$?*
3. *Given a single-valued finite-crossing 2NFT A_1 and a 3-phase finite-crossing 2DPCMT A_2, is $R(A_1) = R(A_2)$?*

References

1. Culik, K., Karhumaki, J.: The equivalence of finite valued transducers (on HDTOL languages) is decidable. Theoret. Comput. Sci. 47, 71–84 (1986)
2. Ginsburg, S., Greibach, S.: Deterministic context-free languages. Inform. and Control 9, 620–648 (1966)
3. Griffiths, T.: The unsolvability of the equivalence problem for Λ-free nondeterministic generalized sequential machines. J. Assoc. Comput. Mach. 15, 409–413 (1968)
4. Gurari, E.: The equivalence problem for deterministic two-way sequential transducers is decidable. SIAM J. Comput. 11, 448–452 (1982)
5. Gurari, E., Ibarra, O.H.: The complexity of decision problems for finite-turn multicounter machines. J. Comput. Syst. Sci. 22, 220–229 (1981)
6. Gurari, E., Ibarra, O.H.: A note on finite-valued and finitely ambiguous transducers. Math. Systems Theory 16, 61–66 (1983)
7. Hopcroft, J., Ullman, J.: Introduction to Automata Theory, Languages, and Computation. Addison-Wesley, Reading (1979)
8. Ibarra, O.H.: Reversal-bounded multicounter machines and their decision problems. J. Assoc. Comput. Mach. 25, 116–133 (1978)
9. Ibarra, O.H.: The unsolvability of the equivalence problem for ε-free NGSM's with unary input (output) alphabet and applications. SIAM J. Computing 7, 524–532 (1978)
10. Ibarra, O.H.: On the universe, disjointness, and containment problems for simple machines. Inf. Comput. 208, 1273–1282 (2010)
11. Ibarra, O.H., Seki, S.: Characterizations of bounded semilinear languages by one-way and two-way deterministic machines. In: Proc. 13th Int. Conf. on Automata and Formal Languages (to appear, 2011)
12. Weber, A.: Decomposing finite-valued trasnsducers and deciding their equivalence. SIAM. J. on Computing 22, 175–202 (1993)

There Does Not Exist a Minimal Full Trio with Respect to Bounded Context-Free Languages

Juha Kortelainen[1] and Tuukka Salmi[2]

[1] Department of Information Processing Science, University of Oulu, Finland
[2] Elektrobit Corporation, Oulu

Abstract. We solve an old conjecture of Autebert et al. [1] saying that there does not exist any minimal full trio with respect to bounded context-free languages[1].

1 Introduction

Small subfamilies of context-free languages were quite extensively studied some decades ago [1-4, 8, 10]. It was, and still is, believed that the structure of almost regular languages can illustrate the properties of other context-free and even of certain nonalgebraic languages. The notion of a *full trio* or *cone* (i.e., a language family closed under morphism, inverse morphism and intersection with regular languages) is in these considerations quite natural: we wish to keep the structure of the (sub)families adequately simple but still nontrivial. The concept of smallness is connected to set inclusion: given language families $\mathcal{L}_1$ and $\mathcal{L}_2$, we say that $\mathcal{L}_1$ is *smaller than* $\mathcal{L}_2$ if $\mathcal{L}_1 \subsetneq \mathcal{L}_2$. Obviously the family of all regular languages $\mathcal{L}_{REG}$ is a full trio contained in any other trio. To avoid this negligible case, we define a *minimal full trio* to be a smallest nontrivial full trio, i.e., a full trio containing nonregular languages. The following important conjecture has been stated by Autebert et al. [3] (see also [1]):

Conjecture 1. There does not exist a minimal full trio in the family $\mathcal{L}_{CF}$ of context-free languages.

The claim above appeared to be difficult to prove, so it was reasonable to restrict the concept of minimality to fixed, in our case, bounded families of languages. Since a minimal full trio is always generated by a single language, we say that a full trio $\mathcal{L}$ is *minimal with respect to a language family* $\mathcal{K}$ if $\mathcal{L} \cap \mathcal{K}$ contains nonregular languages and any nonregular language in $\mathcal{L} \cap \mathcal{K}$ generates $\mathcal{L}$. Recall that, for each nonnegative integer n, a language L is *n-bounded* if there exist words $w_1, w_2, \ldots, w_n$ such that $L \subseteq w_1^* w_2^* \cdots w_n^*$. The language L is *bounded* if it is n-bounded for some n. In this paper we give a proof of the old claim of Autebert et al [1]:

Conjecture 2. There does not exist any minimal full trio with respect to bounded context-free languages.

[1] The main results of this paper appear originally in the dissertation [2].

G. Mauri and A. Leporati (Eds.): DLT 2011, LNCS 6795, pp. 312–323, 2011.

Let $\mathbb{N}$ be the set of all nonnegative integers. Define the bounded languages $S_<$, $S_>$, and $S_{\neq}$ as follows: $S_< = \{a^m b^n | m, n \in \mathbb{N}, m < n\}$, $S_> = \{a^m b^n | m, n \in \mathbb{N}, m > n\}$, and $S_{\neq} = \{a^m b^n | m, n \in \mathbb{N}, m \neq n\}$. For each language L, let $\mathcal{T}(L)$ be the full trio generated by L, i.e., the smallest family of languages containing L and closed under morphism, inverse morphism and intersection with regular sets. Berstel and Boasson have verified in [4] that the families $\mathcal{T}(S_<)$, $\mathcal{T}(S_>)$, and $\mathcal{T}(S_{\neq})$ are the (only) three distinct minimal full trios with respect to 2-bounded context-free languages. Moreover, they showed that none of them is a minimal full trio with respect to k-bounded context-free languages for $k > 2$.

The rest of the paper is organized as follows. In Section 2 some basic concepts are defined and results given. The third section is divided into four subsections and contains the sequence of lemmata and theorems needed to verify the main result. The final section contains some conclusive remarks , we also introduce an open problem and a topic for further research.

2 Preliminaries

In this section we introduce definitions, notations and some auxiliary results needed in the sequel. The reader is referred to theb old classics [5] and [6] for any unexplained notation.

Let $\mathbb{N}_+$ the set of all positive natural numbers. Suppose that $n \in \mathbb{N}_+$, $r \in \mathbb{N}$ and $c, p_1, p_2, \ldots, p_r \in \mathbb{N}^n$. Denote by $L(c; p_1, p_2, \ldots, p_r)$ the linear set with constant (vector) c and periods $p_1, p_2, \ldots, p_r$. We say that the linear set $L(c; p_1, p_2, \ldots, p_r)$ is *proper* if the periods $p_1, p_2, \ldots, p_r$ are linearly independent over $\mathbb{Q}$, the field of rationals. Recall that a set is semilinear if it is a finite union of linear sets. By the Eilenberg-Schützenberger theorem [7] (see also [8]) every semilinear set is a finite union of disjoint proper linear sets.

A linear set $L(c; p_1, p_2, \ldots, p_r) \subseteq \mathbb{N}^n$ is *stratified* if

(i) for each the $i \in \{1, 2, ..., r\}$, the vector p_i contains at most two nonzero coordinates; and

(ii) for all $d, t \in \{1, 2, ..., r\}$, and $i, j, k, l \in \{1, 2, ..., n\}$ such that $i < j < k < l$ the vectors $p_d = (p_{d1}, p_{d2}, ..., p_{dn})$ and $p_t = (p_{t1}, p_{t2}, ..., p_{tn})$ satisfy the equality $p_{di} p_{tj} p_{dk} p_{tl} = 0$.

Let $A = L(c; p_1, p_2, \ldots, p_r) \subseteq \mathbb{N}^n$ be a linear set. The *integer closure* of the set A is the set $\mathrm{intc}(A) = \{c + l_1 p_1 + l_2 p_2 + \cdots + l_r p_r | l_i \geq 0, l_i \in \mathbb{Q}\} \cap \mathbb{N}^n$.

Let the set $A \subseteq \mathbb{N}^n$ be linear and $T \subseteq \mathbb{N}^n$ an arbitrary set. We say that A is *structured by* T if $A = \mathrm{intc}(A) \cap T$. A semilinear set S is *semistructured by* T if there exist $m \in \mathbb{N}_+$ and linear sets $S_1, S_2, ..., S_m$ such that

$$S = \bigcup_{i=1}^{m} S_i = \bigcup_{i=1}^{m} (\mathrm{intc}(S_i) \cap T) .$$

Let Σ be an alphabet. For each word $w \in \Sigma^*$ and symbol $a \in \Sigma$, let $|w|$ denote the length of w, and $|w|_a$ the number of occurrences of a in w. The empty word is denoted by ϵ. Let $L \subseteq \Sigma^*$ be a language. If there exist $k \in \mathbb{N}_+$

distinct letters $a_1, a_2, \ldots, a_k \in \Sigma$ such that $L \subseteq a_1^* a_2^* \cdots a_k^*$, then L is *strictly k-bounded*. The language L is *strictly bounded* if it is strictly k-bounded for some $k \in \mathbb{N}_+$. We shall denote the set of strictly k-bounded languages by $\mathcal{B}_k$. Given that $L \subseteq a_1^* a_2^* \cdots a_k^*$, define $\overline{L} = a_1^* a_2^* \cdots a_k^* \setminus L$. Denote by $\mathcal{L}_{REG}$ ($\mathcal{L}_{CF}$, resp.) the set of all regular (context-free, resp.) languages.

Let $n \in \mathbb{N}_+$, let $\Sigma_n = \{a_1, a_2, \ldots, a_n\}$. Define the *Parikh-map* $\Psi_n : \Sigma_n^* \to \mathbb{N}^n$ by the equality $\Psi_n(w) = (|w|_{a_1}, |w|_{a_2}, \ldots, |w|_{a_n})$. Let $L \subseteq \Sigma_n^*$, and $\Psi_n(L) = \{\Psi_n(w) \mid w \in L\}$. The set $\Psi_n(L)$ is the *Parikh-image* of a language L. When n is understood, we write Ψ instead of of Ψ_n.

Ginsburg ([5], Lemma 5.4.2) gives a necessary and sufficient condition for a strictly bounded language to be context-free: A strictly bounded language is context-free if and only if its Parikh-image is a finite union of stratified linear sets.

Assume $k \in \mathbb{N}_+$ and let $L, K \subseteq a_1^* a_2^* \cdots a_k^*$ be languages with semilinear Parikh-images. Then the language L is *(semi)structured by* $K \subseteq a_1^* a_2^* \cdots a_k^*$ if the set $\Psi_k(L)$ is (semi)structured by $\Psi_k(K)$.

Let Σ and Δ be alphabets and $L \subseteq \Sigma^*$ and $L' \subseteq \Delta^*$ languages. By an important theorem of Nivat (for a proof, see [9]), the language L' is in the full trio $\mathcal{T}(L)$ if and only if there exist an alphabet Γ, a regular language $R \subseteq \Gamma^*$ and alphabetic morphisms $h : \Gamma^* \to \Delta^*$ and $g : \Gamma^* \to \Sigma^*$ such that $L' = h(g^{-1}(L) \cap R)$.

Let $k \in \mathbb{N}$. Denote by $\mathcal{L}_k$ the family of all languages L such that there exist $n \in \mathbb{N}_+$ and words x_{ip}, w_{iq}, $i = 1, 2, \ldots, n$, $p = 0, 1, \ldots k$, and $q = 1, 2, \ldots, k$, such that $L \subseteq \bigcup_{i=0}^{n} x_{i0} w_{i1}^* x_{i1} w_{i2}^* x_{i2} \cdots w_{ik}^* x_{ik}$.

Let $\mathcal{R}_k = \mathcal{L}_k \cap \mathcal{L}_{REG}$. By the properties of bounded regular sets, the language R is in $\mathcal{R}_k$ if and only if R is a finite union of languages of the form $x_0 w_1^* x_1 w_2^* x_2 \cdots w_k^* x_k$ where $x_0, w_1, x_1, w_2, x_2, \ldots, w_k, x_k$ are words.

3 Full Trios Generated by Almost Regular Bounded Context-Free Languages

For a language L over an alphabet Σ, let $\Sigma_L = \{a \in \Sigma \mid \exists w \in L : a \text{ occurs in } w\}$. For each $k \in \mathbb{N}_+$, define

$$\mathcal{C}_k = \{L \in \mathcal{L}_{CF} \mid \forall x_0, w_1, x_1, w_2, x_2, \ldots, w_k, x_k \in \Sigma_L^* : \\ L \cap x_0 w_1^* x_1 w_2^* \cdots w_k^* x_k \in \mathcal{L}_{REG}\} .$$

and $\mathcal{C}_\infty = \cap_{i=1}^{\infty} \mathcal{C}_i$. Clearly $\mathcal{C}_{k+1} \subseteq \mathcal{C}_k$ for all $k \in \mathbb{N}_+$. It follows directly from the definition that $\mathcal{L}_{REG} \subseteq \mathcal{C}_\infty$ and $\mathcal{C}_k \subseteq \mathcal{L}_{CF}$ for all $k \in \mathbb{N}_+$.

We shall proceed by the following sequence of arguments. For all $k \in \mathbb{N}_+$:

1. the sets $\mathcal{C}_k \setminus \mathcal{C}_{k+1}$ (and the set $\mathcal{C}_\infty \setminus \mathcal{L}_{REG}$ as well) are nonempty;
2. any bounded nonregular context-free language is contained in $\mathcal{C}_k \setminus \mathcal{C}_{k+1}$ for some k;
3. any language in $\mathcal{C}_k \setminus \mathcal{C}_{k+1}$ can be transformed into a language in $\mathcal{C}_{k+1} \setminus \mathcal{C}_{k+2}$ using only trio-operations; and
4. the family $\mathcal{C}_k$ is a full trio.

Note that items 1 and 4 imply that, as a nonempty (albeit infinite) intersection of full trios, the family $\mathcal{C}_\infty$ is a full trio, too. It should now be clear that picking any language in the chain $\mathcal{C}_1 \setminus \mathcal{C}_2$, $\mathcal{C}_2 \setminus \mathcal{C}_3$, ... and applying appropriate trio-operations to L, we can always proceed downwards as far as we wish, but never upwards. Our main result becomes thus proved.

The majority of the proofs of results presented are omitted because of lack of space and included to the full version of the paper.

3.1 Examples and Basic Properties of Languages in $\mathcal{C}_k \setminus \mathcal{C}_{k+1}$, $k \in \mathbb{N}_+$

Let $k \in \mathbb{N}_+$ and $\theta_1, \theta_2, \ldots, \theta_k \in \{<, >, \neq\}$. For each $i \in \{1, 2, \ldots, k\}$, define the language

$$S_i(\theta_1, \theta_2, \ldots, \theta_k) = \{a_1^{n_1} a_2^{n_2} \cdots a_k^{n_k} \mid n_1, n_2, \ldots, n_k \in \mathbb{N} \text{ and } \vee_{j=1}^{k} n_i \theta_j n_j\}.$$

Example 1. Let us consider the language $S_1(\neq, >) = \{a_1^{n_1} a_2^{n_2} | n_1 > n_2\}$. Certainly $S_1(\neq, >) = S_2(<, \neq)$. For example, $S_1(\neq, >) \cap a_1^* a_2 = a_1 a_1^+ a_2 \in \mathcal{L}_{REG}$ and $S_1(\neq, >) \cap a_1 a_2^* = a_1 \in \mathcal{L}_{REG}$. It is clear that $S_1(\neq, >) \in \mathcal{C}_1 \setminus \mathcal{C}_2$.

Example 2. Let us consider the language

$$S_1(\neq, \neq, \neq) = \{a_1^{n_1} a_2^{n_2} a_3^{n_3} | n_1 \neq n_2 \text{ or } n_1 \neq n_3\} .$$

Surely $S_1(\neq, \neq, \neq) = S_2(\neq, \neq, \neq) = S_3(\neq, \neq, \neq) = a_1^* a_2^* a_3^* \setminus \{a_1^n a_2^n a_3^n | n \in \mathbb{N}\}$. Now, for instance, $S_1(\neq, \neq, \neq) \cap a_1^* a_2^* a_3 = a_1^* a_2^* a_3 \setminus \{a_1 a_2 a_3\} \in \mathcal{L}_{REG}$. It is easily seen that $S_1(\neq, \neq, \neq) \in \mathcal{C}_2 \setminus \mathcal{C}_3$. Thus it is quite obvious that $S_1(\neq, \neq, \neq, \neq) \in \mathcal{C}_3 \setminus \mathcal{C}_4$ and furthermore $S_1(\underbrace{\neq, \neq, \ldots, \neq}_{k+1 \text{ times}}) \in \mathcal{C}_k \setminus \mathcal{C}_{k+1}$ for all $k \in \mathbb{N}_+$.

Goldstine defined a nonregular context-free language

$$\begin{aligned} L_G &= \left\{ a_2^{i_1} a_1 a_2^{i_2} a_1 \cdots a_2^{i_p} a_1 \mid p \in \mathbb{N}_+, \quad \exists j \in \{1, 2, \ldots, p\} : i_j \neq j \right\} \\ &= (a_2^* a_1)^+ \setminus \{a_2 a_1 a_2^2 a_1 a_2^3 a_1 \cdots a_2^p a_1 | p \in \mathbb{N}_+\}. \end{aligned}$$

and showed that all bounded languages in the full trio $\mathcal{T}(L_G)$ are regular [10]. Thus $L_G \in \mathcal{C}_\infty \setminus \mathcal{L}_{REG}$. On the other hand, certainly $L \cap x_0 w_1^* x_1$ is in $\mathcal{L}_{REG}$ for each L in $\mathcal{L}_{CF}$ and words $x_0, w_1, x_1 \in \Sigma_L^*$. Thus $\mathcal{C}_1 = \mathcal{L}_{CF}$ and

$$\mathcal{L}_{CF} = \mathcal{C}_1 \supsetneq \mathcal{C}_2 \supsetneq \cdots \supsetneq \mathcal{C}_\infty \supsetneq \mathcal{L}_{REG} .$$

Theorem 1. *Let $n \in \mathbb{N}_+$ and $L \subseteq a_1^* a_2^* \cdots a_n^*$. Given $k \in \mathbb{N}_+$, with $k \leq n$, the language L is in $\mathcal{C}_k$ if and only if $L \cap a_1^{\alpha_1} a_2^{\alpha_2} \cdots a_n^{\alpha_n} \in \mathcal{L}_{REG}$ for all $\alpha_1, \alpha_2, \ldots, \alpha_n \in \mathbb{N} \cup \{*\}$, where $|\{i \in \{1, 2, \ldots, n\} | \alpha_i = *\}| = k$.*

Our next theorem gives a necessary and sufficient condition for the language $S_i(\theta_1, \theta_2, \ldots, \theta_{k+1})$ (where $k \in \mathbb{N}_+$ and $i \in \{1, 2, ..., k+1\}$) to belong to the family $\mathcal{C}_k \setminus \mathcal{C}_{k+1}$.

Theorem 2. *Let $k \geq 2$ be an integer and $i \in \{1, 2, \dots, k+1\}$. Then the language $S_i(\theta_1, \theta_2, \dots, \theta_{k+1})$ is in $\mathcal{C}_k \setminus \mathcal{C}_{k+1}$ if and only if $\theta_j \in \{>, \neq\}$ for all $j \in \{1, 2, \dots, k+1\}$ such that $j \neq i$.*

The next result says that if we are searching for a language generating a minimal full trio inside the family $\mathcal{C}_k \setminus \mathcal{C}_{k+1}$, we may always restrict to look it inside the family of strictly $(k+1)$-bounded languages.

Theorem 3. *Let $k \in \mathbb{N}_+$ and $L \in \mathcal{C}_k \setminus \mathcal{C}_{k+1}$. Then there exists a language $L_1 \in \mathcal{T}(L)$ such that $L_1 \in \mathcal{B}_{k+1} \cap (\mathcal{C}_k \setminus \mathcal{C}_{k+1})$.*

Proof. (sketch) Since $L \notin \mathcal{C}_{k+1}$, we have $x_0, x_1, \dots, x_{k+1} \in \Sigma^*$ and $w_1, w_2, \dots, w_{k+1} \in \Sigma^*$ such that

$$L' = L \cap x_0 w_1^* x_1 w_2^* \cdots x_k w_{k+1}^* x_{k+1} \notin \mathcal{L}_{REG}.$$

Hence, $L' \notin \mathcal{C}_{k+1}$. Since $L \in \mathcal{C}_k$, we have $L' \in \mathcal{C}_k \setminus \mathcal{C}_{k+1}$.

Applying full trio-operations on L', one first finds a $(k+1)$-bounded language $L'' \subseteq w_1^* w_2^* \cdots w_{k+1}^*$ in $\mathcal{C}_k \setminus \mathcal{C}_{k+1}$, and then a strictly $(k+1)$-bounded language $L_1 \subseteq a_1^* a_2^* \cdots a_{k+1}^*$ in $\mathcal{C}_k \setminus \mathcal{C}_{k+1}$. □

3.2 Simplifying the Structure of Bounded Languages in $\mathcal{C}_k \setminus \mathcal{C}_{k+1}$, $k \in \mathbb{N}_+$

General strictly $(k+1)$-bounded languages provide still too much redundant data for deeper analysis of the family $\mathcal{C}_k \setminus \mathcal{C}_{k+1}$. We shall see that every nonregular strictly bounded context-free language can be intersected with a regular strictly bounded language R so that the result is nonregular and semistructured by R. The inner structure of these (nonregular and semistructured) languages is simple, but they possess all the important features of languages in $\mathcal{C}_k \setminus \mathcal{C}_{k+1}$.

Note that the Parikh image of a strictly bounded regular language may always be represented as a semilinear set whose periods are parallel with a coordinate axis. In [11] it is proved that for each integer $k \in \mathbb{N}_+$ and proper linear set $S \subseteq \mathbb{N}_+^k$ there exists a regular language $R \subseteq a_1^* a_2^* \cdots a_k^*$ such that S is structured by $\Psi^{-1}(R)$.

Lemma 1. *Let $k \in \mathbb{N}_+$ and $L \subseteq a_1^* a_2^* \cdots a_k^*$ be a language such that $\Psi(L)$ is a semilinear set and L is semistructured by a regular language $R \subseteq a_1^* a_2^* \cdots a_k^*$. Let $R' \subseteq a_1^* a_2^* \cdots a_k^*$ be another regular language. Then $L \cap R'$ is semistructured by $R \cap R'$.*

Proof. Let $n \in \mathbb{N}_+$ and $L_1, L_2, \dots, L_n$ be linear components of L such that $L = \bigcup_{i=1}^n L_i = \bigcup_{i=1}^n (\text{intc}(L_i) \cap R)$. For each $i \in \{1, 2, \dots, n\}$, let $n(i) \in \mathbb{N}_+$ and $L_{i,1}, L_{i,2}, \dots, L_{i,n(i)}$ be linear components of $L_i \cap R'$. Then for each $i \in \{1, 2, \dots, n\}$, we have

$$\bigcup_{j=1}^{n(i)} \text{intc}(L_{i,j}) \subseteq \text{intc}(L_i).$$

By the above relation and the fact that L is semistructured by R, we have

$$\bigcup_{i=1}^{n}\bigcup_{j=1}^{n(i)} \mathrm{intc}(L_{i,j}) \cap R \cap R' \subseteq \bigcup_{i=1}^{n} \mathrm{intc}(L_i) \cap R \cap R' = L \cap R'.$$

On the other hand, since $L \subseteq R$ and $L\cap R' = \bigcup_{i,j} L_{i,j} \subseteq \bigcup_{i,j} \mathrm{intc}(L_{i,j})$, we have

$$L \cap R' = \bigcup_{i,j} \mathrm{intc}(L_{i,j}) \cap R \cap R'.$$

□

The next lemma is slightly modified version of Lemma 3 in [11]. This result and Theorem 3 will imply that if we are searching for a language generating a minimal full trio inside the family $\mathcal{C}_k \setminus \mathcal{C}_{k+1}$, we may always restrict to look it inside the family of strictly $(k+1)$-bounded languages that are semistructured by some regular language.

Lemma 2. *Let $k \in \mathbb{N}_+$ and $L \subseteq a_1^*a_2^* \cdots a_k^*$ be a language such that $\Psi(L)$ is a semilinear set. Then there exist $n \in \mathbb{N}_+$ and regular languages $R_1, R_2, \ldots, R_n \subseteq a_1^*a_2^* \cdots a_k^*$ such that*

1. *$\Psi(R_i)$ is a linear set for all $i \in \{1, 2, \ldots, n\}$;*
2. *$L \cap R_i$ is semistructured by R_i for all $i \in \{1, 2, \ldots, n\}$;*
3. *$L = \bigcup_{i=1}^{n}(L \cap R_i)$.*

Let us have an example of the construction of Lemma 2

Example 3. Let $L = \{a_1^{2n_1}a_2^{n_2} | n_1 < n_2\} \cup \{a_1^{4n_1}a_2^{n_2} | n_1 > n_2\}$. Then the languages

$$L_1 = L \cap (a_1^2)^*(a_2)^* \cap (a_1^4)^*a_2^* = \{a_1^{4n_1}a_2^{n_2} | 2n_1 < n_2 \vee n_1 > n_2\}$$

and

$$L_2 = L \cap (a_1^2)^*(a_2)^* \cap \overline{(a_1^4)^*a_2^*} = \{a_1^{4n_1+2}a_2^{n_2} | 2n_1 + 1 < n_2\}$$

are semistructured by $(a_1^2)^*(a_2)^* \cap (a_1^4)^*a_2^* = (a_1^4)^*a_2^*$ and $(a_1^2)^*(a_2)^* \cap \overline{(a_1^4)^*a_2^*} = a_1^2(a_1^4)^*a_2^*$, respectively. Moreover, $L = L_1 \cup L_2$.

We next state a geometrically obvious lemma.

Lemma 3. *Let $k \in \mathbb{N}_+$ and $L \subset a_1^*a_2^* \ldots a_k^*$ be a semistructured language by a language $K \subset a_1^*a_2^* \ldots a_k^*$. Then either L or $K \setminus L$ does not contain as a subset any language from the family $\mathcal{R}_k \setminus \mathcal{R}_{k-1}$.*

We can reformulate the previous result as follows.

Lemma 4. *Let $k \in \mathbb{N}_+$, $j \in \{1, 2, \ldots, k+1\}$ and $L \subset a_1^*a_2^* \cdots a_{k+1}^*$ be a semistructured language by a regular language $R \subset a_1^*a_2^* \cdots a_{k+1}^*$ such that $\Psi(R)$ is linear and*

$$L \cap a_1^*a_2^* \cdots a_{j-1}^*a_j^m a_{j+1}^* a_{j+2}^* \cdots a_{k+1}^* \in \mathcal{L}_{REG}$$

for all $m \in \mathbb{N}$. *If for each* $m_0 \in \mathbb{N}$, *there exists* $m \geq m_0$ *such that*

$$L \cap a_1^* a_2^* \cdots a_{j-1}^* a_j^m a_{j+1}^* a_{j+2}^* \cdots a_{k+1}^* \in \mathcal{R}_k \setminus \mathcal{R}_{k-1}, \tag{1}$$

then there exists $m_0' \in \mathbb{N}$ *such thuntitled folderat*

$$(R \setminus L) \cap a_1^* a_2^* \cdots a_{j-1}^* a_j^m a_{j+1}^* a_{j+2}^* \cdots a_{k+1}^* \in \mathcal{R}_{k-1}$$

for all $m \geq m_0'$.

Let us generalize the definition of the family $\mathcal{C}_k$ by setting

$$\hat{\mathcal{C}}_k = \{L \in \Sigma^* | L \cap x_0 w_1^* x_1 w_2^* \cdots x_{k-1} w_k^* x_k \in \mathcal{L}_{REG} \quad \forall x_i, w_i \in \Sigma_L^*\}$$

for all $k \in \mathbb{N}_+$.

The following result gives us information of crucial importance about languages in $L \in \hat{\mathcal{C}}_k \setminus \hat{\mathcal{C}}_{k+1}$ for which $\Psi(L)$ is a semilinear set; we need to characterize *any* presentation of $\Psi(L)$ as a finite union of linear sets.

Theorem 4. *Let* $k \in \mathbb{N}_+$ *and* $L \subset a_1^* a_2^* \cdots a_{k+1}^*$ *be a language such that* $\Psi(L)$ *is a semilinear set and* $L \in \hat{\mathcal{C}}_k \setminus \hat{\mathcal{C}}_{k+1}$. *Let* J *be the set of all* $j \in \{1, 2, \ldots, k+1\}$ *for which there exist* $m_0 \in \mathbb{N}$ *such that*

$$L \cap a_1^* a_2^* \cdots a_{j_i-1}^* a_{j_i}^m a_{j_i+1}^* a_{j_i+2}^* \cdots a_{k+1}^* \in \mathcal{R}_{k-1} \tag{2}$$

for all $m \geq m_0$. *Then* $\Psi(L)$ *contains a period with at least* $\min\{k+1,\ |J|+1\}$ *nonzero coordinates in every representation of* $\Psi(L)$ *as a finite union of linear sets.*

Let us now give three simple examples of Theorem 4.

Example 4. Let $L = \{a_1^n a_2^n a_3^n | n \in \mathbb{N}\}$. We have $L \cap a_1^m a_2^* a_3^* = L \cap a_1^* a_2^m a_3^* = L \cap a_1^* a_2^* a_3^m = \{a_1^m a_2^m a_3^m\} \in \mathcal{R}_0 \subset \mathcal{R}_1$ for all $m \in \mathbb{N}$. Therefore we have $L \in \hat{\mathcal{C}}_2 \setminus \hat{\mathcal{C}}_3$ and $l = 3$. Theorem 4 states that $\Psi(L)$ contains a period with $\min\{3, 4\} = 3$ nonzero coordinates in every representation of $\Psi(L)$ as a finite union of linear sets.

Example 5. Let $L = \{a_1^{n_1} a_2^{n_2} a_3^{n_3} | n_1 < n_2 \wedge n_1 < n_3 \in \mathbb{N}\}$. We have $L \cap a_1^m a_2^* a_3^* \in \mathcal{R}_2 \setminus \mathcal{R}_1$ and $L \cap a_1^* a_2^m a_3^*, L \cap a_1^* a_2^* a_3^m \in \mathcal{R}_1$ for all $m \in \mathbb{N}$. Therefore we have $L \in \hat{\mathcal{C}}_2 \setminus \hat{\mathcal{C}}_3$ and $l = 2$. Theorem 4 states that $\Psi(L)$ contains a period with $\min\{3, 3\} = 3$ nonzero coordinates in every representation of $\Psi(L)$ as a finite union of linear sets.

Example 6. Let

$$L = \{a_1^{n_1} a_2^{n_2} a_3^{n_3} a_4^{n_4} \mid (n_1 < n_2 \wedge n_1 < n_3 \wedge n_1 < n_4) \\ \vee (n_4 < n_1 \wedge n_4 < n_2 \wedge n_4 < n_3)\}.$$

It can be easily seen that $L \cap a_1^m a_2^* a_3^* a_4^*, L \cap a_1^* a_2^* a_3^* a_4^m \in \mathcal{R}_3 \setminus \mathcal{R}_2$ and $L \cap a_1^* a_2^m a_3^* a_4^*, L \cap a_1^* a_2^* a_3^m a_4^* \in \mathcal{R}_2 \setminus \mathcal{R}_1$ for all $m \in \mathbb{N}$. Therefore $L \in \hat{\mathcal{C}}_3 \setminus \hat{\mathcal{C}}_4$ and

$l = 2$. Theorem 4 states that $\Psi(L)$ contains a period with at least $\min\{3,5\} = 3$ nonzero coordinates in every representation of $\Psi(L)$ as a finite union of linear sets. It is quite obvious that $\Psi(L)$ contains a period with 4 nonzero coordinates in every representation of $\Psi(L)$ as a finite union of linear sets. Hence the statement of the theorem is not optimal in that sense. It is also obvious that $\Psi(L)$ has representations such that $\Psi(L)$ does not contain a period with exactly 3 nonzero coordinates. Thus the expression "at least" is necessary in the result.

The next theorem states that a strictly bounded language $L \in \mathcal{C}_k \setminus \mathcal{C}_{k+1}$ that is semistructured by a regular language $R \subseteq a_1^* a_2^* \cdots a_{k+1}^*$ is always big, i.e. it contains almost the whole language R as a subset. Since by Theorems 3 and 2, every language $L \in \mathcal{C}_k \setminus \mathcal{C}_{k+1}$ can be transformed to a language of this form with the trio-operations, this theorem gives us important knowledge of the structure of weak languages[2] in the family $\mathcal{C}_k \setminus \mathcal{C}_{k+1}$. The theorem and its variation (Theorem 6) play crucial role in this paper; they are later used for proving that (1) $\mathcal{C}_k$ is closed under morphism and (2) we can transform any language from the family $\mathcal{C}_k \setminus \mathcal{C}_{k+1}$ into the family $\mathcal{C}_{k+1} \setminus \mathcal{C}_{k+2}$ with the full trio-operations.

Theorem 5. *Let $k \geq 2$ be an integer and $L \in \mathcal{C}_k \setminus \mathcal{C}_{k+1}$ a semistructured language by a regular language $R \subseteq a_1^* a_2^* \cdots a_{k+1}^*$ such that $\Psi(R)$ is linear. Then there exist $m_0 \in \mathbb{N}$ and $j_0 \in \{1, 2, \ldots, k+1\}$ for which*

$$(R \setminus L) \cap a_1^* a_2^* \cdots a_{j-1}^* a_j^m a_{j+1}^* a_{j+2}^* \cdots a_{k+1}^* \in \mathcal{R}_{k-1}$$

for all $m \geq m_0$ and $j \in \{1, 2, \ldots, k+1\}$ such that $j \neq j_0$.

Corollary 1. *Let $k \geq 2$ be an integer and $L \in \mathcal{C}_k \setminus \mathcal{C}_{k+1}$ a semistructured language by a regular language $R \subset a_1^* a_2^* \cdots a_{k+1}^*$ such that $\Psi(R)$ is linear. Then $R \setminus L$ does not contain as a subset any language from the family $\mathcal{R}_{k+1} \setminus \mathcal{R}_k$.*

Let us now consider two examples of Theorem 5.

Example 7. Consider the language

$$S_1(\neq, \neq, \neq) = \{a_1^{n_1} a_2^{n_2} a_3^{n_3} | n_1 \neq n_2 \vee n_1 \neq n_3\}.$$

We have $\overline{S_1(\neq, \neq, \neq)} \cap a_1^* a_2^* a_3^* = \{a_1^{n_1} a_2^{n_2} a_3^{n_3} | n_1 = n_2 = n_3\}$. Thus all the languages $\overline{S_1(\neq, \neq, \neq)} \cap a_1^m a_2^* a_3^*$, $\overline{S_1(\neq, \neq, \neq)} \cap a_1^* a_2^m a_3^*$, $\overline{S_1(\neq, \neq, \neq)} \cap a_1^* a_2^* a_3^m$ are singletons, and the index j_0 of Theorem 5 may be chosen freely.

Example 8. Let

$$S_1(\neq, >, >) = \{a_1^{n_1} a_2^{n_2} a_3^{n_3} | n_1 > n_2 \vee n_1 > n_3\}.$$

We have $\overline{S_1(\neq, >, >)} \cap a_1^* a_2^* a_3^* = \{a_1^{n_1} a_2^{n_2} a_3^{n_3} | n_1 \leq n_2 \wedge n_1 \leq n_3\}$. Thus in this case the language $\overline{S_1(\neq, >, >)} \cap a_1^m a_2^* a_3^* = a_1^m a_2^m a_2^* a_3^m a_3^*$ is in $\in \mathcal{R}_2 \setminus \mathcal{R}_1$ and the languages $\overline{S_1(\neq, >, >)} \cap a_1^* a_2^m a_3^*$, $\overline{S_1(\neq, >, >)} \cap a_1^* a_2^* a_3^m$ are in $\mathcal{R}_1$. Hence the index j_0 of Theorem 5 is one.

[2] By the informal term 'weak languages', we mean languages which generate a full trio containing only a small set of nonregular languages.

Next we state a variation of Theorem 5.

Theorem 6. *Let $k \geq 2$ be an integer and $L \in \mathcal{C}_k \setminus \mathcal{C}_{k+1}$ such that $L \subseteq a_1^* a_2^* \cdots a_{k+1}^*$. Then there exist $j_0 \in \{1, 2, \ldots, k+1\}$ and $R \in \mathcal{L}_{REG}$ such that $L \cup R \in \mathcal{C}_k \setminus \mathcal{C}_{k+1}$ and*

$$\overline{L \cup R} \cap a_1^* a_2^* \cdots a_{j-1}^* a_j^m a_{j+1}^* a_{j+2}^* \cdots a_{k+1}^* \in \mathcal{R}_{k-1}$$

for all $m \in \mathbb{N}$ and $j \in \{1, 2, \ldots, k+1\}$ such that $j \neq j_0$.

Theorem 7. *Let $k \in \mathbb{N}_+$ and $L \in \mathcal{C}_k \setminus \mathcal{C}_{k+1}$ be such that $L \subseteq a_1^* a_2^* \cdots a_{k+1}^*$. Then $\Psi(a_1^* a_2^* \cdots a_{k+1}^* \setminus L)$ contains a period with $k+1$ nonzero coordinates in every representation of $\Psi(a_1^* a_2^* \cdots a_{k+1}^* \setminus L)$ as a finite union of linear sets.*

Corollary 2. *Let $k \geq 2$ be an integer and $L \in \mathcal{B}_{k+1} \cap (\mathcal{C}_k \setminus \mathcal{C}_{k+1})$. Then $\overline{L}$ is not context-free.*

Let $L \in \mathcal{C}_k \setminus \mathcal{C}_{k+1}$ be a strictly $(k+1)$-bounded semistructured language by a regular language R. In Corollary 1 we saw that $R \setminus L$ does not contain any language as a subset from the family $\mathcal{R}_{k+1} \setminus \mathcal{R}_k$. However, the next theorem states that $R \setminus L$ contains always a language from the family $\mathcal{L}_{k+1} \setminus \mathcal{L}_k$.

Theorem 8. *Let $k \in \mathbb{N}_+$. If $L \in \mathcal{C}_k \setminus \mathcal{C}_{k+1}$ is a semistructured language by a regular language $R \subset a_1^* a_2^* \cdots a_{k+1}^*$, then $R \setminus L \in \mathcal{L}_{k+1} \setminus \mathcal{L}_k$.*

So far all our example languages $S_i(\theta_1, \theta_2, \ldots, \theta_{k+1})$ have had a property such that $\Psi(S_i(\theta_1, \theta_2, \ldots, \theta_{k+1}))$ has a representation such that each linear component of $\Psi(S_i(\theta_1, \theta_2, \ldots, \theta_{k+1}))$ contains only one period with two nonzero coordinates. One could ask whether this is a global property for all the strictly $(k+1)$-bounded languages in the family $\mathcal{C}_k \setminus \mathcal{C}_{k+1}$. However, this is not the case as seen in the next example.

Example 9. Consider the language

$$L = \{a_1^{n_1} a_2^{n_2} a_3^{n_3} | n_1 > 2n_2 \vee n_1 > 2n_3 \vee n_1 = n_2 + n_3\}.$$

Then $L \in \mathcal{C}_2 \setminus \mathcal{C}_3$. Moreover L contains a linear component with periods $(1, 1, 0)$ and $(1, 0, 1)$ in the most natural representation of $\Phi(L)$. It seems also quite obvious that there exists a linear component with periods $(n_1, n_2, 0)$ and $(n_3, 0, n_4)$, where $n_i > 0$, in every representation of $\Phi(L)$.

3.3 Propagation in the Chain

The following theorem expresses that we can proceed in a controlled way proceed downwards in the chain $\mathcal{C}_1 \setminus \mathcal{C}_2$, $\mathcal{C}_2 \setminus \mathcal{C}_3$, ... by applying full trio-operations.

Theorem 9. *Let $k \in \mathbb{N}_+$. For each language $L \in \mathcal{C}_k \setminus \mathcal{C}_{k+1}$, there exists a strictly $(k+1)$-bounded language $L' \in \mathcal{T}(L)$ such that $L' \in \mathcal{C}_{k+1} \setminus \mathcal{C}_{k+2}$.*

Theorem 10. *A full trio $\mathcal{L}$ is minimal with respect to the $(k+1)$-bounded context-free languages if and only if it is minimal respect to $\mathcal{C}_k \setminus \mathcal{C}_{k+1}$.*

3.4 Closure Properties of the Language Families $\mathcal{C}_k$, $k \in \mathbb{N}_+$

In this section we will prove that the family $\mathcal{C}_k$ is a full trio for each $k \in \mathbb{N}_+$. This will be done in two phases. First we will prove that $\mathcal{C}_k$ is closed under morphism, inverse morphism and intersection with regular languages. Closure under morphism is cosiderable harder to prove than the other full trio-operations and it is verified in the separate theorem. Finally, we are able to prove Conjecture 2.

Theorem 11. *The language family $\mathcal{C}_k$ is closed under inverse morphism and intersection with regular languages for each $k \in \mathbb{N}_+$.*

Proof. Let us prove the claim concerning the families $\mathcal{C}_k$, where $k \in \mathbb{N}_+$. Therefore the claim concerning the family $\mathcal{C}_\infty$ follows from the fact that $\mathcal{C}_\infty$ is the intersection of the families $\mathcal{C}_k$.

Let $L_1, L_2 \in \mathcal{C}_k$, $x_0, x_1, \ldots, x_k \in \Sigma a^*$ and $w_1, w_2, \ldots, w_k \in \Sigma^*$ be arbitrary. Denote $R = x_0 w_1^* x_1 w_2^* \cdots x_{k-1} w_k^* x_k$. Hence $L_1 \cap R \in \mathcal{L}_{REG}$ and $L_2 \cap R \in \mathcal{L}_{REG}$.

Let us first prove closure under intersection with regular languages. Let R_1 be a regular language. Since $(L \cap R_1) \cap R = (L \cap R) \cap R_1 \in \mathcal{L}_{REG}$, we have $L \cap R_1 \in \mathcal{C}_k$.

Let us next consider closure under inverse morphism. Let $L \in \mathcal{C}_k$ and $h : \Sigma^* \to \Sigma_L^*$ be a morphism. Since $R \subset h^{-1}(h(R))$, we have

$$\begin{aligned} & h^{-1}(L) \cap R \\ &= h^{-1}(L) \cap h^{-1}(h(R)) \cap R \\ &= h^{-1}(L \cap h(R)) \cap R. \end{aligned}$$

Furthermore, since $L \cap h(R) \in \mathcal{L}_{REG}$, we have $h^{-1}(L) \cap R \in \mathcal{L}_{REG}$. □

Corollary 3. *The language family $\mathcal{C}_\infty$ is closed under inverse morphism and intersection with regular languages.*

It is quite straightforward to see that Theorem 11 holds also for a greater family

$$\hat{\mathcal{C}}_k = \{L \in \Sigma^* | L \cap x_0 w_1^* x_1 w_2^* \cdots x_{k-1} w_k^* x_k \in \mathcal{L}_{REG} \quad \forall x_i, w_i \in \Sigma_L^*\} .$$

It is easily seen that the family $\hat{\mathcal{C}}_k$ is even closed under intersection. However, the assumption that the languages considered are context-free is necessary for closure under morphism, as seen in the example below.

Example 10. Let $L = \{a_1^n a_2^n a_3^n | n \in \mathbb{N}\}$ and $h : \{a_1, a_2, a_3\}^* \to \{a_1, a_2\}^*$ be the projection. Obviously, $L \in \hat{\mathcal{C}}_2$ but $h(L) = \{a_1^n a_2^n | n \in \mathbb{N}\} \notin \hat{\mathcal{C}}_2$.

Before proving closure under morphism, we need two auxiliary results for noting that it suffices to prove closure under morphism for strictly bounded languages. In [12], Blattner and Latteux show that for each context-free language $L \subset \Sigma^*$, there exists a bounded context-free language $L' \subseteq \Sigma^*$ such that $\Psi(L') = \Psi(L)$ and $L' \subseteq L$.

Lemma 5. *Let $k \in \mathbb{N}_+$ and $L \in \mathcal{C}_k$ a language over the alphabet Σ. Let Δ be an alphabet and $h_1 : \Sigma^* \to \Delta^*$ a morphism such that $h_1(L)$ is strictly bounded. Then there exist a strictly bounded language $L' \in \mathcal{C}_k$ and a morphism $h_2 : \Sigma^* \to \Delta^*$ such that $h_2(L') = h_1(L)$.*

One more technical result is needed to prove the closure under morphism.

Lemma 6. *Let $L \subset \Delta^*$, $L' \subseteq \Sigma^*$ be languages and $h : \Sigma^* \to \Delta^*$ a morphism. Then $h(h^{-1}(\overline{L}) \cap L') = h(\overline{h^{-1}(L)} \cap L')$.*

Theorem 12. *The language family $\mathcal{C}_k$ is closed under morphism for each $k \in \mathbb{N}_+$.*

Corollary 4. *The language family $\mathcal{C}_\infty$ is closed under morphism.*

Combining Theorems 11 and 12, we get the following result.

Theorem 13. *The language family $\mathcal{C}_k$ is a full trio for each $k \in \mathbb{N}_+$. Moreover, the language family $\mathcal{C}_\infty$ is a full trio.*

We are finally able to state our last theorem, the main result of this research:

Theorem 14. *There does not exist a minimal full trio with respect to the family of bounded context-free languages.*

Proof. Let L be an arbitrary nonregular bounded context-free language. Then $L \in \mathcal{C}_k \backslash \mathcal{C}_{k+1}$ for some $k \in \mathbb{N}_+$. By Theorem 9, there exists a nonregular bounded language $L' \in \mathcal{T}(L)$ such that $L' \in \mathcal{C}_{k+1} \setminus \mathcal{C}_{k+2}$. Hence $\mathcal{T}(L') \subseteq \mathcal{T}(L)$. On the other hand, by Theorem 13, the language L is not in $\mathcal{T}(L')$. Thus $\mathcal{T}(L') \subsetneq \mathcal{T}(L)$. Hence $\mathcal{T}(L)$ cannot not be minimal with respect to the bounded context-free languages. Since L was chosen to be an arbitrary nonregular bounded context-free language, the proof is complete. □

4 Conclusions

The Conjecture 1 still remains open. On the other hand, the family of context-free languages L such that $\mathcal{T}(L)$ does not contain a nonregular bounded language is fairly lean. These languages are a good topic for further investigation.

It can be shown that the language familiy $\mathcal{C}_k$ is closed under concatenation and Kleene*, i.e., it is a full AFL for each $k \in \mathbb{N}_+$. This certainly implies that $\mathcal{C}_\infty$ ia a full AFL, too. In [2], it is also proven, with great help of Michel Latteux, that $\mathcal{C}_k$ is also closed under substitution for each $k \in \mathbb{N}_+$.

It should be noted that as a corollary of Theorem 14, none of the languages $S_i(\theta_1, \theta_2, \ldots, \theta_{k+1})$ can be minimal with respect to the bounded context-free languages. Autebert et al. have stated a conjecture that the full trios generated by the languages $S_1(\underbrace{\neq, \neq, \ldots, \neq}_{k+1 \text{ times}})$ would be minimal with respect to the $(k+1)$-bounded context-free languages [1]. It is reasonable to expand the question of the conjecture a little bit.

Open Problem 1. *Let $k \geq 2$ be an integer, $i \in \{1, 2, \ldots, k+1\}$ and $\theta_j \in \{>, \neq\}$ for all $j \in \{1, 2, \ldots, k+1\}$. Are the full trios generated by the languages $S_i(\theta_1, \theta_2, \ldots, \theta_{k+1})$ minimal with respect to the family of $(k+1)$-bounded context-free languages?*

Research Problem 2. *For each $k \in \mathbb{N}_+$, characterize the full trios that are minimal with respect to $(k+1)$-bounded context-free languages.*

References

1. Autebert, J.-M., Beauquier, J., Boasson, L., Latteux, M.: Very small families of algebraic nonrational languages. In: Formal Language Theory, Perspectives and Open Problems, pp. 89–107. Academic Press, Orland (1980)
2. Salmi, T.: Very Small Families Generated by Bounded and Unbounded Context-Free Languages. PhD thesis, University of Oulu, Department of Mathematical Sciences (2009)
3. Autebert, J.M., Beauquier, J., Boasson, L., Nivat, M.: Quelques problèmes ouverts en théorie des langages algébriques. RAIRO Informatique Theorique/Theoretical Informatics 13(4), 363–378 (1979)
4. Berstel, J., Boasson, L.: Une suite décroissante de cônes rationnels. In: Proceedings of the 2nd Colloquium on Automata, Languages and Programming, pp. 383–397. Springer, London (1974)
5. Ginsburg, S.: The Mathematical Theory of Context-Free Languages. McGraw-Hill, Inc., New York (1966)
6. Ginsburg, S.: Algebraic and Automata-Theoretic Properties of Formal Languages. Elsevier Science Inc., New York (1975)
7. Eilenberg, S., Schützenberger, M.P.: Rational sets in commutative monoids. Journal of Algebra 13, 173–191 (1969)
8. Ito, R.: Every semilinear set is a finite union of disjoint linear sets. Journal of Computer and System Sciences 3(2), 221–231 (1969)
9. Berstel, J.: Transductions and Context-Free Languages. Teubner Studienbücher, Stuttgart (1979)
10. Goldstine, J.: Substitution and bounded languages. Journal of Computer and System Sciences 6(1), 9–29 (1972)
11. Kortelainen, J.: Every commutative quasirational language is regular. In: Proceedings of the 12th Colloquium on Automata, Languages and Programming, pp. 348–355. Springer, London (1985)
12. Blattner, M., Latteux, M.: Parikh-bounded languages. In: Proceedings of the 9th Colloquium on Automata, Languages and Programming, pp. 316–323. Springer, London (1981)

Describing Periodicity in Two-Way Deterministic Finite Automata Using Transformation Semigroups

Michal Kunc[1,⋆] and Alexander Okhotin[2,3,⋆⋆]

[1] Masaryk University, Czech Republic
kunc@math.muni.cz
[2] Department of Mathematics, University of Turku, Finland
[3] Academy of Finland
alexander.okhotin@utu.fi

Abstract. A framework for the study of periodic behaviour of two-way deterministic finite automata (2DFA) is developed. Computations of 2DFAs are represented by a two-way analogue of transformation semigroups, every element of which describes the behaviour of a 2DFA on a certain string x. A subsemigroup generated by this element represents the behaviour on strings in x^+. The main contribution of this paper is a description of all such monogenic subsemigroups up to isomorphism. This characterization is then used to show that transforming an n-state 2DFA over a one-letter alphabet to an equivalent sweeping 2DFA requires exactly $n+1$ states, and transforming it to a one-way automaton requires exactly $\max_{0 \leqslant \ell \leqslant n} G(n-\ell) + \ell + 1$ states, where $G(k)$ is the maximum order of a permutation of k elements.

1 Introduction

Two-way deterministic finite automata (2DFA) were introduced in the famous paper by Rabin and Scott [15] alongside the one-way nondeterministic automata (1NFA). Both kinds of automata recognize the same language family as the one-way deterministic finite automata (1DFA). However, they are substantially different in terms of succinctness of description, and the number of states needed to represent a language by one type of finite automata is sometimes much greater than for another type.

While the methods for determining the number of states in one-way automata, both deterministic and nondeterministic, are well-known, and the main descriptional complexity questions [6] have been researched to extinction, the succinctness issues of two-way automata have proved to be truly challenging. The question of whether 2DFAs can simulate their nondeterministic counterparts (2NFA) with only a polynomial blowup has attracted a lot of attention due to its relation

⋆ Supported by the project MSM0021622409 of the Ministry of Education of the Czech Republic and by Grant 201/09/1313 of the Grant Agency of the Czech Republic.

⋆⋆ Supported by the Academy of Finland under grant 134860.

G. Mauri and A. Leporati (Eds.): DLT 2011, LNCS 6795, pp. 324–336, 2011.

to the *L vs. NL* problem in the complexity theory [2,16], yet no definite answers could be found. Even such a basic question as the precise number of states in a 1DFA needed to simulate an n-state 2DFA had not been settled for almost half a century: the $(n+1)^{n+1}$ upper bound by Shepherdson [17] was approached by a relatively close $(\frac{n-5}{2})^{\frac{n-5}{2}}$ lower bound by Moore [13], but only a few years ago the exact value $n(n^n - (n-1)^n)$ was finally determined by Kapoutsis [8]. Simulations of 2NFAs by simpler automata, first studied by Vardi [19], were also determined precisely by Kapoutsis [8]. The complexity of operations on 2DFAs has recently been investigated by Jirásková and Okhotin [7].

In spite of some individual results on the succinctness of 2DFAs, what exactly can a 2DFA with a given number of states do, remains a mystery. This paper undertakes to study the behaviour of 2DFAs in one particular case: on a concatenation of multiple instances of the same string, on which an automaton may first keep track of the number of instances, but eventually loses count and follows a periodic trajectory, repeatedly passing through the same sequence of states. The paper develops a general framework for reasoning about such computations, which describes both the periodic and the non-periodic behaviour in a unified way.

Consider the well-known algebraic representation of a 1DFA by a monoid of *partial transformations* of its set of states [14]. A generalization of this concept to semigroups of *two-way transformations* representing bi-directional motion was given in the work of Birget [3], who used it to represent 2DFAs of a special form. Recalling this notion, this paper applies it to arbitrary 2DFAs in Section 2. Each element of a two-way transformation semigroup defines the behaviour of some 2DFA on some nonempty string x, that is, for the 2DFA entering x from a given side in a given state, the two-way transformation specifies the result of its computation: whether the 2DFA ever leaves the string, and if it does, then from which side and in which state it emerges. The behaviour of the 2DFA on all strings in x^+ is then represented by the subsemigroup generated by the two-way transformation corresponding to x.

The properties of such *monogenic subsemigroups* of the semigroup of two-way transformations are gradually worked out in Section 3. For each two-way transformation on n states, an ordinary transformation on $n+1$ states is constructed, so that the subsemigroups generated by these transformations are isomorphic. The final result is the precise characterization of monogenic semigroups of two-way transformations on n states: their index ℓ (that is, the starting point of the periodicity) and period $\mathrm{lcm}(p_1, \ldots, p_k)$ satisfy the inequality $p_1 + \cdots + p_k + \ell \leqslant n+1$.

The most direct application of this framework is to the state complexity of 2DFAs over a one-letter alphabet. The first study of unary 2DFAs was undertaken by Chrobak [4], who has sketched an argument that an n-state 2DFA over a unary alphabet can be simulated by a $\Theta(G(n))$-state 1DFA, where $G(n) = e^{(1+o(1))\sqrt{n \ln n}}$ is the maximal order of a permutation on n elements, known as *Landau's function* [9]. Further work in this direction was done by Mereghetti and Pighizzini [10] and by Geffert, Mereghetti and Pighizzini [5], who similarly estimated the 2NFA–1DFA tradeoff. Their approach lies with considering only the periodic part of the language and using a general upper bound on the starting point of its periodicity, and thus leads only to asymptotic succinctness tradeoffs. No exact values of succinctness tradeoffs between unary two-way and one-way automata are known up to date. The first such results are obtained in this paper.

Based on the analysis of monogenic two-way transformation semigroups, the use of the states by a unary 2DFA is explained as follows. It is in fact found that 2DFAs can do just two things using fewer states than 1DFAs:

1. count divisibility separately for several numbers, where an equivalent 1DFA would need to count modulo the least common multiple of those numbers;
2. when counting up to a finite bound ℓ, they can count one step less than one would expect, and then use one of the cycles to distinguish the string of length ℓ from longer strings.

In Section 4, this understanding is used to show that every 2DFA over a unary alphabet can be transformed to an equivalent sweeping 2DFA with at most one extra state, thus more or less confirming the unproved claim by Chrobak [4], who stated that this can be done without increasing the number of states. The next Section 5 considers the standard question of converting an n-state unary 2DFA to an equivalent 1DFA. Chrobak's [4] asymptotic estimation $\Theta(G(n))$ is hereby improved to the precise expression, which is $\max_{0 \leqslant \ell \leqslant n} G(n-\ell) + \ell + 1$. The same function applies to the 2DFA to 1NFA transformation.

2 Two-Way Transformation Semigroups

A (partial) 1DFA is a quintuple $\mathcal{A} = (\Sigma, Q, q_1, \delta, F)$, where Σ is a finite alphabet, Q is a finite set of states, of which $q_1 \in Q$ is the initial state and $F \subseteq Q$ are the accepting states, and $\delta \colon Q \times \Sigma \to Q$ is a (partially defined) transition function. For every string $w \in \Sigma^*$, if $\mathcal{A}$ reads the string w beginning in a state $q \in Q$, it can either finish reading it in a state q', or reach an undefined transition. This is represented by a (partial) function $\delta_w^{\mathcal{A}} \colon Q \to Q$, called the *behaviour of* $\mathcal{A}$ *on* $w \in \Sigma^*$. The language recognized by a 1DFA is $L(\mathcal{A}) = \{\, w \mid \delta_w^{\mathcal{A}}(q_1) \in F \,\}$.

Behaviours of 1DFAs on different strings can be analyzed within the monoid $\mathcal{PT}_Q$ of partial functions from Q to Q (also known as *partial transformations of* Q) with the function composition $\circ$ as the product. The computations of the 1DFA $\mathcal{A}$ are characterized by its *transition monoid*, which is the submonoid of $\mathcal{PT}_Q$ consisting of the automaton's behaviours $\delta_w^{\mathcal{A}}$ on all strings $w \in \Sigma^*$. This submonoid is generated by the behaviours on letters, and the function composition takes the form $\delta_v^{\mathcal{A}} \circ \delta_u^{\mathcal{A}} = \delta_{uv}^{\mathcal{A}}$. A detailed introduction was given by Perrin [14]. In this section, this construction of a monoid associated with an automaton is generalized to the case of bi-directional motion.

Consider 2DFAs operating on strings $\vdash w \dashv$ bounded by end-markers, which begin their computation at $\vdash$, and accept by going beyond either of the markers. Thus, a 2DFA is defined as a quadruple $\mathcal{A} = (\Sigma, Q, q_1, \delta)$, in which Σ is a finite alphabet with $\vdash, \dashv \notin \Sigma$, Q is a finite set of states, $q_1 \in Q$ is the initial state and

$\delta\colon Q \times (\Sigma \cup \{\vdash, \dashv\}) \to Q \times \{-1, +1\}$ is a partially defined transition function: for $\mathcal{A}$ in a state q observing a symbol a, the value $\delta(q, a)$ indicates the next state and the direction of motion. Consider the behaviour of a 2DFA on any nonempty string w. It enters the string in a certain state either from its first or from its last symbol. Then the automaton may either loop inside the string, or reach an undefined transition, or eventually leave the string by going to the left of its first symbol or to the right of its last symbol in a certain new state.

This behaviour is represented similarly to the behaviour of a 1DFA, using additional notation for the right and the left sides of the string. Entering the leftmost symbol of w in a state q is described using the notation $\overrightarrow{q}$. Then, leaving w by going to the right of its last symbol in a state r is denoted by $\overrightarrow{r}$, because this means entering the string to the right of w on its first symbol. Symmetrically, the notation $\overleftarrow{q}$ represents entering a string from the right in a state q, and if such a computation results in leaving w by going to the left beyond its first symbol in a state r, this is denoted by $\overleftarrow{r}$.

Thus the behaviour of a 2DFA with the set of states Q on any string can be represented as a partial transformation of a set $N = \overrightarrow{Q} \cup \overleftarrow{Q}$, where $\overrightarrow{Q} = \{\, \overrightarrow{q} \mid q \in Q \,\}$ and $\overleftarrow{Q} = \{\, \overleftarrow{q} \mid q \in Q \,\}$. Transformations of this kind were introduced by Birget [3], who used them to describe the behaviour of a restricted case of 2DFAs. These transformations will be called *two-way transformations* on Q, denoted by symbols f and g, and depicted by oriented graphs, such as in Figure 1. Let $f_w^{\mathcal{A}} : N \to N$ denote the behaviour of a 2DFA $\mathcal{A}$ on a string $w \in (\Sigma \cup \{\vdash, \dashv\})^*$. Then, the language recognized by $\mathcal{A}$ is $L(\mathcal{A}) = \{\, w \in \Sigma^* \mid f_{\vdash w \dashv}^{\mathcal{A}}(\overrightarrow{q_1}) \text{ is defined} \,\}$.

A natural question is whether all two-way transformations may occur as behaviours of some automata on some strings, and the answer is negative. Let f be a behaviour of some automaton on a string $w \in \Sigma^+$, let $\overrightarrow{q} \in \overrightarrow{Q}$ and assume $f(\overrightarrow{q}) = \overrightarrow{r} \in \overrightarrow{Q}$. Then the computation of the automaton going through w to the state r beyond w should pass through the last symbol of w in some state p, from where the automaton went to the right. Accordingly, there should exist a state p with $f(\overleftarrow{p}) = \overrightarrow{r}$.

A symmetric condition applies to elements of $\overleftarrow{Q}$, and the entire necessary condition can be succinctly stated as follows. For all $\alpha, \beta \in N$, if both α and β belong to $\overrightarrow{Q}$ or both belong to $\overleftarrow{Q}$, this is denoted by $\alpha \sim \beta$ and represents entering strings in the same direction. Then

$$\forall \alpha \in N\colon f(\alpha) \sim \alpha \implies \exists \beta \in N\colon \beta \nsim \alpha \ \wedge\ f(\beta) = f(\alpha). \tag{1}$$

Denote by $\mathcal{TT}_Q$ the set of two-way transformations on Q satisfying this condition.

Two-way transformations that occur as behaviours of some automata on one-symbol strings $w = a \in \Sigma$ must satisfy a stronger condition:

$$f_a^{\mathcal{A}}(\overrightarrow{q}) = f_a^{\mathcal{A}}(\overleftarrow{q}) \quad \text{(for all } q \in Q) \tag{2}$$

And conversely, if a two-way transformation satisfies condition (2), then it is a behaviour of some letter in some 2DFA. Such two-way transformations shall be called *distinguished*.

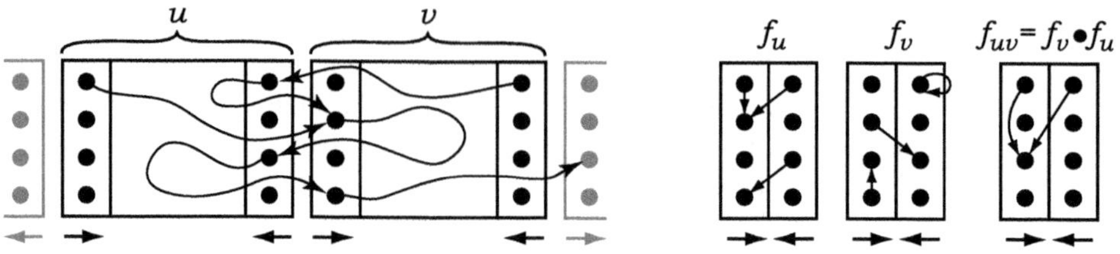

Fig. 1. Composition of behaviours on u and on v as a product $f_v \bullet f_u$

Once the behaviour of a 2DFA on some strings $u, v \in \Sigma^+$ is known to be f and g, respectively, its behaviour on their concatenation uv can be inferred from f and g, as depicted in Figure 1. It is calculated as a certain *product* $g \bullet f$ of two-way transformations $f, g\colon N \to N$, defined by Birget [3]. This product faithfully represents the behaviour of 2DFAs on the concatenation of two strings, that is, $f_{uv}^{\mathcal{A}} = f_v^{\mathcal{A}} \bullet f_u^{\mathcal{A}}$ for any strings $u, v \in (\Sigma \cup \{\vdash, \dashv\})^+$ and for any 2DFA $\mathcal{A}$.

It can be shown that $\mathcal{TT}_Q$ equipped with the product $\bullet$ is a semigroup. This semigroup is generated by distinguished two-way transformations, and therefore, elements of $\mathcal{TT}_Q$ are precisely behaviours of some 2DFA on some string. More precisely, every element of $\mathcal{TT}_Q$ can be obtained as a product of two distinguished transformations. Note that $\mathcal{TT}_Q$ is not a monoid for the lack of an identity element. Though there exists an identity two-way transformation defined by $e(\alpha) = \alpha$, it does not satisfy condition (1). For this reason, the semigroup representation of two-way automata considers only their computations on nonempty inputs. The empty string can be reintroduced later, when turning from semigroups back to automata.

There is the following formal connection between computations of 2DFAs and semigroups $\mathcal{TT}_Q$:

Proposition 1. *Let $\mathcal{A} = (\Sigma, Q, q_1, \delta)$ be a 2DFA, and consider the subsemigroup of $\mathcal{TT}_Q$ generated by distinguished two-way transformations $f_a^{\mathcal{A}}$, for $a \in \Sigma \cup \{\vdash, \dashv\}$. Then*

$$L(\mathcal{A}) = \{\, a_1 \dots a_\ell \mid (f_{\dashv}^{\mathcal{A}} \bullet f_{a_\ell}^{\mathcal{A}} \bullet \dots \bullet f_{a_1}^{\mathcal{A}} \bullet f_{\vdash}^{\mathcal{A}})(\overrightarrow{q_1}) \text{ is defined} \,\}.$$

In particular, for a given fixed automaton $\mathcal{A}$, the membership of a string $a_1 \dots a_\ell \in \Sigma^+$ in $L(\mathcal{A})$ depends only on the element $f_{a_1 \dots a_\ell}^{\mathcal{A}} = f_{a_\ell}^{\mathcal{A}} \bullet \dots \bullet f_{a_1}^{\mathcal{A}}$ of the subsemigroup generated by $\{\, f_a^{\mathcal{A}} \mid a \in \Sigma \,\}$.

3 Monogenic Subsemigroups of $\mathcal{TT}_Q$

Let $f \in \mathcal{TT}_Q$ be any fixed element of the two-way transformation semigroup. This element represents the behaviour of some 2DFA $\mathcal{A}$ on some string w. By Proposition 1, for arbitrary $u, v \in \Sigma^*$, the string $uw^\ell v$ belongs to $L(\mathcal{A})$ if and only if the two-way transformation $f_{v\dashv}^{\mathcal{A}} \bullet (f)^{\bullet\ell} \bullet f_{\vdash u}^{\mathcal{A}}$ is defined on $\overrightarrow{q_1}$, where $f^{\bullet\ell}$

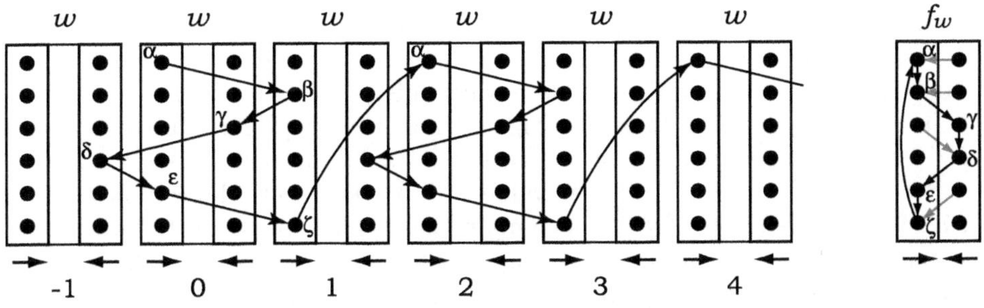

Fig. 2. Computation on a bi-infinite string of ws

stands for $f \bullet \cdots \bullet f$, ℓ times. Beginning with a certain $j \geqslant 1$, two-way transformations in the sequence $\{f^{\bullet \ell}\}_{\ell \geqslant 1}$ repeat periodically. Then, by the above observation, the membership of strings $uw^\ell v$ in $L(\mathcal{A})$ becomes periodic no later than starting from j, and the period divides the period of the powers of f. Therefore, any upper bound on the periodicity of powers $f^{\bullet \ell}$, which is described by the structure of the monogenic subsemigroup generated by f in $\mathcal{TT}_Q$, constitutes a bound on the periodicity of the membership of strings $uw^\ell v$.

In order to describe the structure of a monogenic subsemigroup S generated by some element s in a finite semigroup, one has to determine two numbers. The least positive integer i, for which there exists $j > i$ with $s^i = s^j$, is called the *index* of S, and the least number $p \geqslant 1$ with $s^i = s^{i+p}$ is called the *period* of S. The index and period determine a monogenic semigroup up to isomorphism.

3.1 The Distance Travelled After i Steps

For a string $w \in \Sigma^*$, assume a bi-infinite string of copies of w, numbered by integers. Let $f \in \mathcal{TT}_Q$ be the behaviour of some 2DFA on w. Consider a computation of this 2DFA, that begins by entering copy number 0 from the left in a certain state, which is represented by $\alpha \in \overrightarrow{Q}$. At every j-th step, the automaton proceeds to the neighbouring instance of w: to the right if $f^j(\alpha) \in \overrightarrow{Q}$, and to the left if $f^j(\alpha) \in \overleftarrow{Q}$. Such a computation is illustrated in Figure 2, where $f(\alpha) = \beta$, $f^2(\alpha) = \gamma, \ldots, f^6(\alpha) = \alpha$, etc. Similarly, one can consider computations starting from $\alpha' \in \overleftarrow{Q}$, numbering the instances of w in the reverse direction.

Consider the *distance* travelled in such a computation. In Figure 2, three steps of computation move the head back by one square, while six steps of computation result in moving forward by two squares. This shall be denoted by $\mathrm{d}(\alpha, 3) = -1$ and $\mathrm{d}(\alpha, 6) = 2$, respectively.

Definition 1. *For every $\alpha \in N$ and $i \geqslant 0$, such that $f^i(\alpha)$ is defined, let*

$$\mathrm{d}(\alpha, i) = \left|\{\, j \mid 1 \leqslant j \leqslant i, f^j(\alpha) \sim \alpha \,\}\right| - \left|\{\, j \mid 1 \leqslant j \leqslant i, f^j(\alpha) \nsim \alpha \,\}\right| .$$

In other words, $\mathrm{d}(\alpha, i)$ expresses how far one moves from the original position in the bi-infinite string of fs by means of i steps of the computation represented by the two-way transformation f, where positive numbers mean continuing in the direction of α, while negative numbers mean that the direction was reversed. Observe that $\mathrm{d}(\alpha, i)$ and $\mathrm{d}(\alpha, i+1)$ always differ exactly by one.

In the following, it will be investigated how the graph structure of any $f \in \mathcal{TT}_Q$ determines its behaviour as a two-way transformation.

For any $m \geqslant 1$ and $\alpha \in N$, the value $f^{\bullet m}(\alpha)$ represents the computation on a block of m instances of f. This computation begins on the first or on the last instance of f in this block, depending on whether $\alpha \in \overrightarrow{Q}$ or $\alpha \in \overleftarrow{Q}$. Consider the case of $\alpha \in \overrightarrow{Q}$. Then the computation begins on the first instance of f, and at every j-th step the computation proceeds to the neighbouring instance as described above. Unless $f^{\bullet m}(\alpha)$ is undefined, the computation eventually leaves the block to the right or to the left.

The condition of leaving the block can be defined in terms of d as $\mathrm{d}(\alpha, i) \in \{-1, m\}$ for some i. This is formally established in the following lemma, which handles the cases of $\alpha \in \overrightarrow{Q}$ and $\alpha \in \overleftarrow{Q}$ uniformly.

Lemma 1. *For every $m \in \mathbb{N}$ and $\alpha \in N$, $f^{\bullet m}(\alpha) = f^i(\alpha)$, where $i \in \mathbb{N}$ is the smallest number with $\mathrm{d}(\alpha, i) \notin \{0, \ldots, m-1\}$, or, equivalently, with $\mathrm{d}(\alpha, i) \in \{-1, m\}$. If such an i does not exist, then $f^{\bullet m}(\alpha)$ is undefined.*

3.2 $\tilde{f}$: Moving by f Until Advancing by One Position

Consider the example in Figure 2. What the f-cycle of length 6 does is, starting from α, first going one step forward, then two steps back and finally three steps forward. Provided that there is an extra instance of w in position -1, this results in going 2 steps forward. However, if the instance of w in position 0 is the leftmost one, then this computation would not take place.

In order to deal with computations on long blocks of ws, it is useful to assume that there is an unbounded supply of ws on both sides, and consider the computation starting from α that results in advancing by one position (that is, to the right if $\alpha \in \overrightarrow{Q}$ and to the left if $\alpha \in \overleftarrow{Q}$).

Definition 2. *For every $f \in \mathcal{TT}_Q$, define a partial transformation $\tilde{f} \in \mathcal{PT}_N$ by the rule $\tilde{f}(\alpha) = f^i(\alpha)$, where $i \in \mathbb{N}$ is the smallest number with $\mathrm{d}(\alpha, i) = 1$. If such an i does not exist, $\tilde{f}(\alpha)$ is undefined.*

Returning to Figure 2, $\tilde{f}(\alpha) = f(\alpha) = \beta$, since $\mathrm{d}(\alpha, 1) = 1$. The value of $\tilde{f}(\beta)$ is given by $f^5(\beta) = \alpha$, because $\{\mathrm{d}(\beta, n)\}_{n \geqslant 1} = \{-1, -2, -1, 0, \mathbf{1}, \ldots\}$. For the elements $\gamma, \delta \in \overleftarrow{Q}$, $\tilde{f}$ represents going by one step to the left, and accordingly, $\tilde{f}(\gamma) = \delta$ and $\tilde{f}(\delta)$ is undefined.

Since for every $\alpha \in N$, the last step of the computation defining $\tilde{f}(\alpha)$ is a move in the same direction as α, one has $\tilde{f}(\alpha) \sim \alpha$. Therefore, $\tilde{f}$ is formed of two separate partial transformations on each of the sets $\overrightarrow{Q}$ and $\overleftarrow{Q}$.

While the partial transformation $\tilde{f}$ corresponds to advancing by one position, the m-th power of $\tilde{f}$ represents advancing by m positions, that is, $\tilde{f}^m(\alpha) = f^i(\alpha)$, where $i \in \mathbb{N}$ is the smallest number with $\mathrm{d}(\alpha, i) = m$.

The first step in describing the subsemigroup generated by f in $\mathcal{TT}_Q$ is to prove that it is isomorphic to the subsemigroup generated by $\tilde{f}$ in $\mathcal{PT}_N$. This amounts to showing the following statement:

Lemma 2. *For all positive integers m and k, $f^{\bullet m} = f^{\bullet k}$ if and only if $\tilde{f}^m = \tilde{f}^k$.*

The forward implication follows immediately from the equality $\tilde{f}^m = \widetilde{f^{\bullet m}}$, which means that making m steps forward over f is the same as making one step forward over a block of m instances of f. The converse is more difficult to verify, since the inequality $f^{\bullet m}(\alpha) \neq f^{\bullet k}(\alpha)$ does not necessarily imply that $\tilde{f}^m(\alpha) \neq \tilde{f}^k(\alpha)$. However, a detailed analysis shows that one can always find some node $\beta \in N$, reachable from α by several applications of f, which satisfies $\tilde{f}^m(\beta) \neq \tilde{f}^k(\beta)$.

3.3 Index and Period of the Subsemigroup Generated by $\tilde{f}$

It is well known how to calculate the index and the period of a subsemigroup of $\mathcal{PT}_N$ generated by a partial transformation h, using the structure of h as a directed graph of out-degree 1, with the set of nodes N and with $h(\alpha) = \beta$ represented by an arc $\alpha \to \beta$. Every node from N belongs either to an h-cycle, or to an h-tail, which is any maximal subset of N consisting of elements not belonging to any h-cycle, and in which for any two elements α and β, either β is reachable from α or vice versa. Note that every h-tail leads either into a cycle, or into a *dead element* where h is not defined. The number of elements of N that belong to a given tail is called the *length* of the tail. Then the index of the subsemigroup generated by h is equal to the length of the longest h-tail (it equals 1 if there is no tail) and its period is equal to the least common multiple of the lengths of all h-cycles.

Therefore, both the index and the period of the subsemigroup generated by f are determined by the lengths of $\tilde{f}$-cycles and the longest $\tilde{f}$-tail. Let C be the set of all nodes from N belonging to $\tilde{f}$-cycles, and fix one of the longest $\tilde{f}$-tails. Let T denote this $\tilde{f}$-tail and denote $D = C \cup T \subseteq N$. Denote the restriction of $\tilde{f}$ to D by $\hat{f} \in \mathcal{PT}_D$.

Then the subsemigroups generated by $\tilde{f}$ and $\hat{f}$ have the same index and period, and hence are isomorphic. Therefore, some information about the index and period of the subsemigroup generated by f can be obtained by finding an upper bound on the size of D with respect to $|Q|$. However, it can be shown that condition (1), which is imposed on f by the fact that it arises as a behaviour of a 2DFA, significantly restricts the possible lengths of cycles and tails of $\tilde{f}$. More precisely, the sum of the lengths of its cycles and the length of the longest tail never exceeds $|Q|+1$, and it can reach $|Q|+1$ only in a very special case, which corresponds to the ability of a 2DFA to use one of its cycles to save one state when counting to a certain bound (an example of such an automaton is given in the following section).

The verification of the inequality $|D| \leqslant |Q| + 1$ is rather simple if f is a distinguished transformation, because it can be easily shown that there exists at most one such state q, that both $\overrightarrow{q}$ and $\overleftarrow{q}$ belong to D. Indeed, assume there is such a $q \in Q$. Since $f(\overrightarrow{q}) = f(\overleftarrow{q})$, at least one of the nodes $\overrightarrow{q}$ and $\overleftarrow{q}$ (say the former one) belongs to an f-tail. As they cannot belong to the same f-tail, the other node $\overleftarrow{q}$ belongs to an f-cycle. Therefore, $\overrightarrow{q}$ is the last node of the unique f-tail, which contains some element of D. This implies that such q is uniquely determined. Moreover, one can also see that if such a q exists, then $\hat{f}$ contains a cycle (the node $\overrightarrow{q}$ cannot belong to the only $\hat{f}$-tail due to $\overrightarrow{q} \not\sim \overleftarrow{q}$) and there exists a node in D (namely, the last node of the $\hat{f}$-tail of $\overrightarrow{q}$), on which $\hat{f}$ is not defined.

This argument can be generalised to arbitrary two-way transformations by proving that if a node γ is in the image of $\hat{f}$, then $\gamma = f(\alpha)$ for some node α satisfying $\alpha \not\sim \gamma$ and not belonging to D. There is only one exception to this rule, namely when γ is the node where the unique f-tail containing elements of D joins an f-cycle. This leads to the following upper bound on the size of D.

Lemma 3. *It holds that $|N| \geqslant 2|D| - 2$. Additionally, if $|N| = 2|D| - 2$, then $\hat{f}$ is undefined on some element of D and contains a cycle.*

3.4 The Main Theorem and Its Implications

Because the subsemigroup generated by f in $\mathcal{TT}_Q$ is isomorphic to the subsemigroup generated by $\hat{f}$ in $\mathcal{PT}_D$, Lemma 3 can be used to describe the structure of all monogenic subsemigroups of $\mathcal{TT}_Q$:

Theorem 1. *For every monogenic subsemigroup S of $\mathcal{TT}_Q$ there exist $k \geqslant 1$ and numbers $p_1, \ldots, p_k \geqslant 1$ and $\ell \geqslant 1$, with $p_1 + \cdots + p_k + \ell \leqslant |Q| + 1$, such that S has index ℓ and period $\mathrm{lcm}(p_1, \ldots, p_k)$. More precisely, if S is generated by $f \in \mathcal{TT}_Q$, then ℓ can be obtained as the length of the longest $\tilde{f}$-tail and numbers $p_1, \ldots, p_k$ as the lengths of all $\tilde{f}$-cycles (if there is no cycle in $\tilde{f}$, let $k = p_1 = 1$).*

Conversely, for any integers $k \geqslant 1$, $p_1, \ldots, p_k \geqslant 1$ and $\ell \geqslant 1$ satisfying $p_1 + \cdots + p_k + \ell \leqslant |Q| + 1$, the semigroup $\mathcal{TT}_Q$ contains a distinguished transformation, which generates a subsemigroup with index ℓ and period $\mathrm{lcm}(p_1, \ldots, p_k)$.

This result provides a lower bound on the size of any such Q, that $\mathcal{TT}_Q$ contains a monogenic subsemigroup with a certain index and period:

Corollary 1. *Let S be a monogenic subsemigroup of $\mathcal{TT}_Q$, which has index ℓ and period p. Let $p = p_1 \cdots p_k$, where $p_1, \ldots, p_k$ are powers of distinct primes, be the prime factorization of p. Then, $|Q|$ must be at least $p_1 + \cdots + p_k + \ell - 1$.*

For a 2DFA $\mathcal{A}$, Proposition 1 guarantees that any strings $u, v \in \Sigma^+$ satisfying $f_u^{\mathcal{A}} = f_v^{\mathcal{A}}$ represent the same element of the syntactic semigroup of $L(\mathcal{A})$. This means that the bounds on the size of monogenic subsemigroups of $\mathcal{TT}_Q$ given in Theorem 1 and Corollary 1 are valid also for monogenic subsemigroups of the syntactic semigroup of any language accepted by an n-state 2DFA.

Consider computations of a 2DFA $\mathcal{A}$ on inputs ux^iv, in which the behaviour on the infix x^i is described by the two-way transformation $(f_x^{\mathcal{A}})^{\bullet i}$. Due to Proposition 1, the membership of a string ux^iv in the language $L(\mathcal{A})$ depends only on the element $f^{\mathcal{A}}_{ux^iv} = f^{\mathcal{A}}_v \bullet (f^{\mathcal{A}}_x)^{\bullet i} \bullet f^{\mathcal{A}}_u$ of $\mathcal{TT}_Q$. Therefore, the periodic behaviour of the set $\{\, i \geqslant 1 \mid ux^iv \in L(\mathcal{A}) \,\}$ depends only on the structure of the subsemigroup generated in $\mathcal{TT}_Q$ by $f^{\mathcal{A}}_x$. More precisely, the period of this set divides the period of the subsemigroup, and its index is bounded by the index of the subsemigroup. This leads to the following consequences of Theorem 1 and Corollary 1.

Corollary 2. *Let $\mathcal{A}$ be an n-state 2DFA over an arbitrary finite alphabet Σ, and let $u, v \in \Sigma^*$ and $x \in \Sigma^+$ be any strings. Then, there exist $k \geqslant 1$ and numbers $p_1, \ldots, p_k \geqslant 1$ and $\ell \geqslant 1$, with $p_1+\cdots+p_k+\ell \leqslant n+1$, such that the set of numbers $\{\, i \geqslant 1 \mid ux^iv \in L(\mathcal{A}) \,\}$ is periodic from ℓ with period $p = \operatorname{lcm}(p_1, \ldots, p_k)$.*

Corollary 3. *Let L be a regular language over an arbitrary finite alphabet Σ, and let $u, v \in \Sigma^*$ and $x \in \Sigma^+$ be any strings. Let the set of numbers $\{\, i \geqslant 1 \mid ux^iv \in L(\mathcal{A}) \,\}$ have period p beginning from ℓ. Let $p = p_1 \cdots p_k$, where $p_1, \ldots, p_k$ are powers of distinct primes, be the prime factorization of p. Then, every 2DFA recognizing L must have at least $p_1 + \cdots + p_k + \ell - 1$ states.*

4 Transformation to Sweeping Automata

A 2DFA is called *sweeping* [18] if in every computation its head changes the direction of motion only on the markers. For an arbitrary alphabet, as independently proved by Berman [1] and by Micali [12], the succinctness blowup from general 2DFAs to sweeping 2DFAs is exponential. For a unary alphabet, Mereghetti and Pighizzini [11] established a transformation of an n-state 2NFA to a sweeping 2NFA with $O(n^2)$ states. Regarding the deterministic case, Chrobak [4] mentioned in passing that "it is easy to show that any unary 2DFA can be substituted by an equivalent sweeping 2DFA without increasing the number of its states". This claim was not substantiated, and the best result known in the literature is the $O(n^2)$ bound for unary 2NFAs due to Mereghetti and Pighizzini [11]. The framework developed in this paper allows finally settling this question:

Theorem 2. *Let $n \geqslant 1$. Then for every unary deterministic two-way automaton $\mathcal{A}$ with n states, there exists an equivalent sweeping deterministic two-way automaton with $n+1$ states. For $n \geqslant 2$, this bound is the best possible.*

In short, the intuition of Chrobak was generally right, though one extra state is needed.

The upper bound is proved by constructing a new sweeping automaton, that simulates the transformation $\tilde{f}$, where f is the behaviour of the original 2DFA on the letter. The construction is straightforward in itself: it produces $p_1+\cdots+p_k+\ell$ states, where $p_1, \ldots, p_k$ are the lengths of the cycles in $\tilde{f}$, and ℓ is the length of its longest tail. The nontrivial part of the argument is the upper bound $n+1$ on this sum, given in Theorem 1.

The lower bound is witnessed by the following language.

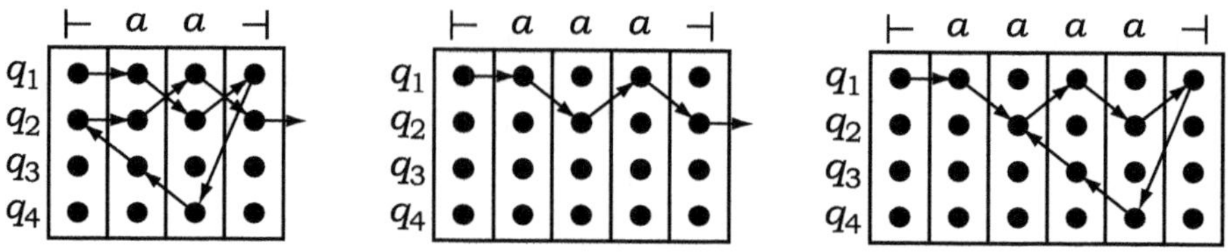

Fig. 3. Computation of the 2DFA for L_n with $n = 4$ in Theorem 2

Example 1. Let $n \geqslant 2$ and consider the language $L_n = a(a^2)^* \cup \{a^{n-2}\}$ if n is even, or $L_n = (a^2)^* \cup \{a^{n-2}\}$ for n odd. Then L_n is recognized by an n-state 2DFA with acceptance only on the right-end marker (that is, with the standard definition). However, every sweeping 2DFA for L_n needs $n + 1$ states.

The obvious automaton for this language is an $(n + 1)$-state 1DFA, and a sweeping 2DFA cannot have fewer states. However, one state can be saved at the expense of losing the sweeping property, as follows. Consider the case of $n = 4$, presented in Figure 3. The automaton begins by traversing the string from left to right, counting modulo 2. Once the right-end marker is reached, the string is either accepted because of its parity, or the automaton proceeds back to the left, counting up to $n - 2$. A sweeping 2DFA would need an extra rejecting state, reached if the string is of length n or greater; the proposed 2DFA can reuse the first two states instead, rejecting by entering an infinite loop.

5 Transformation to One-Way Automata

Now consider the question of the number of states in 1DFAs and 1NFAs needed to represent languages recognized by n-state unary 2DFAs. Chrobak [4] was the first to find out that both tradeoffs are asymptotically equivalent to the function

$$G(n) = \max\{ \operatorname{lcm}\{p_1, \ldots, p_k\} \mid p_1 + \cdots + p_k \leqslant n \},$$

known as *Landau's function* and estimated as $G(n) = e^{(1+o(1))\sqrt{n \ln n}}$ [9].

For a 2DFA over an alphabet $\{a\}$, with $f \in \mathcal{TT}_Q$ representing the behaviour on a, the numbers $p_1, \ldots, p_k$ in the definition of G correspond, according to Theorem 1, to the cycles in $\widetilde{f}$. The length of the tail ℓ in Theorem 1 is actually reflected by the number $n+1-(p_1 + \cdots + p_k)$. The contribution of ℓ to the size of a 1DFA is taken into account in the precise expression for the 2DFA-to-1DFA tradeoff given in the following theorem.

Theorem 3. *Let $n \geqslant 1$. Then for every unary two-way automaton $\mathcal{A}$ with n states, where the transitions on the letter are deterministic, there exists an equivalent complete 1DFA with $\max_{1 \leqslant \ell \leqslant n+1} G(n + 1 - \ell) + \ell$ states. For $n \geqslant 3$, this bound is tight already for the transformation of complete 2DFAs with acceptance on both sides to 1NFAs.*

The upper bound follows from Corollary 2 with $x = a$ and $u = v = \varepsilon$. Proving that this bound is tight requires some efforts. Given $n \geqslant 3$, let $k \geqslant 1$, $\ell \geqslant 1$, and let $p_1, \ldots, p_k \geqslant 2$ be powers of distinct primes, such that $p_1 + \cdots + p_k + (\ell - 1) = n$ and $G(n+1-\ell) = p_1 \cdots p_k = p$. A lower bound $\max_{1 \leqslant \ell \leqslant n+1} G(n+1-\ell) + \ell - 1$ on the 2DFA to 1DFA tradeoff is obtained straightforwardly by constructing an $(n+1)$-state 2DFA for such language as, for instance, $a^{p+\ell-1}(a^p)^*$. To see that one extra state is necessary, one has to embed within such an example the same idea as in the lower bound in Theorem 2. This is achieved in the language

$$L_n = a^{p+\ell-1}(a^p)^* \cup \{\, a^i \mid i \equiv \ell \pmod{p_k} \text{ and } i \equiv \ell - 1 \pmod{p_1 \cdots p_{k-1}} \,\},$$

which is nontrivially recognized by a 2DFA with n states, while every 1DFA for this language obviously requires $p + \ell$ states.

References

1. Berman, P.: A note on sweeping automata. In: de Bakker, J.W., van Leeuwen, J. (eds.) ICALP 1980. LNCS, vol. 85, pp. 91–97. Springer, Heidelberg (1980)
2. Berman, P., Lingas, A.: On complexity of regular languages in terms of finite automata. Report 304, Institute of Computer Science, Polish Academy of Sciences, Warsaw (1977)
3. Birget, J.-C.: Concatenation of inputs in a two-way automaton. Theoretical Computer Science 63(2), 141–156 (1989)
4. Chrobak, M.: Finite automata and unary languages. Theoretical Computer Science 47, 149–158 (1986); Errata 302, 497–498 (2003)
5. Geffert, V., Mereghetti, C., Pighizzini, G.: Converting two-way nondeterministic unary automata into simpler automata. Theoretical Computer Science 295(1-3), 189–203 (2003)
6. Holzer, M., Kutrib, M.: Descriptional and computational complexity of finite automata. In: Dediu, A.H., Ionescu, A.M., Martín-Vide, C. (eds.) LATA 2009. LNCS, vol. 5457, pp. 23–42. Springer, Heidelberg (2009)
7. Jirásková, G., Okhotin, A.: On the state complexity of operations on two-way finite automata. In: Ito, M., Toyama, M. (eds.) DLT 2008. LNCS, vol. 5257, pp. 443–454. Springer, Heidelberg (2008)
8. Kapoutsis, C.A.: Removing bidirectionality from nondeterministic finite automata. In: Jedrzejowicz, J., Szepietowski, A. (eds.) MFCS 2005. LNCS, vol. 3618, pp. 544–555. Springer, Heidelberg (2005)
9. Landau, E.: Uber die Maximalordnung der Permutationen gegebenen Grades (On the maximal order of permutations of a given degree). Archiv der Mathematik und Physik, Ser. 3 5, 92–103 (1903)
10. Mereghetti, C., Pighizzini, G.: Optimal simulations between unary automata. SIAM Journal on Computing 30(6), 1976–1992 (2001)
11. Mereghetti, C., Pighizzini, G.: Two-way automata simulations and unary languages. Journal of Automata, Languages and Combinatorics 5(3), 287–300 (2000)
12. Micali, S.: Two-way deterministic finite automata are exponentially more succinct than sweeping automata. Information Processing Letters 12(2), 103–105 (1981)
13. Moore, F.R.: On the bounds for state-set size in the proofs of equivalence between deterministic, nondeterministic, and two-way finite automata. IEEE Transactions on Computers 20, 1211–1214 (1971)

14. Perrin, D.: Finite Automata. In: van Leeuwen, J. (ed.) Handbook of Theoretical Computer Science vol. B, pp. 1–57. MIT Press, Cambridge (1990)
15. Rabin, M.O., Scott, D.: Finite automata and their decision problems. IBM Journal of Research and Development 3, 114–125 (1959)
16. Sakoda, W.J., Sipser, M.: Nondeterminism and the size of two way finite automata. In: STOC 1978, pp. 275–286 (1978)
17. Shepherdson, J.C.: The reduction of two-way automata to one-way automata. IBM Journal of Research and Development 3, 198–200 (1959)
18. Sipser, M.: Lower bounds on the size of sweeping automata. In: STOC 1979, pp. 360–364 (1979)
19. Vardi, M.: A note on the reduction of two-way automata to one-way automata. Information Processing Letters 30(5), 261–264 (1989)

Deciding Networks of Evolutionary Processors

Florin Manea*

Otto-von-Guericke-Universität Magdeburg, Fakultät für Informatik
PSF 4120, D-39016 Magdeburg, Germany
manea@iws.cs.uni-magdeburg.de

Abstract. In this paper we discuss the usage of Accepting Networks of Evolutionary Processors (ANEPs for short) as deciding devices. In this context we define a new halting condition for this model, which seems more coherent with the rest of the theory than the previous such definition, and show that all the computability results reported so far remain valid in the new framework. Moreover, we give a direct and efficient simulation of an arbitrary ANEP by a complete ANEP, thus, showing that the efficiency of deciding a language by ANEPs is not influenced by the network's topology. Finally, we obtain a surprising characterization of $\mathbf{P}^{\mathbf{NP}[\log]}$ as the class of languages that can be decided in polynomial time by ANEPs.

1 Introduction

The accepting networks of evolutionary processors (ANEPs, for short) are a bio-inspired computational mode, introduced in [1], and having its roots in [2,3]. An ANEP consists in a graph having in each node a processor, which is able to perform very simple operations, namely point mutations in a DNA sequence (insertion, deletion or substitution of nucleotides); these processors are called evolutionary processor. Furthermore, each node contains data, which are organized in the form of multisets of words, each word appearing in an arbitrarily large number of copies, and all copies are processed in parallel such that all the possible events that can take place do actually take place. Following the biological motivation, each node may be viewed as a cell containing genetic information encoded in DNA sequences, which may evolve by local evolutionary events, i.e., point mutations; moreover, each node is specialized just for one of these evolutionary operations.

The computation of an ANEP is conducted as follows. Initially, only one special node, the input node, contains a certain word, the input word. Further, the words are processed in alternative evolutionary and communication steps. In an evolutionary step, the words found in each node are rewritten according to the rules of that node. In a communication step, the words of a node are communicated to the other nodes, as permitted by some filtering condition associated

* Also at: *Faculty of Mathematics and Computer Science, University of Bucharest, Str. Academiei 14, RO-010014 Bucharest, Romania* (flmanea@fmi.unibuc.ro). The work of Florin Manea is supported by the *Alexander von Humboldt Foundation.*

G. Mauri and A. Leporati (Eds.): DLT 2011, LNCS 6795, pp. 337–349, 2011.

with both the sending and the receiving node. The classical definition assumes that a computation halts and accepts, when a word enters a special node of the network, the output node, or halts and rejects, when the words contained in each node do not change consecutive evolutionary or communication steps.

In this paper we propose a new halting condition for such a computation. Namely, a computation halts (and is called finite) as soon as at least one word enters the output node. The input word is accepted if the computation of the ANEP on this word is finite, and at least one word that is found in the output node at the end of the computation contains a special symbol; otherwise, the computation is either infinite, or, when it is finite, rejecting. The motivations behind this new halting condition are discussed in Section 3.

Results on ANEPs, seen as formal languages accepting devices, were surveyed recently in [4]. In this paper we see how these results change when the new halting condition is used. While the computational power of the ANEPs remains the same, the time complexity results are not preserved. To this end, we obtain a surprising characterization of $\mathbf{P}^{\mathbf{NP}[\log]}$ as the class of languages that can be decided in polynomial time by ANEPs.

We also show that an arbitrary ANEP can be simulated efficiently by an ANEP with complete underlying graph. This answers a question from [5] and shows that one cannot expect to decide a language faster, using ANEPs with special topology, than when complete ANEPs are used.

For the full proofs of the results presented in this paper see [10].

2 Basic Definitions

We start by summarizing the notions used throughout the paper; for all unexplained notions the reader is referred to [6]. An *alphabet* is a finite and nonempty set of symbols. The cardinality of a finite set A is written card(A). Any sequence of symbols from an alphabet V is called *word* over V. The set of all words over V is denoted by V^* and the empty word is denoted by λ. The length of a word x is denoted by $|x|$ while $alph(x)$ denotes the minimal alphabet W such that $x \in W^*$. For a word $x \in W^*$, x^r denotes the reversal of the word.

In the following, we introduce a series of rewriting operations, called *evolutionary operations* as they may be viewed as linguistic formulations of local gene mutations. We say that a rule $a \to b$, with $a, b \in V \cup \{\lambda\}$ is a *substitution rule* if both a and b are not λ; it is a *deletion rule* if $a \neq \lambda$ and $b = \lambda$; it is an *insertion rule* if $a = \lambda$ and $b \neq \lambda$. The sets of all substitution, deletion, and insertion rules over an alphabet V are denoted by Sub_V, Del_V, and Ins_V, respectively.

Given a rule σ as above and a word $w \in V^*$, we define the following *actions* of σ on w:

If $\sigma \equiv a \to b \in Sub_V$, then $\sigma^*(w) = \begin{cases} \{ubv \mid \exists u, v \in V^* \ (w = uav)\}, \\ \{w \mid w \text{ contains no } a\}, \end{cases}$

If $\sigma \equiv a \to \lambda \in Del_V$, then: $\sigma^*(w) = \begin{cases} \{uv \mid \exists u, v \in V^* \ (w = uav)\}, \\ \{w \mid w \text{ contains no } a\}, \end{cases}$

$$\sigma^r(w) = \begin{cases} \{u\}, w = ua, \\ \{w\}, \text{ otherwise,} \end{cases} \quad \sigma^l(w) = \begin{cases} \{v\}, w = av, \\ \{w\}, \text{ otherwise,} \end{cases}$$

If $\sigma \equiv \lambda \to a \in Ins_V$, then
$\sigma^*(w) = \{uav \mid \exists u, v \in V^* \ (w = uv)\}$, $\sigma^r(w) = \{wa\}$, $\sigma^l(w) = \{aw\}$.

We say that $\alpha \in \{*, l, r\}$ expresses the way of applying a deletion or insertion rule to a word, namely at any position ($\alpha = *$), in the left ($\alpha = l$), or in the right ($\alpha = r$) end of the word, respectively. For every rule σ, action $\alpha \in \{*, l, r\}$, and $L \subseteq V^*$, we define the *α-action of σ on L* by $\sigma^\alpha(L) = \bigcup_{w \in L} \sigma^\alpha(w)$. Given a finite set of rules M, we define the *α-action of M* on the word w and the language L by $M^\alpha(w) = \bigcup_{\sigma \in M} \sigma^\alpha(w)$ and $M^\alpha(L) = \bigcup_{w \in L} M^\alpha(w)$, respectively.

For two disjoint subsets P and F of an alphabet V and a word w over V, we define the predicates:

$$\varphi^{(s)}(w; P, F) \equiv \ P \subseteq alph(w) \qquad \wedge \ F \cap alph(w) = \emptyset$$
$$\varphi^{(w)}(w; P, F) \equiv alph(w) \cap P \neq \emptyset \ \wedge \ F \cap alph(w) = \emptyset.$$

The construction of these predicates is based on *random-context conditions* defined by the two sets P (*permitting contexts/symbols*) and F (*forbidding contexts/symbols*). Informally, the first condition requires that all permitting symbols are present in w and no forbidding symbol is present in w, while the second one is a weaker variant of the first, requiring that at least one permitting symbol appears in w and no forbidding symbol is present in w.

For every language $L \subseteq V^*$ and $\beta \in \{(s), (w)\}$, we define:

$$\varphi^\beta(L, P, F) = \{w \in L \mid \varphi^\beta(w; P, F)\}.$$

An *evolutionary processor over* V is a tuple (M, PI, FI, PO, FO), where:

– M is a set of substitution, deletion or insertion rules over the alphabet V, the set of rules of the processor. Formally ($M \subseteq Sub_V$) or ($M \subseteq Del_V$) or ($M \subseteq Ins_V$). A processor is "specialized" in one type of operations only.

– $PI, FI \subseteq V$ are the *input* permitting, respectively forbidding, contexts of the processor, while $PO, FO \subseteq V$ are the *output* permitting, respectively forbidding, contexts of the processor. Informally, the permitting input (output) contexts are the set of symbols that should be present in a word, when it enters (respectively, leaves) the processor, while the forbidding contexts are the set of symbols that should not be present in a word in order to enter (respectively, leave) the processor.

We denote the set of evolutionary processors over V by EP_V.

Next we define the central notion of our paper, namely Accepting Networks of Evolutionary Processors (ANEPs for short). Our definition is slightly different from the one that was used in literature so far (see [4] for a survey) by the usage of a special accepting symbol μ.

An *accepting hybrid network of evolutionary processors* (ANEP for short) is a 9-tuple $\Gamma = (V, U, \mu, G, \mathcal{N}, \alpha, \beta, x_I, x_O)$, where:

– V and U are the input and network alphabets, respectively, $V \subseteq U$; the symbol $\mu \in U \setminus V$ is a distinguished symbol, called accepting symbol.

– $G = (X_G, E_G)$ is a directed graph, with the set of nodes X_G and the set of edges $E_G \subseteq X_G \times X_G$. The graph G is called the *underlying graph* of the network, and card(X_G) is the size of Γ.

– $\mathcal{N} : X_G \longrightarrow EP_U$ is a mapping which associates with each node $x \in X_G$ the evolutionary processor $\mathcal{N}(x) = (M_x, PI_x, FI_x, PO_x, FO_x)$.

– $\alpha : X_G \longrightarrow \{*, l, r\}$; $\alpha(x)$ gives the action mode of the rules of node x on the words existing in that node.
– $\beta : X_G \longrightarrow \{(s), (w)\}$ defines the type of the *input/output filters* of a node. More precisely, for every node, $x \in X_G$, the following filters are defined:

$$\text{input filter: } \rho_x(\cdot) = \varphi^{\beta(x)}(\cdot; PI_x, FI_x),$$
$$\text{output filter: } \tau_x(\cdot) = \varphi^{\beta(x)}(\cdot; PO_x, FO_x).$$

That is, $\rho_x(w)$ (respectively, $\tau_x(w)$) indicates whether or not the word w can pass the input (respectively, output) filter of x.
– x_I and $x_O \in X_G$ are the *input node*, and, respectively, the *output node* of Γ.

An ANEP is said to be complete if the underlying graph G has the edges $E_G = \{(x, y) \mid x \neq y \text{ and } x, y \in X_G\}$.

A *configuration* of an ANEP Γ is a mapping $C : X_G \longrightarrow 2^{V^*}$, associating a set of words with every node of the graph. A configuration may be understood as the sets of words which are present in any node at a given moment; it can change either by an *evolutionary step* or by a *communication step.*

When changing by an evolutionary step each component $C(x)$ of the configuration C is changed in accordance to the set of evolutionary rules M_x, of node x, and $\alpha(x)$, the way these rules should be applied. Formally, the configuration C' is obtained in *one evolutionary step* from the configuration C, written as $C \Longrightarrow C'$, if and only if $C'(x) = M_x^{\alpha(x)}(C(x))$, for all $x \in X_G$.

When changing by a communication step, each node-processor $x \in X_G$ sends one copy of each word it contains, and is able to pass its output filter, to all the node-processors connected to x, and receives all the words sent by all the other node processor connected with x, provided that they can pass its input filter. Formally, the configuration C' is obtained in *one communication step* from configuration C, written as $C \vdash C'$, if and only if $C'(x) = (C(x) - \tau_x(C(x))) \cup \bigcup_{(y,x) \in E_G} \rho_x(\tau_y(C(y)))$, for all $x \in X_G$. Note that the words which leave a node are eliminated from that node; if such a word cannot pass the input filter of any node, it is lost.

The computation of Γ on the input word $w \in V^*$ is a (potential infinite) sequence of configurations $C_0^{(w)}, C_1^{(w)}, C_2^{(w)}, \ldots$. The initial configuration $C_0^{(w)}$ is defined by $C_0^{(w)}(x_I) = \{w\}$ and $C_0^{(w)}(x) = \emptyset$ for all $x \in X_G$, $x \neq x_I$. Further, $C_{2i}^{(w)} \Longrightarrow C_{2i+1}^{(w)}$ and $C_{2i+1}^{(w)} \vdash C_{2i+2}^{(w)}$, for all $i \geq 0$.

The acceptance symbol μ is important when defining the accepting computations of an ANEP. A computation as above halts when it reaches a configuration C_t^w in which the output node x_O contains a word; we say that Γ halts on the input word w. We distinguish two situations:

i. There exists $u \in C_t^w(x_O)$, such that u contains the symbol μ. In this case, the computation is an *accepting computation*, and Γ accepts w.
ii. Any word $u \in C_t^w(x_O)$ does not contain the symbol μ. In this case, the computation is a *rejecting computation*, and Γ rejects w.

The *language accepted* by Γ is $L_a(\Gamma) = \{w \in V^* \mid$ the computation of Γ on w is an accepting one$\}$. We say that an ANEP Γ decides the language $L \subseteq V^*$, and write $L(\Gamma) = L$ if and only if $L_a(\Gamma) = L$ and Γ halts on all input words.

Let Γ be an ANEP deciding the language L.
The *time complexity* of the finite computation $C_0^w, C_1^w, C_2^w, \ldots C_m^w$ of Γ on $w \in L$ is denoted by $Time_\Gamma(w)$ and equals m. The time complexity of Γ is the function $Time_\Gamma(n) = \max\{Time_\Gamma(x) \mid |x| = n\}$. We say that Γ works in polynomial time, if there exists a polynomial function f such that $f(n) \geq Time_\Gamma(n)$.

For a function $f : \mathbf{N} \longrightarrow \mathbf{N}$ we define $\mathbf{Time}_{ANEP}(f(n)) = \{L \mid$ there exists the ANEP Γ, such that $L(\Gamma) = L, Time_\Gamma(n) \leq f(n)$, for all $n \in I\!N\}$. Moreover, we write $\mathbf{PTime}_{ANEP} = \bigcup_{k \geq 0} \mathbf{Time}_{ANEP}(n^k)$.

One can define in a similar manner space and length complexity classes ([4]).

3 The New Halting Condition and Computability Results

The single difference between the initial definition of ANEPs ([1]) and ours is the presence and usage of the symbol μ. The way a computation of an ANEP is conducted remains basically the same, but the halting conditions (both in the case of acceptance and rejection) are essentially different.

But let us first recall the definition of a halting ANEP-computation from the literature (see the seminal work [1], the survey [4], and the references therein). A computation halts and accepts if there exists a configuration in which the set of words existing in the output node is non-empty. A computation halts and rejects if there exist two identical configurations obtained either in consecutive evolutionary steps or in consecutive communication steps. The language accepted by the ANEP Γ is $L_a(\Gamma) = \{w \in V^* \mid$ the computation of Γ on w accepts$\}$. Also, it was said that an ANEP Γ decides the language $L \subseteq V^*$ iff $L_a(\Gamma) = L$ and L halts on every input.

The reasons that made us switch to the new definition are related mainly to the rejecting computations.

First, the rejecting condition did not seem coherent with the other conditions verified in a ANEP. The filters verify the existence or absence of several symbols in the communicated words; the application of a rule by a processor consists (in the more complicated cases of substitution and deletion rules) in looking for the occurrences of a symbol in the words of that node (if it contains any), and replacing an arbitrary such occurrence with a symbol or with λ. The rejecting condition assumed a very different process: one checked whether the configurations of all the nodes, in two consecutive steps of the same kind, were equal.

Also, the processes executed by an ANEP are localized: filters are associated with nodes, rules are associated with nodes, and the accepting condition concerned only one node, the output node. In the case of a rejecting computation the definition took us to a global level: we looked at all the words present at a given moment in the network. The condition seemed an artificial formalization of the case when the network enters in an infinite loop, and the computation should halt. However, only infinite loops in which the configurations are repeated in consecutive steps were detected. Although avoiding infinite loops seems to us a good-practice in programming (regardless of the computational model), we do not see a reason for it to be ruled out from the definition.

Nevertheless, verifying the equality between two configurations required to memorize, at any moment, all the words from the last two configurations. Thus, an additional memory-device was needed, and this was not (explicitly) part of an ANEP. This affected the self-containment of the definition.

The new halting conditions seem to overcome these problems. The computation halts as soon as a word enters in a special node. Although it seems to be also a condition of a different nature from the ones that are checked in an ANEP, it is natural to think that before each processing step a processor checks whether it contains some words, and then it looks for the places where the rules can be applied. Further, the decision of a computation is taken according to a test in which we check for the existence of a symbol in the words of a node; this seems coherent with the rest of the definition of a ANEP. Moreover, we do not use any auxiliary devices (as it was the memory we needed in the former case).

It only remains to be settled in which measure the results already reported for ANEPs ([4]) still hold, with respect to the new definition.

Accepting a language. All the ANEP constructions proposed in the literature (for instance, in [7,8,9]), where one was interested only in accepting a language by complete ANEPs, can be still be used. However, we must modify such an ANEP in order to work properly in the new setting: the former output node becomes an insertion node where the symbol μ is inserted in the words that were accepted inside, and then we add a new output node, in which all the words containing μ are allowed to enter. Thus, one can construct, for a recursively enumerable language, an ANEP accepting it, w.r.t. the new definition.

Deciding a language. In [7] one shows that the class of languages decided by an ANEP, with respect to the classical halting condition, is the class of recursive languages. The proof was based on simulating, in parallel, all the possible computations of a nondeterministic Turing machine; the words communicated in the network were encodings of the Turing machine configurations. As we have already mentioned, these proofs can be used, as long as we are not interested in deciding the language, but only in accepting it. However, any recursive language can be decided by a deterministic Turing machine that for each input either enters a single final state and accepts, or enters a single blocking state and rejects. The ANEP simulating a Turing machine, presented in [7], can be easily modified to decide a recursive language L: we simulate the deterministic Turing machine deciding L and allow in the former output node all the words that contain the final state or the blocking state; if a word containing the final state enters the output node, then the network accepts, otherwise, it rejects. In conclusion, the languages decided by ANEPs, w.r.t. the new definition, are the recursive languages.

Computational Complexity. The results regarding polynomial space complexity or polynomial deterministic time complexity reported in [7] can be also proved by simulating deterministic Turing machines by ANEPs and vice versa, so they remain valid when the new acceptance/rejection conditions are used.

However, the results on time-efficient simulations of nondeterministic machines (e.g., the characterization of **NP** [7]) are not preserved, as we show in Section 5.

4 Complete ANEPs

It is worth mentioning that most of the computability and computational complexity results reported so far in literature deal with complete ANEPs or with ANEPs with a restricted topology (see [4] and the references therein). More precisely, there are many results stating that particular types of networks are accepting all the recursively enumerable languages or perform a series of tasks efficiently (e.g. solving NP-complete problems, or simulating efficiently different types of universal devices).

A natural question arises: in which measure is the topology of the networks important, with respect to the computational power or the efficiency of the computations? Such a result would be interesting if we consider that sometimes it is easier to construct an ANEP with a particular topology, solving a given problem efficiently, than to construct a complete one (thus, it is simpler to solve a problem by a non-uniform approach than by an uniform one). For instance, in [5] it was left as an open problem to see if the reported results still hold for complete ANEPs.

A first answer is immediate, but unsatisfactory. Complete ANEPs can simulate (with respect to the former halting conditions) nondeterministic Turing machines, and nondeterministic Turing machines can simulate ANEPs of any kind (see [7]). So one can construct a complete ANEP simulating an arbitrary ANEP via the simulation by Turing machines. However, such a simulation is not time-efficient since the Turing machine simulates a computation of t steps of the ANEP on an input word of length n in time $\mathcal{O}(\max(t^2, tn))$; this approach is also complicated due to the construction of the intermediate Turing machine. Such an approach leads to complete ANEPs that solve quite inefficiently a given problem (see, for instance, Example 1).

In the following we propose a new answer to the above question: we can accept (respectively, decide) with a complete ANEP any language accepted (respectively, decided) by an ANEP with an arbitrary underlying graph, within the same computing time. The new halting conditions play an important role. As we have already explained, they are not relevant in the case when we are interested only in accepting languages: if we use these conditions, we still obtain that given an arbitrary ANEP one can construct a complete ANEP accepting, as efficiently as the arbitrary ANEP, the same language. They come into play in the case of deciding languages. Basically, the proof of our result consists in simulating an ANEP by a complete ANEP. Two consecutive steps of the initial ANEP are simulated in exactly 54 consecutive steps of the complete ANEP. In the classical setting, the initial ANEP rejected when the configurations obtained in two consecutive steps of the same kind were identical; but these configurations do not occur in the complete ANEP consecutively so the new network would not reject, but enter in an infinite cycle. Thus, such a simulation would not preserve the halting property of a computation. However, when the new conditions are used the halting property is preserved canonically.

The proof of the announced results is based on the following two Lemmas.

Lemma 1. *Given an ANEP* $\Gamma = (V, U, \mu, G, \mathcal{N}, \alpha, \beta, In, Out)$*, one can construct an ANEP* $\Gamma' = (V, U, \mu, G', \mathcal{N}', \alpha', \beta', In', Out')$ *such that* Γ' *accepts (decides) the same language as* Γ *accepts (respectively, decides), each node of* Γ' *has at most one rule and* In' *has no rules. Moreover, two consecutive steps of* Γ *(an evolutionary and a communication step) are simulated in exactly* 6 *consecutive steps (*3 *evolutionary and* 3 *communication steps) of* Γ'*.* □

Lemma 2. *Given an ANEP* $\Gamma = (V, U, \mu, G, \mathcal{N}, \alpha, \beta, In, Out)$*, such that all the processors* Γ *have at most one rule and* In *has no rules, one can construct a complete ANEP* $\Gamma' = (V, U', \mu, G', \mathcal{N}', \alpha', \beta', In', Out')$ *such that* Γ' *accepts (decides) the same language as* Γ *accepts (respectively, decides). Two consecutive steps of* Γ *are simulated in exactly* 18 *consecutive steps of* Γ'*.*

Proof. Let $U_1 = \{\#_x, \#_x^\circ, \#_x', \#_x'', \#_x^b, \#_{x,y} \mid b \in U, x, y \in X_G, (x,y) \in E_G\}$.

The complete network Γ' simulates the computation of Γ using the following strategy. We try to construct for each node x, of Γ, a subnetwork $s(x)$, of Γ', that simulates the computation of the processor $\mathcal{N}(x)$; we denote by $\mathrm{set}(s(x))$ the nodes of the subnetwork $s(x)$. The underlying graph of Γ' is complete and has the nodes $\bigcup_{x \in X_G} \mathrm{set}(s(x))$. All the words processed in the new network have a special symbol from U_1. The symbols of U_1 that encode one processor of the initial network indicate the nodes whose actions must be simulated at that point, thus which of the subnetworks should act on the word. The symbols that encode two nodes indicate a possible way to communicate the word containing it between the subnetworks of Γ'. The symbol $\#_{In}$ is inserted in the input word at the beginning of the computation, so we should start by simulating the input node of Γ. Further, the way the computation is conducted, described below, and the way symbols of U_1 are inserted, deleted or modified, in the processed words enable us to simulate, in parallel, all the possible derivations of the input word in Γ, and ensure that the subnetworks act independently.

The alphabet of Γ' is defined as $U' = U \cup \{b', b'' \mid b \in U\} \cup \{\#^{(i)} \mid 1 \le i \le 8\} \cup U_1$. In the following, we define the rest of the network.

We have to analyze more cases, according to the type of the node, the way the operations are applied and the type of the filters.

Assume that the node x verifies $\mathcal{N}(x) = (\emptyset, PI, FI, PO, FO)$. Then, we have $\mathrm{set}(s(x)) = \{x_0, x_1\}$. In $s(x)$ we only change the symbol $\#_x$ into a new symbol $\#_y$, indicating that the word can now go towards the nodes of the subnetwork $s(y)$, and cannot enter the nodes of any other subnetwork. The only trick is that we must do this change in nine steps, instead of a single rewriting step. The rest of the word is left unchanged, as it was also the case in the node x of the initial network, where the whole word stayed unchanged. In the case of the input node In of Γ, the only difference is that we add a new node In_0', which is an insertion node, where $\#_{In}'$ is inserted. The input node of the network is In_0'. This node does not allow any word to enter, and allows all the words to exit.

The subnetwork associated with the output node Out has only the node Out_0, which is the output node of the new network.

We analyze next the case of substitution nodes. For a node x with $\mathcal{N}(x) = (\{a \to b\}, PI, FI, PO, FO)$, $a \in FO$, $\alpha(x) = s$ and $\beta(x) = *$, we have $\mathrm{set}(s(x)) = \{x_0, x_1, x_2, x_3, x_4, x_5, x_6\}$. The simulation implemented by the subnetwork $s(x)$ is, in this case, more involved. Let us assume that w is a word that was sent towards the node x in a communication step of Γ, and the word $w_1\#_x w_2$, with $w_1 w_2 = w$, was communicated in the network Γ'. Our construction ensures that the word w can enter x if and only if $w_1\#_x w_2$ can also enter x_0 (and no other node of Γ'). In the node x we obtain from w a word w' by substituting several a symbols with b symbols (actually, we either substitute exactly one occurrence of an a, if $a \notin FO$, all of them, if $a \in FO$, or none, if w contains no a) and w' leaves the node if it verifies the output conditions. In the network Γ' the word is processed as follows. In x_0 it becomes $w_1\#'_x w_2$ and is sent out. It can only enter x_1, if it contains at least one a, or x_2, otherwise. In the first case it becomes $w'_1\#'_x w'_2$, by substituting exactly one a symbol with a b' symbol. In the second case, the word becomes $w_1\#''_x w_2$ (only the symbol encoding the current node is changed, and the rest of the word remains the same because it contains no a). In the both cases, the obtained words enter x_5 where we obtain, in 5 processing steps, all the words $w_3\#_{x,y} w_4$, for a node y such that $(x, y) \in E_G$ and $w_3 w_4 = w'$, where w' was obtained from w by substituting at most one a symbol with a b symbol. Such words can only be communicated x_6, if they w' verifies the output filters of x. In x_6 they are transformed into $w_3\#_y w_4$, and can go to the nodes of the subnetwork associated with the node y of Γ. There is one more thing to be analyzed: the words that leave x_2 and contain a, in the case when $a \in FO$. These words can go to x_3, where they become $w'_1\#^\circ_x w'_2$ and, further, can only enter x_4 (but only if they do not contain any more forbidden output symbols of x, or their primed copies). In this node, b' becomes b and $\#^\circ_x$ becomes $\#'_x$, after 6 processing steps are made, and the obtained words are sent back to x_2, where another processing step of x is simulated. It is not hard to see now that the action of the node x in Γ was correctly simulated by the subnetwork $s(x)$ in Γ'; more precisely one processing and one communication step of Γ are simulated in 9 processing and 9 communication steps of Γ'.

If $a \notin FO$ then we simply delete the nodes x_3 and x_4 from $s(x)$. The cases of other types of substitution nodes can be treated in a similar fashion.

Now, we present the simulation of the deletion nodes. We can have left, right or arbitrary deletion rules and strong or weak filters. However, all the cases are based on the same general idea so we discuss here only the case of left deletion nodes, with strong filters.

Let us assume that the node x verifies $\mathcal{N}(x) = (\{a \to \lambda\}, PI, FI, PO, FO)$, $\alpha(x) = s$ and $\beta(x) = l$. First, we will assume that $a \in FO$. In this case, we have $\mathrm{set}(s(x)) = \{x_0, x_1, x_2, x_3, x_5\} \cup \{x_b, x^b \mid b \in U\}$. Let us assume that w is a word that was sent towards the node x in a communication step of Γ, and the word $w_1\#_x w_2$, with $w_1 w_2 = w$, was communicated in the network Γ'. If the word w can pass the input filters of x then $w_1\#_x w_2$ can also enter x_0, and vice versa. In the node x we obtain from w a word w' by deleting all the a symbols from the left end of the word, and this word leaves the node if it verifies the output conditions. In the network Γ' the word is processed as follows. In x_0 it becomes $w_1\#_x w_2\#'_x$ and is sent out. It can only enter x_1 where it becomes $w\#'_x$. Now it can only go to x_2. Here it is transformed into $w'_1 b' w'_2 \#'_x$, for all $b \in U$ and $w'_1, w'_2 \in U^*$ such that $w'_1 b w'_2 = w$. Now these word enter x_3, and the network obtains from them the words $w'_1 b' w'_2 \#^c_x$, with $c \in U$. From these, only the words $w'_1 b' w'_2 \#^b_x$ are further processed. More precisely, the node x_b permits these words to enter, and transforms them into $w'_2\#^b_x$, if and only if $w'_1 = \lambda$. Next, the obtained words can only go to node x^b. If $b \neq a$ it means that we simulated a deletion that should not have happened, so we remake the word into $b'' w'_2 \#^b_x$; otherwise, the deletion was correct, and we get in x^a the word $w'_2\#''_x$. In the first case, the words can enter x_5 (if they do not contain any of the forbidden output symbols of node x) where they are transformed into $w'\#_{x,y}$, with $(x,y) \in E_G$, in two processing steps, and sent out; finally, they can enter only the node x_6, but if and only if w' verifies the permitting output filters of x, and here are transformed into $w'\#_y$ and go out in the network (and can only enter the nodes of the subnetwork constructed for the node y). In the second case, the word $w'_2\#''_x$ can either enter x_5, and be processed as above, or, if it still contains a symbols, which are forbidden output symbols for x, it goes to node x_4; in this node they are transformed into $w'_2\#'_x$ (in 5 steps), go back to node x_2, and the whole process described above is repeated. It is not hard to see now that the action of the node x in Γ was correctly simulated by the subnetwork $s(x)$ in Γ'; more precisely, one processing and one communication step of Γ are simulated in 9 processing and 9 communication steps of Γ'.

In the case when $a \notin FO$ we just have to delete the node x_4 from $s(x)$.

Finally, the simulation of the insertion nodes is based on a strategy that follows the ideas used for the other types of nodes.

From the way the subnetworks $s(x)$, with $x \in X_G$, work we see that the following statements are equivalent:

i. w is a word that entered the node x in Γ, was transformed by this node into w' in one step, and w' was communicated to node y (when w' cannot exit x we assume that $y = x$);

ii. the word $w_1\#_x w_2$, with $w_1 w_2 = w$, entered the node x_0, from the subnetwork $s(x)$ of Γ', and it was transformed, by the nodes of $s(x)$, into $w'_1\#_y w'_2$, with $w'_1 w'_2 = w'$, in exactly 9 evolutionary and 9 communication steps.

Note that Γ' accepts a word w if and only if a word $w_1\#_{Out} w_2$, with $w_1 w_2 \in U^*$, can be derived from it; but such a word can be derived only in a number of steps divisible by 9.

According to the above, the computation of Γ on w ends (and accepts) in t steps if and only if the computation of Γ' on w ends (and, respectively, accepts) in $9t$ steps. Therefore, $L(\Gamma) = L(\Gamma')$. □

Theorem 1. *Given an ANEP $\Gamma = (V, U, \mu, G, \mathcal{N}, \alpha, \beta, In, Out)$, one can construct a complete ANEP $\Gamma' = (V, U', \mu, G', \mathcal{N}', \alpha', \beta', In', Out')$ such that Γ' accepts (decides) the same language as Γ accepts (respectively, decides). Two consecutive steps of Γ are simulated in exactly* 54 *consecutive steps of Γ'.* □

Compared to the afore mentioned simulation via Turing machines, we also propose a 2-steps construction, but the only computational model we use is the ANEP model. Also, the both steps rely on the same idea: we replace the nodes of the initial network with a group of nodes that simulate its job; this makes the construction simpler to apply, and easier to follow. Moreover, the complete network simulates in linear time the arbitrary network.

The result in Theorem 1 has a series of consequences. First, it provides a normal topology for ANEPs, allowing us to specify all the results and definitions in an uniform manner, without taking into account the particular topology of the network. Further, the number of nodes in the complete networks is greater only by a constant factor than the number of nodes of the simulated one. Our simulation preserves the computational properties of the initial ANEP, thus, complete networks can be used to prove lower bounds: the most time-efficient ANEP-based solution of a problem can be implemented on a complete ANEP.

The proof of Lemma 2 is an example on how one can design an ANEP by putting together multiple subnetworks; this approach seems close to that of procedural programming. Theorem 1 shows that building a greater ANEP from smaller subnetworks is an approach that can be used without being afraid that such a solution is no longer uniform (i.e., the network has very specific properties, and cannot be transformed efficiently to a general network, such as a complete one). Moreover, if an ANEP constructed from subnetworks works efficiently, so will do the complete variant of that ANEP.

5 Computational Complexity

We recall from [11] the following definition:

Definition 1. *Let M be a nondeterministic polynomial Turing machine and w be a word over the input alphabet of M. The word w is accepted by M with respect to the shortest computations if one of the possible shortest computations of M on w is accepting; w is rejected by M w.r.t. the shortest computations if all the possible shortest computations of M on w are rejecting. We denote by $L_{sc}(M)$ the language decided by M w.r.t. the shortest computations and by $PTime_{sc}$ the class of all the languages decided in this manner.*

It is not hard to see that the class of languages decided w.r.t. shortest computations by nondeterministic polynomial Turing machines with a single tape equals $PTime_{sc}$. In [11] the following result is shown:

Theorem 2. *$PTime_{sc} = \mathbf{P}^{\mathbf{NP}[\log]}$.*[1]

Further, one can show that ANEPs simulate efficiently Turing machines that decide w.r.t. shortest computations, and vice versa.

Theorem 3. i. *For an ANEP Γ, deciding a language L in polynomial time, there exists a nondeterministic polynomial single-tape Turing machine M, deciding L w.r.t. shortest computations.*
ii. *For a nondeterministic polynomial single-tape Turing machine M, deciding a language L w.r.t. shortest computations, there exists a (complete) ANEP Γ, deciding L in polynomial time.* □

As in the case of the former halting conditions, the machine M simulates a computation of t steps of Γ, on an input word of length n, in $\mathcal{O}(\max(t^2, tn))$ time; on the other hand, Γ simulates the computations of M in linear time.

As a consequence of Theorem 3 we obtain the following Theorem.

Theorem 4. $\mathbf{PTime}_{ANEP} = \mathbf{P}^{\mathbf{NP}[\log]}$. □

This result seems interesting to us as we are not aware of any other characterization of the class $\mathbf{P}^{\mathbf{NP}[\log]}$ by computational complexity classes defined for bio-inspired computing models.

Finally, we show that, in fact, one can design complete ANEPs working faster than nondeterministic Turing machines. This shows that solving a problem by nondeterministic Turing machines and then simulating such machines by NEPs does not lead to an optimal ANEP-based solution to that problem.

Example 1. Let $L = \{a^n b \mid n \in I\!N, n \geq 1\}$. Any Turing machine, deciding L in the classical way, w.r.t. shortest computations or using oracles, with an arbitrary number of tapes, makes at least a linear number of moves before stopping on an input word. But L can be accepted in constant time by a complete ANEP.

References

1. Margenstern, M., Mitrana, V., Jesús Pérez-Jímenez, M.J.: Accepting Hybrid Networks of Evolutionary Processors. In: Ferretti, C., Mauri, G., Zandron, C. (eds.) DNA 2004. LNCS, vol. 3384, pp. 235–246. Springer, Heidelberg (2005)
2. Hillis, W.D.: The Connection Machine. MIT Press, Cambridge (1986)
3. Csuhaj-Varjú, E., Salomaa, A.: Networks of Parallel Language Processors. In: Păun, G., Salomaa, A. (eds.) New Trends in Formal Languages. LNCS, vol. 1218, pp. 299–318. Springer, Heidelberg (1997)
4. Manea, F., Martín-Vide, C., Mitrana, V.: Accepting networks of evolutionary word and picture processors: A survey. In: Martín-Vide, C. (ed.) Scientific Applications of Language Methods, pp. 525–560. World Scientific, Singapore (2010)
5. Bottoni, P., Labella, A., Manea, F., Mitrana, V., Sempere, J.M.: Filter Position in Networks of Evolutionary Processors Does Not Matter: A Direct Proof. In: Deaton, R., Suyama, A. (eds.) DNA 15. LNCS, vol. 5877, pp. 1–11. Springer, Heidelberg (2009)

[1] $\mathbf{P}^{\mathbf{NP}[\log]}$ is the class of problems solvable by a deterministic polynomial machine, that can make $\mathcal{O}(\log n)$ queries to an **NP** oracle (where n is the length of the input).

6. Rozenberg, G., Salomaa, A.: Handbook of Formal Languages. Springer, New York (1997)
7. Manea, F., Margenstern, M., Mitrana, V., Perez-Jimenez, M.J.: A New Characterization of NP, P and PSPACE with Accepting Hybrid Networks of Evolutionary Processors. Theor. Comp. Sys. 46(2), 174–192 (2010)
8. Loos, R., Manea, F., Mitrana, V.: Small Universal Accepting Hybrid Networks of Evolutionary Processors. Acta Inf. 47(2), 133–146 (2010)
9. Alhazov, A., Csuhaj-Varjú, E., Martín-Vide, C., Rogozhin, Y.: On the size of computationally complete hybrid networks of evolutionary processors. Theor. Comput. Sci. 410(35), 3188–3197 (2009)
10. Manea, F.: Deciding networks of evolutionary processors. Technical report (2010), `http://theo.cs.uni-magdeburg.de/pubs/preprints/pp-afl-2011-05.pdf`
11. Manea, F.: Deciding according to the shortest computations. In: Normann, D. (ed.) CiE 2011. LNCS, vol. 6735, pp. 191–200. Springer, Heidelberg (2011)

From Linear Partitions to Parallelogram Polyominoes

Roberto Mantaci[1] and Paolo Massazza[2,⋆]

[1] LIAFA, CNRS UMR 7089, Université Paris Diderot - Paris 7,
Case 7014, 75205 Paris Cedex 13, France
mantaci@liafa.jussieu.fr
[2] Università degli Studi dell'Insubria, Dipartimento di Informatica e Comunicazione,
Via Mazzini 5, 21100 Varese, Italy
paolo.massazza@uninsubria.it

Abstract. We provide a bijection between parallelogram polyominoes and suitable pairs of linear partitions. This lets us design a CAT (Constant Amortized Time) algorithm for generating all parallelogram polyominoes of size n using $O(\sqrt{n})$ space.

Keywords: Polyominoes, Exhaustive generation, CAT algorithms.

1 Introduction

As combinatorial objects, polyominoes are simply defined as finite connected sets of edge-to-edge adjacent square unit cells in the cartesian two-dimensional plane. Polyominoes and their applications appear more and more often in several contexts, with interesting overlaps and interactions betweens them.

In enumerative combinatorics, where several families of polyominoes have been characterized and enumerated according to several characteristics (area, perimeter, ...), providing in some cases their generating functions [4,8];

In bijective combinatorics; a polyomino P can be described by a pair of appropriate paths in the $\mathbb{N} \times \mathbb{N}$ lattice. This consideration relates polyominoes to lattice paths theory [1,10,17];

In two-dimensional language theory and image treatment, where a polyomino is considered as a two-dimensional word and appropriate families of polyominoes turn out to be two-dimensional languages with specific properties [5,6], or where the problem of its reconstruction from partial informations is considered [2,12];

In tiling theory, where tile shapes can be described by polyominoes [11,16].

Here we are interested in particular polyominoes called *Parallelogram Polyominoes*. This class consists of all polyominoes bounded by two paths in the $\mathbb{N}\times\mathbb{N}$ lattice composed of north and east steps and intersecting each other only in the initial and the final points. This class has first been studied by Polya [18], who counted the number of parallelogram polyominoes with respect to the perimeter,

⋆ Partially supported by Project M.I.U.R. PRIN 2007–2009: Mathematical aspects and forthcoming applications of automata and formal languages.

G. Mauri and A. Leporati (Eds.): DLT 2011, LNCS 6795, pp. 350–361, 2011.

and provided a (non closed) formula for the generating function by perimeter and area. A nice bijection between parallelogram polyominoes of perimeter $2n+2$ and Dick words of length $2n$ was given in [8] and used in [7] to produce an equation for the generating function according to the area, the width and the number of left path corners. The relation between parallelogram polyominoes and two-dimensional languages has also been studied. In particular, in [6] it is shown that parallelogram polyominoes can be represented by two-dimensional words of a tiling recognizable language. This class of polyominoes has also been investigated from the exhaustive generation point of view. More precisely, in [3] the ECO method has been applied to the recursive generation of parallelogram polyominoes with semi-perimeter n. This method has also been used later to generate and enumerate some classes of convex polyominoes, see [1,9].

In this paper, we highlight a natural bijection between parallelogram polyominoes having size (or area) equal to n and pairs (s,t) of integer partitions, where t is a partition of an integer m and s a partition of the integer $m+n$.

Using this bijection, we provide an algorithm for the exhaustive generation of all parallelogram polyominoes of a given area. This algorithm runs in constant amortized time (CAT), that is, if N is the number of all parallelogram polyominoes of area n, our algorithm generates them all in time kN where k is a constant. Its space complexity is $O(\sqrt{n})$.

The algorithm provided is based on an idea of the second author and R. Radicioni, who in their article [14] describe a CAT algorithm to generate particular linear partitions of an integer n called ice piles. The algorithm is based on the preorder traversal of a spanning tree for the Hasse diagram of the partial order defined by the dominating relation between two partitions.

The article is divided into two main sections and ends with a short conclusion presenting some possible developments of this work.

In section 2 we provide the definitions, the order structure of the set of partitions and the combinatorial results allowing to define the bijection between parallelogram polyominoes and pairs of partitions. In section 3 we describe the algorithm by providing its pseudo-code and analyze its complexity.

2 Preliminaries

A linear partition of n is a non-increasing sequence of nonnegative integers with sum n. We indicate by $\mathrm{LP}(n)$ the set of the linear partitions of n.
For any two integers x,p with $p>0$, we denote by $x^{[p]}$ the sequence $(\underbrace{x,\ldots,x}_{p})$ and by $\cdot$ the *catenation product* of sequences. The *length* of $s=(s_1,\ldots,s_l)\in\mathrm{LP}(n)$ is $l(s)=l$, the *height* is $h(s)=s_1$ and the *size* (or *weight*) is $w(s)=\sum s_i=n$. Moreover, we define $\Delta(s)=\sum_{i<l}(s_i-s_{i+1})=h(s)-s_l$. Note that s can be univocally written as $s=s_1^{[m_1]}\cdot s_{j_2}^{[m_2]}\cdot\ldots\cdot s_{j_k}^{[m_k]}$ with $s_1>s_{j_2}>\cdots>s_{j_k}$ and $m_i>0$. The following notations are useful when dealing with sequences, $s_{<i}=(s_1,\ldots,s_{i-1})$ ($s_{\le i}=(s_1,\ldots,s_i)$) and $s_{>i}=(s_{i+1},\ldots,s_l)$ ($s_{\ge i}=(s_i,\ldots,s_l)$).

The set LP(n) can be ordered with respect to the *negative lexicographic* order:

Definition 1. *Let* $s = (s_1, \ldots, s_l)$ *and* $t = (t_1, \ldots, t_m)$ *belong to* $LP(n)$. *Then,* $s <_{nlex} t$ *if and only if there is* i *such that* $s_{<i} = t_{<i}$ *and* $s_i > t_i$.

Covering Partitions On the set of integer partitions we define a binary relation $\frown$ that is of particular interest for generating parallelogram polyominoes.

Definition 2. *Let* $s \in LP(n_1)$ *and* $t \in LP(n_2)$. *Then,* s *strongly covers* t, *denoted as* $s \frown t$, *if and only if* $l = l(s) = l(t)$, $s_1 > t_1$ *and* $s_j > t_{j-1}$ *for all* j *with* $1 < j \leq l$.

From the previous definition we immediately get:

Lemma 1. *Let* $s \in LP(n_1)$, $t \in LP(n_2)$ *and* $s \frown t$. *Then* $n_1 \geq n_2 + l(t) + \Delta(t)$.

Proof. By Definition 2 one has

$$n_1 = s_1 + \sum_{i=2}^{l(s)} s_i \geq t_1 + 1 + \sum_{i=1}^{l(t)-1} (t_i + 1) = t_{l(t)} + \Delta(t) + l(t) + \sum_{i=1}^{l(t)-1} t_i = n_2 + l(t) + \Delta(t).$$

□

Figure 1 shows the linear partition s of smallest weight that strongly covers a linear partition t. It clearly follows that $w(s) = w(t) + l(t) + \Delta(t)$.

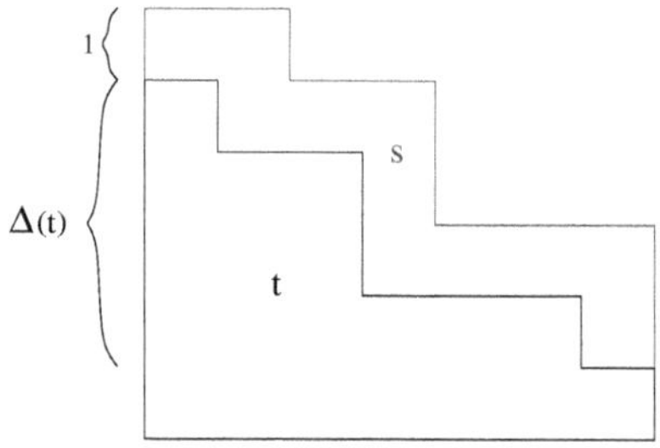

Fig. 1. The smallest covering partition

Given $t \in \mathrm{LP}(n_2)$, the set of linear partitions of n_1 that strongly cover t is denoted by $\mathrm{LP}(n_1|t) = \{s \in \mathrm{LP}(n_1) | s \frown t\}$. Observe that for all $v \in \mathrm{LP}(n_1|t)$ one has $l(v) = l(t)$ and that $\mathrm{LP}(n_1) = \bigcup_{r=1\ldots n_1} \mathrm{LP}(n_1|0^{[r]})$. Definitions 1 and 2 let us easily characterize $\hat{s} = \min_{<_{\mathrm{nlex}}}(\mathrm{LP}(n_1|t))$.

Lemma 2. *Let* $t = (t_1, \ldots, t_l) \in LP(n_2)$. *Then, for* $n_1 \geq n_2 + l + \Delta(t)$ *one has*

$$\hat{s} = (t_1 + 1 + n_1 - n_2 - \Delta(t) - l, t_1 + 1, t_2 + 1, \ldots, t_{l-1} + 1).$$

We can interpret the elements of $\mathrm{LP}(n_1|t)$ as states of a simple discrete dynamical system which simulates the movements occurring into a heap of $n_1 - n_2$ grains stacked into a (two-dimensional) silo of radius $l(t)$ and whose bottom is shaped like the profile of t. This system has an evolution rule called *Move* which corresponds to moving one grain from column i of s to the first column k, with $k > i$, such that $s_i > s_{i+1}$ and $s_i - s_k \geq 2$. Figuratively, the rightmost grain of a *plateau* can slide on the plateau on its right and find place at the bottom of the next *cliff*. More formally, one has:

Definition 3. *Given $t \in LP(n_2)$, $s \in LP(n_1|t)$ and an integer i, let*

$$k = \begin{cases} i+1 \text{ if } s_i - s_{i+1} \geq 2 \\ \quad j \quad \text{if } s_i - s_j \geq 2 \wedge s_{i+1} = s_{j-1} = s_i - 1 \\ \quad \bot \quad \text{otherwise} \end{cases}$$

Then

$$Move(s,t,i) = \begin{cases} s_{<i} \cdot (s_i - 1, s_{i+1}, \ldots, s_{k-1}, s_k + 1) \cdot s_{>k} & \text{if } k \neq \bot, s_i - 1 > t_{i-1} \\ \bot & \text{otherwise} \end{cases}$$

The initial state of the dynamical system is $\hat{s}$, corresponding to the configuration where all the grains are in column 1, with the exception of those that create a covering layer of height 1 at the bottom of the silo. Figure 2 shows s and $\mathrm{Move}(s,t,2)$ for $t = (3,3,3,2,1,1,1)$. Note that $\mathrm{Move}(s,t,i) = \bot$ for $i = 3,4,6,7$. We write $s \overset{i,t}{\Rightarrow} s'$ if $s' = \mathrm{Move}(s,t,i)$ and, more generally, $s \overset{\star}{\Rightarrow} v$ if there is a

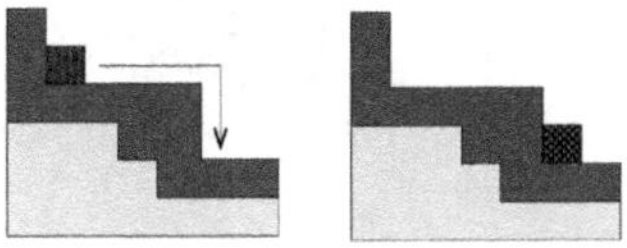

Fig. 2. $s = (6,5,4,4,4,2,2)$ and $\mathrm{Move}(s,t,2) = (6,4,4,4,4,3,2)$

sequence of moves leading from s to v. The dynamical system has a set of states denoted by $G(\hat{s}|t) = \{v \in \mathrm{LP}(n_1|t) | \hat{s} \overset{\star}{\Rightarrow} v\}$, which turns out to be equal to $\mathrm{LP}(n_1|t)$, that is, all partitions of n_1 covering t can be generated starting from the minimum in the neglex order.

Lemma 3. *Let $t = (t_1, \ldots, t_l) \in LP(n_2)$. Then, for any $n_1 \geq n_2 + l + \Delta(t)$ one has $LP(n_1|t) = G(\hat{s}|t)$.*

Proof. Suppose $G(\hat{s}|t) \subsetneq \mathrm{LP}(n_1|t)$ and let $s' = \min_{<_{\mathrm{nlex}}}(\mathrm{LP}(n_1|t) \setminus G(\hat{s}|t))$. Let i be the largest integer such that $s'_i > t_{i-1} + 1$. If $i = 1$ then $s'_{>1} = \hat{s}_{>1}$ and so $s' = \hat{s}$. Thus, let $j_1 = \min\{k \leq i | s'_k = s'_i\}$, $j_2 = \min\{k < j_1 | s'_k = s'_{j_1 - 1}\}$ and set either $j = j_1$ (if $j_1 < i$) or $j = j_2$ (if $j_1 = i$). Now, consider $s'' = s'_{<j} \cdot (s'_j + 1, s'_{j+1}, \ldots, s'_{i-1}, s'_i - 1) \cdot s'_{>i}$ and note that $s'' \in \mathrm{LP}(n_1|t)$ and $s'' <_{\mathrm{nlex}} s'$. This implies $s'' \in G(\hat{s}|t)$. Lastly, from $s'' \overset{j,t}{\Rightarrow} s'$ we get $s' \in G(\hat{s}|t)$. □

The *set of moves* of $s \in \text{LP}(n_1|t)$ is defined as the set of integers $\text{M}(s|t) = \{i|\text{Move}(s,t,i) \neq \perp\}$. $\text{LP}(n_1|t)$ admits exactly one element v such that $\text{M}(v|t) = \emptyset$, called *fixed point* and denoted as $\text{FP}(n_1|t)$. In particular, $\text{LP}(n_1|t)$ turns out to be a lattice with respect to the relation induced by $\Rightarrow$, with the top and the bottom given by $\hat{s}$ and $\text{FP}(n_1|t)$, respectively.

We denote by $s^{\downarrow}$ the linear partition in $\text{LP}(n_1+1|t)$ given by $s^{\downarrow} = s_{<i} \cdot (s_i + 1) \cdot s_{>i}$, where i is the largest integer such that $i \leq l(s)$ and $s_{i-1} > s_i$ ($i = 1$ if $l(s) = 1$). The following lemma provides a characterization of $\text{FP}(n_1|t)$.

Lemma 4. *Let $t = (t_1, \ldots, t_l) \in LP(n_2)$ and $n_1 \geq n_2 + l + \Delta(t)$. Then*

$$v = FP(n_1|t) = \begin{cases} (t_1+1, t_1+1, t_2+1, \ldots, t_{l-1}+1) & \text{if } n_1 = n_2 + l + \Delta(t) \\ FP(n_1-1|t)^{\downarrow} & \text{otherwise} \end{cases}$$

Proof. The case $n_1 = n_2 + l + \Delta(t)$ follows by noting that $v \in \text{LP}(n_1|t)$ and $\text{Move}(v,t,i) = \perp$ for all i. Otherwise, let $u = \text{FP}(n_1-1|t)$ and let i be the only integer such that $u_i \neq u_i^{\downarrow}$. Since u is a fixed point, it is sufficient to prove that $\text{Move}(u^{\downarrow}, t, i) = \perp$. This follows from $u_{>i} = u^{\downarrow}_{>i}$ and $u_i = u_{i+1} = \ldots = u_l$. □

Figure 3 shows the linear partition $t = (5,4,4,2,2,2,1)$ together with the two fixed points $s = \text{FP}(31|t)$ and $r = \text{FP}(36|t)$ (note that $r \not\Rightarrow s$).

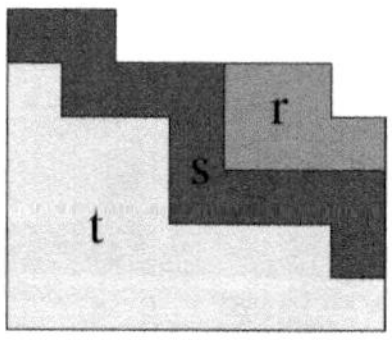

Fig. 3. $s = \text{FP}(31|t)$ and $r = \text{FP}(36|t)$

We associate with each linear partition in $\text{LP}(n_1|t)$ its *grand ancestor*, that is, a particular element of $\text{LP}(n_1|t)$ playing an important role in the process of generating $\text{LP}(n_1|t)$. To enumerate the elements in neglex order, we should apply to the current partition s the rightmost move, in order to preserve the longest possible prefix. However, whenever such move is also yet to be done in another partition s' previously generated (and hence $s' <_{neglex} s$), the move needs to be applied not to s, but to s' that is still waiting to make that move. s' is the grand ancestor of s, it is the smallest partition for which $\max(\text{M}(s|t))$ is still a valid move. Proper grand ancestors correspond to branching nodes in the implicit spanning tree that is being traversed to generate the partitions.

Definition 4. *Let $s \in LP(n_1|t)$. The* grand ancestor *of s is the linear partition $A(s|t) = \min_{<_{nlex}}\{v \in LP(n_1|t) | s_{\leq i} = v_{\leq i}, i = \max(M(s|t)) \in M(v|t)\}$.*

The grand ancestor of a linear partition is characterized in

Lemma 5. *Let $s \in LP(n_1|t)$, $i = \max(M(s|t))$ and $q = \sum_{j>i}(s_j - t_{j-1} - 1)$. If p is the largest integer such that $d = \sum_{j=i+1}^{i+p}(s_i - 1 - t_{j-1} - 1) \leq q$, then $A(s|t)$ is the linear partition*

$$s' = s_{\leq i} \cdot (s_i - 1)^{[p]} \cdot (t_{i+p} + 1 + q - d, t_{i+p+1} + 1, \ldots, t_{l-1} + 1).$$

Proof. First, note that $s' \in \mathrm{LP}(n_1|t)$, $i \in \mathrm{M}(s'|t)$ and $s'_{\leq i} = s_{\leq i}$. Thus we have only to prove that s' is the smallest linear partition with such properties. In fact, suppose that $z \in \mathrm{LP}(n_1|t)$ satisfies $z <_{\mathrm{nlex}} s'$, with $z_{\leq i} = s_{\leq i}$ and $i \in \mathrm{M}(z|t)$. Let j be the smallest integer such that $z_j > s'_j$ (obviously $j > i+p$). Then, one has $\sum_{e>j} z_e < \sum_{e>j} s'_e = \sum_{e>j}(t_{e-1} + 1)$. This implies the existence of an index $\hat{\jmath}$ such that $z_{\hat{\jmath}} < t_{\hat{\jmath}-1} + 1$ and so $z \not\succ t$. □

Example 1. Let $t = (5, 4, 4, 3, 2, 2, 2, 1, 1, 1, 1)$ and $s = (8, 8, 8, 8, 6, 5, 5, 4, 4, 4, 4)$. Then, $A(s|t) = (8, 8, 8, 8, 6, 5, 5, 5, 5, 4, 2)$ (see Figure 4). Note how $A(s|t)$ has the same prefix as s up to position $\max(\mathrm{M}(s|t))$, whereas the remaining suffix has the leftmost entries as large as possible, while ensuring the covering condition.

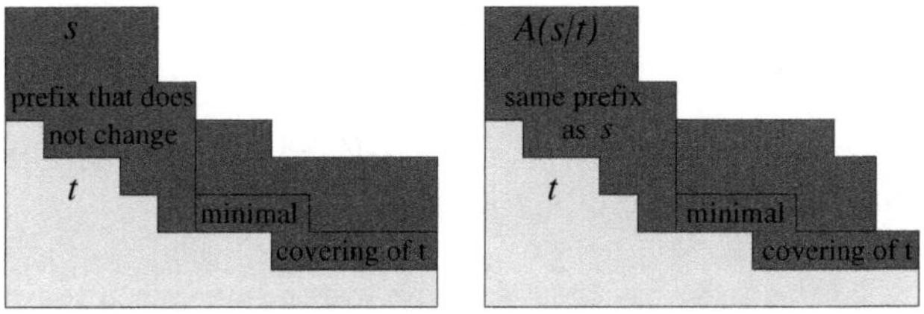

Fig. 4. A linear partition and its grand ancestor (right)

When n_1 and t are clear from the context, we denote by $s^{(e)}$ the eth linear partition in $\mathrm{LP}(n_1|t)$. We can obtain a spanning tree associated with the lattice $\mathrm{LP}(n_1|t)$ by defining, for $e > 1$, $\mathrm{A}(s^{(e-1)}|t)$ as the father of $s^{(e)}$ ($\hat{s} = s^{(1)}$ is the root). The following lemma shows how we can implicitly traverse such tree in order to generate the sequence of elements in $\mathrm{LP}(n_1|t)$.

Lemma 6. *Let $t \in LP(n_2)$ and $n_1 \geq n_2 + l(t) + \Delta(t)$. Then, for any $e > 0$, we have*

$$A(s^{(e)}|t) \overset{i_e}{\Rightarrow} s^{(e+1)}, \qquad \text{with } i_e = \max(M(s^{(e)}|t)).$$

Proof. It is sufficient to follows the proof of [14, Lemma 2.16], where the same result is stated with respect to the states of a dynamical system associated with the Ice Pile Model. We have only to take into account the (slightly) different definition of the evolution rule (the function *Move*). □

We define a function Next which, having as inputs $s^{(e)}, t$, determines first $i = \max(\mathrm{M}(s^{(e)}|t))$ and $s' = \mathrm{A}(s^{(e)}|t)$, and then outputs $s^{(e+1)} = \mathrm{Move}(s', t, i)$ ($\mathrm{Next}(s^{(e),t}) = \perp$ if $s^{(e)} = \mathrm{FP}(n_1|t)$). Hence, Algorithm 1 generates $\mathrm{LP}(n_1|t)$.

Theorem 1. LinPartGen(n_1, t) *generates* $LP(n_1|t)$ *in time* $O(\sharp LP(n_1|t))$ *using* $O(\sqrt{n_1})$ *space.*

Proof. (Outline) The space requirement follows from section 3.1 and by implementing $M(s|t)$ as a stack of links to nodes (in the list of s) where moves occur. With respect to the time, the result follows by reasoning as in [14, Lemma 4.2] and showing that, by Lemma 5, $s' = A(s^{(e)}|t)$ can be computed in time $O(d_e)$, where d_e is the distance between s' and $s^{(e)}$ in the spanning tree of $LP(n_1|t)$. □

Algorithm 1. Exhaustive Generation of $LP(n_1|t)$.

```
1: PROCEDURE LINPARTGEN(n1, t)
2: s:=MINLINPART(n1, t); {s = min_<nlex (LP(n1|t))}
3: while M(s|t) ≠ ∅ do
4:     s:=NEXT(s, t);
5: end while
```

Similarly, we can easily generate the set $\{v \in LP(n_1|t) | s \overset{\star}{\Rightarrow} v\}$ for any $s \in LP(n_1|t)$. Then, one has:

Corollary 1. *For any* $s \in LP(n_1|t)$ *the set* $V = \{v \in LP(n_1|t) | s \overset{\star}{\Rightarrow} v\}$ *can be generated in time* $O(\sharp V)$ *using* $O(\sqrt{n_1})$ *space.*

Parallelogram Polyominoes A parallelogram polyomino is a polyomino whose boundary consists of two north-east lattice paths which are non-crossing (apart the starting and ending points). Without loss of generality, we suppose that the starting point of such paths is $(1, 0)$. Thus, a polyomino P can be specified by a pair of paths (p, q) where $p = (1, 0), (1, 1), \ldots, (x-1, y), (x, y)$ and $q = (1, 0), (2, 0), \ldots, (x, y-1), (x, y)$. The *height* of P is y, the *width* is $x - 1$, while its *size* (or *area*) is the number of unit squares within the boundary identified by p and q. The following theorem provides a useful bijection between polyominoes of size n and suitable pairs of linear partitions.

Theorem 2. *The set of parallelogram polyominoes of size n is in bijective correspondence with the set of pairs of linear partitions* (s, t) *such that*

- $1 \le l = l(t) \le n$;
- $t_l = 1$;
- $l + \Delta(t) \le n$;
- $t \in LP(m)$ *with* $l \le m \le (l-1)(\Delta(t)+1)+1$;
- $s \in LP(m+n|t)$.

Proof. First, let us see how a polyomino P with area n univocally identifies a pair (s, t) of linear partitions. Let $p = (1, 0), (1, 1), (x_1, y_1), \ldots, (x_k, y_k), (x-1, y), (x, y)$ and $q = (1, 0), (2, 0), (x'_1, y'_1), \ldots, (x'_k, y'_k), (x, y-1), (x, y)$ be the two paths defining P. Obviously one has $y + x - 1 \leq n$. Note that there is h such that the first $h + 2$ steps of p

$$(1,0),(1,1),(x_1,y_1),\ldots,,(x_h,y-1),(x_{h+1},y),$$

with $x_h = x_{h+1} \leq x-1$, identify a linear partition $t = (t_1,\ldots,t_l)$ with $l = y \leq n$, $t_1 = x_h$, $t_l = 1$ and $\Delta(t) = x_h - 1$. Thus, one has $l + \Delta(t) = y + x_h - 1 \leq n$ and $t \in \mathrm{LP}(m)$ for a suitable integer m with $l \leq m \leq (l-1)t_1 + 1 = (l-1)(\Delta(t) + 1) + 1$. Similarly, by considering q we can find an index r such that the sequence $(x'_r, 0), (x'_{r+1}, 1), \ldots, (x'_k, y'_k), (x, y-1), (x, y)$ identifies a partition s of size $m+n$ with $s \frown t$ (recall that the two paths are non-crossing), that is, $s \in \mathrm{LP}(m+n|t)$. Lastly, observe that if two different polyominoes of size n are identified by two pairs of paths, say (p,q) and (p',q'), then one necessarily has $p \neq p'$ or $q \neq q'$, and the associated pairs of linear partitions are different.

Now, we show how a pair (s,t) of linear partitions satisfying the given conditions uniquely defines two non-crossing paths p, q which start at $(1,0)$ and end at $(s_1, l(t))$. Indeed, p is the only north-east path passing through the points

- for all j, with $1 \leq j \leq l$, $(t_j, l-j)$;
- if $t_{j-1} > t_j$ $(t_j, l-j+1), (t_j+1, l-j+1), \ldots, (t_{j-1}-1, l-j+1)$;
- $(t_1+1, l), (t_1+2, l), \ldots, (s_1-1, l)$.

Analogously, q is identified by the points

- $(1,0), (2,0), \ldots, (s_l - 1, 0)$;
- for all j, with $1 \leq j \leq l$, $(s_j, l-j)$;
- if $s_{j-1} > s_j$ $(s_j, l-j+1), (s_j+1, l-j+1), \ldots, (s_{j-1}-1, l-j+1)$.

Note that p and q are non-crossing since $s \frown t$. Lastly, the size of the polyomino is just $w(s) - w(t) = n$. Moreover, if $(s,t) \neq (s',t')$ then the pair (p,q) identified by (s,t) is different from the pair (p',q') associated with (s',t') and the two polyominoes are different. □

Example 2. Figure 5 illustrates the parallelogram polyomino of size 27 identified by (s,t) with $t = (8,8,6,5,5,5,2,2,1,1)$ and $s = (9,9,9,8,8,6,6,6,6,3)$. The two north-east paths p,q associated with (s,t) are drawn as dashed and dotted lines, respectively. You can see the two partitions s and t represented in the usual way by turning the picture 90 degrees counterclockwise.

We denote by $\mathrm{PPol}(n)$ the set of parallelogram polyominoes of size n. In the sequel, the generation of $\mathrm{PPol}(n)$ respects the following ordering which can be obtained by considering the bijection in Theorem 2.

Definition 5. *Let (s,t) and (s',t') be two pairs of linear partitions identifying $P, P' \in PPol(n)$, respectively. Then $P < P'$ if and only if*

$$l(s) > l(s')$$

or

$$l(s) = l(s'), t \in LP(m), t' \in LP(m'), m > m'$$

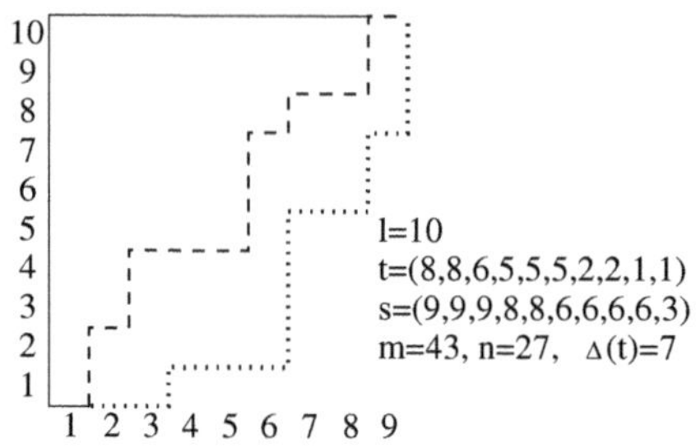

Fig. 5. A parallelogram polyomino identified by two linear partitions

or

$$l(s) = l(s'), t, t' \in LP(m), t <_{nlex} t'$$

or

$$l(s) = l(s'), t = t', s <_{nlex} s'.$$

3 The Algorithm

Here we present an algorithm which solves the following problem:

Problem Parallelogram Polyominoes Generation (PPG)
Input an integer n
Output the ordered sequence of parallelogram polyominoes in PPol(n).

Our approach is that of producing the sequence of parallelogram polyominoes (ordered with respect to Definition 5)

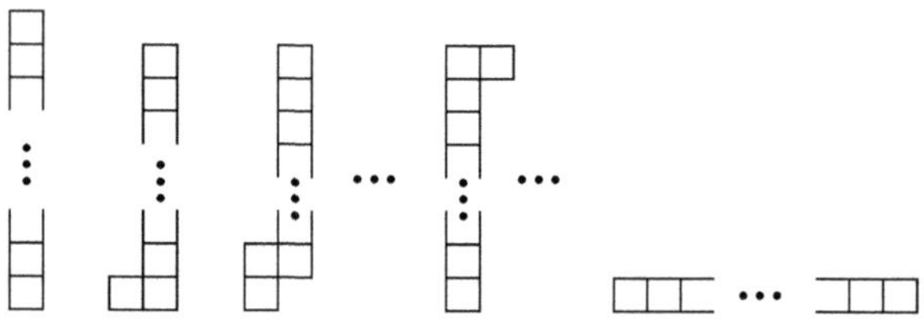

by generating the associated sequence of pairs (s, t) of linear partitions characterized in Theorem 2. Thus, Definition 5 directly leads to Algorithm 2 shown below.

Procedure GenParPol generates the sequence of pairs (s, t) of linear partitions associated with PPol(n). The two loops at lines 2, 3 are used to set the height l of the polyomino, that is, the length of t (from n down to 1) and its size m, respectively. Because of Lemma 1 and Theorem 2, one has $n + m \geq m + l + \Delta(t)$: this means that the largest value for $\Delta(t)$ is $n - l$ and thus m has to satisfy the relation $l \leq m \leq (l-1)(n-l) + l$. The code

Algorithm 2. Generation of parallelogram polyominoes of size n.

```
1: PROCEDURE GENPARPOL(n)
2: for l:=n downto 1 do
3:    for m:=(l − 1)(n − l) + l downto l do
4:       t:= INITLINPART(m, l, n − l + 1);
5:       while M(t, 0) ≠ ∅ do
6:          s:= MINLINPART(m + n, t);
7:          while M(s, t) ≠ ∅ do
8:             s:=NEXT(s, t);
9:          end while
10:         t:=NEXT(t, 0^[l]);
11:      end while
12:   end for
13: end for
```

at lines 4–11 generates all the pairs (s, t) such that $l(t) = l$, $t \in \mathrm{LP}(m)$ and $s \in \mathrm{LP}(m + n|t)$. $\textsc{InitLinPart}(m, l, n - l + 1)$ returns the representation of the smallest linear partition of weight m and length l having height $n - l + 1$, while $\textsc{MinLinPart}(m + n, t)$ returns $\min_{<_{\mathrm{nlex}}}(\mathrm{LP}(m + n|t))$. $\textsc{Next}(s, t)$ returns the linear partition in $\mathrm{LP}(m + n|t)$ which follows s (w.r.t $<_{\mathrm{nlex}}$), while $\textsc{Next}(t, 0^{[l]})$ returns the linear partition of length l following t.

3.1 The Data Structure

In order to efficiently deal with a pair (s, t) of linear partitions with $s \frown t$, we define a data structure consisting of two double-linked lists. The list representing t has as many elements as different values in t, that is, it has $d = \sharp\{i|t_i > t_{i+1}\}$ nodes which contain two integers (a value t_i and the largest k such that $t_k = t_i$),

$$(t_1, k_1), (t_{j_2}, k_2), \ldots, (t_{j_d}, k_d) \quad \approx \quad t_1^{[k_1]} \cdot t_{j_2}^{[k_2]} \cdot \ldots \cdot t_{j_d}^{[k_d]} = t.$$

The list representing s has a similar structure, with the only difference that each node (s_{i_p}, h_p) has a link to the node (t_{j_q}, k_q) in the list of t such that $k_{q-1} < h_p \leq k_q$. Moreover, in order to consider t itself as a covering partition, we add a zero node representing $\mathrm{LP}(0)$. Figure 6 illustrates the representation of the linear partitions s and t drawn in Figure 5. Obviously, the sum of the lengths of the lists representing t and $s \in \mathrm{LP}(n|t)$ is $O(\sqrt{n})$. This data structure is used in Algorithm 1 and its properties are useful to prove Theorem 1. In particular, it lets develop an implementation of function Move running in time $O(1)$, while permitting to update $\mathrm{M}(s|t)$ in time $O(1)$ too. Lastly, one has:

Property 1. Let $t \in \mathrm{LP}(m)$ and $n > l(t) + \Delta(t)$. If k is the length of the list representing $s = \min_{<_{\mathrm{nlex}}}(\mathrm{LP}(m + n|t))$ then:

- $k = 2$ if and only if s is a fixed point;
- $\sharp\{v \in \mathrm{LP}(m + n|t) | s \stackrel{\star}{\Rightarrow} v\} = \Omega(k)$.

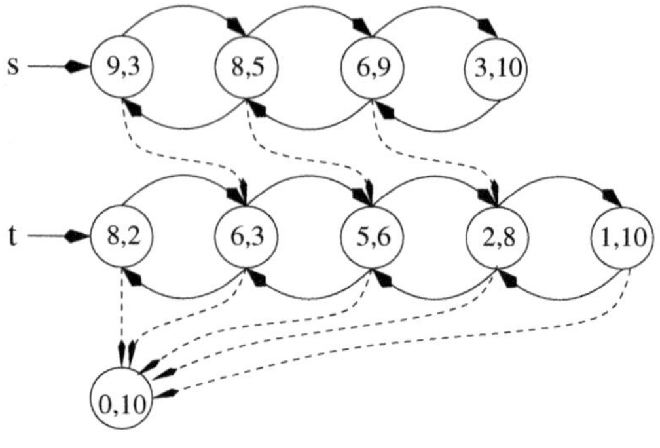

Fig. 6. Representing (s, t)

3.2 Complexity

By considering the data structure in Section 3.1, it is immediate to see that procedure InitLinPart runs in time $O(1)$ since it creates a list of length at most 3. Similarly, MinLinPart$(m+n, t)$ returns a list of length $l(t)$ and so it runs in time $O(l(t))$. Then, we can state:

Theorem 3. GenParPol*(n) runs in time* $O(\sharp PPol(n))$ *and uses* $O(\sqrt{n})$ *space.*

Proof. First, note that the two for-loops at lines 2,3 set $O(n^2)$ different integer pairs (l, m). Moreover, each pair (l, m) is used (lines 4-11) to generate all polyominoes identified by a pair (s, t) with $t \in \mathrm{LP}(m|0^{[l]})$ and $s \in \mathrm{LP}(m+n|t)$. Therefore, it is sufficient to prove that the code consisting of lines 4-11 is CAT. To this aim, by Theorem 1 and Property 1, for each fixed t, the cost of instructions at lines 6-9 is $O(\sharp V)$, where $V = \{v | s \overset{*}{\Rightarrow} v\}$ and $s = \min_{<_{\mathrm{nlex}}}(\mathrm{LP}(m+n|t))$. Similarly, by Corollary 1 the generation of all $t \in \mathrm{LP}(m|0^{[l]})$ with $h(t) \leq n - l + 1$ (the outer while -loop) is CAT. Hence, the set $F(l, m) = \{(s, t) | t \in \mathrm{LP}(m|0^{[l]}), s \in \mathrm{LP}(m+n|t), h(t) \leq n - l + 1\}$ is CAT generated and, since $\sharp \mathrm{PPol}(n) = \sum_{l=1}^{n} \sum_{m=l}^{(l-1)(n-l)+l} \sharp F(l, m)$, the result follows. With respect to the space complexity, we recall that the length of the lists representing s and t is $O(\sqrt{n})$. □

4 Conclusions

It is noteworthy that the idea behind the initial algorithm intended to generate ice piles has proved to be extremely adaptable, allowing to provide CAT algorithms for the exhausting generation of a still growing set of classes of combinatorial objects : (linear) integer partitions, symmetric ice and sand piles [13,15], unimodal sequences, plain partitions, standard Young tableaux and now parallelogram polyominoes. It seems to be possible to modify the algorithms to generates other families of combinatorial objects, such as semi-standard Young tableaux. This raises a more general question in order theory, namely, finding a characterization of posets whose Hasse diagram admits a spanning tree having a natural way to be traversed, defining this way a total order on the elements of the poset. Our approach to generate all the elements of such posets could likely be applied to all of them.

References

1. Barcucci, E., Frosini, A., Rinaldi, S.: Direct-convex polyominoes: ECO method and bijective results. In: Brak, R., Foda, O., Greenhill, C., Guttman, T., Owczarek, A. (eds.) Proceedings of Formal Power Series and Algebraic Combinatorics 2002, Melbourne (2002)
2. Barcucci, E., Del Lungo, A., Nivat, M., Pinzani, R.: Reconstructing convex polyominoes from horizontal and vertical projections. Theoret. Comp. Sci. 155(2), 321–347 (1996)
3. Barcucci, E., Del Lungo, A., Pergola, E., Pinzani, R.: ECO:a methodology for the enumeration of combinatorial objects. J. of Diff. Eq. and App. 5, 435–490 (1999)
4. Bousquet-Mélou, M.: A method for the enumeration of various classes of column-convex polygons. Discrete Math. 154, 1–25 (1996)
5. Castiglione, G., Vaglica, R.: Recognizable Picture Languages and Polyominoes. In: Bozapalidis, S., Rahonis, G. (eds.) CAI 2007. LNCS, vol. 4728, pp. 160–171. Springer, Heidelberg (2007)
6. De Carli, F., Frosini, A., Rinaldi, S., Vuillon, L.: On the Tiling System Recognizability of Various Classes of Convex Polyominoes. Ann. Comb. 13, 169–191 (2009)
7. Delest, M., Dubernard, J.P., Dutour, I.: Parallelogram Polyominoes and Corners. J. Symb. Comp. 20, 503–515 (1995)
8. Delest, M., Viennot, X.G.: Algebraic languages and polyominoes enumeration. Theoret. Comp. Sci. 34, 169–206 (1984)
9. Del Lungo, A., Duchi, E., Frosini, A., Rinaldi, S.: On the generation and enumeration of various classes of convex polyominoes. Electronic Journal of Combinatorics 11, 60 (2004)
10. Del Lungo, A., Mirolli, M., Pinzani, R., Rinaldi, S.: A Bijection for Directed-Convex Polyominoes. In: Proc. of DM-CCG 2001, Discrete Mathematics and Theoretical Computer Science AA, pp. 133–144 (2001)
11. Golomb, S.W.: Checker Boards and Polyominoes. The American Mathematical Monthly 61, 675–682 (1954)
12. Kuba, A., Balogh, E.: Reconstruction of convex 2D discrete sets in polynomial time. Theoret. Comp. Sci. 283(1), 223–242 (2002)
13. Massazza, P.: A CAT algorithm for sand piles. PU.M.A. 19(2-3), 147–158 (2008)
14. Massazza, P., Radicioni, R.: A CAT algorithm for the exhaustive generation of ice piles. RAIRO Theoretical Informatics and Applications 44, 525–543 (2010)
15. Massazza, P., Radicioni, R.: On the Exhaustive Generation of Symmetric Sand Piles. In Proc. of GASCom 2010, Montréal, September 2-4 (2010)
16. Ollinger, N.: Tiling the Plane with a Fixed Number of Polyominoes. In: Dediu, A.H., Ionescu, A.M., Martín-Vide, C. (eds.) LATA 2009. LNCS, vol. 5457, pp. 638–647. Springer, Heidelberg (2009)
17. Pergola, E., Sulanke, R.A.: Schröder Triangles, Paths, and Parallelogram Polyominoes. Journal of Integer Sequences, 1 Article 98.1.7 (1998)
18. Polya, G.: On the number of certain lattice polygons. J. Comb. Theory 6, 102–105 (1969)

On Brzozowski's Conjecture for the Free Burnside Semigroup Satisfying $x^2 = x^3$

Andrey N. Plyushchenko and Arseny M. Shur

Ural State University, Ekaterinburg, Russia

Abstract. In this paper we examine Brzozowski's conjecture for the two-generated free Burnside semigroup satisfying $x^2 = x^3$. The elements of this semigroup are classes of equivalent words, and the conjecture claims that all elements are regular languages. The case of the identity $x^2 = x^3$ is the only one, for which Brzozowski's conjecture is neither proved nor disproved. We prove the conjecture for all the elements containing an overlap-free or an "almost" overlap-free word. In addition, we show that all but finitely many of these elements are "big" languages in terms of growth rate.

Introduction

In order to introduce the notion of a free Burnside semigroup, let us take a finite alphabet Σ, $|\Sigma| = k$, and positive integers n and m. As usual, we write Σ^* for the monoid of all words over Σ (including the empty word λ) and Σ^+ for the semigroup of all non-empty words over Σ. Two words U and V are *neighbours* if and only if one of them can be obtained from the other by replacing some factor of the form Y^n by the factor Y^{n+m}. Then we get the *neighbourhood relation*

$$\pi_{n,m,k} = \{(XY^nZ, XY^{n+m}Z), (XY^{n+m}Z, XY^nZ) \mid X, Y, Z \in \Sigma^*\};$$

its transitive closure is a semigroup congruence, denoted by $\sim_{n,m,k}$. The quotient semigroup $B(n, m, k) = \Sigma^+/\sim_{n,m,k}$ is called the *k-generated free Burnside semigroup satisfying $x^n = x^{n+m}$*. Thus, the elements of free Burnside semigroups are classes of equivalent words. The congruence class of a word U is denoted by $[U]$.

The structure of free Burnside semigroups is far from being completely described. However, a considerable progress was achieved in the 1990's, when many structural properties of such semigroups were discovered. The reader is referred to the survey [10] for the history and the formulations of remarkable results. The most important problems related to free Burnside semigroups are the *finiteness problem* (to determine for each triple (n, m, k), whether the semigroup $B(n, m, k)$ is finite or not) and the *word problem* (to decide for arbitrary pair (U, V) of words, whether or not $U \sim_{n,m,k} V$).

Obviously, the free Burnside semigroups $B(n, m, 1)$ are cyclic and finite, and the word problem for $B(n, m, 1)$ is decidable. Suppose $k > 1$ from now on. Due to Green and Rees [4] and later Kaďourek and Polák [7], the finiteness problem and the word problem for semigroups $B(1, m, k)$ were solved modulo periodic

G. Mauri and A. Leporati (Eds.): DLT 2011, LNCS 6795, pp. 362–373, 2011.

groups (more precisely, they were reduced to the corresponding problems for the groups satisfying $x^m = 1$).

Using Thue-Morse words [18], Brzozowski, Culik, and Gabrielian proved [2] that free Burnside semigroups $B(n, m, k)$ are infinite whenever $n \geq 2$. On the other hand, Brzozowski suggested that these semigroups should possess some finiteness properties. In 1969, he conjectured that all elements of semigroups $B(n, 1, k)$ are regular languages. Later McCammond generalized this conjecture for all semigroups $B(n, m, k)$ with $n \geq 2$. Note that Brzozowski-McCammond's conjecture implies decidability of the word problem: for any given words U and V, one can construct an automaton recognizing the class $[U]$ and try to accept V by this automaton (see [5, 6]).

So, Brzozowski-McCammond's conjecture became the key problem in the area of free Burnside semigroups. Due to de Luca and Varricchio [11], McCammond [12], Guba [5, 6], and do Lago [8, 9], this conjecture was proved for all semigroups $B(n, m, k)$ with $n \geq 3$. However, the conjecture does not hold for semigroups $B(2, m, k)$ with $m \geq 2$ (see [8]). For the case $n = 2, m = 1$, this problem is still open. Note that this particular case was explicitly mentioned by Brzozowski [3] and was considered to be the hardest one to analyze [8]. Recently, the first author [14] showed that Brzozowski's conjecture and the word problem for any semigroup $B(2, 1, k)$ can be reduced to the corresponding problems for the particular semigroup $B(2, 1, 2)$. So, the semigroup $B(2, 1, 2)$ now is the key object in the study of free Burnside semigroups. In what follows, we consider only this semigroup and write π and $\sim$ instead of $\pi_{2,1,2}$ and $\sim_{2,1,2}$, respectively.

In [13], it was proved that each congruence class $[U]$ contains at most one overlap-free word. Later we generalized this result for "almost" overlap-free words and showed that such a word in a given class, if any, can be efficiently found [15, 16]. Thus, we constructed an algorithm that partially solves the word problem for the semigroup $B(2, 1, 2)$. The aim of this paper is to prove the regularity of the congruence class $[U]$ for any almost overlap-free word U. In addition, we show that these classes are rather big in terms of growth rates, so that the particular case, in which Brzozowski's conjecture is now confirmed, is wide enough. Our argument heavily uses the previously developed technique [15, 16].

The text is subdivided into four sections. In Section 1 we introduce the main tools and techniques. Section 2 contains the construction and the analysis of procedure Ancestor, which was introduced in [15, 16]. This procedure is used to describe the congruence classes of almost overlap-free words. Section 3 is devoted to the proof of the main result (Theorem 1). Finally, in the last section we find an exponential lower bound for the combinatorial complexity of the congruence classes containing long almost overlap-free words.

1 Preliminaries

From now on, let $\Sigma = \{a, b\}$. We study only finite words over Σ and use standard notions of prefixes, suffixes, and factors. For a word W, its length is denoted by $|W|$ and its i-th letter is denoted by $W[i]$; thus, $W = W[1] \ldots W[|W|]$.

Negation is an automorphism of Σ^*, defined by the rules $\bar{a} = b$, $\bar{b} = a$. We also put $\overline{\mathcal{L}} = \{\overline{W} \mid W \in \mathcal{L}\}$ for any $\mathcal{L} \subseteq \Sigma^*$.

A word U is called *overlap-free* if it contains no factor of the form $XYXYX$ for any $X \in \Sigma^+, Y \in \Sigma^*$. If U contains no *proper* factor (that is, not equal to U) of the form above, then we call U *almost overlap-free.* The *Thue-Morse morphism* θ of Σ^+ is defined by the equalities $\theta(a) = ab$, $\theta(b) = ba$. This morphism preserves overlap-freeness [18]. Words from the set $\theta(\Sigma^+)$ are called *θ-images.*

Combinatorial complexity of a language $\mathcal{L} \in \Sigma^*$ is the function $c_{\mathcal{L}}(n) = |\Sigma^n \cap \mathcal{L}|$. That is, $c_{\mathcal{L}}(n)$ is the number of words of length n in $\mathcal{L}$. *Growth rate* of a language $\mathcal{L}$ is the number $\alpha(\mathcal{L}) = \limsup_{n\to\infty} \sqrt[n]{c_{\mathcal{L}}(n)}$.

Regular expressions, regular languages, deterministic and *nondeterministic finite automata* (dfa and nfa) are defined in the usual way. For an nfa $\mathcal{A} = (Q, \delta, \Sigma, q_0, T)$, the set of transitions δ is viewed as a ternary relation $\delta \subseteq Q \times \Sigma \times Q$. Thus, an nfa is considered as a labeled digraph. In what follows, $\widetilde{A}$ (resp., $\widetilde{B}$) abbreviates the language $aba(ba)^*$ (resp., $bab(ab)^*$).

Now we proceed with some tools and techniques. As it was shown in [16], any word from the congruence class of an almost overlap-free word can be transformed to a θ-image by performing a sequence of reductions described below. Surprisingly, all these reductions, applied to a pair of words, preserve the congruence $\sim$ under certain conditions.

The simplest reduction operation is r_1*-reduction.* It replaces all factors of the form c^n, where $c \in \Sigma$ and $n > 2$, by c^2. We write $r_1(U)$ for the word obtained from the word U by this operation. We say that U is r_1*-reduced* if $U = r_1(U)$.

Now let U be an r_1-reduced word. Then $r(U)$ is the word obtained from U by performing all possible reductions of the form $W \to aa$, where $W \in a\widetilde{A}a$, and the form $W \to bb$, where $W \in b\widetilde{B}b$. The operation r is called *complete reduction.* Its result is independent of the order of reductions [1]. We call a word U r*-reduced* if $r(U) = U$. Clearly, any θ-image and all its factors are r-reduced. It appears that the completely reduced words are exactly the factors of θ-images.

Proposition 1. ([16]) *A word U is completely reduced if and only if U is a factor of a θ-image.*

In the sequel, we denote the set of all r_1-reduced (r-reduced) words that are equivalent to a given word U by $[U]_{r_1}$ (resp., $[U]_r$). We put $r_1(\mathcal{L}) = \{r_1(U) \mid U \in \mathcal{L}\}$, $r(\mathcal{L}) = \{r(U) \mid U \in \mathcal{L}\}$ for any language $\mathcal{L}$. Clearly, $r_1(U) \sim U$ for any word $U \in \Sigma^*$, therefore $r_1([U]) = [U]_{r_1}$. In contrast to r_1-reduction, the condition $r(U) \sim U$ does not hold in general case. However, under certain restrictions, the words $r(U)$ and U are equivalent, see Proposition 2 below. These restrictions use the notion of $\widetilde{AB}$-whole word introduced in [1]. An r_1-reduced word W is $\widetilde{AB}$*-whole* if any factor $X \in a\widetilde{A}a$ ($X \in b\widetilde{B}b$) occurs in W inside the factor $abXba$ (resp., $baXab$). Obviously, the r-reduced words are $\widetilde{AB}$-whole.

Proposition 2. ([16]) *Suppose that an $\widetilde{AB}$-whole word U has no prefix from the set $(aba)(aba)^*(ab)^2(ab)^*aa$ and no suffix from the set $aa(ba)^*(ba)^2(aba)^*(aba)$, up to negation. Then $U \sim r(U)$.*

We refer to the prefixes and the suffixes mentioned in Proposition 2 as *non-reducible tails.* As it was shown in [16], the property of a word to have non-reducible tails is preserved by the congruence $\sim$. Since the almost overlap-free words have no non-reducible tails, we get the following proposition.

Proposition 3. *Let a word U be equivalent to an almost overlap-free word. Then U has no non-reducible tails.*

Let V be an almost overlap-free word. As it was proved in [1], the property of a word to be $\widetilde{AB}$-whole is preserved by the congruence $\sim$. So, if V is $\widetilde{AB}$-whole, then any word from $[V]_{r_1}$ is $\widetilde{AB}$-whole as well. Now we investigate the case when the word V is not $\widetilde{AB}$-whole. Define

$$\mathcal{S}' = \{aaa, aabaa, aabaab, baabaa, baabaab, aabaabb, \\ bbaabaa, aabaaba, abaabaa, aabaabbaabaa\}$$

and let $\mathcal{S}_1 = \mathcal{S}' \cup \overline{\mathcal{S}'}$. It is easy to see that any word from $\mathcal{S}_1$ is not $\widetilde{AB}$-whole. As it was proved in [17, 13], the word V either belongs to $\mathcal{S}_1$ or has the prefix *aabaabba* and/or the suffix *abbaabaa*, up to negation. For any word $U \in \Sigma^+$, its prefix (suffix) from the set $[aabaabba]_{r_1}$ or $[bbabbaab]_{r_1}$ (resp., $[abbaabaa]_{r_1}$ or $[baabbabb]_{r_1}$) is called its *non-uniform left* (resp., *right*) *tail.* So, there are four sets of equivalent non-uniform tails:

$$[aabaabba]_{r_1} = (aab)^* aabaabba, \quad [abbaabaa]_{r_1} = abbaabaa(baa)^*, \qquad (1)$$

and their negations. *Tail reduction* r_T [13] is the operation that reduces any left non-uniform tail of a word U to the last 7 symbols and any right such tail to the first 7 symbols. (Note that a tail of an almost overlap-free word, if any, has exactly 8 symbols. Hence the operation r_T deletes exactly one letter from such a tail.) If a word U has no non-uniform tails, then $r_T(U) = U$. Also we put $r_T(\mathcal{L}) = \{r_T(U) \mid U \in \mathcal{L}\}$ for any language $\mathcal{L}$.

Proposition 4. ([13]) *Let V be an almost overlap-free word such that $V \notin \mathcal{S}_1$ and let $U \in [V]_{r_1}$. Then*
(1) *if one of the words V and U has a non-uniform tail T, then the other has a tail $T' \sim T$;*
(2) *the words $r_T(V)$ and $r_T(U)$ are $\widetilde{AB}$-whole;*
(3) *$r_T(V) \sim r_T(U)$.*

Note that Proposition 4 includes the case when the word V is $\widetilde{AB}$-whole.

In the sequel, we use some sort of inverses of the reduction operations introduced above. Namely, for any word U, we put

$$r_1^{-1}(U) = \{W \in \Sigma^* \mid r_1(W) = U\},$$
$$r^{-1}(U) = \{W \in \Sigma^* \mid W \text{ is } \widetilde{AB}\text{-whole and has no non-reducible tails, } r(W) = U\},$$
$$r_T^{-1}(U) = \{W \in \Sigma^* \mid W \text{ is } r_1\text{-reduced, } W \sim U, \text{ and } r_T(W) = r_T(U)\}\ .$$

In a natural way, we define $r_1^{-1}(\mathcal{L}) = \bigcup_{U \in \mathcal{L}} r_1^{-1}(U)$, $r^{-1}(\mathcal{L}) = \bigcup_{U \in \mathcal{L}} r^{-1}(U)$, and $r_T^{-1}(\mathcal{L}) = \bigcup_{U \in \mathcal{L}} r_T^{-1}(U)$ for any language $\mathcal{L}$. The next statement immediately follows from the definition of r_1 and Proposition 2.

Proposition 5. *For any word U, (1) $[U]_{r_1} = r_1([U])$ and $[U] = r_1^{-1}([U]_{r_1})$, (2) if U is $\widetilde{AB}$-whole and has no non-reducible tails, then $[U]_r = r([U]_{r_1})$ and $[U]_{r_1} = r^{-1}([U]_r)$.*

By Proposition 1, any r-reduced word U is a factor of a θ-image. Hence,

$$r(U) = cQ_1 \dots Q_k d, \text{ where } Q_1, \dots, Q_k \in \{ab, ba\}, c, d \in \{a, b, \lambda\} \ .$$

This representation is unique if U has the factor aa or bb. If U has no such factors, we additionally require $c = \lambda$ to get a unique representation as well. Now put $\eta(U) = Q_1 \dots Q_k$ in order to transform the word U to a θ-image. We denote $h(U) = c$, $t(U) = d$. Note that the function η does not preserve the congruence $\sim$ in general case. We say that a pair (U_1, U_2) of r-reduced words is *bad* if $U_1 \sim U_2$, but $\eta(U_1) \not\sim \eta(U_2)$. If a pair $(r(W_1), r(W_2))$ is bad and the words W_1, W_2 are $\widetilde{AB}$-whole and have no non-reducible tails, then the pair (W_1, W_2) will be also called *bad*. Otherwise we say that a pair (W_1, W_2) is *good*. The function η and bad pairs were investigated in [16].

2 Procedure Ancestor and Primary Series of Words

The algorithm, constructed in [16] to solve partially the word problem for the semigroup $B(2, 1, 2)$, contains the following essential procedure.

Procedure Ancestor.
Input. A word $U \in \Sigma^+$.
Output. A word $\text{Anc}(U) \in \Sigma^+$, an integer k, length k arrays L, R, h, t of letters.
Step 0. Let $k := 0$.
Step 1. Let $U := r_1(U)$; $k := k + 1$; $L[k] := R[k] := h[k] := t[k] := \lambda$.
Step 2. If $|U| \leq 2$ or $U \sim W$ for some $W \in \mathcal{S}_1$, then Anc $:= U$; stop.
Step 3. If U has a non-uniform left tail, set $L[k] := U[1]$; if U has a non-uniform right tail, set $R[k] := U[|U|]$; let $U' := r_T(U)$.
Step 4. If U' is not $\widetilde{AB}$-whole or U' has a non-reducible tail, then Anc $:= U$; stop.
Step 5. Let $U' := r(U')$.
Step 6. Let $h[k] := h(U')$; $t[k] := t(U')$; $U := \theta^{-1}(\eta(U'))$; GOTO step 1.

Starting with $U_1 = r_1(U)$, procedure Ancestor constructs the sequence of words $U_1, U_2, \dots$ by the rule:

$$U_{k+1} = r_1(\theta^{-1}(\eta(r(r_T(U_k)))))$$

until one of the stop conditions is fulfilled. The sequence $\{U_k\}$ is called the *primary U-series*, its length (that is, the number of words in it) is denoted by

$\ell(U)$. We say that the output word $\mathrm{Anc}(U) = U_{\ell(U)}$ is the *ancestor of* U, and the arrays $L = L_U$, $R = R_U$, $h = h_U$, and $t = t_U$ returned by Procedure Ancestor are *associated with* U. We omit the index U if it is clear from context.

Let V be an almost overlap-free word and let $U \sim V$. We apply procedure Ancestor to the pair (U, V). In view of the definition of r_1 and Propositions 2–4, all steps of this procedure except for Step 6 preserve the congruence $\sim$. The case when the last step preserves $\sim$ as well, is described by the next lemma.

Lemma 1. ([16]) *Let U and V be equivalent words and let $\{U_k\}_{k=1}^{\ell(U)}$, $\{V_k\}_{k=1}^{\ell(V)}$ be the primary U- and V-series respectively. If all pairs $(r_T(V_k), r_T(U_k))$ are good for $k < \min\{\ell(U), \ell(V)\}$, then*
(1) $\ell(U) = \ell(V)$;
(2) $U_k \sim V_k$ for each $k = 1, \ldots, \ell(V)$; in particular, $\mathrm{Anc}(U) \sim \mathrm{Anc}(V)$;
(3) $L_U = L_V$, $R_U = R_V$, $h_U = h_V$, and $t_U = t_V$.

If some pairs $(r_T(V_k), r_T(U_k))$ for $k < \min\{\ell(V), \ell(U)\}$ are bad, we get a more difficult case, described in Lemmas 2, 3 below. For a word U, we define the word $U^{3/2}$ as follows: if $U = YY$ for some $Y \in \Sigma^*$, then $U^{3/2} = YYY$; if U is not a square of any word, we put $U^{3/2} = U$.

Lemma 2. ([16]) *Let $U \sim V$, where V is an almost overlap-free word, and let $\{U_k\}_{k=1}^{\ell(U)}$, $\{V_k\}_{k=1}^{\ell(V)}$ be the primary U- and V-series respectively. Assume that there exists an integer $k' < \min\{\ell(V), \ell(U)\}$ such that all pairs $(r_T(V_k), r_T(U_k))$ are good for $k < k'$ and the pair $(r_T(V_{k'}), r_T(U_{k'}))$ is bad. Then $r_T(V_{k'})$ is a square. Denote by W and $\{W_j\}_{j=1}^{\ell(W)}$ the word $r_T(V_{k'})^{3/2}$ and its primary series respectively. Then*
(1) $\ell(V) - 1 \leq \ell(U) = (k' - 1) + \ell(W) \leq \ell(V)$;
(2) $U_k \sim V_k$ for each $k = 1, \ldots, k'$ and $U_k \sim W_{k-k'+1} \not\sim V_k$ for each $k = k'+1, \ldots, k'-1+\ell(W)$; in particular, $\mathrm{Anc}(U) \sim \mathrm{Anc}(W) \not\sim \mathrm{Anc}(V)$;
(3) $L_U[k] = L_V[k]$, $R_U[k] = R_V[k]$, $h_U[k] = h_V[k]$, and $t_U[k] = t_V[k]$ for all $k \leq \ell(U)$; $h_W[j] = h_V[k'-1+j]$ and $t_W[j] = t_V[k'-1+j]$ for all $j \leq \ell(W)-1$; $L_W[j] = R_W[j] = L_V[k'-1+j] = R_V[k'-1+j] = \lambda$ for all $j \in [2, \ell(W)]$;
(4) The words V_k and W_j are not squares for all $k \in [k'+1, \ell(V)-1]$ and all $j \in [2, \ell(W)-1]$. Moreover, they are r-reduced.

For an almost overlap-free word V, its primary series $\{V_k\}_{k=1}^{\ell(V)}$, define

$$\overline{k} = \max\{k < \ell(V) \mid \mathrm{Anc}(r_T(V_k)^{3/2}) \not\sim \mathrm{Anc}(V)\}$$

if the set in the right-hand side is nonempty. We call the value $\overline{k}$ *critical* (for V). Using Lemma 2 (2, 3), it is not hard to prove that the following holds.

Lemma 3. *Let $U \sim V$, where V is an almost overlap-free word, and let $\{V_k\}_{k=1}^{\ell(V)}$, $\{U_k\}_{k=1}^{\ell(U)}$ be the primary V- and U-series respectively. Then the pair $(r_T(V_k), r_T(U_k))$ is good for any non-critical $k < \ell(V)$.*

Let us define $\mathcal{S}'' = \{aa, aabaabaa, abaabaab, baabaaba\}$ and let $\mathcal{S}_2 = \mathcal{S}'' \cup \overline{\mathcal{S}''}$. The next lemma provides a recursive description of the congruence class $[V]$ for any almost overlap-free word V.

Lemma 4. *Suppose that V is an almost overlap-free word and $\{V_k\}_{k=1}^{\ell(V)}$ is its primary series. For each $k = 1,\dots,\ell(V)-1$, let $m_k = \ell(r_T(V_k)^{3/2})$ and let $\{V_{k,j}^{3/2}\}_{j=1}^{m_k}$ be the primary $r_T(V_k)^{3/2}$-series. Then*
(1) $\mathrm{Anc}(V) \in \mathcal{S}_1 \cup \{a, b, aa, bb, ab, ba\}$;
(2) $[r_T(V_k)]_r = h[k]\theta([V_{k+1}])t[k]$ *for all non-critical* $k \le \ell(V)-1$;
(3) $[r_T(V_{\overline{k}})]_r = h[\overline{k}]\theta([V_{\overline{k}+1}])t[\overline{k}] \cup h[\overline{k}]\theta([V_{\overline{k},2}^{3/2}])t[\overline{k}]$ *for the critical value* $\overline{k}$ *(if such value exists);*
(4) $\mathrm{Anc}(r_T(V_{\overline{k}})^{3/2}) \in \mathcal{S}_2$ *for the critical value* $\overline{k}$ *(if such value exists);*
(5) $[V_{\overline{k},j}^{3/2}]_r = h[\overline{k}-1+j]\,\theta([V_{\overline{k},j+1}^{3/2}])\,t[\overline{k}-1+j]$ *for the critical* $\overline{k}$ *and each* $j = 2,\dots,m_{\overline{k}}-1$ *(if the critical value exists);*
(6) $[V_{\overline{k},j}^{3/2}]_{r_1} = r^{-1}([V_{\overline{k},j}^{3/2}]_r)$ *for the critical* $\overline{k}$ *and each* $j = 2,\dots,m_{\overline{k}}-1$ *(if the critical value exists);*
(7) $[r_T(V_k)]_{r_1} = r^{-1}([r_T(V_k)]_r)$ *for each* $k = 1,\dots,\ell(V)-1$;
(8) $[V_k]_{r_1} = r_T^{-1}(L[k][r_T(V_k)]_{r_1}R[k])$ *for each* $k = 1,\dots,\ell(V)-1$;
(9) $[V_k] = r_1^{-1}([V_k]_{r_1})$, $[V_{k,j}^{3/2}] = r_1^{-1}([V_{k,j}^{3/2}]_{r_1})$ *for each* $k = 1,\dots,\ell(V)-1$ *and each* $j = 2,\dots,m_k-1$.

Proof. Statements 1 and 4 were proved in [16], statements 6, 7, and 9 follow from Proposition 5. Let us prove statements 2 and 3. Clearly, if $U \sim V_{k+1}$ or $U \sim V_{k,2}^{3/2}$ for some $k < \ell(V)$, then the word $h[k]\theta(U)t[k]$ is equivalent to the word $h[k]\theta(V_{k+1})t[k] = r(r_T(V_k))$ or the word $h[k]\theta(V_{k,2}^{3/2})t[k] = r(r_T(V_k)^{3/2})$, respectively. By Proposition 2 and the construction of procedure Ancestor, we have $r(r_T(V_k)^{3/2}) \sim r_T(V_k)^{3/2} \sim r_T(V_k) \sim r(r_T(V_k))$. Since the word $h[k]\theta(U)t[k]$ is r-reduced (as a factor of a θ-image), we conclude that, for any $k < \ell(V)$,

$$h[k]\theta([V_{k+1}])t[k] \cup h[k]\theta([V_{k,2}^{3/2}])t[k] \subseteq [r_T(V_k)]_r\,.$$

Conversely, suppose that $U \in [r_T(V_k)]_r$ for some $k < \ell(V)$, that is, U is an r-reduced word and $U \sim r_T(V_k)$. We aim to prove that $U \in h[k]\theta(U)t[k]$ if k is not critical and $U \in h[k]\theta([V_{k+1}])t[k] \cup h[k]\theta([V_{k,2}^{3/2}])t[k]$ for the critical k. One can easily check that the words V_k are almost overlap-free for all $k \le \ell(V)$. Let $\{U_j\}_{j=1}^{\ell(U)}$ be the primary U-series. Obviously, the primary $r_T(V_k)$-series is $r_T(V_k), V_{k+1},\dots,V_{\ell(V)} = \mathrm{Anc}(V)$. Since the word U is r-reduced, it has no non-uniform tails and $r(r_T(U)) = U$. By Lemma 3, if k is not critical, then the pair $(r_T(V_k), U)$ is good. This implies $U_2 \sim V_{k+1}$ whence $U = h[k]\theta(U_2)t[k] \in h[k]\theta([V_{k+1}])t[k]$, as desired.

Now suppose that k is the critical value for V. Then the value 1 is critical for $r_T(V_k)$. Clearly, if the pair $(r_T(V_k), U)$ is good, we get $U \in h[k]\theta([V_{k+1}])t[k]$, as before. Otherwise we have $U_2 \sim V_{k,2}^{3/2}$ by Lemma 2 whence $U \in h[k]\theta([V_{k,2}^{3/2}])t[k]$, as desired. The proof of statements 2 and 3 is complete. Note that the congruence class $[V_{k,2}^{3/2}]$ coincides with $[V_{k+1}]$ if and only if k is not critical.

Statement 5 can be proved in the same way (as it was shown in [16], if $\overline{k}$ is critical, then any pair $(V_{\overline{k},j}^{3/2}, U)$, where $2 \le j \le m_{\overline{k}} - 1$ and U is an r_1-reduced word, is good).

Finally, we prove statement 8. Suppose that the word U is r_1-reduced and $U \sim V_k$. Since $k < \ell(V)$, we have $V_k \notin \mathcal{S}_1$. Therefore, $r_T(U) \sim r_T(V_k)$ by Proposition 4. Hence we get $L[k]r_T(U)R[k] \in L[k][r_T(V_k)]_{r_1}R[k]$. Consider the arrays L_U and R_U associated with U. From Lemmas 1,2 it follows that $L_V[k] = L_U[1]$ and $R_V[k] = R_U[1]$. So, we have $L_V[k]r_T(U)R_V[k] = L_U[1]r_T(U)R_U[1] \sim U$ by definitions of r_T and non-uniform tails. Moreover, we have $r_T(L_U[1]r_T(U)R_U[1]) = r_T(U)$. According to the definition of r_T^{-1}, we conclude that $U \in r_T^{-1}(L_V[k]\,[r_T(V_k)]_{r_1}\,R_V[k])$.

Conversely, let $U \in r_T^{-1}(L[k]\,[r_T(V_k)]_{r_1}\,R[k])$. This means that U is r_1-reduced and there exists a word $U' \in [r_T(V_k)]_{r_1}$ such that $U \sim L[k]U'R[k]$ and $r_T(U) = r_T(L[k]U'R[k])$. Thus we have $U \sim L[k]U'R[k] \sim L[k]r_T(V_k)R[k] = V_k$, as required. This completes the proof of the lemma. □

By mere construction of regular expressions, one can prove that the following holds.

Proposition 6. ([16]) *For any $V \in \mathcal{S}_1 \cup \mathcal{S}_2$, the class $[V]$ is a regular language.*

3 The Proof of the Main Result

Theorem 1. *For any almost overlap-free word V, the congruence class $[V]$ is a regular language.*

The proof is based on two lemmas.

Lemma 5. *If U is an $\widetilde{AB}$-whole word without non-reducible tails, and $[U]_r$ is a regular language, then the language $[U]_{r_1}$ is regular as well.*

Proof. For any word $U \in \Sigma^*$ and any positive integer k, we define $\mathrm{suff}(U,k)$ to be the k-letter suffix of U if $|U| \geq k$ and to be the word U itself otherwise.

Now suppose that $[U]_r$ is a regular language and $\mathcal{A} = (\Sigma, Q, \delta, q_0, T)$ is a dfa recognizing $[U]_r$. We construct a new dfa $\mathcal{A}' = (\Sigma, Q', \delta', q_0', T')$ such that

$$Q' = \{(q,W) \mid q \in Q, W{\in}\Sigma^*, |W|{\leq}7\},\ q_0' = (q_0, \lambda),\ T' = \{(t,W){\in}Q' \mid t \in T\},$$
$$\text{and}\ \ \delta'((q,W),c) = (\delta'(q,c), \mathrm{suff}(Wc,7)) \text{ for any } (q,W) \in Q',\ c \in \Sigma\ .$$

In fact, we introduce multiple copies of each state of the automaton $\mathcal{A}$ to store the last 7 symbols of the processed input. It can be directly verified that $\delta'(q_0', W) = (\delta(q_0, W), \mathrm{suff}(W,7))$. Hence, the dfa's $\mathcal{A}'$ and $\mathcal{A}$ accept the same language $[U]_r$.

Now we take $\mathcal{A}'$ and add some states and transitions to it, getting an nfa $\mathcal{A}'' = (\Sigma, Q'', \delta'', q_0'', T'')$. Namely, we add 18 new copies of each state $q \in Q$: $Q'' = Q' \cup \{(q, la), (q, lb) \mid q \in Q, l \in [1,9]\}$. The states $(q, 5a)$ and $(q, 5b)$ are terminal whenever $q \in T$, while all other added states are nonterminal. For each state $q \in Q$, we introduce six sets of transitions, called A-sets and B-sets, concerning the states that are copies of q. Three *A-sets* are shown in Fig. 1, and three *B-sets* are defined in a symmetric way.

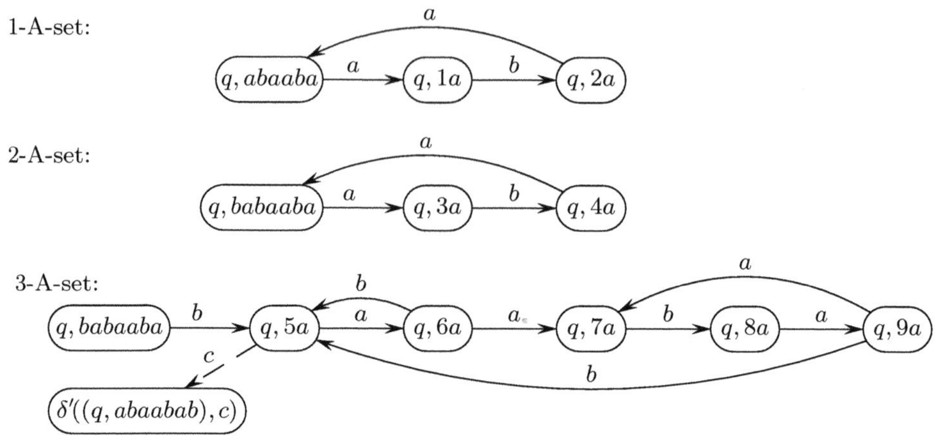

Fig. 1. *A-sets of transitions. The dashed transition is defined for each* $c \in \{a, b\}$.

Note that for each $q \in Q$, the states $(q, 5a)$ and $(q, 5b)$ "duplicate" the states $(q, abaabab)$ and $(q, babbaba)$ respectively. As we will see below, the 3-sets allow us to process the factors from the sets $aba\widetilde{A}^*aba$ and $bab\widetilde{B}^*bab$. The 1-sets and 2-sets deal with the case when the input word begins with *abaaba* and/or ends by *abaaba* (up to negation), in order to provide non-acceptance of the words that have non-reducible tails.

We claim that the automaton $\mathcal{A}''$ recognizes the language $[U]_{r_1}$. Take a word $W \in [U]_r$ and consider the set $r^{-1}(W)$. Any word from this set is obtained from W by a sequence of insertions of the words from $\widetilde{A}$ (between two a's) and the words from $\widetilde{B}$ (between two b's). Since the words from $a\widetilde{A}a$ (resp., $b\widetilde{B}b$) do not contain bb (resp., aa), these two types of insertions can be studied separately. Due to symmetry, we consider only insertions from $\widetilde{A}$. Since any word $V \in r^{-1}(W)$ is $\widetilde{AB}$-whole, the insertions are to be made in the middle of the factors *abaaba*. If this factor is a prefix or a suffix of W, then only the word *aba* can be inserted into it (possibly, several times), because V has no non-reducible tails. Further, W has no factor *aabaa*. As a result, the word V can be obtained from W by replacing, up to negation, the prefix *abaaba* and/or the suffix *abaaba* by a word from $aba(aba)^*aba$ (if such a prefix and/or suffix exists) and replacing all factors *babaabab* by some words from the set $baba(\widetilde{A})^*abab$.

Using 1-sets and 2-sets, nfa $\mathcal{A}''$ accepts any word from the set $aba(aba)^*abaX$ (resp., $Xaba(aba)^*aba$) whenever the word $abaabaX$ (resp., $Xabaaba$) is accepted by $\mathcal{A}'$. Further, if a word $X'babaababX''$ is accepted by dfa $\mathcal{A}'$, then nfa $\mathcal{A}''$ accepts any word from $X'baba(aba)^*ababX''$ (using 2-sets only) and any word from $X'baba(aba)^*(abab(ab)^*a)\widetilde{A}^*ababX''$ (using 2-sets and 3-sets). Hence, nfa $\mathcal{A}''$ accepts any word from the set $r^{-1}([U]_r)$, which equals $[U]_{r_1}$ by Proposition 5 (2).

Now suppose that a word W is accepted by $\mathcal{A}''$ and prove that $W \in [U]_{r_1}$. Let $\alpha = (q''_0, q''_1, \ldots, q''_{|W|})$ be an accepting path labeled by W. Let us replace the state $q''_i = (q, 5a)$ in α by the state $(q, abaabab)$ whenever $q''_{i-1} = (q, babaaba)$

and $W[i] = b$; the symmetric replacement should be also done for the states $(q, 5b)$. After that, we delete all states from $Q'' \setminus Q'$ in the path α. As a result, we get an accepting path α' that belongs to the dfa $\mathcal{A}'$ and is labeled by some word W'. Thus, $W' \in [U]_r$.

The deletion of all states of 1-sets and 2-sets from α induces the reductions of the form $aba(aba)^*aba \to abaaba$ and $bab(bab)^*bab \to babbab$ in the word W. The deletion of all states of 3-sets from α induces the reductions of the form

$$babaAabab \to babaabab, \quad ababBbaba \to ababbaba,$$

where $A \in \widetilde{A}^*$ and $B \in \widetilde{B}^*$, and some reductions of the form

$$abab(ab)^k \to abab \quad \text{and} \quad baba(ba)^k \to baba,$$

where $k \geq 1$. Since these reductions turn W into an r-reduced word, we conclude that W is an $\widetilde{AB}$-whole word and has no non-reducible tails. Then we have $r(W) \sim W' \sim U$, implying $r(W) \in [U]_r$ and finally $W \in r^{-1}([U]_r) = [U]_{r_1}$, as required. □

Lemma 6. *If U is a word such that $[U]_{r_1}$ is a regular language, then the language $[U]$ is regular as well.*

Proof. Suppose that the language $[U]_{r_1}$ is recognized by a dfa $\mathcal{A}=(\Sigma, Q, \delta, q_0, T)$. Similar to the proof of Lemma 5, we define a new dfa $\mathcal{A}' = (\Sigma, Q', \delta', q_0', T')$ as follows:

$$Q' = \{(q, W) \mid q \in Q, W \in \Sigma^*, |W| \leq 2\},\ q_0' = (q_0, \lambda),\ T' = \{(t, W) \in Q' \mid t \in T\},$$
$$\text{and } \delta'((q, W), c) = (\delta'(q, c), \mathrm{suff}(Wc, 2)) \text{ for any } (q, W) \in Q',\ c \in \Sigma\ .$$

Obviously, the automata $\mathcal{A}$ and $\mathcal{A}'$ accept the same language. The second component of each state from Q' stores the last two letters of the processed input. Now we add loops labeled by a (by b) for all states (q, aa) (resp., (q, bb)). Clearly, the resulting automaton recognizes the language $r_1^{-1}([U]_{r_1}) = [U]$. □

Proof (of Theorem 1). Let V be an almost overlap-free word and let $\{V_k\}_{k=1}^{\ell(V)}$ be its primary series. For any $k < \ell(V)$,. let $\{V_{k,j}^{3/2}\}_{j=1}^{m_k}$, where $m_k = \ell(r_T(V_k)^{3/2})$, be the primary $r_T(V_k)^{3/2}$-series.

First, we prove by induction that if the primary V-series has a critical value $\overline{k}$, then the language $[V_{\overline{k},j}^{3/2}]$ is regular for each $j = m_{\overline{k}}, \ldots, 2$. For inductive base, note that $\mathrm{Anc}(r_T(V_{\overline{k}})^{3/2}) \in \mathcal{S}_2$ by Lemma 4. In view of Proposition 6, the congruence class $[\mathrm{Anc}(r_T(V_{\overline{k}})^{3/2})]$ is a regular language. For inductive step, assume that the language $[V_{\overline{k},j+1}^{3/2}]$ is regular for some $j \geq 2$. Since the class of regular languages is closed under product and morphisms, from Lemma 4 (5) we derive that $[V_{\overline{k},j}^{3/2}]_r$ is a regular language. According to Lemmas 5, 6, and 2 (4), the language $[V_{\overline{k},j}^{3/2}]$ is regular as well, and we are done with the inductive step. In particular, we have proved that the congruence class $[V_{\overline{k},2}^{3/2}]$ is a regular language.

Now we prove by induction that the congruence class $[V_k]$ is regular for each $k = \ell(V), \ldots, 1$. For $k = \ell(V)$, this is true in view of Lemmas 4 (1) and Proposition 6. Suppose that the congruence class $[V_{k+1}]$ is a regular language for some $k > 1$ and prove the regularity of $[V_k]$. Since the class of regular languages is closed under union, from Lemma 4 (2, 3) it follows that $[r_T(V_k)]_r$ is a regular language. By Lemma 5, the language $[r_T(V_k)]_{r_1}$ is regular. Hence, $L[k][r_T(V_k)]_{r_1}R[k]$ is a regular language as well. Using (1), one can easily check that

$$r_T^{-1}(L[k][r_T(V_k)]_{r_1}R[k]) = (L[k]L[k]\overline{L[k]})^*L[k][r_T(V_k)]_{r_1}R[k](\overline{R[k]}R[k]R[k])^* \ .$$

Thus, in view of Lemma 4 (8), the language $[V_k]_{r_1}$ is regular. Then $[V_k]$ is a regular language by Lemma 6. By induction, we conclude that the congruence class $[V_1]$ is a regular language. But $[V_1] = [V]_{r_1}$, and one more reference to Lemma 6 completes the proof. □

4 Growth Rates of Congruence Classes

If V is a long almost overlap-free word, then the congruence class $[V]$ is big enough, as the following theorem shows.

Theorem 2. *Let V be an almost overlap-free word such that $|V| > 50$. Then $\alpha([V]) \geq \varphi$, where $\varphi = (1 + \sqrt{5})/2$ is the golden ratio.*

Proof. First, consider the languages $\mathcal{L}_A = (a + \widetilde{A})^*$ and $\mathcal{L}_B = (b + \widetilde{B})^*$. It is easy to show that $\alpha(\mathcal{L}_A) = \alpha(\mathcal{L}_B) = \varphi$.

Now suppose that the word V has the factor *ababbaba*, say, $V = XababbabaZ$. Since $r(r_1(ababBbaba)) = ababbaba$ and the word $r_1(ababBbaba)$ is obviously $\bar{A}\bar{B}$-whole for any $B \in \mathcal{L}_B$, we get $ababBbaba \sim ababbaba$ by Proposition 2. Hence, $Xabab\mathcal{L}_B babaZ \subseteq [V]$. As a result, we obtain, for any $n \geq |V|$ and some constant $C > 0$,

$$c_{[V]}(n) \geq c_{\mathcal{L}_B}(n - |V|) \geq C\varphi^{-|V|} \cdot \varphi^n ,$$

whence $\alpha([V]) \geq \varphi$. The same argument works in the case when V contains *babaabab*. So, in order to prove the theorem, it remains to show that any almost overlap-free word V with $|V| > 50$ has a factor *ababbaba* or *babaabab*. Consider the word

$$\theta^5(a) = abba\ baab\ ba\mathbf{ab}\ \mathbf{abba}\ \mathbf{baab}\ \mathbf{ab}ba\ abba\ baab \ .$$

Its prefix and suffix of length 18 both contain the mentioned factors. The same is true for the word $\theta^5(b)$. Any factor of length ≥ 35 of the infinite *Thue-Morse word* $\theta^\infty(a)$ obviously contains either an 18-letter prefix or an 18-letter suffix of one of the words $\theta^5(a), \theta^5(b)$. Hence, any 35-letter factor of the Thue-Morse word contains *ababbaba* or *babaabab*.

Now apply the main theorem of [17], which says that any overlap-free word U contains a factor of the Thue-Morse word of length $\geq 7|U|/10$. Thus, any overlap-free word of length ≥ 50 contains a factor of the Thue-Morse word of length 35. Since V has an overlap-free factor of length $|V| - 1$ by definition, we have proved the required property. □

References

1. Bakirov, M.F., Sukhanov, F.V.: Thue-Morse words and $\mathcal{D}$-structure of the free Burnside semigroup. Proc. Ural State Univ. Ser. Mathematics and Mechanics 18(3), 5–19 (2000) (Russian)
2. Brzozowski, J., Culik, K., Gabrielian, A.: Classification of non-counting events. J. Computer and System Sci. 5, 41–53 (1971)
3. Brzozowski, J.: Open problems about regular languages. In: Formal language theory: perspectives and open problems, pp. 23–47. Academic Press, New York (1980)
4. Green, J. A., Rees, D.: On semigroups in which $x^r = x$. Proc. Cambridge Phil. Soc. 48, 35–40 (1952)
5. Guba, V.S.: The word problem for the relatively free semigroups satisfying $t^m = t^{m+n}$ with $m \geq 4$ or $m \geq 3$, $n = 1$. Internat. J. Algebra Comput. 3(2), 125–140 (1993)
6. Guba, V.S.: The word problem for the relatively free semigroups satisfying $t^m = t^{m+n}$ with $m \geq 3$. Internat. J. Algebra Comput. 3(3), 335–348 (1993)
7. Kad'ourek, L., Polák, J.: On free semigroups satisfying $x^r \simeq x$. Simon Stevin 64(1), 3–19 (1990)
8. do Lago, A.P.: On the Burnside semigroups $x^n = x^{n+m}$. Internat. J. Algebra Comput. 6(2), 179–227 (1996)
9. do Lago, A.P.: Maximal groups in free Burnside semigroups. In: Lucchesi, C.L., Moura, A.V. (eds.) LATIN 1998. LNCS, vol. 1380, pp. 70–81. Springer, Heidelberg (1998)
10. do Lago, A.P., Simon, I.: Free Burnside semigroups. RAIRO Theoret. Inform. Appl. 35(6), 579–595 (2001)
11. de Luca, A., Varricchio, S.: On non-counting regular classes. In: Proc. ICALP 1990. LNCS, vol. 443, pp. 74–87. Springer, Berlin (1990)
12. McCammond, J.: The solution to the word problem for the relatively free semigroups satisfying $t^a = t^{a+b}$ with $a \geq 6$. Internat. J. Algebra Comput. 1(1), 1–32 (1991)
13. Plyushchenko, A.N.: Overlap-free words and the two-generated free Burnside semigroup satisfying $x^2 = x^3$. Siberian Electron. Math. Rep. 6, 166–181 (2009)
14. Plyushchenko, A.N.: On the word problem for the free Burnside semigroups satisfying $x^2 = x^3$. Russian Math, Iz. Vuz. (2011) (submitted)
15. Plyushchenko, A.N., Shur, A.M.: Almost overlap-free words and the word problem for the free Burnside semigroup satisfying $x^2 = x^3$. In: Proceedings of WORDS 2007, Marseille, France, pp. 245–253 (2007)
16. Plyushchenko, A.N., Shur, A.M.: Almost overlap-free words and the word problem for the free Burnside semigroup satisfying $x^2 = x^3$. Internat. J. Algebra Comput. (submitted), `http://arxiv.org/abs/1102.4315`
17. Shur, A.M.: Overlap-free words and Thue-Morse sequences. Internat. J. Algebra Comput. 6(3), 353–367 (1996)
18. Thue, A.: Uber die gegenseitige Lage gleicher Teile gewisser Zeichentreihen. Norske Videnskabssellskabets Skrifter I Mat. Nat. Kl. 1, 1–67 (1912)

Never Minimal Automata and the Rainbow Bipartite Subgraph Problem

Emanuele Rodaro and Pedro V. Silva

Departamento de Matemática, Faculdade de Ciências
Universidade do Porto, 4169-007 Porto, Portugal
{emanuele.rodaro,pvsilva}@fc.up.pt

Abstract. Never minimal automata, introduced in [4], are strongly connected automata which are not minimal for any choice of their final states. In [4] the authors raise the question whether recognizing such automata is a polynomial time task or not. In this paper, we show that the complement of this problem is equivalent to the problem of checking whether or not in an edge-colored graph there is a bipartite subgraph whose edges are colored using all the colors. We prove that this graph theoretic problem is **NP**-complete, showing that checking the property of never-minimality is unlikely a polynomial time task.

1 Introduction

Let $\mathscr{A} = \langle Q, \Sigma, \delta \rangle$ be a deterministic (not necessarily complete) finite-state automaton (DFA). The action of the transition function δ can naturally be extended to the free monoid Σ^*. This extension will still be denoted by δ. For convenience for each $v \in \Sigma^*$ and $q \in Q$ we will write $q \,.\, v = \delta(q, v)$ and put $S \,.\, v = \{q \,.\, v \mid q \in S\}$ for any $S \subseteq Q$. The set $q^{-1}F$, with $q \in Q, F \subseteq Q$, denotes, as usual, the set $\{w \in \Sigma^* : q \,.\, w \in F\}$.
An equivalence relation σ on a set Q is said to saturate a subset $F \subseteq Q$ if F is union of equivalence classes of σ. A congruence σ of the automaton $\mathscr{A}$ is an equivalence relation on Q such that if $q\sigma q'$ then $(q \,.\, u)\sigma(q' \,.\, u)$ for all $u \in \Sigma^*$. By $\mathcal{M}(\mathscr{A})$ we denote the set of minimal (non-trivial) congruences of $\mathscr{A}$. In [4] the authors introduce some classes of automata with extremal conditions. The class of uniformly minimal automata is formed by strongly connected automata which are minimal for any choice of the final states. In [4] the authors provide a characterization in terms of the state-pair graph which leads to a polynomial time algorithm to decide whether a given DFA is uniformly minimal. This class has also interesting connections with *multi-entry automata* and *symbolic dynamics*. The other interesting case introduced in [4] is the opposite extremal case of *never-minimal automata* which is considered in our paper.

Definition 1. *Except for the last section, we restrict our attention to strongly connected automata. We say that a DFA $\mathscr{A} = \langle Q, \Sigma, \delta \rangle$ is never-minimal if and only if for any $F \subseteq Q$ and $i \in Q$ the automaton $\mathscr{A}_{i,F} = \langle Q, \Sigma, \delta, i, F \rangle$ is not minimal.*

G. Mauri and A. Leporati (Eds.): DLT 2011, LNCS 6795, pp. 374–385, 2011.

In [4] the authors exhibit an infinite sequence of never-minimal automata and raise the problem of characterizing such property in order to give a polynomial time algorithm for recognizing such automata. Formally `NEVER-MINIMAL` is the problem that given as input a strongly connected DFA $\mathscr{A}$ checks whether or not $\mathscr{A}$ is never-minimal. In this paper we prove that co-`NEVER-MINIMAL` is **NP**-complete showing that `NEVER-MINIMAL` is unlikely in **P**.
The paper is organized as follows. In Section 2 we introduce the concept of a syntactic graph which is useful in characterizing never-minimal automata. In Section 3 we introduce some graph theoretic problems which are proved to be **NP**-complete. These graph theoretic problems turn out to be equivalent to the `DISJUNCTIVE SET` problem already considered in [1] in which the authors show the **NP**-completeness. However our reduction gives the **NP**-completeness for a smaller class. In Section 4 we prove that co-`NEVER-MINIMAL` is equivalent to the `DISJUNCTIVE SET` problem showing that the former problem is also **NP**-complete. In Section 5 we explore some connections with the `SYNTACTIC MONOID` problem (cf. [1]) and we end with Section 6 where we pose some open problems.

2 The Syntactic Graphs

In this paper we deal with graphs which are simple undirected and without loops. Given a symmetric, reflexive relation $R \subseteq V \times V$, there is a natural way to associate to R a graph $\mathcal{G}(R) = (V, E)$. Namely for each pair of distinct elements x, y we say that $\{x, y\} \in E$ if $(x, y) \in R$. Conversely a graph G gives rise to a symmetric reflexive relation $\mathcal{R}(G)$ in the obvious way. We say that a family $\mathfrak{R}$ of equivalence relations on a set V is *orthogonal* (or *pairwise separating* in [1]) if for any pair $R, R' \in \mathfrak{R}$ of distinct relations, $R \cap R' = 1_V$, where 1_V is the identity relation on V. In [4] the authors introduce the state-pair graphs as a tool to characterize uniformly minimal automata. We introduce an analogous tool which is a slight generalization of these graphs. This will be useful to characterize never-minimal automata. For any pair x, y of distinct states we associate an undirected graph $\mathfrak{G}_{x,y}$ called the *syntactic graph* generated by the pair x, y.

Definition 2 (syntactic graph of the pair $\{x, y\}$). *Let $x \neq y$ be two states of the automaton $\mathscr{A} = \langle Q, \Sigma, \delta \rangle$. The syntactic graph of the pair $\{x, y\}$ is the undirected graph $\mathfrak{G}_{x,y} = (Q, E_{x,y})$ having as set of vertices Q and the set of edges $E_{x,y}$ formed by the pairs $\{\alpha, \beta\}$ with $\alpha \neq \beta$ such that there is some $u \in \Sigma^*$ with $\{x, y\} . u = \{\alpha, \beta\}$. We denote by $\Gamma^i_{x,y}$, for $i = 1, \ldots, C(x, y)$, the connected components of $\mathfrak{G}_{x,y}$.*

Given a set $F \subseteq Q$, the *syntactic congruence* $\sim_F$ generated by F is the largest congruence saturating F and it is defined by $x \sim_F y$ if $\forall w \in \Sigma^*\ x . w \in F \Leftrightarrow y . w \in F$ or equivalently if $x^{-1}F = y^{-1}F$. The following proposition characterizes the syntactic congruences $\sim_F$, $F \subseteq Q$ with $x \sim_F y$ in terms of the connected components of the syntactic graph $\mathfrak{G}_{x,y}$. Using the notation of Definition 2 we have the following proposition.

Proposition 1. *Let $x \neq y$ be two states of the automaton $\mathscr{A}$ and let $F \subseteq Q$. Then $\sim_F$ is a syntactic congruence with $x \sim_F y$ if and only if*

$$F = \bigcup_{i \in I} \Gamma^i_{x,y}$$

for some $I \subseteq \{1, \ldots, C(x,y)\}$. Moreover the number of sets $F \subseteq Q$ such that $x \sim_F y$ is $2^{|C(x,y)|}$.

Proof. This is a consequence of Definition 2. □

Therefore, given x, y, the connected components of $\mathfrak{G}_{x,y}$ describe all the possible subsets $F \subseteq Q$ such that x, y are identified via the syntactic congruence $\sim_F$. We have a first characterization of never-minimal automata given in terms of their syntactic graphs.

Theorem 1. *A strongly connected DFA $\mathscr{A} = \langle Q, \Sigma, \delta \rangle$ is never-minimal if and only if for all $F \subseteq Q$ there are a pair of distinct states x, y and a subset $I \subseteq \{1, \ldots, C(x,y)\}$ such that*

$$F = \bigcup_{i \in I} \Gamma^i_{x,y}$$

Proof. Clearly $\mathscr{A}$ is never-minimal if and only if for all $F \subseteq Q$ there is a pair of distinct states x, y such that $x \sim_F y$. By Proposition 1 this is equivalent to say

$$F = \bigcup_{i \in I} \Gamma^i_{x,y}$$

for some $I \subseteq \{1, \ldots, C(x,y)\}$. □

We have the following fact.

Proposition 2. *The equivalence relation $\sigma_{x,y} \subseteq Q \times Q$ defined by $\alpha \sigma_{x,y} \beta$ if $\alpha, \beta \in \Gamma^i_{x,y}$ for some $i \in \{1, \ldots, C(x,y)\}$ is a congruence. Moreover it is the smallest congruence which identifies x with y.*

Proof. Suppose that $\alpha \, . \, a \in \Gamma^j_{x,y}$ for some $\alpha \in \Sigma$ and $j \in \{1, \ldots, C(x,y)\}$. Let $\alpha_1, \alpha_2, \ldots, \alpha_n$ be a path in $\Gamma^i_{x,y}$ connecting $\alpha = \alpha_1$ with $\beta = \alpha_n$. Since $\{\alpha_i, \alpha_{i+1}\} \in E_{x,y}$ then the image of this path through the action $. \, a$ is also a path contained in some connected component which contains also α thus this path is contained in $\Gamma^j_{x,y}$, whence $\beta \, . \, a \in \Gamma^j_{x,y}$. The last statement is also an easy consequence of the definition of syntactic graph. □

Definition 3. *Let $\mathscr{G}$ be the set of all syntactic graphs of the automaton $\mathscr{A}$. We define a preorder in $\mathscr{G}$ by $\mathfrak{G}_{x,y} \preceq \mathfrak{G}_{x',y'}$ if for every $i \in \{1, \ldots, C(x,y)\}$ there is a $j \in \{1, \ldots, C(x',y')\}$ such that $\Gamma^i_{x,y} \subseteq \Gamma^j_{x',y'}$. Equivalently $\mathfrak{G}_{x,y} \preceq \mathfrak{G}_{x',y'}$ if and only if the partition induced by the connected components of $\mathfrak{G}_{x,y}$ is a refinement of the partition induced by the connected components of $\mathfrak{G}_{x',y'}$, i.e. $\sigma_{x,y} \subseteq \sigma_{x',y'}$.*

We have the following lemma.

Lemma 1. *Let x, y be two distinct states of the automaton $\mathscr{A} = \langle Q, \Sigma, \delta \rangle$ and let $\Gamma^i_{x,y}$ be a connected component of $\mathfrak{G}_{x,y}$. For any pair $x', y' \in \Gamma^i_{x,y}$ of distinct states we have $\mathfrak{G}_{x',y'} \preceq \mathfrak{G}_{x,y}$.*

Proof. We remark that the statement of the lemma is equivalent to the following: if x', y' are two distinct states with $(x', y') \in \sigma_{x,y}$ then $\sigma_{x',y'} \subseteq \sigma_{x,y}$. By Proposition 2, $\sigma_{x,y}$ is a congruence, moreover by the definition $x'\sigma_{x,y}y'$. Since $\sigma_{x',y'}$ is the smallest congruence which identifies x' with y', then $\sigma_{x',y'} \subseteq \sigma_{x,y}$, i.e. $\mathfrak{G}_{x',y'} \preceq \mathfrak{G}_{x,y}$. □

We remark that the relation $\equiv$ defined on $(\mathscr{G}, \preceq)$ by $\mathfrak{G}_{x',y'} \equiv \mathfrak{G}_{x,y}$ if $\mathfrak{G}_{x',y'} \preceq \mathfrak{G}_{x,y}$ and $\mathfrak{G}_{x,y} \preceq \mathfrak{G}_{x',y'}$ is an equivalence relation such that $\mathscr{G}/\equiv$ is endowed with an obvious partial order which is isomorphic to the partial order $(\{\sigma_{x,y} : x \neq y\}, \subseteq)$. In view of Proposition 2 it is not difficult to see that $\mathcal{M}(\mathscr{A})$ coincides with the set of minimal elements of $(\{\sigma_{x,y} : x \neq y\}, \subseteq))$. We have the following property.

Proposition 3. *The family of equivalence relations $\mathcal{M}(\mathscr{A})$ is orthogonal. Moreover the automaton $\mathscr{A} = \langle Q, \Sigma, \delta \rangle$ is never-minimal if and only if for any non-empty subset $F \subseteq Q$ there is a $\sigma \in \mathcal{M}(\mathscr{A})$ such that F is union of equivalence classes of σ.*

Proof. The intersection of two minimal non-trivial congruences is the identity congruence, thus $\mathcal{M}(\mathscr{A})$ is an orthogonal family. The last statement is a consequence of Theorem 1 and the definition of $\mathcal{M}(\mathscr{A})$. □

3 The Rainbow Bipartite Subgraph Problem

In this section we introduce some graph theoretic problems and we study their computational complexity class. In this paper a colored graph is a pair (G, φ) where $G = (V, E)$ is a graph and φ is a function, called coloring, from the set of edges E to a set $C = \{1, \ldots, N\}$ of colors. This definition can be extended to the case of list colored graphs (G, φ), where the list coloring is a function $\varphi : E \to 2^C$. For each $i \in C$ by $G(i)$ we denote the maximal subgraph of G formed by the edges whose lists contain i. We call the subgraphs $G(i) = (V, E_{G(i)}), i = 1, \ldots, n$, the *maximal monochromatic components* of (G, φ). It is clear that a list coloring φ is completely described by all the maximal monochromatic components $\{G(i)\}_{i \in C}$. Namely $\varphi(\{\alpha, \beta\}) = \{i \in C : \{\alpha, \beta\} \in E_{G(i)}\}$.

We say that a coloring φ is *splittable* if there is a partition of the set of vertices of V into two sets V_1, V_2 such that for any $i \in \{1, \ldots, n\}$ there is an edge $\{v_1, v_2\}$ with $v_1 \in V_1, v_2 \in V_2$ such that $\varphi(\{v_1, v_2\}) = i$. In this case we say that the partition V_1, V_2 splits φ. A more graph-theoretic way to see this property is via the concept of rainbow subgraphs (cf. [2]), i.e. subgraphs (in the weakest sense) such that all the edges are colored differently. Indeed a coloring φ is splittable if there is a bipartite rainbow subgraph of G colored using all the colors $\{1, \ldots, n\}$. Everything extends naturally to the case of the list coloring. Indeed we say that a list coloring φ is *list splittable* (for short splittable) if there is a partition of V into two sets V_1, V_2 such that for any $i \in \{1, \ldots, N\}$ there is an edge $\{v_1, v_2\}$

with $v_1 \in V_1, v_2 \in V_2$ such that $i \in \varphi(\{v_1, v_2\})$. Also in this case a similar characterization holds in terms of bipartite subgraphs. Indeed a list coloring φ is splittable if there is a bipartite subgraph such that each color is contained in the color list of some edge. Extending the definition of n-bounded coloring (cf. [2]) from colored graphs to list colored graphs, we say that a list coloring φ of a graph G is n-bounded if each color appears at most n times in the lists of colors associated to the edges.

With the previous definitions it makes sense defining the problem `SPLITTABLE`. This is the problem of determining, given a colored graph (G, φ), whether φ is splittable. Analogously `LIST-SPLITTABLE` is the problem of determining, given a list colored graph (G, φ) whether φ is list splittable. The n-`SPLITTABLE` problem is the sub-problem of checking, given an n-bounded colored graph (G, φ), whether φ is splittable. n-`LIST-SPLITTABLE` is defined analogously. We say that the (list) coloring φ on a graph $G = (V, E)$ is *anti-incidence* if for all pairs of incident edges $\{v, v_1\}, \{v, v_2\}$, $\varphi(\{v, v_1\}) \cap \varphi(\{v, v_2\}) = \emptyset$.

Although `LIST-SPLITTABLE` may appear a more difficult problem with respect to `SPLITTABLE`, the following proposition shows that this is not the case.

Proposition 4. *There is a reduction η from `LIST-SPLITTABLE` to `SPLITTABLE` bringing an n-bounded list colored graph (G, φ) into an n-bounded colored graph (G', φ'). Moreover we can find a reduction η' which brings an n-bounded colored graph (G, φ) into an n-bounded colored graph (G'', φ'') such that φ'' is anti-incidence.*

Proof. Given an instance (G, φ) with $G = (V, E)$ of `LIST-SPLITTABLE`, where φ is a list coloring on the set of colors $C = \{1, \ldots, N\}$, we reduce it to an instance (G', φ') with $G' = (V', E')$ of `SPLITTABLE`. Starting from (G, φ) we iteratively apply the following construction. For each edge $\{v, v'\}$ of a list colored graph (H, ψ) with $H = (Y, T)$ such that $\psi(\{v, v'\}) = \{i_0, \ldots, i_k\}$ with $k \geq 1$, we add $k+1$ new vertices $\overline{v}, v_1, \ldots, v_k$, $2k+1$ new edges and $k+1$ new colors $c_0, \ldots, c_k$. For each $j \in \{1, \ldots, k\}$ we add an edge $\{v', v_j\}$ colored by $\psi'(\{v', v_j\}) = i_j$ and take $\psi'(\{v', v\}) = i_0$. Putting $v_0 = v$ we also add for each $j \in \{0, \ldots, k\}$ edges $\{v_j, \overline{v}\}$ colored by $\psi'(\{v_j, \overline{v}\}) = c_j$ (see Fig. 1). Leaving $\psi = \psi'$ for the other edges, we obtain a new list colored graph (H', ψ') with $H' = (Y', T')$ such that $|T'| - |T| \leq 2N - 1$ and the number of added colors is upper bounded by N.

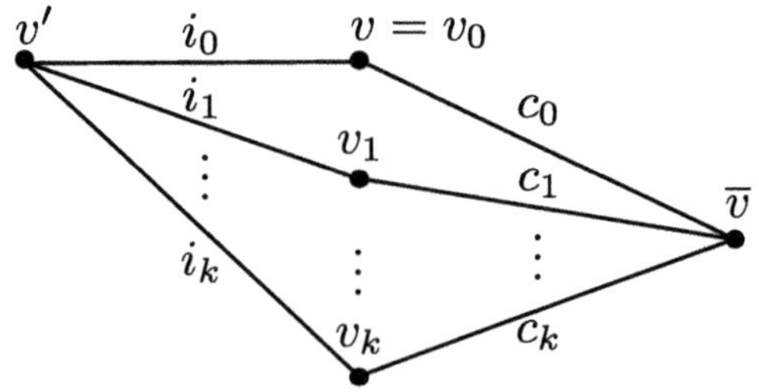

Fig. 1. One step of the iterated construction of Proposition 4

It is clear that after at most $|E|$ iterations we get a colored graph (G', φ') with $|E'| \leq 2N|E|$ whose number of colors is upper bounded by $N|E| + N$. Then to prove that η is a reduction we have to show that φ is list splittable iff φ' is splittable. To prove it, it is enough to show that the splittability property is preserved in each step of the previous iteration. Indeed suppose that ψ' is list splittable in H' and let Y_1', Y_2' be the associated partition. We observe that, since the edges $\{v_j, \overline{v}\}$ for $j \in \{0, \dots, k\}$ are the only edges in H' colored by $c_0, \dots, c_k$, then $\overline{v} \in Y_l'$ for some $l \in \{1, 2\}$ iff $v_0, \dots, v_k \in Y_{3-l}'$. We claim that $Y \cap Y_1', Y \cap Y_2'$ is a partition that splits ψ. Indeed let $c \in \{1, \dots, N\}$. Since ψ' is splittable there is an edge $\{w, z\}$ with $c \in \psi'(\{w, z\})$ and $w \in Y_l', z \in Y_{3-l}'$. If $w, z \in Y$ we are done, hence we may assume without loss of generality that $w = v' \in Y$ and z belongs to the added vertices $\{v_1, \dots, v_k\}$, say $z = v_j$. Since $w = v' \in Y_l' \cap Y$ and $z = v_j \in Y_{3-l}'$, it follows from the previous remark that $\overline{v} \in Y_l'$ and $v \in Y_{3-l}' \cap Y$. Since $c \in \psi(\{v, v'\})$, we get that $Y \cap Y_1', Y \cap Y_2'$ splits ψ.

Conversely suppose that Y_1, Y_2 splits the list coloring ψ. Suppose that $\{v, v'\} \in T$ with $v \in Y_l, v' \in Y_k$ for some $k, l \in \{1, 2\}$. We put all the added vertices $\{v_1, \dots, v_k\}$ in Y_l' and $\overline{v}$ in Y_{3-l}'. In this way all the added colors are splitted by Y_l', Y_{3-l}'. Thus, by the construction of the graph H', we get that Y_l', Y_{3-l}' splits ψ'.

It is easy to check that if φ is an n-bounded coloring then also φ' is n-bounded.

The last statement can be obtained in a similar way. Indeed starting from the previous colored graph (G', φ'), to obtain (G'', φ'') it is enough to apply the following construction iteratively. Suppose that the colored graph (H, ψ) has a vertex v such that there are two incident edges $\{v, v_1\}, \{v, v_2\}$ with $\varphi(\{v, v_1\}) = \varphi(\{v, v_2\}) = i$. Then build a new colored graph (H', ψ') erasing from H the edge $\{v, v_2\}$ and adding two new vertices $\overline{v}, v'$ and the following three new edges: $\{v', v_2\}$, colored by $\psi'(\{v', v_2\}) = i$, and the edges $\{\overline{v}, v\}, \{\overline{v}, v'\}$ colored by two new colors $\psi'(\{\overline{v}, v\}) = c$, $\psi'(\{\overline{v}, v'\}) = c'$. Since c, c' are new colors it is easy to see that in a splitting, v, v' belong to the same component of the partition. Therefore ψ' is spittable iff ψ is splittable. Since in each iteration we reduce the number of incident edges having the same colors, the number of iterations is upper bounded by $N(|V'| - 2)$ where N is the number of colors of (G', φ') and $G' = (V', E')$. Thus $|E''| \leq 2N(|V'| - 2)$ and so η'' is a reduction. □

Since `SPLITTABLE` is a sub-problem of `LIST-SPLITTABLE`, from Proposition 4 we have that actually `SPLITTABLE` and `LIST-SPLITTABLE` are equivalent problems. It is easy to see that both 1-`LIST-SPLITTABLE` and 1-`SPLITTABLE` coincide with the problem of checking if a graph is bipartite and so they belong to the computational class **P**. The following theorem shows that things change radically when we consider 2-`SPLITTABLE`, indeed we have the following:

Theorem 2. *2-`SPLITTABLE` is **NP**-complete.*

Proof. The problem is clearly in **NP**. To prove the completeness we reduce `NAESAT` to 2-`SPLITTABLE`. `NAESAT` is the problem of checking, given a boolean formula in CNF

$$\mathcal{F} = \bigwedge_{i=1}^{m} \mathcal{C}_i$$

where in each clause C_i there are three literals, whether there is a truth assignment such that in no clause all three literals are equal in truth value (neither all true nor all false). This is a well know **NP**-complete problem [3].

Suppose that the boolean formula $\mathcal{F}$ is on the set of variables $X = \{x_1, \ldots, x_n\}$. We build a graph $G = (V, E)$ and a 2-bounded list coloring φ in the following way. The set of vertices is $V = X \cup \{\neg x : x \in X\}$. The set of colors $C = \{1, \ldots, m, t_1, \ldots, t_n\}$ corresponds to the set of clauses and the set of variables. The set of edges contains all pairs $\{x, \neg x\}$ for $x \in X$ and $t_j \in \varphi(\{x_j, \neg x_j\})$ for all $j \in \{1, \ldots, n\}$. Moreover for each color $i \in \{1, \ldots, m\}$ suppose that the clause $\mathcal{C}_i = l_1^i \vee l_2^i \vee l_3^i$. Then $\{l_1^i, l_2^i\}, \{l_2^i, l_3^i\} \in E$ and $i \in \varphi(\{l_1^i, l_2^i\}), i \in \varphi(\{l_2^i, l_3^i\})$. Clearly (G, φ) is a 2-bounded list colored graph. Let us prove that φ is list splittable if and only if there is a truth assignment such that in no clause all the three literals are all equal in truth value.

Suppose that φ is list splittable, and consider the corresponding partition of V into two disjoint sets V_1, V_2. Since for each $i = 1, \ldots, n$, t_i is contained only in the list coloring of the the edge $\{x_i, \neg x_i\}$, it is clear that if a literal $l \in V_1$ then $\neg l \in V_2$. Therefore there is a truth assignment that makes (for instance) all the literals of V_1 true and all the literals of V_2 false. Since φ is list splittable, for each clause $\mathcal{C}_i = l_1^i \vee l_2^i \vee l_3^i$ there is one edge $\{\alpha, \beta\}$ among $\{l_1^i, l_2^i\}, \{l_2^i, l_3^i\}$ such that $\alpha \in V_1, \beta \in V_2$. Hence in each clause there is a literal which is true and one which is false.

Conversely suppose that there is a truth assignment such that in no clause all the three literals are equal in truth value. Let V_1 be the set of literals in V that are true and $V_2 = V \setminus V_1$ be the one that are false. Clearly if $l \in V_1$ then $\neg l \in V_2$ thus the colors $t_i, i = 1, \ldots, n$ which are contained only in the edges $\{x_i, \neg x_i\}$ are clearly splitted. Moreover in each clause $\mathcal{C}_i = \{l_1^i, l_2^i, l_3^i\}$ there are two literals say $\alpha, \beta \in \mathcal{C}_i$ which are respectively true and false, whence $\alpha \in V_1, \beta \in V_2$. Since $\{l_1^i, l_2^i\}, \{l_2^i, l_3^i\} \in E$ it is not difficult to see that there is an edge $\{\alpha', \beta'\} \in \{\{l_1^i, l_2^i\}, \{l_2^i, l_3^i\}\}$ with $\alpha' \in V_1, \beta' \in V_2$ and so the color i is splitted. Since this holds for all the colors $i \in \{1, \ldots, m\}$ we can conclude that V_1, V_2 splits φ.

By now we have reduced `NAESAT` to `LIST-SPLITTABLE` with a 2-bounded list coloring graph. Hence by Proposition 4 we can reduce `NAESAT` to 2-`SPLITTABLE`. □

Let (G, φ) be a list colored graph with $G = (V, E)$ and $\varphi : E \to 2^C$ for some set of colors C. We say that φ is *chromatic-transitive* if for any $i \in C$ the connected components of $G(i)$ are complete subgraphs. Equivalently iff the associated relation $\mathcal{R}(G(i))$ is an equivalence relation on V. Therefore we can define the chromatic-transitive closure of (G, φ) as the list colored graph $(\overline{G}, \overline{\varphi})$ with vertex set V and whose maximal monochromatic components are

$$\left\{ \mathcal{G}\left(\overline{\mathcal{R}(G(i))}^{tr} \right) \right\}_{i \in C}$$

where $\cdot^{tr}$ is the transitive closure operator. The definition of chromatic-transitive closure is interesting under the following proposition.

Proposition 5. *Let (G,φ) be a list colored graph. Then (G,φ) is list splittable if and only if $(\overline{G},\overline{\varphi})$ is list splittable.*

Proof. Since G is a subgraph of $\overline{G}$ and $\overline{\varphi}$ is an extension of φ, then if (G,φ) is list splittable then also $(\overline{G},\overline{\varphi})$ is list splittable. Conversely, assume that $(\overline{G},\overline{\varphi})$ is list splittable and let V_1, V_2 be a corresponding partition of V. Given a color $i \in C$, we have an edge $\{v_0, w\}$ in $\overline{G}(i)$ with $v_0 \in V_1$ and $w \in V_2$. Then we have a sequence of edges $\{v_0, v_1\}, \ldots, \{v_{m-1}, v_m\}$ in $G(i)$ with $v_m = w$. Since $v_0 \in V_1$ and $w \in V_2$, there is a $j \in \{1, \ldots, m\}$ such that $(v_{j-1}, v_j) \in V_1 \times V_2 \cup V_2 \times V_1$ and so (G,φ) is list splittable by considering the partition V_1, V_2. □

In view of this proposition and the fact that a list coloring is determined by its maximal monochromatic components we define the problem `SEPARATING`. Given a set $\mathfrak{R}$ of equivalence relations on a set V as input, `SEPARATING` is the problem of checking whether or not there is a set $F \subseteq V$ which is not saturated by any equivalence relation of $\mathfrak{R}$, i.e. for any $\sigma \in \mathfrak{R}$, F is not union of equivalence classes of σ. The following lemma is an easy consequence of the definitions.

Lemma 2. *Let $\mathfrak{R} = \{\sigma_1, \ldots, \sigma_k\}$ be a family of equivalence relations on a set V and let (G,φ) be the associated list colored graph:*

$$G = \mathcal{G}(\cup_{\sigma\in\mathfrak{R}}\sigma),\ \varphi(\{x,y\}) = \{i \in \{1,\ldots,k\} : (x,y) \in \sigma_i\}$$

There is a set $F \subseteq V$ which is not saturated by any $\sigma \in \mathfrak{R}$ if and only if (G,φ) is splittable.

By Proposition 5 and Lemma 2 it immediately follows that `LIST-SPLITTABLE` and `SEPARATING` are equivalent problems. The problem `ORTHOGONAL-SEPARATING` is the analogous problem but with the difference that the input $\mathfrak{R}$ is a family of orthogonal equivalence relations. This problem has been introduced in [1] under the name `DISJUNCTIVE SET` as a clue of the possible **NP**-completeness of the `SYNTACTIC MONOID` problem. The same article provides a proof of the **NP**-completeness of `DISJUNCTIVE SET`, even when the instances of the problem are restricted to families of the following kind:

1. (Corollary 2.5 [1]) Orthogonal families $\mathfrak{R}$ such that each equivalence relation $\sigma \in \mathfrak{R}$ has at most three non-singleton classes and these non-singleton classes have exactly two elements.
2. (Corollary 2.4 [1]) Orthogonal families $\mathfrak{R}$ such that each equivalence relation $\sigma \in \mathfrak{R}$ has exactly one non-singleton class and this non-singleton class has at most four elements.

We remark that `ORTHOGONAL-SEPARATING` restricted to the case (2) with families $\mathfrak{R}$ such that each equivalence relation $\sigma \in \mathfrak{R}$ has exactly one non-singleton class and this non-singleton class has at most two elements elements is equivalent to 1-`SPLITTABLE`, i.e. to check whether or not the associated colored graph as in Lemma 2 is bipartite. Therefore in this case ORTHOGONAL-SEPARATING is in **P**. The same occurs if we restrict it to the case (1) allowing the equivalence relations to have at most one non-trivial equivalence class. The following theorem establishes the exact borderline for which ORTHOGONAL-SEPARATING turns out to be **NP**-complete.

Theorem 3. *ORTHOGONAL-SEPARATING is still **NP**-complete if we assume*

1. *Orthogonal families $\mathfrak{R}$ such that each equivalence relation $\sigma \in \mathfrak{R}$ has at most two non-singleton classes and these non-singleton classes have exactly two elements.*
2. *Orthogonal families $\mathfrak{R}$ such that each equivalence relation $\sigma \in \mathfrak{R}$ has exactly one non-singleton class and this non-singleton class has at most three elements.*

Proof. We prove the statement (1). Case (2) can be obtained analogously by the structure of the obtained graph of Theorem 2, Lemma 2 and a similar technique of Proposition 4 to pass from a list coloring to a coloring.

By Theorem 2, 2-SPLITTABLE is **NP**-complete. Thus given a 2-bounded coloring graph (G, φ) with $G = (V, E)$ and $\varphi : E \to C$, by Proposition 4 we can suppose without loss of generality that φ is anti-incidence. Therefore for each color $i \in C$, $\overline{\mathcal{R}(G(i))}^{tr} = \mathcal{R}(G(i))$ and so the transitive closure $(\overline{G}, \overline{\varphi})$ is equal to (G, φ). Moreover since (G, φ) is a colored graph, the associated family of equivalence relations $\mathfrak{R} = \{\mathcal{R}(G(i)) : i \in C\}$ is orthogonal. By Lemma 2, $\mathfrak{R}$ is in ORTHOGONAL-SEPARATING if and only if φ splits. Since φ is a 2-bounded coloring and $(\overline{G}, \overline{\varphi}) = (G, \varphi)$, it is also evident that the family $\mathfrak{R} = \{\mathcal{R}(G(i)) : i \in C\}$ is formed by equivalence classes composed by at most two elements and there are at most two equivalence classes which are not singletons. □

4 NP-Completeness of co-NEVER-MINIMAL

In this section we show that, given an orthogonal family $\mathfrak{R}$ of equivalence relations, we can always build a strongly connected DFA $\mathscr{A}$ having $\mathfrak{R}$ as the set of minimal non-trivial congruences of $\mathscr{A}$. Indeed we have the following theorem.

Theorem 4. *Let $\mathfrak{R} = \{\sigma_1, \ldots, \sigma_m\}$ be an orthogonal family of equivalence relations on a set Q and let $G = \mathcal{G}(\cup_{\sigma \in \mathfrak{R}} \sigma)$ with $G = (Q, E)$. There is a strongly connected DFA $\mathscr{A}_{\mathfrak{R}} = \langle Q, \Sigma, \delta \rangle$ with*

$$|\Sigma| \leq 3 \sum_{\sigma \in \mathfrak{R}} |Q/\sigma| + \binom{|Q|}{2} - |E|$$

such that $\mathcal{M}(\mathscr{A}_{\mathfrak{R}}) = \mathfrak{R}$.

Proof. For each $\sigma \in \mathfrak{R}$ suppose that in Q/σ there are $[q_0]_\sigma, \ldots, [q_{t-1}]_\sigma$ non singleton classes and $[q_t]_\sigma, \ldots, [q_n]_\sigma$ singleton classes, where $n = |Q/\sigma|$. Putting $[q_i]_\sigma = \{q_i^0, \ldots, q_i^{n_i - 1}\}$ for $i = 0, \ldots, t-1$, we define an alphabet

$$\Sigma_\sigma = \{a(\sigma)_0, b(\sigma)_0, \ldots, a(\sigma)_{t-1}, b(\sigma)_{t-1}, c(\sigma)_0, \ldots c(\sigma)_{t-1}, d(\sigma)_{t-1}, \ldots, d(\sigma)_n\}$$

Notice that $|\Sigma_\sigma| = 2t + (n+1) \leq 3n$ if $t < n$, otherwise $|\Sigma_\sigma| = 3n$. The action of Σ_σ on Q is defined by the following rules

$$\delta(q_i^j, a(\sigma)_i) = q_i^{j+1 \bmod n_i},$$

$$\delta(q_i^0, b(\sigma)_i) = q_i^0, \delta(q_i^{n_i-1}, b(\sigma)_i) = q_i^0, \delta(q_i^s, b(\sigma)_i) = q_i^{s+1} \text{ for } s = 1, \ldots, n_i - 2,$$

$$\delta(q_i^0, c(\sigma)_i) = q_{i+1 \bmod t}^0,\ \delta(q_i^1, c(\sigma)_i) = q_{i+1 \bmod t}^1,$$

$$\delta(q_{t-1}^0, d(\sigma)_{t-1}) = q_t, \delta(q_n, d(\sigma)_n) = q_{t-1}^0,\ \delta(q_i, d(\sigma)_i) = q_{i+1} \text{ for } t \leq i < n.$$

The alphabet $\Sigma = \cup_{\sigma \in \mathfrak{R}} \Sigma_\sigma \cup \Sigma'$ is the disjoint union of the alphabets $\Sigma_\sigma, \sigma \in \mathfrak{R}$ and an alphabet Σ' used to satisfy the minimality condition $\mathcal{M}(\mathscr{A}_\mathfrak{R}) = \mathfrak{R}$. The action of Σ' is defined in the following way. For any pair p, q of distinct elements such that (p, q) does not belong to $\cup_{\sigma \in \mathfrak{R}} \sigma$ we have to satisfy the condition $\sigma \subseteq \sigma_{p,q}$ for some $\sigma \in \mathfrak{R}$. Therefore, if $G = (Q, E) = \mathcal{G}(\cup_{\sigma \in \mathfrak{R}} \sigma)$, we define $\Sigma' = \{a(p,q) : \{p,q\} \notin E\}$. Regarding the action we first fix two states $\overline{q}, \overline{q}'$ such that $(\overline{q}, \overline{q}') \in \overline{\sigma}$ for some $\overline{\sigma} \in \mathfrak{R}$ and we put

$$\delta(p, a(p,q)) = \overline{q},\ \delta(q, a(p,q)) = \overline{q}'.$$

It is not difficult to see that $\mathscr{A}$ is strongly connected (in particular it is strongly connected even if we restrict the action to Σ_σ for any $\sigma \in \mathfrak{R}$). Moreover the action δ is transitive on each $\sigma \in \mathfrak{R}$ in the sense that if $(x,y), (x',y') \in \sigma$ are two distinct pairs with $x \neq y$ and $x' \neq y'$, then there is a word $w \in \Sigma_\sigma^*$ such that $\delta(\{x,y\}, w) = \{x', y'\}$. Moreover since $\mathfrak{R}$ is orthogonal the action of any letter in $\Sigma \setminus \Sigma_\sigma$ brings any pair of distinct elements $\{x,y\}$ with $(x,y) \in \sigma$ into a singleton. Thus for each $\sigma \in \mathfrak{R}$ and for each pair $(x,y) \in \sigma$ with $x \neq y$, we have $\sigma_{x,y} = \sigma$. The minimality condition is also satisfied since for any $\{p,q\} \notin E$ we have $\mathfrak{G}_{\overline{q},\overline{q}'} \preceq \mathfrak{G}_{p,q}$, i.e. $\overline{\sigma} \subseteq \sigma_{p,q}$, hence $\mathcal{M}(\mathscr{A}_\mathfrak{R}) = \mathfrak{R}$. Moreover, since $|\Sigma_\sigma| \leq 3|Q/\sigma|$ for all $\sigma \in \mathfrak{R}$ and $|\Sigma'| = \binom{|Q|}{2} - |E|$ the bound for $|\Sigma|$ of the statement immediately follows. □

This theorem gives also a way to build never-minimal automata. Indeed to build a never-minimal DFA, by Proposition 3, it is enough to consider an orthogonal family $\mathfrak{R}$ which is not in `ORTHOGONAL-SEPARATING`, and then apply the construction of Theorem 4 to $\mathfrak{R}$. A very simple class of orthogonal families which are not `ORTHOGONAL-SEPARATING` are the families obtained from graphs which are not bipartite. Indeed consider a non-bipartite graph $G = (V, E)$, color it using the identity map $1_E : E \to E$ where E is now the set of colors. Therefore the set $\mathfrak{R}_G = \{\mathcal{R}(G(e)) : e \in E\}$ is clearly an orthogonal family of equivalence relations which saturates all the subsets of V since, by Lemma 2, $(G, 1_E)$ is not splittable. We also remark that condition C_3 defined in [4] is translated to the condition that the colored graph associated to the DFA which satisfies C_3 contains a rainbow triangle and no other edge is colored using colors of this triangle, whence this graph is clearly non-splittable and so the automaton is never-minimal. We conclude the section with the following consequences of Theorem 4.

Theorem 5. *co-`NEVER-MINIMAL` is equivalent to `ORTHOGONAL-SEPARATING`.*

Proof. By Proposition 3, $\mathscr{A} = \langle Q, \Sigma, \delta \rangle$ is not never-minimal if and only if $\mathcal{M}(\mathscr{A})$ is a family of orthogonal relations belonging to ORTHOGONAL-SEPARATING. It is straightforward to check that computing $\mathcal{M}(\mathscr{A})$ is a polynomial time and space task since the construction of the syntactic graphs is polynomial and there are at most $\binom{|Q|}{2}$ such graphs. Thus co-NEVER-MINIMAL is reducible to ORTHOGON AL-SEPARATING. Conversely given an orthogonal family $\mathfrak{R} = \{\sigma_1, \ldots, \sigma_m\}$ of equivalence relations on a set Q by Theorem 4 we can build in polynomial time an automaton $\mathscr{A}_{\mathfrak{R}} = \langle Q, \Sigma, \delta \rangle$ with $\mathcal{M}(\mathscr{A}_{\mathfrak{R}}) = \mathfrak{R}$. By Proposition 3, $\mathfrak{R}$ is in ORTHOGONAL-SEPARATING if and only if $\mathscr{A}_{\mathfrak{R}}$ is not never-minimal. □

From Theorems 3, 5 we have also the following consequence.

Corollary 1. *co-NEVER-MINIMAL is **NP**-complete.*

5 Connections with the Syntactic Monoid Problem

As already mentioned in [4], NEVER-MINIMAL is related to the SYNTACTIC MONOID problem (cf. [1]). A finite monoid $(M, \cdot)$ is called syntactic if there is a $P \subseteq M$ such that the congruence $\sim_P$ on M defined by

$$x \sim_P y \Leftrightarrow \forall a, b \in M(axb \in P \Leftrightarrow ayb \in P)$$

is the identity congruence on M. If a monoid M is the transition monoid of a DFA $\mathscr{A}$, and it is not syntactic, then $\mathscr{A}$ is never-minimal. However the problem of the positioning among the complexity classes of this problem is still open.

As a concluding result we give a characterization of the monoids which are not syntactic in terms of never-minimal automata. This result places the SYNTACTIC MONOID problem as a subproblem of NEVER-MINIMAL problem when the class of automata considered are the two-side Cayley automata associated to a monoid.

We consider complete automata here, hence an automaton $\langle Q, \Sigma, \delta, i, F \rangle$ is minimal if and only if every vertex is accessible from the initial vertex i and the Nerode equivalence defined by $p \mathcal{N} q$ if $p^{-1}F = q^{-1}F$ is the identity relation on Q. Let $(M, \cdot)$ be a finite monoid and let $A \subseteq M$ be a set of generators for M. The *two-side Cayley automaton* of M is the automaton $\hat{\Gamma}_A(M) = \langle M, A \cup A', \delta \rangle$ where $A' = \{a' : a \in A\}$ is a disjoint copy of A and $\circ'$ is an involution in $(A \cup A')^*$ such that $(uv)' = v'u'$ for all $u, v \in (A \cup A')^*$. The action δ is defined by:

$$\forall a \in A,\ \delta(u, a) = u \cdot a, \quad \forall a \in A',\ \delta(u, a') = a \cdot u$$

We have the following characterization.

Theorem 6. *M is not syntactic if and only if $\hat{\Gamma}_A(M)$ is never-minimal.*

Proof. Assume first that M is syntactic. Then there exists some $P \subseteq M$ such that $\sim_P = 1_M$. We claim that $(\hat{\Gamma}_A(M), 1, P)$ is a minimal automaton. Since every vertex is accessible from 1, it remains to show that the Nerode equivalence is trivial. Let $u, v \in M$ be different. Then $(x, y) \notin \sim_P$, hence we may assume without loss of generality that $xuy \in P$ and $xvy \notin P$ for some $x, y \in M$. It follows that $\delta(u, x'y) = xuy \in P$ and $\delta(v, x'y) = xvy \notin P$, hence $(u, v) \notin \mathcal{N}$ as required.

Conversely suppose that $(\hat{\Gamma}_A(M), i, P)$ is a minimal automaton. We claim that $\sim_P = 1_M$. Indeed, take $u, v \in M$ distinct. Since $(u, v) \notin \mathcal{N}$, we may assume that $\delta(u, w) \in P$ and $\delta(v, w) \notin P$ for some $w \in (A \cup A')^*$. Now there exist $x, y \in M$ such that $\delta(u, w) = xuy$ and $\delta(v, w) = xvy$, hence $u \not\sim_P v$ and so $\sim_P = 1_M$. Therefore M is syntactic. □

6 Conclusions and Open Problems

In Theorem 4 it is shown how to build, from an orthogonal family $\mathfrak{R}$ of equivalence relations on a set of n elements, an automaton $\mathcal{A}_{\mathfrak{R}}$ with $\mathcal{M}(\mathcal{A}_{\mathfrak{R}}) = \mathfrak{R}$ over an alphabet of dimension $O(n^2)$. This is a key fact to prove the **NP**-completeness of co-`NEVER-MINIMAL`. In Theorem 6 the `SYNTACTIC MONOID` problem on a monoid M is reduced to check whether or not an automaton over an alphabet of $O(|M|)$ elements is never-minimal. Therefore, to have a more precise clue about the complexity class of the `SYNTACTIC MONOID` problem, it seems natural to ask if Theorem 4 still holds also in the case of automata over a constant or linear growing alphabets. This leads also to the problem of finding the complexity class of co-`NEVER-MINIMAL` in the class of automata with a constant or linear growing alphabets.

Acknowledgments

Both authors acknowledge the support of the Centro de Matemática da Universidade do Porto financed by FCT through the programmes POCTI and POSI, with Portuguese and European Community structural funds. The first-named author acknowledges also the support of the FCT project SFRH/BPD/65428/2009.

References

1. Goralcik, P., Koubek, V.: On the disjunctive set problem. Theoretical Computer Science 204, 99–118 (1998)
2. Kano, M., Li, X.: Monochromatic and heterochromatic subgraphs in edge-colored graphs-a survey. Graphs and Combinatorics 24, 237–263 (2008)
3. Papadimitriou, C.H.: Computational Complexity. Addison-Wesley Longman, Amsterdam (1995)
4. Restivo, A., Vaglica, R.: Automata with extremal minimality conditions. In: Gao, Y., Lu, H., Seki, S., Yu, S. (eds.) DLT 2010. LNCS, vol. 6224, pp. 399–410. Springer, Heidelberg (2010)

Boolean Algebras of Regular Languages

Victor Selivanov[1] and Anton Konovalov[1]

[1] A.P. Ershov Institute of Informatics Systems
Siberian Division Russian Academy of Sciences
and Novosibirsk State University
vseliv@iis.nsk.su
[2] Institute of Informatics Systems
Siberian Division Russian Academy of Sciences
Jack@sibmail.ru

Abstract. We characterize up to isomorphism the Boolean algebras of regular languages and of regular aperiodic languages, and show decidability of classes of regular languages related to these characterizations.

Keywords: Boolean algebra, Frechét ideal, regular language, aperiodic language.

1 Introduction

Boolean algebras (BA's) are of principal importance for several branches of mathematics. Accordingly, characterization of naturally arising BA's attracts attention of many researchers. As examples we mention characterizations of natural BA's in logic and computability theory [Han75, LPS02, Se91, Se92, Se03].

In automata theory, people consider many classes of languages which form BA's but apparently there was so far no attempt to characterize those BA's up to isomorphism. At the same time, such characterizations could be of some interest because they provide new information on well-known classes of regular languages. Due to Stone duality, this contributes to understanding of the corresponding Stone spaces which are closely related to the profinite topology [Pip97, GGP08].

In this note, we start investigations of such questions and characterize some fundamental BA's of regular languages up to isomorphism.

If $\mathbb{B}$ is a BA and α an ordinal, let $F_\alpha(\mathbb{B})$ be the α-th iterated Frechét ideal of $\mathbb{B}$, $\mathbb{B}^{(\alpha)} = \mathbb{B}/F_\alpha(\mathbb{B})$ is the α-th Frechét derivative of $\mathbb{B}$ and $\mathbb{B}' = \mathbb{B}^{(1)}$. (See [Go96] for a detailed treatment, some definitions are recalled in the next section.)

For a finite alphabet Σ, let $\mathbb{R}_\Sigma$ (resp. $\mathbb{A}_\Sigma$) denote the BA of all regular (resp. of all regular aperiodic) languages over Σ. The main result of this paper looks as follows:

Theorem 1. *1. For any alphabet Σ, $\mathbb{R}_\Sigma$ is an atomic BA with infinitely many atoms, and $\mathbb{R}'_\Sigma$ is a countable atomless BA.*
2. For a unary alphabet Σ, $\mathbb{A}_\Sigma$ is an atomic BA with infinitely many atoms, and $\mathbb{A}'_\Sigma$ is a two-element BA.

G. Mauri and A. Leporati (Eds.): DLT 2011, LNCS 6795, pp. 386–396, 2011.

3. *For any alphabet Σ with at least two symbols, $F_0(\mathbb{A}_\Sigma) \subset F_1(\mathbb{A}_\Sigma) \subset \cdots \subset F_\omega(\mathbb{A}_\Sigma) = F_{\omega+1}(\mathbb{A}_\Sigma)$, for each $n < \omega$ the BA $\mathbb{A}_\Sigma^{(n)}$ is atomic with infinitely many atoms, and $\mathcal{A}_\Sigma^{(\omega)}$ is a countable atomless BA.*

From the well-known facts on BA's (see Chapter 1 of [Go96]) it follows that the assertions 1 — 3 characterize the corresponding BA's up to isomorphism. Moreover, our proofs imply decidability of the classes of regular languages related to the main theorem.

While the assertions 1 and 2 above are in fact easy exercises, the assertion 3 is less trivial. The classes of languages $F_k(\mathbb{A}_\Sigma)$ turn out to be closely related to subclasses of sparse regular languages investigated in [SY92] (see also the surveys [Yu97, Pin1?]).

In Sections 2 and 3 we provide the necessary background on BA's and regular languages, respectively. In Section 4 we characterize the BA $\mathbb{R}_\Sigma$ and some other classes of regular languages. In Section 5 we characterize the BA $\mathbb{A}_\Sigma$ and some other classes of aperiodic languages, and we conclude in Section 6.

2 Preliminaries on Boolean Algebras

In this section we briefly recall some notions and facts on BA's used in the sequel. We assume the reader to be familiar with basic notions related to BA's like ideal of a BA, quotient-algebra of a BA modulo a given ideal, and canonical homomorphism of a BA onto its quotient-algebra (for detailed treatments of BA's see e.g. [Go96, Si64]). BA's are considered in the signature $\{\cup, \cap, \bar{}, 0, 1\}$.

It is often useful to extend the notion of a BA to the notion of Ershov algebra (EA for short). EA's, known also as relatively complemented distributive lattices with 0, are considered in the signature $\{\cup, \cap, \setminus, 0\}$. Up to isomorphism, EA's are the ideals of BA's with the natural interpretation of $\cup, \cap, \setminus, 0$. EA is a BA iff it has a greatest element. For any element a of a EA $\mathbb{A} = (A; \cup, \cap, \setminus, 0)$, the set $\hat{a} = \{x \mid x \leq a\}$ with the induced operations is a BA. Here $\leq$ is the partial order on A such that $x \leq y$ iff $x \cap y = x$. If $a \in A$ and I is an ideal of $\mathbb{A}$, a/I denotes the correspondent element of the quotient-algebra $\mathbb{A}/I$.

Recall that an element a of a EA $\mathbb{A}$ is an *atom* if $a \neq 0$ and $x < a$ implies $x = 0$. A EA $\mathbb{A}$ is called *atomless* if it has no atom, and it is called *atomic* if below any non-zero element there is an atom. The ideal of a EA $\mathbb{A}$ generated by atoms is called *Frechét ideal* of $\mathbb{A}$. It consists of all finite unions (including the empty union which coincides with 0) of atoms and it is denoted by $F(\mathbb{A})$. The quotient-algebra $\mathbb{A}/F(\mathbb{A})$ is called the *Frechét derivative* of $\mathbb{A}$ and is denoted by $\mathbb{A}'$. The following result is well known [Go96]:

Proposition 1. *For any countable atomic EA's $\mathbb{A}$ and $\mathbb{B}$, $\mathbb{A} \simeq \mathbb{B}$ iff $\mathbb{A}' \simeq \mathbb{B}'$.*

Define the transfinite sequence $\{F_\alpha(\mathbb{A})\}$ of *iterated Frechét ideals* of a EA $\mathbb{A}$ as follows: $F_0(\mathbb{A}) = \{0\}$, $F_{\beta+1}(\mathbb{A}) = \{x \mid x/F_\beta(\mathbb{A}) \in F(\mathbb{A}^{(\beta)})\}$ where $\mathbb{A}^{(\beta)} = \mathbb{A}/F_\beta(\mathbb{A})$, and $F_\alpha(\mathbb{A}) = \bigcup_{\beta<\alpha} F_\beta(\mathbb{A})$ for a limit ordinal α. This sequence is ascending under inclusion, and $F_\alpha(\mathbb{A}) = F_{\alpha+1}(\mathbb{A})$ for some α; the least ordinal

α with this property is called the *ordinal type of* $\mathbb{A}$ and is denoted $\sigma(\mathbb{A})$. Note that if $\mathbb{A}$ is a BA then the ordinal $\sigma(\mathbb{A})$ is not limit (i.e., it is either zero or a successor).

A EA $\mathbb{A}$ is called *superatomic* if $\mathbb{A}^{(\alpha)}$ is singleton for some α. Relate to any such EA its *superatomicity type* as follows: if $\mathbb{A}^{(\gamma)}$ is not a BA for each $\gamma < \sigma(\mathbb{A})$ then the type is $(\sigma(\mathbb{A}), \sigma(\mathbb{A}), 0)$ (in this case $\sigma(\mathbb{A})$ is a limit ordinal), otherwise the type is $(\sigma(\mathbb{A}) - 1, \beta, k)$ where β is the least ordinal for which $\mathbb{A}^{(\beta)}$ is a BA and k is the number of atoms in the finite BA $\mathbb{A}^{(\sigma(\mathbb{A})-1)}$ (in this case $\sigma(\mathbb{A})$ is a successor ordinal). The next characterization of countable superatomic EA's is due to Yu.L. Ershov (cf. Proposition 1.8.4 of [Go96]).

Proposition 2. *For any countable superatomic EA's $\mathbb{A}$ and $\mathbb{B}$, $\mathbb{A} \simeq \mathbb{B}$ iff $\mathbb{A}$ and $\mathbb{B}$ have the same superatomicity type.*

It is well known (see Proposition 1.8.3 of [Go96]) that there are exactly two isomorphism types of countable atomless EA's which are determined by the presence or absence of the greatest element.

There is also a nice characterization of isomorphism types of arbitrary countable BA's $\mathbb{A}$ (due to J. Ketonen [Ke78]), which was extended to countable EA's by Yu.L. Ershov [Er79], in terms of the superatomic EA $F_{\sigma(\mathbb{A})}(\mathbb{A})$ and the atomless BA $\mathbb{A}^{(\sigma(\mathbb{A}))}$. Since we do not need this rather technical result in full generality, we formulate only the following very particular case relevant to this paper.

Proposition 3. *There is a unique, up to isomorphism, countable BA $\mathbb{B}$ with the properties specified (for the case of BA $\mathbb{A}_\Sigma$) in item 3 of the main result in Introduction.*

We conclude this section by mentioning the close relationship of BA's to linear orderings. Relate to any linear ordering $\mathbb{L} = (L; \leq)$ the BA $\mathbb{B}_L$ generated (in the BA of subsets of L) by the sets $\{x \mid a \leq x\}$ and $\{x \mid a \leq x < b\}$ where $a, b \in L$. It is well known (see e.g. Section 1.6 of [Go96]) that any countable BA is isomorphic to the BA $\mathbb{B}_L$ for some countable ordering L with a least element. For instance, for a unary alphabet Σ we have BA $\mathbb{A}_\Sigma \simeq \mathbb{B}_\omega$ because the last BA satisfies the conditions of item 2 of Theorem 1, $\mathbb{B}_{1+\eta}$ (where $1+\eta$ is a countable dense linear ordering with a least element and without a greatest element) is a countable atomless BA, $\mathbb{R}_\Sigma \simeq \mathbb{B}_{\omega\cdot(1+\eta)}$ (where $\omega \cdot (1+\eta)$ is obtained from $1+\eta$ by replacing any point with a copy of ω) because the last BA satisfies the conditions of item 1 of Theorem 1, and for a non-unary alphabet Σ we have $\mathbb{A}_\Sigma \simeq \mathbb{B}_{\omega^\omega\cdot(1+\eta)}$ (where $\omega^\omega = sup\{\omega, \omega^2, \omega^3, \ldots\}$) because the last BA satisfies the conditions of item 3 of Theorem 1.

3 Preliminaries on Regular Languages

Here we briefly recall some notation, notions and facts on regular languages used in the sequel. Let Σ^* denote the set of words over Σ including the empty word ε. For any $n < \omega$, let Σ^n be the set of words over Σ of length n. The length of a word w is denoted by $|w|$. Let $\sqsubseteq$ be the prefix partial order on Σ^*, i.e. $u \sqsubseteq v$ iff $ux = v$ for some $x \in \Sigma^*$.

We freely use some standard definitions and notation on words, regular languages and finite automata like regular expressions or deterministic finite automata (DFA) denoted as $\mathcal{A} = (Q, \Sigma, f, q_0, F)$ (for a more systematic treatment see e.g. [Str94, PP04, Th96, Yu97]). The automaton $\mathcal{A}$ recognizes the language $L_\mathcal{A} = \{u \in \Sigma^* \mid q_0 \cdot u \in F\}$ where $q \cdot u = f(q, u)$ is the state reached by $\mathcal{A}$ when it reads the word u started from a state q. Let $\mathcal{R}_\Sigma$ denote the class of regular languages over Σ (i.e., languages recognized by DFA's or, equivalently, denoted by regular expressions) For $L \in \mathcal{R}_\Sigma$, $\mathcal{M}_L$ denotes the minimal DFA that recognizes L; the minimal DFA is unique up to isomorphism.

The following fact is the simplest form of *pumping lemma* (see e.g. [Yu97] for more details).

Proposition 4. *For any language $L \in \mathcal{R}_\Sigma$ there is $k < \omega$ such that any word $w \in L$ with $|w| \geq k$ may be factorized as $w = xyz$ where $|xy| \leq k$, $|y| \geq 1$ and $xy^nz \in L$ for each $n < \omega$.*

For $k < \omega$, a language $L \in \mathcal{R}_\Sigma$ is called *k-sparse* if the function $p_L(n) = |\Sigma^n \cap L|$ is $O(n^k)$ (cf. [SY92, Yu97]). Let $\mathcal{S}_k$ denote the class of regular k-sparse languages over Σ. Note that $\mathcal{S}_0 \subset \mathcal{S}_1 \subset \cdots$. Languages from $\mathcal{S}_\omega = \bigcup_{k<\omega} \mathcal{S}_k$ are called *sparse languages.* 0-sparse languages are also known as *slender languages.* We will need the following characterizations of k-sparse languages in terms of regular expressions (cf. [SY92, Yu97, Pin1?]).

Proposition 5. *For any $L \in \mathcal{R}_\Sigma$ and $k < \omega$, $L \in \mathcal{S}_k$ iff L is a finite union of languages $xy_0^*z_0 \cdots y_k^*z_k$ where $x, y_i, z_i \in \Sigma^*$.*

There is also a nice characterization of sparse languages in terms of (graphs of) their minimal automata (cf. [Pin1?]). We say that a DFA $\mathcal{A}$ *has an ω-pattern* if there are $u, v_1, v_2, w \in \Sigma^*$ such that v_1, v_2 are $\sqsubseteq$-incomparable (i.e., $v_1(i) \neq v_2(i)$ for some $i < |v_1|, |v_2|$), $q_0 \cdot u = q_0 \cdot uv_1 = q_0 \cdot uv_2$ and $q_0 \cdot uv_1w \in F$.

Proposition 6. *For any $L \in \mathcal{R}_\Sigma$, $L \in \mathcal{S}_\omega$ iff the minimal DFA of L has no ω-pattern.*

For $q \in Q$ and $u \in \Sigma^*$, let (q, u) denote the path in $\mathcal{A}$ along u started at q, and let $Q(q, u) = \{q \cdot v \mid v \sqsubseteq u\}$. We say that a path (q, u) *meets* a set of states $G \subseteq Q$ if $Q(q, u) \cap G \neq \emptyset$. A path (q, u) is a *cycle of $\mathcal{A}$* if u is nonempty and $q \cdot u = q$. A cycle is *simple* if it has no repeated vertices other that the starting and ending vertices. Let $C_\mathcal{A}$ be the set of all $Q(q, u)$ where (q, u) is a cycle of $\mathcal{A}$. Define the preorder $\leq$ on $C_\mathcal{A}$ as follows: $G \leq H$ if there is a path of $\mathcal{A}$ starting in G and ending in H. Let $\equiv$ denote the equivalence relation on $C_\mathcal{A}$ induced by $\leq$.

Note that if $G \equiv H$ then $K \equiv G$ for some $K \supseteq G \cup H$, i.e. any element $[G]$ of the quotient-set $C_\mathcal{A}/\equiv$ has a greatest set under inclusion; these greatest sets are called *strongly connected components* (SCC's) of $\mathcal{A}$ and they may serve as canonical representatives for the equivalence classes $[G]$. The next result is known (cf. [Pin1?]) and easy to check:

Proposition 7. *For any sparse language $L = L_{\mathcal{A}} \in \mathcal{R}_\Sigma$ and any $u \in L$, if the path (q_0, u) meets a SCC of $\mathcal{A}$ then $[G] = \{G\}$ for some $G \in C_{\mathcal{A}}$, and $G = Q(s, v)$ for some simple cycle of $\mathcal{A}$.*

We also recall the following classical fact of automata theory. A DFA $\mathcal{A}$ is called *counter-free* if $q \cdot u^n = q$ implies $q \cdot u = q$, for all $q \in Q$, nonempty word $u \in \Sigma^*$ and $n > 0$.

Proposition 8. *For any $L \in \mathcal{R}_\Sigma$ the following conditions are equivalent:*
1. *There is $n < \omega$ such that $xy^nz \in L$ iff $xy^{n+1}z \in L$ for all $x, y, z \in \Sigma^*$.*
2. *The minimal DFA of L is counter-free.*

Languages satisfying the conditions in the last proposition are called *aperiodic.* By *aperiodicity index* of an aperiodic language L we mean the least number n satisfying the condition 1 above. There are several other important characterizations of the class $\mathcal{A}_\Sigma$ of aperiodic languages but in this paper we use only those mentioned above. The class $\mathcal{A}_\Sigma$ is closed under the Boolean operations.

4 Boolean Algebra $\mathbb{R}_\Sigma$

First we characterize the BA $\mathbb{R}_\Sigma$ formed by the class $\mathcal{R}_\Sigma$ of regular languages over Σ. Thus, we prove item 1 of the main theorem from Introduction.

Proposition 9.
1. *For any alphabet Σ, $\mathbb{R}_\Sigma$ is an atomic BA with infinitely many atoms, and $\mathbb{R}'_\Sigma$ is a countable atomless BA.*
2. *The class $F(\mathbb{R}_\Sigma)$ of regular languages is decidable.*

Proof. 1. Obviously, the atoms of $\mathbb{R}_\Sigma$ are exactly the singleton languages, hence the first assertion holds. For the second assertion, it suffices to show that for any infinite language $L \in \mathcal{R}_\Sigma$ there is an infinite regular language $M \subseteq L$ such that $L \setminus M$ is infinite. By Proposition 4, there are $x, y, z \in \Sigma^*$ such that $|y| \geq 1$ and $xy^*z \subseteq L$. Then the language $M = x(yy)^*z$ has the desired properties.

2. The assertion is well-known because $F(\mathbb{R}_\Sigma)$ is the class of finite languages. □

According to Proposition 1 and Proposition 1.8.3 of [Go96]), the item 1 of Proposition 9 characterizes the BA $\mathbb{R}_\Sigma$ up to isomorphism. This immediately implies

Corollary 1. *For any alphabet Σ, the BA $\mathbb{R}_\Sigma$ is isomorphic to the BA of computable subsets of ω.*

We conclude this section with some remarks on sparse languages.

Lemma 1. *Let $L \in \mathcal{R}_\Sigma$,let the minimal DFA $\mathcal{M}_L$ has no ω-pattern and let $F_0 < \cdots < F_{k+1}$ be SCC's of $\mathcal{M}_L$ such that a final state of $\mathcal{M}_L$ is reachable from F_{k+1}. Then there exist $x, y_i, z_i \in \Sigma^*$ and $s_i \in F_i$ such that $xy_0^*z_0 \cdots y_{k+1}^*z_{k+1} \subseteq L$, $F_i = Q(s_i, y_i)$ for each $i \leq k+1$, (s_i, y_i) is a simple cycle of $\mathcal{M}_L$, $q_0 \cdot x \in F_0$, $q_0 \cdot xz_0 \in F_1, \ldots, q_0 \cdot xz_0 \cdots z_k \in F_{k+1}$, $xz_0 \cdots z_{k+1} \in F$, the words z_i are nonempty, and for each $i \leq k+1$ the first letters in y_i, z_i are distinct.*

Proof. Choose a path u in $\mathcal{M}_L$ that meets all SCC's $F_0, \ldots, F_{k+1}$. Then existence of the desired objects (except maybe the property that the first letters in y_i, z_i are distinct) follows from Proposition 7. The last property may be achieved by "shifting" the cycles (s_i, y_i) if needed. □

The next result extends the corresponding folklore fact on slender languages in [Pin1?].

Proposition 10. *For any $L \in \mathcal{R}_\Sigma$ and $k < \omega$, $L \in \mathcal{S}_k$ iff the minimal DFA $\mathcal{M}_L$ has no ω-pattern and there is no chain $F_0 < \cdots < F_{k+1}$ of SCC's of $\mathcal{M}_L$ such that a final state of $\mathcal{M}_L$ is reachable from F_{k+1}.*

Proof. Let $L \in \mathcal{S}_k$, then $\mathcal{M}_L$ has no ω-pattern by Proposition 6. Suppose, for a contradiction, that a chain as above exists. Consider the language $xy_0^*z_0 \cdots y_{k+1}^*z_{k+1}$ where the words from the previous lemma are used. By Lemmas 1,2 in [SY92] we get $xy_0^*z_0 \cdots y_{k+1}^*z_{k+1} \notin \mathcal{S}_k$ which is a contradiction.

For the opposite direction, assume that there is no chain of length $k+1$ as in the formulation. There are only finitely many of such chains of length $\leq k$. By Proposition 7, L is a finite union of languages of the form $xy_0^*z_0 \cdots y_t^*z_t$, $t \leq k$ constructed as in the last paragraph. By Proposition 5 $L \in \mathcal{S}_k$. □

As an immediate consequence, we obtain

Proposition 11. *The classes of regular languages $\mathcal{S}_\omega, \mathcal{S}_0, \mathcal{S}_1, \ldots$ are ideals of the BA $\mathbb{R}_\Sigma$ and they are decidable.*

Proof. The first assertion is obvious. The decidability of $\mathcal{S}_\omega$ follows from Proposition 6. The decidability of $\mathcal{S}_k$ for $k < \omega$ follows from Proposition 10. □

We finish this section with remarks on some EA's related to sparse languages. For any $k \leq \omega$, let $\mathbb{S}_k$ be the EA formed $\mathcal{S}_k$ and let $\mathbb{T}_k$ be the BA formed by the sets in $\mathcal{S}_k$ and their complements, with the usual set-theoretic operations.

Proposition 12. *1. The EA's $\mathbb{S}_\omega\mathbb{S}_0, \mathbb{S}_1, \ldots$ are atomic with infinitely many atoms, and their first Frechét derivatives are countable atomless EA's without greatest elements.*
2. The BA's $\mathbb{T}_\omega\mathbb{T}_0, \mathbb{T}_1, \ldots$ are atomic with infinitely many atoms, and their first Frechét derivatives are countable atomless BA's.
3. The quotient-algebra $\mathbb{R}_\Sigma/\mathcal{S}_\omega$ is a countable atomless BA.

Proof. Items 1 and 2 are checked in a straightforward way, as in the proof of Proposition 9.

3. We have to show that for any non-sparse language $L \in \mathcal{R}_\Sigma$ there is a non-sparse regular language $M \subseteq L$ such that $L \setminus M$ is not sparse. By Proposition 6, the minimal automaton for L has an ω-pattern with some words u, v_1, v_2, w as in Section 3. Let us take $M = uv_1(v_1+v_2)^*w$, then $uv_2(v_1+v_2)^*w \subseteq L \setminus M$. Since both languages $uv_1(v_1+v_2)^*w$ and $uv_2(v_1+v_2)^*w$ are not sparse by Proposition 6, this completes the proof. □

By Proposition 1, the EA's in items 1 and 2 above are characterized up to isomorphism, so we have

Corollary 2. *1. The EA's* $\mathbb{S}_\omega\mathbb{S}_0, \mathbb{S}_1, \dots$ *are pairwise isomorphic.*
2. The BA's $\mathbb{T}_\omega\mathbb{T}_0, \mathbb{T}_1, \dots$ *are pairwise isomorphic.*

5 Boolean Algebra $\mathbb{A}_\Sigma$

First we characterize the BA $\mathbb{A}_\Sigma$ formed by the class $\mathcal{A}_\Sigma$ of aperiodic regular languages over a unary alphabet Σ. Thus, we prove item 2 of the main theorem from Introduction.

Proposition 13. *For any unary alphabet* $\Sigma = \{a\}$, $\mathbb{A}_\Sigma$ *is an atomic BA with infinitely many atoms, and* $\mathbb{A}'_\Sigma$ *is a 2-element BA.*

Proof. Obviously, the atoms of $\mathbb{A}_\Sigma$ are exactly the singleton languages, hence the first assertion holds. For the second assertion, it suffices to show that any infinite aperiodic language $L \in \mathcal{A}_\Sigma$ is cofinite. By Proposition 4, there are $x, y, z \in \Sigma^*$ such that $|y| \geq 1$ and $xy^*z \subseteq L$. In other words, $a^m \in L$ for infinitely many m. By Proposition 8, $a^m \in L$ for all $m \geq n$ where n is the aperiodicity index of L. □

According to Proposition 1, the assertion above characterizes the BA $\mathbb{A}_\Sigma$ up to isomorphism. This immediately implies

Corollary 3. *For any unary alphabet* Σ*, the BA* $\mathbb{A}_\Sigma$ *is isomorphic to the BA of finite and cofinite subsets of* ω*.*

For the sequel we need the following lemma which follows easily from Proposition 8 and Theorem 2.1 in [CK97]. A non-empty word $u \in \Sigma^*$ is called *primitive* if $v^n = u$ implies $n = 1$ (and hence $v = u$).

Lemma 2. *For any* $u \in \Sigma^*$, $u^* \in \mathcal{A}_\Sigma$ *iff* u *is either empty or primitive.*

The next two lemmas relate the iterated Frechét ideals of $\mathbb{A}_\Sigma$ to sparse languages.

Lemma 3. *For any* $k < \omega$ *and* $x, y_i, z_i \in \Sigma^*$ *with* $y_1^*, \cdots, y_k^* \in \mathcal{A}_\Sigma$*, the element* $xy_1^*z_1 \cdots y_k^*z_{/}F_k(\mathbb{A}_\Sigma)$ *is either zero or an atom of the BA* $\mathbb{A}_\Sigma^{(k)}$*. If* L *is a finite union of such languages* $xy_1^*z_1 \cdots y_k^*z_k$ *then* $L \in F_{k+1}(\mathbb{A}_\Sigma)$*.*

Proof. Since the second assertion follows from the first one, it suffices to check the first assertion by induction on k. For $k = 0$ the assertion is obvious because $\{x\}$ is an atom of $\mathbb{A}_\Sigma = \mathbb{A}_\Sigma^{(0)}$.

Let $k = 1$. The case $y_1 = \varepsilon$ is trivial, so assume y_1 to be non-empty (hence, primitive by Lemma 2). Then $L = xy_1^*z_1$ is infinite and we have to show that if A is an infinite aperiodic subset of L then $L \setminus A$ is finite. We have $xy_1^mz_1 \in A$ for infinitely many m. By Proposition 8, $xy_1^mz_1 \in A$ for all $m \geq n$ where n is the aperiodicity index of A. Therefore, $L \setminus A$ is finite.

Let now $k \geq 2$. For any $n < \omega$, set $L_n = xy_1^ny_1^*z_1 \cdots y_k^ny_k^*z_k$, then $L = L_0 \supset L_1 \supset L_2 \supset \cdots$ where $L = xy_1^*z_1 \cdots y_k^*z_k$. It suffices to show that for any aperiodic language $A \subseteq L$ at least one of languages $A, L \setminus A$ is in $F_k(\mathbb{A}_\Sigma)$. We distinguish two cases.

Case 1. $L_n \subseteq \overline{A}$ for some n. Then $A = L \setminus \overline{A} \subseteq L \setminus L_n \in F_k(\mathbb{A}_\Sigma)$ by induction on k. Thus, $A \in F_2(\mathbb{A}_\Sigma)$.

Case 2. $L_n \not\subseteq \overline{A}$ for all n, i.e. for any n there are $m_1, \ldots, m_k \geq n$ such that $xy_1^{m_1} z_1 \cdots y_k^{m_k} z_k \in A$. By Proposition 8, $xy_1^{m_1} z_1 \cdots y_k^{m_k} z_k \in A$ for all $m_1, \ldots, m_k \geq m$ where m is the aperiodicity index of A. Then $L_m \subseteq A$, hence $L \setminus A \subseteq L \setminus L_m \in F_k(\mathbb{A}_\Sigma)$, hence $L \setminus A \in F_k(\mathbb{A}_\Sigma)$. □

Lemma 4. *Let $u, v_1, v_2, w \in \Sigma^*$ be such that the words v_1, v_2 are primitive and $\sqsubseteq$-incomparable. Then $uv_1^*w \notin F_1(\mathbb{A}_\Sigma)$, $uv_1^*v_2v_1^*w \notin F_2(\mathbb{A}_\Sigma)$, $uv_1^*v_2v_1^*v_2v_1^*w \notin F_3(\mathbb{A}_\Sigma)$ and so on.*

Proof. The assertion $uv_1^*w \notin F_1(\mathbb{A}_\Sigma)$ is clear because the language uv_1^*w is infinite while $F_1(\mathbb{A}_\Sigma)$ is the class of finite languages.

The language $uv_1^*v_2v_1^*w$ is a disjoint union of languages $K_n = uv_1^n v_2 v_1^* w$, and, for each n, $K_n/F_1(\mathbb{A}_\Sigma)$ is an atom of $\mathbb{A}'_\Sigma$ by the previous lemma. Therefore, $uv_1^*v_2v_1^*w \notin F_2(\mathbb{A}_\Sigma)$. Continuing in this manner, we derive the desired assertions. □

Lemma 5. *For any $L \in F_\omega(\mathbb{A}_\Sigma)$, the minimal DFA of L has no ω-pattern.*

Proof. By contraposition, assume that the minimal DFA of L has an ω-pattern with some words u, v_1, v_2, w as in Section 3. By Proposition 8, the words v_1, v_2 are primitive. We have to show that $L \notin F_\omega(\mathbb{A}_\Sigma)$. Since $u(v_1+v_2)^*w \subseteq L$, it suffices to show that $uv_1^*w \notin F_1(\mathbb{A}_\Sigma)$, $uv_1^*v_2v_1^*w \notin F_2(\mathbb{A}_\Sigma)$, $uv_1^*v_2v_1^*v_2v_1^*w \notin F_3(\mathbb{A}_\Sigma)$ and so on. But this holds by the previous lemma. □

We are ready to provide useful characterizations of the iterated Frechét ideals of $\mathbb{A}_\Sigma$.

Theorem 2. *For any $L \in \mathcal{R}_\Sigma$ the following conditions are equivalent:*

1. *$L \in F_\omega(\mathbb{A}_\Sigma)$.*
2. *The minimal DFA of L is counter-free and has no ω-pattern.*
3. *$L \in \mathcal{A}_\Sigma \cap \mathcal{S}_\omega$.*
4. *L is a finite union of languages $xy_0^*z_0 \cdots y_k^*z_k$ where $k < \omega$, $x, y_i, z_i \in \Sigma^*$ and $y_0^*, \cdots, y_k^* \in \mathcal{A}_\Sigma$.*

Proof. 1→2. Follows from the previous lemma and Proposition 8.

2↔3. Follows from Propositions 8 and 6.

2→4. The desired representation of L follows from Lemmas 1 and 2.

4→1. Follows from Lemma 3. □

Corollary 4. *The class of regular languages $F_\omega(\mathbb{A}_\Sigma)$ is decidable.*

Next we characterize $F_k(\mathbb{A}_\Sigma)$ for $k < \omega$. Note that $F_0(\mathbb{A}_\Sigma) = \{\emptyset\}$ and $F_1(\mathbb{A}_\Sigma)$ is the class of finite languages over Σ.

Theorem 3. *For any $k < \omega$ and $L \in \mathcal{R}_\Sigma$ the following conditions are equivalent:*

1. *$L \in F_{k+2}(\mathbb{A}_\Sigma)$.*
2. *The minimal DFA of L is counter-free, has no ω-pattern, and there is no chain $F_0 < \cdots < F_{k+1}$ of SCC's of $\mathcal{M}_L$ such that a final state of $\mathcal{M}_L$ is reachable from F_{k+1}.*
3. *$L \in \mathcal{A}_\Sigma \cap \mathcal{S}_k$.*
4. *L is a finite union of languages $xy_0^*z_0 \cdots y_k^*z_k$ where $x, y_i, z_i \in \Sigma^*$ and $y_0^*, \cdots, y_k^* \in \mathcal{A}_\Sigma$.*

Proof. 1→2. Follows from Proposition 8, Lemma 5 and the proofs of Lemma 5 and Proposition 10.

2↔3. Follows from Propositions 8 and 10.

2→4. The desired representation of L follows from Lemmas 1 and 2.

4→1. Follows from Lemma 3. □

Corollary 5. *For any $k < \omega$, the class of regular languages $F_k(\mathbb{A}_\Sigma)$ is decidable.*

Next we prove the item 3 of the main theorem in Introduction.

Theorem 4. *Item 3 of the main theorem in Introduction holds.*

Proof. First we check that $F_k(\mathbb{A}_\Sigma) \subset F_{k+1}(\mathbb{A}_\Sigma)$ for each $k < \omega$. For $k = 0$ the inclusion is trivial. Let $a, b \in \Sigma$, $a \neq b$. Let $y = ab$ and $z = aab$, then y, z are primitive $\sqsubseteq$-incomparable words. It suffices to show that $y^* \in F_2(\mathbb{A}_\Sigma) \setminus F_1(\mathbb{A}_\Sigma)$, $y^*zy^* \in F_3(\mathbb{A}_\Sigma) \setminus F_2(\mathbb{A}_\Sigma)$, $y^*zy^*zy^* \in F_4(\mathbb{A}_\Sigma) \setminus F_3(\mathbb{A}_\Sigma)$, and so on. By Theorem 2 and Lemma 3, $y^* \in F_2(\mathbb{A}_\Sigma)$, $y^*zy^* \in F_3(\mathbb{A}_\Sigma)$, $y^*zy^*zy^* \in F_4(\mathbb{A}_\Sigma)$, and so on. By Lemma 4, $y^* \not\in F_1(\mathbb{A}_\Sigma)$, $y^*zy^* \not\in F_2(\mathbb{A}_\Sigma)$, $y^*zy^*zy^* \not\in F_3(\mathbb{A}_\Sigma)$, and so on.

By Lemmas 3 — 5, elements $y^*/F_1(\mathbb{A}_\Sigma), y^*zy^*/F_2(\mathbb{A}_\Sigma), \ldots$ are atoms respectively in $\mathbb{A}_\Sigma^{(1)}, \mathbb{A}_\Sigma^{(2)}, \ldots$, and, for each $n < \omega$, the same applies to the elements $z^ny^*/F_1(\mathbb{A}_\Sigma), z^ny^*zy^*/F_2(\mathbb{A}_\Sigma), \ldots$. Since the languages z^ny^* (as well as the languages $z^ny^*zy^*$ and so on) are pairwise disjoint for distinct n, the BA's $\mathbb{A}_\Sigma^{(1)}, \mathbb{A}_\Sigma^{(2)}, \ldots$ have infinitely many atoms (as well as the BA $\mathbb{A}_\Sigma^{(0)} = \mathbb{A}_\Sigma$).

Next we check that the BA $\mathbb{A}_\Sigma^{(k)}$ is atomic for each $k < \omega$. For $k < 2$ this is again obvious, so it remains to show that for any $k < \omega$ and $L \in \mathcal{A}_\Sigma \setminus F_{k+2}(\mathbb{A}_\Sigma)$ there is an aperiodic language $A \subseteq L$ such that $A/F_{k+2}(\mathbb{A}_\Sigma)$ is an atom of $\mathbb{A}_\Sigma^{(k+2)}$. We distinguish the cases $L \not\in F_\omega(\mathbb{A}_\Sigma)$ and $L \in F_\omega(\mathbb{A}_\Sigma)$.

In the first case, by Propositions 6 and 8 the minimal DFA of L has an ω-pattern, hence A exists by Theorem 2 and Lemma 4. In the second case, by Theorem 2 and Lemma 1 there are SCC's $F_0 < \cdots < F_{k+1}$ and the words specified there. The words $z_0, \ldots, z_k$ are non-empty and the first letters in z_i, y_i are distinct for each $i \leq k$. Then the language $A = xy_0^*z_0 \cdots y_k^*z_k$ has the desired property.

It remains to show that for any $L \in \mathcal{A}_\Sigma \setminus F_\omega(\mathbb{A}_\Sigma)$ there is an aperiodic language $M \subseteq L$ such that $M, L \setminus M \not\in F_\omega(\mathbb{A}_\Sigma)$. By Proposition 8 and Theorem 2, the minimal DFA $\mathcal{M}_L$ of L is counter-free and has an ω-pattern with some words u, v_1, v_2, w as in Section 3. Since $\mathcal{M}_L$ is counter-free, $v_1^*, v_2^* \in \mathcal{A}_\Sigma$. By the proof of Proposition 12, we can take $M = uv_1(v_1 + v_2)^*w$. □

From the results above we immediately obtain characterizations of the EA's $\mathbb{F}_k(\mathbb{A}_\Sigma)$ formed by the ideals $F_k(\mathbb{A}_\Sigma)$ $(k \leq \omega)$ and of the BA's $\mathbb{G}_k(\mathbb{A}_\Sigma)$ formed by the languages in $F_k(\mathbb{A}_\Sigma)$ and their complements.

Corollary 6. *Let Σ be an alphabet with at least two letters.*

1. *For any $k \leq \omega$, $\mathbb{F}_k(\mathbb{A}_\Sigma)$ is a superatomic EA of superatomicity type $(k, k, 0)$.*
2. *For any $k \leq \omega$, $\mathbb{G}_k(\mathbb{A}_\Sigma)$ is a superatomic BA of superatomicity type $(k, 0, 1)$.*

6 Conclusion

We hope that results of this paper demonstrate that characterizing of other well-known BA's (and, more generally, of lattices) of regular languages and ω-languages may be also of interest. In particular, it is instructive to characterize the BA's formed by levels of the dot-depth hierarchy.

After presenting the main theorem at the Mal'tsev Meeting-2010 in Novosibirsk we learned from Andrea Sorbi that he with his colleagues independently started an investigation of natural BA's in formal language theory, and they independently obtained the assertion 1 of the main result in Introduction. According to Andrea, this is the only intersection of their (yet unpublished) results with ours.

Acknowledgement. We are grateful Sergey Goncharov for useful hints on BA's and to Jean-Eric Pin for a useful discussion of slender and sparse languages and for providing a part of his yet unpublished manuscript [Pin1?].

References

[CK97] Choffrut, C., Karhumäki, J.: Combinatorics of Words. In: Handbook of Formal Languages. Springer, Berlin (1997)

[Er79] Ershov, Y.L.: Relatively complemented distributive lattices. Algebra and Logic 18(6), 680–722 (1979) (Russian, there is an English translation)

[GGP08] Gehrke, M., Grigorieff, S., Pin, J.-É.: Duality and equational theory of regular languages. In: Aceto, L., Damgård, I., Goldberg, L.A., Halldórsson, M.M., Ingólfsdóttir, A., Walukiewicz, I. (eds.) ICALP 2008, Part II. LNCS, vol. 5126, pp. 246–257. Springer, Heidelberg (2008)

[Go96] Goncharov, S.S.: Countable Boolean Algebras and Decidability. Plenum, New York (1996)

[Han75] Hanf, W.: The boolean algebra of logic. Bull. Amer. Math. Soc. 20(4), 456–502 (1975)

[Ke78] Ketonen, J.: The structure of countable Boolean algebras. Annals of Mathematics 108, 41–89 (1978)

[LPS02] Lempp, S., Peretyat'kin, M., Solomon, R.: The Lindenbaum algebra of the theory of the class of all finite models. Journal of Mathematical Logic 2(2), 145–225 (2002)

[Pin1?] Pin, J.-E.: Unpublished manuscript on regular languages

[Pip97] Pippenger, N.: Regular languages and Stone duality. Theory of Computing Systems 30(2), 121–134 (1997)

[PP04] Perrin, D., Pin, J.-E.: Infinite Words. Pure and Applied Mathematics, vol. 141. Elsevier, Amsterdam (2004)

[Se91] Selivanov, V.L.: Universal Boolean algebras with applications. In: Abstracts of Int. Conf. in Algebra, Novosibirsk, p. 127 (1991) (in Russian)

[Se92] Selivanov, V.L.: Hierarchies, Numerations, Index Sets. Handwritten Notes, 290 pp (1992)

[Se03] Selivanov, V.L.: Positive structures. In: Barry Cooper, S., Goncharov, S.S. (eds.) Computability and Models, Perspectives East and West, pp. 321–350. Kluwer Academic/Plenum Publishers, New York (2003)

[Si64] Sikorski, R.: Boolean Algebras. Springer, Berlin (1964)

[Str94] Straubing, H.: Finite automata, formal logic and circuit complexity. Birkhäuser, Boston (1994)

[SY92] Szilard, A., Yu, S., Zhang, K., Shallit, J.: Characterizing Regular Languages with Polynomial Densities. In: Havel, I.M., Koubek, V. (eds.) MFCS 1992. LNCS, vol. 629, Springer, Heidelberg (1992)

[Th96] Thomas, W.: Languages, automata and logic. In: Handbook of Formal Language Theory, vol. B, pp. 133–191 (1996)

[Yu97] Yu, S.: Regular Languages. In: Rozenberg, G., Salomaa, A. (eds.) A chapter of Handbook of Formal Languages. Springer, Heidelberg (1997)

Fife's Theorem Revisited

Jeffrey Shallit

University of Waterloo, Waterloo, ON N2L 3G1 Canada
shallit@cs.uwaterloo.ca

Abstract. We give another proof of a theorem of Fife — understood broadly as providing a finite automaton that gives a complete description of all infinite binary overlap-free words. Our proof is significantly simpler than those in the literature. As an application we give a complete characterization of the overlap-free words that are 2-automatic.

1 Introduction

Repetitions in words is a well-researched topic. Among the various themes studied, the binary overlap-free words play an important role, both historically and as an example exhibiting interesting structure. Here by an *overlap* we mean a word of the form $axaxa$, where a is a single letter and x is a (possibly empty) word.

It is easy to see that neither the finite nor the infinite binary overlap-free words form a regular language. Nevertheless, in 1980, Earl Fife [7] proved a theorem characterizing the infinite binary overlap-free words as encodings of paths in a finite automaton. His theorem was rather complicated to state and the proof was difficult. Berstel [3] later simplified the exposition, and both Carpi [5] and Cassaigne [6] gave an analogous analysis for the case of finite words. Also see [4].

In this note we show how to use the factorization theorem of Restivo and Salemi [10] to give an alternate (and, we hope, significantly simpler) proof of Fife's theorem — here understood in the general sense of providing a finite automaton whose paths encode all infinite binary overlap-free words.

2 Notation

Let Σ be a finite alphabet. We let Σ^* denote the set of all finite words over Σ and Σ^ω denote the set of all (right-) infinite words over Σ. We say y is a *factor* of a word w if there exist words x, z such that $w = xyz$.

If x is a finite word, then x^ω represents the infinite word $xxx\cdots$.

As mentioned above, an *overlap* is a word of the form $axaxa$, where $a \in \Sigma$ and $x \in \Sigma^*$. An example of an overlap in English is the word `alfalfa`. A finite or infinite word is *overlap-free* if it contains no finite factor that is an overlap.

From now on we fix $\Sigma = \{0, 1\}$. The most famous infinite binary overlap-free word is $\mathbf{t}$, the Thue-Morse word, defined as the fixed point, starting with 0, of the Thue-Morse morphism μ, which maps 0 to 01 and 1 to 10. We have

$$\mathbf{t} = t_0 t_1 t_2 \cdots = 0110100110010110 \cdots .$$

G. Mauri and A. Leporati (Eds.): DLT 2011, LNCS 6795, pp. 397–405, 2011.

The morphism μ has a second fixed point, $\overline{\mathbf{t}} = \mu^\omega(1)$, which is obtained from $\mathbf{t}$ by applying the complementation coding defined by $\overline{0} = 1$ and $\overline{1} = 0$.

We let $\mathcal{O}$ denote the set of (right-) infinite binary overlap-free words.

We now recall the infinite version of the factorization theorem of Restivo and Salemi [10] as stated in [1, Lemma 3].

Theorem 1. *Let* $\mathbf{x} \in \mathcal{O}$, *and let* $P = \{p_0, p_1, p_2, p_3, p_4\}$, *where* $p_0 = \epsilon$, $p_1 = 0$, $p_2 = 00$, $p_3 = 1$, *and* $p_4 = 11$. *Then there exists* $\mathbf{y} \in \mathcal{O}$ *and* $p \in P$ *such that* $\mathbf{x} = p\mu(\mathbf{y})$. *Furthermore, this factorization is unique, and* p *is uniquely determined by inspecting the first 5 letters of* $\mathbf{x}$.

We can now iterate the factorization theorem to get

Corollary 1. *Every infinite overlap-free word* $\mathbf{x}$ *can be written uniquely in the form*

$$\mathbf{x} = p_{i_1}\mu(p_{i_2}\mu(p_{i_3}\mu(\cdots))) \tag{1}$$

with $i_j \in \{0,1,2,3,4\}$ *for* $j \geq 1$, *subject to the understanding that if there exists* c *such that* $i_j = 0$ *for* $j \geq c$, *then we also need to specify whether the "tail" of the expansion represents* $\mu^\omega(0) = \mathbf{t}$ *or* $\mu^\omega(1) = \overline{\mathbf{t}}$. *Furthermore, every truncated expansion*

$$p_{i_1}\mu(p_{i_2}\mu(p_{i_3}\mu(\cdots p_{i_{n-1}}\mu(p_{i_n})\cdots)))$$

is a prefix of $\mathbf{x}$, *with the understanding that if* $i_n = 0$, *then we need to replace* 0 *with either* 1 *(if the "tail" represents* $\mathbf{t}$*) or* 3 *(if the "tail" represents* $\overline{\mathbf{t}}$*).*

Proof. The form (1) is unique, since each p_i is uniquely determined by the first 5 characters of the associated word.

Thus, we can associate each infinite binary overlap-free word $\mathbf{x}$ with the essentially unique infinite sequence of indices $\mathbf{i} := (i_j)_{j\geq 0}$ coding elements in P, as specified by (1). If $\mathbf{i}$ ends in 0^ω, then we need an additional element (either 1 or 3) to disambiguate between $\mathbf{t}$ and $\overline{\mathbf{t}}$ as the "tail". In our notation, we separate this additional element with a semicolon so that, for example, the string $000\cdots;1$ represents $\mathbf{t}$ and $000\cdots;3$ represents $\overline{\mathbf{t}}$.

Other sequences of interest include $203000\cdots;1$, which codes $001001\overline{\mathbf{t}}$, the lexicographically least infinite word, and $2(31)^\omega$, which codes the word having, in the i'th position, the number of 0's in the binary expansion of i.

Of course, not every possible sequence of $(i_j)_{j\geq 1}$ of indices corresponds to an infinite overlap-free word. For example, every infinite word coded by $21\cdots$ represents $00\mu(0\mu(\ldots))$ and hence begins with 000 and has an overlap. Our goal is to characterize precisely, using a finite automaton, those infinite sequences corresponding to overlap-free words.

We recall some basic facts about overlap-free words.

Lemma 1. *Let* $a \in \Sigma$. *Then*

(a) $\mathbf{x} \in \mathcal{O} \iff \mu(\mathbf{x}) \in \mathcal{O}$;
(b) $a\,\mu(\mathbf{x}) \in \mathcal{O} \iff \overline{a}\,\mathbf{x} \in \mathcal{O}$;

(c) $a\,a\,\mu(\mathbf{x}) \in \mathcal{O} \iff \overline{a}\,\mathbf{x} \in \mathcal{O}$ *and* $\mathbf{x}$ *begins* $\overline{a}\,a\,\overline{a}$.

Proof. See, for example, [1].

We now define 11 subsets of $\mathcal{O}$:

$$
\begin{aligned}
A &= \mathcal{O} \\
B &= \{\mathbf{x} \in \Sigma^\omega \ : \ 1\mathbf{x} \in \mathcal{O}\} \\
C &= \{\mathbf{x} \in \Sigma^\omega \ : \ 1\mathbf{x} \in \mathcal{O} \text{ and } \mathbf{x} \text{ begins with } 101\} \\
D &= \{\mathbf{x} \in \Sigma^\omega \ : \ 0\mathbf{x} \in \mathcal{O}\} \\
E &= \{\mathbf{x} \in \Sigma^\omega \ : \ 0\mathbf{x} \in \mathcal{O} \text{ and } \mathbf{x} \text{ begins with } 010\} \\
F &= \{\mathbf{x} \in \Sigma^\omega \ : \ 0\mathbf{x} \in \mathcal{O} \text{ and } \mathbf{x} \text{ begins with } 11\} \\
G &= \{\mathbf{x} \in \Sigma^\omega \ : \ 0\mathbf{x} \in \mathcal{O} \text{ and } \mathbf{x} \text{ begins with } 1\} \\
H &= \{\mathbf{x} \in \Sigma^\omega \ : \ 1\mathbf{x} \in \mathcal{O} \text{ and } \mathbf{x} \text{ begins with } 1\} \\
I &= \{\mathbf{x} \in \Sigma^\omega \ : \ 1\mathbf{x} \in \mathcal{O} \text{ and } \mathbf{x} \text{ begins with } 00\} \\
J &= \{\mathbf{x} \in \Sigma^\omega \ : \ 1\mathbf{x} \in \mathcal{O} \text{ and } \mathbf{x} \text{ begins with } 0\} \\
K &= \{\mathbf{x} \in \Sigma^\omega \ : \ 0\mathbf{x} \in \mathcal{O} \text{ and } \mathbf{x} \text{ begins with } 0\}
\end{aligned}
$$

Next, we describe the relationships between these classes:

Lemma 2. *Let* $\mathbf{x}$ *be an infinite binary word. Then*

$$\mathbf{x} \in A \iff \mu(\mathbf{x}) \in A \tag{2}$$
$$\mathbf{x} \in B \iff 0\mu(\mathbf{x}) \in A \tag{3}$$
$$\mathbf{x} \in C \iff 00\mu(\mathbf{x}) \in A \tag{4}$$
$$\mathbf{x} \in D \iff 1\mu(\mathbf{x}) \in A \tag{5}$$
$$\mathbf{x} \in E \iff 11\mu(\mathbf{x}) \in A \tag{6}$$
$$\mathbf{x} \in D \iff \mu(\mathbf{x}) \in B \tag{7}$$
$$\mathbf{x} \in B \iff 0\mu(\mathbf{x}) \in B \tag{8}$$
$$\mathbf{x} \in E \iff 1\mu(\mathbf{x}) \in B \tag{9}$$
$$\mathbf{x} \in B \iff \mu(\mathbf{x}) \in D \tag{10}$$
$$\mathbf{x} \in D \iff 1\mu(\mathbf{x}) \in D \tag{11}$$
$$\mathbf{x} \in C \iff 0\mu(\mathbf{x}) \in D \tag{12}$$
$$\mathbf{x} \in I \iff \mu(\mathbf{x}) \in E \tag{13}$$
$$\mathbf{x} \in C \iff 0\mu(\mathbf{x}) \in E \tag{14}$$
$$\mathbf{x} \in F \iff \mu(\mathbf{x}) \in C \tag{15}$$
$$\mathbf{x} \in E \iff 1\mu(\mathbf{x}) \in C \tag{16}$$
$$\mathbf{x} \in J \iff 0\mu(\mathbf{x}) \in I \tag{17}$$
$$\mathbf{x} \in G \iff 1\mu(\mathbf{x}) \in F \tag{18}$$
$$\mathbf{x} \in K \iff \mu(\mathbf{x}) \in J \tag{19}$$
$$\mathbf{x} \in J \iff \mu(\mathbf{x}) \in K \tag{20}$$

$$\begin{aligned}
\mathbf{x} \in B &\iff 0\mu(\mathbf{x}) \in J && (21)\\
\mathbf{x} \in C &\iff 0\mu(\mathbf{x}) \in K && (22)\\
\mathbf{x} \in H &\iff \mu(\mathbf{x}) \in G && (23)\\
\mathbf{x} \in G &\iff \mu(\mathbf{x}) \in H && (24)\\
\mathbf{x} \in D &\iff 1\mu(\mathbf{x}) \in G && (25)\\
\mathbf{x} \in E &\iff 1\mu(\mathbf{x}) \in H && (26)
\end{aligned}$$

Proof.

(2): Follows immediately from Lemma 1 (a).

(3), (5), (7), (10): Follow immediately from Lemma 1 (b).

(4), (6), (9), (12): Follow immediately from Lemma 1 (c).

(8): $0\mu(\mathbf{x}) \in B \iff 10\mu(\mathbf{x}) = \mu(1\,\mathbf{x}) \in \mathcal{O} \iff 1\,\mathbf{x} \in \mathcal{O}$.

(11): Just like (8).

(13): $\mu(\mathbf{x}) \in E \iff (0\mu(\mathbf{x}) \in \mathcal{O}$ and $\mu(\mathbf{x})$ begins with 010) $\iff$ ($1\mathbf{x} \in \mathcal{O}$ and $\mathbf{x}$ begins with 00).

(15): Just like (13).

(14): $0\mu(\mathbf{x}) \in E \iff (00\mu(\mathbf{x}) \in \mathcal{O}$ and $0\mu(\mathbf{x})$ begins with 010) $\iff$ ($1\mathbf{x} \in \mathcal{O}$ and $\mathbf{x}$ begins with 101).

(16): Just like (14).

(17): $0\mu(\mathbf{x}) \in I \iff (10\mu(\mathbf{x}) \in \mathcal{O}$ and $0\mu(\mathbf{x})$ begins with 00) $\iff$ ($\mu(1\mathbf{x}) \in \mathcal{O}$ and $\mathbf{x}$ begins with 0) $\iff$ ($1\mathbf{x} \in \mathcal{O}$ and $\mathbf{x}$ begins with 0).

(18): Just like (17).

(19): $\mu(\mathbf{x}) \in J \iff (1\mu(\mathbf{x}) \in \mathcal{O}$ and $\mu(\mathbf{x})$ begins with 0) $\iff$ ($0\mathbf{x} \in \mathcal{O}$ and $\mathbf{x}$ begins with 0).

(23), (20), (24): Just like (19).

(21): $0\mu(\mathbf{x}) \in J \iff (10\mu(\mathbf{x}) \in \mathcal{O}$ and $0\mu(\mathbf{x})$ begins with 0) $\iff \mu(1\mathbf{x}) \in \mathcal{O} \iff 1\mathbf{x} \in \mathcal{O}$.

(25): Just like (21).

(22): $0\mu(\mathbf{x}) \in K \iff (00\mu(\mathbf{x}) \in \mathcal{O}$ and $0\mu(\mathbf{x})$ begins with 0) $\iff$ ($1\mathbf{x} \in \mathcal{O}$ and $\mathbf{x}$ begins with 101).

(26): Just like (22).

We can now use the result of the previous lemma to create an 11-state automaton that accepts all infinite sequences $(i_j)_{j\geq 1}$ over $\Delta := \{0, 1, 2, 3, 4\}$ such that $p_{i_1}\mu(p_{i_2}\mu(p_{i_3}\mu(\cdots)))$ is overlap-free. Each state represents one of the sets $A, B, \ldots, K$ defined above, and the transitions are given by Lemma 2.

Of course, we also need to verify that transitions not shown correspond to the empty set of infinite words. For example, a transition out of B on the symbol 2 would correspond to the set $\{\mathbf{x} : 100\mu(\mathbf{x}) \in \mathcal{O}\}$. But if $\mathbf{x}$ begins with 0, then $100\mu(\mathbf{x}) = 10001\cdots$ contains the overlap 000 as a factor, whereas if $\mathbf{x}$ begins with 10, then $100\mu(\mathbf{x}) = 1001001\cdots$ contains the overlap 1001001 as a factor, and if $\mathbf{x}$ begins with 11, then $100\mu(\mathbf{x}) = 1001010\cdots$ contains 01010 as a factor. Similarly, we can (somewhat tediously) verify that all other transitions not given in Figure 1 correspond to the empty set:

$$
\begin{aligned}
\delta(B,4) &= \{\mathbf{x} \in \Sigma^\omega \ : \ 111\mu(\mathbf{x}) \in \mathcal{O}\} = \emptyset \\
\delta(D,2) &= \{\mathbf{x} \in \Sigma^\omega \ : \ 000\mu(\mathbf{x}) \in \mathcal{O}\} = \emptyset \\
\delta(D,4) &= \{\mathbf{x} \in \Sigma^\omega \ : \ 011\mu(\mathbf{x}) \in \mathcal{O}\} = \emptyset \\
\delta(C,1) &= \{\mathbf{x} \in \Sigma^\omega \ : \ 10\mu(\mathbf{x}) \in \mathcal{O} \text{ and } 0\mu(\mathbf{x}) \text{ begins with } 101\} = \emptyset \\
\delta(C,2) &= \{\mathbf{x} \in \Sigma^\omega \ : \ 100\mu(\mathbf{x}) \in \mathcal{O} \text{ and } 00\mu(\mathbf{x}) \text{ begins with } 101\} = \emptyset \\
\delta(C,4) &= \{\mathbf{x} \in \Sigma^\omega \ : \ 111\mu(\mathbf{x}) \in \mathcal{O} \text{ and } 11\mu(\mathbf{x}) \text{ begins with } 101\} = \emptyset \\
\delta(E,2) &= \{\mathbf{x} \in \Sigma^\omega \ : \ 000\mu(\mathbf{x}) \in \mathcal{O} \text{ and } 00\mu(\mathbf{x}) \text{ begins with } 010\} = \emptyset \\
\delta(E,3) &= \{\mathbf{x} \in \Sigma^\omega \ : \ 01\mu(\mathbf{x}) \in \mathcal{O} \text{ and } 1\mu(\mathbf{x}) \text{ begins with } 010\} = \emptyset \\
\delta(E,4) &= \{\mathbf{x} \in \Sigma^\omega \ : \ 011\mu(\mathbf{x}) \in \mathcal{O} \text{ and } 11\mu(\mathbf{x}) \text{ begins with } 010\} = \emptyset \\
\delta(F,0) &= \{\mathbf{x} \in \Sigma^\omega \ : \ 0\mu(\mathbf{x}) \in \mathcal{O} \text{ and } \mu(\mathbf{x}) \text{ begins with } 11\} = \emptyset \\
\delta(F,1) &= \{\mathbf{x} \in \Sigma^\omega \ : \ 00\mu(\mathbf{x}) \in \mathcal{O} \text{ and } 0\mu(\mathbf{x}) \text{ begins with } 11\} = \emptyset \\
\delta(F,2) &= \{\mathbf{x} \in \Sigma^\omega \ : \ 000\mu(\mathbf{x}) \in \mathcal{O} \text{ and } 00\mu(\mathbf{x}) \text{ begins with } 11\} = \emptyset \\
\delta(F,4) &= \{\mathbf{x} \in \Sigma^\omega \ : \ 011\mu(\mathbf{x}) \in \mathcal{O} \text{ and } 11\mu(\mathbf{x}) \text{ begins with } 11\} = \emptyset \\
\delta(J,2) &= \{\mathbf{x} \in \Sigma^\omega \ : \ 100\mu(\mathbf{x}) \in \mathcal{O} \text{ and } 00\mu(\mathbf{x}) \text{ begins with } 0\} = \emptyset \\
\delta(J,3) &= \{\mathbf{x} \in \Sigma^\omega \ : \ 11\mu(\mathbf{x}) \in \mathcal{O} \text{ and } 1\mu(\mathbf{x}) \text{ begins with } 0\} = \emptyset \\
\delta(J,4) &= \{\mathbf{x} \in \Sigma^\omega \ : \ 111\mu(\mathbf{x}) \in \mathcal{O} \text{ and } 11\mu(\mathbf{x}) \text{ begins with } 0\} = \emptyset \\
\delta(K,2) &= \{\mathbf{x} \in \Sigma^\omega \ : \ 000\mu(\mathbf{x}) \in \mathcal{O} \text{ and } 00\mu(\mathbf{x}) \text{ begins with } 0\} = \emptyset \\
\delta(K,3) &= \{\mathbf{x} \in \Sigma^\omega \ : \ 01\mu(\mathbf{x}) \in \mathcal{O} \text{ and } 1\mu(\mathbf{x}) \text{ begins with } 0\} = \emptyset \\
\delta(K,4) &= \{\mathbf{x} \in \Sigma^\omega \ : \ 011\mu(\mathbf{x} \in \mathcal{O}) \text{ and } 11\mu(\mathbf{x}) \text{ begins with } 0\} = \emptyset
\end{aligned}
$$

The proof of most of these is immediate. (We have not listed $\delta(I,a)$ for $a \in \{0,2,3,4\}$, nor $\delta(G,a)$ for $a \in \{1,2,4\}$, nor $\delta(H,a)$ for $a \in \{1,2,4\}$, as these are symmetric with other cases.) The only one that requires some thought is $\delta(F,4)$:

- If $\mathbf{x}$ begins 00, then $011\mu(\mathbf{x}) = 0110101\cdots$, which has 10101 as a factor.
- If $\mathbf{x}$ begins 01, then $011\mu(\mathbf{x}) = 0110110\cdots$, which has 0110110 as a factor.
- If $\mathbf{x}$ begins 1, then $011\mu(\mathbf{x}) = 01110\cdots$, which has 111 as a factor.

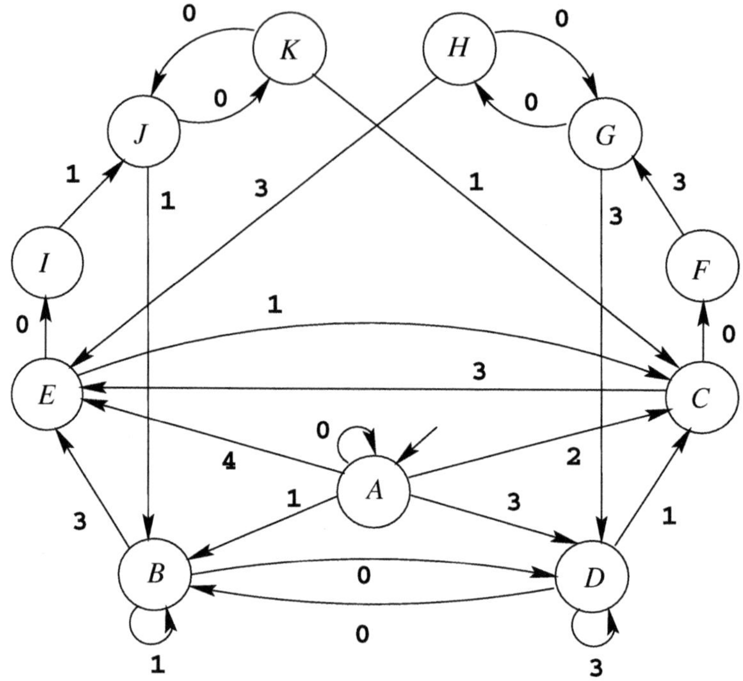

Fig. 1. Automaton coding infinite binary overlap-free words

From Lemma 2 and the results above, we get

Theorem 2. *Every infinite binary overlap-free word* $\mathbf{x}$ *is encoded by an infinite path, starting in A, through the automaton in Figure 1.*

Every infinite path through the automaton not ending in 0^ω *codes a unique infinite binary overlap-free word* $\mathbf{x}$. *If a path* $\mathbf{i}$ *ends in* 0^ω *and this suffix corresponds to a cycle on state A or a cycle between states B and D, then* $\mathbf{x}$ *is coded by either* $\mathbf{i};1$ *or* $\mathbf{i};3$. *If a path* $\mathbf{i}$ *ends in* 0^ω *and this suffix corresponds to a cycle between states J and K, then* $\mathbf{x}$ *is coded by* $\mathbf{i};1$. *If a path* $\mathbf{i}$ *ends in* 0^ω *and this suffix corresponds to a cycle between states G and H, then* $\mathbf{x}$ *is coded by* $\mathbf{i};3$.

Corollary 2. *Each of the 11 sets* $A, B, \ldots, K$ *is uncountable.*

Proof. We prove this for K, with the proof for the other sets being similar. Elements in the set K correspond to those infinite paths leaving the state K in Figure 1. It therefore suffices to produce uncountably many distinct paths leaving K. One way to do this, for example, is by $\{13010, 1301000\}^\omega$.

3 The Lexicographically Least Overlap-Free Word

We now recover a theorem of [1]:

Theorem 3. *The lexicographically least infinite binary overlap-free word is* $001001\overline{\mathbf{t}}$.

Proof. Let $\mathbf{x}$ be the lexicographically least infinite word, and let $\mathbf{y}$ be its code. Then $\mathbf{y}[1]$ must be 2, since any other choice codes a word that starts with 01 or something lexicographically greater. Once $\mathbf{y}[1] = 2$ is chosen, the next two symbols must be $\mathbf{y}[2..3] = 03$. Now we are in state G. We argue that the lexicographically least string that follows causes us to alternate between states G and H on 0, producing $100\cdots$. For otherwise our only choices are 30, 31, or (if we are in G) 33 as the next two symbols, and all of these code a word lexicographically greater than 100. Hence $\mathbf{y} = 203\,0^\omega; 1$ is the code for the lexicographically least sequence, and this codes $001001\overline{\mathbf{t}}$.

4 Automatic Infinite Binary Overlap-Free Words

As a consequence of Theorem 2, we can give a complete description of the infinite binary overlap-free words that are 2-automatic [2]. Recall that an infinite word $(a_n)_{n\geq 0}$ is k-automatic if there exists a deterministic finite automaton with output that, on input n expressed in base k, produces an output associated with the state last visited that is equal to a_n.

Theorem 4. *An infinite binary overlap-free word is 2-automatic if and only if its code is both specified by the DFA given above in Figure 1, and is ultimately periodic.*

First, we need two lemmas:

Lemma 3. *An infinite binary word* $\mathbf{x} = a_0a_1a_2\cdots$ *is 2-automatic if and only if* $\mu(\mathbf{x})$ *is 2-automatic.*

Proof. For one direction, we use the fact that the class of k-automatic sequences is closed under uniform morphisms ([2, Theorem 6.8.3]). So if $\mathbf{x}$ is 2-automatic, so is $\mu(\mathbf{x})$.

For the other, we use the well-known characterization of automatic sequences in terms of the k-kernel [2, Theorem 6.6.2]: a sequence $(c_n)_{n\geq 0}$ is k-automatic if and only if its k-kernel defined by

$$\{(c_{k^e n+i})_{n\geq 0} \ : \ e \geq 0 \text{ and } 0 \leq i < k^e\}$$

is finite. Furthermore, each sequence in the k-kernel is k-automatic.

Now if $\mathbf{y} = \mu(\mathbf{x}) = b_0b_1b_2\cdots$, then $b_{2n} = a_n$. So one of the sequences in the 2-kernel of $\mathbf{y}$ is $\mathbf{x}$, and if $\mathbf{y}$ is 2-automatic, then so is $\mathbf{x}$.

Now we can prove Theorem 4.

Proof. Suppose the code of $\mathbf{x}$ is ultimately periodic. Then we can write its code as yz^ω for some finite words y and z. Since the class of 2-automatic sequences is closed under appending a finite prefix [2, Corollary 6.8.5], by Lemma 3, it suffices to show that the word coded by z^ω is 2-automatic.

The word z^ω codes an overlap-free word $\mathbf{w}$ satisfying $\mathbf{w} = t\varphi(\mathbf{w})$, where t is a finite word and φ is a power of μ. If t is empty the result is clear. Otherwise, by iteration, we get that

$$\mathbf{w} = t\varphi(t)\varphi^2(t)\cdots. \tag{27}$$

The 2-kernel of a sequence is obtained by repeated 2-*decimation*, that is, recursively splitting a sequence into its even- and odd-indexed terms. When we apply 2-decimation to $\mu^k(t)$, where t is a finite word, we get $\mu^{k-1}(t)$ and $\mu^{k-1}(\overline{t})$. These words are both of even length, provided k is at least 1. Hence iteratively applying 2-decimation to $\mathbf{w}$, as given in (27), shows that if $\varphi = \mu^k$, then the 2-kernel of $\mathbf{w}$ is contained in

$$S := \{u\mu^i(v)\mu^{i+k}(v)\mu^{i+2k}(v)\cdots \ : \ |u| \leq |t| \text{ and } v \in \{t, \overline{t}\} \text{ and } 1 \leq i \leq k\},$$

which is a finite set.

On the other hand, suppose the code for $\mathbf{x}$ is not ultimately periodic. Then we show that the 2-kernel is infinite. To see this, note that the code for $\mathbf{x}$ contains a 2 or 4 only at the beginning, so we can assume without loss of generality that the code for $\mathbf{x}$ contains only the letters $0, 1, 3$. Now it is easy to see that if the code for $\mathbf{x}$ is $a\mathbf{y}$ for some letter $a \in \{0, 1, 3\}$ and infinite string $\mathbf{y} \in \{0, 1, 3\}^\omega$, then one of the sequences in the 2-kernel (obtained by taking either the odd- or even-indexed terms) is either coded by $\mathbf{y}$ or its complement is coded by $\mathbf{y}$. Since the code for $\mathbf{x}$ is not ultimately periodic, there are infinitely many distinct sequences in the orbit of the code for $\mathbf{x}$, under the shift. (By the orbit of $\mathbf{y}$ we mean the set of sequences of the form $\mathbf{y}[i..\infty]$ for $i \geq 1$.) Now infinitely many of these sequences correspond to a sequence in the 2-kernel, or its complement. Hence $\mathbf{x}$ is not 2-automatic.

5 Remarks

According to a theorem of Karhumäki and the author [8], there is a similar factorization theorem for all exponents α with $2 < \alpha \leq \frac{7}{3}$. Recently we have proven similar results for $\alpha = \frac{7}{3}$ [9].

I am grateful to the referees for a careful reading of the manuscript.

References

1. Allouche, J.-P., Currie, J., Shallit, J.: Extremal infinite overlap-free binary words. Electronic J. Combinatorics 5(1), R27 (1998) (electronic), `http://www.combinatorics.org/Volume_5/Abstracts/v5i1r27.html`
2. Allouche, J.-P., Shallit, J.: Automatic Sequences: Theory, Applications, Generalizations. Cambridge University Press, Cambridge (2003)
3. Berstel, J.: A rewriting of Fife's theorem about overlap-free words. In: Karhumäki, J., Rozenberg, G., Maurer, H. (eds.) Results and Trends in Theoretical Computer Science. LNCS, vol. 812, pp. 19–29. Springer, Heidelberg (1994)
4. Blondel, V.D., Cassaigne, J., Jungers, R.M.: On the number of α-power-free binary words for $2 < \alpha \leq 7/3$. Theoret. Comput. Sci. 410, 2823–2833 (2009)

5. Carpi, A.: Overlap-free words and finite automata. Theoret. Comput. Sci. 115, 243–260 (1993)
6. Cassaigne, J.: Counting overlap-free binary words. In: Enjalbert, P., Wagner, K.W., Finkel, A. (eds.) STACS 1993. LNCS, vol. 665, pp. 216–225. Springer, Heidelberg (1993)
7. Fife, E.D.: Binary sequences which contain no *BBb*. Trans. Amer. Math. Soc. 261, 115–136 (1980)
8. Karhumäki, J., Shallit, J.: Polynomial versus exponential growth in repetition-free binary words. J. Combin. Theory. Ser. A 105, 335–347 (2004)
9. Rampersad, N., Shallit, J., Shur, A.: Fife's theorem for $\frac{7}{3}$-powers (2011) (preprint)
10. Restivo, A., Salemi, S.: Overlap free words on two symbols. In: Perrin, D., Nivat, M. (eds.) Automata on Infinite Words. LNCS, vol. 192, pp. 198–206. Springer, Heidelberg (1985)

Infinite Words Rich and Almost Rich in Generalized Palindromes

Edita Pelantová and Štěpán Starosta

Department of Mathematics, FNSPE, Czech Technical University in Prague, Trojanova 13, 120 00 Praha 2, Czech Republic

Abstract. We focus on Θ-rich and almost Θ-rich words over a finite alphabet $\mathcal{A}$, where Θ is an involutive antimorphism over $\mathcal{A}^*$. We show that any recurrent almost Θ-rich word $\mathbf{u}$ is an image of a recurrent Θ'-rich word under a suitable morphism, where Θ' is again an involutive antimorphism. Moreover, if the word $\mathbf{u}$ is uniformly recurrent, we show that Θ' can be set to the reversal mapping. We also treat one special case of almost Θ-rich words. We show that every Θ-standard word with seed is an image of an Arnoux-Rauzy word.

Keywords: palindrome, pseudopalindrome, palindromic defect, richness.

1 Introduction

In this paper we deal with infinite words over a finite alphabet $\mathcal{A}$. Given a word $\mathbf{u} \in \mathcal{A}^{\mathbb{N}}$ we are interested whether its language $\mathcal{L}(\mathbf{u})$ is saturated, in a certain sense, by generalized palindromes, here called Θ-palindromes. We use the symbol Θ for an involutive antimorphism, i.e., a mapping $\Theta : \mathcal{A}^* \mapsto \mathcal{A}^*$ such that $\Theta^2 = \mathrm{Id}$ and $\Theta(uv) = \Theta(v)\Theta(u)$ for all $u, v \in \mathcal{A}^*$. Fixed points of Θ are called Θ-palindromes.

A strong impulse for study of palindromes recently came from outside of mathematics. Physicists discovered a role of classical palindromes in the description of the spectrum of Schrödinger operators with aperiodic potentials, see [14]. In genetics, the so called Watson-Crick palindromes play, for instance, an important role in the description of unwanted bindings of nucleotides in a DNA strand, see [15]. In our terminology, the Watson-Crick palindromes are Θ-palindromes where the involutive antimorphism Θ acts on a quaternary alphabet and has no fixed point of length one.

The most common antimorphism used in combinatorics on words is the reversal mapping. We denote it by R. The reversal mapping associates to every word $w = w_1w_2\ldots w_n$ its mirror image $R(w) = w_nw_{n-1}\ldots w_1$. In the case $w = R(w)$, we sometimes say that w is a palindrome or classical palindrome instead of R-palindrome.

The set of distinct Θ-palindromes occurring in a finite word w is denoted $\mathrm{Pal}_\Theta(w)$. Since the empty word ε is a Θ-palindrome for any Θ, we have a simple lower bound $\#\mathrm{Pal}_\Theta(w) \geq 1$.

G. Mauri and A. Leporati (Eds.): DLT 2011, LNCS 6795, pp. 406–416, 2011.

In 2001, Droubay et al. gave in [12] an upper bound for the reversal mapping R. They deduced that $\#\mathrm{Pal}_R(w) \leq |w| + 1$, where $|w|$ denotes the length of the word w. In [4], Blondin Massé et al. studied involutive antimorphisms with no fixed points of length 1. For such Θ they decreased the upper bound, in particular, they showed that $\#\mathrm{Pal}_\Theta(w) \leq |w|$ for all non-empty word w. In [18], the upper bound is more precise. The following estimate is valid for any involutive antimorphism Θ:

$$\#\mathrm{Pal}_\Theta(w) \leq |w| + 1 - \gamma_\Theta(w), \tag{1}$$

where $\gamma_\Theta(w) := \#\big\{\{a, \Theta(a)\} \mid a \in \mathcal{A}, a \text{ occurs in } w, \text{and } a \neq \Theta(a)\big\}$. Let us note that if $\Theta = R$, then $\gamma_\Theta(w) = 0$, and the upper bound in (1) is the same as for classical palindromes.

According to the terminology for classical palindromes introduced in [13] and for Θ-palindromes in [18], we say that a finite word w is *Θ-rich* if the equality in (1) holds. An infinite word $\mathbf{u} \in \mathcal{A}^\mathbb{N}$ is *Θ-rich* if every factor $w \in \mathcal{L}(\mathbf{u})$ is Θ-rich. In [5], the authors introduced the *palindromic defect* of a finite word w as the difference between the upper bound $|w| + 1$ and the actual number of distinct palindromic factors. We define analogously the *Θ-palindromic defect* of w as

$$D_\Theta(w) := |w| + 1 - \gamma_\Theta(w) - \#\mathrm{Pal}_\Theta(w).$$

We define for an infinite word $\mathbf{u}$ its *Θ-palindromic defect* as

$$D_\Theta(\mathbf{u}) = \sup\{D_\Theta(w) \mid w \in \mathcal{L}(\mathbf{u})\}.$$

Words with finite Θ-palindromic defect are referred to as *almost Θ-rich*. The notion of almost richness for classical palindromes was introduced and studied in [13].

In [7], it is shown that rich words (i.e. R-rich words) can be characterized using an inequality shown in [2] for infinite words with languages closed under reversal. Results of both mentioned papers were generalized in [18] for an arbitrary involutive antimorphism. In particular, it is shown that if an infinite word has its language closed under Θ, the following inequality holds

$$\mathcal{C}(n+1) - \mathcal{C}(n) + 2 \geq \mathcal{P}_\Theta(n) + \mathcal{P}_\Theta(n+1) \text{ for all } n \geq 1, \tag{2}$$

where $\mathcal{C}(n)$ is the *factor complexity* defined by $\mathcal{C}(n) := \#\{w \in \mathcal{L}(\mathbf{u}) \mid n = |w|\}$ and $\mathcal{P}_\Theta(n)$ is the *Θ-palindromic complexity* defined by $\mathcal{P}_\Theta(n) := \#\{w \in \mathcal{L}(\mathbf{u}) \mid w = \Theta(w) \text{ and } n = |w|\}$. The gap between the left-hand side and the right-hand side in (2) decides about Θ-richness. Let us therefore denote by $T_\Theta(n)$ the quantity

$$T_\Theta(n) := \mathcal{C}(n+1) - \mathcal{C}(n) + 2 - \mathcal{P}_\Theta(n+1) - \mathcal{P}_\Theta(n).$$

In [18], it is also shown that an infinite word with language closed under Θ is Θ-rich if and only if

$$T_\Theta(n) = 0 \text{ for all } n \geq 1.$$

The list of infinite words which are R-rich is quite extensive. See for instance [2,9,11,13]. Examples of Θ-rich words can be found in the class of words called Θ-episturmian words. A condition when such a word is Θ-rich can be found in [18]. In [1], the authors also deal with Θ-episturmian words (they are called pseudopalindromic in the paper). However, the result of Theorem 2 in [1] is valid only for the subset of Θ-rich Θ-episturmian words, not for all Θ-episturmian words as stated in the paper.

Fewer examples of words with finite non-zero palindromic defect are known. Periodic words with finite non-zero R-defect can be found in [5], aperiodic ones are studied in [13] and [3]. To our knowledge, examples of words with $0 < D_\Theta(\mathbf{u}) < +\infty$ and $\Theta \neq R$ have not yet been explicitly exhibited. As we will show, such examples are Θ-standard words with seed defined in [10] and thus also their subset, standard Θ-episturmian words, which can be constructed from standard episturmian words, see [8].

The main aim of this paper is to show that among words with finite Θ-palindromic defect, Θ-rich words, i.e. words with $D_\Theta(\mathbf{u}) = 0$, play an important role. We will show the following theorems.

Theorem 1. *Let $\Theta_1 : \mathcal{A}^* \mapsto \mathcal{A}^*$ be an involutive antimorphism. Let $\mathbf{u} \in \mathcal{A}^\mathbb{N}$ be a recurrent infinite word such that $D_{\Theta_1}(\mathbf{u}) < +\infty$. Then there exist an involutive antimorphism $\Theta_2 : \mathcal{B}^* \mapsto \mathcal{B}^*$, a morphism $\varphi : \mathcal{B}^* \mapsto \mathcal{A}^*$ and an infinite recurrent word $\mathbf{v} \in \mathcal{B}^\mathbb{N}$ such that*

$$\mathbf{u} = \varphi(\mathbf{v}) \text{ and } \mathbf{v} \text{ is } \Theta_2\text{-rich.}$$

A stronger statement can be shown if uniform recurrency is assumed.

Theorem 2. *Let $\Theta : \mathcal{A}^* \mapsto \mathcal{A}^*$ be an involutive antimorphism. Let $\mathbf{u} \in \mathcal{A}^\mathbb{N}$ be a uniformly recurrent infinite word such that $D_\Theta(\mathbf{u}) < +\infty$. Then there exist a morphism $\varphi : \mathcal{B}^* \mapsto \mathcal{A}^*$ and an infinite uniformly recurrent word $\mathbf{v} \in \mathcal{B}^\mathbb{N}$ such that*

$$\mathbf{u} = \varphi(\mathbf{v}) \text{ and } \mathbf{v} \text{ is } R\text{-rich.}$$

One can conclude that rich words, using the classical notion of palindrome, play somewhat a more important role than Θ-rich words for an arbitrary $\Theta \neq R$.

The proofs of the two stated theorems do not provide any relation between the size of the alphabet $\mathcal{B}$ of the word $\mathbf{v}$ and the size of the original alphabet $\mathcal{A}$. In the following special case, the size of $\mathcal{B}$ can be bounded. Moreover, the word $\mathbf{v}$ is more specific, namely it is Arnoux-Rauzy. Let us recall that an infinite word $\mathbf{v}$ is an Arnoux-Rauzy word if for every n we have $\mathcal{C}(n) = (\#\mathcal{A} - 1)n + 1$ and there is exactly one factor $w \in \mathcal{L}(\mathbf{v})$ of length n which can be extended to the left in more than one way, i.e., is left special. Ternary Arnoux-Rauzy words were first metioned in [17].

Theorem 3. *Let $\Theta : \mathcal{A}^* \mapsto \mathcal{A}^*$ be an involutive antimorphism and $\mathbf{u} \in \mathcal{A}^\mathbb{N}$ be a Θ-standard word with seed. Then there exist an Arnoux-Rauzy word $\mathbf{v} \in \mathcal{B}^\mathbb{N}$ and a morphism $\varphi : \mathcal{B}^* \mapsto \mathcal{A}^*$ such that*

$$\mathbf{u} = \varphi(\mathbf{v}) \text{ and } \#\mathcal{B} \leq \#\mathcal{A}.$$

One of the reviewers of this paper pointed out to us the fact that the last theorem is in fact a restatement in a weak form of Theorem 1 in [6]. We keep it here with a proof for the sake of completeness in the context of Θ-richness.

All three mentioned theorems present an almost Θ_1-rich word as an image of a Θ_2-rich word by a suitable morphism. The opposite question when a morphic image of a Θ_1-rich word is almost Θ_2-rich is not tackled here. In [13], a type of morphisms preserving the set of almost R-rich words is studied.

2 Properties of Words with Finite Θ-Defect

We consider mainly infinite words $\mathbf{u} = (u_n)_{n \in \mathbb{N}} \in \mathcal{A}^{\mathbb{N}}$ having their language $\mathcal{L}(\mathbf{u})$ closed under a given involutive antimorphism Θ. In other words, for any factor $w \in \mathcal{L}(\mathbf{u})$ we have $\Theta(w) \in \mathcal{L}(\mathbf{u})$.

For any factor $w \in \mathcal{L}(\mathbf{u})$ there exists an index i such that w is a prefix of the infinite word $u_i u_{i+1} u_{i+2} \ldots$. Such an index is called an *occurrence* of w in $\mathbf{u}$. If each factor of $\mathbf{u}$ has infinitely many occurrences in $\mathbf{u}$, the infinite word $\mathbf{u}$ is said to be *recurrent*. It is easy to see that if the language of $\mathbf{u}$ is closed under Θ, then $\mathbf{u}$ is recurrent. For a recurrent infinite word $\mathbf{u}$, we may define the notion of a *complete return word* of any $w \in \mathcal{L}(\mathbf{u})$. It is a factor $v \in \mathcal{L}(\mathbf{u})$ such that w is a prefix and a suffix of v and w occurs in v exactly twice. By a *return word* of a factor w we mean a word $q \in \mathcal{L}(\mathbf{u})$ such that qw is a complete return word of w. If every factor w of a recurrent word $\mathbf{u}$ has only finitely many return words, then the infinite word $\mathbf{u}$ is called *uniformly recurrent.*

An important role for the description of languages closed under Θ is played by the so-called super reduced Rauzy graphs $G_n(\mathbf{u})$, introduced in [7]. Before defining them, we introduce some necessary notions.

We say that a factor $w \in \mathcal{L}(\mathbf{u})$ is left special (LS) if w has at least two left extensions, i.e., if there exist two letters $a, b \in \mathcal{A}$, $a \neq b$, such that $aw, bw \in \mathcal{L}(\mathbf{u})$. A right special (RS) factor is defined analogously. If a factor is LS and RS, we refer to it as bispecial. The fact that $\mathcal{L}(\mathbf{u})$ is closed under Θ assures the following relation: a factor w is LS if and only if the factor $\Theta(w)$ is RS.

An *n-simple path* e is a factor of $\mathbf{u}$ of length at least $n+1$ such that the only special (right or left) factors of length n occurring in e are its prefix and suffix of length n. If w is the prefix of e of length n and v is the suffix of e of length n, we say that the n-simple path e begins with w and ends with v. We denote by $G_n(\mathbf{u})$ an undirected graph whose set of vertices is formed by unordered pairs $\{w, \Theta(w)\}$ such that $w \in \mathcal{L}(\mathbf{u})$, $|w| = n$, and w is RS or LS. We connect two vertices $\{w, \Theta(w)\}$ and $\{v, \Theta(v)\}$ by an unordered pair $\{e, \Theta(e)\}$ if e or $\Theta(e)$ is an n-simple path beginning with w or $\Theta(w)$ and ending with v or $\Theta(v)$. Note that the graph $G_n(\mathbf{u})$ may have multiple edges and loops.

As first shown for classical palindromes in [7], the super reduced Rauzy graph $G_n(\mathbf{u})$ can be used to detect the equality in (2). Let us cite Corollary 7 from [18].

Proposition 4. *Let $n \in \mathbb{N}$ and $\mathcal{L}(\mathbf{u})$ be closed under Θ. Then $T_\Theta(n) = 0$ if and only if*

1. *all n-simple paths forming a loop in $G_n(\mathbf{u})$ are Θ-palindromes and*
2. *the graph obtained from $G_n(\mathbf{u})$ by removing all loops is a tree.*

Analogous to the case of the reversal mapping, one can see from the definition of Θ-defect that an infinite word $\mathbf{u}$ has finite Θ-defect if and only if there exists an integer H such that every prefix p of $\mathbf{u}$ of length greater than H has a unioccurrent Θ-palindromic suffix, i.e., a suffix occurring exactly once in p. We use this fact to prove the following lemma, which generalizes Proposition 4.4. in [13].

Lemma 5. *Let $\mathbf{u}$ be a recurrent infinite word with finite Θ-defect. Then $\mathcal{L}(\mathbf{u})$ is closed under Θ.*

Proof. Let H be an integer such that every prefix of $\mathbf{u}$ of length greater than H has a unioccurrent Θ-palindromic suffix. Suppose that w is a factor of $\mathbf{u}$ such that $\Theta(w) \notin \mathcal{L}(\mathbf{u})$. Since $\mathbf{u}$ is recurrent, we can find two consecutive occurrences i and j of the factor w such that $i, j > H$ and $i < j$. Denote p the prefix of $\mathbf{u}$ ending with w occurring at j, i.e., $|p| = j + |w|$. Since $|p| > H$, there exists a unioccurrent Θ-palindromic suffix of p. Denote s to be such a suffix. If $|s| \leq |w|$, then s is a factor of w and thus occurs at least twice in p - a contradiction with the unioccurrence of s. If $|s| > |w|$, the w is a factor of s which is a Θ-palindrome and thus contains $\Theta(w)$ as well - a contradiction with the assumption that $\Theta(w) \notin \mathcal{L}(\mathbf{u})$.

In [3], various properties are shown for words with finite R-palindromic defect. These properties and their proofs are valid even if we replace the antimorphism R by an arbitrary Θ.

Proposition 6. *Let $\mathbf{u}$ be a recurrent infinite word such that $D_\Theta(\mathbf{u}) < +\infty$. Then there exists a positive integer H such that*

- *every prefix of $\mathbf{u}$ longer than H has a unioccurent Θ-palindromic suffix;*
- *for any factor $w \in \mathcal{L}(\mathbf{u})$ such that $|w| > H$, occurrences of w and $\Theta(w)$ in the word $\mathbf{u}$ alternate;*
- *for any $w \in \mathcal{L}(\mathbf{u})$ such that $|w| > H$, every factor $v \in \mathcal{L}(\mathbf{u})$ beginning with w, ending with $\Theta(w)$, and with no other occurrences of w or $\Theta(w)$ is a Θ-palindrome;*
- *$T_\Theta(n) = 0$ for any integer $n > H$.*

The main difference for an arbitrary Θ is that there can be non-Θ-palindromic letters. However, this can be dealt with by a good choice of the constant H, and one can then follow the proofs for $\Theta = R$ in [3]. Therefore, we do not give a proof of the previous proposition.

Let us recall that in case of the reversal mapping and $D_R(\mathbf{u}) = 0$, the listed properties are generalizations of properties already shown in [13] and [11]. In this case we have $H = 0$.

As already mentioned, the first property stated in the previous proposition, in fact, characterizes words with finite Θ-defect. We do not know whether this is the case of the remaining properties. If we restrict our attention to uniformly recurrent words, only then several characterizations of words with finite Θ-defect can be shown. The next proposition states two of them that we use in what follows. Again, the proposition is based on the work done in [3] for $\Theta = R$. No modifications besides replacing R by Θ in its proof are needed, therefore, we omit it.

Proposition 7. *Let* $\mathbf{u}$ *be a uniformly recurrent infinite word with language closed under* Θ*. The following statements are equivalent.*

- $D_\Theta(\mathbf{u}) < +\infty$;
- *there exists a positive integer* K *such that for any* Θ*-palindrome* $w \in \mathcal{L}(\mathbf{u})$ *of length* $|w| \geq K$*, all complete return words of* w *are* Θ*-palindromes;*
- *there exists a positive integer* H *such that for any* $w \in \mathcal{L}(\mathbf{u})$*,* $|w| > H$*, the longest* Θ*-palindromic suffix of* w *is unioccurrent in* w*.*

A Θ-standard word with seed is an infinite word defined by using Θ-palindromic closure, for details see [10]. Construction of such word $\mathbf{u}$ guarantees that $\mathbf{u}$ is uniformly recurrent (cf. Proposition 3.5. in [10]). The authors of [10] showed (Proposition 4.8) that any complete return word of a sufficiently long Θ-palindromic factor is a Θ-palindrome as well. Therefore, Θ-standard words with seed serve as an example of almost Θ-rich words.

Corollary 8. *Let* $\mathbf{u}$ *be a* Θ*-standard word with seed. Then* $D_\Theta(\mathbf{u}) < +\infty$*.*

3 Proofs

In this section we give proofs of all three theorems stated in Introduction. Although Theorem 2 seems to be only a refinement of Theorem 1, constructions of the morphisms φ in their proofs differ substantially. It is caused by stronger properties we may exploit for a uniformly recurrent word.

Proof (Proof of Theorem 1). Recall that according to Theorem 5 the language $\mathcal{L}(\mathbf{u})$ is closed under Θ_1.

If $\mathbf{u}$ is an eventually periodic word with language closed under Θ_1, then $\mathbf{u}$ is purely periodic. Any purely periodic word is a morphic image of a word $\mathbf{v}$ over one-letter alphabet under the morphism which assigns to this letter the period of $\mathbf{u}$. Therefore we may assume without loss of generality that $\mathbf{u}$ is not eventually periodic.

Since $D_{\Theta_1}(\mathbf{u}) < +\infty$, according to Theorems 4 and 6, there exists $H \in \mathbb{N}$ such that

1. $\forall w \in \mathcal{L}(\mathbf{u})$, $|w| > H$, occurrences of w and $\Theta_1(w)$ alternate;
2. $\forall w \in \mathcal{L}(\mathbf{u})$, $|w| > H$, every factor beginning with w, ending with $\Theta_1(w)$ and with no other occurrences of w or $\Theta_1(w)$ is a Θ_1-palindrome;
3. $\forall n \geq H$, every loop in $G_n(\mathbf{u})$ is a Θ_1-palindrome and the graph obtained from $G_n(\mathbf{u})$ by removing all loops is a tree.

Fix $n > H$. If an edge $\{b, \Theta_1(b)\}$ in $G_n(\mathbf{u})$ is a loop, then, according to Property 3, we have $b = \Theta_1(b)$. If the edge $\{b, \Theta_1(b)\}$ connects two distinct vertices $\{w_1, \Theta_1(w_1)\}$ and $\{w_2, \Theta_1(w_2)\}$, then there exist exactly two n-simple paths b and $\Theta_1(b)$ such that without loss of generality the n-simple path b begins with w_1 and ends with w_2 and the n-simple path $\Theta_1(b)$ begins with $\Theta_1(w_2)$ and ends with $\Theta_1(w_1)$.

We assign to every n-simple path b a new symbol $[b]$, i.e., we define the alphabet $\mathcal{B}$ as

$$\mathcal{B} := \{[b] \mid b \in \mathcal{L}(\mathbf{u}) \text{ is an } n\text{-simple path}\}$$

and on this alphabet we define an involutive antimorphism $\Theta_2 : \mathcal{B}^* \mapsto \mathcal{B}^*$ in the following way:

$$\Theta_2([b]) := [\Theta_1(b)].$$

We are now going to construct a suitable infinite word $\mathbf{v} \in \mathcal{B}^{\mathbb{N}}$. Let $(s_i)_{i\in\mathbb{N}}$ denote a strictly increasing sequence of indices such that s_i is an occurrence of a LS or RS factor of length n and every LS and RS factor of length n occurs at some index s_i. We define $\mathbf{v} = (v_i)_{i\in\mathbb{N}}$ by the formula

$$v_i = [b] \quad \text{if} \quad b = u_{s_i} u_{s_i+1} u_{s_i+2} \ldots u_{s_{i+1}+n-1}.$$

This construction can be done for any $n > H$. Since infinitely many prefixes of $\mathbf{u}$ are LS or RS factors, we can choose $n > H$ such that the prefix of $\mathbf{u}$ of length n is LS or RS, i.e., $s_0 = 0$.

According to Proposition 12 in [18], to prove that $\mathbf{v}$ is Θ_2-rich we need to show the following:

(i) for every non-empty factor $w \in \mathcal{L}(\mathbf{v})$, any factor v beginning with w and ending with $\Theta_2(w)$, with no other occurrences of w or $\Theta_2(w)$, is a Θ_2-palindrome;

(ii) for every letter $[b] \in \mathcal{B}$ such that $[b] \neq \Theta_2([b])$, the occurrences of $[b]$ and $\Theta_2([b])$ in the word $\mathbf{v}$ alternate.

Let us first verify (i). Let e and f be factors of $\mathbf{v}$ such that e is a prefix of f and $\Theta_2(e)$ is a suffix of f and there are no other occurrences of e or $\Theta_2(e)$ in f. In that case there exist integers $r \leq k$ such that $f = [b_1][b_2]\ldots[b_k]$ and $e = [b_1][b_2]\ldots[b_r]$. The case $r = k$ is trivial. Suppose $r < k$. Since $\mathbf{v}$ is defined as a coding of consecutive occurrences of n-simple paths in $\mathbf{u}$, factor f codes a certain segment of the word $\mathbf{u}$. Let us denote that segment $F = u_j \ldots u_l$ where $j = s_t$ for some $t \in \mathbb{N}$ and $l = s_{t+k-1} + n - 1$. Factor e codes in the same way a factor $E = u_j \ldots u_h$ where $h = s_{t+r-1} + n - 1$.

Due to the definition of Θ_2, the fact that e is a prefix of f and $\Theta_2(e)$ is a suffix of f ensures that E is a prefix of F and $\Theta(E)$ is a suffix of F. Suppose f is not a Θ_2-palindrome. This implies that F is not a Θ_1-palindrome which contradicts Property 3.

Let us now verify (ii). Consider $[b] \in \mathcal{B}$ such that $[b] \neq \Theta_2([b])$. Moving along the infinite word $\mathbf{u} = u_0u_1u_2\ldots$ from the left to the right with a window of width n corresponds to a walk in the graph $G_n(\mathbf{u})$. The pair b and $\Theta_1(b)$ of

n-simple paths in $\mathbf{u}$ represents an edge in $G_n(\mathbf{u})$ connecting two distinct vertices. Moreover, moving along the n-simple path b and moving along $\Theta_1(b)$ can be viewed as traversing that edge in opposite directions. Since the graph obtained from $G_n(\mathbf{u})$ by removing all loops is a tree, the only way to traverse an edge is alternately in one direction and in the other. Thus, the occurrences of letters $[b]$ and $\Theta_2([b])$ in $\mathbf{v}$ alternate.

We have shown that $\mathbf{v}$ is Θ_2-rich. It is now obvious how to define a morphism $\varphi : \mathcal{B}^* \mapsto \mathcal{A}^*$. If an n-simple path b equals $b = u_{s_i}u_{s_i+1}\dots u_{s_{i+1}+n-1}$, then we set $\varphi([b]) := u_{s_i}u_{s_i+1}\dots u_{s_{i+1}-1}$.

Proof (Proof of Theorem 2). Recall again that according to Theorem 5 the language $\mathcal{L}(\mathbf{u})$ is closed under Θ.

Next, we show that infinitely many Θ-palindromes are also prefixes of $\mathbf{u}$. Consider an integer H whose existence is guaranteed by Proposition 6 and denote by w a prefix of $\mathbf{u}$ longer than H. Since occurrences of factors w and $\Theta(w)$ in $\mathbf{u}$ alternate, according to the same proposition, the prefix of $\mathbf{u}$ ending with the first occurrence of $\Theta(w)$ is a Θ-palindrome.

Let us denote by p a Θ-palindromic prefix of $\mathbf{u}$ of length $|p| > K$ where K is the constant from Proposition 7. All complete return words of p are Θ-palindromes. Since $\mathbf{u}$ is uniformly recurrent, there exist only finite number of complete return words to p. Let $r^{(1)}, r^{(2)}, \dots, r^{(M)}$ be the list of all these complete return words. Any complete return word $r^{(i)}$ has the form $q^{(i)}p = r^{(i)}$ for some factor $q^{(i)}$, usually called return word of p. Since $r^{(i)}$ and p are Θ-palindromes, we have

$$p\Theta(q^{(i)}) = q^{(i)}p \quad \text{for any return word } \ q^{(i)}. \tag{3}$$

Let us define a new alphabet $\mathcal{B} = \{1, 2, \dots, M\}$ and morphism $\varphi : \mathcal{B}^* \to \mathcal{A}^*$ by the prescription

$$\varphi(i) = q^{(i)}, \quad \text{for } \ i = 1, 2, \dots, M\,.$$

First, we shall check the validity of the relation

$$\Theta\big(\varphi(w)p\big) = \varphi\big(R(w)\big)p \quad \text{for any } \ w \in \mathcal{B}^*\,. \tag{4}$$

Let $w = i_1 i_2 \dots i_n$. Then $\Theta\big(\varphi(i_1 i_2 \dots i_n)p\big)$ equals to

$$\Theta(p)\Theta\big(\varphi(i_n)\big)\Theta\big(\varphi(i_{n-1})\big)\dots\Theta\big(\varphi(i_1)\big) = p\Theta\big(q^{(i_n)}\big)\Theta\big(q^{(i_{n-1})}\big)\dots\Theta\big(q^{(i_1)}\big)$$

and we may apply gradually n times the equality (3) to rewrite the right-hand side as

$$q^{(i_n)}q^{(i_{n-1})}\dots q^{(i_1)}p = \varphi(i_n)\varphi(i_{n-1})\dots\varphi(i_1)p = \varphi\big(R(i_1 i_2 \dots i_n)\big)p.$$

This proves the relation (4).

An important property of the morphism φ is its injectivity. Indeed, in accordance with the definition, the number of occurrences of the factor p in $\varphi(w)p$

equals to the number of letters in w plus one. Moreover, each occurrence of p in $\varphi(w)p$ indicates a beginning of an image of a letter under φ. Therefore, $\varphi(w)p = \varphi(v)p$ necessarily implies $w = v$.

Let us finally define the word $\mathbf{v}$. As p is a prefix of $\mathbf{u}$, the word $\mathbf{u}$ can be written as a concatenation of return words $q^{(i)}$ and thus we can determine a sequence $\mathbf{v} = (v_n) \in \mathcal{B}^{\mathbb{N}}$ such that

$$\mathbf{u} = q^{(v_0)}q^{(v_1)}q^{(v_2)}\ldots$$

Directly from the definition of $\mathbf{v}$ we have $\mathbf{u} = \varphi(\mathbf{v})$. Since $\mathbf{u}$ is uniformly recurrent, the word $\mathbf{v}$ is uniformly recurrent as well. To prove that $\mathbf{v}$ is an R-rich word, we shall show that any complete return word of any R-palindrome in the word $\mathbf{v}$ is an R-palindrome as well. According to Theorem 2.14 in [13], this implies the R-richness of $\mathbf{v}$.

Let s be an R-palindrome in $\mathbf{v}$ and w its complete return word. Then $\varphi(w)p$ has precisely two occurrences of the factor $\varphi(s)p$. Since s is an R-palindrome, we have according to (4) that $\varphi(s)p$ is a Θ-palindrome of length $|\varphi(s)p| \geq |p| > K$. Therefore $\varphi(w)p$ is a complete return word of a long enough Θ-palindrome and according to our assumption $\varphi(w)p$ is a Θ-palindrome as well. Therefore by using (4) we have

$$\varphi(w)p = \Theta\big(\varphi(w)p\big) = \varphi\big(R(w)\big)p$$

and injectivity of φ gives $w = R(w)$, as we claimed.

Theorem 6.1 in [8] states that every standard Θ-episturmian word is an image of a standard episturmian word. Again, the role of R can be perceived as more important. Also, compared to Theorem 2, it may be seen as a special case since Θ-episturmian words, according to Theorem 8, have finite Θ-defect.

Proof (Proof of Theorem 3).

If $\mathbf{u}$ is periodic, then the claim is trivial. Suppose $\mathbf{u}$ is aperiodic.

We are going to repeat the proof of Theorem 2 with a more specific choice of p. Theorem 4.4 in [10] implies that there exists $L \in \mathbb{N}$ such that any LS factor of $\mathbf{u}$ longer than L is a prefix of $\mathbf{u}$. Without loss of generality, we may assume that the constant L is already chosen in such a way that all prefixes of $\mathbf{u}$ longer than L have the same left extensions. Let us denote their number by M. According to the same theorem, infinitely many prefixes of $\mathbf{u}$ are Θ-palindromes and thus bispecial factors as well.

According to Corollary 8, $\mathbf{u}$ has finite Θ-palindromic defect. Let K be the constant from Theorem 7. Altogether, there exists a bispecial factor p, $|p| > \max\{L, K\}$, such that it is a prefix of $\mathbf{u}$ and a Θ-palindrome. Since p is longer than K, all complete return words to p are Θ-palindromes. As p is the unique left special factor of length $|p|$ in $\mathbf{u}$, its return words (i.e., complete return words after erasing the suffix p) end with distinct letters. It means that there are exactly M return words of p, denoted again $q^{(i)}$. Let us recall that by M we denoted the number of left extensions of some factor, therefore $M \leq \#\mathcal{A}$.

The construction of the word $\mathbf{v}$ and the definition of the morphism φ over the alphabet $\mathcal{B} = \{1, 2, \ldots, M\}$ can be done in exactly the same way as in the proof of Theorem 2. It remains to show that $\mathbf{v}$ is an Arnoux-Rauzy word.

According to Theorem 2 we know that $\mathbf{v}$ is R-rich and uniformly recurrent. Applying Theorem 5 we deduce that the language $\mathcal{L}(\mathbf{v})$ is closed under reversal.

Suppose there exist $v, w \in \mathcal{L}(\mathbf{v})$, two LS factors such that $|v| = |w|$ and $v \neq w$. Since the words $q^{(i)}$ end with distinct letters, it is clear that $\varphi(w)p$ is a LS factor of $\mathbf{u}$ and it has the same number of left extensions as w. The same holds for $\varphi(v)p$. Since both these factors have their length greater than or equal to $|p| > L$ and are both LS, one must be prefix of another. Let without loss of generality $\varphi(w)p$ be a prefix of $\varphi(v)p$, i.e., $\varphi(v)p = \varphi(ww')p$. The injectivity of φ implies $w' = \varepsilon$ and thus $v = w$ – a contradiction.

Remark 9. Note that the proof of Theorem 3 is in fact a combination of methods used in the preceding proofs of Theorems 1 and 2 in the sense that the set of complete return words $r^{(i)}$ of the factor p and the set of $|p|$-simple paths in $\mathbf{u}$ coincide.

4 Conclusion

All presented results concern infinite words whose language is closed under one antimorphism. In particular, we proved that any uniformly recurrent Θ-rich word $\mathbf{u}$ is a morphic image of an R-rich word, or equivalently, that the reversal mapping R is more important than other antimorphisms. The question whether this statement is valid even in case when $\mathbf{u}$ is not uniformly recurrent is still open.

Infinite words whose languages are invariant under more antimorphisms are not treated at all in the paper. The famous Thue-Morse word belongs among such words. Recently the authors proved that words with a larger group of symmetries cannot be Θ-rich for any antimorphisms Θ from the group. Therefore, a new definition of richness which respects all symmetries presented in infinite word is suggested, see [16].

Acknowledgement

We would like to thank the reviewers of this paper for their fruitful remarks. We acknowledge financial support by the Czech Science Foundation grant GAČR 201/09/0584, by the grants MSM6840770039 and LC06002 of the Ministry of Education, Youth, and Sports of the Czech Republic, and by the grant of the Grant Agency of the Czech Technical University in Prague grant No. SGS11/162/OHK4/3T/14.

References

1. Anne, V., Zamboni, L.Q., Zorca, I.: Palindromes and pseudo-palindromes in episturmian and pseudo-episturmian infinite words. In: Brlek, S., Reutenauer, C. (eds.) Words 2005, vol. (36), pp. 91–100. LACIM (2005)

2. Baláži, P., Masáková, Z., Pelantová, E.: Factor versus palindromic complexity of uniformly recurrent infinite words. Theoret. Comput. Sci. 380(3), 266–275 (2007)
3. Balková, L., Pelantová, E., Starosta, Š.: Infinite words with finite defect. To appear in Adv. Appl. Math., (2011), preprint available at http://arxiv.org/abs/1009.5105
4. Blondin Massé, A., Brlek, S., Garon, A., Labbé, S.: Combinatorial properties of f-palindromes in the Thue-Morse sequence. Pure Math. Appl. 19(2-3), 39–52 (2008)
5. Brlek, S., Hamel, S., Nivat, M., Reutenauer, C.: On the palindromic complexity of infinite words. Internat. J. Found. Comput. 15(2), 293–306 (2004)
6. Bucci, M., De Luca, A.: On a family of morphic images of arnoux-rauzy words. In: Dediu, A.H., Ionescu, A.M., Martín-Vide, C. (eds.) LATA 2009. LNCS, vol. 5457, pp. 259–266. Springer, Heidelberg (2009)
7. Bucci, M., De Luca, A., Glen, A., Zamboni, L.Q.: A connection between palindromic and factor complexity using return words. Adv. in Appl. Math. 42(1), 60–74 (2009)
8. Bucci, M., de Luca, A., De Luca, A.: Characteristic morphisms of generalized episturmian words. Theor. Comput. Sci. 410, 2840–2859 (2009)
9. Bucci, M., de Luca, A., De Luca, A.: Rich and periodic-like words. In: Diekert, V., Nowotka, D. (eds.) DLT 2009. LNCS, vol. 5583, pp. 145–155. Springer, Heidelberg (2009)
10. Bucci, M., de Luca, A., De Luca, A., Zamboni, L.Q.: On different generalizations of episturmian words. Theoret. Comput. Sci. 393(1-3), 23–36 (2008)
11. Bucci, M., de Luca, A., De Luca, A., Zamboni, L.Q.: On theta-episturmian words. European J. Combin. 30(2), 473–479 (2009)
12. Droubay, X., Justin, J., Pirillo, G.: Episturmian words and some constructions of de Luca and Rauzy. Theoret. Comput. Sci. 255(1-2), 539–553 (2001)
13. Glen, A., Justin, J., Widmer, S., Zamboni, L.Q.: Palindromic richness. European J. Combin. 30(2), 510–531 (2009)
14. Hof, A., Knill, O., Simon, B.: Singular continuous spectrum for palindromic Schrödinger operators. Comm. Math. Phys. 174, 149–159 (1995)
15. Kari, L., Magalingam, K.: Watson-Crick palindromes in DNA computing. Nat. Comput. (9), 297–316 (2010)
16. Pelantová, E., Starosta, Š.: Languages invariant under more symmetries: overlapping factors versus palindromic richness. (2011), preprint available at http://arxiv.org/abs/1103.4051
17. Rauzy, G.: Suites à termes dans un alphabet fini. Séminaire de Théorie des Nombres de Bordeaux Anné 1982–1983(exposé 25) (1983)
18. Starosta, Š.: On theta-palindromic richness. Theoret. Comput. Sci. 412(12-14), 1111–1121 (2011)

Models of Pushdown Automata with Reset

Nuri Taşdemir and A.C. Cem Say

Boğaziçi University Department of Computer Engineering
P.K. 2 TR-34342 Bebek, Istanbul, Turkey
{nuri.tasdemir,say}@boun.edu.tr
http://www.cmpe.boun.edu.tr

Abstract. We examine various pushdown automaton variants that are architecturally intermediate between the one-way PDA and the two-way PDA (2PDA), where leftward moves of the input head can only reset it to the left end of the tape, and some component of the machine configuration may be "forgotten", that is, reset to its initial value, whenever such a move is performed. Most of these model variants are shown to be equivalent in power to either the 2PDA or the one-way PDA. One exception is the Resettable Pushdown Automaton (RPDA), where the stack contents are lost every time the input is reset, and which we prove to be intermediate in power between the PDA and the 2PDA. We give full characterizations of the classes of languages recognized by both the deterministic and the nondeterministic versions of the RPDA.

Keywords: automata and formal languages, two-way PDAs.

1 Introduction

The differences in computational power caused by restricting the input head of an automaton to a one-way, rather than two-way traversal of the tape content have been studied in great detail for various types of automata [1,11,12,13,16]. One interesting restriction is to allow leftward movements to only "reset" the head to the beginning of the tape [7,8], an action that would cause the machine to "forget" where it was looking at the input, if the input head is read-only, and no further worktape is available, as in the recently introduced probabilistic and quantum finite automata with restart [15].

In this paper, we consider one of the simplest classical computation models where the one-way and two-way versions have different computational power, namely, the pushdown automaton (PDA), and examine various variants that are architecturally intermediate between these two "extremes". None of these variants can move its input head one tape square to the left, but each is endowed with a different instruction that allows it to simultaneously reset some components of its configuration to its initial value. Most of these model variants are shown to be equivalent in power to either the one-way or the two-way PDA. The only exception is what we call the Resettable Pushdown Automaton (RPDA), where the stack contents are also lost every time the input head

G. Mauri and A. Leporati (Eds.): DLT 2011, LNCS 6795, pp. 417–428, 2011.

is reset, and which we prove to be intermediate in power. We prove that general (nondeterministic) RPDAs recognize precisely the languages in the finite intersection closure of the context-free languages (CFLs), whereas the class of languages recognized by deterministic RPDAs equals the Boolean closure of the deterministic CFLs.

The rest of this paper is structured as follows: Sect. 2 covers the necessary background. Our findings about PDAs with various alternative types of reset instructions are presented in Sect. 3. Section 4 is a conclusion.

2 Preliminaries

2.1 Basic Notation

In this paper, Σ denotes the input alphabet not containing the end markers ¢ and \$, and $\tilde{\Sigma} = \Sigma \cup \{¢, \$\}$. The empty string is denoted by ε. Σ_ε stands for $\Sigma \cup \{\varepsilon\}$. $W \in \Sigma^*$ represents an input string. $W(i)$ is the ith character of W, where i is a natural number, and $W(0)$ is the first character of W. $|W|$ is the length (number of symbols) of W . $W(i, j)$ denotes the substring of W containing characters from index i to j, including i, and excluding j. Γ is the finite stack alphabet. Q is the finite set of states. p, q, s represent states in Q. a, b, c represent characters in Σ. x, y, z represent characters in Γ. $Z \in \Gamma^*$ represents the stack content. $d \in \{-1, 0, 1\}$ denotes the direction of the input head. $\mathfrak{C}_*(\mathbb{L})$ is the smallest family of languages containing the class $\mathbb{L}$ and closed under operation $*$.

2.2 Basic Computation Models

We provide the standard definitions of the one- and two-way PDAs in this subsection.

Pushdown Automata. In simple terms, a pushdown automaton (PDA) is a finite state machine augmented with a last-in first-out infinite storage. A formal definition is given below. In the literature, there are several alternative definitions with minor variations, but all of them recognize the same language family, the context-free languages (CFLs).

A pushdown automaton is a 7-tuple, $(Q, \Sigma, \Gamma, \delta, q_0, z_0, F)$ [4], where

1. Q is a finite nonempty set (of states),
2. Σ is a finite nonempty set (of input symbols),
3. Γ is a finite nonempty set (of stack symbols),
4. δ is any function from $(Q \times \Sigma_\varepsilon \times \Gamma)$ into finite subsets of $Q \times \Gamma^*$,
5. $q_0 \in Q$ is the initial state,
6. $z_0 \in \Gamma$ is the initial stack symbol, and
7. $F \subseteq Q$ is the set of accept states.

The PDA is a nondeterministic model by definition, and its deterministic counterpart is strictly less powerful. A PDA is deterministic if it satisfies the following conditions:

1. $\forall q \in Q \forall a \in \Sigma_\varepsilon \forall x \in \Gamma[|\delta(q,a,x)| \leq 1]$
2. $\forall q \in Q \forall x \in \Gamma[(|\delta(q,\varepsilon,x)| = 1) \Rightarrow (\forall a \in \Sigma[|\delta(q,a,x)| = 0])]]$
3. $\forall q \in F \forall x \in \Gamma[(|\delta(q,\varepsilon,x)| = 0)]$
4. $\forall q \in Q \forall a \in \Sigma_\varepsilon[(|\delta(q,a,z_0)| \neq 0) \Rightarrow (\exists q' \in Q \exists Z \in \Gamma^*[(q', z_0 Z) \in \delta(q,a,z_0)]])]$

The first two conditions guarantee that there is at most one applicable instruction in every situation. The third condition guarantees that we do not have to choose between accepting the string or continuing the computation. And the last condition guarantees the perpetual existence of z_0 at the bottom of the stack. Therefore until the end of input there exists one and only one computation path.

The languages recognized by deterministic PDAs (DPDAs) are called the deterministic CFLs (DCFLs).

Two-Way Pushdown Automata. Gray et al. introduced the two-way PDA model in [3]. A two-way pushdown automaton (2PDA) is basically a PDA which can move its reading head leftward, as well as rightward, on its input. The standard PDAs of the previous subsection can be thought of as one-way machines.

A 2PDA M is a 9-tuple $(Q, \Sigma, ¢, \$, \Gamma, \delta, q_0, z_0, F)$, where

1. Q is a finite nonempty set (of states),
2. Σ is a finite nonempty set (of input symbols),
3. $¢ \notin \Sigma$ and $\$ \notin \Sigma$ are the left and right input end-markers,
4. Γ is a finite nonempty set (of stack symbols),
5. δ is any function from $Q \times \tilde{\Sigma} \times \Gamma$ into finite subsets of $\{-1,0,1\} \times Q \times \Gamma^*$,
6. $q_0 \in Q$ is the initial state,
7. $z_0 \in \Gamma$ is the initial stack symbol, and
8. $F \subseteq Q$ is the set of accept states.

The collection of its current state, contents of the stack and input tapes, and the position of the reading head on the input tape make up the instantaneous configuration of a 2PDA. We can represent a configuration as a 3-tuple of the form (q, A, Z), where $A \in ((\upharpoonleft ¢\Sigma^*\$) \cup (¢\Sigma^* \upharpoonleft \Sigma^*\$) \cup (¢\Sigma^*\$ \upharpoonleft))$. A represents the contents of the input tape, with $\upharpoonleft$ marking the place of the reading head. A 2PDA starts its computation in the initial state q_0, with z_0 in its stack and with input in the form of $¢W\$$ where $W \in \Sigma^*$. The reading head of the 2PDA is initially on the ¢ symbol. Therefore, the initial configuration of a 2PDA is of the form $(q_0, \upharpoonleft ¢\Sigma^*\$, z_0)$.

The configurations of a 2PDA evolve according to its transition function δ. If $(d, q', Z) \in \delta(q, a, z)$, then any configuration where the state is q, the scanned input symbol is a[1], and the topmost stack symbol is z will yield a new configuration that is obtained by

- changing the state to q',
- moving the input head to the left if $d = -1$, to the right if $d = 1$, and leaving it on the same place if $d = 0$, and

[1] The transition will take place regardless of the input tape content if $a = \varepsilon$.

– popping z from the stack, and then pushing Z. (If nonempty, the first symbol of Z is written first, and so its last symbol of Z will be on top.)

For convenience and without loss of generality we assume that the 2PDA does not attempt to move to the left when it is scanning ¢ on the input tape.

If configuration c_1 yields c_2, we denote this by $c_1 \vdash c_2$. $\vdash^*$ is the reflexive and transitive closure of $\vdash$. We say that 2PDA M accepts string W iff $\exists q \in F \exists Z \in \Gamma^*[(q_0, \upharpoonleft ¢W\$, z_0) \vdash^* (q, ¢W\$ \upharpoonleft, Z)]$. Then $L(M)$, the language recognized by 2PDA M, is equal to $\{W \in \Sigma^* | M \text{ accepts } W\}$.

We call a 2PDA deterministic if it satisfies the following two conditions:

1. $\forall(q, a, z) \in Q \times \tilde{\Sigma} \times \Gamma[|\delta(q, a, z)| \leq 1]$ (Therefore we will use $\delta(q, a, z) = (d, q', Z)$ instead of $\delta(q, a, z) = \{(d, q', Z)\}$)
2. $\forall q \in Q \forall a \in \tilde{\Sigma}[(\delta(q, a, z_0) = (d, q', Z)) \Rightarrow (Z \in z_0\Gamma^*)]$

3 PDA with Reset States

We enumerate the intermediate PDA variants that will be considered in our study in the following manner: The instantaneous configuration of a pushdown automaton has three variable components; input head position, stack contents, and internal state. In each of the considered variants, we will augment the one-way PDA model with an instruction which resets a different subset of these configuration components to their initial values simultaneously. This will be realized by including special states (called *reset states*) in the machine description. The associated reset operation is performed whenever such a reset state is entered, after which point the computation proceeds as usual from the newly attained configuration.

Let us now start going through our PDA variants. There are seven nonempty subsets of our set ({state, stack, input}) of potentially resettable components. The four variants which have reset instructions corresponding to the subsets {state}, {stack}, {state,stack}, and {state, stack, input} are easily seen to be equivalent in power to the one-way PDA: We can of course simulate a {state}-reset simply with a transition to the initial state. A {stack}-reset can be simulated by emptying the stack symbol by symbol. A {state,stack}-reset can be handled by combining the two aforementioned operations. Any accepting computation path of a machine that visits {state, stack, input}-reset states has a final segment that goes from the starting configuration to an accepting one without visiting any reset states, so the language recognized by any such machine is context-free.

The remaining types of reset combinations require detailed analysis. In Sect. 3.1, we show that resetting the input head and the stack simultaneously yields machines, which we call Resettable Pushdown Automata (RPDAs), that are more powerful than PDAs, but weaker than two-way PDAs. The remaining two variants are demonstrated to be equivalent to the 2PDA in Sect. 3.2.

3.1 Resettable Pushdown Automata

Nondeterministic Version. Briefly put, a RPDA is just a restricted 2PDA that cannot move its input head just one square to the left. The only way the input head can go leftward is by jumping all the way to the left end of the tape, and each such jump is accompanied by the complete loss of the current stack contents. A formal definition follows.

Definition 1. *A* ***resettable pushdown automaton*** *is a 9-tuple* $(Q, \Sigma, \$, \Gamma, \delta, q_0, z_0, F, R)$*, where* Q, Σ, Γ, F *and* R *are all finite sets, and*

1. Q *is a finite nonempty set of states,*
2. Σ *is a finite nonempty set of input symbols,*
3. $\$ \notin \Sigma$ *is the input end-marker,*
4. Γ *is a finite nonempty set of stack symbols,*
5. δ *is any function from* $Q \times \Sigma_{\varepsilon,\$} \times \Gamma$ *into finite subsets of* $Q \times \Gamma^*$,
6. $q_0 \in Q$ *is the initial state,*
7. $z_0 \in \Gamma$ *is the initial stack symbol,*
8. $F \subseteq Q$ *is the set of accept states, and*
9. $R \subseteq Q$ *is the set of reset states.*

Note that $\Sigma_{\varepsilon,\$} = \Sigma \cup \{\varepsilon, \$\}$ and $\Gamma_\varepsilon = \Gamma \cup \{\varepsilon\}$. The condition that states that all legal moves should leave z_0 at the bottom of the stack is same as in Sect. 2.2.

A resettable pushdown automaton $M = (Q, \Sigma, \$, \Gamma, \delta, q_0, z_0, F, R)$ computes as follows: It accepts string W if and only if there exist sequences of states p_0, p_1, ..., $p_m \in Q$, natural numbers k_0, k_1, ..., $k_m \in N$, strings Z_0, Z_1, ..., $Z_m \in \Gamma^*$, and five-tuples of $Q \times \Sigma_{\varepsilon,\$} \times \Gamma_\varepsilon \times Q \times \Gamma^*$ t_0, t_1, ..., t_{m-1} $(t = (s, a, y, s', Y))$ that satisfy the nine conditions listed below when it runs on input $W\$$. Here, p_i is the state the machine is in at step i. k_i represents the location of the input reading head. And the string Z_i represents the stack content. The five-tuples represent valid transitions from one step to the next.

1. $p_0 = q_0$, $k_0 = 0$ and $Z_0 = z_0$ (This condition signifies that M starts out properly, in the initial state with input reading head on the left end, and with only z_0 in the stack.)
2. For each t_i, $(s'_i, Y_i) \in \delta(s_i, a_i, y_i)$ and $p_i = s_i$ and $p_{i+1} = s'_i$
3. If $a_i \neq \varepsilon$ then $a_i = W(k_i)$
4. $y_i = Z_i(|Z_i| - 1)$
5. If $p_{i+1} \in R$ then $k_{i+1} = 0$ and $Z_{i+1} = z_0$
6. If $p_{i+1} \notin R$ and $a_i \neq \varepsilon$ then $k_{i+1} = k_i + 1$
7. If $p_{i+1} \notin R$ and $a_i = \varepsilon$ then $k_{i+1} = k_i$
8. If $p_{i+1} \notin R$ then $Z_{i+1} = Z_i(0, |Z_i| - 1) + Y_i$
9. $p_m \in F$ and $k_m = |W|$ (This condition states that an accept state comes into effect only when the input head is at the right end-marker)

Theorem 1. *The class of languages recognizable by (nondeterministic) RPDAs is the finite-intersection closure of the context-free languages.*

Proof. This requires a two-way proof. First we will show that every language which can be obtained by intersecting finitely many CFLs can be recognized by an RPDA. We will then show that every language which can be recognized by an RPDA can be written as the intersection of finitely many CFLs.

Part I. $L = L_1 \cap L_2 \ldots L_k$ where L_i is a CFL for $i = 1\ldots k$. Say that $M_i = (Q_i, \Sigma_i, \Gamma_i, \delta, q_{i,0}, F_i)$ is a PDA for the language L_i, where $q_{i,j}$ is the jth state of Q_i. Now we will construct an RPDA $M = (Q, \Sigma, \$, \Gamma, \delta, q_0, F, R)$ using the M_i. Basically we will run the input first on M_1. If it accepts, then we will reset the input head and the stack and pass control to M_2. This will continue all the way to M_k if all the submachines accept the input. The formal construction of the RPDA is described below.

- $Q = \bigcup(Q_i) \cup \{r_1, r_2 \ldots r_{k-1}\}(i \neq j \Rightarrow Q_i \cap Q_j = \emptyset)$
- $\Sigma = \bigcup(\Sigma_i) \quad \Gamma = \bigcup(\Gamma_i) \quad q_0 = q_{1,0} \quad F = F_k$
- $R = \{r_1, r_2 \ldots r_{k-1}\}(\forall i[Q_i \cap R = \emptyset])$
- $\delta(q,a,z) = \begin{cases} \delta_i(q_{i,j}, a, z) & \text{if } q = q_{i,j} \in Q_i,\ a \in \Sigma_i \cup \{\varepsilon\} \text{ and } z \in \Gamma_i \\ \{(r_i, \varepsilon)\} & \text{if } q \in F_i,\ a = \$ \text{ and } z \in \Gamma_i \\ \{(q_{i+1,0}, z)\} & \text{if } q = r_i \in R,\ a = \varepsilon \text{ and } z \in \Gamma_i \\ \emptyset & \text{otherwise} \end{cases}$

Part II. Now we will show that every language L which can be recognized by an RPDA can be written as the intersection of finitely many CFLs. We will construct finitely many PDAs from the RPDA for L, and write L as the finite intersection of some context-free languages.

We have a RPDA $M = (Q, \Sigma, \$, \Gamma, \delta, q_0, z_0, F, R)$ which recognizes L. $k = |R|$, and $r_i \in R$ is the ith reset state. We define PDA $M_0 = (Q, \Sigma, \Gamma, \delta_0, q_0, F)$ and $M_i = (Q, \Sigma, \Gamma, \delta_i, r_i, F)$ for $i = 1\ldots k$. Note that r_i acts as a start state for M_i. We will construct M_i such that it will accept exactly the set of strings which are accepted by M without the need of any further reset after r_i, assuming that they reach r_i somehow. We will also define $M_i^j = (Q, \Sigma, \Gamma, \delta_i^j, p_i, \{r_j\})$ for $i = 0\ldots k$, $j = 1\ldots k$ and $i \neq j$ and $p_i = r_i$ for $i > 0$ and $p_i = q_0$ for $i = 0$. And M_i^j will represent a passage between r_i and r_j. It will accept exactly the set of strings which reach reset state r_j without the need of any further reset after reset at r_i, assuming that they reach r_i somehow. (Or for the case $i = 0$, the set of strings which reach reset state r_j without the need of any extra reset after starting computation from q_0.)

δ_0 of M_0 and δ_i of the M_i are the same.

$$\delta_0(q,a,z) = \delta_i(q,a,z) = \delta(q,a,z) - \{(q',Z) | q' \in R, Z \in \Gamma^*\}$$

However M_i^j has a different transition function.

$$\delta_i^j(q,a,z) = \begin{cases} \delta(q,a,z) - \{(q',Z) | q' \in (R - \{r_j\}), Z \in \Gamma^*\} & \text{if } q \in (Q - \{r_j\}) \\ \{(r_j, z)\} & \text{if } q = r_j \end{cases}$$

Let us say that the languages recognized by M_i and M_i^j are L_i and L_i^j, respectively. Note that, $L_0^i \cap L_i^j \cap L_j$ is a subset of L and it consists of exactly the strings which are accepted by M, after resetting at r_i, then at r_j, and finally reaching an accept state at the end of the input. Also we should note that if a string is accepted by M, it may do so by visiting every reset state at most once. We will now express L using these languages, but first let us define P_i^k as the set of permutations of the set $\{1, 2, \ldots, k\}$ with i elements. Then we will define C_i as

$$C_i = \bigcup_{p \in P_i^k} (L_0^{p(1)} \cap L_{p(1)}^{p(2)} \cap \ldots L_{p(i)})$$

Now, we can write

$$L = L_0 \cup \bigcup_{i=1}^{k} C_i$$

Using the fact that union distributes over intersection, we obtain an expression depicting intersections of languages which are themselves unions of CFLs. Since the unions of CFLs are also CFLs, L is the intersection of finitely many CFLs. □

We should note that Okhotin proposes conjunctive grammars in [10]. This grammar recognizes intersection closure of CFLs and more. It adds intersection operation to the CFG. Though this intersection operation occurs not on strings of the language but parts of the strings. This intersection can be thought as a reset operation which only clears the stack and the restores the input up to a specific point. This is the essence of the extra recognition power.

Deterministic Version. We will now characterize the languages recognized by deterministic RPDAs (DRPDAs). These are defined by restricting the RPDA model in the same way one restricts the PDA to obtain the DPDA.

After analyzing the two constructions given in the proof for Theorem 1, it can be seen that those constructions obtain nondeterministic machines from nondeterministic ones, and deterministic machines from deterministic ones.

Corollary 1. *Every language in the finite intersection closure of DCFLs is recognizable by DRPDAs.*

Proof. Due to the Part I of the proof for Theorem 1, such a DRPDA can be constructed. □

However, we cannot say that the class of languages recognized by DRPDAs is exactly the finite intersection closure of the DCFLs. Since the class of DCFLs is not closed under union, Part II of the proof of Theorem 1 no longer stands. Instead, we are going to show that the class of languages recognized by DRPDAs is exactly the finite Boolean closure of the DCFLs.

Now we will prove some lemmata in order to prove the main theorem.

Lemma 1. *The class of languages recognizable by DRPDAs is closed under the finite union operation.*

Proof. Let us say that L_1 and L_2 are any two languages recognizable by DRPDAs and $M_1 = (Q_1, \Sigma_1, \$, \Gamma_1, \delta_1, q_{1,0}, z_0, F_1, R_1)$ and $M_2 = (Q_2, \Sigma_2, \$, \Gamma_2, \delta_2, q_{2,0}, z_0, F_2, R_2)$ are their respective DRPDA recognizers. We will show that $L = L_1 \cup L_2$ is DRPDA recognizable by constructing a DRPDA M for it.

For convenience and without loss of generality, we assume that q_0 is the initial state, and q_i for $i = 1 \ldots |R|$ are reset states. And through preprocessing of DRPDA we make sure that DRPDA consumes its input in finite time. This can be done via the process given in Lemma 12.1 in [5].

In the constructed machine, computations of M_1 and M_2 are done in a merged way. M_1 starts computation. If it accepts, the combined machine accepts, otherwise (upon termination of computation of M_1 or reaching a reset state,) computation continues with M_2. The copy of M_2 that is simulated at this point "knows" which reset state M_1 was left in, and is able to transit to the correct state of M_1 when it is time to pass control back to M_1. This method handles the cases where one of the machines is stuck in a loop of resets. Formally, the combined machine is[2]

$M = (Q, \Sigma, \$, \Gamma, \delta, q_{1,0,0}, z_0, F, R)$ where

- $Q = (\bigcup_{i=0}^{i=|R_2|} Q_{1,i}) \cup (\bigcup_{i=0}^{i=|R_1|} Q_{2,i}) \cup R$ where $Q_{i,j}$ is j^{th} copy of Q_i
- $\Sigma = \Sigma_1 \cup \Sigma_2 \qquad \Gamma = \Gamma_1 \cup \Gamma_2$
- $\delta(q, a, z) = \begin{cases} \delta_i(q_k, a, z) & \text{if } q = q_{i,j,k} \in Q_{i,j},\ q_k \notin R_i \text{ and } a \neq \$ \\ (s_{i',j,0}, z) & \text{if } q = q_{i,j,k} \in Q_{i,j},\ q_k \notin R_i \text{ and } a = \$ \\ (s_{i',j,k}, z) & \text{if } q = q_{i,j,k} \in Q_{i,j} \text{ and } q_k \in R_i \\ (q_{i,k,l}, Z) & \text{if } q = s_{i,j,k},\ z = z_0 \text{ and } (q_{i,l}, Z) = \delta_i(q_{i,j}, a, z_0) \\ \emptyset & \text{otherwise} \end{cases}$

  where $i' = mod(i, 2) + 1$
- $F = \{q | q = q_{i,j,k} \in Q \text{ and } q_{i,j} \in F_i\}$
- $R = \{s_{i,j,k} | i \in \{1, 2\},\ j \in \{1 \ldots |R_i|\} \text{ and } k \in \{1 \ldots |R_{mod(i,2)+1}|\}\}$ □

Lemma 2. *The set of languages recognizable by DRPDAs is* $\mathfrak{C}_{\cup}(\mathfrak{C}_{\cap}(\text{DCFL}))$.

Proof. This requires a two-way proof. First we will show that every language which can be expressed by finitely many unions of finitely many intersections of DCFLs can be recognized by a DRPDA. And after that we will show that every language which can be recognized by a DRPDA can be written as such a union of intersections of DCFLs.

Part I. Due to Corollary 1, we know that languages that are intersections of finitely many DCFLs can be recognized by a DRPDA. And from Lemma 1, we know that the set of languages recognizable by DRPDAs is closed under finite union. Therefore, a language which can be obtained by taking finitely many unions of finitely many intersections of DCFLs can be recognized by a DRPDA.

Part II. Due to part II of Theorem 1, a language which can be recognized by a DRPDA can be written in the form of finitely many unions of finitely many intersections of DCFLs. □

[2] We do not use set notation in the description of the values of δ for DRPDAs, since there is at most one possible member. However we continue to use $\emptyset$ for expressing the cases where there are no possible continuations.

Theorem 2. *The set of languages recognizable by DRPDAs is the finite Boolean closure of the deterministic context free languages.*

Proof. From [14], we know that

$$\mathfrak{C}_{\mathbf{Bool}}(\text{DCFL}) = \mathfrak{C}_{\cap}(\mathfrak{C}_{\cup}(\text{DCFL})) = \mathfrak{C}_{\cup}(\mathfrak{C}_{\cap}(\text{DCFL}))$$

This is due to the closure of DCFLs under the complementation operation and the distribution property of $\cap$ over $\cup$ and vice versa.

Therefore, due to Lemma 2, Theorem 2 is obvious. □

We can now use the known relationships [14] between CFL, $\mathfrak{C}_{\cap}$(CFL), and $\mathfrak{C}_{\mathbf{Bool}}$(DCFL) to compare the computational powers of the machine variants in question. The context-free language $\{wcx|w, x \in \{a,b\}^* \text{ and } w \neq x\}$ is not in $\mathfrak{C}_{\mathbf{Bool}}$(DCFL). It is also well-known that the non-CFL $\{a^n b^n c^n | n > 0\}$ is in $\mathfrak{C}_{\mathbf{Bool}}$(DCFL), and so we conclude that one-way nondeterministic PDAs and DRPDAs are incomparable in power, and general nondeterministic RPDAs are strictly more powerful than DRPDAs. Finally, the language $\{wcw|w \in \{a,b\}^*\}$ is not in $\mathfrak{C}_{\cap}$(CFL), and therefore cannot be recognized by an RPDA, whereas it can be recognized by a deterministic 2PDA. It is easy to see that any RPDA can be simulated easily by a nondeterministic 2PDA.

3.2 Variants Equivalent to the 2PDA

We will now prove that forcing a 2PDA to jump all the way back to the beginning of the input and simultaneously switch to the initial state whenever it requires the input head to move left does not affect its computational power. The machine can store the information that is apparently lost during the reset in the stack, which is left intact in this type of reset. Just for this section, we will use the name MPDA (Modified PDA) for PDAs with {input,state}-reset states.

Theorem 3. *PDAs with {input,state}-reset states are equivalent in power to 2PDAs.*

Proof. It is easy to see that a 2PDA can simulate an {input,state}-reset by just moving the input head to the left end square by square, and then switching to the initial state. Now let us prove that any 2PDA can be simulated by an MPDA. The difficulty here is to simulate the behavior of 2PDA when its input head stays at the same place, or moves to the left. (Staying on the same place is a problem because a MPDA moves to the right automatically whenever it reads a symbol from input.) To simulate such moves, the MPDA must use its reset capability, but before that, it must store the information that would be lost by the reset (the internal state and the input head position) on the stack. It is simple to represent the state by a stack symbol, but how can the MPDA store the input head position (say, i) in its stack?

This problem is solved (see Fig. 1) by first pushing $m = n - i$ symbols, where n is the overall input length, on the stack. The MPDA then resets, and pops these symbols to help move the input head to the mth position. From there, the input head moves once again to the right end, guiding us to push exactly the requied i symbols on the stack. The machine then resets again, and is guided to the proper configuration by making use of the information in the stack.

The formal construction follows.

If we have a 2PDA T (Q_T, Σ, ¢, \$, Γ_T, δ_T, z_0, q_0, F), then we can simulate T with MPDA M (Q_M, Σ, \$, Γ_M, δ_M, z_0, q_s, F, R).

- $Q_M = Q_T \cup Q_T^1 \cup \{q_s, q_r, q_A, q_B, q_C, q_D, q_E, q_F, q_G, q_H, q_I\}$
- $R = \{q_r\}$
- $\Gamma_M = \Gamma_T \cup Q_T \cup \{q_B, q_D, q_F, q_H, \alpha\}$ $(\Gamma_T \cap (Q_T \cup \{q_B, q_D, q_F, q_H, \alpha\}) = \emptyset)$
- $$\delta_M(q,a,z) = \begin{cases} \{(q_0^1, z_0)\} & \text{if } q = q_s,\ a = \varepsilon \text{ and } z = z_0 \\ \{(q_B, \varepsilon)\} & \text{if } q = q_s,\ a = \varepsilon \text{ and } z = q_B \\ \{(q_D, \varepsilon)\} & \text{if } q = q_s,\ a = \varepsilon \text{ and } z = q_D \\ \{(q_F, \varepsilon)\} & \text{if } q = q_s,\ a = \varepsilon \text{ and } z = q_F \\ \{(q_H, \varepsilon)\} & \text{if } q = q_s,\ a = \varepsilon \text{ and } z = q_H \\ \mathfrak{T} \text{ (is given below)} & \text{if } q \in Q_T \\ \mathfrak{U} \text{ (is given below)} & \text{if } q = q_i^1 \text{ and } a = \varepsilon \\ \{(q_A, z\alpha)\} & \text{if } q = q_A \text{ and } a \in \Sigma \\ \{(q_r, zq_B)\} & \text{if } q = q_A \text{ and } a = \$ \\ \{(q_B, \varepsilon)\} & \text{if } q = q_B,\ a \in \Sigma \text{ and } z = \alpha \\ \{(q_C, z)\} & \text{if } q = q_B,\ a = \varepsilon \text{ and } z \in Q_T \\ \{(q_C, z\alpha)\} & \text{if } q = q_C \text{ and } a \in \Sigma \\ \{(q_r, zq_D)\} & \text{if } q = q_C \text{ and } a = \$ \\ \{(q_D, \varepsilon)\} & \text{if } q = q_D,\ a \in \Sigma \text{ and } z = \alpha \\ \{(q_i, \varepsilon)\} & \text{if } q = q_D,\ a = \varepsilon \text{ and } z = q_i \\ \{(q_E, z\alpha)\} & \text{if } q = q_E \text{ and } a \in \Sigma \\ \{(q_r, zq_F)\} & \text{if } q = q_E \text{ and } a = \$ \\ \{(q_F, \varepsilon)\} & \text{if } q = q_F,\ a \in \Sigma \text{ and } z = \alpha \\ \{(q_G, z)\} & \text{if } q = q_F,\ a = \varepsilon \text{ and } z \in Q_T \\ \{(q_G, z\alpha)\} & \text{if } q = q_G \text{ and } a \in \Sigma \\ \{(q_r, zq_H)\} & \text{if } q = q_G \text{ and } a = \$ \\ \{(q_i^1, \varepsilon)\} & \text{if } q = q_H,\ a = \varepsilon \text{ and } z = q_i \\ \{(q_I, \varepsilon)\} & \text{if } q = q_H,\ a = \varepsilon \text{ and } z = \alpha \\ \{(q_I, \varepsilon)\} & \text{if } q = q_I,\ a \in \Sigma \text{ and } z = \alpha \\ \{(q_i, \varepsilon)\} & \text{if } q = q_I,\ a = \varepsilon \text{ and } z = q_i \\ \emptyset & \text{otherwise} \end{cases}$$

$$\text{where } \mathfrak{T} = \begin{cases} \{(q_i, Z) | (q_i, Z, 1) \in \delta_T(q, a, z)\} \cup \\ \{(q_A, Zq_i\alpha) | a \in \Sigma \text{ and } (q_i, Z, 0) \in \delta_T(q, a, z)\} \cup \\ \{(q_r, Zq_iq_B) | a = \$ \text{ and } (q_i, Z, 0) \in \delta_T(q, a, z)\} \cup \\ \{(q_E, Zq_i\alpha) | a \in \Sigma \text{ and } (q_i, Z, -1) \in \delta_T(q, a, z)\} \cup \\ \{(q_r, Zq_iq_F) | a = \$ \text{ and } (q_i, Z, -1) \in \delta_T(q, a, z)\} \end{cases}$$

and $\mathfrak{U} = \{(q_j^1, Z) | (q_j, Z, 0) \in \delta_T(q_i, ¢, z)\} \cup \{(q_j, Z) | (q_j, Z, 1) \in \delta_T(q_i, ¢, z)\}$

□

The corresponding proof for the version with {input}-reset states is similar.

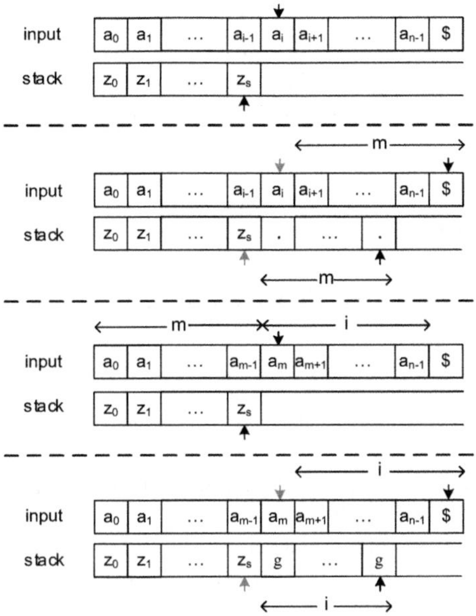

Fig. 1. Using the stack to store the input head position

4 Conclusion

Figure 2 illustrates the comparative powers of the machines examined in this paper. Note that we do not know exactly where to place the deterministic 2PDA, whose relationships with both one-way and two-way nondeterministic PDAs have formed open questions for many decades, in this figure. One question that arises from the work reported here is

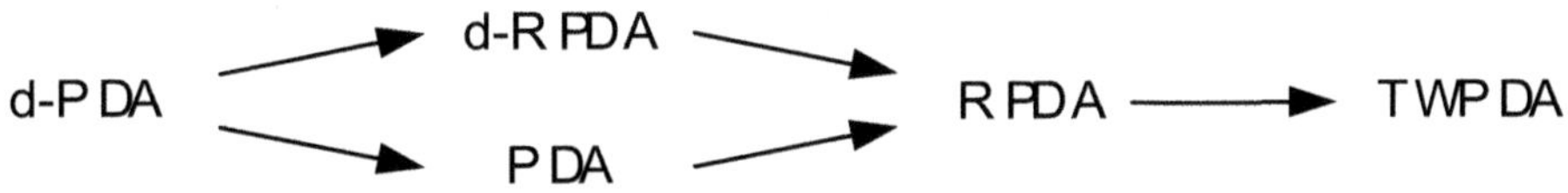

Fig. 2. Recognition powers

Open Problem 1. Does limiting the maximum number of reset operations allowed during RPDA computation result in a language class hierarchy? That is, if $\mathcal{L}_n$ is the class of languages recognized by an RPDA whose reset states stop resetting and behave as non-resetting states after a total of n reset operations, is $\mathcal{L}_n \subsetneq \mathcal{L}_{n+1}$ for $n \geqslant 0$?

Hierarchy of intersections of CFL is investigated in [9]. This work may play an important role in solving Open Problem 1.

We plan to extend this study to probabilistic [6] and quantum [2] versions of PDAs. It would also be interesting to examine a similar setup to our Section 3.1 in the context of sublinear-space Turing machines whose read-only input tape head can only move leftward by resetting to the left end, and losing the contents of the worktape.

Acknowledgement

The authors wish to thank Oscar Ibarra for his helpful answers to their questions about 2PDAs.

References

1. Freivalds, R., Karpinski, M.: Lower space bounds for randomized computation. In: ICALP 1994, pp. 580–592 (1994)
2. Golovkins, M.: Quantum pushdown automata. In: Jeffery, K., Hlaváč, V., Wiedermann, J. (eds.) SOFSEM 2000. LNCS, vol. 1963, pp. 336–346. Springer, Heidelberg (2000)
3. Gray, J.N., Harrison, M.A., Ibarra, O.H.: Two-way pushdown automata. Information and Control 11(1-2), 30–70 (1967)
4. Hoogeboom, H.J., Engelfriet, J.: Pushdown automata. In: Formal Languages and Applications. SFSC, ch. 6, vol. 148, pp. 117–138. Springer, Berlin (2004)
5. Hopcroft, J.E., Ullman, J.D.: Formal languages and their relation to automata. Addison-Wesley Longman Publishing Co., Boston (1969)
6. Hromkovic, J., Schnitger, G.: On probabilistic pushdown automata. Information and Computation 208(8), 982–995 (2010)
7. Jančar, P., Mráz, F., Plátek, M., Procházka, M., Vogel, J.: Deleting automata with a restart operation. In: Bozapalidis, S. (ed.) Proceedings of DLT 1997, Greece, pp. 191–202 (1997)
8. Jančar, P., Mráz, F., Plátek, M., Vogel, J.: Restarting automata. LNCS, vol. 965, pp. 283–292 (1995)
9. Liu, L.Y., Weiner, P.: An infinite hierarchy of intersections of context-free languages. Mathematical Systems Theory 7, 185–192 (1973)
10. Okhotin, A.: Conjunctive grammars. J. A. Lang. Comb. 6, 519–535 (2001)
11. Shepherdson, J.C.: The reduction of two-way automata to one-way automata. IBM Journal of Research and Development 3, 198–200 (1959)
12. Stearns, R.E., Hartmanis, J., Lewis II, P.M.: Hierarchies of memory limited computations. In: SWCT 1965, pp. 179–190 (1965)
13. Szepietowski, A.: Turing Machines with Sublogarithmic Space. Springer, Heidelberg (1994)
14. Wotschke, D.: Nondeterminism and boolean operations in PDA's. Journal of Computer and System Sciences 16, 456–461 (1978)
15. Yakaryılmaz, A., Cem Say, A.C.: Succinctness of two-way probabilistic and quantum finite automata. DMTCS 12(4), 19–40 (2010)
16. Yakaryılmaz, A., Cem Say, A.C.: Unbounded-error quantum computation with small space bounds. Technical Report arXiv:1007.3624 (2010)

Towards Dual Approaches for Learning Context-Free Grammars Based on Syntactic Concept Lattices

Ryo Yoshinaka*

ERATO MINATO Discrete Structure Manipulation System Project,
Japan Science and Technology Agency
ry@ist.hokudai.ac.jp

Abstract. Recent studies on grammatical inference have demonstrated the benefits of "distributional learning" for learning context-free and context-sensitive languages. Distributional learning models and exploits the relation between strings and contexts in the language of the learning target. There are two main approaches. One, which we call *primal*, constructs nonterminals whose language is characterized by strings. The other, which we call *dual*, uses contexts to characterize the language of a nonterminal of the conjecture grammar. This paper demonstrates and discusses the duality of those approaches by presenting some powerful learning algorithms along the way.

1 Introduction

Recent studies on grammatical inference have demonstrated how powerful the idea of "distributional learning" is for learning context-free and context-sensitive languages. Distributional learning algorithms exploit information on combinations of strings and contexts that form grammatical sentences. Clark [6] has discussed two main approaches in distributional learning, which are called *primal* and *dual*, respectively. In fact, many concrete algorithms of distributional learning proposed so far can be classified into those two approaches. The primal approach is taken in [4, 8, 13, 14, 15, 16] and the dual approach is taken in [12, 2, 5, 7]. Among those, let us compare the learning of *congruential* CFGs by Clark [4] and that of *c-deterministic* CFGs by Shirakawa and Yokomori [12] as examples of those two approaches. A CFG G is said to be congruential if

$$X \overset{*}{\underset{G}{\Rightarrow}} w,\ X \overset{*}{\underset{G}{\Rightarrow}} w' \text{ and } uwv \in \mathcal{L}(G) \text{ implies } uw'v \in \mathcal{L}(G)$$

for all nonterminal symbols X, where $\mathcal{L}(G)$ denotes the language of G. A CFG G is said to be c-deterministic if

$$S \overset{*}{\underset{G}{\Rightarrow}} uXv \text{ implies } \mathcal{L}(G, X) = \{\, w \mid uwv \in \mathcal{L}(G) \,\}$$

* The author is concurrently working in Hokkaido University.

G. Mauri and A. Leporati (Eds.): DLT 2011, LNCS 6795, pp. 429–440, 2011.

for all nonterminal symbols X, where $\mathcal{L}(G,X)$ denotes the language of X. Roughly speaking, the former requires that one string w (actually any of $\mathcal{L}(G,X)$) characterizes the context set of strings in $\mathcal{L}(G,X)$: whatever occurs as a context of w, it is a context of all other strings in $\mathcal{L}(G,X)$ as well. The latter requires that one context $\langle u,v\rangle$ is enough to characterize the strings in $\mathcal{L}(G,X)$: whatever occurs in that context, it is a member of $\mathcal{L}(G,X)$. In Clark's learning algorithm for congruential CFGs, nonterminals are assigned strings (or string sets) and the conjecture grammar is designed so that the language of a nonterminal is characterized by the assigned string(s) along the way consistent with the congruentiality. The *primal* approach uses strings to determine the language that should be generated by a nonterminal.[1] Contexts are rather auxiliary but play an important role to determine production rules by which nonterminals generate the desired languages. The *dual* approach takes the opposite way. The conjecture grammar is constructed so that the language generated by a nonterminal is characterized by contexts assigned to that nonterminal. Shirakawa and Yokomori's algorithm takes the dual approach, which favors the property of c-determinism.

Clark [2] has proposed to use the concept lattice based on the Galois connection between strings and contexts in an arbitrary language L and named them *syntactic concept lattices*. He has proposed variants of powerful learning algorithms based on syntactic concept lattices taking the dual approach [2,5,7]. Although the concept of syntactic concept lattices is neutral, no learning algorithms based on syntactic concept lattices that takes the primal approach has been proposed so far. Actually the duality between strings and contexts is not so perfect, as we can concatenate strings to generate strings, while we cannot do the same on contexts.

This paper aims to give a clearer view on the duality of distributional learning by proposing concrete algorithms for learning CFGs, which include an algorithm based on the syntactic concept lattice that takes the primal approach. We tackle the question to what extent we have the symmetry in the dual approaches and where it will be violated. After preliminaries, Section 3 reviews Clark's syntactic concept lattices and discusses how the dual approaches would work on them. We give some concrete distributional learning algorithms of the primal and of the dual approach in Section 4. We then conclude the paper.

2 Preliminaries

Let Σ be a nonempty finite set of letters. We denote the empty string by λ. Any element of $\Sigma^* \times \Sigma^*$ is called a *context*. $\mathbb{Z}$ denotes the set of integers and $\mathbb{N}$ is the set of non-negative integers. This paper is concerned with the learning of CFGs, but we limit our attention to CFGs in a variant of the Chomsky normal form. A CFG is a tuple $G = \langle \Sigma, V, P, I\rangle$, where Σ is the set of terminal symbols, V is the set of nonterminal symbols, $P \subseteq V \times (VV \cup \Sigma \cup \{\lambda\})$ is the set of rules,

[1] Clark [6] uses a more general term "primitives" in stead of "nonterminals". In fact distributional lattice grammars [2] directly use sets of strings and contexts as primitive objects in generation rules instead of nonterminals.

and $I \subseteq V$ is the set of initial symbols. The derivation relation of G is denoted by $\Rightarrow_G^*$. The language generated by each nonterminal symbol X is denoted by $\mathcal{L}(G, X) = \{ w \mid X \Rightarrow_G^* w \}$ and the language of G is $\mathcal{L}(G) = \bigcup_{X \in I} \mathcal{L}(G, X)$.

Our learning paradigm is *identification in the limit from positive data and membership queries.* A *positive presentation* of a language L_* over Σ is an infinite sequence $w_1, w_2, \dots \in \Sigma^*$ such that $L_* = \{ w_i \mid i \geq 1 \}$. A learner is given a positive presentation of the language $L_* = \mathcal{L}(G_*)$ of the target grammar G_* and each time a new example w_i is given, it outputs a grammar G_i computed from $w_1, \dots, w_i$ with the aid of a *membership oracle.* One may query the oracle whether an arbitrary string w is in L_*, and the oracle answers in constant time. We say that a learning algorithm *identifies G_* in the limit from positive data and membership queries* if for any positive presentation $w_1, w_2, \dots$ of $\mathcal{L}(G_*)$, there is an integer n such that $G_n = G_m$ for all $m \geq n$ and $\mathcal{L}(G_n) = \mathcal{L}(G_*)$. Trivially every grammar admits a successful learning algorithm. An algorithm should learn a rich class of grammars in a uniform way. We say that a learning algorithm *identifies a class $\mathbb{G}$ of grammars in the limit from positive data and membership queries* iff it identifies all $G \in \mathbb{G}$. We remark that as we have membership queries, learning algorithms based on exhaustive enumeration will work, hence a learner should have further favorable properties in terms of efficiency.

3 Syntactic Concept Lattices

3.1 Overview

To make this paper self-contained, this section reviews the syntactic concept lattices proposed by Clark [2, 3], but we do not have enough space to discuss them in detail. The readers are referred to his original papers for further details.

For a context $\langle x, z \rangle$ and a string $y \in \Sigma^*$, we define a binary operation $\odot$ by $\langle x, z \rangle \odot y = xyz$, which is naturally extended to sets of contexts and strings as $C \odot S = \{ xyz \mid \langle x, z \rangle \in C,\ y \in S \}$ for $C \subseteq \Sigma^* \times \Sigma^*$ and $S \subseteq \Sigma^*$. We write $C \odot S_1 S_2$ to mean $C \odot (S_1 S_2)$.

Let us fix a language $L \subseteq \Sigma^*$. A pair $\langle S, C \rangle$ is called a *biclique*[2] on L if $C \odot S \subseteq L$. We denote the set of contexts for a string set S by S' and the set of strings admitted by a context set C by C':

$$S' = \{ \langle x, z \rangle \in \Sigma^* \times \Sigma^* \mid xyz \in L \text{ for all } y \in S \},$$
$$C' = \{ y \in \Sigma^* \mid xyz \in L \text{ for all } \langle x, z \rangle \in C \}.$$

In other words, S' and C' are the maximum sets such that $\langle S, S' \rangle$ and $\langle C', C \rangle$ are bicliques on L, respectively. Hence $\langle S'', S' \rangle$ and $\langle C', C'' \rangle$ are maximal bicliques on L induced from S and C, respectively. We call such maximal bicliques *concepts.* A biclique $\langle S, C \rangle$ is a concept iff $S' = C$ and $C' = S$. When two sets S_1 and S_2 induce the same concept, i.e., $S_1' = S_2'$, we write $S_1 \equiv_L S_2$. Those notions depend

[2] We may think of $\langle S, C \rangle$ as a biclique of the bipartite graph with the node set $\Sigma^* \cup \Sigma^* \times \Sigma^*$ where $w \in \Sigma^*$ and $\langle u, v \rangle \in \Sigma^* \times \Sigma^*$ are adjacent iff $uwv \in L$.

on L, but our notation usually suppresses L for legibility, provided that it is understood from the context. It is useful to note that for any $S_1, S_2, S_3, S_4 \subseteq \Sigma^*$,

$$\begin{cases} S_1' = S_1''' \text{ and hence } S_1 \equiv S_1'', \\ \text{if } S_1 \equiv S_3 \text{ and } S_2 \equiv S_4, \text{ then } S_1 S_2 \equiv S_3 S_2 \equiv S_3 S_4. \end{cases} \tag{1}$$

Let us denote the set of all concepts of L by $\mathfrak{B}(L)$. It is easy to see that for any two concepts $\langle S_1, C_1 \rangle$ and $\langle S_2, C_2 \rangle$ of $\mathfrak{B}(L)$, we have $S_1 \subseteq S_2$ iff $C_1 \supseteq C_2$. Let us write $\langle S_1, C_1 \rangle \leq \langle S_2, C_2 \rangle$ if $S_1 \subseteq S_2$. With this partial order $\leq$, $\mathfrak{B}(L)$ forms a lattice. Among the concepts in $\mathfrak{B}(L)$, the one induced from the singleton of the empty context $\Lambda = \{\langle \lambda, \lambda \rangle\}$ is special: $\langle \Lambda', \Lambda'' \rangle = \langle L, L' \rangle$, which is induced by L as well. It is not hard to see that $w \in L$ iff $\langle \{w\}'', \{w\}' \rangle \leq \langle L, L' \rangle$. Two concepts $\langle S_1, C_1 \rangle$ and $\langle S_2, C_2 \rangle$ are concatenated by

$$\langle S_1, C_1 \rangle \circ \langle S_2, C_2 \rangle = \langle (S_1 S_2)'', (S_1 S_2)' \rangle.$$

Example 1. Let $L = \{\, a^n b^n \mid n \geq 0 \,\}$. Then

$$\begin{aligned} \mathfrak{B}(L) &= \{\bot, \top, E\} \cup \{\, A_k \mid k \geq 1 \,\} \cup \{\, B_k \mid k \geq 1 \,\} \cup \{\, D_k \mid k \in \mathbb{Z} \,\}, \text{ where} \\ \bot &= \langle \varnothing, \Sigma^* \times \Sigma^* \rangle,\ \top = \langle \Sigma^*, \varnothing \rangle,\ E = \langle \{\lambda\}, \{\, \langle u, v \rangle \mid uv \in L \,\} \rangle, \\ A_k &= \langle \{a^k\}, \{\, \langle a^i, a^j b^{i+j+k} \rangle \mid i, j \geq 0 \,\} \rangle,\ B_k = \langle \{b^k\}, \{\, \langle a^{i+j+k} b^i, b^j \rangle \mid i, j \geq 0 \,\} \rangle, \\ D_k &= \langle \{\, a^i b^j \mid i, j \geq 1,\ i - j = k \,\}, \{\, \langle a^i, b^j \rangle \mid i, j \geq 0,\ j - i = k \,\} \rangle. \end{aligned}$$

We note that $D_0 = \langle \Lambda', \Lambda'' \rangle = \langle L, L' \rangle$, $E \leq D_0$, and $\bot \leq X \leq \top$ for any $X \in \mathfrak{B}(L)$. The following table summarizes the concatenation operation over $\mathfrak{B}(L)$:

$\circ$	$\bot$	$\top$	E	A_l	B_m	D_n
$\bot$	$\bot$	$\bot$	$\bot$	$\bot$	$\bot$	$\bot$
$\top$	$\bot$	$\top$	$\top$	$\top$	$\top$	$\top$
E	$\bot$	$\top$	E	A_l	B_m	D_n
A_i	$\bot$	$\top$	A_i	A_{i+l}	D_{i-m}	D_{i+n}
B_j	$\bot$	$\top$	B_j	$\top$	B_{j+m}	$\top$
D_k	$\bot$	$\top$	D_k	$\top$	D_{k-m}	$\top$

where i, j, l, m are positive integers and k, n are integers.

3.2 Grammatical Representation of Syntactic Concept Lattices

This paper assumes L to be context-free, and we try to capture the structure of $\mathfrak{B}(L)$ in terms of CFGs. It seems quite a natural idea to treat concepts as nonterminals and we would like the language of a nonterminal $\langle S, C \rangle$ to be S. If we were allowed to have infinitely many nonterminals and rules, the following rules are compatible with that idea:

- $\langle S, C \rangle \to w$ for $w \in S$,
- $\langle S, C \rangle \to \langle S_1, C_1 \rangle \langle S_2, C_2 \rangle$ if $\langle S, C \rangle \geq \langle S_1, C_1 \rangle \circ \langle S_2, C_2 \rangle$,

with the initial symbol $\langle L, L' \rangle$. This grammar correctly characterizes every string in the sense that the grammar derives w from $\langle S, C \rangle$ iff $w \in S$. Yet such a grammar is impossible, because it is infinite in two respects. First, unless the language is regular, the grammar has infinitely many nonterminals. Second, each nonterminal is represented by a pair of (possibly) infinite sets.

Actually the first point is not a real problem as long as we assume that the target language L is context-free. It is enough to have concepts $\langle S_X, C_X \rangle = \langle \mathcal{L}(G, X)'', \mathcal{L}(G, X)' \rangle$ induced from $\mathcal{L}(G, X)$ for each nonterminal X of a CFG G where $L = \mathcal{L}(G)$, if what we need to know is whether $w \in L$ rather than whether $w \in S$ for an arbitrary concept $\langle S, C \rangle$. In general, it is not necessarily the case that $\mathcal{L}(G, X) = S_X$, but the fact $\mathcal{L}(G, X) \equiv S_X$ guarantees the grammar constructed in this way still defines the same language. Moreover, we can restrict w to be a letter in Σ in the rule of the form $\langle S, C \rangle \to w$. For $L = \{ a^n b^n \mid n \geq 0 \}$ in Example 1, the following five rules suffice:

$$A_1 \to a, \quad B_1 \to b, \quad D_0 \to \lambda, \quad D_0 \to A_1 D_{-1}, \quad D_{-1} \to D_0 B_1,$$

where D_0 is the initial symbol of our grammar. We now have lost the knowledge about what the concatenation of A_1 and A_1 produces, for example, but it does not matter to decide whether a given string is generated by D_0.

The second problem is crucial for learning algorithms. We need to construct nonterminals and rules from finitely many strings and contexts. Suppose that we have a finite set $K \subseteq \Sigma^*$ of strings such that $\lambda \in K$ and a finite set $F \subseteq \Sigma^* \times \Sigma^*$ of contexts such that $\langle \lambda, \lambda \rangle \in F$. We use finite bicliques $\langle S, C \rangle \subseteq K \times F$ as finite imitations of concepts. There are two interpretations of $\langle S, C \rangle$ as a concept. One focuses on S and assumes that $\langle S, C \rangle$ is a finite approximation of $\langle S'', S' \rangle$ and the other presumes $\langle C', C'' \rangle$. In general, $\langle C', S' \rangle$ may not be a biclique. We call the former idea *primal* and the latter *dual*. Suppose that $\langle S, C \rangle$, $\langle S_1, C_1 \rangle$, $\langle S_2, C_2 \rangle$ are finite bicliques. According to the former interpretation, we should have a rule

$$\langle S, C \rangle \to \langle S_1, C_1 \rangle \langle S_2, C_2 \rangle \quad \text{if } \langle S'', S' \rangle \geq \langle S_1'', S_1' \rangle \circ \langle S_2'', S_2' \rangle. \tag{2}$$

We cannot compute the polar map $(\cdot)'$, but, with the aid of the membership oracle, the following restricted polar maps $(\cdot)^{(F)}$ and $(\cdot)^{(K)}$ are computable:

$$S^{(F)} = \{ \langle x, z \rangle \in F \mid xyz \in L \text{ for all } y \in S \},$$
$$C^{(K)} = \{ y \in K \mid xyz \in L \text{ for all } \langle x, z \rangle \in C \}.$$

Because we always have $S_1 S_2 \equiv (S_1 S_2)'' \equiv S_1'' S_2''$ by (1), the condition (2) is approximated by $S^{(F)} \odot S_1 S_2 \subseteq L$, where $(\cdot)^{(F)}$ is substituted for $(\cdot)'$. In the dual approach, we should have $\langle S, C \rangle \to \langle S_1, C_1 \rangle \langle S_2, C_2 \rangle$ if $\langle C', C'' \rangle \geq \langle C_1', C_1'' \rangle \circ \langle C_2', C_2'' \rangle$, which is approximated as the condition $C \odot C_1^{(K)} C_2^{(K)} \subseteq L$. In that way one can construct a CFG as a finitely describable imitation of $\mathfrak{B}(L)$.

We want such a finite approximation to work effectively and efficiently. Thus we target CFGs such that the concept associated with a nonterminal can be approximated by a relatively small set.

Definition 1. For a positive integer k, we say that a CFG G has the k-FKP *(finite kernel property)* if each nonterminal X admits a finite set $S_X \subseteq \Sigma^*$ such that $|S_X| \leq k$ and $S_X \equiv_{\mathcal{L}(G)} \mathcal{L}(G, X)$. We call such a set S_X a *k-kernel* of X.

Definition 2. For a positive integer f, we say that a CFG H has the f-FCP *(finite context property)* if each nonterminal X admits a finite set $C_X \subseteq \Sigma^* \times \Sigma^*$ such that $|C_X| \leq f$ and $C'_X \equiv_{\mathcal{L}(H)} \mathcal{L}(H, X)$. We call such a set C_X an *f-context* of X.

The notion of the FKP by Clark et al. [9] corresponds to the 1-FKP in this paper. The 1-FKP is not too strong a restriction as all regular languages and the Dyck language over $m \in \mathbb{N}$ pairs of parentheses, for example, are generated by CFGs with the 1-FKP. The following examples show that the k-FKP properly generalizes Clark et al.'s original definition of the FKP.

Example 2. The palindrome language $L_{\text{pal}} = \{ w \in \Sigma^* \mid w^R = w \}$ with $|\Sigma| \geq 2$, where $\lambda^R = \lambda$ and $(wz)^R = z(w^R)$ for $z \in \Sigma$ and $w \in \Sigma^*$, is generated by a CFG with the 2-FKP, but not by a CFG with the 1-FKP. Let G consist of the following rules:

$$I_0 \to z,\ I_0 \to X_z Y_z,\ X_z \to Y_z I_0,\ Y_z \to z \text{ for each } z \in \Sigma, \text{ and } I_0 \to \lambda,$$

where I_0 is the initial symbol of G. Each nonterminal has the following 2-kernel, respectively:

$$I_0 : \{a, b\},\ X_z : \{za, zb\},\ Y_z : \{z\},$$

where $a, b \in \Sigma$ are arbitrary distinct terminals. To see that $\{a, b\} \equiv \mathcal{L}(G, I_0)$, i.e., $\{a, b\}' = \mathcal{L}(G, I_0)'$, it suffices to show $\{a, b\}' \subseteq \mathcal{L}(G, I_0)'$, because the opposite inclusion $\{a, b\}' \supseteq \mathcal{L}(G, I_0)'$ immediately follows from the fact $\{a, b\} \subseteq \mathcal{L}(G, I_0)$. Suppose that $\langle u, v\rangle \in \{a, b\}'$, i.e., $\langle u, v\rangle \odot \{a, b\} \subseteq L_{\text{pal}}$. If $|u| < |v|$, by $uav = v^R a u^R \in L_{\text{pal}}$, one has $v = xau^R$ for some $x \in \Sigma^*$, whereas the fact $ubv = v^R b u^R \in L_{\text{pal}}$ implies $v = ybu^R$ for some $y \in \Sigma^*$, which is a contradiction. The opposite assumption $|u| > |v|$ leads a contradiction as well. Hence $|u| = |v|$ and $v = u^R$. It is clear that $\langle u, v\rangle \odot \mathcal{L}(G, I_0) \subseteq L_{\text{pal}}$, i.e., $\langle u, v\rangle \in \mathcal{L}(G, I_0)'$. Similarly one can check that $\{za, zb\}$ is a 2-kernel of X_z for all $z \in X_z$. Trivially $\{z\}$ is a 2-kernel of Y_z.

On the other hand, the palindrome language cannot be generated by any CFG with the 1-FKP. If a CFG generates L_{pal}, it must have an initial nonterminal symbol I_1 that generates at least two palindromes of the form $a^i bx$ and $a^j by$ for some $x, y \in \Sigma^*$ and $i, j \in \mathbb{N}$ with $i \neq j$. Suppose that $\{w\}$ could be a 1-kernel of I_1. By $\langle \lambda, w^R\rangle \in \{w\}'$, it holds that $a^i bxw^R \in L_{\text{pal}}$. If $|w| > i$, w must have the form $a^i bz$ for some $z \in \Sigma^*$, but $\langle \lambda, w^R\rangle \odot a^j by = a^j byz^R ba^i$ is not a palindrome, since $i \neq j$. If $|w| \leq i$, we have $w \in \{a\}^*$. Then $\langle \lambda, a^{i+1}\rangle \in \{w\}'$, but $\langle \lambda, a^{i+1}\rangle \odot a^i bx = a^i bxa^{i+1}$ is not a palindrome. Thus $\{w\}' \neq \mathcal{L}(G, I_1)'$. No CFG with the 1-FKP generates L_{pal}.

Other examples that distinguish the expressive power of the 2-FKP and the 1-FKP are $L_m = \{ a^n b^{kn} \mid n \geq 0 \text{ and } 1 \leq k \leq m \}$ for every $m \geq 2$. Furthermore, the language $M_k = \{ a_1^{n_1} a_2^{n_2} \dots a_k^{n_k} \mid n_i = n_j \text{ for some } i \neq j \}$ separates the k-FKP from the $(k-1)$-FKP for $k \geq 3$. The hierarchy formed by the k-FKP is proper.

4 Learning of Context-Free Grammars

When we can work only on finite sets $K \subseteq \Sigma^*$ and $F \subseteq \Sigma^* \times \Sigma^*$, it is a natural idea to take maximal bicliques with respect to K and F in order to get closer to concepts in $\mathfrak{B}(L)$. Clark's [5] algorithm takes as nonterminals only maximal bicliques. Just like the ideal polar map $(\cdot)'$, every maximal biclique can be induced by $S \subseteq K$ and $C \subseteq F$ as $\langle S^{(F)(K)}, S^{(F)} \rangle$ and $\langle C^{(K)}, C^{(K)(F)} \rangle$, respectively. This paper, however, does not take this idea. Actually the restricted polar maps $(\cdot)^{(F)}$ and $(\cdot)^{(K)}$ do not demonstrate properties as nice as the ideal one $(\cdot)'$. While $S'' \equiv S$ always holds, it is not necessarily the case that $S^{(F)(K)} \equiv S$. This means there can be a concept that is approximated only by a non-maximal biclique.

The algorithm presented in Section 4.1 takes the primal approach, where we have bicliques of the form $\langle S, S^{(F)} \rangle$ as nonterminals, in which S may be non-maximal and $S^{(F)}$ is maximal. Section 4.2 presents an algorithm dual to the preceding subsection.

4.1 Primal Approach

In what follows, we fix a target language L_* which is generated by a CFG G_* with the k-FKP, and use the polar maps and other notations with respect to L_*.

Before we present the overall view of our learning algorithm, we describe how we construct a CFG from finite sets $K \subseteq \mathrm{Sub}(D)$ and $F \subseteq \mathrm{Con}(D)$ which are computed from given positive data set D in some way, where

$$\mathrm{Sub}(D) = \{\, y \in \Sigma^* \mid xyz \in D \text{ for some } \langle x, z \rangle \in \Sigma^* \times \Sigma^* \,\},$$
$$\mathrm{Con}(D) = \{\, \langle x, z \rangle \in \Sigma^* \times \Sigma^* \mid xyz \in D \text{ for some } y \in \Sigma^* \,\}.$$

We define a CFG $\mathcal{G}^k(K, F) = \langle \Sigma, V_K, P_{K,F}, I_K \rangle$ as follows. The set of nonterminal symbols is

$$V_K = \{\, [\![S]\!] \mid S \subseteq K \text{ and } |S| \le k \,\}.$$

The intended meaning of $[\![S]\!]$ is the biclique $\langle S, S^{(F)} \rangle$, which can be seen as an approximation of the concept $\langle S'', S' \rangle$. The initial symbols are those $[\![S]\!]$ such that $S \subseteq L_*$. We have two types of rules:

- $[\![S]\!] \to a$ with $a \in \Sigma \cup \{\lambda\}$ if $S^{(F)} \odot a \subseteq L_*$;
- $[\![S]\!] \to [\![S_1]\!][\![S_2]\!]$ if $S^{(F)} \odot S_1 S_2 \subseteq L_*$,

which are constructed with the aid of the membership oracle. The grammar $\mathcal{G}^k(K, F)$ shows the following beautiful monotonicity.

Lemma 1. *If $J \subseteq K$ then $\mathcal{L}(\mathcal{G}^k(J, F)) \subseteq \mathcal{L}(\mathcal{G}^k(K, F))$.*

Proof. Every rule of $\mathcal{G}^k(J, F)$ is also a rule of $\mathcal{G}^k(K, F)$. □

Lemma 2. *If $E \subseteq F$ then $\mathcal{L}(\mathcal{G}^k(K, F)) \subseteq \mathcal{L}(\mathcal{G}^k(K, E))$.*

Proof. Every rule of $\mathcal{G}^k(K, F)$ is also a rule of $\mathcal{G}^k(K, E)$. □

We want a rule $[\![S]\!] \to [\![S_1]\!][\![S_2]\!]$ to be *correct* in the sense that $\langle S'', S'\rangle \geq \langle S_1'', S_1'\rangle\langle S_2'', S_2'\rangle$, which is equivalent to $S' \odot S_1''S_2'' \subseteq L_*$. We approximate S', S_1'' and S_2'' with $S^{(F)}$, S_1 and S_2, respectively. We know that $S_1'' \equiv S_1$ and $S_2'' \equiv S_2$, while it is not certain that $S' \equiv S^{(F)}$. If a rule $[\![S]\!] \to [\![S_1]\!][\![S_2]\!]$ is not correct, there is $\langle x, z\rangle \in S'$ such that $\langle x, z\rangle \odot S_1''S_2'' \not\subseteq L_*$. If $\langle x, z\rangle$ is in F, then such an incorrect rule is not constructed. That is, for every triple $[\![S]\!], [\![S_1]\!], [\![S_2]\!]$ of nonterminals that forms an incorrect rule, one context is enough to prevent the algorithm from constructing the incorrect rule. Similarly we say that $[\![S]\!] \to a$ is *correct* if $S' \odot a \subseteq L_*$. If $[\![S]\!] \to a$ is not correct, having one context in F is enough to suppressing this rule. We remark that whenever $\mathcal{G}^k(K, E)$ has no incorrect rules and $E \subseteq F$, $\mathcal{G}^k(K, F)$ has no incorrect rules.

Lemma 3. *For every K, there is a finite set $F \subseteq \mathrm{Con}(L_*)$ consisting of at most $|V_K|^3 + |V_K|(|\Sigma|+1)$ contexts such that $\mathcal{G}^k(K, F)$ has no incorrect rules. In such a case, we have $\mathcal{L}(\mathcal{G}^k(K, F)) \subseteq L_*$.*

Proof. Let $\hat{G} = \mathcal{G}^k(K, F)$. We prove by induction that $[\![S]\!] \Rightarrow^*_{\hat{G}} w$ implies $S' \odot w \subseteq L_*$. Particularly when $[\![S]\!]$ is an initial symbol, $S \subseteq L_*$ implies that $\langle \lambda, \lambda\rangle \in S'$ and thus $w \in L_*$. Suppose that $[\![S]\!] \Rightarrow^*_{\hat{G}} a$ for some $a \in \Sigma \cup \{\lambda\}$. Since the rule $[\![S]\!] \to a$ is correct, we have $S' \odot a \subseteq L_*$. Suppose that $[\![S]\!] \Rightarrow_{\hat{G}} [\![S_1]\!][\![S_2]\!] \Rightarrow^*_{\hat{G}} w_1 w_2$ where $w_i \in \mathcal{L}(\hat{G}, [\![S_i]\!])$ for $i = 1, 2$. By the induction hypothesis, we have $S_i' \odot w_i \subseteq L_*$, which implies $w_i \in S_i''$ for $i = 1, 2$. Since $[\![S]\!] \to [\![S_1]\!][\![S_2]\!]$ is correct, we have $S' \odot S_1''S_2'' \subseteq L_*$. Hence $S' \odot w_1 w_2 \subseteq S' \odot S_1''S_2'' \subseteq L_*$. □

Lemma 4. *Let L_* be generated by a* CFG *G_* with the k-*FKP*. Then $L_* \subseteq \mathcal{L}(\mathcal{G}^k(K, F))$ if K includes k-kernels S_X of all nonterminals X of G_*.*

Proof. Let $\hat{G} = \mathcal{G}^k(K, F)$. We show that each nonterminal X of G_* is simulated by $[\![S_X]\!]$ in $\hat{G}$. Suppose that $X \to a$ is a rule of G_*. By $a \in \mathcal{L}(G, X) \equiv S_X$, we have $S_X \odot a \in L_*$. $\hat{G}$ has the rule $[\![S_X]\!] \Rightarrow_{\hat{G}} a$. Suppose that $X \to YZ$ is a rule of G_*. Since $S_Y S_Z \equiv \mathcal{L}(G_*, Y)\mathcal{L}(G_*, Z) \subseteq \mathcal{L}(G_*, X) \equiv S_X''$, we have $S_X^{(F)} \odot S_Y S_Z \subseteq S_X' \odot S_Y S_Z \subseteq L$. Thus $\hat{G}$ has the rule $[\![S_X]\!] \to [\![S_Y]\!][\![S_Z]\!]$. If X is an initial symbol of G_*, then $S_X \equiv \mathcal{L}(G, X) \subseteq L_*$, which implies that $[\![S_X]\!]$ is an initial symbol of $\hat{G}$. □

Due to the nice properties shown through Lemmas 1 to 4, our learning strategy is quite simple. Whenever we get a positive example that is not generated by our current conjecture, we expand K. On the other hand, to suppress incorrect rules, we keep expanding F.

Theorem 1. *Algorithm 1 identifies* CFGs *with the k-*FKP *in the limit from positive data and membership queries.*

Proof. Let $D_n = \{w_1, \dots, w_n\}$. Lemma 4 ensures that Algorithm 1 does not update K infinitely many times, because k-kernels of nonterminals of G_* are finite subsets of $\mathrm{Sub}(L_*)$ and there is a point in time when the positive examples cover all the k-kernels of G_*. Let $K_{m_0} = \mathrm{Sub}(D_{m_0})$ be the limit. There is a

Algorithm 1. Learning CFGs with k-FKP

Data: A positive presentation $w_1, w_2, \ldots$ of L_*; membership oracle $\mathcal{O}$;
Result: A sequence of CFGs $G_1, G_2, \ldots$;
let $D := \varnothing$; $K := \varnothing$; $F := \varnothing$; $\hat{G} := \mathcal{G}^k(K, F)$;
for $n = 1, 2, \ldots$ **do**
  let $D := D \cup \{w_n\}$; $F := \mathrm{Con}(D)$;
  if $D \not\subseteq \mathcal{L}(\hat{G})$ **then**
    let $K := \mathrm{Sub}(D)$;
  end if
  output $\hat{G} = \mathcal{G}^k(K, F)$ as G_n;
end for

point n_0 such that $\mathcal{G}^k(K_{m_0}, F_{n_0})$ has no incorrect rules for $F_{n_0} = \mathrm{Con}(D_{n_0})$. By Lemma 3, we see that for any $n \geq \max\{m_0, n_0\}$, Algorithm 1 outputs $\hat{G}_n = \mathcal{G}^k(K_{m_0}, F_{n_0})$, which generates the learning target L_*. □

We remark on the efficiency of our algorithm. It is easy to see that the description sizes of K and F are bounded by the square of that of D, and V_K consists of at most $(|K| + 1)^k$ nonterminals. We need an at most polynomial number of membership queries to determine rules among those nonterminals. All in all, Algorithm 1 updates its conjecture in polynomial time in $\|D\|$. Moreover, we do not need too much data. To get k-kernels of all nonterminals, $k|V_*|$ examples are enough, where V_* is the set of nonterminals of the target CFG G_*. To suppress incorrect rules, $\mathrm{O}(|V_K|^3)$ contexts are enough by Lemma 3.

4.2 Dual Approach

Clark [5] has proposed a learning algorithm for CFGs with the f-FCP that takes maximal bicliques as nonterminals. This subsection presents an even simpler learning algorithm for the same class. Our algorithm demonstrates the exact symmetry to the one presented in the previous subsection. Due to the simplicity of the definition of nonterminals, proofs of lemmas that support the correctness of the algorithm are much easier than Clark's.

For two finite sets $K \subseteq \Sigma^*$ and $F \subseteq \Sigma^* \times \Sigma^*$, we define a CFG $\mathcal{H}^f(F, K)$ as follows. The set of nonterminal symbols is

$$V_F = \{\, [\![C]\!] \subseteq F \mid C \subseteq F \text{ and } |C| \leq f \,\}.$$

The initial symbols are those $[\![C]\!]$ such that $\langle \lambda, \lambda \rangle \in C$. The intended meaning of $[\![C]\!]$ is the biclique $\langle C^{(K)}, C \rangle$, which can be seen as an approximation of the concept $\langle C', C'' \rangle$. We have rules of the following two types:

- $[\![C]\!] \to a$ with $a \in \Sigma \cup \{\lambda\}$ if $C \odot a \subseteq L_*$;
- $[\![C]\!] \to [\![C_1]\!][\![C_2]\!]$ if $C \odot C_1^{(K)} C_2^{(K)} \subseteq L_*$.

We again have monotonicity lemmas, which work in the direction opposite to Lemmas 1 and 2.

Lemma 5. *If $E \subseteq F$ then $\mathcal{L}(\mathcal{H}^f(E, K)) \subseteq \mathcal{L}(\mathcal{H}^f(F, K))$.*

Lemma 6. *If $J \subseteq K$ then $\mathcal{L}(\mathcal{H}^f(F, K)) \subseteq \mathcal{L}(\mathcal{H}^f(F, J))$.*

We say that a rule $[\![C]\!] \to [\![C_1]\!][\![C_2]\!]$ is *correct* if $C'' \odot C_1' C_2' \subseteq L_*$. If $[\![C]\!] \to [\![C_1]\!][\![C_2]\!]$ is not correct, there are $w_i \in C_i'$ for $i = 1, 2$ such that $C'' \odot w_1 w_2 \equiv C \odot w_1 w_2 \nsubseteq L_*$. If $w_1, w_2 \in K$, such an incorrect rule is suppressed.

Lemma 7. *Every F admits a finite set $K \subseteq \mathrm{Sub}(L_*)$ consisting of at most $2|V_F|^3$ strings such that all rules of $\hat{H} = \mathcal{H}^f(F, K)$ are correct. In such a case, $\mathcal{L}(\hat{H}) \subseteq L_*$.*

Proof. One can prove by induction that $[\![C]\!] \Rightarrow^*_{\hat{H}} w$ implies $C \odot w \subseteq L_*$. □

Lemma 8. *Let L_* be generated by a* CFG *H_* with the f-*FCP*. If F includes f-contexts of all nonterminals of H_*, then $L_* \subseteq \mathcal{L}(\mathcal{H}^f(F, K))$.*

Proof. Each nonterminal X of H_* is simulated by $[\![C_X]\!]$ in $\mathcal{H}^f(F, K)$. □

Algorithm 2. Learning CFGs with f-FCP

Data: A positive presentation $w_1, w_2, \dots$ of L_*; membership oracle $\mathcal{O}$;
Result: A sequence of CFGs $H_1, H_2, \dots$
let $D := \varnothing$; $F := \varnothing$; $K := \varnothing$; $\hat{H} := \mathcal{H}^f(F, K)$;
for $n = 1, 2, \dots$ **do**
  let $D := D \cup \{w_n\}$; $K := \mathrm{Sub}(D)$;
  if $D \nsubseteq \mathcal{L}(\hat{H})$ **then**
    let $F := \mathrm{Con}(D)$;
  end if
  output $\hat{H} = \mathcal{H}^f(F, K)$ as H_n;
end for

Now it is clear that Algorithm 2 efficiently learns CFGs with the f-FCP just like Algorithm 1 learns CFGs with the k-FKP.

Theorem 2. *Algorithm 2 identifies* CFGs *with the f-*FCP *in the limit from positive data and membership queries.*

4.3 Learning with the Maximal Bicliques

In the previous two subsections, we have presented two learning algorithms that have the same rule construction schemes with different sets of nonterminals, or bicliques. It is interesting to see that those rule construction schemes show the opposite monotonicity on K and F just depending on the choice of the bicliques we use as nonterminals. On the other hand, if we take maximal bicliques as nonterminals, we lose the monotonicity of both directions.

For $\hat{G} = \mathcal{G}^k(K, F)$ with $k \geq |K|$, it is easy to see that $[\![S]\!] \Rightarrow^*_{\hat{G}} w$ iff $[\![S^{(F)(K)}]\!] \Rightarrow^*_{\hat{G}} w$ for any $S \subseteq K$ and $w \in \Sigma^*$. This suggests that we may merge $[\![S_1]\!]$

and $[\![S_2]\!]$ such that $S_1^{(F)} = S_2^{(F)}$ into $[\![S_1^{(F)(K)}]\!]$. That is, maximal bicliques are sufficient for constructing a reasonable conjecture. Yet this does not imply that we may consider only nonterminals of the form $[\![S^{(F)(K)}]\!]$ *before* constructing rules, though one may merge such $[\![S_1]\!]$ and $[\![S_2]\!]$ *after* all rules are determined. This observation leads us to the following construction scheme of a conjecture:

- $V = \{\, [\![S^{(F)(K)}]\!] \mid S \subseteq K \text{ and } |S| \leq k \,\}$,
- $[\![S]\!] \to a$ with $a \in \Sigma \cup \{\lambda\}$ if $S^{(F)} \odot a \subseteq L_*$, i.e., $a \in S$,
- $[\![S]\!] \to [\![S_1]\!][\![S_2]\!]$ if there are $T_i \subseteq S_i$ such that $|T_i| \leq k$, $S_i = T_i^{(F)(K)}$ for $i = 1, 2$ and $S^{(F)} \odot T_1 T_2 \subseteq L_*$.

The rule construction scheme looks a little more complex, but it is essentially the same as the one presented in Section 4.1. The new learning algorithm based on this grammar construction still identifies CFGs with the k-FKP and this modification reduces the size of the output grammar.

On the other hand, for $\hat{H} = \mathcal{H}^f(F, K)$, it does not holds that $[\![C]\!] \Rightarrow^*_{\hat{H}} w$ iff $[\![C^{(K)(F)}]\!] \Rightarrow^*_{\hat{H}} w$. Thus the idea from the previous paragraph does not apply to the dual approach. The grammar constructed by Clark's learning algorithm [5] for CFGs with the f-FCP can be described by our notation as follows:

- $V = \{\, [\![C^{(K)(F)}]\!] \mid C \subseteq F \text{ and } |C| \leq f \,\}$,
- $[\![C]\!] \to a$ with $a \in \Sigma \cup \{\lambda\}$ if $C \odot a \subseteq L_*$,
- $[\![C]\!] \to [\![C_1]\!][\![C_2]\!]$ if $C \odot (C_1^{(K)} C_2^{(K)})^{(F)(K)} \subseteq L_*$.

This does not look symmetric to the one in the previous paragraph.

5 Concluding Remarks

Algorithms 1 and 2 demonstrate a clear symmetry of the dual approaches of distributional learning. On the other hand, when we work on maximal bicliques, which can be seen as a neutral choice of nonterminals, we do not have such a perfect symmetry.

The original paper [9] that proposes the notion of the 1-FKP works on *Binary Feature Grammars*, which are more expressive than CFGs. One can generalize the discussion of [9] with the k-FKP.

We remark the analogy between CFGs with the 1-FCP and residual finite state automata (RFSAs [10]) by translating strings to suffixes and contexts to prefixes. Each state of an RFSA corresponds to the residual language $x^{-1}L = \{\, y \mid xy \in L \,\}$ for some $x \in \Sigma^*$, while each nonterminal of a CFGs with 1-FCP corresponds to the language $\{\langle x, z\rangle\}' = \{\, y \in \Sigma^* \mid xyz \in L \,\}$ for some $\langle x, z\rangle$. In fact Denis et al. [11] and Bollig et al. [1] have proposed efficient learning algorithms for RFSAs. Analogously to CFGs with f-FCP, one may think of learning of f-RFSAs for $f \geq 1$, whose states correspond to $S^{-1}L = \{\, y \mid xy \in L \text{ for all } x \in S \,\}$ for some S with $|S| \leq f$, which could be even more succinct representations of regular languages.

Acknowledgement

The author is grateful to Alexander Clark and Anna Kasprzik for helpful discussions. This work was supported in part by MEXT KAKENHI (B-20700124).

References

1. Bollig, B., Habermehl, P., Kern, C., Leucker, M.: Angluin-style learning of NFA. In: Boutilier, C. (ed.) IJCAI, pp. 1004–1009 (2009)
2. Clark, A.: A learnable representation for syntax using residuated lattices. In: Proceedings of the 14th Conference on Formal Grammar, Bordeaux, France (2009)
3. Clark, A.: Three learnable models for the description of language. In: Dediu, A.-H., Fernau, H., Martín-Vide, C. (eds.) LATA 2010. LNCS, vol. 6031, pp. 16–31. Springer, Heidelberg (2010)
4. Clark, A.: Distributional learning of some context-free languages with a minimally adequate teacher. In: [17], pp. 24–37 (2010)
5. Clark, A.: Learning context free grammars with the syntactic concept lattice. In: [17], pp. 38–51 (2010)
6. Clark, A.: Towards general algorithms for grammatical inference. In: Hutter, M., Stephan, F., Vovk, V., Zeugmann, T. (eds.) ALT 2010. LNCS, vol. 6331, pp. 11–30. Springer, Heidelberg (2010)
7. Clark, A.: Efficient, correct, unsupervised learning of context-sensitive languages. In: Proceedings of CoNLL. Association for Computational Linguistics, Uppsala (2010)
8. Clark, A., Eyraud, R.: Polynomial identification in the limit of substitutable context-free languages. Journal of Machine Learning Research 8, 1725–1745 (2007)
9. Clark, A., Eyraud, R., Habrard, A.: Using contextual representations to efficiently learn context-free languages. Journal of Machine Learning Research 11, 2707–2744 (2010)
10. Denis, F., Lemay, A., Terlutte, A.: Residual finite state automata. Fundam. Inform. 51(4), 339–368 (2002)
11. Denis, F., Lemay, A., Terlutte, A.: Learning regular languages using RFSAs. Theor. Comput. Sci. 313(2), 267–294 (2004)
12. Shirakawa, H., Yokomori, T.: Polynomial-time MAT learning of c-deterministic context-free grammars. Transaction of Information Processing Society of Japan 34, 380–390 (1993)
13. Yoshinaka, R.: Identification in the limit of k,l-substitutable context-free languages. In: Clark, A., Coste, F., Miclet, L. (eds.) ICGI 2008. LNCS (LNAI), vol. 5278, pp. 266–279. Springer, Heidelberg (2008)
14. Yoshinaka, R.: Learning mildly context-sensitive languages with multidimensional substitutability from positive data. In: Gavaldà, R., Lugosi, G., Zeugmann, T., Zilles, S. (eds.) ALT 2009. LNCS, vol. 5809, pp. 278–292. Springer, Heidelberg (2009)
15. Yoshinaka, R.: Polynomial-time identification of multiple context-free languages from positive data and membership queries. In: [17], pp. 230–244 (2010)
16. Yoshinaka, R., Clark, A.: Polynomial time learning of some multiple context-free languages with a minimally adequate teacher. In: Proceedings of the 15th Conference on Formal Grammar, Copenhagen, Denmark (2010)
17. Sempere, J.M., García, P. (eds.): ICGI 2010. LNCS, vol. 6339. Springer, Heidelberg (2010)

On Highly Repetitive and Power Free Words

Narad Rampersad[1] and Elise Vaslet[2]

[1] Department of Mathematics, University of Liège, Belgium
narad.rampersad@gmail.com
[2] Institut de Mathématiques de Luminy, Université Aix-Marseille II, France
vaslet@iml.univ-mrs.fr

Abstract. Answering a question of Richomme, Currie and Rampersad proved that 7/3 is the infimum of the real numbers $\alpha > 2$ such that there exists an infinite binary word that avoids α-powers but is highly 2-repetitive, i.e., contains arbitrarily large squares beginning at every position. In this paper, we prove similar statements about β-repetitive words, for some other β's, on the binary and the ternary alphabets.

1 Introduction

In this paper we study words that are non-repetitive, in the sense that they avoid α-powers for some α, but yet are highly repetitive, in the sense that for some β close to α, they contain infinitely many β-powers starting at every position. First we recall some basic definitions. A finite, non-empty word v can be written as $v = p^k e$, where $k \geq 1$, the word e is a prefix of p, and the length of p is minimal. We say that v has *period* p, *excess* e, and *exponent* $\text{expo}(v) = |v|/|p|$. A word with exponent α is called an α-power. An α^+-*power* is a word that is a β-power for some $\beta > \alpha$. A 2-power is called a *square*; a 2^+-power is called an *overlap*. A word is α-*free* if none of its factors is a β-power for any $\beta \geq \alpha$. A word is α^+-*free* if none of its factors is an α^+-power.

Thue [TH2] proved that there exist infinite overlap-free binary words. Dekking [DEK] later showed that any infinite overlap-free binary word must contain arbitrarily large squares. Currie, Rampersad, and Shallit [CRS] constructed 7/3-free words containing infinitely many overlaps. Their motivation came from the following result of Shur [SHU]: Any bi-infinite 7/3-free binary word is overlap-free. The analogue of this surprising result is not true for one-sided infinite words: There are infinite 7/3-free binary words that are not overlap-free. The result of Currie et al. shows that in fact the infinite 7/3-free binary words can be significantly different from the infinite overlap-free binary words.

Currie and Rampersad [CR] continued this line of study in order to respond to the following question of Richomme [RIC]: What is the infimum of the real numbers $\alpha > 2$ such that there exists an infinite word that avoids α-powers but contains arbitrarily large squares beginning at every position? They showed that over the binary alphabet the answer to Richomme's question is $\alpha = 7/3$. In this paper we show that if instead of requiring arbitrarily large squares at every position, we only ask for arbitrarily large β-powers at every position for some

G. Mauri and A. Leporati (Eds.): DLT 2011, LNCS 6795, pp. 441–451, 2011.

fixed $\beta < 2$, then the answer of 7/3 for Richomme's question can be replaced by 2. We also answer the analogous question for the ternary alphabet.

We also mention here the related work of Saari [SAA], who also studied infinite words containing squares (not necessarily arbitrarily large) beginning at every position. He called such words *squareful*, but he imposed the additional condition that the word contain only finitely many distinct minimal squares.

2 Preliminary Definitions and Results

In this paper, A_k will denote the k-letter alphabet, for an integer k.

Let α be a real number.

Definition 1. *A finite word is an α-repetition (resp. α^+-repetition) if it is a β-power for some $\beta \geq \alpha$ (resp. $\beta > \alpha$).*

Definition 2. *A word (finite or infinite) is highly α-repetitive (resp. highly α^+-repetitive) if it contains infinitely many α-repetitions (resp. α^+-repetitions) beginning at every position.*

Definition 3. *A word (finite or infinite) is α-free (resp. α^+-free) if it contains no α-repetition (resp. α^+-repetition).*

Definition 4. *A morphism $\mu : A^* \rightarrow B^*$ is α-free (resp. α^+-free) if for any word $w \in A^*$, the following equivalence holds:*

$$w \text{ is } \alpha\text{-free} \Leftrightarrow \mu(w) \text{ is } \alpha\text{-free}.$$

We now define three particular morphisms which we will use in the paper because of their α-freeness for certain α.

The Thue-Morse morphism is defined by

$$\mu_{TM} : \begin{cases} 0 \mapsto 01 \\ 1 \mapsto 10. \end{cases}$$

Proposition 1 ([TH2]). *μ_{TM} is a 2^+-free morphism.*

Dejean's morphism is defined by

$$\mu_D : \begin{cases} a \mapsto abc\ acb\ cab\ c\ bac\ bca\ cba \\ b \mapsto \pi(\mu(a)) = bca\ bac\ abc\ a\ cba\ cab\ acb \\ c \mapsto \pi^2(\mu(a)) = cab\ cba\ bca\ b\ acb\ abc\ bac, \end{cases}$$

where π is the morphism that cyclically permutes the alphabet $\{a, b, c\}$.

Proposition 2 ([DEJ]). *μ_D is a $7/4^+$-free morphism.*

The following morphism was given by Brandenburg in [BRA]:

$$\mu_B : \begin{cases} a \mapsto abacbabcbac \\ b \mapsto abacbcacbac \\ c \mapsto abcbabcacbc. \end{cases}$$

Proposition 3 ([BRA]). *μ_B is a 2-free morphism.*

3 Highly Repetitive Binary Words

The results of Currie and Rampersad in [CR, Theorems 4 and 5] state that for any $\alpha > 7/3$, there exists an infinite binary word which is both α-free and highly 2-repetitive, and that such a word does not exist if $\alpha \leq 7/3$. The following proposition gives a similar result in the case of highly β-repetitive words, with $\beta < 2$: for any $\alpha > 2$ and any $\beta < 2$, there exists an infinite binary word which is both α-free and highly β-repetitive.

Theorem 1. *For any real number $\beta < 2$, there exists an infinite binary word which is both 2^+-free and highly β-repetitive.*

Proof. Let $\beta < 2$ be a real number. There exists an integer $m \geq 1$ such that $\beta \leq 2 - \frac{1}{2^m} < 2$. Let $r = 2 - \frac{1}{2^m}$, and let us construct a binary infinite word with infinitely many r-powers at every position, and which is 2^+-free. We recall that μ_{TM} is the Thue-Morse morphism.

We remark that $\mu_{TM}^m(11)$ is a factor of $\mu_{TM}^{m+3}(1)$. Indeed, $\mu_{TM}^3(1) = 10010110$, and so $\mu_{TM}^{m+3}(1) = \mu_{TM}^m(10010110) = u\mu_{TM}^m(11)\mu_{TM}^m(0)$, where $u = \mu_{TM}^m(10010)$. Let us define $q = m + 3$. We define the following sequence of finite words:

$$\begin{aligned} A_0 &= \mu_{TM}^m(11) \\ A_1 &= (u)^{-1}\mu_{TM}^q(A_0) \\ &\vdots \\ A_n &= (u)^{-1}\mu_{TM}^q(A_{n-1}), \end{aligned}$$

where we denote by $(x)^{-1}y$ the word obtained by removing the prefix x from the word y (here, the word $\mu_{TM}^q(A_n)$ always has u as a prefix, since A_n always has 1 as a prefix). As 11 and μ_{TM} are 2^+-free, then for any $n \geq 0$, A_n is 2^+-free.

We show that for any $n \geq 0$, A_n is a prefix of A_{n+1}. This is clearly true for $n = 0$. If A_{n-1} is a prefix of A_n, then we can write $A_n = A_{n-1}S$, where S is a binary word. Then, $A_{n+1} = (u)^{-1}\mu_{TM}^q(A_{n-1}S) = A_n\mu_{TM}^q(S)$. Thus, the sequence A_n converges to an infinite limit word we denote by w, and this limit is 2^+-free. We now prove that w contains infinitely many r-powers beginning at every position. It can easily be seen that for any $n \geq 1$,

$$A_n = (u)^{-1}[\mu_{TM}^q(u)]^{-1} \cdots [\mu_{TM}^{q(n-1)}(u)]^{-1}\mu_{TM}^{qn}(A_0).$$

Let us denote by p_n the word $\mu_{TM}^{q(n-1)}(u)\cdots\mu_{TM}^{q}(u)u$. Then we have $A_n = (p_n)^{-1}\mu_{TM}^{qn}(A_0)$, since for any words u and v, $(uv)^{-1} = (v)^{-1}(u)^{-1}$. Moreover, we can see that p_n is a prefix of $\mu_{TM}^{qn}(1)$. Indeed,

$$\begin{aligned}
|p_n| &= |u| \sum_{i=0}^{n-1} |\mu_{TM}(1)|^{qi} \\
&= |u| \sum_{i=0}^{n-1} 2^{qi} \\
&< 2^{qn} = |\mu_{TM}^{qn}(1)|,
\end{aligned}$$

since $|u| \leq 2^q - 1$. So, we can write $\mu_{TM}^{qn}(1) = p_n s_n$, where s_n is a binary word. For any $n \geq 1$, $A_n = (p_n)^{-1}\mu_{TM}^{qn}(A_0) = (p_n)^{-1}\mu_{TM}^{qn}(\mu_{TM}^{m}(11))$. We know that $\mu_{TM}^{m}(1)$ begins with the letter 1, and let us denote its last letter by $a \in \{0,1\}$. We can write $\mu_{TM}^{m}(1) = 1xa$, where $x \in \{0,1\}^*$. We can also write $\mu_{TM}^{qn}(a) = \tilde{p_n}\tilde{s_n}$, with $\tilde{p_n}$ and $\tilde{s_n}$ two binary words such that $|\tilde{p_n}| = |p_n|$ and $|\tilde{s_n}| = |s_n|$. Then,

$$A_n = s_n \mu_{TM}^{qn}(x)\tilde{p_n}\tilde{s_n} p_n s_n \mu_{TM}^{qn}(x)\tilde{p_n}\tilde{s_n}.$$

It follows that at all positions $0 \leq j \leq |s_n|$, there begins a repetition with period a conjugate of $s_n\mu_{TM}^{qn}(x)\tilde{p_n}\tilde{s_n}p_n$, and of length

$$\begin{aligned}
&|s_n \mu_{TM}^{qn}(x)\tilde{p_n}\tilde{s_n} p_n s_n \mu_{TM}^{qn}(x)\tilde{p_n}| \\
&= 2 \cdot |x| \cdot 2^{qn} + 3 \cdot 2^{qn} \\
&= 2^{qn}(2^{m+1} - 1),
\end{aligned}$$

since $|x| = |\mu_{TM}^{m}(1)| - 2 = 2^m - 2$. These repetitions have exponents $\frac{2^{qn}(2^{m+1}-1)}{2^{qn}|x|+2\cdot 2^{qn}} = \frac{2^{m+1}-1}{2^m} = r$. As n can be taken arbitrarily large, the result follows. □

Moreover, this result is optimal. Indeed, on the one hand, it is easy to see that for $\alpha \leq 2$, there is no infinite binary α-free word. On the other hand, Currie and Rampersad proved the following result.

Proposition 4 ([CR]). *If w is an infinite overlap-free binary word, then there is a position i such that w does not contain a square beginning at position i.*

4 Highly Repetitive Ternary Words

In this part, we will state some similar results in the case of the ternary alphabet. We recall that the repetition threshold of the ternary alphabet is 7/4, so for any $\alpha \leq 7/4$, there is no infinite α-free ternary word.

Theorem 2. *For any real number $\beta < 7/4$, there exists an infinite ternary word that is both $7/4^+$-free and highly β-repetitive.*

Proof. For any real number $\beta < 7/4$, there exists an integer $m \geq 1$ such that $\beta \leq 7/4 - \frac{1}{19^{m+2}} < 7/4$. Let $r = 7/4 - \frac{1}{19^{m+2}}$, and let us construct a $7/4^+$-free word which contains infinitely many r-powers at every position. We recall that μ_D is Dejean's morphism.

We remark that $\mu_D^m(abcbabc)$ is a factor of $\mu_D^{m+2}(a)$. Indeed,

$$\mu_D^2(a) = \mu_D(a)\mu_D(b)cabcbabcabacbabcbac\mu_D(acbcabcbacbcacba),$$

and so

$$\mu_D^{m+2}(a) = u\mu_D^m(abcbabc)\mu_D^m(abacbabcbac)\mu_D^{m+1}(acbcabcbacbcacba),$$

where $u = \mu_D^{m+1}(ab)\mu_D^m(c)$. Let us define $q = m + 2$.

We define the following sequence of finite words:

$$\begin{aligned} A_0 &= \mu_D^m(abcbabc) \\ A_1 &= (u)^{-1}\mu_D^q(A_0) \\ &\vdots \\ A_n &= (u)^{-1}\mu_D^q(A_{n-1}). \end{aligned}$$

(Here, the word $\mu_D^q(A_n)$ always has u as a prefix, since A_n always has a as a prefix). As $abcbabc$ and μ_D are $7/4^+$-free, then for any $n \geq 0$, A_n is $7/4^+$-free.

Similar arguments as in the proof of Theorem 1 prove that the sequence A_n converges to an infinite limit word we denote by w, and this limit is $7/4^+$-free. We now prove that w contains infinitely many α-powers beginning at every position. Again with similar arguments as in the proof of Theorem 1, we can see that for any $n \geq 1$,

$$A_n = (p_n)^{-1}\mu_D^{qn}(A_0),$$

where p_n is the word $\mu_D^{q(n-1)}(u)\cdots\mu_D^q(u)u$. Moreover, we can see that p_n is a prefix of $\mu_D^{qn}(a)$. Indeed,

$$\begin{aligned} |p_n| &= |u|\sum_{i=0}^{n-1}|\mu_D(a)|^{qi} \\ &= |u|\sum_{i=0}^{n-1}19^{qi} \\ &< 19^{qn} = |\mu_D^{qn}(a)|. \end{aligned}$$

So, we can write $\mu_D^{qn}(a) = p_n s_n$, where s_n is a ternary word. For any $n \geq 1$, $A_n = (p_n)^{-1}\mu_D^{qn}(A_0) = (p_n)^{-1}\mu_D^{qn}(\mu_D^m(abcbabc))$. We know that $\mu_D^m(a)$ begins with the letter a, and let us denote its last letter by $x \in \{a, b, c\}$. We can write $\mu_D^m(a) = aXx$, where $X \in \{a, b, c\}^*$. Similarly, let us write $\mu_D^m(c) = cYy$, where $Y \in \{a, b, c\}^*$, and $y \in \{a, b, c\}$. We can also write $\mu_D^{qn}(y) = \tilde{p_n}\tilde{s_n}$, with $\tilde{p_n}$ and $\tilde{s_n}$ two ternary words such that $|\tilde{p_n}| = |p_n|$ and $|\tilde{s_n}| = |s_n|$. Then,

$$A_n = s_n\mu_D^{qn}(Xx\mu_D^m(b)cY)\tilde{p_n}\tilde{s_n}\mu_D^{qn+m}(b)p_n s_n\mu_D^{qn}(Xx\mu_D^m(b)cY)\tilde{p_n}\tilde{s_n}.$$

It follows that at all positions $0 \leq j \leq |s_n|$, there begins a repetition with period a conjugate of $s_n \mu_D^{qn}(Xx\mu_D^m(b)cY)\tilde{p_n}\tilde{s_n}\mu_D^{qn+m}(b)p_n$, and of length

$$\begin{aligned} l &= |s_n \mu_D^{qn}(Xx\mu_D^m(b)cY)\tilde{p_n}\tilde{s_n}\mu_D^{qn+m}(b)p_n s_n \mu_D^{qn}(Xx\mu_D^m(b)cY)\tilde{p_n}| \\ &= 2 \cdot |Xx\mu_D^m(b)cY| \cdot 19^{qn} + 19^{qn+m} + 3 \cdot 19^{qn} \\ &= 19^{qn} \cdot (7 \cdot 19^m - 1), \end{aligned}$$

since $|X| = |Y| = |\mu_D^m(a)| - 2 = 19^m - 2$. These repetitions have exponent

$$\frac{19^{qn}(7 \cdot 19^m - 1)}{19^{qn}|Xx\mu_D^m(b)cY| + 19^{qn+m} + 2 \cdot 19^{qn}} = \frac{7 \cdot 19^m - 1}{4 \cdot 19^m} = \alpha.$$

As n can be taken arbitrarily large, the result follows. □

Now, we will consider the problem for highly 7/4-repetitive words. This would be Richomme's question for the ternary alphabet : what is the infimum of the real numbers α such that there exists an infinite ternary word which is both α-free and highly 7/4-repetitive?

Proposition 5. *There is no $7/4^+$-free ternary word which is highly 7/4-repetitive.*

Proof. Let us suppose there exists such a word w. Then, any factor of w is *left-special* (i.e., has two left extensions). Indeed, let u be a factor of w, and let i denote its position in w. w is *recurrent* (every factor occurs infinitely often) because it is highly repetitive, so we can suppose $i \geq 1$. There is a long enough 7/4-repetition v beginning at position i. We denote its period by p. Then, u is repeated in the excess of v : pu is a factor of v. Now, let $x_1 = w_{i-1}$ and $x_2 = p_{|p|}$. It is clear that $x_1 \neq x_2$, since otherwise, $x_1 v$ would be a $7/4^+$-repetition, which is impossible since w is $7/4^+$-free. So u has two left extensions. Moreover, as the factors aa, bb and cc are clearly forbidden, we can see that w contains the factor $abab$ (by extending b on the left). But $abab$ is a square, which contradicts the fact that w is $7/4^+$-free. □

Theorem 3. *There exists an infinite ternary word which is both 2-free and highly 7/4-repetitive.*

Proof. We use the morphism of Brandenburg, which is a square-free morphism :

$$\mu_B : \begin{cases} a \mapsto abacbabcbac \\ b \mapsto abacbcacbac \\ c \mapsto abcbabcacbc. \end{cases}$$

We use it to give a similar construction as in the proofs of Theorems 1 and 2. We consider the sequence:

$$\begin{aligned} A_0 &= cbac\mu_B(cbc)abac = cbacabcbabcacbcabacbcacbacabcbabcacbcabac \\ A_1 &= (u)^{-1}\mu_B^3(A_0) \\ &\vdots \\ A_n &= (u)^{-1}\mu_B^3(A_{n-1}), \end{aligned}$$

where $u = \mu_B^2(a)\mu_B(ab)abacbab$. The sequence A_n converges to a limit we denote by w, which is a square-free word because μ_B is square-free. For any $n \geq 1$, if we denote by p_n the word $\mu_B^{3(n-1)}(u) \cdots \mu_B^3(u)u$, we can see that $A_n = (p_n)^{-1}\mu_B^{3n}(A_0)$. But p_n is a prefix of $\mu_B^{3n}(c)$, since

$$\begin{aligned} |p_n| &= |u| \sum_{i=0}^{n-1} |\mu_B(a)|^{3i} \\ &= |u| \sum_{i=0}^{n-1} 11^{3i} \\ &< 11^{3n} = |\mu_B^{3n}(c)|. \end{aligned}$$

So, we can write $\mu_B^{3n}(c) = p_n s_n$, where s_n is a ternary word. Then, for any $n \geq 1$,

$$\begin{aligned} A_n &= (p_n)^{-1}\mu_B^{3n}(A_0) \\ &= s_n \mu_B^{3n}(bacabcbabcacbcaba) p_n s_n \mu_B^{3n}(bca) p_n s_n \mu_B^{3n}(bacabcbabcacbcaba) p_n s_n. \end{aligned}$$

It follows that at all positions $0 \leq j \leq |s_n|$, there begins a repetition with period a conjugate of $s_n \mu_B^{3n}(bacabcbabcacbcaba) p_n s_n \mu_B^{3n}(bca) p_n$, and of length

$$\begin{aligned} l &= |s_n \mu_B^{3n}(bacabcbabcacbcaba) p_n s_n \mu_B^{3n}(bca) p_n s_n \mu_B^{3n}(bacabcbabcacbcaba) p_n| \\ &= 40 \cdot 11^{3n}. \end{aligned}$$

These repetitions have exponents $40/22 = 20/11 \geq 7/4$. As n can be taken arbitrarily large, the result follows. □

Remark 1. In fact, with this proof, we have the result: There exists an infinite 2-free ternary word which is highly 20/11-repetitive.

So the problem of finding the infimum of the real numbers α such that there exists an infinite ternary α^+-free word which is highly 7/4-repetitive is open. We proved that this infimum is between 7/4 and 2.

5 Weaker Results on Ternary Words

In the following section, we will prove some weaker results for words over the ternary alphabet. First, instead of considering highly repetitive words, we consider words with powers at every positions. Then, we will discuss words containing infinitely many powers, but not necessary at every position. These questions were asked at the end of [CR] by Currie and Rampersad.

The following results concern Dejean's morphism μ_D and its behaviour regarding repetitions and synchronizing properties, and were proved in [VAS]. If v is a word, we denote by δ the application that removes the first letter of v (and so, $\delta^t(v)$ removes the first t letters of v). Moreover, we denote by $\mathfrak{F}_D$ the set $\mathrm{Fact}(\mu_D(A_3^*))$ of factors of $\mu_D(A_3^*)$.

Lemma 1 ([VAS]). *Let* $\alpha \in]7/4, 2[$ *be given. Let* s, t *be natural numbers such that* $\mu_D^s(b) = xabcbabcy$, *with* $|x| = t$. *Let* $\beta = 2 - \frac{t}{4 \cdot 19^s}$. *Suppose that* $7/4 < \beta < \alpha$, *and that* $abcbabcv \in \mathfrak{F}_D$ *is* α*-free. Consider the word* $w = \delta^t \mu_D^s(babcbabcv)$. *Then, we have:*

1. w *has a prefix with exponent* β.
2. *If* $abcbabcv$ *has a factor with exponent* γ *and period* p, *then,* w *has a factor with exponent* γ *and a period of length* $19^s|p|$.
3. w *is* α*-free.*

Definition 5 ([VAS]). *A real number* $\beta < \alpha$ *is said to be obtainable if* β *can be written as* $\beta = 2 - \frac{t}{4 \cdot 19^s}$, *where the natural numbers* s *and* t *verify:*

* $s \geq 3$
* *the word* $\delta^t(\mu_D^s(b))$ *begins with abcbabc.*

We note that for any given $s \geq 3$, it is possible to choose t such that

* $7/4 < \beta = 2 - \frac{t}{4 \cdot 19^s} < \alpha$
* $|\alpha - \beta| \leq \frac{19^2}{4 \cdot 19^s}$.

Indeed, $\mu_D^2(a)$, $\mu_D^2(b)$, and $\mu_D^2(c)$ have length 19^2, and each has *abcbabc* as a factor. Therefore, choosing a large enough s, we can always find some obtainable real numbers β arbitrarily close to α.

Theorem 4. *For every real number* $\alpha > 7/4$, *there exists an infinite* α*-free ternary word that contains* $7/4$*-powers beginning at every position.*

Proof. Theorem 3.4 in [CR] established the result for any $\alpha > 2$. Let us consider $7/4 < \alpha \leq 2$. Let β be an obtainable number (see Definition 5), that is, it can be written as:

$$\beta = 2 - \frac{t}{4 \cdot 19^s},$$

where $s \geq 3$, and $\delta^t \mu_D^s(b)$ begins with *abcbabc*.

For any word v in $\mathfrak{F}_D$, we denote by $\Phi(v)$ the word $\delta^t \mu_D^s(bv)$ (we recall that $\delta(u)$, for some word u, removes the first letter of u) and we consider the sequence:

$$\begin{aligned} v_1 &= \Phi(abcbabc) = \delta^t \mu_D^s(babcbabc) \\ v_2 &= \Phi(v_1) = \delta^t \mu_D^s(b\delta^t \mu_D^s(babcbabc)) \\ &\vdots \\ v_n &= \Phi(v_{n-1}) \\ &\vdots \end{aligned}$$

Let $w = \lim v_n$ (possible because each v_i is clearly a prefix of v_{i+1}). By iteration of Lemma 1, as *abcbabc* is α-free, each v_i is α-free, and then w is α-free.

For any $n \geq 0$, w begins with a prefix of the form

$$\delta^t \mu_D^s(b) \mu_D^s(\delta^t \mu_D^s(b)) \cdots \mu_D^{ns}(\delta^t \mu_D^s(b)) \mu_D^{(n+1)s}(abcbabc).$$

Thus, w contains the word $\mu_D^{ns}(\delta^t(\mu_D^s(b))\mu_D^{(n+1)s}(abcbabc)$ at position

$$\begin{aligned}
p_n &= |\delta^t\mu_D^s(b)\mu_D^s(\delta^t\mu_D^s(b))\cdots\mu_D^{(n-1)s}(\delta^t\mu_D^s(b))| \\
&= \sum_{i=1}^{n}|\mu_D^{is}(b)| - \sum_{i=0}^{n-1} t\cdot 19^{is} \\
&= (19^s - t)\sum_{i=0}^{n-1} 19^{is} \\
&= (19^s - t)\frac{19^{ns}-1}{19^s-1}
\end{aligned}$$

Moreover, the word $\mu_D^{ns}(\delta^t\mu_D^s(b))\mu_D^{(n+1)s}(abcbabc)$ contains 7/4-powers at every position between 1 and $q_n = |\mu_D^{ns}(\delta^t\mu_D^s(b))| = (19^s - t)\cdot 19^{ns}$. So, in w, there is a 7/4-repetition starting at every position j, for every $j \in [p_n, p_n + q_n - 1]$. Since $p_{n+1} = p_n + q_n$, every position j in $\mathbb{N}$ is reached. □

Theorem 5. *For every real number $\alpha \geq 7/4$ there exists a real number β, with $7/4 \leq \beta < \alpha$, arbitrarily close to α, such that there is an infinite β^+-free ternary word containing infinitely many β-powers.*

Proof. Theorem 3.4 in [CR] established the result for any $\alpha > 2$. Let us consider $7/4 < \alpha \leq 2$. Let β be an obtainable number, that is, it can be written as:

$$\beta = 2 - \frac{t}{4\cdot 19^s},$$

where $s \geq 3$, and $\delta^t\mu_D^s(b)$ begins with $abcbabc$.

Similarly to the proof of Theorem 4, we consider the sequence:

$$\begin{aligned}
v_1 &= \Phi(abcbabc) = \delta^t\mu_D^s(babcbabc) \\
v_2 &= \Phi(v_1) = \delta^t\mu_D^s(b\delta^t\mu_D^s(babcbabc)) \\
&\vdots \\
v_n &= \Phi(v_{n-1}) \\
&\vdots
\end{aligned}$$

Let $w = \lim v_n$. w is α-free, and it is β^+-free, since $\beta < \alpha$. Moreover, for any $n \geq 1$, w contains the word $\mu_D^{ns}(\delta^t\mu_D^s(babcbabc))$, which has length $19^{ns}(8\cdot 19^s - t)$ and exponent β. □

6 Conclusion

We propose the following definition. For an integer k, consider the set

$$\mathfrak{S}_k = \{(\alpha,\beta) \in \mathbb{R}^2 : \text{there exists a word in } A_k^\omega \text{ which is } \alpha\text{-free and highly } \beta\text{-repetitive}\}.$$

The combined results of Currie and Rampersad [CR], and of Sections 3 and 4 of this paper, give a partial desciption of $\mathfrak{S}_k$ for $k = 2$ and $k = 3$. This is summarized in Figures 1 and 2.

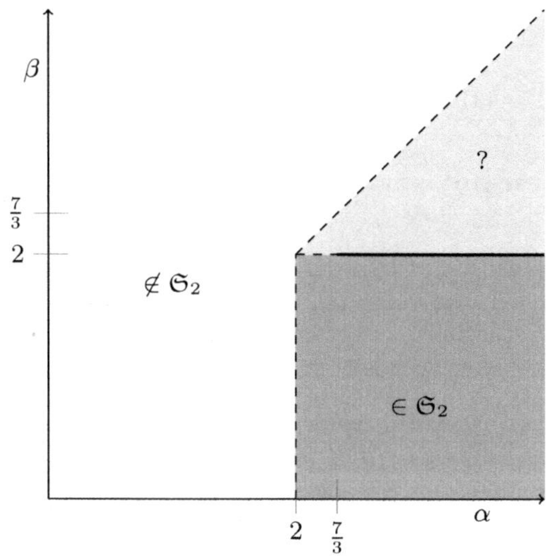

Fig. 1. $k = 2$

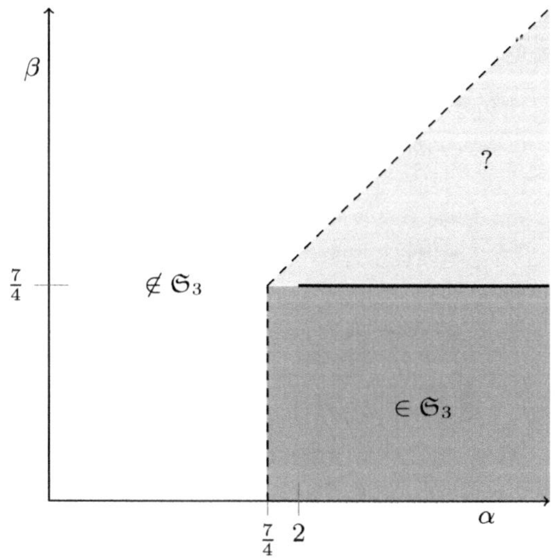

Fig. 2. $k = 3$

References

[BER] Berstel, J.: Axel Thue's papers on repetitions in words: a translation. In: Publications du LACIM, vol. 20 (1994)

[BRA] Brandenburg, F.J.: Uniformly growing k-th powerfree homomorphisms. Theoret. Comput. Sci. 23, 69–82 (1983)

[CAR] Carpi, A.: On Dejean's conjecture over large alphabets. Theor. Comput. Sci. 385, 137–151 (2007)

[CUR] Currie, J., Rampersad, N.: For each $\alpha > 2$ there is an infinite binary word with critical exponent α. Electron. J. Combinatorics 15, #N34 (2008)

[CURR] Currie, J., Rampersad, N.: Dejean's conjecture holds for $n \geq 30$. Theoret. Comput. Sci. 410, 2885–2888 (2009)

[CR] Currie, J., Rampersad, N.: Infinite words containing squares at every position. Theor. Inform. Appl. 44, 113–124 (2010)

[CRS] Currie, J., Rampersad, N., Shallit, J.: Binary words containing infinitely many overlaps. Electron. J. Combinatorics 13, #R82 (2006)

[DEJ] Dejean, F.: Sur un théoréme de Thue. J. Combin. Theory Ser. A 13 (1972)

[DEK] Dekking, F.M.: On repetitions in binary sequences. J. Comb. Theory Ser. A 20, 292–299 (1976)

[KRI] Krieger, D., Shallit, J.: Every real number greater than 1 is a critical exponent. Theoret. Comput. Sci. 381 (2007)

[MIG] Mignosi, F., Pirillo, G.: Repetitions in the Fibonacci infinite word. RAIRO Inform. Theor. Appl. 26 (1992)

[PAN] Pansiot, J.-J.: A propos d'une conjecture de F. Dejean sur les répétitions dans les mots. Discrete Appl. Math. 7 (1984)

[RIC] Richomme, G.: Personal communication (2005)

[SAA] Saari, K.: Everywhere α-repetitive sequences and Sturmian words. Europ. J. Combin. 31, 177–192 (2010)

[SHU] Shur, A.M.: The structure of the set of cube-free $\mathbb{Z}$-words in a two-letter alphabet. Izv. Ross. Akad. Nauk Ser. Mat. 64, 201–224 (2000); English translation in Izv. Math. 64, 847–871 (2000)

[TH1] Thue, A.: Uber unendliche Zeichenreihen. Norske Vid. Selsk. Skr. I. Mat. Nat. Kl. Christiana (7) (1906)

[TH2] Thue, A.: Uber die gegenseitige Lage gleicher Teile gewisser Zeichenreihen. Norske Vid. Selsk. Skr. I. Mat. Nat. Kl. Christiana (10) (1912)

[VAS] Vaslet, E.: Critical exponents of words over 3 letters (submitted)

A Sufficient Condition for Erasing Productions to Be Avoidable

Georg Zetzsche

Fachbereich Informatik, Technische Universität Kaiserslautern,
Postfach 3049, 67653 Kaiserslautern, Germany
zetzsche@cs.uni-kl.de

Abstract. In each grammar model, it is an important question whether erasing productions are necessary to generate all languages. Using the concept of grammars with control languages by Salomaa, which offers a uniform treatment of a variety of grammar models, we present a condition on the class of control languages that guarantees that erasing productions are avoidable in the resulting grammar model. On the one hand, this generalizes the previous result that in Petri net controlled grammars, erasing productions can be eliminated. On the other hand, it allows us to infer that the same is true for vector grammars.

1 Introduction

Since it had become evident that the classical grammar models of the Chomsky Hierarchy either lack the ability to generate many important languages or do not admit sufficient means of analysis, many grammar models have been proposed with the aim of extending the generative power of context-free languages while retaining the decidability of relevant questions. These extensions are commonly referred to as *grammars with regulated rewriting*. In most cases, these models consist of context-free productions and some mechanism that restricts the set of valid derivations.

For every such grammar model, it is an important question whether each grammar has an equivalent that does not utilize erasing productions. This is due to the fact that in the absence of erasing productions, a given word cannot be derived from a sentential form of length greater than that of the word. Thus, in the search for a derivation of a given word, the set of possible intermediate sentential forms is finite. Another reason for studying the generative capacity of erasing productions is that sometimes, the erasing and non-erasing variant correspond to language classes whose relation is of independent interest.

For some grammar models, the generative capacity of erasing productions has been determined. Aho obtained a normal form result for *indexed grammars* in the same article that introduced this grammar model [1]. Furthermore, Fernau and Stiebe proved that for *valence grammars* over commutative monoids (and thus *unordered vector grammars*) analogs of the Chomsky and Greibach normal forms are available [7]. In both cases, the normal forms imply that erasing productions can be eliminated. Moreover, it has recently been shown that in

G. Mauri and A. Leporati (Eds.): DLT 2011, LNCS 6795, pp. 452–463, 2011.

permitting random context grammars, erasing productions can be avoided [10]. Finally, in *Petri net controlled grammars*, erasing productions can be eliminated as well [11].

For *matrix grammars*, a partial result and reformulations have been obtained [10,11], but the problem is still open[1]. In addition, to the knowledge of the author, it is open wether erasing productions are necessary in *forbidding random context grammars*. Up to date, it was also not known whether in *vector grammars*, erasing productions increase the generative capacity.

This work provides a general condition with respect to *grammars with control languages* that guarantees that erasing productions are avoidable. The concept of grammars with control languages, introduced by Salomaa [9], unifies some of the grammar models with regulated rewriting [2]. Such a grammar consists of context-free productions and a control language whose words specify the sequences of productions that correspond to valid derivations. On the one hand, our sufficient condition is fulfilled for Petri net controlled grammars, which allows us to recover the result from [11]. On the other hand, it applies to vector grammars, which means that erasing productions do not increase their capacity.

2 Basic Notions

Let Σ be a fixed countable set of abstract symbols, the finite subsets of which are called *alphabets*. For an alphabet X, we will write X^* for the set of words over X. The empty word is denoted by λ. In particular, $\emptyset^* = \{\lambda\}$. We will regard every $x \in X$ as an element of X^*, namely the word consisting of only one occurence of x. For a symbol $x \in X$ and a word $w \in X^*$, let $|w|_x$ be the number of occurrences of x in w. For a subset $Y \subseteq X$, let $|w|_Y := \sum_{x \in Y} |w|_x$. By $|w|$, we will refer to the length of w. Given an alphabet X, subsets of X^* are called *languages*. We define the *shuffle* $L_1 \sqcup\!\sqcup L_2$ of two languages $L_1, L_2 \subseteq X^*$ to be the set of all words $w \in X^*$ such that $w = u_1 v_1 \cdots u_n v_n$ for some $u_i, v_i \in X^*$, $1 \leq i \leq n$, with $u_1 \cdots u_n \in L_1$, $v_1 \cdots v_n \in L_2$. The *shuffle closure* $L^{\sqcup\!\sqcup}$ of a language L is defined as $L^{\sqcup\!\sqcup} := \bigcap_K K$, where K ranges over those $K \subseteq X^*$ such that $\lambda \in K$ and $K \sqcup\!\sqcup L \subseteq K$. Equivalently, $L^{\sqcup\!\sqcup}$ is the smallest language in X^* that contains λ and contains $u \sqcup\!\sqcup v$ whenever $u \in K$ and $v \in L$. The *Dyck language over a and b*, $a, b \in X$, $a \neq b$, is the set $\{ab\}^{\sqcup\!\sqcup}$. Equivalently, it is the set of all words $w \in \{a, b\}^*$ such that $|w|_a = |w|_b$ and $|p|_b \leq |p|_a$ for every prefix p of w. A language $L \in X^*$ is called a *Dyck language* if there are symbols $a, b \in X$, $a \neq b$, such that $L = \{a, b\}^{\sqcup\!\sqcup}$. A *languages class* is a set of languages. A language class $\mathcal{C}$ is said to be *closed under shuffling with Dyck languages* if $L \sqcup\!\sqcup D$ is in $\mathcal{C}$ whenever L is in $\mathcal{C}$ and D is a Dyck language. For a language class $\mathcal{C}$, we denote by $\mathbf{D}(\mathcal{C})$ the smallest language class that contains $\mathcal{C}$ and is closed under shuffling with Dyck languages.

---

[1] In [4, Theorem 2.1, p. 106], it is claimed that erasing productions cannot be avoided. However, none of the given references contains a proof for this claim.

[2] Control sets on grammars were also considered by Ginsburg and Spanier [8]. They, however, require all derivations to be leftmost.

For a set A, let $\wp(A)$ denote the set of subsets of A. Given alphabets X and Y, a *letter substitution* is a map $\alpha : X \to \wp(Y)$. For $u = u_1 \cdots u_n$, $u_1, \ldots, u_n \in X$, $\alpha(u)$ is the set of words $v \in Y^*$ such that $v = v_1 \cdots v_n$, $v_1, \ldots, v_n \in Y$, and $v_i \in \alpha(u_i)$, $1 \le i \le n$. The map is extended to languages by means of $\alpha(L) = \bigcup_{w \in L} \alpha(w)$. A class of languages is said to be *closed under letter substitutions* if $\alpha(L)$ is in $\mathcal{C}$ whenever L is in $\mathcal{C}$ and α is a letter substitution. The smallest language class that contains $\mathcal{C}$ and that is closed under letter substitutions is denoted by $\mathbf{S}(\mathcal{C})$.

For an alphabet X and a subset $Y \subseteq X$, we have the homomorphisms $\delta_Y : X^* \to (X \setminus Y)^*$ and $\pi_Y : X^* \to Y^*$, defined by $\pi_Y(y) = y$ for $y \in Y$ and $\pi_Y(x) = \lambda$ for $x \in X \setminus Y$ and $\delta_Y = \pi_{X \setminus Y}$.

Furthermore, we write $X^\oplus$ for the set of *multisets* over the set X, that is, $X^\oplus$ is the set of mappings $\alpha : X \to \mathbb{N}$. The operation $+$ on $X^\oplus$ is defined by $(\alpha + \beta)(x) := \alpha(x) + \beta(x)$ for all $x \in X$. Together with the neutral element $\mathbf{0}$, defined by $\mathbf{0}(x) := 0$ for every $x \in X$, $X^\oplus$ is a (commutative) monoid. We write $\alpha \sqsubseteq \beta$ if $\alpha(x) \le \beta(x)$ for every $x \in X$. As in the case of words, we will regard X as a subset of $X^\oplus$ by identifying each $x \in X$ with $\mu_x \in X^\oplus$, which is defined by $\mu_x(x) := 1$ and $\mu_x(y) := 0$ for $y \in X$, $y \neq x$. For a multiset $\mu \in X^\oplus$, let $|\mu| := \sum_{x \in X} \mu(x)$. Here, $|\mu|$ is called the *size* of μ. For $\alpha \sqsubseteq \beta$, let $(\beta - \alpha)(x) := \beta(x) - \alpha(x)$. The *Parikh mapping* is the mapping $\Psi : X^* \to X^\oplus$ defined by $\Psi(w)(x) := |w|_x$ for all $w \in X^*$ and $x \in X$.

A *finite automaton* is a tuple $A = (Q, X, E, q_0, F)$, where Q is a finite set of *states*, X is an alphabet, E is a finite subset of $Q \times X^* \times Q$ called the set of *edges*, an *initial state* $q_0 \in Q$ and a set $F \subseteq Q$ of *final states*. The binary relation $\to_A$ on $X^* \times Q$ is defined by $(u, p) \to_A (v, q)$ iff there is an edge (p, w, q) such that $v = uw$. For $p, q \in Q$, the set $L_{p,q}(A)$ is then

$$L_{p,q}(A) := \{w \in X^* \mid (\lambda, p) \to_A^* (w, q)\}.$$

The language *accepted by A* is $L(A) := \bigcup_{q \in F} L_{q_0,q}(A)$. A language is said to be *regular* if it is accepted by a finite automaton.

A *controlled grammar* is a tuple $G = (N, T, P, S, \Lambda, \rho, C)$, in which

- N and T are disjoint alphabets, called the *nonterminal* and *terminal* symbols, respectively,
- $P \subseteq N \times (N \cup T)^*$ is a finite set of *productions*,
- $S \in N$ is the *start symbol*,
- Λ is an alphabet of *labels*,
- $\rho : \Lambda \to P$ assigns a production to each label, and
- $C \subseteq \Lambda^*$ is a non-empty language, called the *control language*.

For a controlled grammar G, a *configuration* is a pair $(u, c) \in (N \cup T)^* \times \Lambda^*$. The *derivation relation* is a binary relation on the set of configurations and is defined as follows. For $u, v \in (N \cup T)^*$ and $c, d \in \Lambda^*$, let $(u, c) \Longrightarrow_G (v, d)$ iff there are words $r, s \in (N \cup T)^*$ and a label $t \in \Lambda$ such that $u = rAs$, $v = rws$, and $d = ct$, where $\rho(t) = A \to w$. The *language generated by G* is then

$$L(G) := \{w \in T^* \mid \exists c \in C : (S, \lambda) \Longrightarrow_G^* (w, c)\}.$$

A production $A \to w$ in P is called *erasing* if $w = \lambda$ and otherwise *non-erasing*.

We now come to the definition of non-erasing controlled grammars. If we would define them to be those grammars that contain no erasing productions, we would forbid them to generate the empty word. This problem is usually addressed by allowing non-erasing grammars to have a single erasing production $S \to \lambda$ such that S does not appear on any right side. With this definition, however, it would still be impossible to generate the empty word when the control language at hand does not provide a control word of length one. Therefore, we relax this condition even further and allow a non-erasing grammar to have productions $S_i \to S_{i+1}$, $1 \leq i \leq n-1$, and $S_n \to \lambda$ such that $S = S_1$ and none of the symbols S_i appears on any other right side. This definition enables the generation of λ by control languages without short words and still retains desirable properties of non-erasing grammars. That is, it is still possible to conclude that from a sentential form of length $\geq n$, no word of length $< n$ can be derived, with the only exception of the unique derivation $S \Longrightarrow_G^* \lambda$. Furthermore, if a language L is generated by a non-erasing grammar with control language, then it can be written as either $L = K$ or $L = K \cup \{\lambda\}$, where K is generated by a grammar with the same control language that has no erasing productions at all (for a partial converse, see Lemma 2).

Thus, we call a controlled grammar $G = (N, T, P, S, \Lambda, \rho, C)$ *non-erasing* if either

- P does not contain any erasing productions or
- there is a subset $\{S_1, \ldots, S_n\} \subseteq N$ and productions $S_i \to S_{i+1}$ for $1 \leq i \leq n-1$ and $S_n \to \lambda$ such that $S = S_1$ and no S_i, $1 \leq i \leq n$, appears on the right side of any other production than these.

Let $\mathcal{C}$ be a language class. The class of languages generated by arbitrary and non-erasing controlled grammars with a control language in $\mathcal{C}$ is denoted by $\mathbf{CF}^\lambda(\mathcal{C})$ and $\mathbf{CF}(\mathcal{C})$, respectively.

3 Control Languages and Erasing Productions

It is a well-known fact that in context-free grammars, erasing productions can be avoided. This is due to the fact that, in a derivation, all nonterminals are rewritten independently. That is, it has no effect on the other nonterminals how a particular occurrence is rewritten. Therefore, in order to avoid erasing productions, one can introduce productions that, instead of creating and then deleting instances of nonterminals, prevent them from being generated.

However, in the case of grammars with some additional control mechanism, preventing the generation of nonterminals could interfere with the rest of the derivation. Specifically, the process of generating and then deleting instances of nonterminals could be necessary to obey the control mechanism. Thus, in order to eliminate erasing productions in grammars with regulated rewriting, the strategy above cannot be applied.

Here, we will make assumptions about the control mechanism that allow us to use it as a storage for the part of the sentential form that will eventually be rewritten to the empty word. Specifically, we require a closure property that enables us to transform one control language into another one. This new language poses the same restriction on the set of derivations, except that it allows additional labels that correspond to operations on counters (*increase* and *decrease*) such that the words in the control language describe valid sequences of these operations (i.e., the counters are never negative and are zero in the end). Thus, the new control language constitutes a control mechanism that (in addition to the control executed by the original language) acts like a storage (of natural numbers). This will allow us to transfer occurrences of later deleted nonterminals back and forth between the sentential form and the storage. The latter will then be used to reduce the number of such later deleted symbols in the sentential form below a bound that only depends on the grammar: After generating a later deleted symbol, it can be moved to the storage. When it is to be rewritten, it is brought back to the sentential form so that the production can be applied and the inserted right hand side can be moved to the storage, etc.

Thus, in each sentential form, there is only a bounded number of symbols that will eventually be rewritten to the empty word. These can then be accounted for by "attaching" multisets to nonterminals that can hold a finite number of occurrences. Attaching means that the new set of nonterminals will consist of pairs of old (terminal or nonterminal) symbols and multisets. The idea to seperate later deleted symbols from other symbols, reduce the number of their occurrences and then account for the remaining occurrences via attached multisets has also been used in [10] to obtain an analogous result for permitting random context grammars. However, in the latter work, the technique used for the reduction step is completely different from the one used here.

Some of our proofs involve the construction of a controlled grammar $G' = (N', T, P', S', \Lambda', \rho', C')$ from a grammar $G = (N, T, P, S, \Lambda, \rho, C)$ such that $C' = \alpha(C)$, where α is a letter substitution. This will be the case when for every production p in G, there is a set of productions in G' that in some way correspond to p, i.e., we are given a relation $R \subseteq P' \times P$, the *simulation (relation)*, such that $(p', p) \in R$ whenever p' corresponds to p. The substitution α is then defined such that for a label $t \in \Lambda$, the set $\alpha(t) \subseteq \Lambda'$ consists of labels of those productions that correspond to $\rho(t)$. Thus, we define

$$\Lambda' := \{\ell_{t,p} \mid t \in \Lambda,\ p \in P',\ (p, \rho(t)) \in R\},$$

where the $\ell_{t,p}$ are new symbols. That is, for every pair $(t, p) \in \Lambda \times P'$ such that p corresponds to the production of t, we have a unique label in Λ'. The maps $\rho' : \Lambda' \to P'$ and $\alpha : \Lambda \to \wp(\Lambda')$ are then defined as $\rho'(\ell_{t,p}) := p$ and $\alpha(t) := \{\ell_{t,p} \in \Lambda' \mid p \in P'\}$. The triple $(\Lambda', \rho', \alpha)$ will be called the *triple induced by R*.

The first step in the proof of our main result is to establish a normal form for grammars with control languages. An analogous concept was used in [10] for permitting random context grammars. This normal form requires the set of nonterminals to be partitioned into two subsets. One of these subsets contains only symbols from which no terminal word but the empty one can be derived. The other set contains only symbols from which only non-empty words can be derived. Thus, this normal form allows us to determine whether an occurrence will be rewritten to the empty word just by looking at its symbol.

A controlled grammar $G = (N, T, P, S, \Lambda, \rho, C)$ is in *erasing normal form* if there is a subset $\Delta \subseteq N$ such that for each production $A \to w \in P$ we have that $A \in \Delta$ if and only if $w \in \Delta^*$. Since a symbol $A \in \Delta$ can only be directly rewritten to words in Δ^*, the only terminal word derivable from A is the empty word. Furthermore, in every word derivable from a symbol $A \in N \setminus \Delta$, there occurs at least one symbol outside of Δ. Thus, symbols outside of Δ cannot derive the empty word. If $G = (N, T, P, S, \Lambda, \rho, C)$ is in erasing normal form with $\Delta \subseteq N$, we also write $G = (N, \Delta, T, P, S, \Lambda, \rho, C)$.

Lemma 1. *Let G be a controlled grammar with control language C such that $\lambda \notin L(G)$. Then, there is an equivalent grammar with control language $\alpha(C)$ in erasing normal form, such that α is a letter substitution.*

Proof. Let $G = (N, T, P, S, \Lambda, \rho, C)$ be a controlled grammar. Let $\bar{A}$ be a new symbol for each $A \in N$ and let $\Delta := \{\bar{A} \mid A \in N\}$ and $N' := N \cup \Delta$. Furthermore, let $\varphi : (N' \cup T)^* \to (N \cup T)^*$ and $\psi : N^* \to \Delta^*$ be the homomorphisms defined by $\varphi(\bar{A}) = A$ for $A \in N$ and $\varphi(x) = x$ for $x \in N \cup T$ and $\psi(A) = \bar{A}$ for $A \in N$. Let $P = \{A_i \to w_i \mid 1 \leq i \leq n\}$ such that $w_i \in N^*$ iff $i \leq m$ for some $m \leq n$. For each i, $1 \leq i \leq n$, we include the productions $A_i \to u$ for each $u \in (N' \cup T)^* \setminus \Delta^*$ such that $\varphi(u) = w_i$ and let $A_i \to u$ simulate $A_i \to w_i$. Furthermore, for each i, $1 \leq i \leq m$, we add $\psi(A_i) \to \psi(w_i)$ and let it simulate $A_i \to w_i$.

This defines a set of productions P' and a simulation $R \subseteq P' \times P$. Let $(\Lambda', \rho', \alpha)$ be the triple induced by R. The grammar $G' = (N', \Delta, T, P', S', \Lambda', \rho', C')$ with $S' := S$ and $C' = \alpha(C)$ is in erasing normal form. This is due to the fact that in the productions $A \to u$, $u \in (N' \cup T)^* \setminus \Delta^*$, none of the sides is in Δ^* and in the productions $\psi(A) \to \psi(w)$, both sides are. Thus, it remains to be shown that $L(G') = L(G)$.

Since for every production $A \to w$ in P', the production $\varphi(A) \to \varphi(w)$ exists in P and is simulated by $A \to w$, it follows immediately that $L(G') \subseteq L(G)$.

In order to prove that $L(G) \subseteq L(G')$, one can show by induction on n that for every derivation $(w_1, c_1) \Longrightarrow_G \cdots \Longrightarrow_G (w_n, c_n)$, where $c_1 = \lambda$, $w_n \in T^*$, there is a derivation $(w'_1, c'_1) \Longrightarrow_{G'} \cdots \Longrightarrow_{G'} (w'_n, c'_n)$ such that $\varphi(w'_i) = w_i$ and $c'_i \in \alpha(c_i)$ for $1 \leq i \leq n$. Then, every derivation $(S, \lambda) \Longrightarrow_G^* (w, c)$, $w \in T^+$, yields a derivation $(A, \lambda) \Longrightarrow_{G'}^* (w', c')$ with $\varphi(A) = S$, $\varphi(w') = w$, and $c' \in \alpha(C) = C'$. Since $w \in T^+$, this means that $w' = w$ and that $A = S$ or $A = \bar{S}$. However, $A = \bar{S}$ would imply $w = \lambda$ and thus we have $A = S$, and hence $w \in L(G')$. Since $L(G) \subseteq T^+$, this proves $L(G) \subseteq L(G')$. □

The following lemmas can be proven using standard techniques. Therefore, we omitted them in order to meet the space restrictions.

Lemma 2. *For each L in $\mathbf{CF}(\mathcal{C})$, the language $L \cup \{\lambda\}$ is contained in $\mathbf{CF}(\mathbf{S}(\mathcal{C}))$.*

Lemma 3. *Let L be in $\mathbf{CF}^\lambda(\mathcal{C})$ and K be a regular language. Then, $L \cap K$ is in $\mathbf{CF}^\lambda(\mathbf{S}(\mathcal{C}))$.*

We now come to the key lemma in the proof of our main result. It shows that shuffling with Dyck-languages allows us to reduce the number of Δ-symbols in the sentential forms below a fixed bound.

A controlled grammar $G = (N, \Delta, T, P, S, \Lambda, \rho, C)$ in erasing normal form is called *Δ-bounded* iff there is a constant $k \in \mathbb{N}$ such that each word $w \in L(G)$ has a derivation $S = w_1 \Longrightarrow_G \cdots \Longrightarrow_G w_n = w$ such that $|w_i|_\Delta \leq k$ for $1 \leq i \leq n$.

Lemma 4. *For each language L in $\mathbf{CF}^\lambda(\mathcal{C})$ with $\lambda \notin L$, there is a Δ-bounded controlled grammar with a control language in $\mathbf{S}(\mathbf{D}(\mathcal{C}))$.*

Proof. Let $G = (N, \Delta, T, P, S, \Lambda, \rho, C)$ be a controlled grammar for L in erasing normal form. We shall construct the grammar $G' = (N, \Delta, T, P', \Lambda', \rho', C')$, for $L = L(G)$, where C' is in $\mathbf{S}(\mathbf{D}(\mathcal{C}))$, and prove that it is Δ-bounded.

For each $A \in \Delta$ and $B \in \Delta$, let x_A, $\bar{x}_A$, and $\bar{x}_{A,B}$ be new symbols and $D_A \subseteq \{x_A, \bar{x}_A\}^*$ be the Dyck language over x_A and $\bar{x}_A$. Furthermore, let $\Gamma := \{x_A, \bar{x}_{A,B} \mid A, B \in \Delta\}$. With these, let $\Lambda' := \Lambda \cup \Gamma$.

The new set of productions is $P' := P \cup Q$, where Q consists of the productions $A \to \lambda$ and $B \to BA$ for each $A, B \in \Delta$. The assignment of labels is given by $\rho'(t) := \rho(t)$ for $t \in \Lambda$ and

$$\rho'(x_A) := A \to \lambda, \qquad \rho'(\bar{x}_{A,B}) := B \to BA$$

for $A, B \in \Delta$. Thus, the production with label x_A removes an occurrence of A from the sentential form and the production with label $\bar{x}_{A,B}$ creates an occurrence of A by using B as a left hand side. For $A \in \Delta$, the homomorphism $\varphi_A : \Lambda'^* \to \{x_A, \bar{x}_A\}^*$ is defined by $\varphi_A(t) = \lambda$ for $t \in \Lambda$, $\varphi_A(x_A) = x_A$, $\varphi_A(x_{B,C}) = \varphi_A(x_B) = \lambda$ for $B, C \in \Delta$, $B \neq A$, and $\varphi_A(x_{A,B}) = x_A$ for $B \in \Delta$. The letter substitution α is given by

$$\alpha(t) = \begin{cases} \{t\} & \text{if } t \in \Lambda \cup \{x_A \mid A \in \Delta\}, \\ \{\bar{x}_{A,B} \mid B \in \Delta\} & \text{if } t = \bar{x}_A \text{ for some } A \in \Delta. \end{cases}$$

Finally, let

$$C' := \alpha\left(C \sqcup\!\sqcup \coprod_{A \in \Delta} D_A\right).$$

It remains to be shown that $L(G') = L(G)$ and that G' is Δ-bounded. The constant for Δ-boundedness will be $k := 1 + \max\{|w|_\Delta \mid A \to w \in P\}$. In order to describe the relationship between configurations in G and in G', we define the following multisets. For a word $w \in (T \cup N)^*$, let $\mu(w) := \Psi(\pi_\Delta(w))$. Thus, $\mu(w)$ contains the number of occurrences of each Δ-symbol in w. Furthermore, when for $c \in \Lambda'^*$ the word $\varphi_A(c) \in \{x_A, \bar{x}_A\}^*$ is a prefix of a word in D_A, then we have $|\varphi_A(c)|_{\bar{x}_A} \leq |\varphi_A(c)|_{x_A}$ and can define $\nu(c)(A) := |\varphi_A(c)|_{x_A} - |\varphi_A(c)|_{\bar{x}_A}$ to obtain $\nu(c) \in \Delta^\oplus$.

We say that a configuration (w', c'), $w' \in (N \cup T)^*$, $c' \in C'$, *represents* a configuration (w, c), $w \in (N \cup T)^*$, $c \in C$, iff

1. for each $A \in \Delta$, $\varphi_A(c')$ is a prefix of a word in D_A.
2. $\delta_\Delta(w') = \delta_\Delta(w)$,
3. $\delta_\Gamma(c') = c$, and
4. $\mu(w) = \mu(w') + \nu(c')$.

First, we claim that whenever there is a derivation $(S, \lambda) = (w_1, c_1) \Longrightarrow_G \cdots \Longrightarrow_G (w_n, c_n)$, then there is a derivation $(S, \lambda) = (w_1', c_1) \Longrightarrow_{G'} \cdots (w_m', c_m')$ such that (w_m', c_m') represents (w_n, c_n) and

$$|\mu(w_m')| \leq 1, \quad \text{if } \nu(c_m') \neq \mathbf{0}, \text{ then } |\mu(w_m')| = 1. \tag{$*$}$$

and $|w_i'|_\Delta \leq k$ for each $1 \leq i \leq m$.

We prove this by induction on n. The claim clearly holds for $n = 1$, so let $n > 1$ and assume that there is a derivation $(w_1', c_1') \Longrightarrow_{G'} \cdots \Longrightarrow_{G'} (w_m', c_m')$ such that $|w_i'|_\Delta \leq k$ and (w_m', c_m') respresents (w_{n-1}, c_{n-1}) and $(*)$ holds. Let $A \to w$ be the production applied in $(w_{n-1}, c_{n-1}) \Longrightarrow_G (w_n, c_n)$. We distinguish the following cases.

- $A \in \Delta$, $w \in \Delta^*$, $\mu(w_m')(A) = 1$ and $w = \lambda$. In order to make sure that $(*)$ is fulfilled we might have to introduce another nonterminal from Δ. If $\nu(c_m') \neq \mathbf{0}$, choose a $B \in \Delta$ with $\nu(c_m')(B) \geq 1$ and apply the production $A \to AB$ with label $\bar{x}_{B,A}$. Then, apply the production $A \to \lambda$.
- $A \in \Delta$, $w \in \Delta^*$, $\mu(w_m')(A) = 1$ and $w \neq \lambda$. Then, write $w = Bw'$ with $B \in \Delta$, $w' \in \Delta^*$. In this case, we have to bring the Δ-occurrences introduced by $A \to w$ into the storage of the control mechanism, except the occurrence of B, which is needed to fulfill $(*)$: Apply $A \to w$ and then $|w'|_E$ times the production $E \to \lambda$ with label x_E for each $E \in \Delta$.
- $A \in \Delta$, $w \in \Delta^*$, and $\mu(w_m')(A) = 0$. Then, by condition 4, we have $\nu(c_m')(A) \geq 1$ and by $(*)$, we have $\mu(w_m')(B) \geq 1$ for some $B \in \Delta$. First we bring A into the sentential form and apply $B \to BA$ with label $\bar{x}_{A,B}$. Then, we apply $A \to w$ and then bring its right side into the storage: Apply $|w|_E$ times the production $E \to \lambda$ with label x_E for each $E \in \Delta$.
- $A \notin \Delta$, $w \notin \Delta^*$, $|w|_\Delta > 0$, and $\mu(w_m') = \mathbf{0}$. The production $A \to w$ introduces Δ-occurrences into a sentential form that does not contain any up to that point. We want to keep one of those occurrences in the sentential form and thus write $\mu(w) = B + \gamma$ for some $B \in \Delta$ and $\gamma \in \Delta^\oplus$. Apply $A \to w$ and then, apply $\gamma(E)$ times the production $E \to \lambda$ with label x_E for each $E \in \Delta$.
- $A \notin \Delta$, $w \notin \Delta^*$, $|w|_\Delta > 0$, and $\mu(w_m') \neq \mathbf{0}$. There already is one occurrence of a Δ-symbol. Thus, we can apply $A \to w$ and then transfer all Δ-occurrences in w to the storage: Apply $|w|_E$ times the production $E \to \lambda$ with label x_E.
- $A \notin \Delta$, $w \notin \Delta^*$, and $|w|_\Delta = 0$. We just apply $A \to w$ and leave the storage mechanism unused.

Note that, in all cases, there are at most k occurrences of Δ-symbols in the sentential form at any time. Furthermore, after applying the mentioned productions, we reach a pair $(w_{m+\ell}', c_{m+\ell}')$ that represents (w_n, c_n). This proves our claim.

In particular, we have $L(G) \subseteq L(G')$. Indeed, if $(S, \lambda) \Longrightarrow_G^* (w, c)$ for some $c \in C$, then the derivation $(S, \lambda) \Longrightarrow_{G'}^* (w', c')$ where (w', c') represents (w, c) has the following properties. By condition 2, we have $w' = w$, since both are in T^*. Condition 4 then implies $\nu(c') = \mathbf{0}$ and thus c' is contained in C' by condition 1. This proves $L(G) \subseteq L(G')$ and that all words in $L(G)$ can be derived using at most k occurrences of Δ-symbols in each sentential form.

Conversely, let $(u', c') \Longrightarrow_{G'} (v', d')$ and let (u', c') represent (u, c). If the production in $(u', c') \Longrightarrow_{G'} (v', d')$ is in P, then we have $(u, c) \Longrightarrow_G (v, d)$ such that (v', d') represents (v, d). If the production is one added to obtain P', then (v', d') also represents (u, c). Thus, when $(S, \lambda) \Longrightarrow_{G'}^* (w', c')$ for some $w' \in T^*$ and $c' \in C'$, then we have $(S, \lambda) \Longrightarrow_G^* (w, c)$ such that (w', c') represents (w, c). This means in particular that $w = w'$, $c \in C$, and therefore $w \in L(G)$. Thus, $L(G') \subseteq L(G)$, which concludes the proof. □

We are now ready to prove the main result of the article.

Theorem 1. *For any language class $\mathcal{C}$, we have $\mathbf{CF}^\lambda(\mathcal{C}) \subseteq \mathbf{CF}(\mathbf{S}(\mathbf{D}(\mathcal{C})))$.*

Proof. Let L be in $\mathbf{CF}^\lambda(\mathcal{C})$. Then, we can assume that $\lambda \notin L$, since otherwise, we have $L \cap T^+ \in \mathbf{CF}^\lambda(\mathbf{S}(\mathcal{C}))$ by Lemma 3 and the theorem yields $L \cap T^+ \in \mathbf{CF}(\mathbf{SDS}(\mathcal{C}))$ and thus $L = (L \cap T^+) \cup \{\lambda\} \in \mathbf{CF}(\mathbf{SSDS}(\mathcal{C}))$ by Lemma 2. Since $\mathbf{SSDS}(\mathcal{C}) = \mathbf{SD}(\mathcal{C})$, this means $L \in \mathbf{CF}(\mathbf{S}(\mathbf{D}(\mathcal{C})))$.

Therefore, we can use Lemma 4 to find a (possibly erasing) Δ-bounded grammar $G = (N, \Delta, T, P, S, \Lambda, \rho, C)$ in erasing normal form for L such that $C \in \mathbf{S}(\mathbf{D}(\mathcal{C}))$. We shall construct a non-erasing grammar $G' = (N', T', P', S', \Lambda', \rho', C')$ for L such that $C' = \alpha(C)$ for some letter substitution α. This implies that $L \in \mathbf{CF}(\mathbf{SSD}(\mathcal{C}))$, which together with the fact $\mathbf{SSD} = \mathbf{SD}$ proves the theorem.

For the sake of simplicity, the constructed grammar G' will have a terminal set T' such that there is a bijection $\varphi : T \to T'$ and $L(G') = \varphi(L(G))$. This clearly implies that G' can be modified so as to obtain a grammar that is equivalent to G and is still non-erasing.

In the construction of G', we will prevent the symbols in Δ from being generated. This, however, has to be done in a way that retains their influence on the derivation. Thus, their occurrences will be stored in multisets attached to nonterminals. Since this means we only have a certain amount of space to store these occurrences, we will make use of the fact that G is Δ-bounded.

Let G be Δ-bounded with the constant $k \in \mathbb{N}$. The set of nonterminals is

$$\begin{aligned} N' :=& \{(x, \mu) \mid x \in N \setminus \Delta,\ \mu \in \Delta^\oplus,\ |\mu| \leq k\} \\ &\cup \{(x, \mu) \mid x \in T,\ \mu \in \Delta^\oplus \setminus \{\mathbf{0}\},\ |\mu| \leq k\}. \end{aligned}$$

and the terminal symbols are $T' := \{(x, \mathbf{0}) \mid x \in T\}$ and $\varphi : T \to T'$ is defined by $\varphi(x) = (x, \mathbf{0})$. The homomorphism $\iota : (T \cup N \setminus \Delta)^* \to (T' \cup N')^*$ is given by $\iota(x) := (x, \mathbf{0})$ for $x \in T \cup N \setminus \Delta$. The set of productions P' is defined as follows. For each production $A \to w \in P$, we distinguish two cases.

– Suppose $A \notin \Delta$ and $w \notin \Delta^*$. Then $\delta_\Delta(w) \neq \lambda$. Hence, write $\delta_\Delta(w) = w_1 w'$, $w_1 \in T \cup N \setminus \Delta$, where $w' \in (T \cup N \setminus \Delta)^*$. We include the production

$$(A, \mu) \to (w_1, \mu + \Psi(\pi_\Delta(w)))\iota(w')$$

for each $\mu \in \Delta^\oplus$ with $|\mu| \leq k$ and $|\mu + \Psi(\pi_\Delta(w))| \leq k$ and let it simulate $A \to w$.
– Suppose $A \in \Delta$ and $w \in \Delta^*$. Then, we include

$$(x, \mu) \to (x, \mu - A + \Psi(w))$$

for each $x \in T \cup N \setminus \Delta$ and each $\mu \in \Delta^\oplus$ such that $|\mu| \leq k$, $\mu(A) \geq 1$, and $|\mu - A + \Psi(w)| \leq k$. Again, these productions will all simulate $A \to w$.

We have thus defined a simulation $R \subseteq P' \times P$. Let $(\Lambda', \rho', \alpha)$ be the triple induced by R. Together with $S' := (S, \mathbf{0})$, this defines the grammar G'.

In order to prove $L(G') = \varphi(L(G))$, we have the homomorphisms

$$(T \cup N)^* \xrightarrow{\beta} (T \cup N \setminus \Delta)^* \times \Delta^\oplus \xleftarrow{\gamma} (T' \cup N')^*,$$

with $\beta(x) = (x, \mathbf{0})$ for $x \in T \cup N \setminus \Delta$, $\beta(x) = (\lambda, x)$ for $x \in \Delta$, and $\gamma((x, \mu)) = (x, \mu)$ for $(x, \mu) \in T' \cup N'$. We say that a configuration $(w', c') \in (T' \cup N')^* \times \Lambda'^*$ of G' *represents* a configuration $(w, c) \in (T \cup N)^* \times \Lambda^*$ of G iff $\beta(w) = \gamma(w')$ and $c' \in \alpha(c)$.

The central argument for the equality $L(G') = \varphi(L(G))$ is illustrated in the following diagrams. Here, a dotted line means that the lower configuration represents the upper configuration. Furthermore, the framed configuration is claimed to exist.

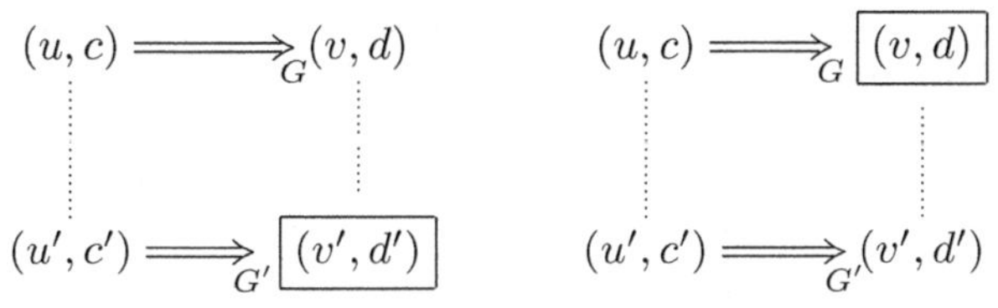

Let (u, c), (v, d) be configurations in G such that $|u|_\Delta \leq k$ and $|v|_\Delta \leq k$. Furthermore, let (u', c') be a configuration in G' such that $(u, c) \Longrightarrow_G (v, d)$ and (u', c') represents (u, c). Then, there clearly exists a configuration (v', d') that represents (v, d) such that $(u', c') \Longrightarrow_{G'} (v', d')$. On the other hand, for configurations (u', c'), (v', d') in G' and (u, c) in G such that (u', c') represents (u, c), there exists a configuration (v, d) in G represented by (v', d') such that $(u, c) \Longrightarrow_G (v, d)$. This immediately implies that for every derivation $(S, \lambda) = (w_1, c_1) \Longrightarrow_G \cdots \Longrightarrow_G (w_n, c_n)$ with $w_n \in T^*$ and $|w_i|_\Delta \leq k$ for $1 \leq i \leq n$, there is a derivation $(S', \lambda) = (w'_1, c'_1) \Longrightarrow_{G'} \cdots \Longrightarrow_{G'} (w'_n, c'_n)$ such that (w'_i, c'_i) represents (w_i, c_i) for $1 \leq i \leq n$. In particular, $w'_n = \varphi(w_n)$ and thus $\varphi(L(G)) \subseteq L(G')$. Analogously, one obtains $L(G') \subseteq \varphi(L(G))$. □

Corollary 1. *Let $\mathcal{C}$ be closed under letter substitutions and shuffling with Dyck languages. Then,* $\mathbf{CF}^\lambda(\mathcal{C}) = \mathbf{CF}(\mathcal{C})$.

4 Applications

We will now apply the main result to concrete grammar models.

A *labeled Petri net* is a tuple $N = (X, P, T, \partial_0, \partial_1, \sigma, \mu_0, F)$, where X is a finite alphabet, P is a finite set of *places*, T is a finite set of *transitions*, $\partial_0, \partial_1 : T^\oplus \to P^\oplus$ are homomorphisms (where $\partial_0(t)$ and $\partial_1(t)$ specifies the *pre-* or *post-multiset*, respectively, for each transition $t \in T$), $\sigma : T \to X \cup \{\lambda\}$ is the *labeling function*, $\mu_0 \in P^\oplus$ is the *initial marking*, and $F \subseteq P^\oplus$ is a finite set of *final markings*.

The binary relation $\to_N$ on $X^* \times P^\oplus$ is defined by $(w, \mu) \to_N (w', \mu')$ iff there exists a $t \in T$ such that $w' = w\sigma(t)$, $\partial_0(t) \sqsubseteq \mu$ and $\mu' = (\mu - \partial_0(t)) + \partial_1(t)$. The *language generated by* N is given by

$$L(N) := \{w \in X^* \mid \exists \mu \in F : (\lambda, \mu_0) \to_N^* (w, \mu)\}.$$

A language is called *Petri net language* if it is generated by a labeled Petri net. A *Petri net controlled grammar* is a grammar with a Petri net language as control language. These grammars have been introduced by Dassow and Turaev [6] and we will denote the class of languages generated by arbitrary and non-erasing Petri net controlled grammars by $\mathbf{PN}^\lambda$ and $\mathbf{PN}$, respectively. Since it is a well-known fact that Petri net languages contain the Dyck languages and are closed against shuffling and letter substitutions, Corollary 1 implies the following, which has been shown directly in [11].

Theorem 2. $\mathbf{PN}^\lambda = \mathbf{PN}$.

A *vector grammar* is a grammar with a control language of the form $V^{⧢}$, where V is a finite set of words. The class of languages generated by arbitrary and non-erasing vector grammars is denoted by $\mathbf{V}^\lambda$ and $\mathbf{V}$, respectively. Vector grammars were introduced by Cremers and Mayer [2] and were originally dubbed *generalized vector grammars* (we follow the terminology of [3]). Up to date, it was not known whether vector grammars with and without erasing productions have the same generative capacity. Here, this question can be answered positively using Corollary 1. In order to do this, we just have to convince ourselves that the class of languages of the form $V^{⧢}$ is closed against letter substitutions and shuffling with Dyck languages. The former follows from the fact that $\alpha(V^{⧢}) = \alpha(V)^{⧢}$ for a letter substitution α. The latter can be seen by noting that $V^{⧢} ⧢ D = (V \cup \{ab\})^{⧢}$ if D is the Dyck language over a and b. This proves the following theorem.

Theorem 3. $\mathbf{V}^\lambda = \mathbf{V}$.

While this result is of interest itself, it has other noteworthy consequences: A *matrix grammar* is a controlled grammar with a control language of the form M^*, where M is finite set of words. We denote the class of languages generated by arbitrary and non-erasing matrix grammars by $\mathbf{MAT}^\lambda$ and $\mathbf{MAT}$, respectively. As mentioned above, it is a longstanding open problem whether every matrix grammar has a non-erasing equivalent. It is a classic result that vector grammars are equivalent to matrix grammars, when in both models, erasing productions are allowed (see [3, Theorem 2.1.2]). Combining this with Theorem 3 yields that $\mathbf{V} = \mathbf{MAT}^{\lambda}$, which in turn means that we have a new reformulation for the question about erasing productions in matrix grammars: *Are non-erasing matrix grammars as powerful as non-erasing vector grammars?* A restatement of a similar form has been presented in [11] and is a consequence of Theorem 2. Both restatements have the advantage of avoiding any reference to erasing and thus providing a different perspective on the problem. The problem on matrix grammars was also linked to the problem of whether permitting random context grammars are strictly less powerful than non-erasing matrix grammars [10].

Another consequence is that there are several subclasses of $\mathbf{PN}^{\lambda}$, introduced by Dassow and Turaev [5], that were shown by them to lie between $\mathbf{V}$ and $\mathbf{V}^{\lambda}$. By Theorem 3, all these classes conincide.

Acknowledgements. I would like to thank Reiner Hüchting and Matthias Jantzen for valuable comments, which have improved the presentation of the paper. Furthermore, I'm grateful to one of the anonymous referees, whose careful review helped correct many typing errors.

References

1. Aho, A.V.: Indexed grammars—an extension of context-free grammars. Journal of the ACM 15, 647–671 (1968)
2. Cremers, A.B., Mayer, O.: On matrix languages. Information and Control 23(1), 86–96 (1973)
3. Dassow, J., Păun, G.: Regulated rewriting in formal language theory. Springer, Berlin (1989)
4. Dassow, J., Păun, G., Salomaa, A.: Grammars with controlled derivations. In: Rozenberg, G., Salomaa, A. (eds.) Handbook of Formal Languages, vol. 2, pp. 101–154. Springer, Berlin (1997)
5. Dassow, J., Turaev, S.: Petri net controlled grammars: the case of special petri nets. Journal of Universal Computer Science 15(14), 2808–2835 (2009)
6. Dassow, J., Turaev, S.: Petri net controlled grammars: the power of labeling and final markings. Romanian Journal of Information Science and Technology 12(2), 191–207 (2009)
7. Fernau, H., Stiebe, R.: Sequential grammars and automata with valences. Theoretical Computer Science 276, 377–405 (2002)
8. Ginsburg, S., Spanier, E.H.: Control sets on grammars. Mathematical Systems Theory 2(2), 159–177 (1968)
9. Salomaa, A.: On grammars with restricted use of productions. Annales Academiæ Scientiarum Fennicæ. Series A I. Mathematica 454 (1969)
10. Zetzsche, G.: On erasing productions in random context grammars. In: Abramsky, S., Gavoille, C., Kirchner, C., Meyer auf der Heide, F., Spirakis, P.G. (eds.) ICALP 2010. LNCS, vol. 6199, pp. 175–186. Springer, Heidelberg (2010)
11. Zetzsche, G.: Toward understanding the generative capacity of erasing rules in matrix grammars. International Journal of Foundations of Computer Science 22, 411–426 (2011)

Encoding Centered Polyominoes by Means of a Regular Language

Daniela Battaglino[1], Jean Marc Fedou[2], Andrea Frosini[3], and Simone Rinaldi[1]

[1] Dipartimento di Matematica e Informatica
Università di Siena, Siena, Italy
battaglino3@unisi.it, rinaldi@unisi.it

[2] Departement d'Informatique
UNS, Nice, France
Jean-Marc.Fedou@unice.fr

[3] Dipartimento di Sistemi e Informatica
Università di Firenze, Firenze, Italy
andrea.frosini@unifi.it

In [3] the authors proposed a classification of convex polyominoes based on the number of changes of direction in the paths connecting any two cells of a polyomino. More precisely, a convex polyomino is k-convex if every pair of its cells can be connected by a monotone path with at most k changes of direction. In 1-convex (also called L-convex) polyominoes, any two cells can be connected by a path with at most one change of direction. The polyomino in Fig. 1 (a) is L-convex, while the polyominoes (b), (c) are not, but are 2-convex.

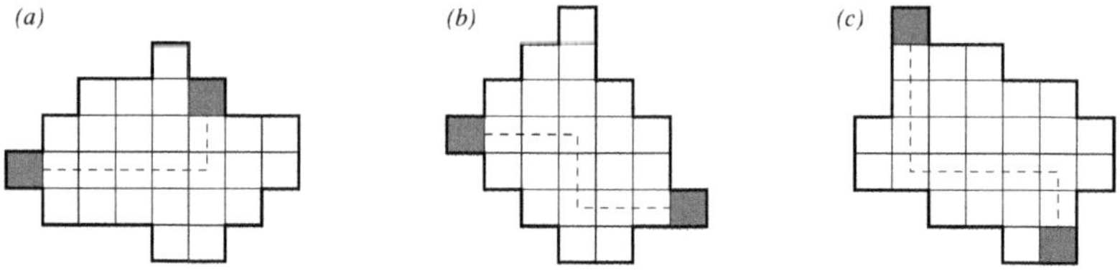

Fig. 1. (a) an L-convex polyomino, and a monotone path with a single change of direction joining two of its cells; (b) a Z-convex but not L-convex polyomino; (c) a centered polyomino (not L-convex)

In [1] it was proved that the number l_n of L-convex polyominoes with semi-perimeter $n+2$ satisfies the recurrence relation $l_n = 4l_{n-1} - 2l_{n-2}$, for $n \geq 3$, with $l_0 = 1, l_1 = 2, l_2 = 7$. Successively, in [2], the authors proved the same result by encoding L-convex polyominoes by words of a regular language of four letters. Then we have considered 2-convex polyominoes (also called Z-convex polyominoes), which do not inherit the most interesting properties of L-convex polyominoes: the generating function of Z-convex polyominoes with respect to the semi-perimeter is algebraic [4] (not rational) and Z-convex polyominoes are not uniquely determined by their orthogonal projections.

G. Mauri and A. Leporati (Eds.): DLT 2011, LNCS 6795, pp. 464–465, 2011.

Here, we consider the class of *horizontally centered* (or simply *centered*) convex polyominoes. A convex polyomino is said to be centered if it contains at least one row touching both the left and the right side of its minimal bounding rectangle (see Fig. 1 (*c*)). We observe that any L-convex polyomino is centered, and, more importantly for us, any centered polyomino is Z-convex, while the converse statement does not hold. Figure 1 (*c*) shows a centered polyomino which is not L-convex, and Figure 1 (*b*) a Z-convex polyomino which is not centered. Centered convex polyominoes can also be described as made of two stack polyominoes glued together at their basis. This class of objects was first considered in [4], where – by means of purely analytical techniques – the authors proved that the number c_n of centered convex polyominoes having semi-perimeter equal to $n+2$ satisfies the recurrence relation:

$$c_n = 6c_{n-1} - 8c_{n-2}, \qquad n \geq 3, \text{ with } c_0 = 1, c_1 = 2, c_2 = 7. \tag{1}$$

Comparing the asymptotic behavior of the three classes, the number of Z-convex polyominoes with semi-perimeter $n+2$ grows asymptotically like $n \cdot 4^n$, while the number of L-convex polyominoes grows only like $(2+\sqrt{2})^n$, and the number of centered polyominoes grows like 4^n. The reason for the interest in centered polyominoes is that they effectively constitute a bridge between L-convexity and Z-convexity. In fact, they have a rational generating function and they can be easily reconstructed by a tomographical point of view (like L-convex polyominoes), but their asymptotic growth is dominated by the factor 4^n (like Z-convex polyominoes). Moreover, until now, they resisted standard decomposition technics leading to rational generating functions.

Our main goal is to give a combinatorial explanation to (1). In practice, we tackle the problem of proving the rationality of centered polyominoes by encoding them by means of a regular language, of a six letter alphabet, extending the language encoding L-convex polyominoes in [2]. The computation of the generating function of centered polyominoes can easily be obtained from the regular expression of the language encoding centered polyominoes, and we re-obtain the formula in (1). To reach our goal, we need introduce an auxiliary class of objects, the *bi-colored stack* polyominoes. Then, the encoding of centered polyominoes as words of a regular language consists of two steps: first we map each centered polyomino into a *bi-colored stack*, then we represent each bi-colored stack as a word of a non-ambiguous regular language.

References

1. Castiglione, G., Frosini, A., Restivo, A., Rinaldi, S.: Enumeration of L-convex polyominoes by rows and columns. Theoret. Comput. Sci. 347, 336–352 (2005)
2. Castiglione, G., Frosini, A., Munarini, E., Restivo, A., Rinaldi, S.: Combinatorial aspects of L-convex polyominoes. European J. Combin. 28, 1724–1741 (2007)
3. Castiglione, G., Restivo, A.: Reconstruction of L-convex Polyominoes. Electron. Notes Discrete Math. 12 (2003)
4. Duchi, E., Rinaldi, S., Schaeffer, G.: The number of Z-convex polyominoes. Advances in Applied Math. 40, 54–72 (2008)

Computational Aspects of Asynchronous Cellular Automata*

Jérôme Chandesris[2], Alberto Dennunzio[2], Enrico Formenti[2,**], and Luca Manzoni[1]

[1] Università degli Studi di Milano–Bicocca, Dipartimento di Informatica, Sistemistica e Comunicazione, Viale Sarca 336, 20126 Milano, Italy
luca.manzoni@disco.unimib.it

[2] Université Nice-Sophia Antipolis, Laboratoire I3S, 2000 Route des Colles, 06903 Sophia Antipolis, France
{enrico.formenti,alberto.dennunzio}@unice.fr, chandesris@i3s.unice.fr

Cellular Automata (CA) are a computational model widely used in many scientific fields. A CA consists of identical finite automata arranged over a regular lattice (*i.e.* every configuration of a CA is an element of $A^{\mathbb{Z}}$ where A is a finite set of local states). Each automaton updates its state on the basis of its own state and the one of its neighbors according to a local rule. All updates are synchronous.

Even if CA has been successfully used to model natural systems [3], the synchronicity of the updates can be inadequate to simulate many natural processes that are inherently asynchronous (*e.g.* [9]). Asynchronous CA (ACA) has been introduced to overcame this limitation. Many models for ACA has been proposed (purely asynchronous [5], α-asynchronous [8], *etc.*). This work focuses on fully asynchronous CA (*i.e.* two cells are never updated simultaneously). An (fully) ACA is defined with the same components of a classical CA (a finite set A, a local rule λ and a radius r) with the addition of a sequence $(\theta_t)_{t>0}$ with cell positions $\theta_t \in \mathbb{Z}$ to be updated. For every time $t \in \mathbb{N}$ only the cell in position θ_t is updated.

ACA are capable of complex computational tasks. Indeed, both CA and ACA are Turing universal (see [7,5,6,4,10]). However, the computational cost of simulating an universal Turing machine (TM) has never been taken into account. This is the main concern of the present paper. We analyze two simulation schemes: *strict* and *scattered strict* simulation (see [2] for precise definitions and results).

Strict simulation requires that the ACA exactly reproduces the tape of the TM under simulation allowing cells of the ACA to store only a limited additional information. In this case the simulation can be correctly completed by *universal* updating sequences θ *i.e.* such that every $z \in \mathbb{Z}$ appears infinitely often inside it.

* This work has been partially supported by the French National Research Agency project EMC (ANR-09-BLAN-0164).

** Corresponding author.

G. Mauri and A. Leporati (Eds.): DLT 2011, LNCS 6795, pp. 466–468, 2011.

In fact, for every updating sequence θ and for any TM $\mathcal{M}$, there exists an ACA that strictly simulates $\mathcal{M}$ iff θ is universal. To allow a bound on the simulation time a specific set of sequences must be chosen. In particular, the sequences that touches all the cells between $-n$ and n for larger and larger n allow the simulation to be performed in $O\left(T(n)^2\right)$ if the TM halts in $T(n)$ time when started on inputs of size n. Even when the ACA is designed with a specific sequence (*i.e.* the simulation will not be carried on by all universal sequences) the time can be reduced only by a multiplicative factor.

Scattered strict simulation is similar to strict simulation but a subset of the cells of the simulating ACA is allowed to be "inactive". Given two positions on the TM tape the corresponding cells of the automaton can be non-consecutive (*i.e.* they are separated by a set of "inactive" cells). The possibility to build an ACA that simulates a TM in this way depends on the properties of the sequence of updates. Given an updating sequence θ, for any TM there exists an ACA that scatter strictly simulates it iff there exists a syndetic subset of θ on which θ is universal. Also in this case no bound can be given on the simulation time if a specific set of sequences has not been chosen. For particular sequences the running time of the simulation can be show to be $O\left(T(n)^2\right)$, as before $T(n)$ is the halting time of the TM on inputs of size n. This means that a scattered simulation could require only more space but not necessarily more time.

The simulation of TM could be considered artificial, but some models of real life processes (*e.g.* [1]) are actually disguised ACA with the update sequence given by some stochastic process. When this process is a 1D random walk the updating sequence will be almost surely universal, allowing computation to be performed. The authors are currently investigating the bound on time complexity of the simulation when the updating sequence is generated by a stochastic process.

References

1. Amar, P., Bernot, G., Norris, V.: Hsim: a simulation programme to study large assemblies of proteins. Journal of Biological Physics and Chemistry 4, 79–84 (2004)
2. Chandesris, J., Dennunzio, A., Formenti, E., Manzoni, L.: Computational aspects of asynchronous cellular automata (2010), `http://arxiv.org/find/all/1/all:+239813/0/1/0/all/0/1`
3. Chopard, B.: Modelling physical systems by cellular automata. In: Rozenberg, G., et al. (ed.) Handbook of Natural Computing: Theory, Experiments, and Applications. Springer, Heidelberg (to appear, 2011)
4. Lee, J., Adachi, S., Peper, F., Mashiko, S.: Delay-insensitive computation in asynchronous cellular automata. J. Comput. Syst. Sci. 70, 201–220 (2005)
5. Nakamura, K.: Asynchronous cellular automata and their computational ability. Systems, Computers, Control 5, 58–66 (1974)
6. Nehaniv, C.L.: Evolution in asynchronous cellular automata. Artificial Life VIII, 65–73 (2002)

7. Ollinger, N.: Universalities in cellular automata. In: Rozenberg, G., et al. (ed.) Handbook of Natural Computing: Theory, Experiments, and Applications, Springer, Heidelberg (to appear, 2011)
8. Regnault, D., Schabanel, N., Thierry, E.: Progresses in the analysis of stochastic 2d cellular automata: A study of asynchronous 2d minority. Theoretical Computer Science 410, 4844–4855 (2009)
9. Schönfisch, B., de Roos, A.: Synchronous and asynchronous updating in cellular automata. BioSystems 51, 123–143 (1999)
10. Worsch, T.: A note on (intrinsically?) universal asynchronous cellular automata (2010) (preprint)

Short 3-Collapsing Words over a 2-Letter Alphabet

Alessandra Cherubini[1], Achille Frigeri[1], and Brunetto Piochi[2]

[1] Politecnico di Milano {alessandra.cherubini,achille.frigeri}@polimi.it
[2] Università di Firenze
piochi@math.unifi.it

Let $\mathcal{A} = (Q, \Sigma, \delta)$ be a finite deterministic complete automaton. $\mathcal{A}$ is called k-compressible if there is a word $w \in \Sigma^+$ such that the image of the state set Q under the action of w has at most size $|Q| - k$, in such case the word w is called k-compressing for $\mathcal{A}$. A word $w \in \Sigma^+$ is k-collapsing if it is k-compressing for each k-compressible automaton of the alphabet Σ and it is k-synchronizing if it is k-compressing for each k-compressible automaton with $k+1$ states (see [1,2] for details).

For each alphabet Σ and $k \geq 1$, k-collapsing words always exist [4], and are k-full, i.e contain as factors each word of length k on Σ. Let $|\Sigma| = t$, $c(k,t)$ denotes the length of the shortest k-collapsing words on the alphabet Σ. Exact values of $c(k,t)$ are known only for $k = 2$ and $t = 2, 3$; moreover the shortest 3-synchronizing words for $|\Sigma| = 2$ (i.e., the word $s_{3,2} = ab^2aba^3b^2a^2babab^2a^2b^3aba^2ba^2b^2a$ and its dual $\bar{s}_{3,2}$) have length 33 [1], hence by the construction in [3], one gets $33 \leq c(3,2) \leq 154$, and more precisely, since we prove that neither $s_{3,2}$ nor $\bar{s}_{3,2}$ are not 3-collapsing (see Fig. 1), $34 \leq c(3,2) \leq 154$. Nevertheless, we observe that any 3-compressible 5-states automaton over a two letter alphabet, is 3-compressed either by $s_{3,2}$ or by $\bar{s}_{3,2}$.

We are interested in 3-compressible *proper* automata, i.e, 3-compressible automata such that no word of length 3 is 3-compressing for them. So identifying each letter of the alphabet with its action on Q we can assume that each letter acts as a permutation or as a transformation α of one of the following types (different letters represent different states):

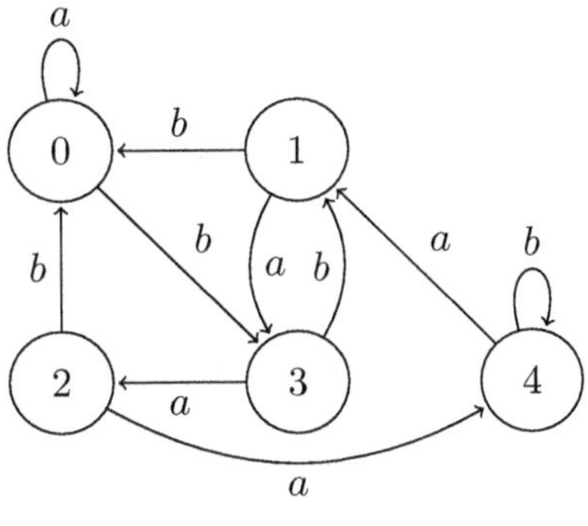

Fig. 1. An automaton which is not 3-compressed by $s_{3,2}$: $\delta(Q, s_{3,2}) = \{0,1,3\}$, $\delta(Q, \bar{s}_{3,2}) = \{3\}$.

1. $[x, y, z]/x, y$;
2. $[x, y][z, v]/x, z$;
3. $[x, y]/x$;
4. $[x, y]/z$ with $z\alpha = x$;

where this notation means that the states in the same square brackets are the unique states identified by α, the state after the slash do not belong to $\text{Im}(\alpha)$.

G. Mauri and A. Leporati (Eds.): DLT 2011, LNCS 6795, pp. 469–471, 2011.

We say that $\mathcal{A}$ is a $(\mathbf{i.},\mathbf{j.})$-*automaton*, $1 \le i, j \le 4$, if a letter is of type **i.** and the other letter is of type **j.**, and we say that $\mathcal{A}$ is a $(\mathbf{i.},\mathbf{p})$-*automaton*, with $1 \le i \le 4$, if a letter is of type **i.** and the other letter is a permutation. Then we can prove that:

Lemma 1. *A 3-compressible* $(\mathbf{i.},\mathbf{j.})$*-automaton, with* $i \in \{1,2\}$ *and* $j \in \{1,2,4\}$ *on the alphabet* $\{a,b\}$*, is not proper.*

Considering exhaustively the case of $(\mathbf{i.},\mathbf{p})$-automata, we give a necessary and sufficient condition for such automata to be not 3-compressible. In the proof of such proposition, we produce for each proper 3-compressible $(\mathbf{i.},\mathbf{p})$-automaton a short 3-compressing word, leading to the following corollary:

Corollary 1. *Let* $\mathcal{A}$ *be a 3-compressible proper automaton with input alphabet* $\Sigma = \{a,b\}$ *in which* b *acts as a permutation. Then:*

1. *if letter* a *is of type* **1.**, *then the word* ab^2a *3-compress the automaton;*
2. *if letter* a *is of type* **2.**, *then either the words* ab^2a *or* ab^3a *3-compress the automaton;*
3. *if letter* a *is of type* **3.**, *then one of the following words 3-compresses the automaton:*

$$ababa,\ abab^2a,\ aba^2ba,\ abab^2aba,\ ab^2ab^2a,\ ab^2a^2b^2a,$$
$$ab^2abab^2a,\ ab^2aba,\ ab^3aba,\ abab^3a,\ ab^3ab^3a;$$

4. *if letter* a *is of type* **4.**, *then one of the following words 3-compress the automaton:*

$$a^2ba,\ a^2b^2a,\ aba^2,\ a^2b^2a^2,\ ababa,\ a^2ba^2,\ ab^2a^2,\ ab^3a^2,\ a^2b^3a,\ a^2b^3a^2.$$

One can also prove again via an exhaustive search, or proving that for each proper 3-compressible $(\mathbf{i.},\mathbf{j})$-automaton $\mathcal{A}$ there exists an associated $(\mathbf{i.},\mathbf{p})$-automaton $\mathcal{A}'$ such that a word 3-compressing $\mathcal{A}'$ also 3-compresses $\mathcal{A}$, that at least a word in Corollary 1 or a dual of one of such words 3-compresses all proper 3-compressible automata of the remaining cases (i.e., $(\mathbf{1.},\mathbf{3})$-, $(\mathbf{2.},\mathbf{3})$-, $(\mathbf{3.},\mathbf{3})$-, $(\mathbf{3.},\mathbf{4})$-, and $(\mathbf{4.},\mathbf{4})$-automata). It follows that the word

$$a^2b^3a^3b^2ab^2abab^2aba^3ba^3bab^3ab^3aba^2baba^2ba^2b^2a^2b^2a$$

which has length 55 and is a shortest word having all the above words and their duals as factors, is 3-collapsing, improving the known upper bound. We also recall that a short 3-collapsing word, can be used in the procedure arising from ([3], Theorem 3.5) to obtain shorter k-collapsing words for $k \ge 4$, e.g., one have $c(4,2) \le 1803$ and $c(5,2) \le 113847$, instead of the known upper bound of 4872 and 307194, respectively [3].

References

1. Ananichev, D.S., Petrov, I.V., Volkov, M.V.: Collapsing words: A progress report. In: De Felice, C., Restivo, A. (eds.) DLT 2005. LNCS, vol. 3572, pp. 11–21. Springer, Heidelberg (2005)
2. Cherubini, A.: Synchronizing and collapsing words. Milan J. Math. 75, 305–321 (2007)
3. Margolis, S.W., Pin, J.-E., Volkov, M.V.: Words guaranteeing minimum image. Internat. J. Foundations Comp. Sci. 15, 259–276 (2004)
4. Sauer, N., Stone, M.G.: Composing functions to reduce image size. Ars. Combinatoria 1, 171–176 (1991)

A Cascade Decomposition of Weighted Finite Transition Systems*

Manfred Droste[1], Ingmar Meinecke[1], Branimir Šešelja[2], and Andreja Tepavčević[2,**]

[1] Institut für Informatik, Universität Leipzig,
D-04109 Leipzig, Germany
{droste,meinecke}@informatik.uni-leipzig.de

[2] Department of Mathematics and Informatics, University of Novi Sad,
Trg Dositeja Obradovica 4, 21000 Novi Sad, Serbia
{seselja,andreja}@dmi.uns.ac.rs

Abstract. We consider weighted finite transition systems with weights from naturally ordered semirings. Such semirings comprise distributive lattices as well as the natural numbers with ordinary addition and multiplication, and the max-plus-semiring. For these systems we explore the concepts of covering and cascade product. We show a cascade decomposition result for such weighted finite transition systems using special partitions of the state set of the system. This extends a classical result of automata theory to the weighted setting.

Synthesis and analysis of automata have been a central topic of computer science since its beginning. By Kleene's famous result [6], combinations of very simple automata have the full power of all automata. On a related strand, cascade and wreath products of transition systems and semigroups were investigated, leading to the fundamental theorem of Krohn-Rhodes [7], cf. [2,9,11]; see also Holcombe [5] for various applications of cascade products. In our work, we have investigated coverings by cascade products for the class of weighted automata.

In weighted automata, the transitions may carry weights modeling, e.g., the resources used during the execution of the transition. Already Schützenberger [10] extended Kleene's fundamental result to this model of quantitative automata whose theory quickly developed, cf. [4] and recently [3]. Here, the weights are taken from a semiring; multiplication is used to define the weights of paths and addition gives the total weight of all paths realizing a given word.

In our work, assuming that the semiring carries a partial order compatible with the semiring operations, we introduce a notion of *covering* for weighted transition systems (WTS); if a WTS M' covers M, then a subset of the states of M' can be mapped onto the states of M such that the weights of the transitions of M' bound the weight of the corresponding transition of M. This extends the corresponding notion for classical (unweighted) transition systems, and it is related to (but different from) other notions of

* This research was partially supported by the DAAD-Serbia project "Weighted Automata over Semirings and Lattices".

** The research of the last two authors is supported by Serbian Ministry of Science and Tech. Develop., Grant No. 174014.

G. Mauri and A. Leporati (Eds.): DLT 2011, LNCS 6795, pp. 472–473, 2011.

coverings from the literature, cf., e.g. [1]. We also extend the classical notion of cascade products to weighted transition systems, and we derive that cascade products preserve the covering relation. Next, we turn to admissible equivalence relations for WTS; they naturally lead to a quotient WTS covered by the original system.

For our main results, we assume that the naturally ordered semiring S satisfies the condition that $s \leq s^2$ for all $s \in S$. This assumption is satisfied by many important semirings including the semiring of natural numbers, the max-plus-semiring, all distributive lattices, and non-commutative semirings of formal languages, but not by e.g. the min-plus-semiring. Examples show that our main results do not hold without this condition on S. Assuming the condition, we show that we can cover any given WTS by a sequence of cascade products of smaller WTS. Furthermore, in case there are two admissible equivalence relations which are orthogonal, the original WTS can be covered by a direct product of the two corresponding quotient WTS. We obtain the classical results for unweighted transition systems as an immediate consequence by choosing S as the Boolean semiring $(\{0,1\}, \vee, \wedge)$.

In Mordeson-Malik [8], such a cascade decomposition result was claimed for the fuzzy semiring $([0,1], \vee, \wedge)$ and a stronger notion of covering with equality of weights (instead of bounds by weights). We give a counter-example showing that the proposed proof method does not work for such *strong* coverings.

To the best of our knowledge, this is the first paper dealing with cascade products of general semiring weighted transition systems. It remains open whether suitable wreath products could also be defined leading (ideally) to a weighted Krohn-Rhodes theory, and whether the applications of cascade products for e.g. neural networks (cf. Holcombe [5]) could also be developed for a quantitative setting.

References

1. Béal, M.-P., Lombardy, S., Sakarovitch, J.: On the equivalence of $\mathbb{Z}$-automata. In: Caires, L., Italiano, G.F., Monteiro, L., Palamidessi, C., Yung, M. (eds.) ICALP 2005. LNCS, vol. 3580, pp. 397–409. Springer, Heidelberg (2005)
2. Dömösi, P., Nehaniv, C.L.: Algebraic Theory of Automata Networks. In: SIAM Monographs on Discrete Mathematics and Applications, vol. 11. Society for Industrial and Applied Mathematics, Philadelphia (2004)
3. Droste, M., Kuich, W., Vogler, H. (eds.): Handbook of Weighted Automata. EATCS Monographs in Theoretical Computer Science. Springer, Heidelberg (2009)
4. Eilenberg, S.: Automata, Languages, and Machines, vol. A. Academic Press, London (1974)
5. Holcombe, W.M.L.: Algebraic Automata Theory. Cambridge University Press, Cambridge (1982)
6. Kleene, S.: Representations of events in nerve nets and finite automata. In: Shannon, C., McCarthy, J. (eds.) Automata Studies, pp. 3–42. Princeton University Press, Princeton (1956)
7. Krohn, K., Rhodes, J.L.: Algebraic theory of machines, I. Prime decomposition theorem for finite semigroups and machines. Trans. Amer. Math. Soc. 116, 450–464 (1965)
8. Mordeson, J.N., Malik, D.S.: Fuzzy Automata and Languages – Theory and Applications. Computational Mathematics Series. Chapman & Hall, CRC (2002)
9. Rhodes, J.L., Steinberg, B.: The q-Theory of Finite Semigroups. Springer, Heidelberg (2008)
10. Schützenberger, M.: On the definition of a family of automata. Information and Control 4, 245–270 (1961)
11. Straubing, H.: Finite Automata, Formal Logic, and Circuit Complexity. Birkäuser, Basel (1994)

Morphic Characterizations in Terms of Insertion Systems with a Context of Length One

Kaoru Fujioka

Office for Strategic Research Planning, Kyushu University,
6-10-1 Hakozaki Higashi-ku Fukuoka-shi, 812-8581, Japan
kaoru@tcslab.csce.kyushu-u.ac.jp

Representing a class of languages through operations on its subclasses is a traditional issue within formal language theory. Among the variety of representation theorems for context-free languages, Chomsky-Schützenberger theorem is unique in that it consists of Dyck languages, regular languages, and simple operations. In this work, we obtain some characterizations and representation theorems of context-free languages and regular languages in Chomsky hierarchy by insertion systems, strictly locally testable languages, and morphisms in the framework of Chomsky-Schützenberger theorem.

For insertion systems, we focus on the one of weight $(i,1)$ with $i \geq 1$, that is, the insertion operation is controlled by a context of length 1 and inserts a string of length i. In the general case, for $i, j \geq 0$, let INS_i^j be the class of all languages generated by insertion systems of weight (i', j') with $i' \leq i$ and $j' \leq j$.

On the other hand, for each $k \geq 1$, a strictly k-testable language is prescribed by a triplet of finite sets consisting of strings of length k. It has already been known the proper inclusion $LOC(1) \subset LOC(2) \subset \cdots \subset LOC(k) \subset \cdots \subset REG$, where REG is the class of regular languages and $LOC(k)$ is the class of strictly k-testable languages for $k \geq 1$.

For languages L_1, L_2, and a morphism h, we use the following notation: $h(L_1 \cap L_2) = \{h(w) \mid w \in L_1 \cap L_2\}$. For language classes $\mathcal{L}_1$ and $\mathcal{L}_2$, we introduce the following language class:

$$H(\mathcal{L}_1 \cap \mathcal{L}_2) = \{h(L_1 \cap L_2) \mid h \text{ is a morphism}, L_i \in \mathcal{L}_i \ (i = 1, 2)\}.$$

Within the framework of this notation, we consider characterizations and representation theorems of language families in Chomsky hierarchy.

In our previous research, using insertion systems of weight $(i,0)$ for $i \geq 1$, in which no insertion operation can be controlled by contexts, we proved the following concerning the class of regular languages denoted by REG and the class of context-free languages denoted by CF [1]:

- $REG = H(INS_1^0 \cap LOC(k)))$ with $k \geq 2$.
- $H(INS_1^0 \cap LOC(1)) \subset REG \subset H(INS_i^0 \cap LOC(k))$ with $i, k \geq 2$.
- $CF = H(INS_i^0 \cap LOC(k))$ with $i, k \geq 2$.

In this work, first of all, we consider the generative power of insertion systems of weight $(i,1)$ with $i \geq 1$. As is shown in [3], INS_i^1 is known to be a proper

G. Mauri and A. Leporati (Eds.): DLT 2011, LNCS 6795, pp. 474–475, 2011.

subset of the class of context-free languages, CF. We prove that INS_i^1 with $i \geq 1$ is incomparable with the class of regular languages, REG.

With the help of strictly 1-testable languages and simple operations, we can show the inclusion $INS_i^1 \subseteq H(INS_i^1 \cap LOC(1))$. The inclusion can be proved easily if we consider the fact $V^* \in LOC(1)$ for any alphabet V.

Furthermore, we show a characterization of context-free languages using insertion systems of weight $(i, 1)$ with $i \geq 1$ and strictly 1-testable languages. The inclusion $H(INS_i^1 \cap LOC(1)) \subseteq CF$ $(i \geq 1)$ can be derived directly from the fact $INS_i^1 \subset CF$ [3], $LOC(1) \subset REG$ [2], and the closure property of context-free languages. We prove the proper inclusion $H(INS_i^1 \cap LOC(1)) \subset CF$ $(i \geq 1)$ by showing that for a context-free language $L = \{a^n b a^n \mid n \geq 1\}$, there are no insertion system γ of weight $(i, 1)$, strictly 1-testable language R, and morphism h such that $L = h(L(\gamma) \cap R)$.

In contrast to this, we prove the representation theorem for the class of context-free languages with insertion systems of weight $(1, 1)$ and strictly 2-testable languages, that is, $H(INS_1^1 \cap LOC(2)) = CF$.

To show the representation theorem, from Chomsky-Schützenberger theorem $CF = H(Dyck \cap REG)$, we consider the inclusion $H(INS_1^1 \cap LOC(2)) \supseteq H(Dyck \cap REG)$, where $Dyck$ is the class of Dyck languages. For any language $L = h(D \cap L(G))$ with Dyck language D, regular grammar G, and morphism h, we construct an insertion system γ of weight $(1, 1)$, a strictly 2-testable language R, and a morphism h' and prove that $h'(L(\gamma) \cap R) = L$ holds.

The converse inclusion $H(INS_1^1 \cap LOC(2)) \subseteq CF$ can be derived directly from the fact that $INS_1^1 \subset CF$, $LOC(2) \subset REG$, and the closure property of context-free languages.

From the representation theorem, we have a corollary that for any $i \geq 1$, $k \geq 2$, $H(INS_i^1 \cap LOC(k)) = CF$.

In this work, we contribute to the study of insertion systems controlled by a context of length 1 for new characterizations of context-free and regular languages. Specifically, we showed that

- $H(INS_i^1 \cap LOC(1)) \subset CF$ with $i \geq 1$.
- $H(INS_i^1 \cap LOC(k)) = CF$ with $i \geq 1, k \geq 2$.

Acknowledgments. This work was supported in part by Grants-in-Aid for scientific research from the Ministry of Education, Culture, Sports, Science and Technology of Japan (No. 23740081).

References

1. Fujioka, K.: Morphic characterizations of languages in chomsky hierarchy with insertion and locality. Inf. Comput. 209(3), 397–408 (2011)
2. McNaughton, R., Papert, S.A.: Counter-Free Automata (M.I.T. research monograph), vol. (65). The MIT Press, Cambridge (1971)
3. Păun, G., Rozenberg, G., Salomaa, A.: DNA Computing. In: New Computing Paradigms. Springer, Heidelberg (1998)

Inference of Residual Finite-State Tree Automata from Membership Queries and Finite Positive Data

Anna Kasprzik

FB IV (Theoretical Computer Science), University of Trier, Germany
kasprzik@informatik.uni-trier.de

Poster – Abstract

The area of Grammatical Inference centers on *learning algorithms*: Algorithms that infer a description (e.g., a grammar or an automaton) for an unknown formal language from given information in finitely many steps. Various conceivable learning settings have been outlined, and based on those a range of algorithms have been developed. One of the language classes studied most extensively with respect to its algorithmical learnability is the class of regular string languages.

Possible sources of information include *membership queries* (MQs) where a learner may query an oracle if a certain element is in the target language L, and *equivalence queries* (EQs) where a learner may ask if the current hypothesis is correct and is given a counterexample if this is not the case. Moreover, a learner can for example be presented with a *positive sample*, i.e., a finite subset of L.

A significant part of the existing algorithms, of which Angluin's well-known LSTAR [1] learning regular string languages via MQs and EQs was one of the first, use the device of an *observation table*. If such a table fulfils certain conditions we can directly derive a DFA from it, and if the information entered into the table is sufficient this DFA is isomorphic to the minimal DFA $\mathcal{A}_L$ for L. The states of $\mathcal{A}_L$ correspond to the equivalence classes of L under the Myhill-Nerode relation.

However, there is a price to pay for the uniqueness of the DFA $\mathcal{A}_L$: In a worst case it can have exponentially many more states than a minimal NFA for the same language, and as for many applications a small number of states is a desirable feature it seems worth tackling the question if there is a way to obtain an NFA instead. In [2], Denis et al. introduce a special case of NFA, so-called *residual finite-state automata* (RFSA), where each state represents a residual language of the language recognized. There is a natural correspondence between the residual languages and the equivalence classes of a language. Thus, contrary to NFA in general, RFSA also have the advantageous property that there is a unique minimal RFSA $\mathcal{R}_L$ for every regular language L which makes them an attractive choice for descriptions in the design of learning algorithms due to their succinctness since $\mathcal{R}_L$ can still be exponentially smaller than $\mathcal{A}_L$.

There are algorithms learning regular string languages from MQs and EQs (e.g., [1]), and from MQs and positive samples (e.g., [3]). Moreover, the scope

G. Mauri and A. Leporati (Eds.): DLT 2011, LNCS 6795, pp. 476–477, 2011.

of interest has been extended from strings to trees – [4] and [5] present algorithms learning regular tree languages from MQs and EQs, and an algorithm learning regular tree languages from MQs and positive samples is given in [6]. The outputs of all algorithms mentioned so far are deterministic finite-state automata. However, in [7] Denis et al. also present an algorithm learning regular string languages from given data using RFSA, and finally Bollig et al. [8] describe an algorithm for the inference of regular string languages from MQs and EQs that in case of success returns an RFSA as well.

The notion of RFSA can equally be extended to trees: *Residual finite-state tree automata* (RFTA) have been defined and studied in [9]. As in the case of strings, for every regular tree language there is a unique minimal RFTA, which moreover can be exponentially more succinct than the corresponding deterministic tree automaton (DFTA). The algorithm in [8] inferring an RFSA from MQs and EQs can be adapted to trees in a rather straightforward manner.

We present a learning algorithm that infers a regular tree language from MQs and a positive sample, and returns an RFTA. The algorithm RESI is of polynomial complexity which benefits further from a technique that is parallel to and slightly improves the one used in [6] for the inference of DFTA.

Potential applications are all those that particularly benefit from a succinct description. One application named in [8] is verification. Other areas specifically related to trees include the extraction of information from semi-structured data (p.ex., `http://mostrare.lille.inria.fr`), or applications in computational linguistics since linguistic structure is often represented in tree form and one can imagine various situations where one might want to derive a succinct description from huge amounts of data contained in treebanks or similar databases.

References

1. Angluin, D.: Learning regular sets from queries and counterexamples. Information and Computation 75(2), 87–106 (1987)
2. Denis, F., Lemay, A., Terlutte, A.: Residual finite state automata. Fundamentae Informaticae 51, 339–368 (2002)
3. Angluin, D.: A note on the number of queries needed to identify regular languages. Information and Control 51(1), 76–87 (1981)
4. Sakakibara, Y.: Learning context-free grammars from structural data in polynomial time. Theoretical Computer Science 76(2-3), 223–242 (1990)
5. Drewes, F., Högberg, J.: Learning a regular tree language from a teacher. In: Ésik, Z., Fülöp, Z. (eds.) DLT 2003. LNCS, vol. 2710, pp. 279–291. Springer, Heidelberg (2003)
6. Besombes, J., Marion, J.-Y.: Learning tree languages from positive examples and membership queries. Theoretical Computer Science 382, 183–197 (2007)
7. Denis, F., Lemay, A., Terlutte, A.: Learning regular languages using RFSA. In: Abe, N., Khardon, R., Zeugmann, T. (eds.) ALT 2001. LNCS (LNAI), vol. 2225, pp. 348–363. Springer, Heidelberg (2001)
8. Bollig, B., Habermehl, P., Kern, C., Leucker, M.: Angluin-style learning of NFA. In: Online Proceedings of IJCAI, vol 21 (2009)
9. Carme, J., Gilleron, R., Lemay, A., Terlutte, A., Tommasi, M.: Residual finite tree automata. In: Ésik, Z., Fülöp, Z. (eds.) DLT 2003. LNCS, vol. 2710, pp. 171–182. Springer, Heidelberg (2003)

On the Representability of Line Graphs

Sergey Kitaev[1,2,⋆], Pavel Salimov[1,⋆⋆], Christopher Severs[1,⋆⋆⋆], and Henning Úlfarsson[1,†]

[1] Reykjavik University, School of Computer Science, Menntavegi 1, 101 Reykjavik, Iceland
[2] University of Strathclyde, Department of Computer and Information Sciences, Livingstone Tower, 26 Richmond Street, Glasgow G1 1XH, UK

1 Introduction

A graph $G = (V, E)$ is representable if there exists a word W over the alphabet V such that letters x and y alternate in W if and only if $(x, y) \in E$ for each $x \neq y$. Such a W is called a *word-representant* of G. Note that in this paper we use the term graph to mean a finite, simple graph, even though the definition of representable is applicable to more general graphs.

The notion of a representable graph comes from algebra, where it was used by Kitaev and Seif to study the growth of the free spectrum of the well known *Perkins semigroup* [5]. There are also connections between representable graphs and robotic scheduling as described by Graham and Zang in [1]. Moreover, representable graphs are a generalization of *circle graphs*, which was shown by Halldórsson, Kitaev and Pyatkin in [2], and thus they are interesting from a graph theoretical point of view. Finally, representable graphs are interesting from a combinatorics on words point of view as they deal with the study of alternations in words.

Not all graphs are representable. Examples of minimal (with respect to the number of nodes) non-representable graphs given by Kitaev and Pyatkin in [4] are presented in Fig. 1.

It was remarked in [2] that very little is known about the effect of the line graph operation on the representability of a graph. We attempt to shed some light on this subject by showing that the line graph of the smallest known non-representable graph, the wheel on five vertices, W_5, is in fact non-representable. In fact we prove a stronger result, which is that $L(W_n)$ (where $L(G)$ denotes the line graph of G) is non-representable for $n \geqslant 4$. From the non-representability of $L(W_4)$ we are led to a more general theorem regarding line graphs. Our main result is that $L^k(G)$, where G is not a cycle, a path or the claw graph,

⋆ Supported by a grant from Iceland, Liechtenstein and Norway through the EEA Financial Mechanism. Supported and coordinated by Universidad Complutense de Madrid.

⋆⋆ Supported by the Russian Foundation for Basic Research grants no. 10-01-00616, 11-01-00997 and by the Icelandic Research Fund grant no. 090038013.

⋆⋆⋆ Supported by grant no. 090038012 from the Icelandic Research Fund.

† Supported by grant no. 090038011 from the Icelandic Research Fund.

G. Mauri and A. Leporati (Eds.): DLT 2011, LNCS 6795, pp. 478–479, 2011.

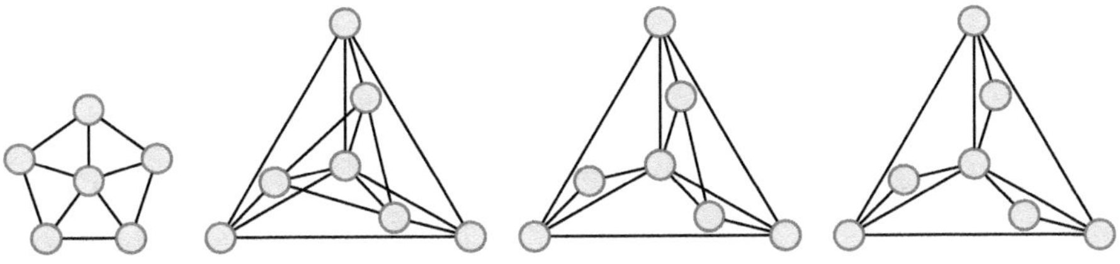

Fig. 1. Minimal non-representable graphs

is guaranteed to be non-representable for $k \geqslant 4$. However an important open question still remains: Is the line graph of a non-representable graph always non-representable.

2 Results

The *wheel graph*, denoted by W_n, is a graph we obtain from a cycle C_n by adding one external vertex adjacent to every other vertex.

A line graph $L(G)$ of a graph G is a graph on the set of edges of G such that in $L(G)$ there is an edge (a, b) if and only if edges a, b are adjacent in G.

Theorem 1. *The line graph $L(W_n)$ is not representable for each $n \geqslant 4$.*

Theorem 2. *The line graph $L(K_n)$ is not representable for each $n \geqslant 5$.*

It was shown by van Rooji and Wilf [6] that iterating the line graph operator on most graphs results in a sequence of graphs which grow without bound. This unbounded growth results in graphs that are non-representable after a small number of iterations of the line graph operator since they contain the line graph of a large enough clique.

Theorem 3. *If a connected graph G is not a path, a cycle, or the claw graph $K_{1,3}$, then $L^n(G)$ is not representable for $n \geqslant 4$.*

References

1. Halldórsson, M., Kitaev, S., Pyatkin, A.: Graphs capturing alternations in words. In: Gao, Y., Lu, H., Seki, S., Yu, S. (eds.) DLT 2010. LNCS, vol. 6224, pp. 436–437. Springer, Heidelberg (2010)
2. Halldórsson, M., Kitaev, S., Pyatkin, A.: On representable graphs, semi-transitive orientations, and the representation numbers, arXiv:0810.0310v1 (math.CO) (2008)
3. Halldórsson, M., Kitaev, S., Pyatkin, A.: Graphs capturing alternations in words. In: Gao, Y., Lu, H., Seki, S., Yu, S. (eds.) DLT 2010. LNCS, vol. 6224, pp. 436–437. Springer, Heidelberg (2010)
4. Kitaev, S., Pyatkin, A.: On representable graphs. Automata, Languages and Combinatorics 13, 1, 45–54 (2008)
5. Kitaev, S., Seif, S.: Word problem of the Perkins semigroup via directed acyclic graphs. Order (2008), doi:10.1007/s11083-008-9083-7
6. van Rooij, A.C.M., Wilf, H.S.: M van Rooij and H.S. Wilf. The interchange graph of a finite graph. Acta Mathematica Academiae Scientiarum Hungaricae 16, 263–269 (1965)

Author Index